MySQL 8.0 实用手册

性能优化、架构设计、运维管理、应用开发与云数据库建设

崔虎龙　胡自贵　编著

本书是一本面向广大 DBA（数据库管理员）的 MySQL 数据库实用手册，分为 10 章，分别介绍了 MySQL 的基础知识和安装部署，体系架构特性，功能架构特性，升级与迁移，性能优化，运维管理，故障分析，安全管理，架构设计与应用开发，以及云数据库建设等内容，让读者对 MySQL 8.0 数据库管理和运维更加得心应手。

本书的目标读者包括：数据库架构师、运维管理人员、开发人员，以及对相关技术感兴趣的人。非专业技术人员也可以从本书中了解 MySQL 8.0 的技术特性，加深对数据库行业的了解。

图书在版编目（CIP）数据

MySQL 8.0 实用手册：性能优化、架构设计、运维管理、应用开发与云数据库建设 / 崔虎龙，胡自贵编著. 北京：机械工业出版社，2025.3. -- ISBN 978-7-111-78146-2

Ⅰ. TP311.132.3-62

中国国家版本馆 CIP 数据核字第 2025YS1147 号

机械工业出版社（北京市百万庄大街 22 号　邮政编码 100037）
策划编辑：王　斌　　　　　　　　　　责任编辑：王　斌　马　超
责任校对：杜丹丹　甘慧彤　马荣华　景　飞　责任印制：常天培
河北虎彩印刷有限公司印刷
2025 年 6 月第 1 版第 1 次印刷
184mm×260mm・34 印张・906 千字
标准书号：ISBN 978-7-111-78146-2
定价：139.00 元

电话服务　　　　　　　　　　　　网络服务
客服电话：010-88361066　　　　　机　工　官　网：www.cmpbook.com
　　　　　010-88379833　　　　　机　工　官　博：weibo.com/cmp1952
　　　　　010-68326294　　　　　金　　书　　网：www.golden-book.com
封底无防伪标均为盗版　　　　　　机工教育服务网：www.cmpedu.com

序　言

MySQL 在数据库领域是一个奇迹，其成功的原因至今引人深思。

当 1995 年 MySQL AB 公司创立时，Monty 肯定无法预见 MySQL 对行业产生的颠覆性影响。而从起点看，**无畏的勇气尤其值得钦佩**。

1996 年，MySQL 第 3 版正式公开发布，因为 MySQL 没有第 2 版，第 1 版仅在内部发布，所以这是其第一个正式版本。

MySQL 的第一个正式版本面对的是怎样的世界？

1996 年，Oracle 7.3 版本已经发布，这是 Oracle 数据库历史上最为经典的版本之一，帮助 Oracle 确立了不可撼动的市场地位。

同样是 1996 年，增加了 SQL 支持的 Postgres95 更名为 PostgreSQL，版本号已经是 6.0。

无论是商业还是开源数据库，市场上看似已经有了最好的选择，MySQL 何以成功？

我想当年 Monty 肯定没有问过这个问题，他对于技术的态度只有两个字：**痴迷**。

在一次访谈中，Monty 提到，"顶尖黑客是万里挑一的，他们奉献了所有能用的时间"。正是因为他为 MySQL 奉献了所有的时间，不断围绕着用户需求快速发布，带领开源社区不断向前，最终成就了一个互联网时代的神话。而彼时，Stonebraker 正在忙于将 Postgres 商业化，并带领 Illustra 团队加入 Informix，那也是 1996 年。

江山代有才人出，一个又一个无畏的勇士，成就了一座又一座的数据库历史丰碑。

云和恩墨的两位专家崔虎龙和胡自贵，长期奋战于数据库工作的一线，最先接触到用户的问题与需求，也因此对数据库技术有着最为直接的领会与体验。**春江水暖鸭先知**，他们对 MySQL 新技术的理解与跟踪，代表了 MySQL 的活力之所在；**不需扬鞭自奋蹄**，他们又是最为勤奋的自我驱动者，因此才有了本书的成型和出版。在此热烈祝贺崔虎龙和胡自贵的新书付梓！

云和恩墨在 MySQL 方面的投入亦有两方面，一方面是通过技术服务帮助用户解决生产中遇到的种种疑难杂症，这是崔虎龙的强项；另一方面是通过 zCloud 等产品设计解决方案，帮助用户构建自治智能的数据库云管平台，这是胡自贵的强项。

现在两位专家联手，我期待本书能够为关心 MySQL 技术的读者带来帮助，并持续促进中文世界 MySQL 的技术应用。

盖国强

云和恩墨（北京）信息技术有限公司创始人

推荐序 1

随着 MySQL 5.7 版本的项目终止（End Of Life），MySQL 8.0 版本无疑成为新部署系统的唯一选择。同时 MySQL 8.0 中带来了大量的新功能：从内核优化器的增强到完整性的 DDL 一致性，从降序索引和索引跳跃扫描的功能到资源组等安全性的全面提升，使得 MySQL 的处理能力进一步提升。MySQL 8.0 很快会变成占有率最高的版本，广大 MySQL DBA 也会越来越多地维护和管理 8.0 版本的 MySQL 数据库。

关于 MySQL 的书籍并不少，但是针对 8.0 版本做深入讲解的很少见。尤其是很多 DBA 已经对 MySQL 有了比较丰富的经验，只是缺少对 8.0 版本新特性和功能的理解，而在面临新版本的时候，只能沿用一些老的方法，无法更好地利用版本提升带来的优势。本书的出版正好可以帮助这些有经验的 DBA，更快地全面了解 MySQL 8.0 的新功能，以及在 8.0 版本环境中运维操作、故障处理和性能优化会发生哪些变化。

本书可以作为 MySQL 8.0 初学者的快速入门手册，因为它对新手很友好，虽然本书的重点放在了新特性的讲解上，但对 MySQL 数据库基本的体系架构、安装部署、日常管理等内容的介绍很全面。除了基本的功能、特性介绍以外，作者根据自己多年一线工作经验，在书中介绍了大量的实战案例：通过真实案例来讲解如何对 MySQL 数据库问题进行深入分析；介绍了 MySQL 数据库的最佳实践部署；针对 8.0 版本的数据库架构设计，以及针对不同的数据库架构对应的开发规范；MySQL 在安全性方面的功能和考虑；MySQL 在云环境中的设计和实践。这些内容对于想要深入学习 MySQL，快速提升自己数据库水平的技术人员非常有帮助。

本书的作者崔虎龙和胡自贵在行业中都是有着超过 15 年工作经验的专家，他们在日常工作中接触了各行各业的客户，全方位地帮助客户解决生产环境的故障，每天都会回答众多的咨询问题。这使得他们对故障的处理驾轻就熟，对各种数据库和应用架构都烂熟于胸，因此才能总结提炼出这本内容翔实、知识覆盖面极广的著作。希望本书可以帮助众多的数据库从业人员，为数据库行业的发展添砖加瓦。

杨廷琨
云和恩墨技术与平台服务部总经理，前 Oracle ACE

推荐序 2

在数据库的发展史中，MySQL 无疑是近十年最受欢迎的数据库之一，无论是在国外还是在国内，MySQL 都拥有着非凡的吸引力和众多的使用者。现今，MySQL 不但仍然拥有全球领先的热度，同时也拥有大量由其衍生而来的数据库品类，因此，众多数据库技术爱好者与用户都有着学习与掌握 MySQL 的需求。

随着 MySQL 数据库的发展与创新，MySQL 8.0 已经成为众多用户的必选产品，相对于早期版本，MySQL 8.0 拥有大量新的特性与功能，因此，无论是初学者，还是有经验的 DBA，都需要对 MySQL 8.0 有个全面的认识。

本书的作者崔虎龙是一名有着多年 MySQL 一线实战经验的资深"MySQLer"，从最早的版本开始接触 MySQL 数据库，经历了十余年的实战打磨，为众多企业进行了上百场 MySQL 技术培训与分享。他通过将大量 MySQL 实战经验与理论相结合，与同样富有经验的资深数据库技术专家胡自贵一起，最终完成了本书。本书内容从易到难，讲解深入浅出，既能够让初学者快速获取 MySQL 的安装部署要点，掌握体系架构理论，又能够通过数据迁移、运维实战让用户快速上手，具备使用、维护、优化、故障处理的能力。本书也为 MySQL 的开发者、设计师提供安全防范、应用规范、云上 MySQL 的规则与理念，是一本具备"实战、实用、实际"特点的 MySQL 技术教材。本书非常值得广大数据库同仁借鉴学习。

李轶楠
云和恩墨培训产品部总经理，前 Oracle ACE

推荐序 3

MySQL 依然是当前最具影响力、应用最为广泛的数据库之一。本书面向运维、源于实战、聚焦问题，围绕 MySQL 安装部署、升级迁移、性能优化、故障处置及安全防护等一系列专项主题，展开深入阐述。本书中 MySQL 云数据库建设部分更是结合云原生、容器化、DevOps、资源池等关键技术，从架构设计的角度，给出了 DBaaS 平台的参考模型，使得本书的写作理念和内容得到进一步的突破与充实。

作为一名数据库从业人员，在日常工作中经常会遇到一时无法解决的难点问题，不少人或许第一反应就是借助搜索引擎寻找答案，长期以来，得到的几乎都是快餐式、碎片化的知识点导致知识结构不系统、不完整，这肯定是不理想的。而本书在很大程度上实现了对 MySQL 知识结构的梳理与重构，可以帮助广大数据库从业者实现对 MySQL 数据库的体系化认知，从而极大提升业务能力以应对纷繁复杂的业务场景，各式各样的技术难点。

有感于本书前言所述，"为了让广大同行更好地了解和运维 MySQL 这个重要的数据库，为 MySQL 开源社区的建设尽一份绵薄之力，我们把自己多年积累的、针对 MySQL 数据库的开发运维工作经验毫无保留地通过本书展现给大家。"由此衷心感谢两位作者的情怀与坚持，能够在紧张繁忙的工作之余，用心完成书稿的编写。

创作不易，开卷有益，希望本书的出版能够帮助更多的数据库技术人员。

安新亚

云和恩墨技术市场部总经理，北京华胜天成科技股份有限公司原总工程师

前　言

　　MySQL 是当前最受欢迎的关系数据库之一，应用十分广泛。MySQL 的每个版本都有着自己的生命周期，目前最新版本为 8.0。2022 年，MySQL 5.6 Extended 的支持到期，经历了升级至 5.7 版本的艰苦历程；2023 年 10 月，MySQL 5.7 版本生命周期已尽，同时 MySQL 8.0 版本周期又延长至 2026 年。在 MySQL 5.7 版本上新增诸多特性的 MySQL 8.0 日益成为 MySQL 的主要版本。

　　为了让广大同行更好地了解和运维 MySQL 这个重要的数据库，为 MySQL 开源社区的建设尽一份绵薄之力，我们把自己多年积累的、针对 MySQL 数据库的开发运维工作经验毫无保留地通过本书展现给大家。

　　本书从知识结构上分为以下六大部分。

　　第一部分（第 1~3 章）详细介绍 MySQL 8.0 的各种安装部署方式、体系结构和功能，并针对这些内容进行了实例展示。

　　第二部分（第 4 章）介绍如何将数据库迁移到 MySQL 8.0，包含迁移的方法、迁移工具和迁移注意事项。

　　第三部分（第 5~7 章）介绍 MySQL 的性能优化、运维管理、故障分析。以实际运行中的 MySQL 案例为基础，说明如何分析性能问题、优化数据库性能的关键点，如何有效地管理 MySQL，以及碰到故障时如何应对，应该怎样避免故障发生。

　　第四部分（第 8 章）为安全方面的内容，介绍如何利用社区版提供的特点，做好 MySQL 的安全设置，包含数据加密、SLR 设置、账号密码安全设置、权限最小原则设置方法等内容。

　　第五部分（第 9 章）为 MySQL 8.0 应用开发的内容，介绍 MySQL 架构设计和应用开发的规范。

　　第六部分（第 10 章）介绍 MySQL 的云数据库建设。随着云计算的普及，MySQL 将向云原生架构演进。这一部分内容介绍如何实现云数据库上 MySQL 的管理，包含基础架构承载层设计、资源池设计、一体化运维管理与服务设计、规范化和专业化体系建设。

致谢（崔虎龙）

　　古人云："活到老，学到老。"在几十年的职业生涯里，感谢一起工作的同事，让我学到了很多 MySQL 方面的知识。

　　感谢云和恩墨公司，让我有机会备战在项目交付一线，让我在 MySQL 领域积累了丰富的经验和专业知识。

感谢机械工业出版社的编辑们,在这半年多的时间中始终支持我的写作,他们的鼓励和帮助十分有价值。

谨以此书献给我最亲爱的家人,以及众多热爱 MySQL 的朋友。

致谢(胡自贵)

非常开心能够和崔虎龙一起编写本书,让我有机会结合用户的实际需求场景,把我对 MySQL 数据库的理解,构建 MySQL 云数据库的思路和大家分享。

感谢云和恩墨公司给了我这个舞台,让我始终奋斗在面向客户的第一线,能够结合不同用户的需求场景制定出有针对性的解决方案。感谢机械工业出版社给予我出版图书、分享个人经验的机会。感谢所有关心、支持、信任我的朋友,你们是我坚持写作的动力源泉。

谨以此书献给我最亲爱的妻子和孩子们,因为有你们,我才一直充满了前进的动力。

由于作者的水平有限,书中难免会出现一些错误或者不准确的地方,恳请读者批评指正。您可以将书中的错误发到邮箱 cuihulong@163.com ,我们将尽力提供让您满意的解答。如果您有更多的宝贵意见,也欢迎发送邮件至邮箱,我们衷心期待您的真挚反馈。

编　者

目 录

序 言
推荐序 1
推荐序 2
推荐序 3
前 言

第 1 章 MySQL 概述 1
1.1 MySQL 基础知识 1
1.1.1 MySQL 发展历史 1
1.1.2 MySQL 的版本生命周期 1
1.1.3 MySQL 的主要分支 3
1.1.4 MySQL 各版本特点 4
1.1.5 MySQL 开发过程版本标识 7
1.1.6 MySQL 版本选择原则 7
1.2 MySQL 8.0 的安装部署 7
1.2.1 MySQL 的安装部署流程 8
1.2.2 MySQL 8.0 的基本安装部署 9
1.2.3 使用 Shell 脚本批量部署 MySQL 环境 16
1.2.4 定制化 RPM 包安装部署 24
1.2.5 基于 Docker 环境安装 MySQL 29

第 2 章 MySQL 8.0 体系架构特性 34
2.1 数据字典 34
2.1.1 全局事务性数据字典 34
2.1.2 新的 SDI 元数据结构 40
2.1.3 隐藏的数据字典表的可视化功能 45
2.1.4 全新的各类表空间 49
2.1.5 采用锁竞争算法的事务处理机制 56
2.2 日志体系 62
2.2.1 binlog 的新增算法与功能 62
2.2.2 Redo 日志的新增功能 65
2.2.3 慢查询日志的附加信息 70
2.3 引擎 72
2.3.1 InnoDB 引擎底层结构的变化 72
2.3.2 引入数据分析引擎 HeatWave 77

第 3 章 MySQL 8.0 功能架构特性 83
3.1 MySQL 8.0 新增函数和新增集合操作 83
3.1.1 窗口函数 83
3.1.2 集合操作 89
3.2 MySQL 8.0 新增索引类型及特性 93
3.2.1 隐藏索引 94
3.2.2 降序索引 96
3.2.3 函数索引 97
3.2.4 Hash Join 特性 98
3.2.5 Skip Scan Range 特性 101
3.2.6 Anti Join（反连接）特性 103
3.3 复制和高可用性方面的新增功能 106
3.3.1 高可用组复制（MGR）功能 106
3.3.2 异步复制源配置功能 111
3.3.3 MGR 集群容灾功能 116
3.3.4 增强的多源复制功能 126
3.3.5 MySQL Shell 快速创建、纳管副本集和 MGR 集群功能 128
3.3.6 MySQL Router+MGR 方式实现高可用 138
3.3.7 ProxySQL+MGR 方式实现高可用 147

IX

3.4 MySQL 8.0 新增功能 ················ 161
  3.4.1 角色管理 ················ 161
  3.4.2 直方图 ················ 164
  3.4.3 资源组 ················ 169
  3.4.4 优化器提示 ················ 172
  3.4.5 新增的优化器行为标志 ········ 178
  3.4.6 DDL 即时操作 ················ 183
  3.4.7 增强密码机制 ················ 185
  3.4.8 增强的 JSON 功能 ········· 190
  3.4.9 增强的 EXPLAIN 功能 ······ 195
  3.4.10 GIPK 隐藏主键可视化功能 ···· 199
  3.4.11 参数修改持久化功能 ········ 202
  3.4.12 克隆插件功能 ················ 204
  3.4.13 MySQL Shell 的逻辑备份恢复 API 功能 ···················· 212

第 4 章 MySQL 8.0 的升级与迁移 ··· 219
4.1 MySQL 8.0 的版本升级 ············ 219
  4.1.1 MySQL 数据库升级的方法 ··· 219
  4.1.2 MySQL 8.0 升级的注意事项 ··· 221
  4.1.3 MySQL 5.7 升级至 MySQL 8.0 的步骤 ·················· 225
4.2 MySQL 8.0 的迁移 ·················· 229
  4.2.1 MySQL 数据库迁移方案设计 ··· 229
  4.2.2 MySQL 8.0 数据库迁移工具及注意事项 ·················· 232

第 5 章 MySQL 8.0 性能优化 ········ 236
5.1 MySQL 8.0 性能优化概述 ········ 236
  5.1.1 性能优化的作用与方法 ········ 236
  5.1.2 性能分析需要收集的 11 类信息 ······················ 239
  5.1.3 导致性能突发事件的十大原因 ···· 242
  5.1.4 性能监控指标 ················ 244
5.2 MySQL 8.0 性能优化的关键点 ··· 246
  5.2.1 数据库配置优化 ··············· 246
  5.2.2 库、表、字段和索引的设计优化 ······················ 250
  5.2.3 SQL 语句优化 ················ 252
5.3 MySQL 8.0 性能优化实践 ········ 253
  5.3.1 SQL 语句执行性能的指标——QRTi ······················ 253
  5.3.2 通过 events_statements_summary_by_digest 表发现问题 SQL 语句 ··· 255
  5.3.3 使用 statement_analysis 视图分析 SQL 语句 ·················· 258
  5.3.4 通过分析 sys 库的存储过程排查性能问题 ·················· 260
  5.3.5 通过监控 InnoDB 存储引擎进行性能优化 ·················· 268
  5.3.6 问题 SQL 语句优化命令行 ····· 277
  5.3.7 定位导致 CPU 使用率高的问题 ··· 282

第 6 章 MySQL 8.0 的运维管理 ····· 288
6.1 MySQL 8.0 运维管理概述 ········ 288
  6.1.1 数据库运维管理的作用 ········ 288
  6.1.2 数据库运维管理的主要工作 ··· 288
6.2 MySQL 8.0 运维管理的关键点 ··· 290
  6.2.1 高频使用的运维管理操作 ···· 290
  6.2.2 运维管理中的高危操作 ········ 306
  6.2.3 运维管理中常用的官方工具 ··· 309
  6.2.4 运维管理中常用的周边工具 ··· 312
6.3 MySQL 8.0 运维实践 ·············· 319
  6.3.1 binlog 文件查看和解析 ········ 319
  6.3.2 利用 setup_actors 命令进行资源使用统计 ·················· 326
  6.3.3 数据库备份和恢复实践 ········ 329
  6.3.4 数据库热数据加载设置 ········ 335
  6.3.5 Query Rewrite 插件的使用 ··· 338
  6.3.6 控制 InnoDB 的并发线程 ···· 344
  6.3.7 备份中全局读锁 FTWRL 对数据库的影响 ·················· 347
  6.3.8 如何快速删除大量数据 ········ 352

第 7 章 MySQL 8.0 故障分析 ········ 355
7.1 MySQL 8.0 故障分析概述 ········ 355
  7.1.1 MySQL 8.0 的故障类型 ······ 355
  7.1.2 MySQL 8.0 故障分析方法 ··· 356
7.2 MySQL 8.0 故障分析关键点 ···· 358
  7.2.1 日志信息 ················ 358
  7.2.2 监控指标 ················ 360
  7.2.3 诊断工具 ················ 362
  7.2.4 SQL 语句 ················ 366
7.3 MySQL 8.0 典型故障分析实践 ··· 367
  7.3.1 导致服务器 OOM 的故障分析 ··· 367

7.3.2 导致 Got an error reading communication packet 提示的故障分析 ………… 373
7.3.3 导致服务器信号量不足的故障分析 ………………………………… 379
7.3.4 Undo 日志无法清理导致阻塞数据库的故障分析 ……………………… 386
7.3.5 导致服务器 CPU 的 sys 使用率过高的故障分析 …………………… 388

第 8 章 MySQL 8.0 安全管理 ………… 394
8.1 MySQL 8.0 的安全管理概述 ………… 394
　8.1.1 MySQL 安全管理的作用 …… 395
　8.1.2 MySQL 权限管理的作用 …… 395
8.2 MySQL 8.0 的安全管理关键点 ……… 396
　8.2.1 安全管理制度的执行和管理 … 396
　8.2.2 建立数据库审计制度 ………… 397
　8.2.3 敏感数据加密 ………………… 398
8.3 MySQL 8.0 的安全管理实践 ………… 399
　8.3.1 密码插件的使用 ……………… 399
　8.3.2 数据加密功能的使用 ………… 405
　8.3.3 SSL 安全的设置 ……………… 412
　8.3.4 用户数据库访问权限的设置 … 421

第 9 章 MySQL 8.0 架构设计与应用开发 ………………………………… 432
9.1 MySQL 8.0 架构设计 ………………… 432
　9.1.1 架构设计的原则 ……………… 432
　9.1.2 架构设计实践 1：读写分离方案 …………………………… 433
　9.1.3 架构设计实践 2：库内分库分表方案 …………………………… 437
9.2 MySQL 8.0 应用开发 ………………… 441
　9.2.1 MySQL 8.0 应用开发的概念 ……… 441
　9.2.2 MySQL 8.0 常用的开发规范 ……… 442
9.3 MySQL 8.0 应用开发实践 …………… 444
　9.3.1 时间类型的设置 ……………… 444
　9.3.2 BIT 数据类型的使用 ………… 447
　9.3.3 INSERT INTO 语句的使用 … 453
　9.3.4 分区表的使用 ………………… 457
　9.3.5 全文索引的使用 ……………… 469
　9.3.6 自增键的设计 ………………… 478
　9.3.7 外键的设计 …………………… 484
　9.3.8 表主键的设计 ………………… 488
　9.3.9 字符集的设计 ………………… 493
　9.3.10 MySQL 对 InnoDB 存储引擎、列、行格式的限制 ……………… 498

第 10 章 MySQL 8.0 云数据库建设 …… 503
10.1 云数据库的概念和发展趋势 ………… 503
　10.1.1 云数据库的概念 …………… 503
　10.1.2 云数据库的发展趋势 ……… 503
　10.1.3 云数据库面临的挑战 ……… 504
10.2 MySQL 云数据库设计方法 ………… 506
　10.2.1 层次化基础架构承载层设计 … 507
　10.2.2 标准化 MySQL 资源池设计 … 510
　10.2.3 一体化运维管理与服务设计 … 512
　10.2.4 规范化 MySQL 标准体系设计 … 525
　10.2.5 专业化 MySQL 保障体系设计 … 526
10.3 MySQL 云数据库设计方案 ………… 527
　10.3.1 公有云 RDS：公有云厂商的云数据库设计 ……………… 527
　10.3.2 私有云 DBaaS：专业数据库服务厂商云数据库设计 …… 529

第 1 章 MySQL 概述

1.1 MySQL 基础知识

1.1.1 MySQL 发展历史

MySQL 历经 20 多年的发展，现已成为全球最流行的数据库之一，特别是 MySQL 的社区版，由于其快速、稳定、易用和开放源代码等优点，迅速流行起来。MySQL 是目前最受开发者青睐的数据存储解决方案之一。

下面简单回顾一下 MySQL 的发展历史。

第一阶段：创立

MySQL 于 1995 年由 Monty Widenius 和 David Axmark 创立。最初，MySQL 只是用于一些与数据存储有关的小项目。

第二阶段：免费开源

2000 年，MySQL 开始以开源免费的形式发布，公布了自己的源代码，并采用 GPL（GNU General Public License）许可协议。MySQL 成为开源数据库的一个重要选择，也因此收获了广泛的用户群体。

第三阶段：被收购

2008 年，Sun Microsystems 公司收购了 MySQL，开始负责 MySQL 的开发和维护。在这个阶段，MySQL 的功能继续得到改进和增强，包括 InnoDB 存储引擎和查询缓存。

第四阶段：由 Oracle 掌控

2010 年，数据库软件公司 Oracle 收购了 Sun Microsystems，并继续开发和维护 MySQL。Oracle 推出了新版本，包括 MySQL 5.5、MySQL 5.6、MySQL 5.7、MySQL 8.0 和 MySQL 8.1 等，并增加了多种新的功能和特性。

1.1.2 MySQL 的版本生命周期

MySQL 是一种流行的关系数据库管理系统，目前由 Oracle 公司负责开发和维护。截至本书写作时，MySQL 已从早期的 5.0 版本迭代到 8.1 版本。按照版本迭代情况，每个版本的平均生命周期为 8 年。

Oracle 对 MySQL 版本生命周期的支持策略为"From Five Years to Forever（从 5 年到永远）"。目前 Oracle 对 MySQL 版本的支持分为 3 个级别：首要支持、扩展支持和持续支持。

1. **首要支持 Premier Support**

首要支持提供标准的 5 年支持政策，提供的支持包括：
- 主要产品和技术发布；
- 更新、修复、安全警报、数据修复和关键补丁更新；
- 新的税收、法律和监管更新；
- 升级脚本；
- 大多数新的第三方产品/版本认证；
- 新产品的认证。

2. **扩展支持 Extended Support**

扩展支持将支持时长在首要支持的基础上再延长 3~8 年。扩展支持提供的支持包括：
- 主要产品和技术发布；
- 技术支持；
- 更新、修复、安全警报、数据修复和关键补丁更新；
- 新的税收、法律和监管更新；
- 升级脚本；
- 大多数现有第三方产品/版本认证；
- 大多数现有 Oracle 产品的认证。

说明：扩展支持可能不包括某些新的第三方产品/版本的认证。

3. **持续支持 Sustaining Support**

持续支持仅支持重大安全更新。持续支持不包括：
- 新的更新、修复、安全警报、数据修复和关键补丁更新；
- 新的税收、法律和监管更新；
- 升级脚本；
- 第三方新产品/版本认证；
- Oracle 新产品认证；
- 针对严重级别 1 的服务请求，提供 24 小时承诺和响应指南；
- 支持之前发布的修复或更新。

MySQL 8.0 自 2016 年 4 月诞生以来，已进入稳定阶段。2022 年，MySQL 8.0 的生命周期再次进行调整，在之前的基础上延长了两年，标准支持延长到 2026 年 4 月。

按照官方提供的 MySQL 产品的生命周期支持介绍，在标准支持的范围内，MySQL 8.0 会定期提供补丁，进行升级。一旦进入延伸支持阶段，MySQL 仅在认为有必要升级的时候才会提供补丁，通常是为了解决安全性问题。安全性问题可以说是非常严重的问题，一旦碰到，必须进行版本升级。

MySQL 被 Oracle 收购之后，在产品开发方面变得更加规范。产品会遵循事先制定好的生命周期进行开发与维护。例如，MySQL 5.7 在 2023 年 10 月结束其延伸支持，届时 MySQL 将不会提供任何补丁。因此，对于 MySQL 8.0 版本的用户来说，在 2026 年 4 月之前，还会享受到 8.0 版本所带来的新功能和性能提升。图 1-1 所示为目前 MySQL 的版本生命周期。

2023 年 7 月发布的 MySQL 8.1 LTS Innovation 版本类似于 MySQL 团队在某些时候使用的"里程碑版本"，但里程碑版本不被视为"生产就绪"，因为"Innovation"或"Preview"等词汇通常用于描述预览版或包含新特性的测试版，这些版本可能并不适合直接用于生产环境。

Release	GA Date	Premier Support Ends	Extended Support Ends	Sustaining Support Ends
MySQL Database 5.0	Oct 2005	Dec 2011	Not Available	Indefinite
MySQL Database 5.1	Dec 2008	Dec 2013	Not Available	Indefinite
MySQL Database 5.5	Dec 2010	Dec 2015	Dec 2018	Indefinite
MySQL Database 5.6	Feb 2013	Feb 2018	Feb 2021	Indefinite
MySQL Database 5.7	Oct 2015	Oct 2020	Oct 2023	Indefinite
MySQL Database 8.0	Apr 2018	Apr 2025	Apr 2026	Indefinite

图 1-1　MySQL 的版本生命周期

了解了 MySQL 生命周期，可以更好地保证数据库安全、稳定运行，并跟上新技术发展的步伐。

1.1.3　MySQL 的主要分支

MySQL 的主要分支有 Oracle 官方版本的 MySQL、Percona Server、MariaDB，如图 1-2 所示。

截至本书写作时，官网最新的正式发布版本就是 MySQL 8.0，也是在生产环境中推荐使用的一个版本。它无论是在 InnoDB 存储引擎性能和功能的提升上，还是在安全性加固、复制功能、sys schema 库的增强等方面，均有相当出色的表现。

图 1-2　MySQL 的主要分支

Percona Server 是 MySQL 的一个重要分支，它在 InnoDB 存储引擎的基础上，提升了性能和管理性，最后形成了增强版的 XtraDB 引擎，可以更好地发挥服务器硬件的性能。所以 Percona Server 也可以看成增强的 MySQL 与开源的插件（plugin）的结合。由于官方版本的 MySQL 在一些特性的使用上有一定的局限性，即需要收费，因此一些常用的工具包，如 XtraBackup、percona-toolkit 等，成为生产环境中 DBA 的必备"武器"。还有像 XtraDB-Cluster 这种支持多点写入的强同步高可用集群架构，真正实现实时同步的过程，解决了 MySQL 主从复制之间经常出现且令人头疼的延迟问题。Percona 收购了 TokuDB 公司，TokuDB 存储引擎非常优秀。

MariaDB 是由 MySQL 创始人之一 Monty Widenius 创建的，是一款高度兼容的 MySQL 产品，主要由开源社区维护，采用 GPL 授权许可。Oracle 在收购 MySQL 之后，为避免 MySQL 在开源粒度上的下降，于是推出 MariaDB。它不只是 MySQL 的一个替代品，更重要的是，它提高了 MySQL 原有的技术水平并有所创新。它不仅包含 Percona 的 XtraDB 存储引擎，还包含 TokuDB 存储引擎、Spider 水平分片存储引擎等多种存储引擎，并且还有一些复制功能上的新特性，如基于表的并行复制、多源复制（Multi-source Replication）、Galera Cluster（集群）。2015 年，发布了 MariaDB 10.1，产生了独立的版本体系，完全脱离了 MySQL 体系。

上述 3 个分支的对比如下。

1）上述 3 个分支都有免费的开源产品，Percona 公司更偏重于运维工具；MariaDB 更偏向于功能的完善；Oracle MySQL 趋向于 Oracle 数据库一样的功能体系的实现。

2）Oracle 把控原生 MySQL 社区版的发展；Percona 紧跟 MySQL 的功能特性，并做了一些运维工具上的改进；MariaDB 更加开放，功能特性也更强一些。

3）Percona 的 MySQL 版本提供了高性能的 XtraDB 引擎，还提供了 PXC 高可用解决方案，并且附带了 Percona Toolkit 等 DBA 管理工具。MariaDB 从 10.0.9 版本开始使用 XtraDB（名称代号 Aria）来代替 MySQL 的 InnoDB。

社区活跃度：MySQL>MariaDB>Percona。目前，国内用户选择 Oracle MySQL 的比较多。

1.1.4 MySQL 各版本特点

在 Oracle 接管 MySQL 之后，MySQL 继续遵循原有的版本发布周期。每隔几年，MySQL 就会有大的功能版本发布，其间还会发布"仅修复错误"的小版本。MySQL 5.5、MySQL 5.6、MySQL 5.7 和 MySQL 8.0 都是按照这种策略进行维护的。下面介绍各版本的特点。

1. MySQL 5.0 的特点

1）MySQL 5.0 主要增加了商业数据库应该具备的功能，如 Stored procedures、Views、Cursors、Triggers、XA transactions、Event scheduler、Partitioning、Pluggable storage engine API、Row-based replication、Global 级别动态修改、general query log 和 slow query log 的支持。

2）增加 INFORATION_SCHEMA 系统数据库。

3）默认存储引擎更改为 InnoDB。

4）提升了数据库性能和可扩展性。

- 提高了默认线程并发数（innodb_thread_concurrency）。
- 后台输入/输出线程控制（innodb_read_io_threads、innodb_write_io_threads）。
- 主线程输入/输出速率控制（innodb_io_capacity）。

5）提升实用性。

- 半同步复制（Semi-synchronous Replication）。
- 复制 Heartbeat。

6）提升管理性和效率。

- 建立快速索引（Faster Index Creation）。
- 增加 INFORMATION_SCHEMA 表，新表提供了与 InnoDB 压缩和事务处理锁定有关的具体信息。

7）提升可用性。

- 针对 SIGNAL/RESIGNAL 的新 SQL 语法（在 MySQL 中，SIGNAL 和 RESIGNAL 是两个用于在存储过程、函数、触发器或事件中抛出错误或警告的 SQL 语句）。
- 新的表/索引分区选项。MySQL 5.5 将分区表的 RANG 和 LIST 分区范围扩展到了非整数列和日期，并增加了在多个列上分区的能力。

8）改善检测和诊断。MySQL 5.5 引入了一种新的性能架构（performancn_shema），用于监控 MySQL 监控服务器运行时的性能。

2. MySQL 5.6 的特点

MySQL 5.6 版本主要提高了 InnoDB 性能，支持延迟复制。

1）InnoDB 中可以避免大量表打开的时候内存占用过多的问题。

2）InnoDB 的性能加强，如分拆 kernel mutex、flush 操作从主线程分离、多个 purge 线程、大内存优化等。

3）InnoDB 的死锁信息可以记录到 error 日志中，方便分析。

4）支持延迟复制，可以让 slave 与 master 之间控制一个时间间隔，方便特殊情况下的数据恢复。

5）表分区功能增强。

6）行级复制功能加强，可以降低磁盘、内存、网络等资源的开销。

7）binlog 实现 crash-safe。

8）复制事件采用 CRC32 校验，增强 master 与 slave 复制数据的一致性。

3. MySQL 5.7 的特点

MySQL 5.7 版本在安全性、灵活性、易用性等多个方面均有提升。

1）安全性。在 MySQL 5.7 中，有不少安全性相关的改进。

- MySQL 数据库初始化完成以后，会产生一个 root@localhost 用户，但从 MySQL 5.7 开始，root 用户的密码不再是空，而是随机产生一个密码。
- MySQL 官方删除了 test 数据库，默认安装后，MySQL 中是没有 test 数据库的。用户可以创建 test 数据库，也可以对 test 数据库进行权限控制。
- MySQL 5.7 版本提供了更为简单的 SSL 安全访问配置，并且默认连接就采用 SSL 的加密方式。
- 可以为用户设置密码过期策略，一定时间以后，强制用户修改密码。

2）灵活性。MySQL 5.7 具备两个全新功能，即 JSON 和 generated column。

- 随着非结构化数据存储需求的持续增长，以及各种非结构化数据存储数据库的相继产生（如 MongoDB），于是出现了 JSON。
- generated column 字段是 MySQL 5.7 引入的新特性。所谓 generated column，就是数据库中这一列由其他列计算得到。

3）易用性。在 MySQL 5.7 中有很多易用性方面的改进，如客户端快捷键〈Ctrl+C〉的使用，提供了一个系统库（sys）来帮助 DBA 和开发人员使用数据库。

- 在 Linux 中经常使用〈Ctrl+C〉来终止一个命令的运行。在 MySQL 5.7 之前，如果用户输入错误的 SQL 语句，按下〈Ctrl+C〉虽然能够 "结束" SQL 语句的运行，但是，也会退出当前会话，MySQL 5.7 对这一违反直觉的地方进行了改进，不再退出会话。
- MySQL 5.7 可以解释（EXPLAIN）一个正在运行的 SQL，这对于 DBA 分析运行时间较长的语句将会非常有用。
- 在 MySQL 5.7 中，performance_schema 提供了更多监控信息，包括内存使用、MDL 锁、存储过程等。
- sys schema 是 MySQL 5.7.7 中引入的一个系统库，包含一系列视图、函数和存储过程，该项目专注于 MySQL 的易用性。

4）可用性。

- 在线设置复制的过滤规则，不再需要重启 MySQL，只需要停止 SQL thread，修改完成以后，启动 SQL thread 即可。
- 在线修改 buffer pool 的大小。MySQL 5.7 为了支持 online buffer pool resize，引入了 chunk 的概念，每个 chunk 默认是 128MB，当在线修改 buffer pool 的时候，以 chunk 为单位进行扩展或收缩。
- Online DDL。MySQL 5.7 支持在线重命名索引和修改 varchar 的大小，这两项操作在之前的版本中都需要重建索引或表。

- 在线开启 GTID。在之前的版本中，由于不支持在线开启 GTID，因此，如果用户希望将低版本的数据库升级到支持 GTID 的数据库版本，则需要先关闭数据库，再以 GTID 模式启动，导致升级特别麻烦。

5）性能。数据库的性能一直都是用户非常关心的问题，在 MySQL 5.7 中，性能相关的改进非常多，包括临时表相关的性能改进、只读事务的性能优化、连接建立速度的优化和复制性能的改进。

- 临时表相关的性能改进。MySQL 5.7 为了提高临时表相关的性能，对临时表相关的部分进行了大幅修改，包括引入新的临时表空间；对于临时表的 DDL，不持久化相关表定义；对于临时表的 DML，不写 redo、关闭 change buffer 等。临时表只在当前会话中可见，临时表的生命周期是当前连接过程（MySQL "死" 机或重启，则当前连接结束）。
- 只读事务的性能改进。MySQL 5.7 通过避免为只读事务分配事务 ID、不为只读事务分配回滚段、减少锁竞争等多种方式，优化了只读事务的开销，提高了数据库的整体性能。
- 连接建立速度的优化。在 MySQL 5.7 之前，变量的初始化操作（THD、VIO）都是在连接接收线程里面完成的，现在将这些工作下发给工作线程，以减少连接接收线程的工作量，提高连接的处理速度。
- 复制性能的改进。MySQL 的复制延迟一直是令人诟病的问题，MySQL 5.7 版本已经支持 "真正" 的并行复制功能。MySQL 5.7 实现并行复制的思想是：一个组提交的事务都是可以并行回放的，因为这些事务都已进入事务的 prepare 阶段，说明事务之间没有任何冲突（否则就不可能提交）。同时，兼容 MySQL 5.6 基于库的并行复制。

4. MySQL 8.0 的特点

在 MySQL 8.0 版本中，MySQL 进行了许多重要的创新和现有结构的优化。

1）重构数据字典，全部改成 InnoDB 引擎。数据字典的改进会带来诸多方便的特性、原子 DDL 等。同时 INFORMATION_SCHEMA 表已重构为基于数据字典的视图，在此之前，其为临时表。PERFORMANCE_SCHEMA 查询性能得以提升，其已内置多个索引。

2）加入新的索引，包括隐藏索引、函数索引、降序索引、hash 索引等。这些特性有效地提升了 SQL 语句的处理能力。

3）开始使用 UTF-8 编码。从 MySQL 8.0 开始，数据库的默认编码改为 utf8mb4，这个编码包含所有 Emoji 字符和偏僻字，同时增加了 utf8mb4_0900_*的字符集。

4）新增窗口函数（Window Function）。新增一个称为窗口函数的概念，它可以用来实现若干新的查询方式，如 rank()函数。

5）支持加密、添加角色和更强大的密码插件。支持 REDO 和 UNDO，表空间加密，添加角色功能和更强大的密码插件。

6）新的高可用架构 MySQL Group Replication。实现主从复制之外，分布式架构下数据的最终一致性解决方案。

7）SQL 优化。CATS 锁竞争算法、直方图、资源组、EXPLAIN ANALYZE、OPTIMIZER_hint 等功能对于 SQL 语句优化起到了很大的作用。

8）运维设置持久化。持久化、双重密码、快速加列（INSTANT）、克隆、备份锁对运维提供了很大的帮助。

9）其他。Online DDL 增强、JSON 增强、集合操作函数、慢日志附件信息、binlog 附件信息等助力 MySQL 在性能、安全性、可管理性等方面得到全面提升。

1.1.5 MySQL 开发过程版本标识

在 MySQL 开发过程中，会存在下列几种版本标识，它们各自都有一些特点。

1. alpha 版本

alpha 版本表明其中包含大量未被彻底测试的新代码。已知的缺陷在版本说明中描述。在大多数 alpha 版本中，会有新的命令和扩展，也可能有主要代码的更改等。

2. beta 版本

beta 版本意味着版本功能是完整的，并且所有的新代码均被测试，没有增加重要的新特征，应该没有已知的缺陷。当 alpha 版本至少一个月没有出现致命漏洞，并且没有计划增加导致已经实施的功能不稳定的新功能时，则可变为 beta 版本。在之后的 beta 版本、发布版或产品发布中，所有 API、外部可视化结构和 SQL 命令列均不再更改。

3. RC 版本

RC 版本是发布代表（发布代表即以前所称的 gamma 版）；是一个发布了一段时间的 beta 版本，看起来应该运行正常。它只增加了很小的修复。

4. 稳定版本

例如 mysql-8.0.34-linux-glibc2.12-x86_64.tar，意味着该版本已经在很多地方运行了一段时间，并且没有非平台特定的缺陷报告。这时就将其称为一个产品（稳定）或"通用"版本。

5. minimal 版本

相对于完整版，minimal 版本只是不包含测试文件（test file）和调试符号（debug symbol）。同时 minimal 版本在 MySQL crash 时，能够正常产生 core file 文件，该文件无法使用 minimal 版本进行 GDB 调试，需要使用对应完整版进行调试和问题排查。

1.1.6 MySQL 版本选择原则

MySQL 版本更新是比较快的，每个版本都有不同的特点和优势。对于在正式环境中使用的版本，稳定性是非常重要的，因为稳定性不好的版本容易出现崩溃和数据丢失等问题，会给业务带来很大的损失。因此，选择适合自己的 MySQL 版本非常重要。以下是 6 条 MySQL 版本选择原则。

1）选择开源的社区稳定版。
2）选择发布 3 个月以上的稳定版本。
3）要选择发布前后几个月没有大的 bug 的修复版本，而不要选择大量修复 bug 的集中版本，如在 MySQL 5.7.34 之前的版本 bug 列表里，存在很多内存泄漏。
4）要考虑开发人员开发程序使用的 MySQL 版本是否兼容旧的 MySQL 版本。
5）企业非核心业务可采用新版本数据库的稳定版本。
6）向 DBA 高手请教，或者在技术氛围好的群里和大家进行交流，使用高手推荐的好用的稳定版本产品。

1.2 MySQL 8.0 的安装部署

MySQL 8.0 于 2016 年首次发布，其在功能、性能等各方面都已进入稳定阶段。现阶段很多企业都在积极部署 MySQL 8.0。

MySQL 提供多种安装部署方式，如 RPM、TAR 编译包、TAR 源码包。本节将介绍 MySQL 8.0 的 TAR 编译包的安装部署方式以及需要注意的事项。

1.2.1 MySQL 的安装部署流程

MySQL 的安装部署流程如图 1-3 所示，包括 4 个核心阶段。

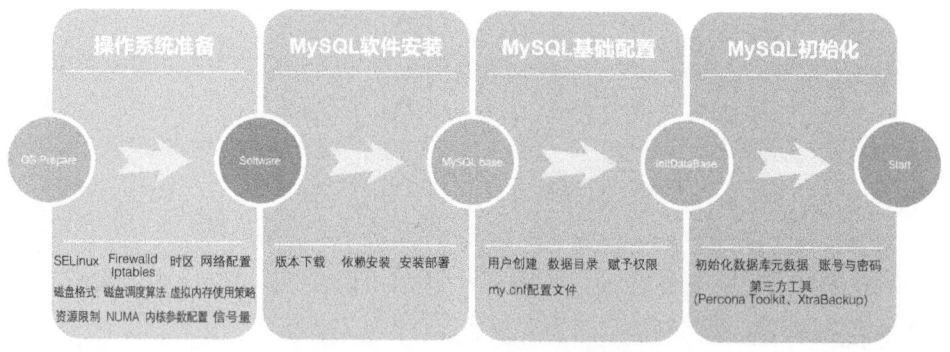

图 1-3 MySQL 的安装部署流程

（1）操作系统准备

为了有效利用操作系统资源，需要对安装环境做如下设置（基于 Linux 系统）。

- SELinux：关闭 SELinux 功能，通过 MySQL 本身进行安全控制。
- Firewalld 和 Iptables：防火墙肯定要进行相关设置或关闭。
- 时区：对于系统来说，时间是非常重要的指标。
- 网络配置：对于高配置机器，网卡的 MTU 值可以提高，建议将私网网卡的 MTU 值增加到 9000，同时启用私网交换机的 Jumbo Frame 属性。
- 磁盘格式：在平均文件较小、并发较小的 I/O 场景中，ext4 的表现和 XFS 差不多，前者略微胜出。当文件较大、并发较大时，XFS 的性能比 ext4 更好，更稳定。从实际使用上来说，一般数据库的文件系统推荐使用 XFS。但 XFS 的恢复比较麻烦，而 ext4 的 fsck 修复成功率较高，而且 ext4 的社区支持比较完备。
- 磁盘调度算法：默认使用的是 CFQ 算法。对于数据库专用服务器，如果为机械磁盘，则建议将磁盘调度算法调整为 deadline 模式；如果为固态硬盘，则建议调整为 noop 模式，以提升 I/O 吞吐量和降低 I/O 响应时间。
- 虚拟内存使用策略：vm.swappiness，以提高 MySQL 对内存的使用效率。
- 资源限制：limits.conf 的 nproc nofile。
- NUMA：非一致性内存访问。
- 内核参数配置：net.ipv4.tcp 相关的参数优化。
- 信号量：对应 InnoDB 引擎长时间信号等待。

（2）MySQL 软件安装

- 版本下载：一定要下载经过兼容性测试的官方版本。
- 依赖安装：MySQL 执行依赖包安装。
- 安装部署：建议使用 tar.gz 包，直接解压缩后即可使用。

(3) MySQL 基础配置
- 用户创建：出于安全考虑。
- 数据目录：便于管理和提升 IO 性能。
- 赋予权限：赋予特定用户相关权限。
- my.cnf 配置文件：根据硬件进行合理的配置。

(4) MySQL 初始化
- 初始化数据库元数据。
- 账号与密码：修改管理员密码，删除不安全账号。
- 第三方工具：Percona Toolkit、XtraBackup 等常用运维工具的安装和调试。

1.2.2 MySQL 8.0 的基本安装部署

1. 安装部署 MySQL 8.0

MySQL 软件下载地址：https://dev.mysql.com/downloads/mysql/。官方下载页面如图 1-4 所示。

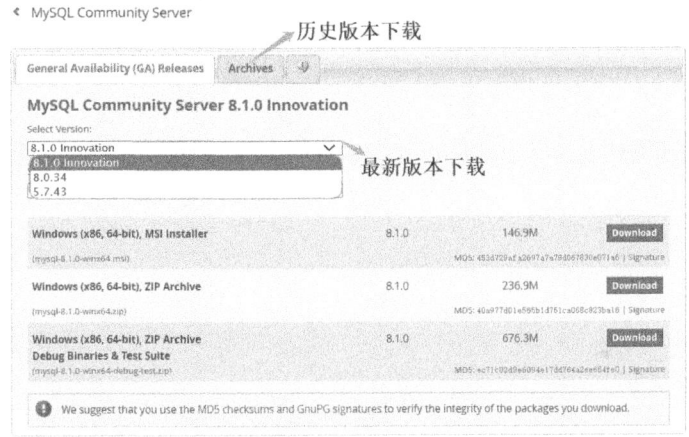

图 1-4 官方下载页面

图 1-5 所示是 MySQL 在 CentOS 8.2 上的安装示例，依次选择 8.0.34→ Linux-Generic→ Linux-

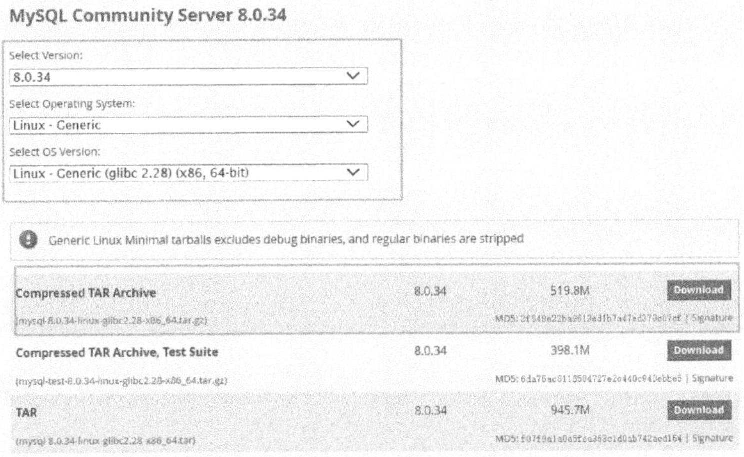

图 1-5 官方版本选择推荐方式

Generic(glibc 2.28)(x86,64-bit),下载 mysql-8.0.34-linux-glibc2.28-x86_64.tar.gz 版本。直接解压缩后就可以使用。

说明:glibc 版本可以通过 ldd --version 命令查看。

安装过程如下。

```
#1.解压缩 TAR 安装包:
shell $>tar -xvf mysql-8.0.34-linux-glibc2.28-x86_64.tar.gz
mysql-8.0.34-linux-glibc2.28-x86_64/bin/
mysql-8.0.34-linux-glibc2.28-x86_64/bin/myisam_ftdump
mysql-8.0.34-linux-glibc2.28-x86_64/bin/myisamchk
...
shell $>mv mysql-8.0.34-linux-glibc2.28-x86_64 mysql8.0.34

#2.数据目录创建
shell $> mkdir -p /opt/data8.0
shell $> chown -R mysql:mysql /opt/data8.0

#3.最简单的 my.cnf 配置文件
shell $> vim my.cnf
[mysqld_safe]
user = mysql
nice = 0

[client]
socket                          = /opt/data8.0/data/mysql.sock
port                            = 3380

[mysqld]
character_set_server            = utf8mb4              #字符集设置
collation_server                = utf8mb4_bin
lower_case_table_names          = 1                    #大小写敏感
default_authentication_plugin   = mysql_native_password #密码策略
transaction_isolation           = READ-COMMITTED       #隔离级别
log_error_verbosity =3                                 #日志记录模式
skip_ssl                                               #SSL 禁用
basedir                         = /opt/idc/mysql8.0    #软件安装目录
datadir                         = /opt/data8.0         #数据目录
socket                          = /opt/data8.0/mysql.sock
pid_file                        = /opt/data8.0/mysql.pid
innodb_buffer_pool_size         = 128M                 #配置最低内存
binlog_format                   = ROW                  #binlog 模式为 ROW
binlog_expire_logs_seconds      = 604800               #binlog 过期时间为 7 天

#4.初始化并启动
shell $> cd /opt/idc/mysql8.0.34
shell $> bin/mysqld --defaults-file=/etc/my.cnf --user=mysql --initialize-insecure

shell $> mysqld_safe -defaults-file=/etc/my.cnf  --user=mysql  &
```

```
#5.登录
shell $> /opt/idc/mysql8.0.34/bin/mysql -uroot -p -S /opt/data8.0/mysql.sock

#按〈Enter〉键就可以直接登录并更改密码
mysql> ALTER USER 'root'@'localhost' IDENTIFIED BY '123456';
mysql> FLUSH PRIVILEGES;
```

到这里，MySQL 8.0 的基本安装部署完成。

2. MySQL 8.0 安装部署中的关键点

MySQL 8.0 安装部署的 4 个关键点说明如下。

（1）SSL 认证设置

安全套接字层（Secure Socket Layer，SSL）及其继任者传输层安全（Transport Layer Security，TLS）都是为网络通信提供安全及数据完整性的协议。SSL 认证设置很重要，因为可通过 SSL 实现数据保护、身份验证，防范恶意行为，从而建立安全可信赖的网络环境。

SSL 协议提供的主要功能如下。

- 数据传输的机密性：利用对称密钥算法对传输的数据进行加密。
- 身份验证机制：基于证书利用数字签名方法对服务器和客户端进行身份验证，其中客户端的身份验证是可选的。
- 消息完整性验证：消息传输过程中使用 MAC 算法来检验消息的完整性。

1）设置 SSL 认证。

在 MySQL 早期版本中，默认关闭 SSL 配置。在 MySQL 5.7 初始化时，只有执行 mysql_ssl_rsa_setup 才会生成 pem 文件（用于验证服务器的身份和建立安全连接的关键文件），但到了后期（5.7 版本以后），都默认生成 SSL 相关文件。使用如下命令行查看在 MySQL 数据目录下生成的 PEM 文件，共 8 个 pem 文件，都是 SSL 密钥相关文件。

```
shell $> ll *.pem
-rw-------  1 mysql mysql 1679 Dec  9  2022 ca-key.pem
-rw-r--r--  1 mysql mysql 1107 Dec  9  2022 ca.pem
-rw-r--r--  1 mysql mysql 1107 Dec  9  2022 client-cert.pem
-rw-------  1 mysql mysql 1679 Dec  9  2022 client-key.pem
-rw-------  1 mysql mysql 1675 Dec  9  2022 private_key.pem
-rw-r--r--  1 mysql mysql  451 Dec  9  2022 public_key.pem
-rw-r--r--  1 mysql mysql 1107 Dec  9  2022 server-cert.pem
-rw-------  1 mysql mysql 1679 Dec  9  2022 server-key.pem
```

2）判断 MySQL 服务是否开启 SSL 认证。

在 MySQL 里，SSL 相关配置也属于配置参数。如图 1-6 所示，可以通过 variables 参数查看是否开启 SSL 认证。这里主要查看 have_openssl 和 have_ssl 参数。

说明：have_openssl 是 MySQL 连接接口 TLS 连接函数；have_ssl 表示服务器是否支持加密连接（从 MySQL 8.0.26 开始，这两个参数被弃用，未来也会删除）。

在 MySQL 8.0 版本中，通过如下命令行进行 SSL 认证的查询：

```
mysql> SHOW GLOBAL STATUS LIKE 'ssl_server%';
+-----------------------------------+----------------------------------+
|Variable_name                      |Value                             |
```

```
+-----------------------+------------------------+
|Ssl_server_not_after   |Dec  6 05:21:22 2032 GMT|
|Ssl_server_not_before  |Dec  9 05:21:22 2022 GMT|
+-----------------------+------------------------+
```

 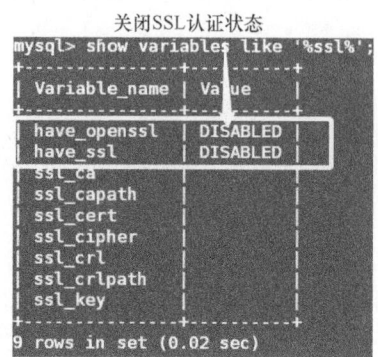

图 1-6 SSL 认证是否开启对比图

说明：ssl_server_not_after，用于指定 MySQL 服务器 SSL 会话证书的有效期限（截止日期）；ssl_server_not_before，表示处于激活状态的 SSL 服务器证书最早可用的日期和时间。

3）通过命令行查看是否采取 SSL 方式。

图 1-7a 所示为本地连接，SSL 是"Not in use"，表示没有使用 SSL 连接。图 1-7b 所示为远程连接，SSL 是"Cipher in use is ECDHE-RSA-AES128-GCM-SHA256"，表示正在使用 SSL 连接。

```
mysql> STATUS;
--------------
mysql Ver 8.0.39 for Linux on x86_64 (MySQL Community Server - GPL)
Connection id:16
Current database:
Current user:root@ localhost
SSL:Not in use
...
```

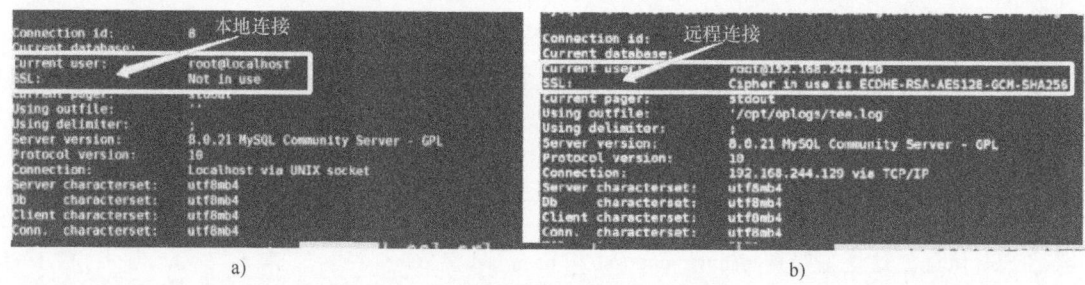

图 1-7 SSL 连接不同状态信息

4）SSL 开启后的常见问题。

如果在开启 SSL 之后，发现无法连接 MySQL 数据库，则会有如下错误提示。在碰到类似问题的时候，首先需要确认是否存在网络丢包问题，然后检查密码是否正确，最后确认 SSL 认证是否开启。

```
[warning] Got error on MySQL select ping: 2013 (Lost connection to MySQL server during query)
ERROR 1045 (28000): Access denied for user 'tester'@'%' (using password: YES)
```

5）SSL 认证对性能的影响。

从诸多测试数据中可以发现，开启 SSL 认证后，数据库的 QPS（每秒查询率）降低了 20% 左右，还是比较影响性能的。从 SSL 实现方式来看，建立连接时需要进行握手、加密、解密等操作。

6）禁用 SSL 认证的方法。

在有些环境下，因为一些应用没有配置 SSL，若单独开启 MySQL 的 SSL 配置，则会导致无法连接。同时，开启 SSL 认证会产生较大的性能影响，所以需要关闭 SSL 认证，如下面在配置文件中进行的 skip_ssl 选项设置。

```
shell $> vim /etc/my.cnf
[mysqld]
skip_ssl
```

7）对于 SSL 在使用方面的建议如下。
- 对于非常敏感、核心的数据，应采用 SSL 方式保证数据安全性。
- 不建议在没有服务器身份验证的情况下建立 SSL 连接。
- 在使用 SSL 的情况下，因为包含加密、解密，所以性能有所下降。此时，需要评估是否能够接受。
- 对于高性能要求的应用，性能和可用性才是关键，建议关闭 SSL 认证。
- 在复制方面，局域网复制可以非 SSL 方式连接，即明文复制，但公网复制建议采用 SSL 连接。

（2）身份认证插件的选择

MySQL 8.0 中的身份认证相较 MySQL 5.7 有不少变化。图 1-8 所示为 MySQL 不同版本密码策略对比。在 MySQL 5.7 中，默认的身份验证插件还是 mysql_native_password。而在 MySQL 8.0 中，caching_sha2_password 是默认的身份验证插件，而不是 mysql_native_password。

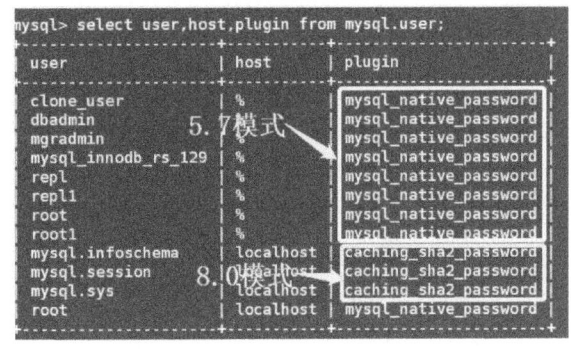

图 1-8　MySQL 不同版本密码策略对比

1）MySQL 目前的稳定版本里提供了以下几类身份验证插件。
- mysql_native_password。执行本地身份验证的插件。在 MySQL 中引入可插入身份验证之前使用的基于密码哈希方法的身份验证。mysql_native_password 插件基于本机密码哈希方法实现身份验证。mysql_old_password 插件基于较旧的（4.1 之前的）密码哈希方法实现本机身份验证（现已弃用）。身份插件可以在服务器启动时在配置文件中设置 --default-authentication-plugin 选项。
- sha256_password。使用 SHA-256 密码哈希执行身份验证的插件。与本地身份验证相比，该插件提供更强大的加密。"sha256" 是指插件用于加密的 256 位摘要长度。"sha2" 更笼统地指 SHA-2 类加密算法，其中 256 位加密是其中的一种实例。
- caching_sha2_password。它是 MySQL 8.0 中的默认身份验证插件，并且提供了 sha256_password 身份验证插件的功能的超集。

sha256_password，实现基本的 SHA-256 身份验证。caching_sha2_password，实现 SHA-256 身份验证（如 sha256_password），但在服务器端使用缓存以获得更好的性能，并具有更广泛的适用范围的附加特性。

说明： 在服务器端，内存中的缓存使之前连接的用户再次连接时能够更快地重新验证身份。无论 MySQL 所连接的 SSL 库是什么，都可以使用基于 RSA 的密码交换，为使用 UNIX 套接字文件和共享内存协议的客户端连接提供了支持。

2）MySQL 连接客户端要求。

如果碰到如下错误，则说明客户端不兼容 MySQL 数据库的身份认证插件。

```
Error: ER_NOT_SUPPORTED_AUTH_MODE: Client does not support authentication
protocol requested by server; consider upgrading MySQL client
```

caching_sha2_password 支持的客户端说明如下。

- caching_sha2_password 支持以下客户端：

➤ libmysqlclientMySQL 8.0（8.0.4 或更高版本）中的客户端库。标准 MySQL 客户端（如 mysql 和 mysqladmin）是基于 libmysqlclient 的，因此它们是兼容的。

➤ ibmysqlclientMySQL 5.7（5.7.23 或更高版本）中的客户端库。标准 MySQL 客户端（如 mysql 和 mysqladmin）是基于 libmysqlclient 的，因此它们也是兼容的。

- MySQL Connector/C++ 1.1.11 或更高版本，或者 MySQL 8.0.7 或更高版本。
- MySQL Connector/J 8.0.9 或更高版本。
- MySQL Connector/NET 8.0.10 或更高版本（通过经典的 MySQL 协议）。
- MySQL Connector/Node.js 8.0.9 或更高版本。
- PHP：X DevAPI PHP 扩展驱动（mysql_xdevapi）支持 caching_sha2_password。
- PHP：PDO_MYSQL 和 ext/mysqli 扩展驱动不支持 caching_sha2_password。
- PHP 版本支持：PHP 7.1.16 之前的版本，以及 PHP 7.2 中 PHP 7.2.4 之前的版本，即使没有使用 caching_sha2_password，也无法连接到 MySQL 服务。

通过设置配置信息和命令行方式密码更改密码插件如下所示。

```
shell $>vim /etc/my.cnf
#配置文件修改完之后,MySQL 服务重新启动
[mysqld]
default_authentication_plugin = mysql_native_password

# MySQL 命令：
mysql> ALTER USER 'root'@'%' IDENTIFIED WITH mysql_native_password BY '******';
mysql> ALTER USER 'root'@'%' IDENTIFIED WITH caching_sha2_password BY '******';
```

说明： 特别是在原系统从 5.7 迁移到 8.0 之后，会发现原先的应用连接不上。

（3）大小写敏感设置

在 MySQL 里，对于大小写敏感对应参数 lower_case_table_names 的说明如下。

- 大小写敏感。表名存储是区分大小写的。
- 不区分大小写。表名存储在磁盘时是小写的，但是比较的时候是不区分大小写的。
- 表名存储在磁盘时是区分大小写的，但是比较的时候是不区分大小写的。

说明： 大小写不可以动态修改，改后必须重启数据库。

如图 1-9 所示，对于不同的操作系统，大小写敏感默认值不同。

> How table and database names are stored on disk and used in MySQL is affected by the `lower_case_table_names` system variable. `lower_case_table_names` can take the values shown in the following table. This variable does *not* affect case sensitivity of trigger identifiers. On Unix, the default value of `lower_case_table_names` is 0. On Windows, the default value is 1. On macOS, the default value is 2.

图 1-9 大小写敏感默认值

lower_case_table_names 的默认值在 UNIX 上是 0，在 Windows 上，是 1，在 macOS 上，是 2。例如，在 MySQL 8.0 安装时，若参数没有更改（lower_case_table_names = 0），直接进行了初始化，现在想要转换为对大小写不敏感，方法如下。

1）将数据库数据和表结构通过 mysqldump 导出。

2）删除原有数据目录，在 my.cnf 中更改 lower_case_table_names = 1，并重启数据库，同时进行初始化。

3）将导出的表结构和数据导入数据库中。

说明： 在 MySQL 5.7 版本中是可以直接进行相应更改的，但后续运行过程中会出现莫名其妙的问题，建议按照上述方法操作。

如图 1-10 所示，在 MySQL 8.0 版本中是无法直接进行相应更改的，会出现错误提示信息。图 1-11 所示为 MySQL 8.0 官方说明。

```
[System] [MY-013576] [InnoDB] InnoDB initialization has started.
[System] [MY-013577] [InnoDB] InnoDB initialization has ended.
[ERROR] [MY-011087] [Server] Different lower_case_table_names settings for server ('0') and data dictionary ('1').
[ERROR] [MY-010020] [Server] Data Dictionary initialization failed.
```

图 1-10 大小写敏感错误提示

> It is prohibited to start the server with a `lower_case_table_names` setting that is different from the setting used when the server was initialized. The restriction is necessary because collations used by various data dictionary table fields are determined by the setting defined when the server is initialized, and restarting the server with a different setting would introduce inconsistencies with respect to how identifiers are ordered and compared.

图 1-11 MySQL 8.0 官方说明

禁止使用与服务器初始化时使用的设置不同的 lower_case_table_names 来启动服务器。这个限制是必要的，因为各种数据字典（MySQL 8.0 存储在单独的表空间 mysql.ibd 中）表字段使用的排序由服务器初始化时定义的设置决定，而使用不同的设置重新启动服务器将导致标识符的排序和比较方式不一致。

所以，在安装 MySQL 时，除了一些特殊情况以外，不能使用默认配置。可如下设置配置参数。

```
shell $>vim /etc/my.cnf
[mysqld]
lower_case_table_names = 1
```

（4）error 日志记录的级别

MySQL 的 error 日志文件包含了数据库服务器运行过程中的各种错误信息和警告。这些信息可以用于排查问题、查看服务运行状态。参数 log_error_verbosity 控制日志输出内容，有错误、警告和信息 3 种类型。

- 1：错误。
- 2：错误、警告。

- 3：错误、警告、信息。

在 MySQL 8.0 中，log_error_verbosity 的默认值是 2。例如，在 log_error_verbosity 值设置为 2 时，输入错误密码也不会记录到日志里。下面是数据库关闭的记录信息，当值设置为 2 时，无具体模块信息。

```
# 设置 log_error_verbosity=2
[System] [MY-013172] [Server] Received SHUTDOWN from user root.Shutting down mysqld (Version: 8.0.34).
[System] [MY-010910] [Server] /opt/idc/mysql8.0/bin/mysqld: Shutdown complete (mysqld 8.0.34)  MySQL Community Server - GPL.
mysqld_safe mysqld from pid file /opt/data8.0/data/mysql.pid ended

# 设置 log_error_verbosity=3
[System] [MY-013172] [Server] Received SHUTDOWN from user root.Shutting down mysqld (Version: 8.0.34).
[Note] [MY-010067] [Server] Giving 0 client threads a chance to die gracefully
[Note] [MY-010117] [Server] Shutting down replica threads
[Note] [MY-010118] [Server] Forcefully disconnecting 0 remaining clients
[Note] [MY-010043] [Server] Event Scheduler: Purging the queue.0 events
[Note] [MY-012330] [InnoDB] FTS optimize thread exiting.
[Note] [MY-010120] [Server]Binlog end
...
[System] [MY-010910] [Server] /opt/idc/mysql8.0/bin/mysqld: Shutdown complete (mysqld 8.0.34)  MySQL Community Server - GPL.
```

强烈建议在 MySQL 配置文件中将 log_error_verbosity 设置成 3，如下所示。

```
shell $>vim /etc/my.cnf
[mysqld]
log_error_verbosity=3
```

（5）总结

虽然 MySQL 8.0 的安装与之前版本基本一致，但需要特别注意 SSL、密码策略、大小写敏感和错误日志输出信息。

1.2.3　使用 Shell 脚本批量部署 MySQL 环境

MySQL 批量（自动化）部署是一项非常重要的任务，可以避免烦琐的命令行操作，降低使用门槛，大大提高工作效率和减少错误。

1. 批量部署包含哪些内容

在一些认知度较高的开源工具软件中，可以实现 MySQL 的一键批量安装部署。然而，这通常需要用户先安装相应的软件，并确保网络访问的畅通无阻，同时还可能需要进行额外的配置。除此之外，用户还需要学习如何使用这些软件。为了简化这一过程，可以采用基础且易于上手的 Linux Shell 脚本来协助用户快速实现 MySQL 的批量部署。

下面是基于 MySQL 8.0.34 编写的 MySQL 批量部署脚本。读者可以按照 MySQL 安装部署的 4 个关键步骤，参考对应的注释内容加以了解。

```
shell $>vim MySQL_AutoSetup.sh
#!/bin/bash
```

```
#####MySQL 8.0.34 数据库自动安装脚本
# Version:      1.0
# Author:       kevinCUI
# Date:         2023-09-01
#####
# MySQL 安装包的绝对路径，去掉.tar.gz
tarGzPath=/opt/idc/
tarGzFile=mysql-8.0.34-linux-glibc2.28-x86_64
# MySQL 安装路径
installPath=/home/mysql/

# my.cnf 配置文件
mysqlcnf=/home/mysql/my.cnf

# MySQL serverid 需要设置唯一的 id,比如 IP 地址+3 位数字(如 192.168.10.1+001)
mysqlServerid=192168101001

# MySQL 密码(不可擅自修改)
defaultPwd=123456

# MySQL 端口
mysqlPort=3306

# MySQL 数据目录
data_default=${installPath}${mysqlPort}
data_datadir=${data_default}/data
data_binlog=${data_default}/binlog
data_dbdata=${data_default}/dbdata
data_logs=${data_default}/logs
data_tmp=${data_default}/tmp
data_undo=${data_default}/undo

# 校验是否为 root 用户
CheckRoot()
{
if [ $(id -u) != "0" ]; then
    echo "Error: You must be root to run this script, please use root to install"
    exit 1
fi
clear
}

# 优化文件最大打开数
DependFile()
{

if [ $( cat /etc/security/limits.conf  |grep "mysql" |wc -l )  -lt 1 ];then
cat >>/etc/security/limits.conf << EOF
```

```
*         soft    nproc   65536
*         hard    nproc   65536
*         soft    nofile  65536
*         hard    nofile  65536
mysql     soft    nproc   65536
mysql     hard    nproc   65536
mysql     soft    nofile  65536
mysql     hard    nofile  65536
EOF

fi

if [ -e /etc/security/limits.d/20-nproc.conf ];then
if [ $( cat /etc/security/limits.d/20-nproc.conf  |grep "mysql" |wc -l )  -lt 1 ] ;then
cat >>/etc/security/limits.d/20-nproc.conf<<EOF
mysql     soft    nproc   unlimited
EOF

fi
fi

if [ -e /etc/security/limits.d/90-nproc.conf ];then
if [ $( cat /etc/security/limits.d/90-nproc.conf  |grep "mysql" |wc -l )  -lt 1 ] ;then
cat >>/etc/security/limits.d/90-nproc.conf<<EOF
mysql     soft    nproc   unlimited
EOF

fi
fi

if [ -e /etc/sysctl.conf ];then
fs_file=$( cat /proc/sys/fs/file-max)
if [ ${fs_file} -lt 65535 ] ;then
sed -i "s/${fs_file}/65535/g" /etc/sysctl.conf
/usr/sbin/sysctl -p

fi
fi
echo -e "\e[31m#1.配置基础资源 \e[0m"
}

#复制tar.gz包
DecompressionTarGz()
{
if [ !-e ${tarGzPath}${tarGzFile}.tar.gz  ];then
    echo -e "\e[31m ${tarGzPath} ${tarGzFile}.tar.gz  不存在!请检查后重新执行脚本 \e[0m"
    exit 1
fi
```

```bash
# 解压缩并重命名到安装目录
if [ !-d ${installPath} ${tarGzFile} ] ;then
    mkdir -p ${installPath}
    tar -xvf ${tarGzPath} ${tarGzFile}.tar.gz -C ${installPath} &> /dev/null
fi

echo -e "\e[31m#2.软件已解压缩 \e[0m"
}

# 添加组合角色
AddMysqlUser()
{
if [ ! $(id -u "mysql") ]; then
    echo "mysql user is not exists for to created"
    /usr/sbin/groupadd mysql
    /usr/sbin/useradd -g mysql -r -s /sbin/nologin -M mysql
fi

echo -e "\e[31m#3.MySQL 启动用户已准备完成 \e[0m"
}

# 创建 MySQL 数据目录
createMysqlFolder()
{
if [ -d ${data_default} ] ;then
    if [ $(du -s ${data_default} | awk 'NR==1{print $1}') -gt 0 ] ;then
        mv ${data_default} ${data_default}"`date +%Y%m%d%H%M`"
    fi
fi

mkdir -p ${data_datadir}
mkdir -p ${data_binlog}
mkdir -p ${data_dbdata}
mkdir -p ${data_logs}
mkdir -p ${data_tmp}
mkdir -p ${data_undo}

# 赋予权限
chown -R mysql:mysql ${data_default}
chmod 700 ${data_tmp}

echo -e "\e[31m#4.mysql 数据目录权限已准备完成 \e[0m"
}

# 创建 my.cnf
MakeMyCnf()
{

if [ -e ${mysqlcnf} ] ;then
```

```
    #mv    ${mysqlcnf}    ${mysqlcnf}"`date +%Y%m%d%H%M`"
    rm ${mysqlcnf}
fi

cat > ${mysqlcnf}<<EOF
[mysqld_safe]
user = mysql
nice = 0

[client]
socket                          = ${data_datadir}/mysql.sock
port                            = ${mysqlPort}

[mysqld]
############# GENERAL #############
skip_ssl
skip-name-resolve
character_set_server            = utf8mb4
collation_server                = utf8mb4_unicode_ci
lower_case_table_names          = 1
port                            = ${mysqlPort}
read_only                       = OFF
transaction_isolation           = READ-COMMITTED
open_files_limit                = 65535
max_connections                 = 2000
expire_logs_days                = 7
default-time_zone               = '+8:00'

####### CACHES AND LIMITS #########
interactive_timeout             = 600
lock_wait_timeout               = 300
max_connect_errors              = 1000000
table_definition_cache          = 2000
table_open_cache                = 2000
table_open_cache_instances      = 8
tmp_table_size                  = 32M
max_heap_table_size             = 64M
sort_buffer_size                = 1M
join_buffer_size                = 1M
sort_buffer_size                = 1M
read_rnd_buffer_size            = 2M

innodb_io_capacity              = 1000
innodb_io_capacity_max          = 2000

max_allowed_packet              = 1024M
slave_max_allowed_packet        = 1024M
slave_pending_jobs_size_max     = 1024M
```

```
############# SAFETY #############
local_infile                    = OFF
skip_name_resolve               = ON
############# LOGGING #############
general_log                     = 0
log_queries_not_using_indexes   = ON
log_slow_admin_statements       = ON
log_warnings                    = 2
long_query_time                 = 1    #1秒慢日志
slow_query_log                  = ON
############# REPLICATION #############
server_id                       = ${mysqlServerid}
binlog_checksum                 = CRC32
binlog_format                   = ROW
binlog_rows_query_log_events    = ON

enforce_gtid_consistency        = ON
gtid_mode                       = ON
log_slave_updates               = ON
master_info_repository          = TABLE
master_verify_checksum          = ON

max_binlog_size                 = 512M
max_binlog_cache_size           = 1024M
binlog_cache_size               = 8M

relay_log_info_repository       = TABLE
skip_slave_start                = ON
slave_net_timeout               = 10
slave_sql_verify_checksum       = ON

sync_binlog                     = 1
sync_master_info                = 1
sync_relay_log                  = 1
sync_relay_log_info             = 1

############# PATH #############
basedir                         = ${installPath}${tarGzFile}
datadir                         = ${data_datadir}
tmpdir                          = ${data_tmp}
socket                          = ${data_datadir}/mysql.sock
pid_file                        = ${data_datadir}/mysql.pid
innodb_data_home_dir            = ${data_dbdata}
log_error                       = ${data_logs}/error.log
general_log_file                = ${data_logs}/general.log
slow_query_log_file             = ${data_logs}/slow.log
log_bin                         = ${data_binlog}/mysql-bin
log_bin_index                   = ${data_binlog}/mysql-bin.index
relay_log                       = ${data_binlog}/relay-log
```

```
relay_log_index                  = ${data_binlog}/relay-log.index

############# INNODB #############
innodb_file_format               = barracuda
innodb_flush_method              = O_DIRECT

innodb_buffer_pool_size          = 1024M
innodb_buffer_pool_instances     = 4
innodb_thread_concurrency        = 0
innodb_flush_log_at_trx_commit   = 1
innodb_support_xa                = ON
innodb_strict_mode               = ON
innodb_data_file_path            = ibdata1:32M;ibdata2:16M:autoextend
innodb_temp_data_file_path       = ibtmp1:1G:autoextend:max:30G
innodb_lock_wait_timeout         = 600
innodb_log_buffer_size           = 8M
innodb_open_files                = 65535
innodb_page_cleaners             = 1
innodb_lru_scan_depth            = 256
innodb_purge_threads             = 4
innodb_read_io_threads           = 4
innodb_write_io_threads          = 4
innodb_print_all_deadlocks       = 1

[mysql]
############# CLIENT #############
max_allowed_packet               = 16M
socket                           = ${data_datadir}/mysql.sock
no-auto-rehash

[mysqldump]
max_allowed_packet               = 16M

EOF
echo -e "\e[31m #5.mysqlcnf 配置完成,【需要按照实际情况更改】\e[0m"
}

# 初始化数据库
InitDataBase()
{
# cd ${installPath} ${tarGzFile}
${installPath} ${tarGzFile}/bin/mysqld --defaults-file=${mysqlcnf}
--basedir=${installPath} ${tarGzFile} --datadir=${data_datadir} --user=mysql
--initialize

${installPath} ${tarGzFile}/bin/mysqld_safe --defaults-file=${mysqlcnf}  --user=mysql  &

echo -e "\e[31m #6.初始化数据库完成并启动服务.\e[0m"
}
```

```bash
#重置密码
ResetPwd()
{
sleep 10s

# 从日志中获取 MySQL 初始密码
pwd=`grep "A temporary password is generated for root@localhost: " ${data_logs}/error.log`
pwd=${pwd##*root@localhost:}

# 防止初始密码中因有特殊字符而出错,所以要拼接单引号
pwd=${pwd// /}
echo ${pwd}
${installPath}${tarGzFile}/bin/mysql -uroot -p ${pwd} -S
${data_datadir}/mysql.sock --connect-expired-password -e "alter user 'root'@'localhost' identified by '${defaultPwd}';"

echo -e "\e[31m #7.已重置数据库密码。登录方式如下: \e[0m"
echo -e "\e[31m ${installPath}${tarGzFile}/bin/mysql -uroot -p -S
${data_datadir}/mysql.sock \e[0m"
}

#重置密码
main()
{
#1.校验是否为 root 用户
CheckRoot

#2.优化文件最大打开数
DependFile

#3.复制 tar.gz 包
DecompressionTarGz

#4.添加组合角色
AddMysqlUser

#5.创建 MySQL 数据目录
createMysqlFolder

#6.创建 my.cnf
MakeMyCnf

#7.初始化数据库
InitDataBase

#8.重置密码
ResetPwd
}

main
```

说明：按照实际需要自行添加或删除脚本内容（如创建数据库、创建用户、导入数据等）。一些内存相关参数需要按照资源情况，自行修改模板内容。

2. 总结

MySQL 一键批量部署 Shell 脚本，可以说自动化地实现了一些常见的部署任务，从而减少了管理员的手动操作，提高了效率。MySQL 的安装部署也需要一些学问，简单粗暴的安装往往会导致后续使用时出现一些问题。考虑得越周全，未来走得越远。

1.2.4 定制化 RPM 包安装部署

在部署 MySQL 时，经常会用到官方提供的 RPM 包进行简易安装。该方式非常快，也非常有效。但该方式默认会安装在根目录下，参数也需要配置。虽然这些看起来无关紧要，但一旦配置不合理，就会导致后续问题不断。为了避免这样的问题，可以按照环境定制化 MySQL 的 RPM 安装包。

Linux 安装软件包分为两大类：
- 二进制类包，包括 RPM 安装包（一般分为 i386 和 x86 等几种）。
- 源码类包，源码包和开发包应该归为此类（.src、.rpm）。

RPM（RedHat Package Manager）有五种基本的操作功能：安装、卸载、升级、查询和验证。RPM 包安装方便，文件也比较小，比较受欢迎。通过官网提供的 rpmbuild 工具，可自行打包 MySQL 软件。下面是 RPM 包的制作过程。

1. RPM 包制作过程

（1）安装 RPM 包

下面先安装 RPM 及其所需的依赖包。因为 RPM 包制作会通过源码编译，所以需要安装比较多的依赖包。

```
# 安装 RPM 的依赖包
shell $> yum install make gcc rpm-build rpmdevtools -y

# MySQL 源码编译依赖包
shell $> yum install make cmake gcc gcc-c++ bison libaio ncurses-devel perl perl-DBI perl-DBD-MySQL perl-Time-HiRes readline-devel numactl zlib-devel curldevel
```

（2）下载源码包

1）下载 MySQL 源码包。下载地址为 https://downloads.mysql.com/archives/community/，选择源码包的选项，如图 1-12 所示。

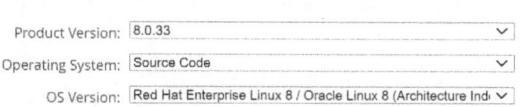

- Product Version：版本。
- Operating System：Source Code。
- OS Version：选择实际环境。

图 1-12 官方源码包选择

2）选择 boost 的 tar.gz 包。在 MySQL 5.6 之后，源码编译都需要依赖 boost_1_59_0.tar.gz 包。图 1-13 所示为 boost 软件下载页面，选择 tar.gz 包（boost 包下载地址：https://www.boost.org/users/history/version_1_59_0.html）。

3）初始化 RPM 所需的目录和源码包存放位置。文件目录

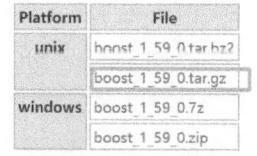

图 1-13 选择 boost 的 tar.gz 包

创建和源码包存放位置如下。

```
shell $> mkdir -p /opt/rpmbuild/{BUILD,RPMS,SOURCES,SPECS,SRPMS}
shell $> ll /opt/rpmbuild/
drwxr-xr-x 2 root root 6 Aug 31 16:00 BUILD
drwxr-xr-x 2 root root 6 Aug 31 16:00 RPMS
drwxr-xr-x 2 root root 6 Aug 31 16:00 SOURCES
drwxr-xr-x 2 root root 6 Aug 31 16:00 SPECS
drwxr-xr-x 2 root root 6 Aug 31 16:00 SRPMS
shell $>ll   /opt/rpmbuild/SOURCES
total 420312
-rw-r--r-- 1 root root      5679 Aug 31 16:06 my.cnf
-rw-r--r-- 1 root root 430390111 Mar 17 00:53 mysql-8.0.33.tar.gz
shell $>ll /opt/rpmbuild/BUILD
total 420312
-rw-r--r-- 1 root root 1654784 Aug 54056899 16:09 boost_1_59_0.tar.gz
```

说明：把下载的源码包放到/opt/rpmbuild/SOURCES 中，同时把 my.cnf 文件也复制到该目录下。需要把 boost_1_59_0.tar.gz 包直接解压缩到 BUILD 目录下，不然检测不到。

（3）配置文件

RPM 编译需要配置 spec 文件。如下编辑 spec 文件内容。

```
shell $>vim mysql8.0.34.spec
Name:         mysql
Version:      8.0.34
Release:      1%{? dist}
License:      GPL
URL:          http://downloads.mysql.com/archives/get/file/mysql-8.0.34.tar.gz
Group:        applications/database
Source:       %{name}-%{version}.tar.gz
BuildRoot:    %(mktemp -ud %{_tmppath}/%{name}-%{version}-%{release}-XXXXXX)
BuildRequires: cmake
Packager:     enmo@enmotech.com
Autoreq:      no
# Source: %{name}-%{version}.tar.gz
prefix: /opt/rpm/mysql-%{version}
Summary: MySQL 8.0.34

%description
The MySQL(TM) software delivers a very fast, multi-threaded, multi-user,
and robust SQL (Structured Query Language) database server.MySQL Server
is intended for mission-critical, heavy-load production systems as well
as for embedding into mass-deployed software.


%define MYSQL_USER mysql
%define MYSQL_GROUP mysql


%prep
%setup -n mysql-%{version}
```

```
%build

#CFLAGS="-O3  -g  -fno-exceptions  -static-libgcc  -fno-omit-frame-pointer
-fno-strict-aliasing"
#CXX=g++
#CXXFLAGS="-O3 -g -fno-exceptions -fno-rtti -static-libgcc -fno-omit-frame-pointer
-fno-strict-aliasing"
#export CFLAGS CXX CXXFLAGS

cmake \
-DCMAKE_INSTALL_PREFIX=%{prefix} \
-DMYSQL_UNIX_ADDR=/data/mysql/mysql.sock \
-DMYSQL_DATADIR=/data/mysql \
-DMYSQL_TCP_PORT=3310 \
-DSYSCONFDIR=/etc \
-DDEFAULT_CHARSET=utf8 \
-DDEFAULT_COLLATION=utf8_general_ci \
-DEXTRA_CHARSETS=all \
-DWITH_ARCHIVE_STORAGE_ENGINE=1 \
-DWITH_BLACKHOLE_STORAGE_ENGINE=1 \
-DWITH_INNOBASE_STORAGE_ENGINE=1 \
-DWITH_FEDERATED_STORAGE_ENGINE=1 \
-DWITH_PARTITION_STORAGE_ENGINE=1 \
-DWITH_PERFSCHEMA_STORAGE_ENGINE=1 \
-DWITH_DEBUG=0 \
-DENABLED_LOCAL_INFILE=1 \
-DWITH_BOOST=../boost_1_59_0  \
-Wno-dev

make -j `cat /proc/cpuinfo | grep processor | wc -l`

%install
rm -rf %{buildroot}
make install DESTDIR=%{buildroot}
cp %{_sourcedir}/my.cnf $RPM_BUILD_ROOT%{prefix}/

%pre
groupadd mysql
useradd -g mysql -s /bin/nologin -M mysql >/dev/null 2>&1

mkdir -p /data
mkdir -p /data/mysql
mkdir -p /data/mysqltmp
mkdir -p /data/dbdata

chown -R mysql:mysql /data
chmod 700 /data/mysqltmp

%post
```

```
/bin/cp %{prefix}/support-files/mysql.server /etc/init.d/mysql
/bin/cp %{prefix}/my.cnf %{_sysconfdir}/my.cnf
chkconfig mysql on
%{prefix}/bin/mysqld              --initialize-insecure              --basedir=%{prefix}
--datadir=/data/mysql --user=mysql
service mysql start
chown -R mysql:mysql /data/mysql
echo "export PATH=.:\$PATH:%{prefix}/bin;">> ~/.bash_profile
source ~/.bash_profile

%preun
service mysql stop
chkconfig --del mysql
userdel -r mysql >/dev/null 2>&1
rm -rf %{prefix} >/dev/null 2>&1
rm -rf /data/mysql >/dev/null 2>&1
rm -rf /etc/init.d/mysql >/dev/null 2>&1

%files
%defattr(-, %{MYSQL_USER}, %{MYSQL_GROUP})
%attr(755, %{MYSQL_USER}, %{MYSQL_GROUP}) %{prefix}/*


%changelog
```

配置 spec 文件里对应参数及其说明见表 1-1。

表 1-1　配置 spec 文件里对应参数及其说明

参　　数	说　　明
Name	源码包的名称，后面可使用%{name}的方式引用
Summary	源码包的内容概要
Version	软件的实际版本号，如 1.0.1 等，后面可使用%{version}引用
Release	发布序列号，如 1linuxing 等，标明第几次打包，后面可使用%{release}引用
Group	软件分组，建议使用标准分组
License	软件授权方式，通常是 GPL
Source	源码包，可以带多个，如 Source1、Source2 等源，后面可以用%{source1}、%{source2}等引用
BuildRoot	安装或编译时使用的"虚拟目录"
URL	软件的主页
Vendor	发行商或打包组织的信息，如 RedFlag Co，Ltd
Disstribution	发行版标识
Prefix: %{_prefix}	生成的 RPM 包前缀名
%build	开始构建包
%install	开始把软件安装到虚拟根目录中
%clean	清理临时文件
%files	定义的那些额外的文件会放入 RPM 包中

（续）

参 数	说 明
%changelog	变更日志
%setu	源码包的安装阶段
%pre rpm	安装前执行的脚本
%post rpm	安装后执行的脚本
%preun rpm	卸载前执行的脚本
%postun rpm	卸载后执行的脚本

（4）执行制作包命令

若所有前置条件均已准备完成，则可执行如下制作包命令：

```
#执行 RPM 包制作命令,需要等待 30~60 分钟
shell $>rpmbuild -bb rpmbuild/SPECS/mysql8.0.34.spec

#完成后在 rpmbuild/RPM 目录下就会有两个 RPM 包
shell $>  ll
total 1616
-rw-r--r-- 1 root root 77264780 Aug 31 17:25 mysql-8.0.34-1.el8.centos.x86 64.rpm
-rw-r--r-- 1 root root 170869956 Aug 31 17:25
mysql-debuginfo-8,0.34-1.el8.centos .x86 64.rpm
```

（5）使用 RPM 包方式安装 MySQL

使用制作完成的 RPM 包安装 MySQL，会出现如下安装信息：

```
shell $>rpm -ivh mysql-8.0.34-1.el8.centos.x86 64.rpm
Preparing...
##########################################[100%]
Updating / installing...
1:mysql-8.0.34-1.el8.centos   ##########################################[100%]
```

2. 包制作过程中需要注意的事项

下面介绍包制作过程中需要注意的 3 个事项。

1）RPM 函数调用无法找到。例如，在使用 rpmbuild 构建 RPM 包时会报如下错误：

```
error: Installed (but unpackaged) file(s) found:
```

解决方法：找到 /usr/lib/rpm/macros 中的 %__check_files /usr/lib/rpm/check-files %{buildroot}，并将它注释掉即可。

2）操作界面。因为制作时间比较长，一般为 30 分钟~2 小时，所以需要操作的 Console 用户界面不能退出或关闭。

3）Warings 信息可以忽略，但是一旦出现错误，就必须解决。另外，在制作过程中，生成的文件全部删除之后，重新执行制作命令。

3. 总结

定制化 MySQL 8.0 的 RPM 安装包，可以实现规范化的部署，避免一些不必要的配置问题，同时实现统一的版本管理环境。

1.2.5 基于 Docker 环境安装 MySQL

Docker 是一种轻量虚拟化的容器技术，提供类似虚拟机的隔离功能（CPU、IO、网络资源隔离）。它通过容器化技术，将应用程序（包含数据库软件）及其依赖的软件包和类库等打包成一个独立的容器，就好像在真实的物理机上运行一样，使其具备高度的可移植性和可复用性。

如图 1-14 所示，数据库硬件技术从传统的物理机技术，发展到传统的虚拟机技术，再发展到目前的容器化技术（如 Docker）。

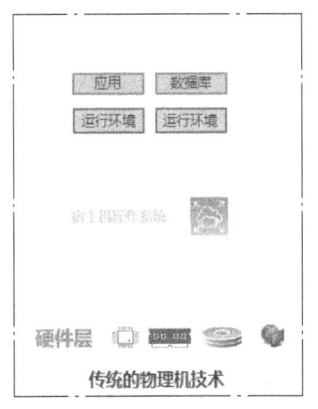

图 1-14　数据库硬件技术的演化过程

表 1-2 为不同运行环境下的指标对比。

表 1-2　不同运行环境下的指标对比

指　标	容　器	虚 拟 机	物 理 机
启动速度	秒级	分钟级	小时级
硬盘使用	MB 级别	GB 级别	GB 或 TB 级别
性能	接近原生	弱于原生	原生
系统支持量	单机支持上千个容器	一般有几个或几十个	单个
隔离性	安全隔离	安全隔离	安全隔离

Docker 展现出来的诸多优点，让很多应用软件在 Docker 中使用起来，MySQL 数据库也逐步踏入其中。虽然在 Docker 中运行 MySQL 具有一些潜在问题和挑战，但总体来看，利大于弊。

下面是在 Docker 中安装 MySQL 的操作步骤。

1. 环境设置

Docker 运行环境必须满足以下内容。

- 必须是 64 位 CPU 架构的计算机，Docker 目前不支持 32 位 CPU。
- 运行 Linux 3.8 或更高版本内核，CentOS 的内核版本不能低于 3.10。
- 内核必须支持一种合适的存储驱动，可以是 Device Manager、AUFS、VFS、BTRFS 和默认的驱动 Device Mapper 中的一个。
- 内核必须支持并开启 CGroup 和命名空间（NameSpace）功能。

检查前提条件。

1）检查系统的内核版本，返回的值大于 3.10 即可。

```
shell $>uname -a
Linux 4.18.0-193.14.2.el8_2.x86_64 #1 SMP Sun Jul 26 03:54:29 UTC 2020 x86_64 x86_64 x86_64 GNU/Linux
```

2）使用 sudo 或 root 权限的用户登入终端。
3）保证包是最新版本的，执行如下命令。

```
shell $> yum update
```

2. 普通方式安装 Docker

在 Linux 环境下使用 yum 命令安装 Docker。

```
shell $>yum -y install docker.io
```

用此命令安装的 Docker 可能不是最新版本。需要安装 Docker 的最新版本，可以使用如下命令：

```
shell $> curl -s https://get.docker.com | sh
# Executing docker install script, commit: c2de0811708b6d9015ed1a2c80f02c9b70c8ce7b
+ sh -c 'yum install -y -q yum-utils'
+ sh -c 'yum-config-manager --add-repo
https://download.docker.com/linux/centos/docker-ce.repo'
...
```

执行如下命令行开启 Docker 并检查版本：

```
shell $> systemctl start docker
shell $> ps -ef | grep docker
root      4101     1  4 15:16 ?        00:00:00 /usr/bin/dockerd -H fd://
--containerd=/run/containerd/containerd.sock
root      4250  2722  0 15:16 pts/1    00:00:00 grep --color=auto docker
shell $> docker --version
Docker version 24.0.5, build ced0996
```

说明：移除 Docker 可使用命令 yum -y remove docker-engine。

3. 在 Docker 里安装 MySQL 8.0

1）拉取镜像文件。

使用 pull 命令拉取 MySQL 8.0 镜像文件。

```
shell $> docker pull docker.io/mysql:8.0.34
8.0.34: Pulling from library/mysql
b193354265ba: Pull complete
14a15c0bb358: Pull complete
02da291ad1e4: Pull complete
9a89a1d664ee: Pull complete
a24ae6513051: Pull complete
5110d0b8df84: Pull complete
71def905d921: Pull complete
c29c4f8eb3c1: Pull complete
769af171cdaa: Pull complete
c1a0ba6abbff: Pull complete
5e7e1ae11403: Pull complete
```

```
Digest: sha256:f0e71f077bb27fe17b1b9551f75d1b35ad4dfe3a33c82412acf19684790f3a30
Status: Downloaded newer image for mysql:8.0.34
docker.io/library/mysql:8.0.34
```

说明：docker pull [OPTIONS] NAME [:TAG]命令的作用是从Docker远程仓库拉取镜像到本地。

通过images命令查看MySQL镜像文件。

```
#查看MySQL 8.0版本的镜像文件
shell $> docker images
REPOSITORY      TAG         IMAGE ID        CREATED         SIZE
mysql           8.0.34      5761fe35fa53    2 weeks ago     577MB
```

说明：docker images [OPTIONS] [REPOSITORY[:TAG]]命令可以用来查看本机中有哪些镜像，也可以验证pull命令是否执行成功。

使用container和ps命令查看容器运行情况。

```
shell $> docker container ls
CONTAINER ID    IMAGE       COMMAND     CREATED     STATUS      PORTS       NAMES
shell $> docker ps
CONTAINER ID    IMAGE       COMMAND     CREATED     STATUS      PORTS       NAMES
```

说明：docker ps [OPTIONS] 命令中的OPTIONS说明如下。

- -a：显示所有的容器，包括未运行的。
- -f：根据条件过滤显示的内容。
- -format：指定返回值的模板文件。
- -l：显示最近创建的容器。
- -n：列出最近创建的n个容器。
- -no-trunc：不截断输出。
- -q：静默模式，只显示容器编号。
- -s：显示总的文件大小。

2）启动MySQL服务。

以下的脚本用于启动MySQL服务和检查服务是否正常启动。

```
#查看镜像的REPOSITORY、TAG信息
shell $> docker images
REPOSITORY      TAG         IMAGE ID        CREATED         SIZE
mysql           8.0.34      5761fe35fa53    2 weeks ago     577MB

#启动MySQL服务,指定启动参数
shell $ > docker run -p 3306:3306 --name mysql8034 -e MYSQL_ROOT_PASSWORD=123456 -d mysql:8.0.34
f14e4e397207e257094e542c6320b133488e3e6532c6ba5974785b41ca512de4

#查看MySQL服务是否启动
shell $> docker container ls
CONTAINER ID    IMAGE           COMMAND                 CREATED         STATUS          PORTS       NAMES
f14e4e397207    mysql:8.0.34    "docker-entrypoint.s…"  9 seconds ago   Up 8 seconds
0.0.0.0:3306->3306/tcp,:::3306->3306/tcp, 33060/tcp   mysql8034
```

说明：docker run [OPTIONS] 命令中的 OPTIONS 说明如下。

- --name：为容器指定一个名称，此处命名为 mysql8034。
- -e：配置信息，此处配置 MySQL 的 root 用户的登录密码。
- -p：端口映射，此处映射主机的 3306 端口到容器的 3306 端口。
- -d：后台运行容器，并返回容器 ID。

登录容器，验证 MySQL 服务是否运行正常。使用的命令行为"docker exec -it 容器 ID bash"。

```
shell $>docker exec -it f14e4e397207  /bin/bash
bash-4.4# mysql -uroot -p ******
mysql:[Warning]Using a password on the command line interface can be insecure.
Welcome to the MySQL monitor.  Commands end with ; or \g.
Your MySQL connection id is 8
Server version: 8.0.34 MySQL Community Server - GPL
Copyright (c) 2000, 2023, Oracle and/or its affiliates.
Oracle is a registered trademark of Oracle Corporation and/or its
affiliates.Other names may be trademarks of their respective
owners.
Type 'help;' or '\h' for help.Type '\c' to clear the current input statement.
mysql>
```

下面的命令行在关闭容器复制之后，通过 docker start 命令启动 MySQL 服务。

```
shell $>docker ps -a
CONTAINER ID   IMAGE         COMMAND              CREATED        STATUS         PORTS    NAMES
f14e4e397207   mysql:8.0.34  "docker-entrypoint.s…"  9 minutes ago  Exited (0) 2
seconds ago              mysql8034
shell $>docker start  f14e4e397207
f14e4e397207
shell $>docker ps
CONTAINER ID   IMAGE         COMMAND              CREATED        STATUS         PORTS    NAMES
f14e4e397207   mysql:8.0.34  "docker-entrypoint.s…"  9 minutes ago  Up 4 seconds
0.0.0.0:3306->3306/tcp,:::3306->3306/tcp, 33060/tcp   mysql8034
```

3）映射 MySQL 服务目录和配置文件。

在主机上创建对应的映射目录。在映射本地目录之后，当 MySQL 容器出现故障时，可通过文件迁移、挂载方式，用原有数据搭建新容器，提供服务。

```
shell $>rm -rf /usr/local/docker/mysql/
shell $>mkdir -p  /usr/local/docker/mysql/conf
shell $>mkdir -p /usr/local/docker/mysql/data
shell $>mkdir -p  /usr/local/docker/mysql/logs
```

在容器的 MySQL 的默认配置中，可以找到 /etc/mysql/my.cnf 文件并配置。可以使用 includedir /etc/mysql/conf 方式加入额外的配置文件。

```
#配置 my.cnf 文件
shell $>vim  /usr/local/docker/mysql/conf/my.cnf
[mysqld]
server_id                              = 1303306
character_set_server                   = utf8mb4
```

```
collation_server                = utf8mb4_unicode_ci
port                            = 3306
transaction_isolation           = READ-COMMITTED
max_connections                 = 1000
datadir                         = /var/lib/mysql
socket                          = /var/run/mysqld/mysqld.sock
pid_file                        = /var/run/mysqld/mysqld.pid
log_error                       = /var/log/mysql/error.log

#赋予权限
shell $>chmod 644   /usr/local/docker/mysql/conf/my.cnf
```

使用配置文件启动 MySQL 服务。

```
shell $> docker run -p 3306:3306 --name mysql \
-v /usr/local/docker/mysql/conf:/etc/mysql \
-v /usr/local/docker/mysql/logs:/var/log/mysql \
-v /usr/local/docker/mysql/data:/var/lib/mysql \
-e MYSQL_ROOT_PASSWORD=123456 \
-d mysql:8.0.34
```

说明：-v 指定主机和容器的目录映射关系，":"前为主机目录，之后为容器目录。

4）实现容器中 MySQL 备份。

对于 MySQL 数据库的备份，可以通过远程访问方式，也可以通过 Docker 参数-c 传递 mysqldump 逻辑备份命令。以下是 mysqldump 备份命令传递方式。

```
shell $>docker exec f14e4e397207sh -c 'execmysqldump -uroot -p123456 -P3309
--single-transaction --master-data --all-databases' > /opt/all-databases.sql
```

4. 总结

通过实践，可以发现，在 Docker 中部署和运维 MySQL 还是比较简单的。经过实际测试，性能损耗约为 10%，主要原因是网络延迟，IO 和 CPU 已经不再是瓶颈。

在 Docker 中，想要 MySQL 更安全、平稳地运行，可对下面几个关键点做进一步的优化。

- Docker 的数据持久化：目录的映射和存储卷（volume）配置方式。
- 资源方面的隔离：命名空间和资源限制（CGroup）。
- 容器服务编排有 3 种常用工具：Google 开源的容器编排平台 Kubernetes，Docker 官方提供的容器编排工具 Docker Swarm，Apache 开源的分布式系统资源管理器 Mesos。
- 网络连接方式的选择。

第 2 章
MySQL 8.0 体系架构特性

2.1 数据字典

2.1.1 全局事务性数据字典

在 MySQL 8.0 版本中使用全新的数据字典（Data Dictionary）结构。MyISAM 系统表全部换成了 InnoDB 表，同时支持数据字典表的原子 DDL 操作。对比 MySQL 5.7 版本，可以说复杂度增加了。这样的数据字典的变化是否会导致性能降低或产生其他未知问题，在实际使用中需要关注。

1. 数据字典结构的变化

数据字典是 MySQL 数据库的重要组成部分，包含表结构、数据库名、表名、字段的数据类型、视图、索引、表字段信息、存储过程、触发器等对象的基础信息。

下面对比 MySQL 5.7 版本，了解 MySQL 8.0 有哪些变化。

图 2-1 所示为 MySQL 5.7 版本数据字典结构。图 2-2 所示为 MySQL 8.0 版本数据字典结构。在 MySQL 8.0 版本中，原先存放于数据字典文件中的信息，全部统一存放到数据库系统表中，即将之前版本的.frm、.trg 和.opt 等文件进行移除。

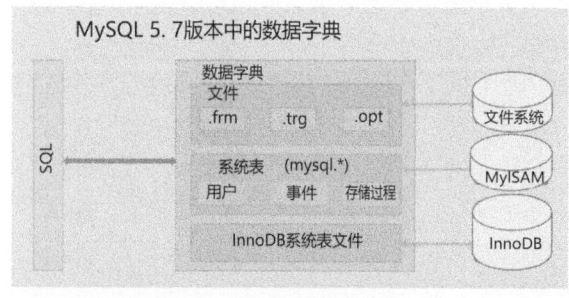

图 2-1　MySQL 5.7 数据字典结构

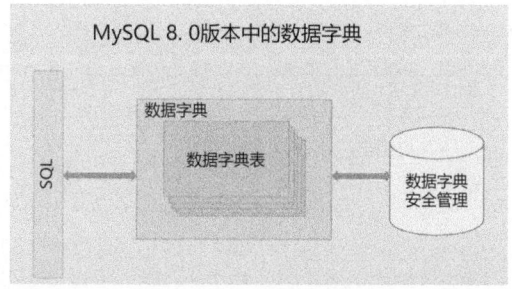

图 2-2　MySQL 8.0 数据字典结构

说明：
- .frm 文件存放表结构信息。
- .opt 文件记录了每个库的一些基本信息，包括库的字符集等信息。
- .trg 文件用于存放触发器的信息。

2. 底层引擎变更

在 MySQL 8.0 中，对系统库 mysql 和 sys 中的表的存储引擎做了改进。原先使用 MyISAM 存储引擎的数据字典表都改为使用 InnoDB 存储引擎表，因此从不支持事务的 MyISAM 存储引擎转变到支持事务的 InnoDB 存储引擎，同时对字典数据的提交、回滚和崩溃恢复提供了功能保护。这为原子 DDL 的实现提供了可能性。

以下是 MySQL 5.7 版本和 MySQL 8.0 版本中数据字典引擎统计的结果集。在 MySQL 8.0 版本中，不存在 MyISAM 引擎系统表。查询表引擎的命令行如下所示。

```
#MySQL 5.7版本查询统计：
mysql> SELECT TABLE_SCHEMA,ENGINE,COUNT(*)
FROM information_schema.tables where table_schema in
('information_schema','mysql','performance_schema','sys') group by
TABLE_SCHEMA,ENGINE;
+--------------------+--------------------+----------+
| TABLE_SCHEMA       | ENGINE             | COUNT(*) |
+--------------------+--------------------+----------+
| information_schema | InnoDB             |       10 |
| information_schema | MEMORY             |       51 |
| mysql              | CSV                |        2 |
| mysql              | InnoDB             |       19 |
| mysql              | MyISAM             |       10 |
| performance_schema | PERFORMANCE_SCHEMA |       88 |
| sys                | NULL               |      100 |
| sys                | InnoDB             |        1 |
+--------------------+--------------------+----------+
8 rows in set (0.06 sec)

#MySQL 8.0版本查询统计：
mysql> SELECT TABLE_SCHEMA,ENGINE,COUNT(*)
FROM information_schema.tables where table_schema in
('information_schema','mysql','performance_schema','sys') group by
TABLE_SCHEMA,ENGINE;
+--------------------+--------------------+----------+
| TABLE_SCHEMA       | ENGINE             | COUNT(*) |
+--------------------+--------------------+----------+
| mysql              | InnoDB             |       37 |
| sys                | NULL               |      100 |
| information_schema | NULL               |       79 |
| mysql              | BLACKHOLE          |        1 |
| performance_schema | PERFORMANCE_SCHEMA |      113 |
| mysql              | CSV                |        2 |
| sys                | InnoDB             |        1 |
+--------------------+--------------------+----------+
7 rows in set (0.01 sec)
```

说明：其中 ENGINE 为 NULL 表示以 x$ 开头的表，主要是 TEMPORARY TABLE 和 VIEW，如 x$waits_global_by_latency、x$user_summary_by_stages 等。

3. 底层文件结构的变化

如图 2-3 所示，在 MySQL 5.7 版本中，每张 MyISAM 存储引擎数据字典表由 3 个物理文件构成（扩展名为.frm、.MYD 和.MYI 的文件）。

```
-rw-r-----  1 mysql mysql      8955 Apr 28 10:13 tables_priv.frm
-rw-r-----  1 mysql mysql      3788 Apr 28 10:13 tables_priv.MYD
-rw-r-----  1 mysql mysql      9216 Apr 28 10:22 tables_priv.MYI
```

图 2-3　MySQL 5.7 数据文件

如图 2-4 所示，在 MySQL 8.0 版本中，所有数据字典表都集中存储在单独的一个表空间的 mysql.ibd 文件中。

```
drwxr-x---  2 mysql mysql       143 May  4 13:03 mysql
-rw-r-----  1 mysql mysql       280 May  4 15:20 mysqld-auto.cnf
-rw-r-----  1 mysql mysql  25165824 May 27 10:55 mysql.ibd
```

图 2-4　MySQL 8.0 数据文件

MySQL 8.0 版本中的数据字典表其体量很小，集中式管理便于管理与维护。

4. 新的索引统计信息的缓存机制

在 MySQL 中，对索引的统计信息记录在 INFORMATION_SCHEMA.STATISTICS 表里。在 MySQL 8.0 版本中，索引的统计信息通过缓存的方式访问，以提高查询的性能。其中 information_schema_stats_expiry 参数定义了缓存的表统计信息过期之前的一段时间。默认时间是 86400 秒（24小时）。如果没有缓存的统计信息或统计信息已过期，那么直接查询表存储引擎。可以使用 ANALYZE table 或 information_schema_stats_expiry=0 保持数据在缓存中。

查看缓存设置参数的命令行如下所示。

```
mysql> SHOW VARIABLES LIKE '%definition%';
+--------------------------------+-------+
|Variable_name                   |Value  |
+--------------------------------+-------+
|schema_definition_cache         |256    |
|stored_program_definition_cache |256    |
|table_definition_cache          |3000   |
|tablespace_definition_cache     |256    |
+--------------------------------+-------+
```

说明：除了索引的统计信息以外，还需要额外的表文件缓存信息。在资源方面，多了额外的内存开销。参数说明如下。

- table_definition_cache：存储表的数量。
- schema_definition_cache：存储库的数量。
- stored_program_definition_cache：存储过程和函数的数量。
- tablespace_definition_cache：存储表空间的数量。

5. 字典表可以使用索引

在 MySQL 8.0 中，mysql 和 information_schema 下的系统表的数据查询，可以有效使用索引。通过对比 MySQL 5.7 和 MySQL 8.0 的系统表的查询执行计划，可以查看是否使用了索引。系统表查询的命令行如下所示。

```
mysql> EXPLAIN SELECT TABLE_NAME FROM INFORMATION_SCHEMA.TABLES  WHERE
TABLE_SCHEMA='test' AND TABLE_NAME='t1';
```

图 2-5 所示为 MySQL 5.7 版本中数据字典表未使用索引，然而在 MySQL 8.0 版本中数据字典表可以使用索引。

图 2-5 MySQL 数据字典表查询执行计划

6. 部分字典表变动

MySQL 5.7 版本中的存储过程和函数基础信息表（mysql.proc、mysql.fun），在 MySQL 8.0 版本中，通过合并汇总到一张表（information_schema.ROUTINES）中，同时用 ROUTINE_TYPE 字段进行类型区分。

7. 非 InnoDB 引擎表的元数据保存到 .sdi 文件中

在 MySQL 8.0 版本中，其他非 InnoDB 引擎表的元数据统一保存到扩展名为 .sdi 的文件中。如图 2-6 所示，peformance_schema 下的引擎为 PERFORMANCE_SCHEMA 的表生成的扩展名为 .sdi 的文件。

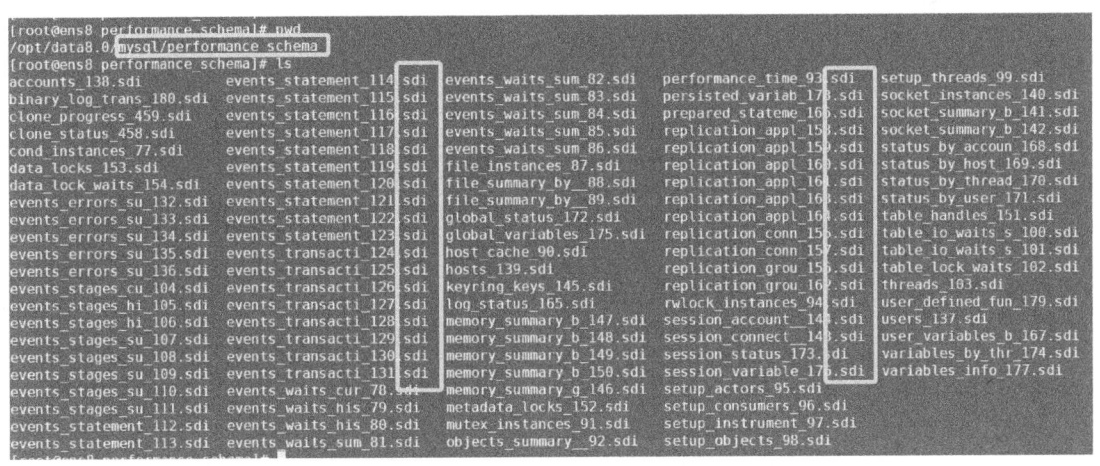

图 2-6 MySQL 8.0 系统中 .sdi 数据文件

8. 实现原子 DDL

原子 DDL 是指将数据字典相关的 DDL 作为单个原子进行操作，MySQL 8.0 支持数据字典的原子操作。为了实现原子 DDL 的重做和回滚机制，MySQL 8.0 将操作的 DDL 日志写到 mysql.innodb_ ddl_log 表中。它是一个隐藏的数据字典表，驻留在 MySQL 中 ibd 数据字典表空间。目前原子 DDL 仅支持 InnoDB 存储引擎的数据字典表，不支持其他引擎表。

innodb_ddl_log 表结构如下所示。

```
#查看 innodb_ddl_log 表结构
mysql> CREATE TABLE `innodb_ddl_log` (
    `id` bigint unsigned NOT NULL AUTO_INCREMENT,
    `thread_id` bigint unsigned NOT NULL,
    `type` int unsigned NOT NULL,
    `space_id` int unsigned DEFAULT NULL,
    `page_no` int unsigned DEFAULT NULL,
    `index_id` bigint unsigned DEFAULT NULL,
    `table_id` bigint unsigned DEFAULT NULL,
    `old_file_path` varchar(512) CHARACTER SET utf8mb3 COLLATE utf8mb3_bin DEFAULT NULL,
    `new_file_path` varchar(512) CHARACTER SET utf8mb3 COLLATE utf8mb3_bin DEFAULT NULL,
    PRIMARY KEY (`id`),
    KEY `thread_id` (`thread_id`)
);
```

说明：在普通模式下，innodb_ddl_log 拒绝访问，需要在 debug 模式下才能访问。每个字段的具体说明如下。

- id：每条 DDL 日志记录的唯一标识符。
- thread_id：每条 DDL 日志记录都与一个 thread_id 相关联，用于重放和删除某个特定 DDL 事务的 DDL 日志。涉及多个数据文件操作的 DDL 事务将会生成多条 DDL 日志记录。
- type：DDL 操作的类型，包括 FREE（删除一棵索引树）、DELETE（删除一个文件）、RENAME（重命名文件）和 DROP（从数据字典表 mysql.innodb_dynamic_metadata 中删除元数据）。
- space_id：表空间 ID。
- page_no：包含分配信息的页面，如索引树的根页面。
- index_id：索引 ID。
- table_id：表 ID。
- old_file_path：旧的表空间文件路径。用于创建或删除表空间文件的 DDL 操作，以及重命名表空间的 DDL 操作。
- new_file_path：新的表空间文件路径。用于重命名表空间文件的 DDL 操作。

原子 DDL 操作步骤如下。

准备：创建所需的对象并将 DDL 日志写入 mysql.innodb_ddl_log 表中。DDL 日志定义了如何前滚和回滚 DDL 操作。

执行：执行 DDL 操作。例如，为 CREATE TABLE 操作执行创建过程。

提交：更新数据字典并提交数据字典事务。

post-DDL：重播并从 mysql.innodb_ddl_log 表中删除 DDL 日志。为确保回滚可以安全执行而不引入不一致性，在此最后阶段执行文件操作（如重命名或删除数据文件）。这一阶段还要从 mysql.innodb_dynamic_metadata 的数据字典表中删除动态元数据，其中包含 DROP TABLE、TRUNCATE 和其他重建表的 DDL 操作。

在 MySQL 错误日志里显示 DDL 操作内容。设置日志输出级别命令行如下所示。

```
# 设置日志记录信息
mysql> SET GLOBAL LOG_ERROR_VERBOSITY=3;
```

```
mysql> SET GLOBAL innodb_print_ddl_logs=1;
mysql> CREATE TABLE t1 (c1 INT) ENGINE =InnoDB;

#mysql_error.log 文件记录内容
shell $> tail -f mysql_error.log
[Note] [MY-012473] [InnoDB] DDL log insert: [DDL record: DELETE SPACE, id=25955, thread_id=
21, space_id=2909, old_file_path=./book/t.ibd]

[Note] [MY-012478] [InnoDB] DDL log delete: 25955
[Note] [MY-012477] [InnoDB] DDL log insert: [DDL record: REMOVE CACHE, id=25956, thread_id=
21, table_id=6843, new_file_path=book/t]

[Note] [MY-012478] [InnoDB] DDL log delete: 25956
[MY-012472] [InnoDB] DDL log insert: [DDL record: FREE, id=25957, thread_id=21, space_id=
2909, index_id=11727, page_no=4]

[MY-012478] [InnoDB] DDL log delete: 25957
[MY-012485] [InnoDB] DDL log post ddl: begin for thread id: 21
[Note] [MY-012486] [InnoDB] DDL log post ddl: end for thread id: 21
```

说明：innodb_print_ddl_logs 输出 DDL 操作日志，LOG_ERROR_VERBOSITY 输出详细日志。

原子 DDL 支持的操作如下。

- DDL 语句：CREATE、ALTER、DROP 对象是库、表空间、表和索引。
- TRUNCATE TABLE。
- non-table DDL：CREATE DROP、ALTER 对象是触发器、视图和函数。
- 账户管理：CREATE、ALTER、DROP、RENAME 对象是用户、角色、授权。

原子 DDL 不支持的操作如下。

- 不是 InnoDB 引擎的 DDL 语句。
- 插件命令：插件加载和删除。
- 组件命令：组件加载和删除。
- 服务命令：服务创建、修改和删除。

9. 其他变化

（1）版本升级

在数据字典表成功升级之后，无法使用旧的服务器二进制文件重新启动服务器。因此，在升级数据字典表之后，不支持将 MySQL 服务器二进制文件降级为以前的 MySQL 版本。

使用 mysqld 启动时可以使用 no-dd-upgrade 选项来防止在启动时自动升级数据字典表。如果指定了参数 no-dd-upgrade，并且发现 MySQL 服务的数据字典版本与存储在数据字典中的版本不同，则启动失败，并出现一个错误提示，表示禁止数据字典升级。

（2）元数据表保护

如图 2-7 所示，在普通模式下访问元数据表时，提示访问拒绝信息。

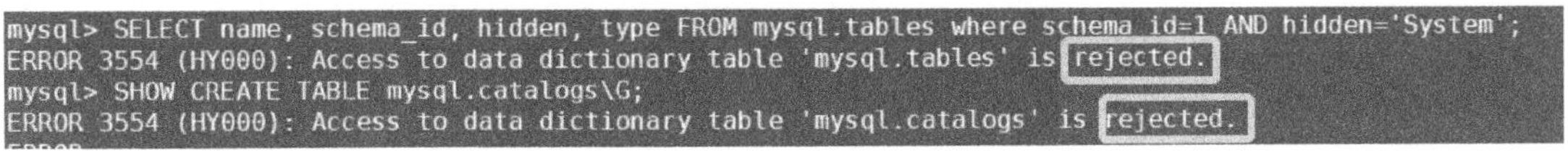

```
mysql> SELECT name, schema_id, hidden, type FROM mysql.tables where schema_id=1 AND hidden='System';
ERROR 3554 (HY000): Access to data dictionary table 'mysql.tables' is rejected.
mysql> SHOW CREATE TABLE mysql.catalogs\G;
ERROR 3554 (HY000): Access to data dictionary table 'mysql.catalogs' is rejected.
```

图 2-7　MySQL 8.0 元数据访问拒绝

可通过 MySQL 源码 CMake 编译时添加-DWITH_DEBUG=1 参数，查看一些隐藏的表的信息和数据。开启隐藏元数据表的命令行如下所示（后面章节会具体介绍）：

```
mysql> SET SESSION debug='+d,skip_dd_table_access_check';
mysql> SELECT name, schema_id, hidden, type FROM mysql.tables where schema_id=1 AND hidden='System';
```

（3）innodb_read_only 参数对数据字典的影响

innodb_read_only 参数是为了防止用户对 InnoDB 存储引擎表创建和删除。从 MySQL 8.0 版本开始，innodb_read_only 参数也适用于数据字典表。通过设置 innodb_read_only 参数，可以防止对系统表的人为操作。

（4）mysqldump 与 mysqlpump 逻辑导出对数据字典的影响

在 MySQL 8.0 版本之前使用逻辑导出命令时，所有对象都会自动导出，而在 MySQL 8.0 版本之后，不再支持，需要额外添加参数。

- 从 MySQL 8.0 版本开始，在使用 all-databases 参数导出数据时，必须添加-routines 和-events 选项，这样才可以导出触发器、存储过程等信息。
- 在 MySQL 8.0 版本中，使用 routines 选项导出时，备份账户需要对所有表有 SELECT 权限（包含系统表）。

（5）底层文件修改限制

在 MySQL 8.0 版本中，不支持在 data 目录下手动创建数据库文件和目录（例如，使用 mkdir、vi），且已无法识别手动创建的数据库文件和目录。数据字典也是一样的。

10. 总结

MySQL 8.0 中数据字典有了全新的改变（存储变化、引擎变化、保证原子性、提高查询性能）。但是它并不是完善的，新版数据字典还是存在一些局限性。例如，DDL 操作会花费更长的时间，因为之前的 DDL 操作直接对.frm 文件进行更改操作，只需要写一个文件，现在需要更新数据字典表，这代表着需要将数据写到存储引擎、read log 和 undo log 中。

2.1.2 新的 SDI 元数据结构

从 MySQL 8.0 版本开始，删除了原来的.frm 文件，采用序列化字典信息；统一使用 InnoDB 存储引擎来存储表的元数据信息。

1. SDI 概述

MySQL 8.0 中的序列化字典信息（Serialized Dictionary Information，SDI）是重新设计数据字典后引入的新产物。它是指表结构的元数据，输出是 JSON（JavaScript 对象表示法）格式的文件。对于 InnoDB 表，SDI 被冗余存储在表空间（临时表空间和 Undo 表空间文件除外）中，而不需要单独的 SDI 文件。对于非 InnoDB 引擎，SDI 则以可读的文件格式存储在磁盘上，命名格式为"表名.sdi"，存放在数据库目录下。

除了临时表空间和 Undo 表空间文件以外，所有 InnoDB 表的数据都保存在 SDI 文件中。SDI 的特性如下。

- SDI 数据通过对表或检查表的 DDL 操作进行更新。
- 使用 DDL 操作期间访问的内部 API 来创建和维护 SDI 记录。
- 当 MySQL 服务器升级到一个新版本时，SDI 数据不会更新。
- 记录数据方式：对于 InnoDB，一条 SDI 记录需要一个索引页，默认大小为 16KB。但实际

SDI 数据进行了压缩以减少存储空间。
- 分区表：对于由多个表空间组成的分区 InnoDB 表，SDI 数据存储在第一个分区的表空间文件中。

注意：SDI 只是元数据的备份，不是元数据本身。数据字典完全存在于 InnoDB 数据字典表空间中。

2. 使用 ibd2sdi 工具查看 SDI 文件

使用官方提供的 ibd2sdi 工具直接读取 SDI 文件，抽取表结构信息。导出的文件格式为 JSON。ibd2sdi 的功能与使用限制如下。
- 支持 file-per-table 表空间文件（*.ibd files）的解析。
- 支持一般表空间文件（*.ibd files）的解析。
- 支持系统表空间文件（ibdata* files）的解析。
- 支持数据字典表空间（mysql.ibd）的解析。
- 不支持临时表空间或 Undo 表空间文件的解析。

ibd2sdi 可以在运行时使用，也可以在服务离线时使用。在进行与 SDI 相关的 DDL 操作、回滚操作和 undo log purge 操作时，可能会出现 ibd2sdi 读取存储在表空间中的 SDI 数据失败的短时间间隔。

ibd2sdi 目前的功能只是单纯地提取字典信息，如图 2-8 所示为 ibd2sdi 解析 ibd 文件的内容（如数据字典版本、数据库版本、文件路径、引擎、表名、列名、列类型、列长度等信息）。

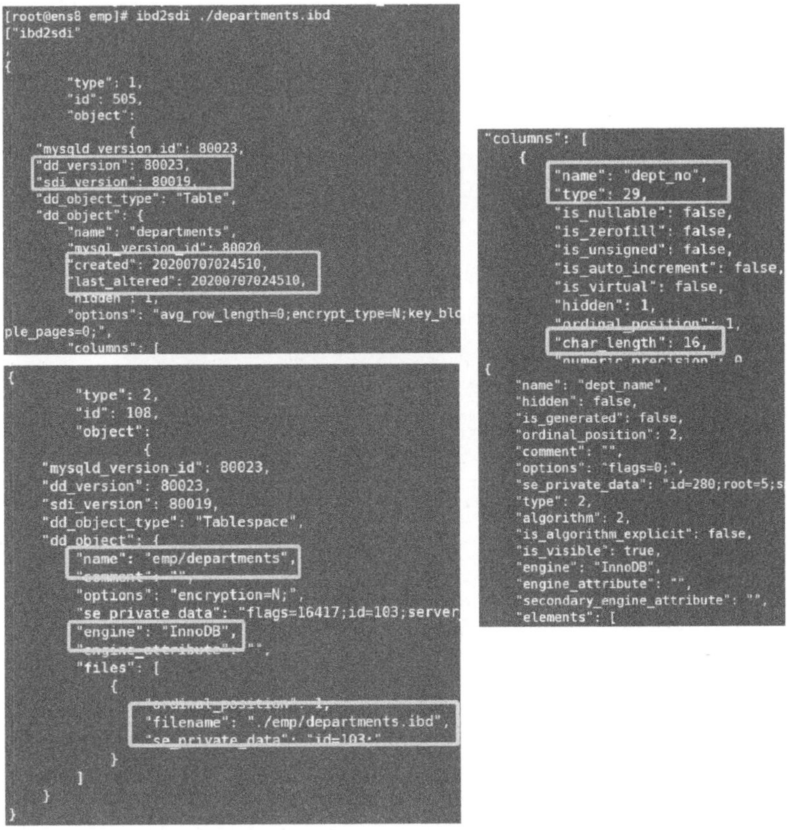

图 2-8　MySQL 8.0 ibd 结构信息

具体信息如下。
- sdi_version→dd_version：从 MySQL 8.0.19 升级到 MySQL 8.0.23。
- created、last_altered：创建时间是最后更新时间。
- 字段、字段长度、数据库引擎和文件路径等。

可以通过参数进行过滤，只保留需要的指标，命令行如下所示。

```
# 查看版本信息
shell $>ibd2sdi -v
ibd2sdi   Ver 8.0.34 for Linux on x86_64 (MySQL Community Server - GPL)

# 指定表.ibd 文件，输出结构信息
shell $>ibd2sdi --skip-data ./db1/a1.ibd
["ibd2sdi"
,
{
    "type": 1,
    "id": 1222
}
,
{
    "type": 2,
    "id": 142
}
]
```

除此之外，对于 MyISAM 引擎表，直接使用 cat 命令读取 SDI 文件，同样会得到基础元数据信息。如图 2-9 所示，可以查看具体表结构信息。

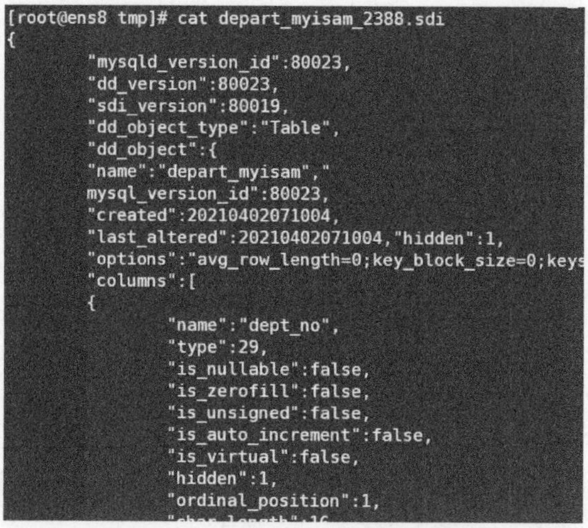

图 2-9　MySQL 8.0 MyISAM 表结构信息

3. 通过 SDI 文件进行数据恢复操作

在数据库运行过程中，难免会出现一些非常极端的情况（如突然断电、异常关闭服务、删除文件时出错、设备损坏等），导致数据文件损坏。下面介绍在 MySQL 8.0 版本中怎样通过 SDI 文件进行数据恢复。需要注意的是，类似数据截断、结构信息不完整的损坏，并不能够保证可以成功恢复。

（1）MyISAM 引擎表恢复操作

MyISAM 引擎表恢复操作具体步骤如图 2-10 所示，先把 .sdi 元数据文件放置到安全目录（secure_file_priv）下，把数据文件放置数据目录（data）下，再通过 IMPORT 命令导入所有文件。

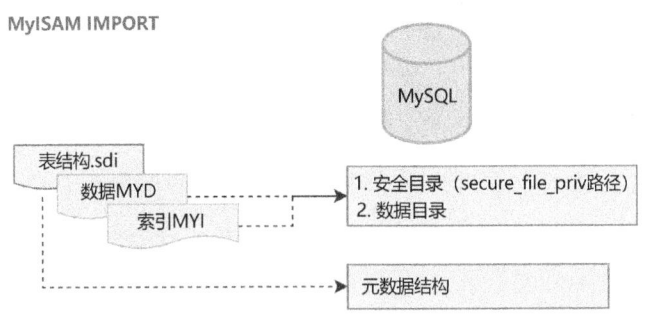

图 2-10　MySQL 8.0 MyISAM 引擎表恢复操作

具体恢复过程如下面示例所示。

```
# 创建表并插入数据
mysql> CREATE TABLE `depart_myisam` (
    `dept_no` char(4) COLLATE utf8mb4_bin NOT NULL,
    `dept_name` varchar(40) COLLATE utf8mb4_bin NOT NULL,
    PRIMARY KEY (`dept_no`)
) ENGINE=MYISAM;
mysql> INSERT INTO  depart_myisam(dept_no,dept_name)  VALUES('001','Beijing');

# 在复制 MyISMA 引擎的 .sdi、.MYD、.MYI 文件之后,进行 DORP 表操作
mysql> DROP TABLE depart_myisam;

# 恢复操作,复制文件
shell $> cp depart_myisam_2389.sdi   /opt/data8.0/tmp
shell $> cp depart_myisam.*   /opt/data8.0/mysql/emp
shell $> chown -R mysql.mysql   ./depart_myisam_2389.sdi
shell $> chown -R mysql.mysql   ./depart_myisam.*

# 导入文件
mysql> IMPORT TABLE FROM  '/opt/data8.0/tmp/depart_myisam_2389.sdi';
```

说明：.sdi 文件可以放在安全目录或数据目录中，.MYD 和 .MYI 文件都需要放到数据目录下。

（2）InnoDB 引擎表空间恢复

InnoDB 引擎表空间恢复的前提是必须开启独立表空间（设置参数 innodb_file_per_table=1，每张表就是一个文件）。通过 ibd2sdi 分析独立表空间。下面会通过 Shell 过滤关键字段信息。解析 SDI 结构的命令行如下所示。

```
shell $>ibd2sdi  --type=1 t1.ibd  --dump-file=/opt/script/json/t1.txt
cat t1.txt | grep -P 'name|column_type_utf8[":]+\K[^"]+'
```

解析出的表的结构信息如图 2-11 所示，列出所有字段名、字符集、类型、长度、索引信息。

说明：在创建新表时，需要之前表的 PRIMARY 信息，其他索引可有可无。

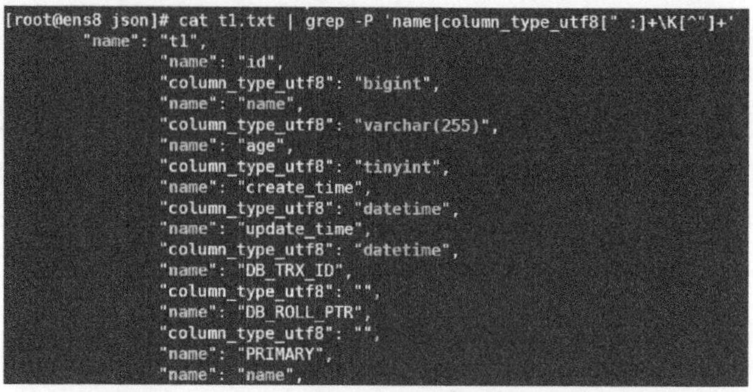

图 2-11　MySQL 8.0 InnoDB 引擎表 ibd 结构

如图 2-12 右侧所示为解析 SDI 获取的新表结构。虽然与原表结构有差异，但是可以进行恢复。

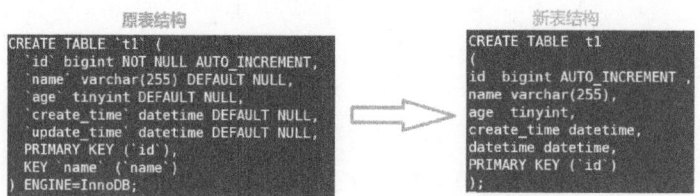

图 2-12　MySQL 8.0 InnoDB 引擎表结构拼接

当获取表结构之后，通过对表执行 DISCARD 或 IMPORT 命令进行恢复。恢复方式如下面示例所示。

```
#1.在目标实例上,丢弃刚刚创建的表的表空间
mysql> ALTER TABLE t1 DISCARD TABLESPACE;

#2.运行 FLUSH TABLES...以暂停要导入的表。当表处于静默状态时,该表上只允许有只读事务
mysql> FLUSH TABLES t1 FOR EXPORT;

#3.复制文件
shell> cp /bak/world/t1.{ibd,cfg}  /datadir/world
shell> chown -R mysql.mysql /datadir/world/t1.*

#4.使用 UNLOCK TABLES 来释放 FLUSH TABLES 获得的锁
mysql> UNLOCK TABLES;

#5.在目标实例上,导入表空间
mysql> ALTER TABLE t1 IMPORT TABLESPACE;
```

说明：如果有原表结构信息，则可以直接使用，没有的情况下只能分析 .ibd 文件。表结构分析工具 ibd2sdi 不像 mysql-utilities 工具那样灵活，目前来说，需要通过 Shell 或 Python 进行二次分析。

4. 总结

MySQL 8.0 版本中新的 SDI 元数据结构信息非常透明，易懂易读。SDI 数据的存在提供了元数据冗余，当 MySQL 数据库"死"机或数据字典变得不可用时，可以使用 ibd2sdi 工具直接从 InnoDB 表空间文件中提取对象元数据，进行单表数据的恢复操作。

2.1.3 隐藏的数据字典表的可视化功能

1. 数据字典表概述

MySQL 的数据字典包含 MySQL 服务运行时所需的信息，考虑到安全问题和维护元数据正确性，避免随意更改，所以无法直接访问。

隐藏的数据字典表如图 2-13 所示。

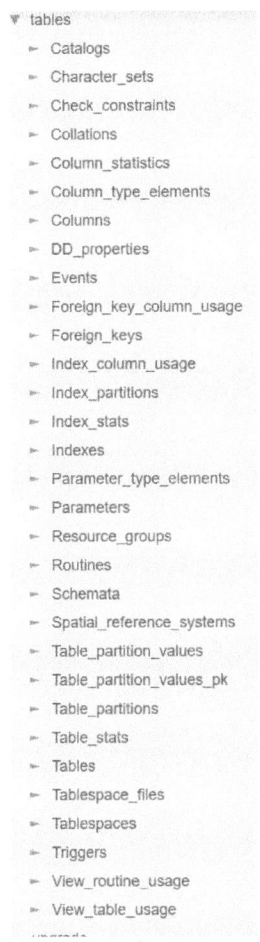

图 2-13　MySQL 8.0 中的隐藏的数据字典表

表 2-1 为隐藏的数据字典表及其说明，可以了解到每个表的具体作用。

表 2-1　隐藏的数据字典表及其说明

隐藏的数据字典表的表名	说　　明
Catalogs	目录信息
Character_sets	可用字符集的信息
Check_constraints	关于表上定义的 CHECK 约束的信息
Collations	关于每个字符集的排序规则的信息

(续)

隐藏的数据字典表的表名	说　　明
Column_statistics	列值的直方图统计信息
Column_type_elements	关于列使用的类型的信息
Columns	关于表中列的信息
DD_properties	标识数据字典属性的表,如它的版本。服务器使用它来确定是否必须将数据字典升级到较新的版本
Events	关于事件调度器调度事件的信息
Foreign_key_column_usage、Foreign_keys	关于外键的信息
Index_column_usage	索引使用的列的信息
Index_partitions	关于索引使用的分区的信息
Index_stats	用于存储在执行 ANALYZE TABLE 时生成的动态索引统计信息
Indexes	关于表索引的信息
Parameter_type_elements	关于存储过程和函数参数的信息
Parameters	存储过程和函数的相关信息
Resource_groups	资源组信息
Schemata	关于库的信息
Spatial_reference_systems	空间数据的可用空间参考系统的信息
Table_partition_values	表分区使用的值的信息
Table_partitions	关于表使用的分区的信息
Table_stats	执行 ANALYZE TABLE 时生成的动态表统计信息
Tables	关于数据库中表的信息
Tablespace_files	表空间使用的文件信息
Tablespaces	活动表空间的信息
Triggers	触发器信息
View_routine_usage	关于视图和视图使用的存储函数之间的依赖关系的信息
View_table_usage	用于跟踪视图与其表之间的依赖关系

2. 访问隐藏的数据字典表

从 8.0 版本开始,MySQL 提供数据字典表的可视化功能。在数据库中直接访问数据字典表会提示错误信息,如下所示。

```
mysql> SELECT name FROM mysql.tables where schema_id=1 AND hidden='System';
ERROR 3554 (HY000): Access to data dictionary table 'mysql.tables' is rejected.
```

在 MySQL 的底层源码中,就对直接访问数据字典表的权限加以限制,不予显示,即在源码 mysql-8.0.34\sql\dd\impl\dictionary_impl.cc 内通过 skip_dd_table_access_check 变量设置,禁止访问,相关代码如下。

```
bool Dictionary_impl::is_dd_table_access_allowed(bool is_dd_internal_thread,
                                                  bool is_ddl_statement,
                                                  const char *schema_name,
```

```
                                         size_t schema_length,
                                         const char *table_name)
const {
    /*
        From WL#6391, we have the following matrix describing access:
        For performance reasons, we first check the schema
    name to shortcut the evaluation.If the table is not in
    the 'mysql' schema, we don't need any further checks.Same for
        checking for internal threads - an internal thread has full
    access.We also allow access if the appropriate debug flag
    is set.
        */
        if (schema_length != MYSQL_SCHEMA_NAME.length ||
            strncmp(schema_name, MYSQL_SCHEMA_NAME.str, MYSQL_SCHEMA_NAME.length) ||
            is_dd_internal_thread ||
    DBUG_EVALUATE_IF("skip_dd_table_access_check", true, false))
            return true;
```

可采用以下方式实现数据字典表的可视化。

通过在编译 MySQL 源码时，设置参数-DWITH_DEBUG=1 来显示数据字典表。另外，可以通过--help 选项来检查是否开启调试服务，命令行如下所示。

```
shell $> mysqld --help | grep debug
mysqld  Ver 8.0.31-debug for Linux on x86_64 (Source distribution)
```

或者，可采用 MySQL 安装包中的 mysqld-debug 执行文件启动调试服务，如下所示。

```
shell $>mysqld-debug --defaults-file=/etc/my8.0.cnf  --user=mysql  &
```

只使用 debug 模式启动 MySQL 服务还无法显示数据字典表，还需要绕过源码中限制访问数据字典表的权限设置。开启访问权限的命令行如下所示。

```
# 绕过权限设置
mysql> SET SESSION debug='+d,skip_dd_table_access_check';
Query OK, 0 rows affected (0.00 sec)
mysql> SELECT @@debug;
+-------------------------------------------+
| @@DEBUG                                   |
+-------------------------------------------+
| d,skip_dd_table_access_check              |
+-------------------------------------------+
1 row in set (0.01 sec)

# 访问数据字典表
mysql> SELECT name FROM mysql.tables where schema_id=1 AND hidden='System';
+-------------------------------------------+
| name                                      |
+-------------------------------------------+
| catalogs                                  |
| character_sets                            |
| check_constraints                         |
```

```
| collations                      |
| column_statistics               |
| column_type_elements            |
| columns                         |
| dd_properties                   |
| events                          |
| foreign_key_column_usage        |
| foreign_keys                    |
| index_column_usage              |
| index_partitions                |
| index_stats                     |
| indexes                         |
| innodb_dynamic_metadata         |
| parameter_type_elements         |
| parameters                      |
| resource_groups                 |
| routines                        |
| schemata                        |
| spatial_reference_systems       |
| table_partition_values          |
| table_partitions                |
| table_stats                     |
| tables                          |
| tablespace_files                |
| tablespaces                     |
| triggers                        |
| view_routine_usage              |
| view_table_usage                |
+---------------------------------+
31 rows in set (0.01 sec)
```

经过上述操作，实现了数据字典表的可视化。接下来可以选择一张元数据表，查看一下它的具体表结构。选择的表是 InnoDB 引擎和 UTF-8 字符集。具体命令行如下所示。

```
mysql> SHOW CREATE TABLE mysql.tables\G;
*************************** 1.row ***************************
       Table: tables
Create Table: CREATE TABLE `tables` (
  `id` bigint unsigned NOT NULL AUTO_INCREMENT,
  `schema_id` bigint unsigned NOT NULL,
  `name` varchar(64) CHARACTER SET utf8 COLLATE utf8_tolower_ci NOT NULL,
  `type` enum('BASE TABLE','VIEW','SYSTEM VIEW') COLLATE utf8_bin NOT NULL,
  `engine` varchar(64) CHARACTER SET utf8 COLLATE utf8_general_ci NOT NULL,
  `mysql_version_id` int unsigned NOT NULL,
  ...
  PRIMARY KEY (`id`),
  UNIQUE KEY `schema_id` (`schema_id`,`name`),
  UNIQUE KEY `engine` (`engine`,`se_private_id`)
) /*!50100 TABLESPACE `mysql` */ ENGINE=InnoDB AUTO_INCREMENT=2412 DEFAULT CHARSET=utf8 COLLATE=utf8_bin STATS_PERSISTENT=0 ROW_FORMAT=DYNAMIC
```

```
1 row in set (0.00 sec)
```

debug 模式下能更改数据字典表数据
```
mysql> UPDATE  mysql.tables SET hidden='Visible' WHERE name='catalogs';
Query OK, 0 rows affected (0.00 sec)
Rows matched: 1  Changed: 0  Warnings:
```

3. 总结

对于开源数据库的研究来说，数据字典表可以显示、修改是非常有意义的。围绕这些基础表进行的操作，可以帮助开发者理解数据库的设计理念；通过访问数据字典表来更改定义和字段，可以实现更多的功能。

另外，数据字典表是数据库系统的根本，十分敏感，必须保护这些表不受 DDL 的影响，同时不被人为更改，因为一旦被更改，就很可能会产生难以承受的灾难性后果，所以数据字典表一直以来都是无法访问，加以隐藏的。

2.1.4　全新的各类表空间

MySQL 数据库的表空间是用来存储数据的逻辑空间，也是存储数据的最大逻辑单元，其下还有段、区、页等逻辑操作单位。MySQL 早期版本中表空间只有一种，而在 MySQL 8.0 版本中，表空间分为以下 5 种，可实现对数据文件的灵活控制。

- 系统表空间（System Tablespaces）。
- 独立表空间（File-Per-Table Tablespaces）。
- 通用表空间（General Tablespaces）。
- 临时表空间（Temporary Tablespaces）。
- 回滚表空间（Undo Tablespaces）。

图 2-14 所示为每个表空间的特性和使用场景。

下面逐一进行介绍。

1. 系统表空间

在 MySQL 8.0 之前的版本中，系统表空间包含 InnoDB 数据字典相关对象的元数据、双写缓存（doublewrite buffer）、改变缓存（change buffer）和回滚日志（undo log）。如果表创建时，不使用独立表空间或通用表空间，那么表数据会存储在系统表空间中，还包含表和索引数据。在 MySQL 8.0 版本中，原系统表空间内只保留了元数据和改变缓存。

设置系统表空间的参数如下所示。

```
# my.cnf 配置文件
[mysqld]
innodb_data_home_dir =/opt/data8.0/dbdata
innodb_data_file_path = ibdata1:16M;ibdata2:16M:autoextend
```

说明：autoextend 属性只能在 innodb_data_file_path 设置的最后一个数据文件中指定。对于系统表空间，对缩减现有系统表空间的大小是不支持的。想要实现更小的系统表空间，唯一的选择是将数据从备份中恢复。

在 MySQL 8.0 里删除数据是不释放空间的，所以 MySQL 8.0 之前版本的系统表空间是非常大的。为了避免大的系统表空间，使用每个表文件的表空间。独立表空间是默认的表空间类型，在创建 InnoDB 表时隐式使用。与系统表空间不同，独立表空间在截断或删除在每个表文件的表空间中创建的表之后，将磁盘空间返回给操作系统。

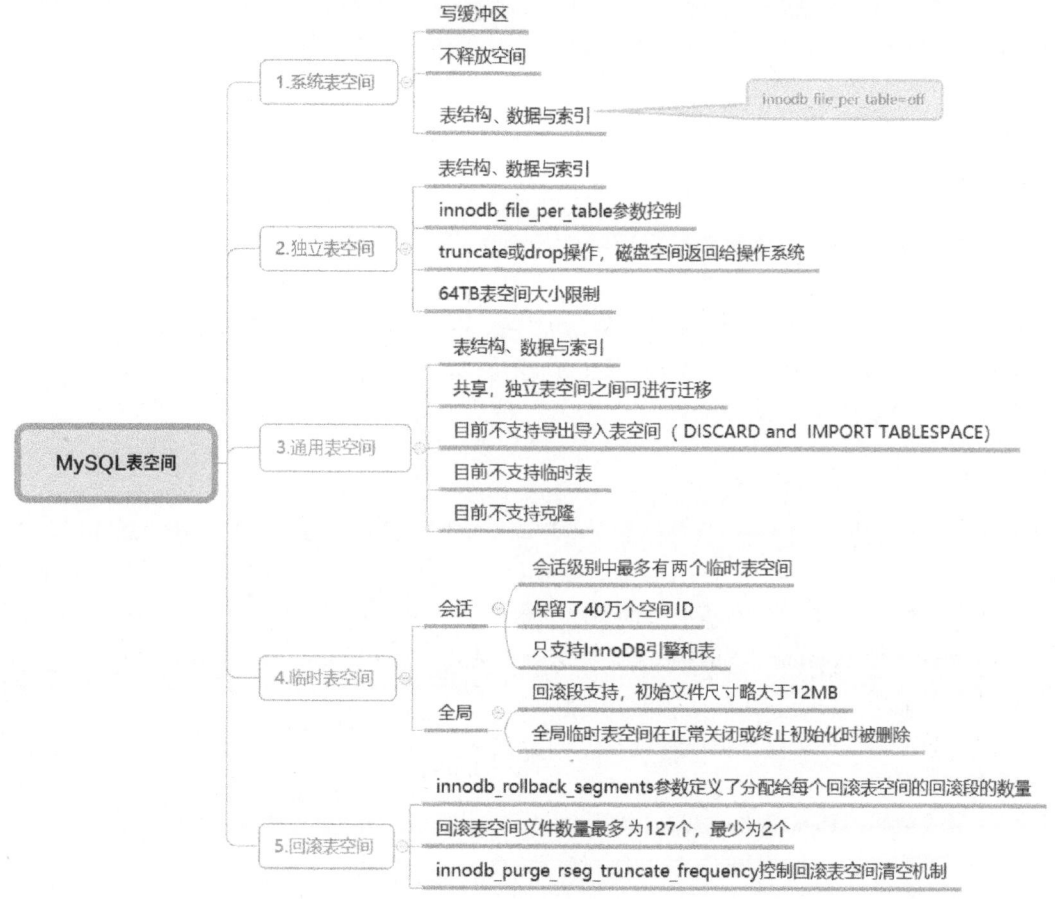

图 2-14　MySQL 8.0 表空间分类

2. 独立表空间

独立表空间包含单个 InnoDB 表的数据和索引，并存储在文件系统中表名对应的数据文件中。设置独立表空间的三种方式如下所示。

```
# my.cnf 配置
[mysqld]
innodb_file_per_table=ON

# 命令行方式设置
mysql> SET GLOBAL innodb_file_per_table=ON;

# 创建表并指定独立表空间
mysql> CREATE TABLE city(ID int)ENGINE=Innodb
TABLESPACE=innodb_file_per_table;
```

独立表空间的优点如下。

- 截断或删除在表上的操作，释放磁盘空间给操作系统。但系统表空间不会释放磁盘空间。

换句话说，系统表空间数据文件的大小不会缩减。
- 在系统表空间的表上执行 ALTER TABLE 操作时，会因为表复制（table-copying）动作，导致表空间所占用的磁盘空间增加，但独立表空间是释放空间。
- 独立表空间在对表执行 TRUNCATE 操作时，性能更好。
- 独立表空间数据文件可以在不同的存储设备上创建，用于 I/O 优化、空间管理或备份。
- 通过 DISCARD 或 IMPORT TABLESPACE 方式迁移独立表空间。
- 独立表空间中创建的表支持与动态和压缩行格式相关的特性，而系统表空间不支持这些特性。
- 存储在独立表空间数据文件中的表可以节省恢复时间，当发生数据损坏时，若备份或二进制日志不可用，或者 MySQL 服务器实例不能重新启动，则成功恢复的概率会上升。独立表空间数据可以通过第三方工具进行分析和抽取。还可以单独恢复独立表空间。
- 每个表文件的表空间，允许通过监视表空间数据文件的大小来监视表的大小。
- 当 innodb_flush_method 设置为 O_DIRECT 时，普通的 Linux 文件系统不允许对单个文件（如系统表空间数据文件）进行并发写操作。但独立表空间可以对多个表空间进行并行处理。因此，当结合此设置使用独立表空间时，可能会有性能上的改进。
- 系统表空间中的表的大小受到 64TB 表空间大小的限制。相比之下，每个独立表空间都有 64TB 的大小限制，这为各个表提供了足够的空间。

独立表空间的缺点如下。
- 对于独立表空间，每个表可能都有未使用的空间，如果管理不当，则可能会导致存储空间的浪费。
- 在 Linux 系统下，fsync 操作会针对单个文件，所以多个表空间的写操作不能组合在一起，这可能导致该操作的总数更高。
- mysqld 必须为每个文件-每个表的表空间保留一个打开的文件句柄，如果有很多独立表空间，就可能会影响性能。
- 当每个表都有自己的数据文件时，需要更多的文件描述符。
- 可能存在碎片，删除表和表扫描性能会下降。
- 当删除独立表空间的表时，会扫描缓冲池，对于大型缓冲池来说，这可能需要长达几秒钟的时间。扫描使用宽的内部锁执行，这可能会延迟其他操作。
- innodb_autoextend_increment 参数定义了增量大小，用于在自动扩展的独立表空间文件满时扩展其大小，但它不适用于独立表空间，因为不管该参数设置如何，这些表空间文件都会自动扩展。初始的每个表文件的表空间扩展量很小，之后的扩展会以 4MB 的增量进行。

3. 通用表空间

通用表空间就是系统表空间的扩展，但只对业务表有效。在独立于 MySQL 数据目录的目录中，数据可以在系统表空间、独立表空间、通用表空间之间转移。这样就会方便迁移数据，特别是在硬盘空间不足的情况下。

通用表空间不是随意放置的，只能在配置的目录下，且由参数 innodb_directories 控制。注意，这个参数拥有只读属性。

创建通用表空间的语法如下所示。

```
CREATE TABLESPACE tablespace_name
    [ADD DATAFILE 'file_name']
```

```
[FILE_BLOCK_SIZE = value]
[ENGINE [=] engine_name]
```

说明：目前只支持 InnoDB 引擎。FILE_BLOCK_SIZE 基于 innodb_page_size 指定默认值，无特殊需求时不需要指定。

配置通用表空间，需要先指定一个目录。通用表空间配置如下所示。

```
#my.cnf 配置参数
[mysqld]
innodb_directories=/opt/data8.0/tmpdata

# 查看路径
mysql> show variables like '%innodb_directories%';
+--------------------+---------------------+
|Variable_name       |Value                |
+--------------------+---------------------+
|innodb_directories  |/opt/data8.0/tmpdata |
+--------------------+---------------------+
1 row in set (0.00 sec)
```

在通用表空间下生成表的方式如下所示。

```
# 指定通用表空间
mysql> CREATE TABLE ext_table(a INT PRIMARY KEY, b CHAR(4)) DATA
DIRECTORY='/opt/data8.0/tmpdata';

# 数据文件生成路径
shell $> cd /opt/data8.0/tmpdata
shell $> ll test/
total 112
-rw-r----- 1 mysql mysql 114688 Aug 31 10:17 ext_table.ibd
```

在通用表空间下引入表，数据目录层次是"数据库文件夹名/数据 ibd 文件"。导入表空间的命令行如下所示。

```
# 导入命令
mysql> CREATE TABLESPACE ext_ts ADD DATAFILE '/opt/data8.0/tmpdata/ext_ts.ibd';
```

通用表空间中其他常用的命令行如下所示。

```
# 查看表空间信息
mysql> SELECT a.NAME AS space_name, b.NAME AS table_name
    FROM INFORMATION_SCHEMA.INNODB_TABLESPACES a,
    INFORMATION_SCHEMA.INNODB_TABLES b
    WHERE a.SPACE=b.SPACE and a.SPACE_TYPE='General';

# 从独立表空间或系统表空间移动到通用表空间中
mysql> ALTER TABLE tbl_name TABLESPACE [=] tablespace_name;

# 从通用表空间移动到系统表空间中
mysql> ALTER TABLE tbl_name TABLESPACE [=] innodb_system;

# 从系统表空间或通用表空间移动到独立表空间中
```

```
mysql> ALTER TABLE tbl_name TABLESPACE [=] innodb_file_per_table;

# 通用表空间重命令
mysql> ALTER TABLESPACE s1 RENAME TO s2;

# 删除通用表空间,但删除之前必须把其中的表移走
mysql> DROP TABLESPACE ts1;
```

通用表空间有下列限制。
- 不能将已存在的表空间更改为通用表空间。
- 不支持创建临时通用表空间。
- 不支持临时表。
- 不支持 DISCARD 和 IMPORT TABLESPACE 操作。
- 可以在 innodb_data_home_dir 和 innodb_directories 目录下生成通用表空间。

4. 临时表空间

在一些常规操作中,MySQL 会使用会话临时表空间和全局临时表空间。

(1) 会话临时表空间

当 InnoDB 被配置为磁盘内部临时表的存储引擎时,会话临时表空间存储了用户创建的临时表和优化器创建的内部临时表(从 MySQL 8.0.16 开始,临时表的存储引擎是 InnoDB,并且由 internal_tmp_disk_storage_engine 参数指定)。

- 每个会话临时表空间最多包含两个临时表空间:一个用于创建的临时表,另一个用于优化器创建的内部临时表。
- 当会话断开连接时,临时表空间将被截断并释放回池中。
- 当服务启动时,将创建一个包含 10 个临时表空间的池。池的大小永远不会缩小,并且表空间会根据需要自动添加到池中。临时表空间池在正常关闭或终止初始化时被删除。如图 2-15 所示,临时表空间为 5 个页大小,扩展名为.ibt。
- 会话临时表空间保留 40 万个空间 ID。由于服务器每次启动时都会重新创建会话临时表空间池,因此在服务关闭时,会话临时表空间的空间 ID 不会持久存在,可能会被重置。

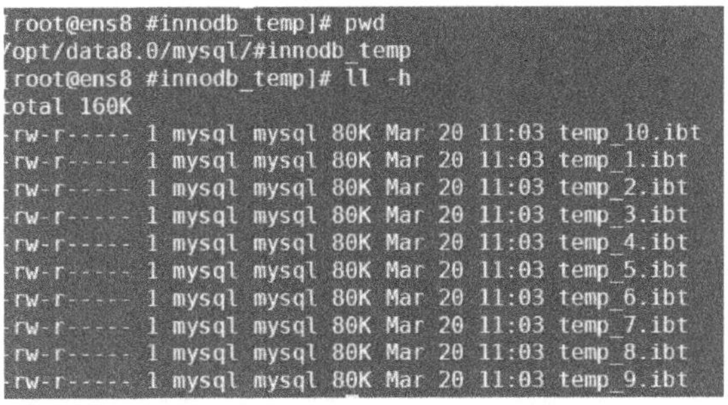

图 2-15 MySQL 8.0 临时表空间

- 参数 innodb_temp_tablespaces_dir 用于设置临时表空间的位置,不支持动态设置。

(2) 全局临时表空间 (ibtmp1)

全局临时表空间存储对用户创建的临时表所做更改的回滚段。

innodb_temp_data_file_path 参数定义了全局临时表空间数据文件的相对路径、名称、大小和属性。如果没有为 innodb_temp_data_file_path 指定值,则默认行为是在 innodb_data_home_dir 目录下创建一个名为 ibtmp1 的自动扩展数据文件。初始文件略大于 12MB。

全局临时表空间在正常关闭服务和终止初始化时被删除,并在每次启动服务时重新创建。全局临时表空间在创建时接收一个动态生成的空间 ID。如果不能创建全局临时表空间,则拒绝启动

服务。如果服务意外停止，则不会删除全局临时表空间。在这种情况下，数据库管理员可以手动删除全局临时表空间或重新启动服务。重新启动服务将自动删除并重新创建全局临时表空间。设置全局临时表空间的方式如下所示。

```
#my.cnf 配置文件
[mysqld]
innodb_temp_data_file_path=ibtmp1:12M:autoextend:max:10G
innodb_temp_tablespaces_dir=/opt/data8.0/temp

# 通过命令行查看大小
mysql> SELECT @@innodb_temp_data_file_path;
+----------------------------------+
| @@innodb_temp_data_file_path     |
+----------------------------------+
| ibtmp1:12M:autoextend            |
+----------------------------------+
```

通过系统表查看临时表空间使用情况，命令行如下所示。

```
# 查看临时表空间路径和使用情况
mysql> SELECT FILE_NAME, TABLESPACE_NAME, ENGINE, INITIAL_SIZE,
TOTAL_EXTENTS * EXTENT_SIZE ASTotalSizeBytes, DATA_FREE, MAXIMUM_SIZE
    FROM INFORMATION_SCHEMA.FILES
    WHERE TABLESPACE_NAME = 'innodb_temporary'\G
*************************** 1.row ***************************
       FILE_NAME: /opt/data8.0/dbdata/ibtmp1
 TABLESPACE_NAME: innodb_temporary
          ENGINE: InnoDB
    INITIAL_SIZE: 134217728
  TotalSizeBytes: 134217728
       DATA_FREE: 126877696
    MAXIMUM_SIZE: 5368709120
1 row in set (0.00 sec)

# 查看临时表空间中每个文件情况
mysql> SELECT * FROM INFORMATION_SCHEMA.INNODB_SESSION_TEMP_TABLESPACES;
+------+------------+------------------------------+-------+----------+-----------+
| ID   | SPACE      | PATH                         | SIZE  | STATE    | PURPOSE   |
+------+------------+------------------------------+-------+----------+-----------+
| 22   | 4243767290 | ./#innodb_temp/temp_10.ibt   | 81920 | ACTIVE   | INTRINSIC |
|  0   | 4243767281 | ./#innodb_temp/temp_1.ibt    | 81920 | INACTIVE | NONE      |
|  0   | 4243767282 | ./#innodb_temp/temp_2.ibt    | 81920 | INACTIVE | NONE      |
|  0   | 4243767283 | ./#innodb_temp/temp_3.ibt    | 81920 | INACTIVE | NONE      |
|  0   | 4243767284 | ./#innodb_temp/temp_4.ibt    | 81920 | INACTIVE | NONE      |
|  0   | 4243767285 | ./#innodb_temp/temp_5.ibt    | 81920 | INACTIVE | NONE      |
|  0   | 4243767286 | ./#innodb_temp/temp_6.ibt    | 81920 | INACTIVE | NONE      |
|  0   | 4243767287 | ./#innodb_temp/temp_7.ibt    | 81920 | INACTIVE | NONE      |
|  0   | 4243767288 | ./#innodb_temp/temp_8.ibt    | 81920 | INACTIVE | NONE      |
|  0   | 4243767289 | ./#innodb_temp/temp_9.ibt    | 81920 | INACTIVE | NONE      |
+------+------------+------------------------------+-------+----------+-----------+
10 rows in set (0.01 sec)
```

```
# 查看临时表情况
mysql> SELECT * FROM INFORMATION_SCHEMA.INNODB_TEMP_TABLE_INFO;
```

5. 回滚表空间

回滚表空间包含 Undo 日志，它是 Undo 日志记录的集合。Undo 日志段包含在回滚段中。innodb_rollback_segments 参数定义了分配给每个回滚表空间的回滚段的数量。查看 Undo 相关参数的命令行如下所示。

```
mysql> SHOW VARIABLES LIKE '%undo%';
+--------------------------------+------------------+
|Variable_name                   |Value             |
+--------------------------------+------------------+
|innodb_max_undo_log_size        |1073741824        |
|innodb_undo_directory           |/opt/data8.0/mysql|
|innodb_undo_log_encrypt         |OFF               |
|innodb_undo_log_truncate        |ON                |
|innodb_undo_tablespaces         |2                 |
+--------------------------------+------------------+
```

说明：innodb_undo_log_truncate 参数指定是否自动截断回滚表空间，需要至少两个活动的回滚表空间，超出 innodb_max_undo_log_size 参数定义的大小限制的回滚表空间将会被截断。

回滚表空间的常规操作的命令行如下所示。

```
# 创建表空间
mysql> CREATE UNDO TABLESPACE tablespace_name ADD DATAFILE 'file_name.ibu';

# 查看回滚表空间信息
mysql> SELECT TABLESPACE_NAME, FILE_NAME FROM INFORMATION_SCHEMA.FILES WHERE FILE_TYPE LIKE 'UNDO LOG';

# 删除表空间
# 回滚表空间在被删除之前必须为空。要清空回滚表空间,必须先将该表空间标记为 INACTIVE,这样该表空间就不再用于为新事务分配回滚段
mysql> ALTER UNDO TABLESPACE tablespace_name SET INACTIVE;   #SET INACTIVE
mysql> DROP UNDO TABLESPACE tablespace_name;

# 清除回滚情况查看
mysql> SELECT NAME, SUBSYSTEM, COMMENT FROM INFORMATION_SCHEMA.INNODB_METRICS WHERE NAME LIKE '%truncate%';

# 回滚运行情况查看
mysql> SHOW STATUS LIKE 'Innodb_undo_tablespaces%';
```

回滚表空间的限制如下。

- 一个 MySQL 实例最多支持 127 个回滚表空间。
- MySQL 实例初始化时创建的默认回滚表空间（innodb_undo_001 和 innodb_undo_002）不能被删除，但是，可以使用 ALTER UNDO TABLESPACE tablespace_name SET INACTIVE 语句使它们处于非活动状态。任何时候，至少需要两个活动的回滚表空间来支持自动截断回滚

表空间。
- innodb_rollback_segments 参数定义了分配给每个回滚表空间和全局临时表空间的回滚段的数量。
- 在 MySQL 服务关闭后,可以手动删除回滚表空间文件,但不建议这样做,因为如果在关闭服务时存在打开的事务,则在服务重启后,关闭的回滚表空间可能会包含活动的回滚日志。
- 加速自动截断回滚表空间。清理线程(Purge thread)负责清空和截断回滚表空间,截断频率由 innodb_purge_rseg_truncate_frequency 参数控制,默认情况下该参数值为 128,即在调用 purge thread 128 次之后,进行清空操作。

当某个回滚表空间被截断时,回滚段将被去激活。其他回滚表空间中的活动回滚段负责平衡整个系统的负载,这可能会导致性能略有下降。性能下降的程度取决于以下几个因素。
- 回滚表空间的数目。
- 回滚日志数量。
- 回滚表空间大小。
- I/O 子系统的速度。
- 现有的长期运行事务。
- 系统负载。

在 MySQL 8.0.21 之前,在回滚表空间截断操作期间会执行以下两个刷新操作。
- 第一个刷新操作是从缓冲池中删除旧的回滚表空间页。
- 第二个刷新操作是将新的回滚表空间的初始页写入磁盘。

在一个繁忙的系统上,如果需要删除大量页面,则第一次刷新操作可能会暂时影响系统性能。
从 MySQL 8.0.21 开始,如果遇到下列情况,则这两个刷新操作都会被删除。
- 最近最少使用时被动释放回滚表空间。
- 完整检查点释放回滚表空间。
- 在截断操作期间,新的回滚页的初始页将被重做记录,而不是刷新到磁盘中。

在日常运维中,如果在 undo_space_number_trunc.log 文件(该日志在 innodb_log_group_home_dir 下)中发现截断操作期间发生系统故障,那么临时日志文件将允许启动进程以识别被截断的回滚表空间,并继续执行该操作。

6. 总结

从表空间的设计上来看,通用表空间能让用户灵活控制数据分布,同时可以解决底层单一硬盘满时迁移的问题。临时表空间能提高处理性能,但对于 MySQL 来说,尽量少用临时表空间。

分离这些表空间确实提高了文件控制的灵活度,但 I/O 的压力肯定会上升。目前,在 MySQL 的使用场景中,很多瓶颈都出在 I/O 上。表空间文件拆分成多个文件,能更有效地处理 MySQL 和 I/O 之间的交互问题。

2.1.5 采用锁竞争算法的事务处理机制

锁竞争问题是关系数据库系统中的核心问题。为了解决锁竞争导致的性能下降问题,在 MySQL 8.0 版本中,引入了 CATS(Contention-Aware Transaction Scheduling,基于竞争感知的事务调度)算法,该算法会计算每个事务阻塞的事务数,然后将锁优先分配给阻塞事务最多的事务。
- 在 MySQL 8.0.20 之前,InnoDB 使用 FIFO(First In First Out,先进先出)算法来调度事务,

该算法会将锁优先分配给最先请求的事务。
- 到 MySQL 8.0.20 版本中，CATS 算法完全取代了 FIFO 算法。

随着 CATS 算法的引入使 FIFO 算法变得多余，所以在后续版本中被移除。在某些情况下，此更改可能会影响事务被授予锁的顺序。

1. CATS 算法介绍

InnoDB 使用 CATS 算法对等待锁的事务进行优先级排序。当多个事务在同一对象上等待锁时，CATS 算法确定哪个事务先收到锁。

CATS 算法通过分配调度权重来对等待的事务进行优先级排序，调度权重是根据事务阻塞的事务数量计算的。例如，如果两个事务正在等待同一对象上的锁，则阻塞最多事务的事务将被分配更大的调度权重。如果权重相等，则高优先级给予等待时间最长的事务。

在论文 Contention-Aware Lock Scheduling for Transactional Databases 中，介绍了几种调度策略，并逐步引申出 CATS 算法。

（1）Number of Locks held

Number of Locks held（NL）是一种在事务拥有诸多锁的情况下，采取 FCFS（First Come First Service，先来先服务）的策略。如图 2-16 所示，通过事务持有锁的数量来判断其优先级。

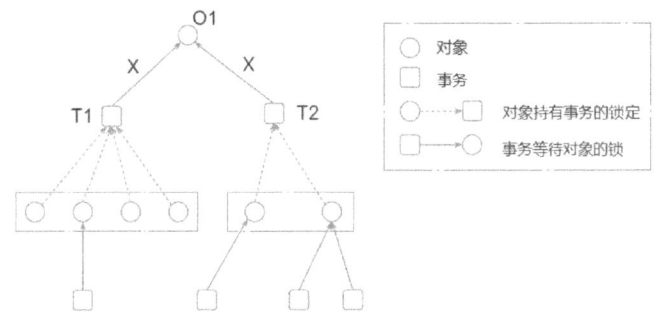

图 2-16　NL 处理方式

NL 处理方式具体解释为：事务 T1 和事务 T2 都在等待对象 O1 的锁，事务 T1 持有的锁数量是 4 个，而事务 T2 持有的锁数量是两个，假如以 "持有锁的数量少" 为标准，那么事务 T2 应该获得锁，但在事务的等待关系中，有 3 个事务等待在 T2 上，而仅有 1 个事务等待在 T1 上。

（2）Number of Locks that Block other Transactions

Number of Locks that Block other Transactions（NLBT）是一种以阻塞其他事务的锁数量来判断优先级的策略。如图 2-17 所示，以阻塞事务的锁数量来判断优先级。

NLBT 处理方式具体解释为：事务 T1 和事务 T2 都在等待对象 O1 的锁，事务 T1 持有的锁中只有一个阻塞了事务 T3，而事务 T2 持有的锁却阻塞了两个事务，假如以等待事务阻塞的事务数量来判断优先级，O1 的锁会被授予 T2，但需要注意的是，事务 T3 阻塞了 3 个其他事务，所以，想提高事务的并发度，最好的选择是将 O1 的锁授予 T1。

（3）Depth of the Dependency Subgraph

Depth of the Dependency Subgraph（DDS）是一种解锁更多的对象，提高并发度的策略。

如图 2-18 所示，以等待事务关系图的深度来判断优先级。

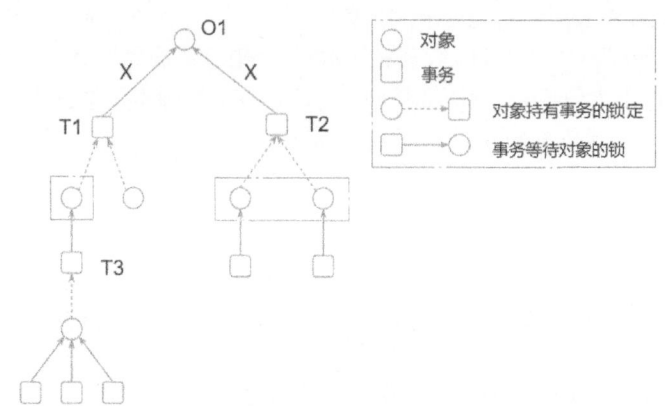

图 2-17　NLBT 的处理方式

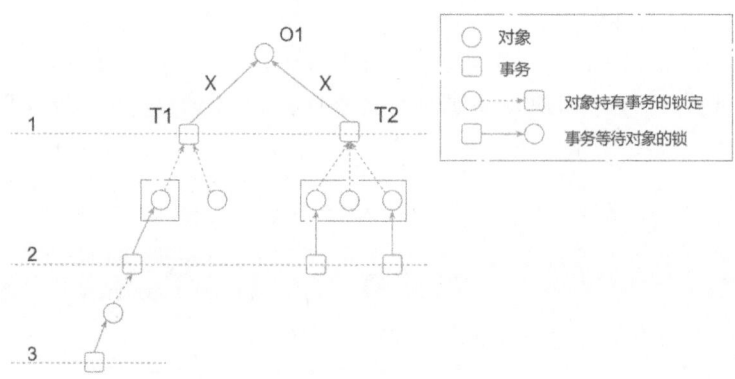

图 2-18　DDS 处理方式

DDS 处理方式具体解释为：事务 T2 同时阻塞两个事务，而事务 T1 有更深的依赖关系，假如将锁授予 T1，势必影响整个数据库的事务并发度。

（4）Largest-Dependency-Set-First

真正的事务等待关系应该表现为有向图，所以计算权重时不应该考虑子树，而是子图。所以，最后提出了一种 Largest-Dependency-Set-First（LDSF）算法。LDSF 是一种根据依赖集的大小锁定的调度算法，根据等待事务所有的等待关系权重来决定锁的调度优先级。LDSF 处理方式如图 2-19 所示。

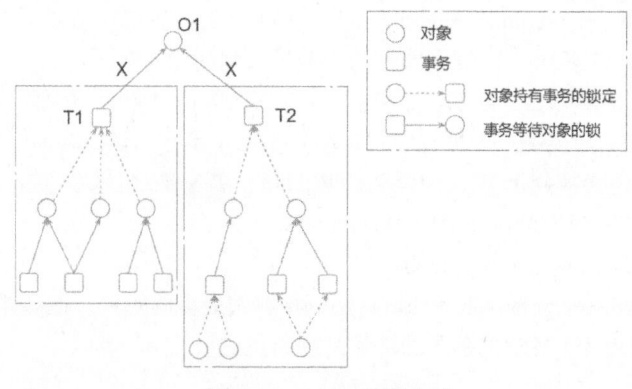

图 2-19　LDSF 处理方式

LDSF 处理方式具体解释为：事务 T1 持有 3 个锁和 4 个事务，事务 T2 持有 5 个锁和 3 个事务。T2 的 3 层子树中依次有两个锁、3 个事务和 3 个锁。所以，可以先释放 T2 最下层子树的 3 个锁，再按照权重释放其他锁。

在 MySQL 8.0 版本中，InnoDB 引擎在原有的事务锁基础上根据 LDSF 实现了 CATS 算法。

2. MySQL 中 CATS 算法监控指标

在 MySQL 8.0 中，可以通过查询 information_schema.INNODB_TRX 表中的 TRX_SCHEDULE_WEIGHT 列来查看事务调度权重。仅为等待事务（waiting transactions）计算权重。等待事务是指那些处于 LOCK WAIT 状态的事务，由 TRX_STATE 列状态而定。

注意，不要将 TRX_STATE 列与 TRX_WEIGHT 列混淆，事务的权重反映的是更改的行数和事务锁定的行数（但不一定是确切的计数）。为了解决死锁问题，InnoDB 选择权重最小的事务作为回滚的"受害者"。

模拟一个语句阻塞场景，如下面示例所示。

```
# 查看事务权重信息
mysql> SELECT trx_id,trx_started,trx_query,trx_weight,trx_schedule_weight
       FROM information_schema.INNODB_TRX;
+---------+---------------------+-----------------------------+------------+---------------------+
| trx_id  | trx_started         | trx_query                   | trx_weight | trx_schedule_weight |
+---------+---------------------+-----------------------------+------------+---------------------+
| 9148699 | 2023-02-24 15:18:46 | update t1 set t0=2 where id=5 |          2 |                   1 |
| 9148698 | 2023-02-24 15:14:16 | update t1 set t0=2 where id=4 |          2 |                   1 |
| 9148697 | 2023-02-24 15:14:07 | NULL                        |          2 |                NULL |
+---------+---------------------+-----------------------------+------------+---------------------+
3 rows in set (0.00 sec)


# 之后可以通过 sys.x$innodb_lock_waits 表找到更详细的对应信息
mysql> SELECT * FROM sys.x$innodb_lock_waits where blocking_trx_id=9148701 limit 1\G;
*************************** 1.row ***************************
                wait_started: 2023-02-24 15:35:39
                    wait_age: 00:05:48
               wait_age_secs: 348
                locked_table: `db1`.`t1`
         locked_table_schema: db1
           locked_table_name: t1
      locked_table_partition: NULL
   locked_table_subpartition: NULL
                locked_index: PRIMARY
                 locked_type: RECORD
              waiting_trx_id: 9148702
         waiting_trx_started: 2023-02-24 15:35:39
             waiting_trx_age: 00:05:48
     waiting_trx_rows_locked: 1
   waiting_trx_rows_modified: 0
                 waiting_pid: 11
               waiting_query: update t1 set t0=2 where id=5
             waiting_lock_id: 140432949921360:2670:4:6:140432881291168
           waiting_lock_mode: X,REC_NOT_GAP
```

```
                blocking_trx_id: 9148701
                   blocking_pid: 14
                 blocking_query: NULL
               blocking_lock_id: 140432949918936:2670:4:6:1404328881272976
             blocking_lock_mode: X,REC_NOT_GAP
              blocking_trx_started: 2023-02-24 15:35:33
                  blocking_trx_age: 00:05:54
         blocking_trx_rows_locked: 7
       blocking_trx_rows_modified: 0
          sql_kill_blocking_query: KILL QUERY 14
     sql_kill_blocking_connection: KILL 14
```

INNODB_METRICS 计数器用于监视代码级事务调度事件，目前有下列 3 种类型。

- lock_rec_release_attempts。试图释放记录锁的次数。一次尝试可能导致零个或多个记录锁被释放，因为在单个结构中可能有零个或多个记录锁。
- lock_rec_grant_attempts。授予记录锁的尝试次数。一次尝试可能导致授予零个或多个记录锁。
- lock_schedule_refreshes。分析等待事件，以及更新事务权重的次数。

查看上述 3 种类型结果的命令行如下所示。

```
mysql> SELECT NAME,SUBSYSTEM,TIME_ENABLED,COUNT,STATUS
       FROM  information_schema.INNODB_METRICS
       WHERE `name` in ('lock_rec_release_attempts',
           'lock_rec_grant_attempts','lock_schedule_refreshes');
+---------------------------+-----------+---------------------+-------+---------+
| NAME                      | SUBSYSTEM | TIME_ENABLED        | COUNT | STATUS  |
+---------------------------+-----------+---------------------+-------+---------+
| lock_rec_release_attempts | lock      | 2023-02-21 15:19:26 |    77 | enabled |
| lock_rec_grant_attempts   | lock      | 2023-02-21 15:19:26 |     0 | enabled |
| lock_schedule_refreshes   | lock      | 2023-02-21 15:19:26 | 66319 | enabled |
+---------------------------+-----------+---------------------+-------+---------+
3 rows in set (0.01 sec)
```

说明：字段说明如下。

- NAME：计数器的唯一名称。
- SUBSYSTEM：指标适用于 InnoDB。
- TIME_ENABLED：计数器启动后经过的时间。
- COUNT：启用计数器以来的计数值。
- STATUS：状态。

3. CAST 算法性能验证

通过 SysBench 工具测试 OLTP-RW 工作负载的性能。测试用例是 8 个表（每个表有 10M 条记录）、128 个客户端测试记录。

- Pareto distribution：帕累托分布，即读写 80/20 分布法则。
- uniformly distributed：读写 50/50 均匀分布法则。

图 2-20 所示为 CAST 算法在 128 线程和 1024 线程下的处理能力对比，其中 msyql-trunk 表示未修改的基线主干、new cats 表示新 CATS 算法、sharding 表示在新 CATS 之上锁定系统碎片。

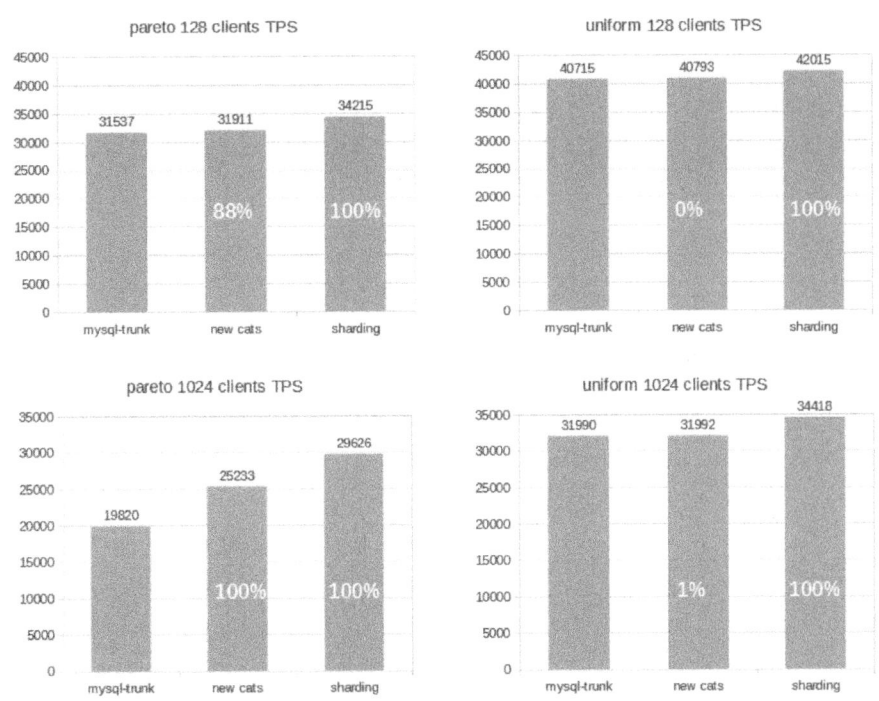

图 2-20　CAST 算法两种线程下处理能力对比

图 2-21 所示为并发数从 1 逐步增加到 1024 过程中，MySQL 的数据字典（dict）、文件（file）、锁（lock）、日志（log）、事务（trx）、等待（wait）、页（page）处理的负载情况。

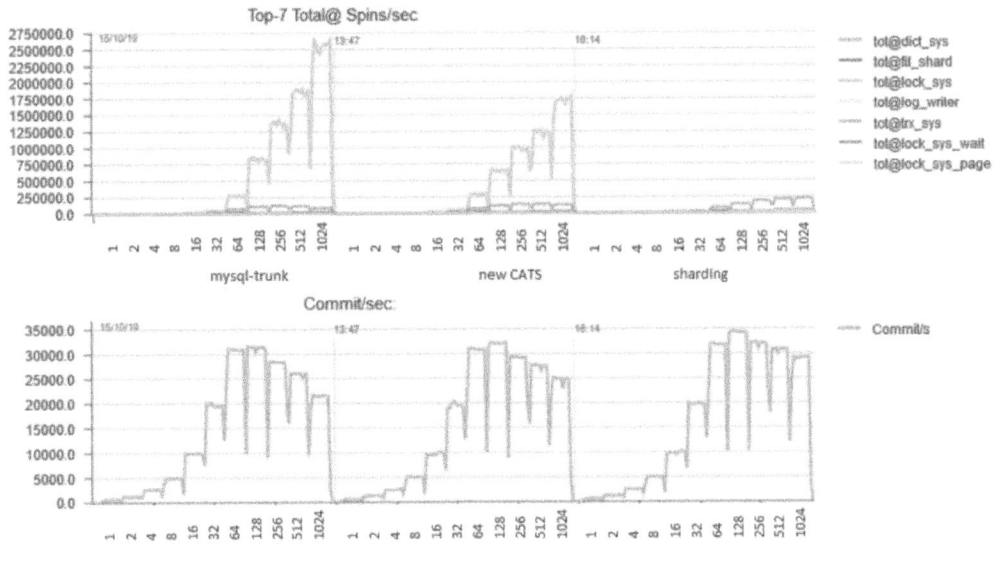

图 2-21　CAST 算法不同并发数下负载情况对比

从上述试验数据来看，事务在争抢锁机制处理方面提升了不少，特别是在数据量小的情况下，效果更明显，另外，在并发数为 1024 的情况下，也有很好的表现。CATS 算法需要更多的指标，赋予对应的权重,有效提高了数据库的事务并发度。

4. 总结

在 MySQL 锁竞争处理上,FIFO 算法在极端情况下无法很好地进行协调,而 CATS 算法有很大的提升。但随着数据量的增加,它们还是存在一定的局限性,可以结合这两种算法,提升 MySQL 8.0 的锁机制处理能力。

2.2 日志体系

2.2.1 binlog 的新增算法与功能

在 MySQL 8.0 中,binlog 有了新的变化,下面具体介绍。

1. binlog 复制回放算法 slave_rows_search_algorithms 参数的变化

当使用基于行的复制格式的副本应用 UPDATE 或 DELETE 操作时,复制进程必须在相关表中搜索匹配的行。用于执行此过程的算法会将表中的一个索引作为首选来执行搜索,如果没有合适的索引,则使用其他算法。

- MySQL 5.7 使用表扫描(默认值:TABLE_SCAN、INDEX_SCAN)。
- MySQL 8.0 使用哈希表(默认值:INDEX_SCAN、HASH_SCAN)。

该回放算法首先评估表定义中的索引,以确定是否有合适的索引可以使用,如果有多种选项,则判断哪个索引最适合此操作。该算法忽略以下类型的索引。

- 全文索引。
- 隐藏索引。
- 计算索引。
- 多值索引。
- 在复制时,更新的表不包含索引字段(与记录 binlog 内容有关,参数对应 binlog_row_image)。

如果记录的语句有可用的索引,则从候选索引中选择一个,复制回放算法优先级顺序见表 2-2。

表 2-2 MySQL 8.0 复制回放算法优先级顺序表

复制回放算法 索引	INDEX_SCAN、HASH_SCAN	INDEX_SCAN、TABLE_SCAN
主键或唯一键	索引扫描	索扫描引
其他索引	哈希扫描	全表扫描
无索引	哈希扫描	全表扫描

可以按照以下原则进行选择。

- 无论什么时候,只要存在主键,就优先选择主键。
- 在无主键且只有一个索引的情况下,选择唯一的索引,其中唯一索引中的每一列都需要赋予 NOT NULL 属性。
- 在无主键且有多个索引的情况下,算法将选择这些索引中最左边的一个。

如果该算法无法找到合适的索引,或者只能找到非唯一的索引或包含空值的索引,则使用哈希表来帮助识别表记录。

该算法创建一个哈希表，其中包含 UPDATE 或 DELETE 操作中的行的键来作为对比哈希值。然后，该算法遍历目标表中的所有记录。对于目标表中的每条记录，它确定该行是否存在于哈希表中。如果在哈希表中找到了该行，则更新目标表中的记录，并从哈希表中删除该行。当目标表中的所有记录都被检查后，该算法验证哈希表现在是否为空。如果哈希表中还有任何不匹配的行，则该算法返回错误 ER_KEY_NOT_FOUND 并停止复制应用程序线程。

可以说，在 MySQL 8.0 主从复制架构的从库回放机制中，主键和唯一键与之前的版本保持一致。对于无主键的情况，MySQL 8.0 采取哈希扫描的方式。哈希扫描利弊兼有，肯定比全表扫描速度更快，但在数据量非常大的情况下，需要构建哈希表，会消耗非常长的时间。

2. binlog 新的日志清理机制

在 MySQL 8.0 中，添加了新的 binlog 日志清理参数。而在 MySQL 8.0 之前的版本中，以天为单位进行清理。

（1）binlog_expire_logs_seconds 参数

该参数表示 binlog 可以秒为单位进行日志清理，之前的参数是以 expire_logs_days（天）为单位进行清理的，因为有时在 1 天之中，可能会生成几百 GB 的 binlog 日志，导致空间不够用，需要以手动或脚本方式清理，现在不用那么麻烦了，更贴合使用现状。

（2）binlog_expire_logs_auto_purge 参数

该参数可设置是否使用 binlog 日志自动清理机制，从 MySQL 8.0.29 开始，可以通过将该参数设置为 OFF 来禁用 binlog 日志的自动清理功能。它的设置优先于 binlog_expire_logs_seconds 参数的任何设置。

从如下代码注释可以看出，expire_logs_days 准备废弃，binlog_expire_logs_auto_purge 和 binlog_expire_logs_seconds 需要配合使用。

```
Warning1287'@@expire_logs_days' is deprecated and will be removed
in a future release.Please use binlog_expire_logs_seconds instead.

# ASSERT: If binlog_expire_logs_auto_purge is set to ON, then the server
#         MUST purge automatically binary log files according to the
#         binlog_expire_logs_seconds setting.
#


#
# ASSERT: If binlog_expire_logs_auto_purge is set to ON and both
#         binlog_expire_logs_seconds and expire_logs_days are set to 0,
#         then binary log files SHALL NOT be automatically purged.
```

3. binlog 加密机制

binlog 记录着对于数据的变更信息，因为 binlog 可以通过工具直接解析，存在一定的数据泄漏风险。从 MySQL 8.0 开始，对 binlog 采用加密机制，在安全性上有一定的增强。相关参数介绍如下。

（1）binlog_encryption 参数

该参数表示为此服务器上的 binlog 和 relay 文件启用加密功能，必须安装并配置一个密钥环插件来提供 MySQL 服务器的密钥环服务。

（2）binlog_rotate_encryption_master_key_at_startup 参数

该参数表示多次加密 binlog 日志。

- ON：每当服务重新启动时，将生成一个新的加密密钥，并将其用作所有后续 binlog 文件和 relay log 的主密钥。
- OFF：再次使用现有的 binlog 主密钥。

4. binlog 记录事件占用空间

binlog 记录事件占用空间是指对于 binlog 记录的事件，最大占用存储空间的软限制（以字节为单位）。通常将存储在 binlog 中的行分组为大小不超过默认值（8192B）的事件，如果单个事件不能拆分，则可以超过最大占用空间值（64 位系统：17179869184GB，32 位系统：4GB）。设置 binlog 记录事件占用空间的参数为 binlog_row_event_max_size，如果事件大小操过这个阈值，就会被截断，无法记录完整事件信息。

5. binlog 列字段记录模式

在 ROW 记录模式下，binlog 日志中记录着元数据的数量信息。主要应对元数据的表结构与源的表结构不同的情况，副本使用元数据来传输数据。同时，外部软件可以使用元数据解码行事件并将数据存储到外部数据库（如数据仓库）中。binlog_row_metadata 参数用于控制记录元数据数量。当设置为默认值 MINIMAL 时，只记录与 SIGNED 标志、列字符集和地理空间类型（Geometry 类型）相关的元数据。当设置为 FULL 时，将记录表的完整元数据，如列名、ENUM 或 SET 字符串值、PRIMARY KEY 信息等。

6. binlog 对于 JSON 数据的处理

JSON 数据格式在可视化方面带来了便利，但数据本身占用的空间过大，若不进行特殊处理，则会浪费存储空间。binlog 里记录的 JSON 事件是逻辑语句，大的 JSON 串，会占有较多空间且传输到从节点的网络资源也会比较多，所以 MySQL 8.0 中，对 JSON 事件记录信息进行了优化。

通过参数 binlog_row_value_options 控制 binlog 记录修改 JSON 文档的一小部分内容，而不是写入完整的 JSON 值。它适用于 UPDATE 语句，使用 JSON_SET 函数、JSON_REPLACE 函数和 JSON_REMOVE 函数的任意序列修改 JSON 列。如果变更语句无法生成部分更新，则使用完整文档。

7. binlog 文件的压缩

对于数据库的变更越多，产生的 binlog 文件会越多，占用空间也会越多，对 binlog 文件压缩可以有效地节省空间。相关参数介绍如下。

（1）binlog_transaction_compression 参数

该参数表示 binlog 启用压缩机制。主节点压缩的 binlog 信息，将以压缩状态传输到从节点并写入 relay log。

（2）binlog_transaction_compression_level_zstd 参数

该参数表示 binlog 内容的压缩级别，随着压缩级别的增加，数据压缩比也会增加，这将减少事务有效负载所需的存储空间和网络带宽。但是，数据压缩所需的工作量也会增加，会占用原始服务器上的时间、CPU 和内存资源。压缩工作量的增加与数据压缩比的增加之间没有线性关系。

8. binlog 精准时间记录

MySQL 复制环节中，作为传输主库事件的 binlog，因为没有准确的执行事件对比时间，会出现延迟误报的情况。MySQL 8.0 添加了供内部复制使用的精准时间。相关记录字段介绍如下。

- original_commit_timestamp：事务在 master 提交 binlog 的时间戳（微秒），该时间戳在每个节

- immediate_commit_timestamp：事务在 slave（包括中继节点）提交 binlog 的时间戳（微秒），该时间戳在 relay log 中与 original_commit_timestamp 一样，在 slave 提交的 binlog 是完成回放的时间戳。

可以通过 mysqlbinlog 解析 binlog 事件，内容如下所示。

```
#221028  8:58:55 server id 129   end_log_pos 276 CRC32 0x17f79afd   GTID
last_committed=0  sequence_number=1  rbr_
only=yes    original_committed_timestamp=1666918735086611
immediate_commit_timestamp=1666918735086611    transaction_length=390
/*!50718 SET TRANSACTION ISOLATION LEVEL READ COMMITTED*//*!*/;
# original_commit_timestamp=1666918735086611 (2022-10-28 08:58:55.086611 CST)
# immediate_commit_timestamp=1666918735086611 (2022-10-28 08:58:55.086611 CST)
/*!80001 SET @@session.original_commit_timestamp=1666918735086611*//*!*/;
/*!80014 SET @@session.original_server_version=80031*//*!*/;
/*!80014 SET @@session.immediate_server_version=80031*//*!*/;
SET @@SESSION.GTID_NEXT= '22228e8c-b0ee-11ec-a2d1-00163e23e2cc:70'/*!*/;
```

说明：这就是将事务原始提交时间写在 binlog 中，提交时间在复制链路上传递，使得 slave 可以计算事务延迟时间，更全面和精准。

9. 总结

MySQL 8.0 的 binlog 新特性，为日常运维和解决问题提供了更大的助力。如果从上述变化中，选择让人受益较大的变化，则必然是以下这 3 个。

- slave_rows_search_algorithms：使用哈希替代传统的扫描表方式，搜索行中的匹配项。
- binlog_expire_logs_seconds：秒级别清理 binlog 日志机制。
- original_commit_timestamp：复制延迟观测新方式。

2.2.2　Redo 日志的新增功能

想要保证关系数据库系统中数据不丢失，可通过 WAL（Write-Ahead Logging）预写 Redo 日志实现。Redo 日志记录着事务里对数据的修改，通过日志文件保证数据的持久性。MySQL 中，Redo 日志是 InnoDB 引擎特有的，其功能如下。

- 保证实现 crash-safe 能力，发生崩溃后也能完整地恢复事务中的数据。
- Redo 日志记录在数据页上的修改内容。
- 循环写。所有 Redo 日志写满后覆盖以重复使用。
- 后台有线程自动刷新落盘，顺序写 I/O。

在 MySQL 中，Redo 日志通过如下参数设置：Redo 日志通过 innodb_log_file_size 和 innodb_log_files_in_group 参数进行调节，Redo 日志数据则存储在数据目录下的 ib_logfile0、ib_logfile1……ib_logfileN 文件当中；由 innodb_log_buffer_size 参数来控制刷新到 Redo 文件的缓存大小；innodb_flush_log_at_trx_commit 参数控制如何将日志缓冲区的内容写入和刷新到磁盘；innodb_flush_log_at_timeout 参数控制日志刷新频率。

在 MySQL 8.0 中，对 Redo 进行了更新。下面了解一下 Redo 新功能。

1. Redo 日志加密机制

MySQL 8.0.16 中，使用 innodb_redo_log_encrypt 参数进行 Redo 日志加密。默认情况下，Redo 日

志加密是禁用的。与表空间数据一样，Redo 日志加密是在将数据写入磁盘时进行的，而解密是在从磁盘读取 Redo 日志数据时进行的。一旦数据读入内存，就处于未加密状态。使用表空间加密密钥对 Redo 日志数据进行加密和解密。

在启用 innodb_redo_log_encrypt 参数时，磁盘上存在的未加密 Redo 日志页面将保持未加密状态，新的 Redo 日志页面将以加密形式写入磁盘。同样，禁用 innodb_redo_log_encrypt 参数时，磁盘上存在的已加密 Redo 日志页面将保持加密状态，新的 Redo 日志页面将以未加密形式写入磁盘。设置命令行如下所示。

```
mysql> SET GLOBAL innodb_redo_log_encrypt=ON;
Query OK, 0 rows affected (0.00 sec)
```

在 MySQL 8.0.30 之前，Redo 日志加密元数据（包括表空间加密密钥）存储在第一个 Redo 日志文件（ib_logfile0）的头中。如果删除此文件，则 Redo 日志加密将被禁用。

从 MySQL 8.0.30 开始，Redo 日志加密元数据，包括表空间加密密钥，存储在最近的检查点 LSN 的 Redo 日志文件的头中。

2. Redo Log Archiving（日志归档）功能

在 MySQL 8.0.17 中，加入了 Redo 日志归档功能，主要是为了解决备份一致性的问题。当备份操作正在进行时，复制 Redo 日志记录的备份实用程序有时可能无法跟上 Redo 日志生成的速度，从而导致 Redo 日志记录由于被覆盖而丢失。

MySQL Enterprise Backup 在备份 MySQL 服务器时使用 Redo 日志归档功能。相应的官方工具还没有提供直接分析 Redo 日志归档文件的功能。Percona Xtrabackup 目前也没有支持 Redo 日志归档的功能。

设置 Redo 日志归档功能如下所示。

```
#设置归档文件
#语法:innodb_redo_log_archive_dirs='label1:directory_path1[;label2:directory_path2;…]'
mysql> SET GLOBAL innodb_redo_log_archive_dirs=
'redo_archiving:/opt/data8.0/redo_archive/';

#启动
mysql> SELECT innodb_redo_log_archive_start('redo_archiving','20220830');
+------------------------------------------------------------+
| innodb_redo_log_archive_start('redo_archiving','20220830') |
+------------------------------------------------------------+
|                                                          0 |
+------------------------------------------------------------+

#停止
mysql> SELECT innodb_redo_log_archive_stop();
+--------------------------------+
| innodb_redo_log_archive_stop() |
+--------------------------------+
|                              0 |
+--------------------------------+
```

下面介绍使用 Redo 日志归档功能时的常见错误。

当设置 Redo 日志归档功能时，提示无权访问错误。需要赋予 INNODB_REDO_LOG_ARCHIVE 权限。赋予权限操作如下所示。

```
#错误日志
ERROR 1227 (42000): Access denied; you need (at least one of) the
INNODB_REDO_LOG_ARCHIVE privilege(s) for this operation

#赋予权限
mysql> GRANT INNODB_REDO_LOG_ARCHIVE  on  *.* to'root'@'localhost';
Query OK, 0 rows affected (0.01 sec)
mysql> FLUSH PRIVILEGES;
Query OK, 0 rows affected (0.00 sec)
```

Redo 日志归档设置提示不存在目录错误,目录必须事先创建。错误信息如下所示。

```
#错误日志
ERROR 3844 (HY000): Redo log archive directory 'directory_path1' does not exist or is not a directory
```

Redo 日志归档设置提示"访问文件路径需要设置成 0700 权限模式"错误。错误信息如下所示。

```
#错误日志
ERROR 3846 (HY000): Redo log archive directory 'directory_path1' is accessible to all OS users
```

3. 禁用 Redo 日志功能

从 MySQL 8.0.21 开始,支持禁用 Redo 日志功能。可以通过避免写入 Redo 和 doublewrite 缓冲来加快数据的加载。一般初期大量灌入数据时可采用此功能,平时不建议使用此功能,因为意外发生或服务停止都可能会导致数据丢失和实例损坏。

禁用 Redo 日志功能,需要对应的权限。命令行如下所示。

```
#启用 Redo 日志功能
mysql> ALTER INSTANCE ENABLE  INNODB REDO_LOG;
ERROR 1227 (42000): Access denied; you need (at least one of) the INNODB_REDO_LOG_ENABLE privilege(s) for this operation

#赋予权限
mysql> GRANT INNODB_REDO_LOG_ENABLE ON *.* to'root'@'localhost';
mysql> ALTER INSTANCE ENABLE  INNODB REDO_LOG;
Query OK, 0 rows affected (0.00 sec)

#查看
mysql> SHOW GLOBAL STATUS LIKE 'Innodb_redo_log_enabled';
+-------------------------------------+------------+
|Variable_name                        |Value       |
+-------------------------------------+------------+
|Innodb_redo_log_enabled              |ON          |
+-------------------------------------+------------+
1 row in set (0.00 sec)

#禁用
mysql> ALTER INSTANCE DISABLE INNODB REDO_LOG;
Query OK, 0 rows affected (0.00 sec)
```

4. Redo 重新定义

MySQL 8.0.30 中重新设计了 Redo 日志物理文件，即拆分成多个 Redo 日志物理文件，替换了原有的模式。在 MySQL 8.0.34 中，启动服务时会提示 Redo 设置参数废弃信息，如下提示。

```
2023-08-19T10:12:23.732750+08:00 0 [Warning] [MY-013907]
[InnoDB] Deprecated configuration parameters innodb_log_file_size and/or
innodb_log_files_in_group have been used to compute
innodb_redo_log_capacity=134217728.Please use innodb_redo_log_capacity instead.
```

从 MySQL 8.0.30 开始，innodb_redo_log_capacity 参数控制 Redo 日志文件占用的磁盘空间，取代了 innodb_log_files_in_group 和 innodb_log_file_size 变量，后两个变量已经被弃用。

可以在启动或运行时使用 SET GLOBAL 语句设置此参数，如将 Redo 日志容量设置为 8GB，命令行如下所示。

```
#my.cnf 配置文件
[mysqld]
innodb_redo_log_capacity = 8589934592

#SQL 命令行设置 Redo 整体大小
mysql> SET GLOBAL innodb_redo_log_capacity = 8589934592;
```

当 MySQL 服务启动时，生成的 Redo 日志文件已经占用物理空间。若在运行时设置，则配置更改将立即生效，但可能需要一段时间才能完全实现新的限制。对于负载高的系统，变更还会存在一定的风险。

- 如果 Redo 日志文件占用的空间小于指定的值，那么从缓冲池中将"脏"页刷新到表空间数据文件的速度就会降低，最终会增加 Redo 日志文件占用的磁盘空间。
- 如果 Redo 日志文件占用的空间超过了指定的值，那么脏页会更频繁地刷新，最终减少 Redo 日志文件占用的磁盘空间。

innodb_log_group_home_dir 参数定义了 InnoDB 日志文件的目录路径。可以使用这个选项将 InnoDB Redo 日志文件放置在与 InnoDB 数据文件不同的物理存储位置，以避免潜在的 I/O 资源冲突。设置 Redo 日志文件路径如下所示。

```
#my.cnf 配置文件
[mysqld]
innodb_log_group_home_dir = /opt/data8.0/redo
```

默认情况下，Redo 日志文件会放在 data 目录的 #innodb_redo 子目录下。如图 2-22 所示，InnoDB 共维护 32 个 Redo 日志文件，每个文件的大小 = 1/32×innodb_redo_log_capacity。

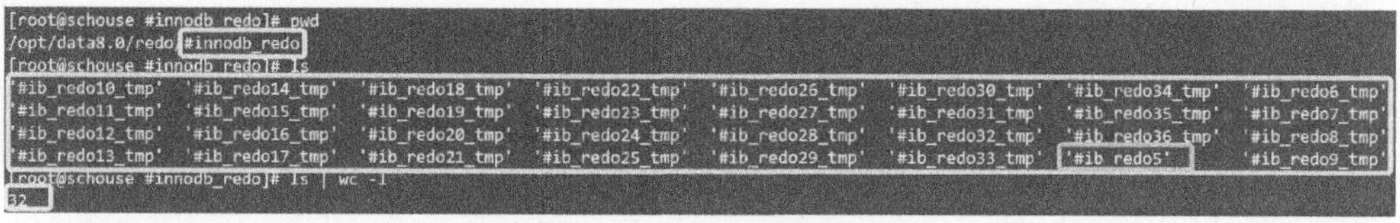

图 2-22 Redo 生成的临时文件

每个 Redo 日志文件都与一个特定的 LSN 值范围相关联（START_LSN 和 END_LSN 值）。查看 LSN 信息的命令行如下所示。

```
mysql> SELECT FILE_ID,FILE_NAME,START_LSN, END_LSN, SIZE_IN_BYTES, IS_FULL, CONSUMER_LEVEL
FROM performance_schema.innodb_redo_log_files\G
*************************** 1.row ***************************
       FILE_ID: 1235
     FILE_NAME: /opt/data8.0/redo/#innodb_redo/#ib_redo1235
     START_LSN: 88783128576
       END_LSN: 88787320832
 SIZE_IN_BYTES: 4194304
       IS_FULL: 0
CONSUMER_LEVEL: 0
1 row in set (0.00 sec)
```

当执行到检查点时，InnoDB 将检查点 LSN 存储在包含该 LSN 的文件头中。在恢复期间，检查所有 Redo 日志文件，并从最新的检查点 LSN 开始恢复 "提供了几个状态变量来对 Redo 日志进行监控和容量调整操作"信息。查看 Redo 日志的状态信息的命令行如下所示。

```
# Redo 大小调整的状态(OK:无调整操作, Resizing down:正在进行调整)
mysql> SHOW STATUS LIKE 'Innodb_redo_log_resize_status';
+--------------------------------+-------+
|Variable_name                   |Value  |
+--------------------------------+-------+
|Innodb_redo_log_resize_status   |OK     |
+--------------------------------+-------+
1 row in set (0.00 sec)

# 当前 Redo 日志的容量限制
mysql> SHOW STATUS LIKE 'Innodb_redo_log_capacity_resized';
+----------------------------------+-----------+
|Variable_name                     |Value      |
+----------------------------------+-----------+
|Innodb_redo_log_capacity_resized  |134217728  |
+----------------------------------+-----------+
1 row in set (0.00 sec)
```

5. 新的参数和状态值

表 2-3 所示为 MySQL 8.0 中 Redo 日志新的参数及其相关说明。

表 2-3 MySQL 8.0 中 Redo 日志新的参数及其相关说明

参　　数	版　　本	类　　型	说　　明
innodb_redo_log_encrypt	8.0.16	variable	Redo 日志数据的加密，与表空间数据一样，当 Redo 日志数据写入磁盘时，Redo 日志数据进行加密；当 Redo 日志数据从磁盘读取时，进行解密
innodb_redo_log_archive_dirs	8.0.17	variable	Redo 日志归档文件目录
Innodb_redo_log_enabled	8.0.21	status	禁用 Redo
innodb_redo_log_capacity	8.0.30	variable	Redo 设置的文件总大小
Innodb_redo_log_read_only	8.0.30	status	表示 Redo 是否只读，目前无设置方式
Innodb_redo_log_uuid	8.0.30	status	Redo UUID

(续)

参　　数	版　本	类　型	说　　明
Innodb_redo_log_checkpoint_lsn	8.0.30	status	Redo 检查点 LSN
Innodb_redo_log_current_lsn	8.0.30	status	Redo 当前检查点 LSN，当前 LSN 表示 Redo 日志中最后写入的位置
Innodb_redo_log_flushed_to_disk_lsn	8.0.30	status	Redo 刷新到磁盘 LSN
Innodb_redo_log_logical_size	8.0.30	status	Redo 正在使用的数据的 LSN 范围
Innodb_redo_log_physical_size	8.0.30	status	Redo 磁盘上所有日志文件（不包括备用 Redo 日志文件）当前消耗的磁盘空间
Innodb_redo_log_capacity_resized	8.0.30	status	Redo 容量调整操作之后，所有日志文件的总容量
Innodb_redo_log_resize_status	8.0.30	status	Redo 大小调整的状态

6. 总结

MySQL 8.0 中的 Redo 具有多样化特点，可以让 DBA 更直观地理解 Redo 的使用情况，也为排查 Redo 相关问题提供了便利。在使用方面，设置完 Redo 大小之后，尽量不要变更。其他建议如下。

- 若使用 MySQL 8.0.30 及以上版本，需要先调整参数配置。
- 对于 MySQL 数据库在线升级，需要谨慎，因为 Redo 的体系变更有可能出现未知的问题，所以需要先做好备份。
- 若没有特殊情况，不建议使用禁用 Redo 功能。
- 目前 Redo 日志归档的意义不大，因为存在 binlog 文件。如果能提供分析 Redo 的工具，那么 MySQL 又多了一种数据恢复手段。

2.2.3 慢查询日志的附加信息

在 MySQL 的 SQL 性能问题排查中，慢查询日志提供了比较完善的 SQL 语句执行情况。通过这些 SQL 语句，能解决性能瓶颈问题。在有些情况下，一个 SQL 语句之所以出现在慢查询日志中，并不是因为它本身执行效率低，而是因为其他因素导致的。

下面介绍 MySQL 8.0 版本的慢查询日志新添加的附加信息。通过这些附加信息，可以更好地判断是否存在性能瓶颈。附加信息包含了 InnoDB 引擎层内部处理的状态值。

1. 慢查询日志的附加信息介绍

MySQL 5.7 中慢查询日志记录信息如下所示。

```
# Time: 2023-06-29T02:52:19.936039Z
# User@Host: root[root] @localhost []  Id:     4
# Query_time: 0.000272  Lock_time: 0.000161 Rows_sent: 3  Rows_examined: 3
SET timestamp=1688007139;
select * from db1.t1;
```

说明：上述记录的慢查询日志中的几个指标介绍如下。

- Query_time：执行时间（秒）。
- Lock_time：锁时间（秒）。
- Rows_sent：发送到客户端的行数。

- Rows_examined：服务层检查的行数（不包括存储引擎内部的任何处理）。

在 MySQL 8.0.14 开始启用 log_slow_extra 参数，它会额外记录每条语句的 InnoDB 性能计数器的值。记录信息如下所示。

```
# Time: 2023-06-29T12:26:25.919007+08:00
# User@Host: root[root] @ localhost [] Id:     7
# Query_time: 0.000243  Lock_time: 0.000003 Rows_sent: 3  Rows_examined: 3
Thread_id: 7Errno: 0 Killed: 0 Bytes_received: 23 Bytes_sent: 168 Read_first: 1
Read_last: 0 Read_key: 1 Read_next: 0 Read_prev: 0 Read_rnd: 0 Read_rnd_next: 4
Sort_merge_passes: 0 Sort_range_count: 0 Sort_rows: 0 Sort_scan_count: 0
Created_tmp_disk_tables: 0 Created_tmp_tables: 0
Start: 2023-06-29T12:26:25.918764+08:00 End: 2023-06-29T12:26:25.919007+08:00
SET timestamp=1688012785;
select * from t1;
```

表 2-4 所示为 MySQL 8.0 慢查询日志性能计数信息。

表 2-4　MySQL 8.0 慢查询日志性能计数信息

性能指标	说明
Thread_id	语句线程 ID
Errno	语句错误号，如果没有发生错误，则为 0
Killed	如果语句非正常终止，则为表示原因的错误号；如果语句正常终止，则为 0
Bytes_received	语句的 Bytes_received 值
Bytes_sent	语句的 Bytes_sent 值
Read_first	语句的 Handler_read_first 值
Read_last	语句的 Handler_read_last 值
Read_key	语句的 Handler_read_key 值
Read_next	语句的 Handler_read_next 值
Read_prev	语句的 Handler_read_prev 值
Read_rnd	语句的 Handler_read_rnd 值
Read_rnd_next	语句的 Handler_read_rnd_next 值
Sort_merge_passes	语句的 Sort_merge_passes 值
Sort_range_count	语句的 Sort_range_count 值
Sort_rows	语句的 Sort_rows 值
Sort_scan_count	语句的 Sort_scan_count 值
Created_tmp_disk_tables	语句的 Created_tmp_disk_tables 值
Created_tmp_tables	语句的 Created_tmp_tables 值
Start	语句执行的开始时间
End	语句执行的结束时间

说明：如果慢查询日志的 SQL 中出现密码信息，则将会由服务重写，而不是以纯文本形式出现。无法解析的语句（如出现语法错误的语句）不会被写入慢查询日志。

2. 慢查询日志的附加信息开启设置

开启慢查询日志的附加信息记录的方式如下所示。

```
#配置文件my.cnf
[mysqld]
log_slow_extra= ON

#以命令行方式设置
mysql> SET PERSIST log_slow_extra=ON;
```

3. 总结

MySQL 8.0 中慢查询日志的附加信息，提供了更详细的语句执行过程，能帮助用户更有效地分析 SQL 语句当时的运行情况。虽然慢查询日志的附加信息的记录会使慢查询日志文件变大，但为分析语句提供了非常重要的信息。在 MySQL 运行环境中，慢查询日志的附加信息记录功能必须开启。

2.3 引擎

2.3.1 InnoDB 引擎底层结构的变化

在数据库系统中，为了加快处理速度，将数据文件加载到内存中，然后读取内存中的数据，并进行处理。MySQL 的 InnoDB 引擎缓冲池，与 Oracle 的 SGA（系统全局内存区）内的共享缓存储区一样，可实现对数据库数据的管理和操作。MySQL 8.0 中 InnoDB 引擎的缓冲池可分为 4 个区域，它们支持不同的内存服务体系。

1. 缓冲池

缓冲池（Buffer Pool）是 MySQL 里保存内存数据的地方。缓冲池里包含实际数据区域和索引数据区域。缓冲池允许直接从内存访问频繁使用的数据。在服务器上，通常物理内存的 50%~80% 分配给缓冲池。

有限的缓冲池大小，无法把大量的数据全部加载到缓冲池中。因此 MySQL 采用经典的 LRU 算法淘汰旧数据。使用 LRU 算法将缓冲池作为列表进行管理。当需要空间向缓冲池添加新页时，将删除最近最少使用的页，并将新页添加到列表的中间。这种中点插入策略将列表分成两个子列表。

图 2-23 所示为缓冲池的结构。

- 头部，最近被访问的新（young）页的子列表。
- 尾部，最近访问次数较少的旧页的子列表。
- 将经常使用的页保留在新子列表中。旧子列表包含较少使用的页。当无缓冲空间可用时，页将被驱逐。
- 缓冲区新、旧子列表默认以 5/8 和 3/8 的比例划分。

想要监控缓冲池的变化，可以通过如下命令行实现。

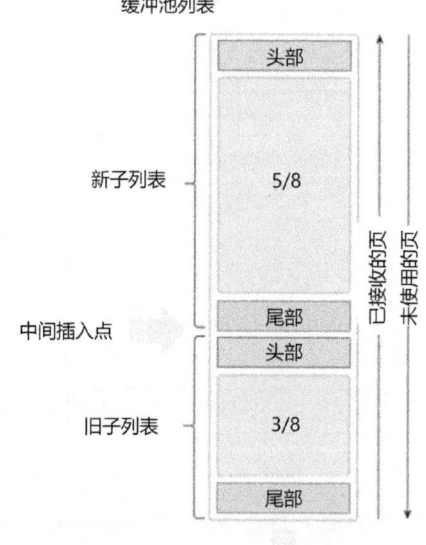

图 2-23 缓冲池结构

通过下列这些状态值，可以分析出缓冲池是否足够满足现有需求。

```
mysql> SHOW ENGINE INNODB STATUS \G;
----------------------
BUFFER POOL AND MEMORY
----------------------
Total large memory allocated 0              # 分配给 InnoDB 的总的内存大小
Dictionary memory allocated 480465          # 分析给 InnoDB 数据字典的内存大小
Buffer pool size      8192                  # 分配给 InnoDB 的总的缓冲池大小,单位为页
Free buffers          6975                  # 数据库中 InnoDB 的缓冲池中空闲页的数量
Database pages        1211                  # 数据库中 InnoDB 的缓冲池中非空闲页的数量
Old database pages    467                   # 旧子列表中的页数量
Modified db pages     12                    # 当前缓冲池中被修改的页数量
Pending reads         4                     # 数据由磁盘读取到缓冲池过程中被挂起的次数
Pending writes: LRU 0, flush list 0, single page 0
                                            # 写入被挂起的次数
                                            # LRU 链表中的页被淘汰出内存,要写入磁盘中
                                            # check point 操作期间页要被写入磁盘中
                                            # 单个页要被写入磁盘中
Pages made young 2, not young 0             # 新页由旧列表移动到新子列表的次数
                                            # 旧页由新子列表移动到旧子列表的次数
0.17 youngs/s, 0.00 non-youngs/s            # youngs/s:平均每秒有多少个页由旧子列表移动到新子列表中
                                            # non-youngs/s:平均每秒有多少个页由新子列表移动到旧子列表中
Pages read 1067, created 144, written 199
                                            # 从缓冲池中读、创建和写的页的总数
90.26 reads/s, 12.18 creates/s, 16.83 writes/s
                                            # 平均每秒从缓冲池中读、创建和写的页的数量
Buffer pool hit rate 941 / 1000, young-making rate 0 / 1000 not 0 / 1000
                                            # 缓冲池的命中率,无限接近于 1
Pages read ahead 0.00/s, evicted without access 0.00/s, Random read ahead 0.00/s
                                            # 每秒预读的次数,每秒淘汰的页的次数,随机预读的次数
LRU len: 1211, unzip_LRU len: 0
I/O sum[0]:cur[0], unzip sum[0]:cur[0]
```

从上述监控的缓冲池变化中就能看出哪些指标会影响缓冲池。

（1）缓冲池大小

理想情况下，将缓冲池的大小设置为实际的最大值，缓冲池越大，InnoDB 越像一个内存数据库，即只需要从磁盘读取一次数据，之后会从内存访问数据。表 2-5 所示为 3 个重要的缓冲池设置参数。

表 2-5　MySQL 8.0 中的缓冲池设置参数

参　　数	说　　明
innodb_buffer_pool_size	缓冲池的大小（以字节为单位），InnoDB 缓存表和索引数据的内存区域
innodb_buffer_pool_chunk_size	缓冲池以块（chunk）为单位进行控制。innodb_buffer_pool_size 应该等于 innodb_buffer_pool_chunk_size×innodb_buffer_pool_instances 的倍数
innodb_buffer_pool_instances	使用哈希函数，将缓冲池划分为多个域，提高并发性，每个缓冲池管理自己的空闲列表、刷新列表、LRU 等结构，并由自己的缓冲池互斥锁保护。单个实例分配域大于 1GB 才生效

说明：MySQL将单个实例分配域设计成大于1GB才生效。为避免潜在的性能问题，块的数量（innodb_buffer_pool_size/innodb_buffer_pool_chunk_size）不应该超过1000。目前官方提供的innodb_buffer_pool_chunk_size没有最大值限制。可以通过算法灵活控制其大小。

（2）新、旧页分配

缓冲池新旧配置基于协调LRU设置值和时间进行替换规则。表2-6所示为新、旧页分配设置参数。

表2-6　MySQL 8.0中缓冲池新、旧页分配设置参数

参　　数	说　　明
innodb_old_blocks_pct	控制LRU链表中"old"旧子链表的百分比。其默认值是37，对应原来的固定比例3/8
innodb_old_blocks_time	防止缓冲池被预读、新、旧页不停切换，可以避免由于表或索引扫描而产生的类似问题。表示第一次访问一个页后的时间窗口（毫秒），在此期间，该页可以被访问而不能被移动到LRU列表的前面

说明：随着工作负载的变化，innodb_old_blocks_time参数的影响比innodb_old_blocks_pct参数更难预测，所以采取默认值即可。

（3）预读设置

InnoDB使用两种预读算法来提高I/O性能：线性预读（linear read-ahead）和随机预读（random read-ahead）。两者的区别在于，线性预读以extent（块）为单位，而随机预读以extent中的页为单位。预读考验的是硬件的I/O处理能力。表2-7所示为预读设置参数及其说明。

表2-7　MySQL 8.0中预读设置参数及其说明

参　　数	说　　明
innodb_read_ahead_threshold	控制是否将下一个extent预读到缓冲池中，如果一个extent中被顺序读取的页超过或者等于该参数变量，则InnoDB将会异步地将下一个extent读取到缓冲池中
innodb_random_read_ahead	随机预读方式则表示当同一个extent中的一些页在缓冲池中被发现时，InnoDB会将该extent中的剩余页一并读到缓冲池中。随机预读方式给InnoDB编码带来了一些不必要的复杂性，同时在性能上也存在不稳定性

说明：因为数据的大小和用途很难预估，在内存足够大、I/O能力足够强的情况下，完全可以全部打开，所以采取默认值即可。

2. Change Buffer

Change Buffer（更改缓冲区）是一种特殊的数据结构，当辅助索引页不在缓冲池中时，它将更改缓冲区到辅助索引页。缓冲区的更改可能来自INSERT、UPDATE或DELETE等操作（DML），当其他读操作将页加载到缓冲区中时，将合并这些更改。图2-24所示为Change Buffer的结构。

与主键不同，二级索引通常不是唯一的，并且对二级索引的插入以相对随机的顺序发生。类似地，删除和更新可能会影响不在索引树中邻接位置的辅助索引页。当受影响的页由其他操作读入缓冲池时，通过合并缓冲区的更改，可以避免大量的随机访问I/O。表2-8所示为Change Buffer设置参数及其说明。

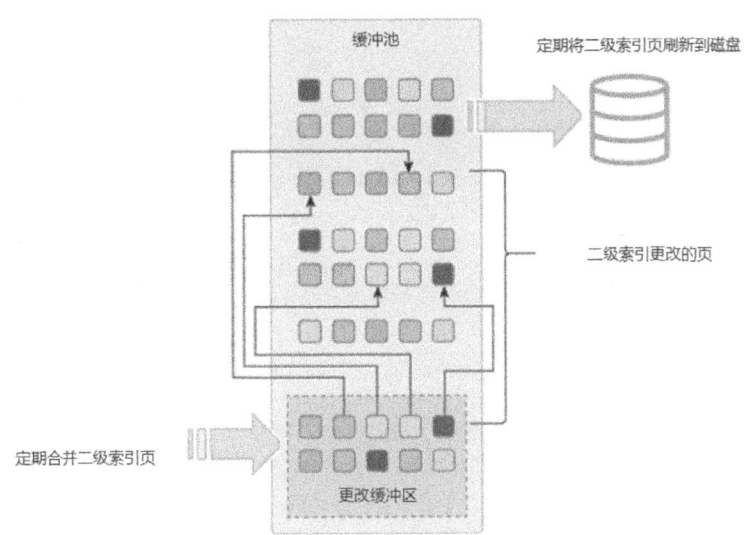

图 2-24　MySQL 8.0 中 Change Buffer 的结构

表 2-8　MySQL 8.0 中 Change Buffer 设置参数及其说明

参　　数	说　　明
innodb_change_buffer_max_size	允许配置更改缓冲区的最大值，占缓冲池总的大小的百分比，这部分内存也包含在 innodb_buffer_pool_size 里
innodb_change_buffering	表示 InnoDB 执行更改缓冲区的程度。允许的选项有 none、inserts、deletes、changes、purges 和 all

说明：随着业务变更和工作负载的变化，缓冲区的数据量大小难测，所以采取默认值即可。查看 Change Buffer 占缓冲池总的大小的百分比的命令行如下所示。

```
mysql> SELECT (SELECT COUNT(*) FROM INFORMATION_SCHEMA.INNODB_BUFFER_PAGE
       WHERE PAGE_TYPE LIKE 'IBUF%') AS change_buffer_pages,
       (SELECT COUNT(*) FROM INFORMATION_SCHEMA.INNODB_BUFFER_PAGE) AS total_pages, (SELECT
((change_buffer_pages/total_pages)*100)) AS change_buffer_page_percentage;
+---------------------+-------------+-------------------------------+
| change_buffer_pages | total_pages | change_buffer_page_percentage |
+---------------------+-------------+-------------------------------+
|                   2 |        8192 |                        0.0244 |
+---------------------+-------------+-------------------------------+
```

说明：如果二级索引包含降序索引列，或者主键包含降序索引列，则不支持更改缓冲区。

3. 自适应哈希索引

自适应哈希索引（Adaptive Hash Index）是 InnoDB 在缓冲池内存中的哈希内存数据库的数据，不会牺牲事务特性或可靠性。哈希索引通过支持对任何元素的直接查找，将索引值转换为一种指针，从而加快查询速度。InnoDB 有一个监视索引搜索的机制。如果 InnoDB 注意到查询可以从建立哈希索引中获益，它就会自动这样做。

目前了解到的自适应哈希索引的自创建条件如下。

- 索引中页访问的模式是相同的，不包含主键。

- 索引命中的页上的记录数要大于该页上总记录数的 1/16。
- 索引中的某个页已经被访问至少 100 次。

满足索引命中总记录数 1/16 的条件就会自动添加到自适应哈希索引中。自适应哈希索引的特性是分区。每个索引都绑定到一个特定的分区，每个分区都由一个单独的锁存器保护。通过查看 INNODB STATUS 信息，可以了解 InnoDB 使用情况，如下所示。

```
mysql> SHOW ENGINE INNODB STATUS \G;
...
-----------------------------------
INSERT BUFFER AND ADAPTIVE HASH INDEX
-----------------------------------
Ibuf: size 10, free list len 174, seg size 176, 8 merges
                    #size 10:已经合并记录页的数量
                    #free list len:插入缓冲区中空闲列表长度
                    #seg size:176 表示当前插入值的长度,大小为 176 * 16KB/1024=2.75MB
                    #merges:合并插入的次数
merged operations:
  insert 0, delete mark 0, delete 0
discarded operations:
  insert 0, delete mark 0, delete 0

#自适应哈希分区表中两个分区在使用
Hash table size 34679, node heap has 2 buffer(s)
Hash table size 34679, node heap has 0 buffer(s)
Hash table size 34679, node heap has 4 buffer(s)

#自适应哈希分区表内第一个分区表中的 114 槽在使用
Hash table size 34679, node heap has 114 buffer(s)
Hash table size 34679, node heap has 0 buffer(s)
Hash table size 34679, node heap has 0 buffer(s)
Hash table size 34679, node heap has 1 buffer(s)
Hash table size 34679, node heap has 0 buffer(s)
0.00 hash searches/s, 0.00 non-hash searches/s
```

表 2-9 所示为控制自适应哈希索引参数列表。

表 2-9　MySQL 8.0 中自适应哈希索引设置参数及其说明

参　数	说　明
innodb_adaptive_hash_index	InnoDB 自适应哈希索引是否启用
innodb_adaptive_hash_index_parts	分区自适应哈希索引搜索系统。每个索引都绑定到一个特定的分区，每个分区由一个单独的锁存器保护。默认设置为 8。最大可设置为 512

说明： 自适应哈希索引默认打开，基本上仅在只读场景下才能提高性能。因为自适应哈希索引本身维护锁机制，其他操作可能会导致锁等待，甚至导致 MySQL hung 或极端场景下的数据损坏。此外，自适应哈希索引会消耗缓冲池空间，虽然用二级索引构建，但索引字段不合理，这可能会影响整体性能。由于很难预测自适应哈希索引功能是否适合特定系统和工作负载，因此建议在 I/O 负载高的场景下禁止。

4. Log Buffer

Log Buffer（日志缓冲区）是存储要写入磁盘上 Redo 文件的数据的内存区域。Log Buffer 的内容会定期刷新到磁盘中。需要保证操作日志落盘和数据的一致性。通过 Log Buffer 的批量处理机制，频繁与磁盘 I/O 交互。表 2-10 所示为 Log Buffer 设置参数列表。

表 2-10　MySQL 8.0 中 Log Buffer 设置参数及其说明

参　　数	说　　明
innodb_log_buffer_size	InnoDB 写入磁盘上日志文件的缓冲区大小（以字节为单位）。默认值是 16MB。大的日志缓冲区使大型事务能够运行，而不需要在事务提交之前将日志写入磁盘
innodb_flush_log_at_trx_commit	控制日志缓冲区的内容如何写入和刷新到磁盘
innodb_flush_log_at_timeout	控制日志缓冲区的内容刷新频率

说明：innodb_log_buffer_size 的最大值可以达到 4GB，可以一次性支持 4GB 的事务。但在内存里寻址是很消耗性能的，同时会占据 InnoDB 整体缓冲池。所以，MySQL 里，还是建议以单个事务方式提交处理。

5. 总结

通过对 MySQL 8.0 中新的 InnoDB 内存结构的理解，可以更合理地设置内存相关参数，并且协助排查诸多问题。

2.3.2　引入数据分析引擎 HeatWave

在轻量级关系数据库领域，MySQL 一直占据着主导地位，但随着数据量的增加，业务的多态化，OLAP 结合的场景越来越多，MySQL 的短板问题逐渐凸显。为了解决类似问题，通常会采用 MySQL 的 binlog 同步机制或 ETL 方式抽取到其他分析平台，再使用 Spark、Impala 等计算引擎做计算，提供 AP 的业务支持。为了在大数据时代继续延伸，MySQL 8.0 为此提供了高性能实时分析计算架构 HeatWave 并且以集群模式提供分析服务。

1. HeatWave 引擎介绍

HeatWave 是一个分布式、可扩展、无共享、内存化、列式存储的查询处理引擎，其设计目的是实现极高的分析性能。

HeatWave 集群包括一个 MySQL 数据库系统节点和两个或更多个 HeatWave 节点。MySQL 数据库系统节点中，HeatWave 插件负责集群管理，将数据加载到 HeatWave 集群，查询调度，并将查询结果返回到 MySQL 数据库系统。其中 HeatWave 节点将数据存储在内存中，并处理分析查询。每个 HeatWave 节点都包含一个 HeatWave 查询处理引擎（RAPID）的实例。并且 HeatWave 集群最多支持 64 个节点。

2. HeatWave 引擎架构

图 2-25 所示为在原生 MySQL 的基础上额外装载插件式 HeatWave 引擎。

（1）查询实现方式

在查询方式上，HeatWave 调用是从连接到 MySQL 数据库系统节点的 MySQL 客户端或应用程序发出的。客户端和应用程序不会直接连接到 HeatWave。支持的查询自动从 MySQL 数据库系统卸载到 HeatWave，以加速处理。结果返回到 MySQL 节点和发出查询请求的 MySQL 客户端或应用程序。MySQL 8.0 中 HeatWave 引擎查询数据方式如图 2-26 所示。

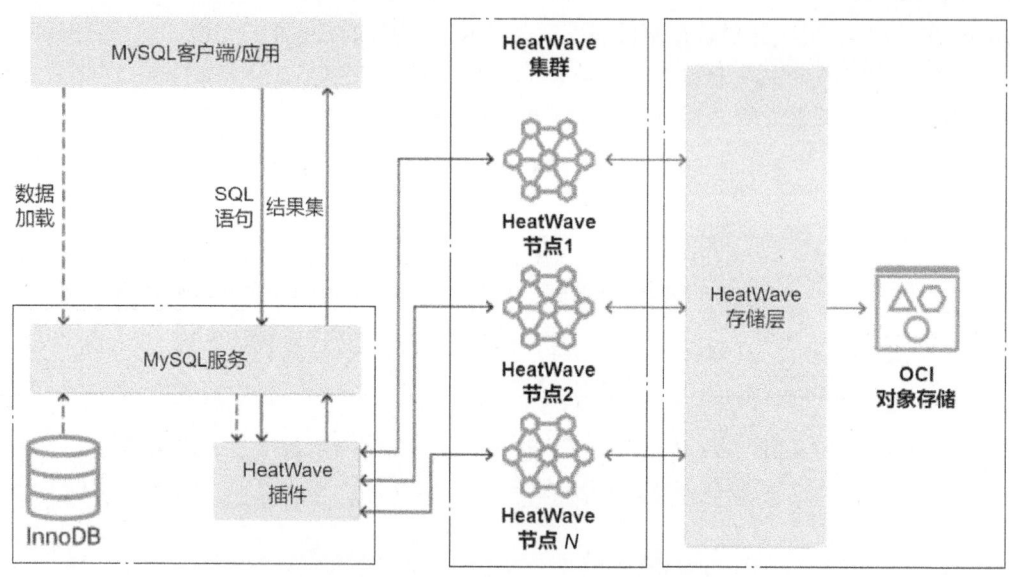

图 2-25　MySQL 8.0 中 HeatWave 引擎架构

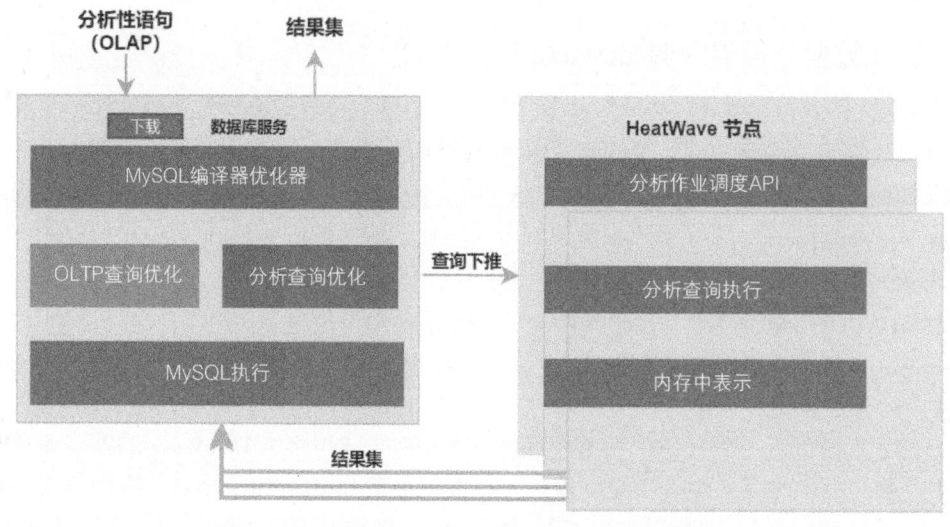

图 2-26　MySQL 8.0 中 HeatWave 引擎查询数据方式

（2）数据加载方式

将数据加载到 HeatWave 中需要在 MySQL 数据库系统上准备表并执行加载命令。准备表包括排除列、定义字符串列编码、数据键以及将表标记为辅助引擎候选表等任务。

Auto Parallel Load 实用程序通过自动化所需的步骤和优化并行加载线程的数量，促进了准备和加载表的过程。

（3）同步方式

当 HeatWave 加载一个表时，数据被共享并分布在 HeatWave 节点之间。加载表后，表上的 DML 操作将自动传播到 HeatWave 节点。同步数据不需要用户操作。

(4）持久化方式

加载到 HeatWave 中的数据（包括传播的更改）由 HeatWave 存储层自动持久化到 OCI 对象存储，以便在 HeatWave 节点或集群故障的情况下快速恢复。

(5）优化方式

如图 2-27 所示，在对数据进行查询之后，使用 HeatWave Advisor 来优化工作负载。Advisor 分析数据和查询历史，以提供字符串列编码和数据键。

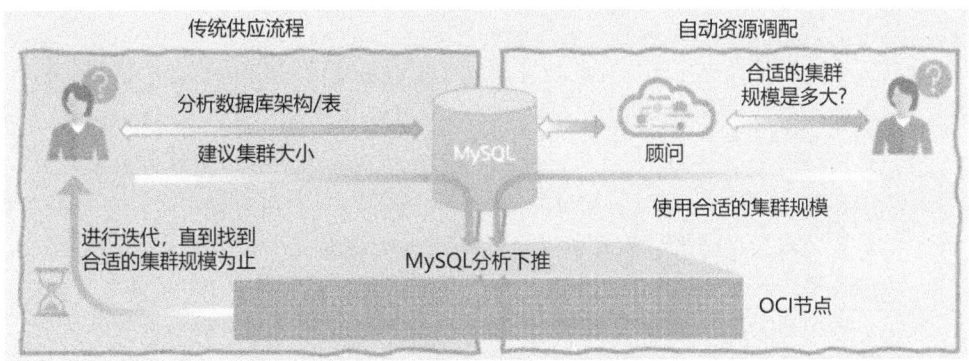

图 2-27　MySQL 8.0 中的 HeatWave Advisor

(6）主要特点

- 内存采用 Hybrid-Columnar 格式。数据以混合列式方式进行保存和处理。
- 基于推送的向量化（Vectorized）查询处理（见图 2-28）。

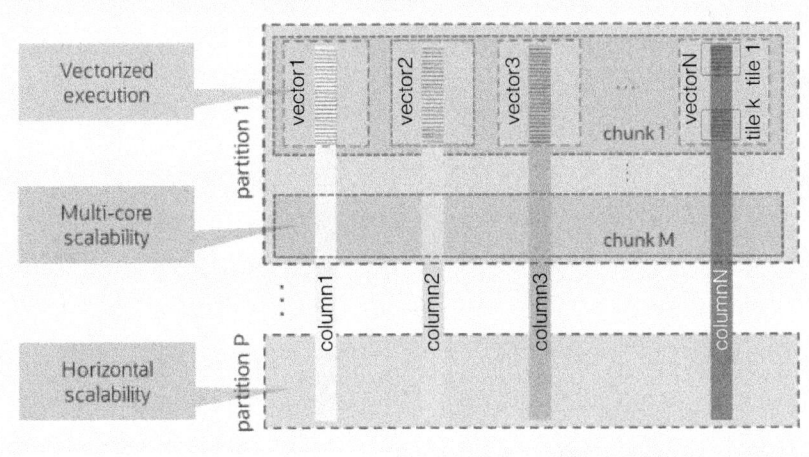

图 2-28　MySQL 8.0 中 HeatWave 引擎向量化查询处理方式

- 大规模并行体系结构。HeatWave 的大规模并行架构是通过节点间和节点内的数据分区实现的。HeatWave 集群中的每个节点和节点中的每个 CPU 核心并行处理分区数据。HeatWave 能够扩展到数千个核心。这种大规模并行架构与高扇形输出、工作负载敏感的分区相结合，可以加速查询处理。MySQL 8.0 HeatWave CPU 处理架构如图 2-29 所示。

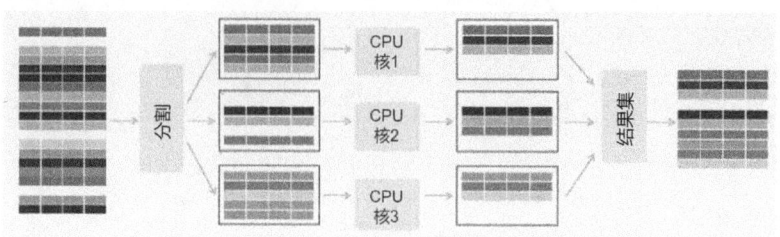

图 2-29　MySQL 8.0 HeatWave CPU 处理架构

- 扩展数据管理。存储层自动将数据保存到 OCI 对象存储中，当恢复故障节点或集群时，数据存储层自动恢复。这个自动的、自我管理的存储层可以伸缩到集群所需的大小，并在后台上独立运行。重新加载数据所需的时间与数据大小或 HeatWave 集群大小无关。
- 插件式 MySQL 集成。HeatWave 被设计为一个可插拔的 MySQL 存储引擎。MySQL 数据库系统上的数据更改会自动实时传播到 HeatWave 节点，这意味着查询总是能够访问最新的数据。更改传播由轻量级算法自动执行。MySQL 查询优化器就会根据查询卸载的先决条件是否满足，透明地决定是否将该查询卸载给 HeatWave 以加速执行。
- 使用语法。语法上继承 MySQL 的基本语法体系，方便、易懂。命令行设置如下所示。

```
# 排除不支持数据类型的列
mysql> CREATE TABLE orders (id INT, description BLOB NOT SECONDARY);

# 关键字定义
mysql> CREATE TABLE orders (date DATE COMMENT 'RAPID_COLUMN=DATA_PLACEMENT_KEY=1');

# HeatWave 查询处理引擎(RAPID)定义为辅助引擎
mysql> CREATE TABLE orders (id INT) SECONDARY_ENGINE = RAPID;
```

- 其他方面。使用 READ COMMITTED 隔离级别读取数据。每张表最多支持 470 列。如果没有主键，则不允许加载表。

3. 性能测试

官网提供了同等条件下（数据量、连接数、并发等），使用传统方式（Amazon 的 RDS、Aurora、Redshift、Snowflake）和 MySQL HeatWave 方式保存数据的性能测试对比。

图 2-30 所示为 4TB 数据量在 Amazon RDS 和 MySQL HeatWave 上计算的耗时和费用对比。在数据量为 4TB 场景下，若在 Amazon RDS 模式下处理，需要 11 小时，但在 HeatWave 下进行同样的处理，只需要 8 秒，同时消耗资源的费用下降为原先的约 2/3。

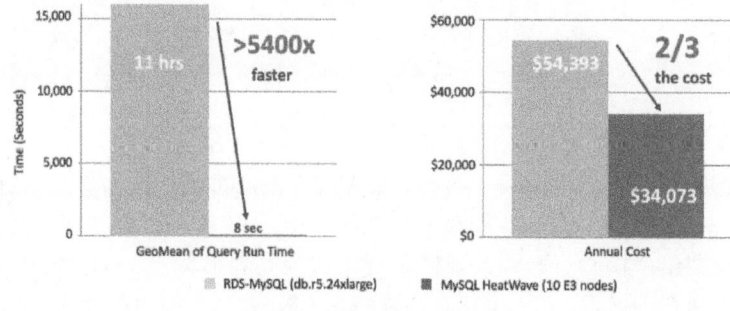

图 2-30　性能对比测试 1

图 2-31 所示为 4TB 数据量在 Aurora 和 MySQL HeatWave 上计算的耗时和费用对比。在数据量为 4TB 场景下，若在 Aurora 模式下处理，需要 2.5 小时，但在 HeatWave 下进行同样的处理，只需要 6.3 秒，同时消耗资源的费用下降为原先的约 1/2。

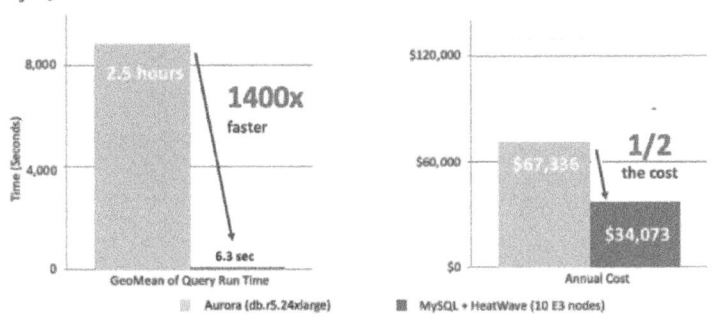

图 2-31　性能对比测试 2

图 2-32 所示为 10TB 数据量在 Amazon Redshift AQUA 和 MySQL HeatWave 上计算的耗时和费用对比。在数据量为 10TB 的场景下，若在 Amazon Redshift AQUA 模式下处理，需要 2380 秒，但在 HeatWave 下同样的处理只需要 346.5 秒完成，同时消耗资源的费用下降为原先的约 1/2。

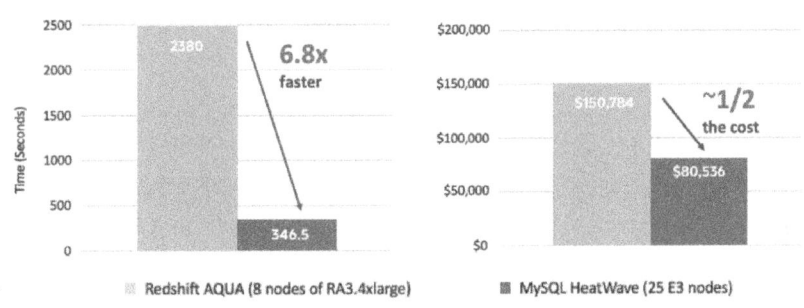

图 2-32　性能对比测试 3

图 2-33 所示为 10TB 数据量在 Snowflake 和 MySQL HeatWave 上计算的耗时和费用对比。在数据量为 10TB 的场景下，若在 Snowflake 模式下处理，需要 2350 秒，但在 HeatWave 下进行同样的处理，只需要 346 秒，同时消耗资源的费用下降为原先的约 1/5。

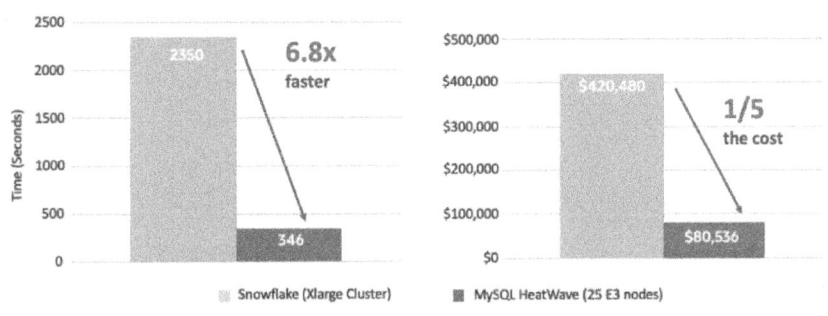

图 2-33　性能对比测试 4

4. 总结

从 MySQL 8.0 的 AP 设计思路和实现上，可以得到如下结论。

- HeatWave 与 MySQL 的结合封装得非常好。
- 无须进行数据同步操作，因为内部自动完成 InnoDB 到 OCI 的持久化。
- AI 功能：可以人工智能方式自动配置集群大小，而不需要人为配置。
- 使用者是无感知的，仍然会通过原有的方式（JDBC/ODBC）连接和使用。
- SQL 语法兼容性基本没变，易用性高。
- MySQL 因其轻量级而存在的本质问题并没有解决，包括 MySQL InnoDB 端的数据量不够多的问题和 MySQL Server 层的负载问题。

HeatWave 也是 MySQL Analytic Engine 服务中的分析执行引擎，令人遗憾的是，目前这一服务仅限于 Oracle Cloud Infrastructure 使用。

第 3 章
MySQL 8.0 功能架构特性

MySQL 8.0 版本在功能、系统表、资源管理、索引、安全、InnoDB 功能、JSON 支持等方面有显著的改进和优化,可以更好地满足用户的需求并提升数据库性能。

本章将系统介绍 MySQL 8.0 版本的功能架构上的特性。

3.1 MySQL 8.0 新增函数和新增集合操作

MySQL 8.0 引入了许多新的功能和改进,其中包括新的函数和集合操作。

3.1.1 窗口函数

从 MySQL 8.0 版本开始支持窗口函数,这个功能在大部分商业数据库和部分开源数据库中早已支持,也称为分析函数。

1. 窗口函数介绍

窗口函数是在满足某种条件的记录集合上执行的特殊函数。窗口的概念非常重要,它可以理解为记录的集合。对于每条记录,都要在此窗口内执行,有些函数对于不同的记录,窗口大小都是固定的,这种窗口属于静态窗口。相反,有些函数会随着不同的记录对应不同的窗口,这种动态变化的窗口称为滑动窗口。

窗口函数和普通聚合函数也很容易混淆。聚合函数是将多条记录聚合为一条。而窗口函数将会执行每条记录,几条记录执行完还是几条。聚合函数也可用于窗口函数中。

窗口函数的语法如下所示。

```
window_spec:
    [window_name][partition_clause][order_clause][frame_clause]
```

- window_name:对于多个 OVER 子句定义指定一个别名,如果 SQL 中涉及的窗口较多,则采用别名可以看起来更清晰易读。
- partition_clause:指示如何将查询行划分为组。给定行的窗口函数结果基于包含该行的分区的行。
- order_clause:ORDER BY 子句指示如何对每个分区中的行进行排序。
- frame_clause:frame 子句指定如何定义这个子集,在分区里面再进一步细分窗口,子句用来定义子集的规则,通常用来作为滑动窗口使用。

frame_clause 的语法如下所示。

```
frame_clause:
    frame_units frame_extent
frame_units:
    {ROWS | RANGE}
```

说明：frame_units 有两种定义方式，分别是 ROWS 和 RANGE。由 ROWS 定义的 frame 是通过开始和结束位置的行号来确定的，而由 RANGE 定义的 frame 则是基于某个值区间内的行来确定的。

2. 非聚合窗口函数介绍

非聚合窗口函数在 SQL 查询中对每一行执行计算，同时考虑与该行相关的其他行。值得注意的是，大多数聚合函数（如 SUM、AVG、MAX、MIN 等）也可以作为窗口函数使用，通过在查询中引入 OVER( ) 子句来实现。

表 3-1 所示为 MySQL 8.0 版本支持的非聚合窗口函数。

表 3-1　MySQL 8.0 非聚合窗口函数

函 数 名	说　　明
CUME_DIST( )	累计分布值
DENSE_RANK( )	当前行在其分区内的排名
FIRST_VALUE( )	函数用于返回当前第一个值
LAG( )	从同一结果集中的当前行访问上一行的数据
LAST_VALUE( )	函数用于返回当前最后一个值
LEAD( )	从同一结果集中的当前行访问后续行的数据
NTH_VALUE( )	函数用于返回有序行的第 n 小的值
NTILE( )	划分为指定数量的组
PERCENT_RANK( )	每个有序分区独立计算函数
RANK( )	排名分配给结果集的分区中的每一行
ROW_NUMBER( )	结果集中的每一行生成序列号行

上述非聚合窗口函数可以按照功能分为以下 5 种类型。
- 分组排序函数：ROW_NUMBER( )、RANK( )、DENSE_RANK( )。
- 分布函数：CUME_DIST( )、PERCENT_RANK( )。
- 头尾中函数：FIRST_VALUE( )、LAST_VALUE( )、NTH_VALUE ( )。
- 前后函数：LAG( )、LEAD( )。
- 分组函数：NTILE( )。

下面对这 5 类函数分别加以说明。

（1）分组排序函数

分组排序函数用于对查询结果进行分组并排序。
- ROW_NUMBER( )：此函数用于为结果集中的每一行分配一个唯一的连续整数编号，这个编号是基于指定的排序顺序在每个分区内生成的。行号的范围从 1 开始，到每个分区的行数结束。
- RANK( )：此函数返回分区中当前行的排名，并带有间隔。对于值相同的行（即对等行），赋予相同的排名。如果存在大小大于 1 的组（即有多行具有相同的值），则此函数不会为这些对等组分配连续的排名。因此，结果是一个不连续的排名。为了确定分区内行的排序顺序，这个函数应该与 ORDER BY 子句一起使用。
- DENSE_RANK()：此函数返回分区内当前行的排名，并且排名之间没有间隔。对于值相同的行（即对等项），被认为地位相等，并获得相同的排名。该函数会为对等组分配连续的排名，因此，即使存在大小大于 1 的组，也不会产生不连续的排名。为了确定分区内行的排序顺序，这个函数应该与 ORDER BY 子句一起使用。

上述分组排序函数的命令行示例如下。

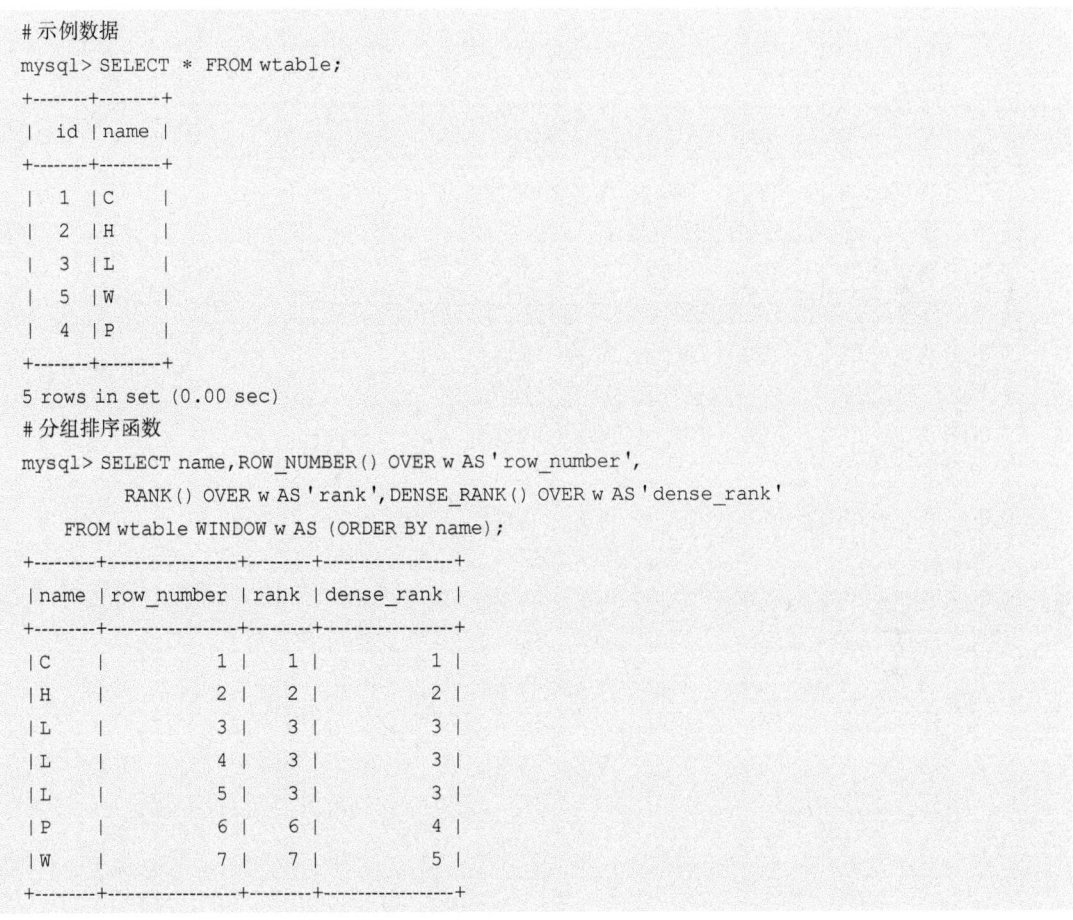

（2）分布函数

分布函数用于计算行在其分区内的相对位置。
- CUME_DIST()：此函数返回一组值中某个值的累计分布，即分区值小于或等于当前行的值的百分比。它表示窗口分区的窗口顺序中在当前行之前或与当前行对等的行数除以窗口分区中的总行数的商。返回值的范围为 0~1。这个函数应该与 ORDER BY 一起使用，将分区行按所需的顺序排序。如果没有 ORDER BY，则所有行都是对等的，值 $N/N=1$，其中 N 是分区大小。
- PERCENT_RANK()：此函数返回当前行在分区中的相对排名百分比，不包括最大值。这个百分比是根据公式 (rank-1)/(total_rows-1) 计算得到的，其中 rank 是当前行的排名（根据 ORDER BY 子句确定），total_rows 是分区中的总行数。因此，返回值范围为 0~1，但不包括 1。

上述分布函数的命令行示例如下。

```
mysql> SELECT id,name,ROW_NUMBER() OVER w AS 'row_number',
          CUME_DIST() OVER w AS 'cume_dist',
          PERCENT_RANK() OVER w AS 'percent_rank'
       FROM wtable WINDOW w AS (ORDER BY id);
+------+------+------------+-----------+--------------+
| id   | name | row_number | cume_dist | percent_rank |
+------+------+------------+-----------+--------------+
|  1   | C    |          1 |       0.2 |            0 |
|  2   | H    |          2 |       0.4 |         0.25 |
|  3   | L    |          3 |       0.6 |          0.5 |
|  4   | P    |          4 |       0.8 |         0.75 |
|  5   | W    |          5 |         1 |            1 |
+------+------+------------+-----------+--------------+
```

说明：ROW_NUMBER：表示排序顺序。

(3) 头尾中函数

头尾中函数是一组用于查找数据集中特定行指定字段值的函数。

- FIRST_VALUE()：从窗口框架的第一行返回值。
- LAST_VALUE()：从窗口框架的最后一行返回值。
- NTH_VALUE()：从窗口框架的第 N 行返回值。如果没有这样的行，则返回值为 NULL。N 必须是一个正整数。

上述头尾中函数的命令行示例如下。

```
# 示例数据
mysql> SELECT * FROM wtable;
+------+------+---------+
| id   | name | subject |
+------+------+---------+
|  1   | C    | C001    |
|  2   | H    | B001    |
|  3   | L    | A001    |
|  5   | W    | C001    |
|  4   | P    | B001    |
|  6   | L    | A001    |
|  7   | L    | B001    |
+------+------+---------+
7 rows in set (0.00 sec)
# 头尾中函数
mysql> SELECT id, subject, name,
          FIRST_VALUE(name) OVER w AS 'first',
          LAST_VALUE(name) OVER w AS 'last',
          NTH_VALUE(name, 2) OVER w AS 'second',
          NTH_VALUE(name, 3) OVER w AS 'fourth'
       FROM wtable WINDOW w AS (PARTITION BY subject
          ORDER BY id ROWS UNBOUNDED PRECEDING);
```

```
+----+---------+------+-------+------+--------+--------+
| id | subject | name | first | last | second | fourth |
+----+---------+------+-------+------+--------+--------+
|  3 | A001    | L    | L     | L    | NULL   | NULL   |
|  6 | A001    | L    | L     | L    | L      | NULL   |
|  2 | B001    | H    | H     | H    | NULL   | NULL   |
|  4 | B001    | P    | H     | P    | P      | NULL   |
|  7 | B001    | L    | H     | L    | P      | L      |
|  1 | C001    | C    | C     | C    | NULL   | NULL   |
|  5 | C001    | W    | C     | W    | W      | NULL   |
+----+---------+------+-------+------+--------+--------+
7 rows in set (0.00 sec)
```

(4) 前后函数

前后函数用于处理在数据窗口范围内的数据。

- LAG()：此函数返回当前行在其分区中向前（即滞后）N 行的值。如果当前行是分区中的前 N 行，或者 N 为负数，则返回默认值。如果未指定 N 或默认值，则 N 默认为 1，若默认值为 NULL，N 必须是一个非负整数。
- LEAD()：此函数返回当前行在其分区中向后（即超前）N 行的值。如果当前行是分区中的最后 N 行，或者 N 为负数，则返回默认值。如果未指定 N 或默认值，则 N 默认为 1，而默认值为 NULL。N 必须是一个非负整数。

上述前后函数的命令行示例如下。

```
mysql> SELECT id,LAG(id, 2,'N') OVER w AS 'lag',
              LEAD(id, 2,'N') OVER w AS 'lead'
       FROM wtable WINDOW w AS (ORDER BY id);
+----+-----+------+
| id | lag | lead |
+----+-----+------+
|  1 | N   | 3    |
|  2 | N   | 4    |
|  3 | 1   | 5    |
|  4 | 2   | 6    |
|  5 | 3   | 7    |
|  6 | 4   | N    |
|  7 | 5   | N    |
+----+-----+------+
7 rows in set (0.00 sec)
```

对于前后函数使用场景，例如，在比较当前订单与后续订单时间间隔时，一个客户想要知道上一次下单与下一次下单之间相隔了多少天，可以使用 LEAD() 函数来求解。LAG() 和 LEAD() 函数允许在同一次查询中取出同一字段的前 N 行（使用 LAG() 函数）和后 N 行（使用 LEAD() 函数）的数据作为独立的列。在实际应用中，与使用 LEFT JOIN 等自连接方式相比，使用 LAG() 和 LEAD() 函数可以使 SQL 查询更加简洁和直观。

(5) 分组函数

分组函数用于对一组值执行计算，并返回单个值。

NTILE()：此函数是将一个分区分成 N 组桶（Bucket），为分区中的每一行分配其桶号，并返回分区中当前行的桶号。如果 N 是 4，则 NTILE( ) 函数将行分成 4 个桶。如果 N 是 100，则将行分成 100 个桶。N 必须是一个正整数。桶号返回值的范围为 1~N。这个函数应该与 ORDER BY 一起使用，将分区行按所需的顺序排序。

上述分组函数的命令行示例如下。

```
mysql> SELECT id,
          ROW_NUMBER()OVER w AS 'row_number',
          NTILE(2)OVER w AS 'ntile2',
          NTILE(4)OVER w AS 'ntile4'
    FROM wtable WINDOW w AS (ORDER BY id);
+------+------------+--------+--------+
| id   | row_number | ntile2 | ntile4 |
+------+------------+--------+--------+
|  1   |     1      |   1    |   1    |
|  2   |     2      |   1    |   1    |
|  3   |     3      |   1    |   2    |
|  4   |     4      |   1    |   2    |
|  5   |     5      |   2    |   3    |
|  6   |     6      |   2    |   3    |
|  7   |     7      |   2    |   4    |
+------+------------+--------+--------+
```

说明：按照需求可以把显示结果分成几部分，如 ntile2 和 ntile4 列等。

3. 窗口函数索引

窗口函数可以正常使用索引来加速数据的检索，但通常还需要对数据进行额外的排序操作，以便按照指定的顺序计算窗口函数的结果。

窗口函数的执行计划的示例如下。

```
mysql> EXPLAIN FORMAT=TREE SELECT id,
          ROW_NUMBER() OVER w AS 'row_number',
          NTILE(2)OVER w AS 'ntile2',
          NTILE(4)OVER w AS 'ntile4'
    FROM wtable WINDOW w AS (ORDER BY id);
+-------------------------------------------------------------------------------------+
| EXPLAIN                                                                             |
+-------------------------------------------------------------------------------------+
-> Window aggregate with buffering: row_number() OVER w,ntile(2) OVER w, ntile(4) OVER w
    -> Sort:wtable.id   (cost=1.5 rows=5)
        -> Index scan on wtable using PRIMARY   (cost=1.5 rows=5)              |
+-------------------------------------------------------------------------------------+1
row in set (0.01 sec)
```

上述 SQL 语句执行计划输出中显示"Index scan on wtable using PRIMARY"，这意味着选择使用 wtable 表的主键索引（PRIMARY）来检索数据，快速地定位到特定的行。

4. 总结

MySQL 8.0 中引入的窗口函数大大扩展了 SQL 查询的能力，使其在数据分析、报告生成等场景下能够更高效地处理数据。不过，这些窗口函数通常需要在数据上进行排序或分组操作，这可能会消耗更多的系统资源，特别是在处理大型数据集时。所以在使用窗口函数时，应该根据具体的业务场景和数据特点来选择合适的策略，以确保查询的性能和效率。

3.1.2 集合操作

在 MySQL 8.0.31 版本中，增加了两种集合操作：INTERSECT 和 EXCEPT。同时新增的 INTERSECT 和 EXCEPT 集合操作都支持 DISTINCT 与 ALL 语句的扩展。结合之前版本支持的集合操作（UNION 和 UNION ALL），下面介绍集合操作和用法。

1. 4 种集合操作介绍

MySQL 8.0 版本支持的集合操作包括 UNION、UNION ALL、INTERSECT 和 EXCEPT。表 3-2 所示为这 4 种集合操作的示例说明。使用图 3-1 中的 A、B 和 C 集合作为示例，可以展示每种集合操作的结果。

表 3-2　MySQL 集合操作

集合操作类型	数　　据	结　　果
UNION	$AC=\{1,2,3\}$，$BC=\{3,4,5,6\}$	$A \cup B=\{1,2,3,4,5,6\}=ABC$
UNION ALL	$AC=\{1,2,3\}$，$BC=\{3,4,5,6\}$	$ALL=\{1,2,3,3,4,5,6\}=ABCC$（包含 C 集合中重复数据）
INTERSECT	$AC=\{1,2,3\}$，$BC=\{3,4,5,6\}$	$A \cap B=\{3\}=C$
EXCEPT	$AC=\{1,2,3\}$，$BC=\{3,4,5,6\}$	$A-B=\{1,2\}=A$

2. 集合操作方式

在 MySQL 8.0 中，UNION、UNION ALL、INTERSECT 和 EXCEPT 集合操作可以用于组合两个或多个 SELECT 语句的结果集。下面将提供一些简单的示例来说明如何在 MySQL 中使用它们。集合操作的命令行示例如下。

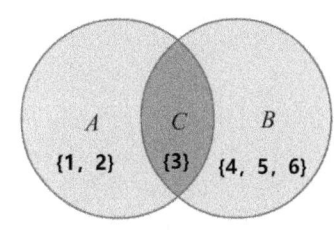

图 3-1　集合交集关系

```
# 模拟数据表和插入数据
mysql> CREATE TABLE `DataSetA` (
  `id` bigint NOT NULL,`name` varchar(10) DEFAULT NULL);
mysql> INSERT INTO DataSetA(id,name) VALUES(1,'A'),(2,'B'),(3,'C');
mysql> CREATE TABLE `DataSetB` (
  `id` bigint NOT NULL,`name` varchar(10) DEFAULT NULL);
mysql> INSERT INTO DataSetB(id,name) VALUES(3,'C'),(4,'D'),(5,'E'),(6,'F');
mysql> CREATE TABLE `DataSetC` (
  `id` bigint NOT NULL,`name` varchar(10) DEFAULT NULL);
mysql> INSERT INTO DataSetC(id,name) VALUES(6,'F'),(7,'G'),(7,'G'),(8,'H');

# 4 种集合操作命令行
mysql> SELECT id,name FROM DataSetA UNION SELECT id,name FROM DataSetB ;
mysql> SELECT id,name FROM DataSetA UNION ALL SELECT id,name FROM DataSetB ;
mysql> SELECT id,name FROM DataSetA INTERSECT SELECT id,name FROM DataSetB ;
mysql> SELECT id,name FROM DataSetA EXCEPT SELECT id,name FROM DataSetB ;
```

图 3-2 所示为 4 种不同的集合操作显示结果集。UNION 会去除重复的数据，UNION ALL 会显示所有元素，INTERSECT 会显示两张表中都存在的值，EXCEPT 会显示属于第一张表但不属于第二张表的数据。

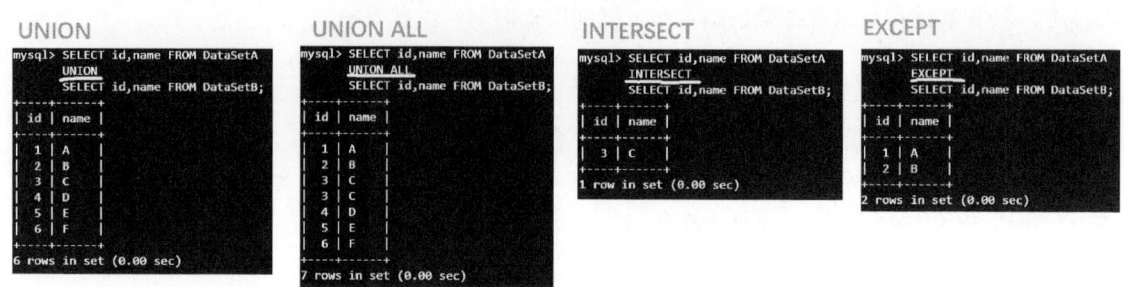

图 3-2　MySQL 8.0 集合操作结果集

3. 查看集合操作的执行计划

当查看集合操作的执行计划时，就会发现，除了 UNION ALL 以外，UNION、INTERSECT 和 EXCEPT 都可能需要通过 Extra 信息中的 Using temporary 来使用临时表进行运算。虽然这可能会对性能产生一定影响，特别是涉及大量数据时，但具体的性能下降程度取决于多种因素。图 3-3 所示为 4 种集合操作的执行计划，进一步揭示了在执行中的性能特点。

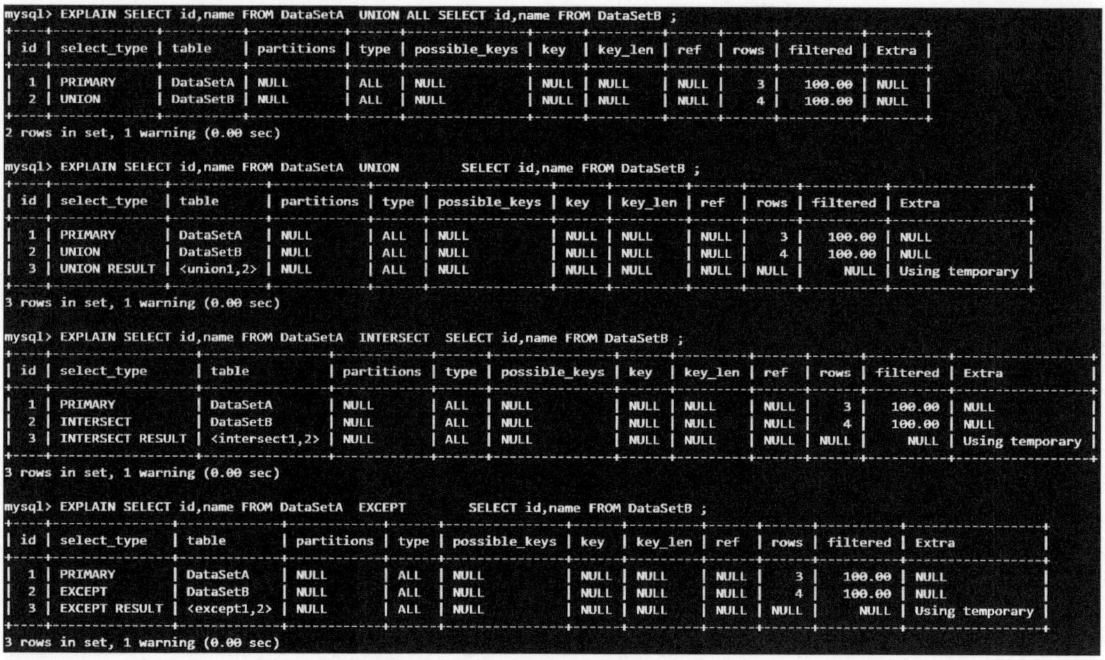

图 3-3　MySQL 8.0 集合操作执行计划

4. 集合函数的 DISTINCT 和 ALL 操作

INTERSECT 和 EXCEPT 集合操作支持 DISTINCT 与 ALL 附加语句的组合使用。如果没有指定附加语句，则默认采取 DISTINCT 的方式。集合操作的附加语句的语法如下所示。

```
query_block INTERSECT [ALL | DISTINCT] query_block
```

附加语句 DISTINCT 和 ALL 的集合操作的命令行示例如下。

```
# DISTINCT 附加语句
mysql> table DataSetC INTERSECT DISTINCT table DataSetC;
+------+---------+
| id   | name    |
+------+---------+
|    6 | F       |
|    7 | G       |
|    8 | H       |
+------+---------+
3 rows in set (0.00 sec)
# ALL 附加语句
mysql> table DataSetC INTERSECT ALL table DataSetC;
+------+---------+
| id   | name    |
+------+---------+
|    6 | F       |
|    7 | G       |
|    7 | G       |
|    8 | H       |
+------+---------+
4 rows in set (0.00 sec)
```

说明：对于 INTERSECT ALL 左表中，任何唯一一行最大支持的重复元素是 4294967295 个。

下面是 EXCEPT ALL 的集合操作的命令行示例。

```
# 模拟数据
mysql> CREATE TABLE `DataSetD` (
  `id` bigint NOT NULL,`name` varchar(10) DEFAULT NULL);
mysql> INSERT INTO DataSetD(id,name) VALUES(1,'A'),(1,'A'),(1,'A'),(3,'C');
mysql> SELECT * FROM DataSetD;
+------+---------+
| id   | name    |
+------+---------+
|    1 | A       |
|    1 | A       |
|    1 | A       |
|    3 | C       |
+------+---------+
3 rows in set (0.00 sec)

mysql> CREATE TABLE `DataSetE` (
  `id` bigint NOT NULL,`name` varchar(10) DEFAULT NULL);
mysql> INSERT INTO DataSetE(id,name) VALUES(1,'A');
mysql> SELECT * FROM DataSetE;
+------+---------+
| id   | name    |
+------+---------+
|    1 | A       |
+------+---------+
1 row in set (0.00 sec)
```

```
# EXCEPT ALL 操作
mysql > table DataSetD EXCEPT ALL table DataSetE;
+------+--------+
| id   | name   |
+------+--------+
|    1 | A      |
|    1 | A      |
|    3 | C      |
+------+--------+
2 rows in set (0.00 sec)
```

说明：在上述示例中进行 EXCEPT ALL 操作时，无法去除所有匹配值（1，A）信息，只能去掉一个重复值，没有全部去掉。

5. 集合操作的数据对比方式

集合操作对比通常直接作用于集合中的元素，先把返回的数据转化成 OBJECT 对象，再进行对比。只要匹配列数量一样，类型就无关紧要。在实际集合操作使用中，确保比较的字段和类型一致。

在集合操作中，必须具有相同的列数量。列数不同时的提示信息如下。

```
mysql> SELECT ID FROM DataSetA
         INTERSECT
       SELECT ID,NAME FROM DataSetB;
ERROR 1222 (21000): The used SELECT statements have a different number of columns
```

在集合操作中，虽然技术上可以对不同类型的元素进行对比，但这种对比通常不会得到有意义的结果，因为类型不匹配可能导致比较的结果不准确或无法解释。数据类型不一样时的集合操作的示例如下。

```
mysql> SELECT ID FROM DataSetA
       UNION
       SELECT NAME FROM DataSetA;
+--------+
| ID     |
+--------+
| 1      |
| 2      |
| 3      |
| A      |
| C      |
+--------+
5 rows in set (0.00 sec)
```

在不同 MySQL 版本中，SQL 语句的写法存在差异。与 MySQL 5.7 相比，在 MySQL 8.0 版本中，集合的解析器规则进行了重构，使其更加一致并减少了重复。

在 MySQL 5.7 和 MySQL 8.0 版本中，集合操作的 LIMIT 写法差异的示例如下。

```
# MySQL 5.7.40 不支持
mysql> (SELECT 1 AS result UNION SELECT 2) LIMIT 1;
ERROR 1064 (42000): You have an error in your SQL syntax
```

```
check the manual that corresponds to your MySQL server version for the right
syntax to use near 'UNION SELECT 2) LIMIT 1' at line 1

# MySQL 8.0.34 支持
mysql> (SELECT 1 AS result UNION SELECT 2) LIMIT 1;
+--------+
| result |
+--------+
|      1 |
+--------+
1 row in set (0.00 sec)
```

在 MySQL 5.7 和 MySQL 8.0 版本中，集合操作的 FOR UPDATE 写法差异的示例如下。

```
# MySQL 5.7.43 中要求比较宽松
mysql> SELECT 1 FOR UPDATE UNION SELECT 1 FOR UPDATE;
+---+
| 1 |
+---+
| 1 |
+---+
| 1 |
+---+
1 row in set (0.00 sec)
mysql> (SELECT 1 FOR UPDATE) UNION (SELECT 1 FOR UPDATE);
+---+
| 1 |
+---+
| 1 |
+---+
| 1 |
+---+
1 row in set (0.00 sec)

# MySQL 8.0.34 中更严谨
mysql> SELECT 1 FOR UPDATE UNION SELECT 1 FOR UPDATE;
ERROR 1064 (42000): You have an error in your SQL syntax
check the manual that corresponds to your MySQL server version for the right
syntax to use near 'UNION SELECT 1 FOR UPDATE' at line 1
mysql> (SELECT 1 FOR UPDATE) UNION (SELECT 1 FOR UPDATE);
+---+
| 1 |
+---+
| 1 |
+---+
| 1 |
+---+
1 row in set (0.00 sec)
```

6. 总结

集合操作实际上是一种聚合操作，它通常需要通过临时表进行处理。由于这种处理可能对性能产生显著影响，因此在生产环境中需要谨慎并合理地使用集合操作，以避免不必要的性能损失。

3.2 MySQL 8.0 新增索引类型及特性

MySQL 8.0 版本在索引方面进行了大量的增强，包括引入新的索引类型和算法，以及对现有索引的优化，这些改进都有助于提高数据库的查询效率和数据处理能力。

3.2.1 隐藏索引

MySQL 8.0 版本支持隐藏索引。其作用是将索引隐藏，避免索引误删除，因为数据库重构索引需要消耗大量的时间和系统资源。

1. 隐藏索引简介

隐藏索引（Invisible Index）又称为不可见索引。隐藏索引实际上是存储在数据库中的索引，但可以根据需要进行隐藏或显示。MySQL 的隐藏索引功能在某些情况下可以修改查询的执行计划，从而提高 SQL 查询的效率。

在 MySQL 执行计划中，隐藏索引不会被查询优化器用于查询路径的选择。同时，主键（Primary Key）不能设置为隐藏，因为主键是表结构的一部分，必须始终存在。使用 SQL 语句创建表结构时，标识隐藏索引的关键字（INVISIBLE）的命令行示例如下。

```
mysql> CREATE TABLE `tb_index` (
  `id` bigint NOT NULL,
  `name` varchar(255) DEFAULT NULL,
  `age` tinyint DEFAULT '10',
  `create_time` datetime DEFAULT NULL,
  `update_time` datetime DEFAULT NULL,
  `addr` varchar(30) DEFAULT NULL,
  `sex` enum('M','F') DEFAULT NULL,
  UNIQUE uni_id (id),
  UNIQUE uni_age(age),
  INDEX idx_addr (addr),
  INDEX idx_name (`name`) INVISIBLE
);
```

查看是否设置隐藏索引的命令行示例如下。

```
# 标识隐藏索引列,Visible=NO
mysql> SHOW INDEX FROM `tb_index` WHERE Visible like 'NO'\G
*************************** 1. row ***************************
        Table: tb_index
   Non_unique: 1
     Key_name: idx_name
 Seq_in_index: 1
  Column_name: name
    Collation: A
  Cardinality: 0
     Sub_part: NULL
       Packed: NULL
         Null: YES
   Index_type: BTREE
      Comment:
Index_comment:
      Visible: NO
   Expression: NULL
1 row in set (0.01 sec)
# 通过 INFORMATION_SCHEMA.STATISTICS 表查看隐藏索引列 IS_VISIBLE
```

```
mysql> SELECT INDEX_NAME, IS_VISIBLE
FROM INFORMATION_SCHEMA.STATISTICS
WHERE TABLE_SCHEMA = 'test' AND TABLE_NAME = 'tb_index';
+------------------+------------------+
| INDEX_NAME       | IS_VISIBLE       |
+------------------+------------------+
| idx_addr         | YES              |
| idx_name         | NO               |
| uni_age          | YES              |
| uni_id           | YES              |
+------------------+------------------+
4 rows in set (0.00 sec)
```

隐藏索引是否可见是通过 ALTER 命令行控制的,命令行示例如下。

```
#隐藏索引
mysql> ALTER TABLE tb_index ALTER INDEX idx_addr INVISIBLE;

#可见索引
mysql> ALTER TABLE tb_index ALTER INDEX idx_addr VISIBLE;
```

在没有设置主键、而表中存在多个唯一键的情况下,第一个唯一键不可以设置不可见属性,但从第二个唯一键开始,可以设置不可见属性。以下是第一个和第二个唯一键设置不可见属性时的提示信息示例。

```
#第一个唯一键设置隐藏属性时,提示无法设置
mysql> ALTER TABLE tb_index ALTER INDEX uni_id INVISIBLE;
ERROR 3522 (HY000): A primary key index cannot be invisible

#第二个唯一键设置隐藏属性时,提示设置成功
mysql> ALTER TABLE tb_index ALTER INDEX uni_age INVISIBLE;
Query OK, 0 rows affected (0.01 sec)
Records: 0  Duplicates: 0  Warnings: 0
```

2. 隐藏索引对执行计划的影响

在 MySQL 的执行计划中,可以通过 optimizer_switch 参数来控制是否使用隐藏索引。use_invisible_indexes 值决定了优化器在构建执行计划时是否使用隐藏索引。如果 use_invisible_indexes 设置为 OFF(默认值),则优化器将不会使用隐藏索引来构建查询的执行计划。相反,如果设置为 ON,即使索引被标记为不可见,优化器在构建执行计划时仍然会使用隐藏索引。隐藏索引的优化器控制的命令行示例如下。

```
#隐藏索引 name 为可见状态时,执行计划仍会使用隐藏索引
mysql> ALTER TABLE tb_index ALTER INDEX idx_name VISIBLE;
mysql> EXPLAIN FORMAT=TREE SELECT * FROM tb_index WHERE name='test';
+-----------------------------------------------------------------------------------+
| EXPLAIN                                                                           |
+-----------------------------------------------------------------------------------+
| -> Index lookup on tb_index using idx_name (name='test')  (cost=1.1 rows=1)       |
+-----------------------------------------------------------------------------------+
1 row in set (0.00 sec)
```

```
#use_invisible_indexes=on,执行计划使用隐藏索引
mysql> SHOW VARIABLES LIKE '%optimizer_switch%'\G
*************************** 1. row ***************************
Variable_name: optimizer_switch
Value: index_merge=on,index_merge_union=on,index_merge_sort_union=on,index_merge_inter
section=on,engine_condition_pushdown=on,index_condition_pushdown=on,mrr=on,mrr_cost_based
=on,block_nested_loop=on,batched_key_access=off,materialization=on,semijoin=on,loosescan=
on,firstmatch=on,duplicateweedout=on,subquery_materialization_cost_based=on,use_index_ex
tensions=on,condition_fanout_filter=on,derived_merge=on,use_invisible_indexes=off,skip_
scan=on,hash_join=on,subquery_to_derived=off,prefer_ordering_index=on,hypergraph_optimizer
=off,derived_condition_pushdown=on
1 row in set (0.00 sec)
mysql> ALTER TABLE tb_index ALTER INDEX idx_name INVISIBLE;
mysql> SET optimizer_switch='use_invisible_indexes=on';
mysql> EXPLAIN FORMAT=TREE SELECT * FROM tb_index WHERE name='test';
+-------------------------------------------------------------------------------------+
| EXPLAIN                                                                             |
+-------------------------------------------------------------------------------------+ | -> Index
lookup on tb_index using idx_name (name='test')  (cost=1.1 rows=1)                   |
+-------------------------------------------------------------------------------------+
1 row in set (0.00 sec)
```

说明：在某些场景下，为了评估特定 SQL 语句在生产环境中的执行效率，可以通过设置 SESSION 级别的 use_invisible_indexes 值，判断优化器是否真正使用了隐藏索引来优化该 SQL 语句。

3. 总结

隐藏索引特性提供了一种方便的方法来测试删除某个索引对查询性能的影响，而无须真正删除索引。这避免了因错误删除而需要重建索引的耗时过程。在大表上删除和重建索引可能会非常耗时，而将其设置为不可见或可见则相对简单快捷。

然而，设置隐藏索引并不会影响增、删、改操作（即 INSERT、DELETE、UPDATE）的性能，因为这些操作仍然需要维护索引。使用隐藏索引时，需要仔细考虑如何使用 EXPLAIN 等工具来测试查询性能，并据此做出决策。

3.2.2 降序索引

在 MySQL 中，索引默认是按照升序排序的。然而，从 MySQL 8.0 版本开始，可以在创建索引时指定列的排序顺序为降序。这允许更好地优化涉及范围查询的特定场景。需要注意的是，即使创建了降序索引，查询的排序方向仍然是由查询本身控制的。

1. 降序索引简介

降序索引是一种特殊的索引类型，它可以按照降序排列索引列的值。也就是说，对于索引定义中的 DESC 语句，不再被忽略，而是按降序存储键值。在 MySQL 8.0 版本前，可以相反的顺序扫描索引，但是会导致性能损失。降序索引可以按前向顺序扫描，效率更高。在多列索引中，当扫描顺序混合了某些列的升序和其他列的降序时，优化器也能使用其中的降序。

降序索引的限制如下。

- 降序索引仅支持 InnoDB 存储引擎。

- 在聚合（MIN()或MAX()）场景中，如果没有使用GROUP BY子语句，则MySQL不会使用降序索引进行优化。
- 降序索引支持BTREE类型索引，不支持哈希索引、全文索引和空间索引。
- 无论是升序还是降序，DISTINCT语句可以利用包含匹配列的索引。
- 所有可用升序索引的数据类型都支持降序索引。

下面是执行计划是否选择多列降序索引的示例，具体命令行如下所示。

```
# 模拟表结构
mysql> CREATE TABLE t (
  c1 INT, c2 INT,
  INDEX idx1 (c1 ASC, c2 ASC),
  INDEX idx2 (c1 ASC, c2 DESC),
  INDEX idx3 (c1 DESC, c2 ASC),
  INDEX idx4 (c1 DESC, c2 DESC)
);
# 多列索引, 查看执行计划的选择
mysql>EXPLAIN FORMAT=TREE SELECT * FROM tORDER BY c1 asc,c2 DESC;
+-----------------------------------------------------------+
|EXPLAIN                                                    |
+-----------------------------------------------------------+
|-> Index scan on tt using idx2   (cost=0.35 rows=1)   |
+-----------------------------------------------------------+
1 row in set (0.00 sec)
```

在上述排序SQL查询语句中，c1是升序，而c2是降序，所以执行计划正确选择了索引idx2。在MySQL 8.0之前，支持GROUP BY子句默认隐式排序的功能，到了MySQL 8.0，不再支持，因此对结果集有排序的需求，就需要显式执行ORDER BY。

2. 总结

索引是提高查询效率的重要手段。在某些组合索引的场景下，适当的排序条件可以利用索引的优势，从而提高查询效率。需要注意的是，索引的创建需要根据实际情况进行，过多或不必要的索引会降低数据库的性能。

3.2.3 函数索引

在MySQL 5.7版本中，为了实现函数索引的功能，采用虚拟列的方式支持建立函数索引，MySQL 8.0.13版本直接支持函数索引，也就是将表达式的值作为索引的内容，而不是列值或列值前缀。

1. 函数索引介绍

函数索引是将函数处理后的结果集作为索引键。

在MySQL 8.0中，如果在时间字段上创建索引，但在查询时通过应用函数来访问该字段，那么无法使用索引加速查询。函数访问索引失效的命令行示例如下。

```
# 引用上面的表 tb_index
mysql> ALTER TABLE tb_index ADD INDEX idx_created (create_time);

# 执行计划进行全表扫描
mysql> EXPLAIN FORMAT=TREE SELECT *
```

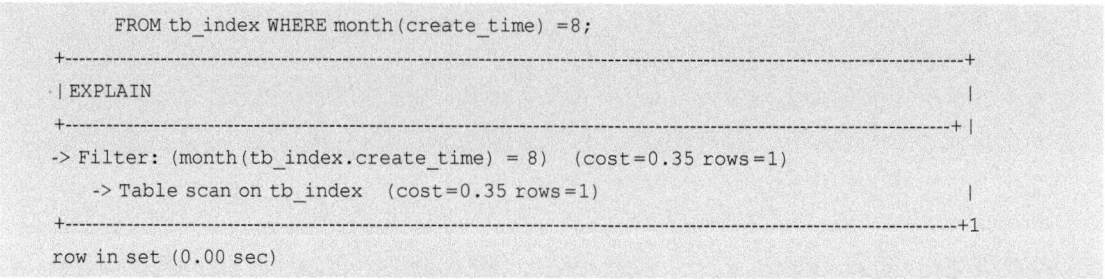

在上述表结构中，为 create_time 列添加 month 函数索引。再次查看执行计划是否使用函数索引。添加函数索引，并查看执行计划的命令行示例如下。

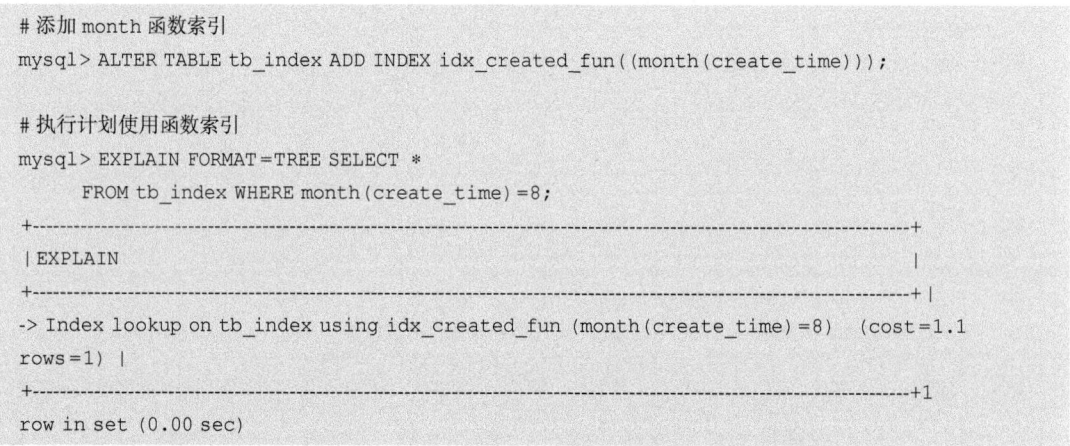

函数索引的实现是基于虚拟列的，因此它受到与虚拟列相同的许多限制。
- 函数索引支持 UNIQUE 选项。然而，主键不能包含函数列。主键仅能使用存储的计算列来实现，而函数索引则是基于虚拟计算列来实现的，不是存储计算列。
- 空间 SPATIAL 索引和全文 FULLTEXT 索引不支持函数索引。
- 如果某个表中没有显式定义主键，那么 InnoDB 存储引擎会自动将第一个 UNIQUE NOT NULL 索引提升为主键。然而，这个提升规则不适用于包含函数列的 UNIQUE NOT NULL 索引。
- 如果字段上存在函数索引，那么需要先删除这些函数索引，然后才能删除该字段。

2. 总结

函数索引的使用条件比较苛刻，必须严格按照索引建立的定义来写，这样才能用到函数索引。

3.2.4 Hash Join 特性

从 MySQL 8.0.18 版本开始，增加了一个新的特性——哈希连接（Hash Join），它适用于等值连接。虽然 Hash Join 特性通常不适用于范围查找、排序等操作，但它可以显著提高等价类型的查询性能，特别是在处理没有索引的表连接操作时。

1. Hash Join 介绍

Hash Join 是一种重要的优化技术，用于加速数据库查询中的表连接操作。这个过程通常涉及在内存中为其中一个表（通常是小表）创建一个哈希表，然后基于哈希值快速查找和连接另一个表（通常是大表）中的相应行。

Hash Join 已经替代之前的 Nested Loop 成为 MySQL 8.0 版本的默认连接方式。以下示例模拟两张表（每个表有 1 万行数据）Join 的场景。模拟场景的 SQL 语句如下所示。

```
# 创建表
mysql> CREATE TABLE `t1` (
  `id` bigint(20) NOT NULL AUTO_INCREMENT,
  `name` varchar(255) DEFAULT NULL,
  `age` tinyint(4) DEFAULT NULL,
  `create_time` datetime DEFAULT NULL,
  `update_time` datetime DEFAULT NULL,
  PRIMARY KEY (`id`),
  key (name)
) ENGINE=InnoDB;
mysql> CREATE TABLE `t2` (
  `id` bigint(20) NOT NULL AUTO_INCREMENT,
  `name` varchar(255) DEFAULT NULL,
  `age` tinyint(4) DEFAULT NULL,
  `create_time` datetime DEFAULT NULL,
  `update_time` datetime DEFAULT NULL,
  PRIMARY KEY (`id`),
  key (name)
) ENGINE=InnoDB;

# 创建批量数据
mysql> delimiter //
DROP PROCEDURE IF EXISTS proc_batch_insert;
CREATE PROCEDURE proc_batch_insert()
BEGIN
DECLARE pre_name BIGINT;
DECLARE ageVal INT;
DECLARE i INT;
SET pre_name=139;
SET ageVal=100;
SET i=1;
WHILE i <= 10000 DO
        INSERT INTO t1(`name`,age,create_time,update_time)
VALUES(CONCAT(pre_name,'@qq.com'),(ageVal+1)*rand()%30,NOW(),NOW());
        INSERT INTO t2(`name`,age,create_time,update_time)
VALUES(CONCAT(pre_name,'@qq.com'),(ageVal+1)*rand()%30,NOW(),NOW());
SET pre_name=pre_name+100;
SET i=i+1;
END WHILE;
END //
mysql> delimiter;

## 执行存储过程
mysql> call proc_ batch_ insert( );
```

使用上述表进行查询,查看执行计划来确认是否使用了 Hash Join,同时与传统的 Nested Loop 的执行时间进行对比。具体执行计划和对比数据如下所示。

```
# 在使用 Hash Join 的情况下,得到结果集耗时 0.19s
mysql> EXPLAIN FORMAT=TREE SELECT COUNT(*) FROM t1 INNER JOIN t2 ON t1.age=t2.age;
```

```
+------------------------------------------------------------------+
| EXPLAIN                                                          |
+------------------------------------------------------------------+
| -> Aggregate: count(0)  (cost=10.7e+6 rows=1)                    |
|    -> Inner hash join (t2.age = t1.age)  (cost=9.69e+6 rows=9.69e+6) |
|       -> Table scan on t2  (cost=0.0126 rows=9884)               |
|       -> Hash                                                    |
|           -> Table scan on t1  (cost=989 rows=9807)              |
+------------------------------------------------------------------+
1 row in set (0.00 sec)
[book]> SELECT COUNT(*) FROM t1 INNER JOIN t2 ON t1.age=t2.age;
+----------+
| COUNT(*) |
+----------+
|  3334479 |
+----------+
1 row in set (0.19 sec)
```

可使用数据库的 optimizer_switch 参数来关闭 Hash Join 的功能，然后在关联查询的字段上添加二级索引。在这种情况下，优化器选择使用索引查找（Index Lookup）来执行查询。具体命令行示例如下。

```
# 得到结果集的时间是 1.02s
mysql>SET optimizer_switch='hash_join=off';
mysql> ALTER TABLE t1 ADD INDEX idx_age(age);
Query OK, 0 rows affected (0.14 sec)

mysql> ALTER TABLE t2 ADD INDEX idx_age(age);
mysql> EXPLAIN FORMAT=TREE SELECT COUNT(*) FROM t1 INNER JOIN t2 ON t1.age=t2.age;
+------------------------------------------------------------------+
| EXPLAIN                                                          |
+------------------------------------------------------------------+
| -> Aggregate: count(0)  (cost=638947 rows=1)                     |
|    -> Nested loop inner join  (cost=326262 rows=3.13e+6)         |
|       -> Filter: (t1.age is not null)  (cost=989 rows=9807)      |
|           -> Index scan on t1 using idx_age  (cost=989 rows=9807)|
|       -> Covering index lookup on t2 using idx_age (age=t1.age)  (cost=1.29 rows=319) |
+------------------------------------------------------------------+
1 row in set (0.00 sec)
[book]> SELECT COUNT(*) FROM t1 INNER JOIN t2 ON t1.age=t2.age;
+----------+
| COUNT(*) |
+----------+
|  3334479 |
+----------+
1 row in set (1.02 sec)
```

从上述结果对比可以看出，Hash Join 执行的性能在添加二级索引并使用 Index Lookup 后得到了显著的提升，性能提升了数倍。然而，使用 Hash Join 也有一些条件和限制，需要用户注意。

- Hash Join 只在没有索引的字段上有效。
- Hash Join 只在等值 Join 条件中有效。
- Hash Join 不能用于 Left Join 和 Right Join。

2. Hash Join 的控制

在 MySQL 中，optimizer_switch 参数的 hash_join 指标用来控制查询优化器的行为。默认情况下，该特性是开启的，以优化查询性能。查看和设置 Hash Join 的命令行示例如下。

```
# 查看 Hash Join
mysql> SHOW VARIABLES LIKE 'optimizer_switch'\G
*************************** 1. row ***************************
Variable_name: optimizer_switch
        Value: index_merge=on,index_merge_union=on,index_merge_sort_union=on,index_merge
_intersection=on,engine_condition_pushdown=on,index_condition_pushdown=on,mrr=on,mrr_cost_
based=on,block_nested_loop=on,batched_key_access=off,materialization=on,semijoin=on,loos-
escan=on,firstmatch=on,duplicateweedout=on,subquery_materialization_cost_based=on,use_
index_extensions=on,condition_fanout_filter=on,derived_merge=on,use_invisible_indexes=off,
skip_scan=on,hash_join=on,subquery_to_derived=off,prefer_ordering_index=on,hypergraph_op-
timizer=off,derived_condition_pushdown=on
1 row in set (0.01 sec)

# 关闭 Hash Join
mysql> SET optimizer_switch='hash_join=off';
```

在 MySQL 8.0.3 版本中，SET_VAR 优化器提示可以用来设置语句级参数（Oracle 和 MariaDB 已支持）。通过 SET_VAR 控制是否使用 Hash Join 的命令行示例如下。

```
mysql> SELECT/*+ set_var(optimizer_switch='hash_join=off')
set_var(join_buffer_size=4M) */ COUNT(*) FROM t1 INNER JOIN t2 ON t1.age=t2.age;

mysql> SELECT/*+ set_var(optimizer_switch='hash_join=off')
set_var(join_buffer_size=4M) */ COUNT(*) FROM t1 INNER JOIN t2 ON t1.age=t2.age;
```

3. 总结

Hash Join 是一种强大的连接策略，在连接小型表时效率非常高。然而，当处理大型表时，将整个表加载到内存中可能会导致性能问题。

3.2.5　Skip Scan Range 特性

从 MySQL 8.0.13 版本开始，MySQL 引入了一个名为"Skip Scan Range"的新功能，这一功能是由 Facebook 贡献的。这个特性主要用于在 InnoDB 存储引擎下优化查询性能，特别是在查询过程中需要扫描大量不必要的索引页时。

1. Skip Scan Range 简介

在 MySQL 8.0 之前，如果有一个组合索引（比如(f1,f2)），并且查询条件只涉及 f2 而没有涉及 f1，那么优化器通常不会使用这个组合索引来扫描数据，因为索引的左前缀（即 f1）没有在查询条件中。然而，在 MySQL 8.0 版本中，查询优化器得到了改进，选择使用该索引进行范围扫描，这种情况下，跳跃式索引（Skip Scan Range）能够更有效地利用索引，包括组合索引。

在利用组合主键索引进行范围查询时，使用 Skip Scan Range 的命令行示例如下。

```
# 模拟表结构和数据
mysql> CREATE TABLE t1 (f1 INT NOT NULL, f2 INT NOT NULL);
mysql> INSERT INTO t1 VALUES (1,1), (1,2), (1,3), (1,4), (1,5), (2,1), (2,2), (2,3), (2,4), (2,5);
mysql> INSERT INTO t1 SELECT f1, f2 + 5 FROM t1;
mysql> INSERT INTO t1 SELECT f1, f2 + 10 FROM t1;
mysql> INSERT INTO t1 SELECT f1, f2 + 20 FROM t1;
mysql> INSERT INTO t1 SELECT f1, f2 + 40 FROM t1;
mysql> ANALYZE TABLE t1;
mysql> SELECT COUNT(*) FROM t1;
+----------+
| COUNT(*) |
+----------+
|      160 |
+----------+
# 添加主键组合索引
mysql> ALTER TABLE t1 ADD PRIMARY KEY(f1, f2);
mysql> EXPLAIN FORMAT=TREE SELECT f1, f2 FROM t1 WHERE f2 > 40;
+-----------------------------------------------------------------------------------------+
| EXPLAIN                                                                                 |
+-----------------------------------------------------------------------------------------+
-> Filter: (t1.f2 > 40)  (cost=15.9 rows=53)
    -> Covering index skip scan on t1 using PRIMARY over 40 < f2  (cost=15.9 rows=53) |
+-----------------------------------------------------------------------------------------+1
row in set (0.00 sec)
```

在上述执行计划中,提示信息"Covering index skip scan on t1 using PRIMARY over 40"说明使用 Skip Scan Range 特性,可避免扫描不必要的索引页或数据页,从而提升了性能,尤其是在索引前缀列区分度比较低时使用。Skip Scan Range 特性适用于以下情况。

1)查询 SQL 语句只引用一个表。

2)查询 SQL 语句不使用 GROUP BY 或 DISTINCT。

3)查询 SQL 语句只引用索引中的列。

4)表 T 至少有一个包含表单关键部分的复合索引 ($[A\_1,\cdots,A\_k],[B\_1,\cdots,B\_m],C,[D\_1,\cdots,D\_n]$)。关键部分 A 和 D 可能是空的,但是 B 和 C 必须是非空的。

- $A\_1,\cdots,A\_k$ 上的谓词必须是等式谓词,它们必须是常数。这包括 IN( )方式查询。
- 查询必须是一个连接查询,如 cond1(key_part1)或 cond2(key_part1)和 cond1(key_part2) 或 cond2(key_part2)。
- 表结构 C 列上必须有一个 range 条件。
- 查询 SQL 语句允许在 D 列上设置条件。D 列上的条件必须与 C 列上的范围条件相结合。

2. Skip Scan Range 的控制

在 MySQL 中,Skip Scan Range 特性通过 optimizer_switch 参数的 skip_scan 指标来控制查询优化器的行为。默认情况下,该特性是开启的,以优化查询性能。控制 Skip Scan Range 的命令行示例如下。

```
# 查看 Skip Scan Range 设置
mysql> SHOW VARIABLES LIKE 'optimizer_switch'\G
```

```
*************************** 1. row ***************************
Variable_name: optimizer_switch
        Value: index_merge=on,index_merge_union=on,index_merge_sort_union=on,index_merge
_intersection=on,engine_condition_pushdown=on,index_condition_pushdown=on,mrr=on,mrr_cost_
based=on,block_nested_loop=on,batched_key_access=off,materialization=on,semijoin=on,loos-
escan=on,firstmatch=on,duplicateweedout=on,subquery_materialization_cost_based=on,use_
index_extensions=on,condition_fanout_filter=on,derived_merge=on,use_invisible_indexes=off,
skip_scan=on,hash_join=on,subquery_to_derived=off,prefer_ordering_index=on,hypergraph_op-
timizer=off,derived_condition_pushdown=on
1 row in set (0.00 sec)

#关闭Skip Scan Range
mysql>SET optimizer_switch='skip_scan=off';
mysql> EXPLAIN FORMAT=TREE SELECT f1, f2 FROM t1 WHERE f2 > 40;
+---------------------------------------------------------------------------------+
| EXPLAIN                                                                         |
+---------------------------------------------------------------------------------+
| -> Filter:
(t1.f2 > 40)    (cost=15.9 rows=53)
    -> Covering index skip scan on t1 using PRIMARY over 40 < f2    (cost=15.9 rows=53)
+---------------------------------------------------------------------------------+
1 row in set
(0.00 sec)
```

3. 总结

Skip Scan Range 功能对使用 MySQL 组合索引查询的 SQL 语句的意义非常重大。在 MySQL 的使用中，一般要求在建立组合索引时将重复值少的字段放在组合索引前面，将重复值多的字段放在组合索引后面，方便 SQL 在使用组合索引时通过前面的字段快速过滤结果。但将索引后面字段作为条件时，无法用到索引。这种情况下，可再次创建合适的索引，但这会导致索引维护成本变高，也有可能导致原先的执行计划变更。

3.2.6 Anti Join（反连接）特性

从 MySQL 8.0.17 版本开始，对任何具有 NOT IN、NOT EXISTS 的子查询语句进行了优化，将其转变为反连接（Anti Join）。

1. Anti Join 介绍

Anti Join 是优化器将 WHERE 条件中的 NOT IN（Subquery）、NOT EXISTS（Subquery）等过程，在内部转化成一个反连接，以便移除里面的子查询（Subquery）。这个优化在某些场景下能够很好地提升性能。

先了解 MySQL 8.0 版本之前 NOT 子句的处理机制。图 3-4 所示为在 MySQL 5.7 版本中 NOT IN 和 NOT EXISTS 条件下的执行计划。

说明：总的来说，在 MySQL 5.7 版本中，对于 NOT IN 和 NOT EXISTS 子查询语句，存在两种运算类型：SUBQUERY 和 DEPENDENT SUBQUERY。其中，SUBQUERY 指的是子查询中的第一个 SELECT 查询，它不依赖于外部查询的结果集；而 DEPENDENT SUBQUERY 则是指子查询中的第一个 SELECT 查询，它依赖于外部查询的结果集。

下面分析执行步骤。

第一步，根据 t_temp 得到一个结果集，其数据量是 rows=1000 行。

```
mysql> EXPLAIN SELECT * FROM t_temp t WHERE id NOT IN (SELECT id FROM r_temp e ) ;
+----+-------------+-------+------------+-------+---------------+---------+---------+------+------+----------+-------------+
| id | select_type | table | partitions | type  | possible_keys | key     | key_len | ref  | rows | filtered | Extra       |
+----+-------------+-------+------------+-------+---------------+---------+---------+------+------+----------+-------------+
|  1 | PRIMARY     | t     | NULL       | ALL   | NULL          | NULL    | NULL    | NULL | 1000 |   100.00 | Using where |
|  2 | SUBQUERY    | e     | NULL       | index | PRIMARY       | name    | 768     | NULL |   12 |   100.00 | Using index |
+----+-------------+-------+------------+-------+---------------+---------+---------+------+------+----------+-------------+
2 rows in set, 1 warning (0.00 sec)

mysql>
mysql>
mysql> EXPLAIN SELECT * FROM t_temp t WHERE NOT EXISTS (SELECT 1 FROM r_temp e where e.id = t.id) ;
+----+--------------------+-------+------------+--------+---------------+---------+---------+----------+------+----------+-------------+
| id | select_type        | table | partitions | type   | possible_keys | key     | key_len | ref      | rows | filtered | Extra       |
+----+--------------------+-------+------------+--------+---------------+---------+---------+----------+------+----------+-------------+
|  1 | PRIMARY            | t     | NULL       | ALL    | NULL          | NULL    | NULL    | NULL     | 1000 |   100.00 | Using where |
|  2 | DEPENDENT SUBQUERY | e     | NULL       | eq_ref | PRIMARY       | PRIMARY | 8       | dbB.t.id |    1 |   100.00 | Using index |
+----+--------------------+-------+------------+--------+---------------+---------+---------+----------+------+----------+-------------+
```

图 3-4　MySQL 5.7 中 NOT IN 和 NOT EXISTS 执行计划

第二步，上面结果集 t_temp 中的每一条记录都将与子表进行 Loop 循环。

在数据量少时，对整体性能没有太大影响，但随着数据量的增加，如第一个结果集中的表达到 100 万行级别，即使这两个步骤中都用到了索引，也会非常慢。DEPENDENT SUBQUERY 会随着外层表的结果集的数据量的增大而导致执行时间增加。如此一来，子查询的执行效率会受制于外层查询的记录数。解决这种问题的方案如下。

- 改成 JOIN 方式。
- 拆分成两个独立查询并顺序执行。
- 临时生成 temp 表，进行二次 JOIN。

在 MySQL 8.0 版本中，对 NOT IN 和 NOT EXISTS 子查询语句转换的 Anti Join 如图 3-5 所示。

```
mysql> EXPLAIN FORMAT=TREE SELECT * FROM t_temp t WHERE t.id NOT IN (SELECT id FROM t_temp e) ;
+----------------------------------------------------------------------------------------------------------------+
| EXPLAIN                                                                                                        |
+----------------------------------------------------------------------------------------------------------------+
| -> Nested loop antijoin  (cost=100201 rows=1e+6)
    -> Table scan on t  (cost=101 rows=1000)
    -> Single-row index lookup on <subquery2> using <auto_distinct_key> (id=t.id)  (cost=201..201 rows=1)
        -> Materialize with deduplication  (cost=201..201 rows=1000)
            -> Filter: (e.id is not null)  (cost=101 rows=1000)
                -> Index scan on e using name  (cost=101 rows=1000)
|
+----------------------------------------------------------------------------------------------------------------+
1 row in set (0.00 sec)

mysql> EXPLAIN FORMAT=TREE SELECT * FROM t_temp t WHERE id NOT IN (SELECT id FROM t_temp e where e.id=t.id) ;
+----------------------------------------------------------------------------------------------------------------+
| EXPLAIN                                                                                                        |
+----------------------------------------------------------------------------------------------------------------+
| -> Nested loop antijoin  (cost=100201 rows=1e+6)
    -> Table scan on t  (cost=101 rows=1000)
    -> Single-row index lookup on <subquery2> using <auto_distinct_key> (id=t.id, id=t.id)  (cost=201..201 rows=1)
        -> Materialize with deduplication  (cost=201..201 rows=1000)
            -> Filter: ((e.id is not null) and (e.id is not null))  (cost=101 rows=1000)
                -> Index scan on e using name  (cost=101 rows=1000)
|
+----------------------------------------------------------------------------------------------------------------+
1 row in set, 1 warning (0.00 sec)
```

图 3-5　Anti Join 的执行

执行计划第一行提示 "ant i join" 采用 Anti Join 策略，同时 select_type 中出现了 Materialize 类型。Materialize 是数据库优化器使用的一种技术，通过物化子查询结果（即生成临时表，通常在内存中，但在需要时也可能使用磁盘）来加速查询。在第一次需要子查询结果时，它会将该结果物化到一个临时表中，后续需要结果时，会再次引用这个临时表。此外，优化器可能会利用哈希索引来快速定位数据，但实际的查找操作可能依赖其他类型的索引。子查询物化在适当的情况下可以显著提高查询性能。

下列情况下子查询将转换为 Anti Join。
- NOT IN (SELECT ... FROM ...)。
- NOT EXISTS (SELECT ... FROM ...)。
- IN (SELECT ... FROM ...) IS NOT TRUE。
- EXISTS (SELECT ... FROM ...) IS NOT TRUE。
- IN (SELECT ... FROM ...) IS FALSE。
- EXISTS (SELECT ... FROM ...) IS FALSE。

在 Anti Join 中，限制 Materialize 物化。
- 内部表达式的类型和外部表达式必须匹配。如果两个表达式都是整数或小数，则优化器可以使用具体化，但如果一个表达式是整数而另一个是小数，则不能使用。
- 内部表达式不能是 BLOB 类型。

2. Anti Join 的控制

在 MySQL 中，Anti Join 通过 optimizer_switch 参数的 semijoin 指标来控制查询优化器的行为。这里存在一个疑问，即为什么与半连接（Semi Join）有关联。

先看一下 Semi Join 和 Anti Join 的关系。Semi Join 和 Anti Join 是处理表间关系时互补的操作，仅目的是相反的。
- Semi Join（或称为 IN/EXISTS 子查询）找出左表中在右表有至少一个匹配的行。
- Anti Join（或称为 NOT IN/NOT EXISTS 子查询）找出左表中在右表没有匹配的行。

所以 MySQL 官方把 Semi Join 和 Anti Join 的指标开关合并。下面是控制 Anti Join 的命令行示例。

```
# 查看优化器 Anti Join (semijoin) 指标,关闭 Semi Join
mysql> SHOW VARIABLES LIKE 'optimizer_switch';
*************************** 1. row ***************************
Variable_name: optimizer_switch
        Value: index_merge=on,index_merge_union=on,index_merge_sort_union=on,index_merge
_intersection=on,engine_condition_pushdown=on,index_condition_pushdown=on,mrr=on,mrr_cost_
based=on,block_nested_loop=on,batched_key_access=off,materialization=on,semijoin=on,loos-
escan=on,firstmatch=on,duplicateweedout=on,subquery_materialization_cost_based=on,use_
index_extensions=on,condition_fanout_filter=on,derived_merge=on,use_invisible_indexes=off,
skip_scan=on,hash_join=on,subquery_to_derived=off,prefer_ordering_index=on,hypergraph_op-
timizer=off,derived_condition_pushdown=on
mysql> SET SESSION optimizer_switch="semijoin=off";

# 执行同一个 SQL 语句,查看执行计划
mysql>EXPLAIN FORMAT=TREE SELECT *
      FROM t_temp t WHERE t.id NOT IN (SELECT id FROM t_temp e);
+-------------------------------------------------------------------------------------+
| EXPLAIN                                                                             |
| +---------------------------------------------------------------------------------+ |
|-> Filter: <in_optimizer>(t.id,<exists>(select #2) is false)  (cost=101 rows=1000)
    -> Table scan on t  (cost=101 rows=1000)
    -> Select #2 (subquery in condition; dependent)
        -> Limit: 1 row(s)  (cost=0.35 rows=1)
            -> Filter: (<cache>(t.id) = e.id)  (cost=0.35 rows=1)
```

```
                     -> Single-row covering index lookup on e using PRIMARY (id=t.id)    (cost=0.35
rows=1) |
  | +------------------------------------------------------------------------------------------+
1 row in set (0.00 sec)
```

最终输出的执行计划中已经没有提示信息"anti join"了。

3. 总结

在 MySQL 中，IN 和 EXISTS 子查询都普遍存在效率低的问题，在 MySQL 8.0 版本中，这些子查询的性能通常已经得到优化。对于 Anti Join，如果查询没有适当优化，则可能会导致大量的内存使用和临时表的创建，尽量合理使用。

3.3 复制和高可用性方面的新增功能

MySQL 8.0 在复制和高可用性方面新增了一些重要的功能，这些功能有助于提升数据库的性能、可靠性和灵活性。

3.3.1 高可用组复制（MGR）功能

MySQL Group Replication（MGR）是 MySQL 8.0 版本中引入的新的高可用解决方案，以插件形式提供。它提供了高可用性和数据一致性保证，尤其适用于需要高可靠性的系统。其特点如下。

- 高一致性：基于 Paxos 的分布式一致性协议（XCom），确保了在组成员间复制数据时的强一致性。
- 高容错性：自动检测机制，只要不是大多数节点都宕机，就可以继续工作，内置防脑裂保护机制。
- 高扩展性：节点的增加与移除会自动更新组成员信息。新节点加入后，自动从其他节点同步增量数据，直到与其他节点所有数据都一致为止。
- 高灵活性：提供单主模式和多主模式。单主模式在主库宕机后能够自动选新主，并且所有写入都在主节点进行。多主模式支持多节点写入。

下面具体介绍 MGR。

1. MGR 演进过程

MGR 目前支持 MySQL 5.7 版本和 MySQL 8.0 版本。其演进过程如下。

- MySQL 主从复制。提供了一种简单的 Primary-Secondary 复制方法。如图 3-6 所示，有一个主节点（Source）和一个或多个从节点（Replica），通过 binlog 进行数据同步，实现异步复制，从而实现高可用。
- 为了解决异步复制中可能的数据丢失问题，MySQL 复制协议中添加了半同步复制机制。半同步复制旨在确保数据在主从节点之间以准实时的方式同步，从而提高了数据的一致性。然而，需要注意的是，半同步复制并不能完全消除从节点复制的延迟，因为从节点仍然需要时间来回放（应用）binlog 的事件。如图 3-7 所示，变更的 SQL 语句必须复制到从节点中，但从节点的回放时间可能因各种因素（如性能、负载和网络延迟）而有所不同。
- MySQL 主从架构都是单点写入，之后将数据同步到从节点。MGR 引入了一种分布式方法来处理复制。Group Replication 是一种可用于实现容错系统的技术。复制组是一组服务节点，每个服务节点都有自己的完整数据副本（不是数据共享方案），并通过原子消息和总订单消息传递相互交互，实现数据的一致性，构建更高级的数据库复制解决方案。如图3-8所示，在保证数据一致性的同时，实现准实时级别的复制。

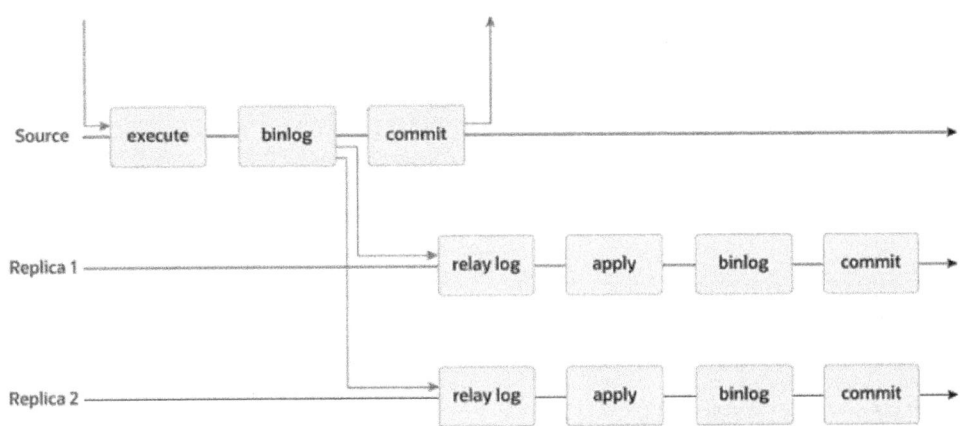

图3-6　MySQL 异步复制

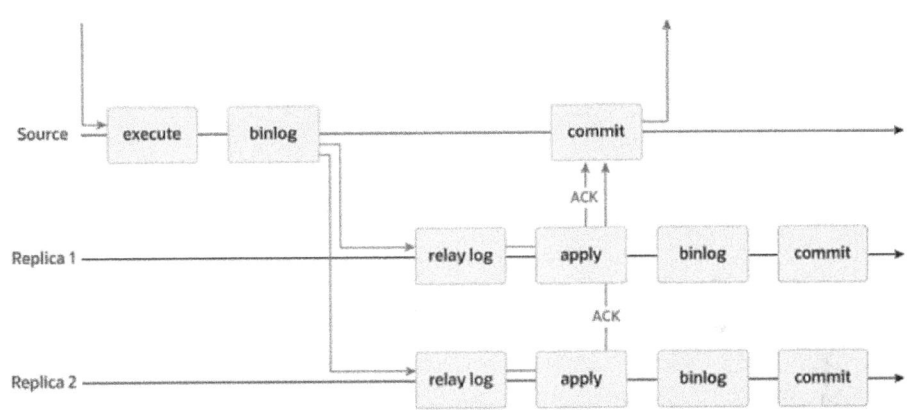

图3-7　MySQL 半同步复制

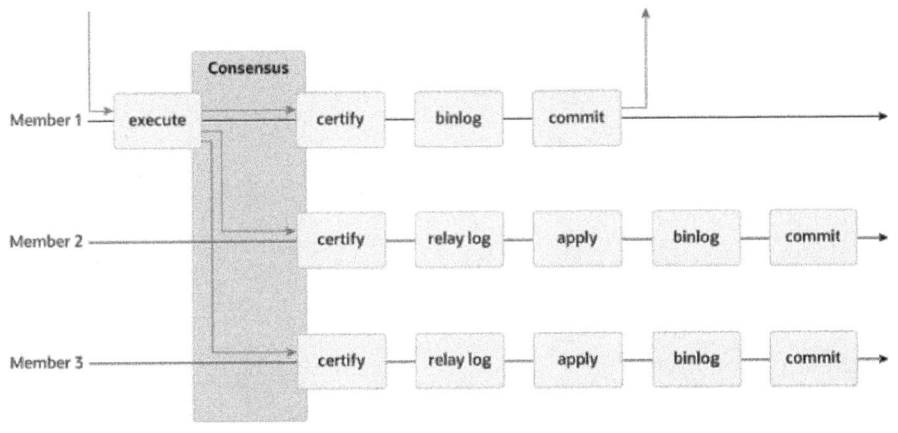

图3-8　MySQL MGR 组复制

2. MGR 插件

MGR 是一个 MySQL 插件，它建立在现有的 MySQL 复制基础设施之上，通过利用 binlog 日志和全局事务标识（GTID）来实现其高可用性、容错性和数据一致性。图 3-9 所示为 MGR 插件的架构图，其中包括各个组件之间的交互和数据流。

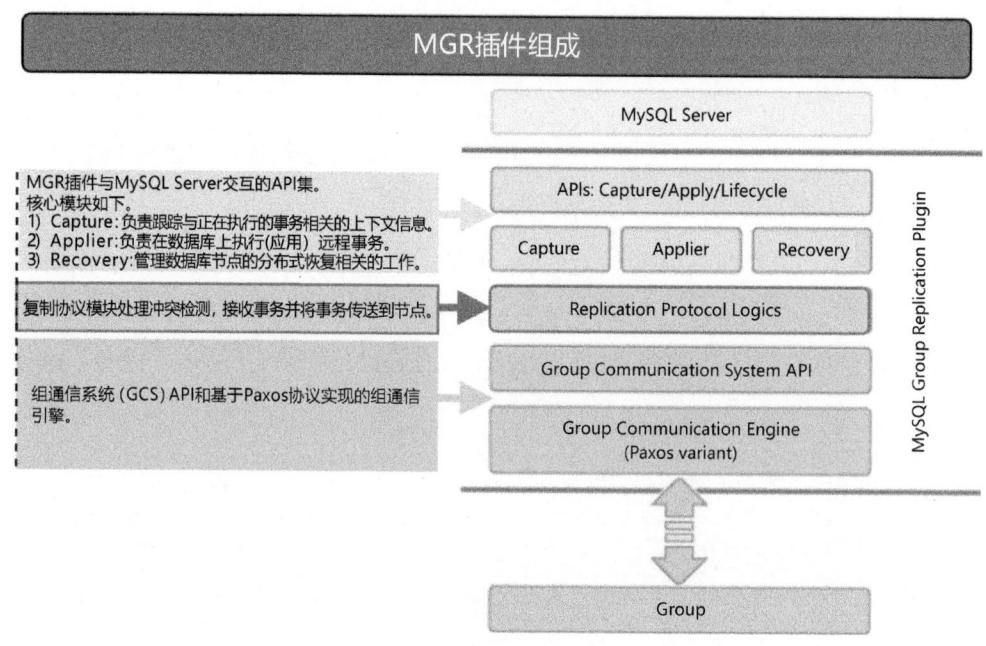

图 3-9　MGR 插件架构图

MGR 插件包括一组用于 Capture、Apply 和 Lifecycle 的 API。这些 API 控制插件如何与 MySQL 服务器交互。例如，有事件通知、服务启动、服务恢复、服务准备接受连接、服务即将提交事务、中止正在进行的事务、在中继日志中排队事务等操作。下面是组件的说明。

- Capture：负责跟踪与正在执行的事务相关的上下文。
- Applier：负责在数据库上执行远程事务。
- Recovery：主要负责管理和协调系统在面临故障或数据丢失时的恢复流程，确保系统的可用性和数据的一致性。
- Replication Protocol Logics：包含复制协议的特定逻辑。它处理冲突检测，接收事务并将其传播到节点。

MGR 插件是基于组通信系统（GCS）API 和基于 Paxos 的组通信引擎（XCom）实现的。

（1）Paxos 算法

Paxos 算法旨在解决分布式系统中的一致性问题，尤其是在可能遇到消息延迟、丢失、重复等异常情况时，确保各个节点（或称为进程）能够就某个值（或称为决议）达成一致，并且这种一致性不会被上述异常所破坏。

Paxos 算法的目的是实现多个节点并发操纵数据，保证在读写过程中数据的一致性。

Paxos 有以下两个组件。

- Proposer：提议发起者，处理客户端请求，将客户端的请求发送到集群中，以便决定这个值是否可以被批准。

- Acceptor：决策者，负责投票决定是否接受提案。

Paxos 有两个原则，一是安全原则，即保证不能做错的事；二是存活原则，只要有多数服务器存活并且彼此间可以通信，最终数据就可以达到一致。

（2）XCom 组件

XCom 组件是 MySQL Group Replication 中用于实现节点间通信的关键部分。它负责在 MySQL 集群中的不同节点之间传递消息和数据，以确保数据的一致性和高可用性。它使用 C 语言编译，可依赖协程实现任务调度，并通过 TCP 连接与 Client 进行交互。

- 单线程驱动，无多线程能力。
- 通信流需要额外一次的 TCP 协议栈。
- XCom 实现了 Batching（批处理）和 Pipelining（流水线化）等优化技术，以提高数据传输效率和吞吐量。

3. MGR 内部处理机制

- MGR 提供分布式状态复制机制，在服务之间具有很强的协调性。该技术的核心是 Paxos 算法。所有节点通过组通信系统（GCS）协议提供支持。它还提供了故障检测机制、组成员服务以及安全且完全有序的消息传递机制。所有这些属性确保数据在组之间一致。
- MGR 内置的组成员服务，可以自动保持组视图的一致性，并在任何给定的时间点对所有服务可用。服务可以离开或加入组，视图也会相应地更新。有时，服务可能会意外地离开组，在这种情况下，故障检测机制会检测到这种情况，并通知组视图已更改。
- 对于 MGR 提交的事务，组中的大多数成员必须就给定事务在全局事务序列中的顺序达成一致。决定提交或终止事务是由每个服务节点单独完成的。如果存在网络分区，导致成员无法达成协议的分裂，则系统不会继续执行此事务，直到此问题得到解决为止。因此，MGR 内部也有一个内置的、自动的防脑裂保护机制。
- MGR 中每个服务节点都可以在任何时候独立执行事务，所有写事务只有在被组允许后才会提交。不同服务节点上并发执行的事务之间可能存在冲突。这种冲突是通过检查和比较两个不同并发事务的写集来检测的，冲突解决之后，事务在所有服务节点上提交。
- Group Replication 能够通过将系统状态复制到一组服务器来创建具有冗余的容错系统。即使一些服务器随后出现故障，只要不是全部或大部分，系统仍然可用。根据发生故障的服务节点的数量，该组可能会降低性能或可扩展性，但它仍然可用。所以故障节点是孤立和独立的。表 3-3 所示为容错节点数。

表 3-3 MySQL MGR 容错节点数

集群节点数	必须存活节点数	允许故障节点数
1	1	0
2	2	0
3	2	1
4	3	1
5	3	2
6	4	2
7	4	3

说明：MGR 建立在 Paxos 分布式算法的实现之上，所以需要大多数节点处于活动状态，这样才能达到允许故障节点数，从而做出决策。这意味着要容忍一定的节点服务无法使用。

- Group Replication 引入流控机制，确保集群内各个节点的延迟尽可能小，避免 Fail Over 和 relay log 回放时间太长。流控机制会检查事务个数、节点写事务数量、调控周期等。
- 在 MGR 管理和使用方面，MySQL 官方推出了 MySQL Shell 工具，它可以和 MySQL Router 中间件配合使用。其中 MySQL Router 可以替换成开源 ProxySQL 中间件。图 3-10 所示为 MySQL Router 中间件结合 MGR 使用的架构图。MySQL Router 用于在 MySQL 和客户端应用程序之间路由连接。它可以提供基于规则的路由、读写分离、故障转移和负载均衡等功能。

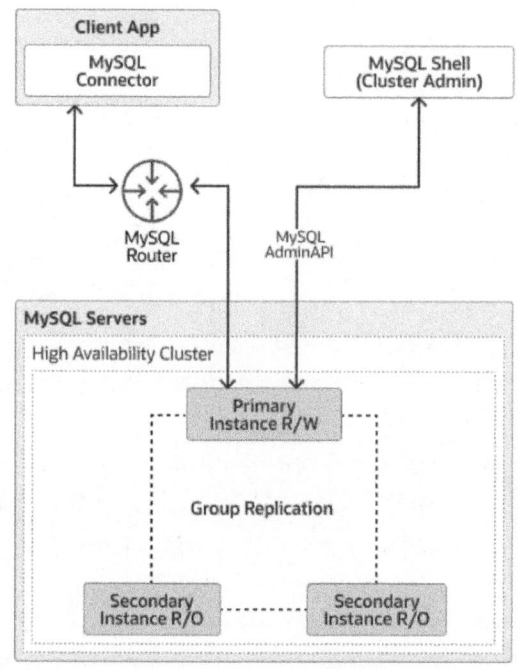

图 3-10　MySQL Router+MGR 架构

- MGR 必须满足一系列使用要求。MGR 必须存储在 InnoDB 存储引擎中。每个表都必须有一个已定义的主键，或者主键的等价物（其中等价物是非空唯一键）。MGR 被设计为部署在服务器实例彼此非常接近的集群环境中（网络延迟和网络带宽都会影响组的性能和稳定性）。

4. MGR 参数设置

想要使用 MGR 集群，必须配置以下参数。
- 保证唯一服务器标识符 server_id。
- binlog 必须开启，同时 binlog_format 的格式为 ROW 模式。
- 副本更新记录 log_replica_updates 必须开启。
- 全局事务标识符 gtid_mode 必须开启。
- 事务写集必须设置 transaction_write_set_extraction=XXHASH64。
- 开启 Multi-threaded 多线程并行应用。
- MySQL 8.0.29 版本及以后版本支持独立 XA 事务。
- 单个 MGR 集群最大支持 9 个 MySQL 节点。

5. MGR 注意事项

在使用 MGR 集群时，需要注意以下事项。

- 当 MGR 节点宕机时，会导致事物锁缺失。建议复制一致性参数（group_replication_consistency）合理选择值。
- 事务隔离级别建议使用 READ_COMMIT。
- 序列化隔离级别：多主模式不支持。
- 并发 DDL 和 DML：多主模式下，不支持一边对一个表进行 DDL，一边进行更新，这样对于 DDL 在其他实例上操作有未检出的风险。
- 外键级联约束：多主模式下，多级外键依赖会引起多级操作，因此可能导致未知冲突，建议设置 group_replication_enforce_update_everywhere_checks=ON。
- 若大事务超过 5s 未提交，则会导致组通信失败。
- 多主模式下，SELECT * FROM FOR UPDATE 会导致死锁，因为 FOR UPDATE 语句产生的锁并非全组共享。
- 部分复制不支持：在 MGR 下，设置部分复制，会过滤事务，导致组事务不一致。
- 在停止复制的情况下，若某个节点执行命令后再启动，则会因为本地有私有事务，而无法加入集群。这时，需要执行 RESET MASTER 命令行重置节点 GTID，并且重新初始化数据。
- 单个 MGR 集群最大支持 9 个 MySQL 节点。

6. 总结

MGR 是主从复制之外，另一个适用于数据库场景的高可用与高扩展解决方案。通过结合分布式协议和常用插件，实现了真正的多主模式。然而，需要特别注意以下事项。

- 目前多主模式存在一些问题，因此在生产环境中建议使用单主模式。多主模式冲突检查机制不太稳定，为了避免潜在的问题，推荐在生产环境中暂时使用单主模式。
- 单个集群建议部署在同一机房内，跨机房部署可能会因为网络心跳无法发送或无法接收通信信息而导致脑裂问题。网络不稳定或延迟可能会导致节点之间的同步问题，从而引发脑裂问题。
- 在选择使用 MGR 的场景时，建议优先考虑读多写少的场景，这样可以更好地利用 MGR 的特性和优势，同时避免在高写入负载下可能出现的性能问题。

3.3.2 异步复制源配置功能

从 MySQL 8.0.22 版本开始，引入了异步复制源配置功能。当某个复制源出现问题（如宕机或复制链路中断）时，被复制的节点会根据配置的权重自动选择一个新的源进行同步。这与 MongoDB 的复制机制相似，但也存在一些区别。

而在 MySQL 8.0.23 版本中，进一步支持了组复制拓扑。该机制会自动监视组成员的更改，并区分主节点和副本节点。当向源列表添加组成员并将其定义为托管组的一部分时，异步连接故障转移机制会相应地更新源列表。只有占多数的在线组成员，才会用于连接和获取状态。如果管理组的最后一个成员离开了该组，则该成员不会被自动删除，以便保留管理组的配置。

1. 切换原理

异步复制源的工作原理如下。

- 当到源的现有连接失败时，副本首先重试相同的连接，通过配置的 SOURCE_CONNECT_

RETRY 和 SOURCE_RETRY_COUNT 参数值进行尝试。当这些尝试结束时，异步连接故障转移机制将接管。
- 异步连接故障转移机制在副本到源的连接失败后被激活，它发出一个 START REPLICA 语句，试图连接到一个新的源。
- 如果复制 I/O 线程由于源停止或网络故障而停止，则连接将中断。例如，当复制线程被 STOP REPLICA 语句停止时，异步连接故障转移机制就会失效。

表 3-4 所示为异步复制源判断故障转移的影响参数。

表 3-4 MySQL 异步复制源判断故障转移的影响参数

参 数	说 明
SLAVE_NET_TIMEOUT	表示 slave 在 SLAVE_NET_TIMEOUT 时间之内没有收到 master 的任何数据（包括 binlog、heartbeat），slave 认为连接断开，会进行重连
SOURCE_CONNECT_RETRY	默认值为 60s。表示重连的时间间隔。在 SLAVE_NET_TIMEOUT 超时后，立刻重连
SOURCE_RETRY_COUNT	表示重连的最大次数。默认值为 86400 次。合适的值是 SOURCE_RETRY_COUNT = 3（重试连接 3 次）和 SOURCE_CONNECT_RETRY = 10（间隔 10s）

2. 配置和监控异步复制源

在主从架构下，配置异步复制源需要用到以下两个复制函数。
- 添加单点复制源信息：asynchronous_connection_failover_add_source。
- 删除单点复制源信息：asynchronous_connection_failover_delete_source。

配置主从复制源的命令行示例如下。

```
# 添加异步复制源
asynchronous_connection_failover_add_source(channel, host, port, #network_namespace, weight)
    mysql> SELECT asynchronous_connection_failover_add_source('channel2','127.0.0.1', 3310, '', 80);
# 删除异步复制源配置
    mysql> SELECT asynchronous_connection_failover_delete_source('channel2', '127.0.0.1', 3310, '');
```

在 MGR 架构下，配置异步复制源需要用到以下两个复制函数。
- 添加 MGR 集群复制源信息：asynchronous_connection_failover_add_managed。
- 删除 MGR 集群复制源信息：asynchronous_connection_failover_delete_managed。

配置 MGR 集群复制源的命令行示例如下。

```
# 命令行示例:asynchronous_connection_failover_add_managed(channel, managed_type, #managed_
# name, host, port, network_namespace, primary_weight, secondary_weight)
    mysql>SELECT asynchronous_connection_failover_add_managed('channel2', 'GroupReplication',
'aaaaaaaa-aaaa-aaaa-aaaa-aaaaaaaaaaaa', '127.0.0.1', 3310, '', 80, 60);
    mysql>SELECT asynchronous_connection_failover_delete_managed('channel2', 'aaaaaaaa-aaaa-
aaaa-aaaa-aaaaaaaaaaaa');
```

说明：对于 MGR，还可以指定管理服务的类型（目前只有"组复制"可用）和管理组的标识符（对于"组复制"，这是系统变量 group_replication_group_name 的值），只需要添加一个组成员，副本将自动添加当前组成员中的其余成员。

复制源的配置信息支持持久化系统表和提供监控信息。源信息存储在 mysql 系统库下的两张表中。

- 单点复制源信息：replication_asynchronous_connection_failover。
- MGR 集群复制源信息：replication_asynchronous_connection_failover_managed。

MySQL 官方提示不允许使用 TRUNCATE 命令行。但实际测试发现，可以使用 TRUNCATE 命令行。

在 Performance Schema 库下，提供了如下两张视图，便于监视线程来跟踪托管组的成员关系并更新源列表。

- 单点异步复制源系统视图名：replication_asynchronous_connection_failover。
- MGR 集群异步复制源系统视图名：replication_asynchronous_connection_failover_managed。

3. 主从/MGR 集群下配置异步复制源

在 MySQL 的高可用架构中，异步复制源的配置可以根据具体的架构类型（如主从复制或 MGR 集群）有所不同。以下是在这两种架构下配置异步复制源的方法。

（1）主从复制源

图 3-11 所示为 MySQL 主从（双主）架构下异步复制源实现方式。虽然主从或双主架构有所不同，但配置方式一样，都是单一配置，需要使用权重（weight）区分优先级。

传统复制源

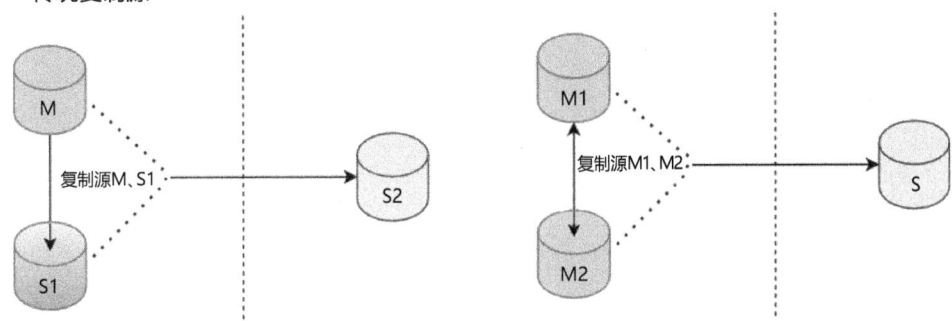

图 3-11　MySQL 主从（双主）复制架构

为了实现副本集的异步复制源配置，MySQL 提供了两个函数 asynchronous_connection_failover_add_source 和 asynchronous_connection_failover_delete_source 来管理源列表，以便添加和删除单个复制源服务器。添加异步复制源信息的命令行示例如下。

```
# 添加两个节点的异步复制源信息
mysql> SELECT asynchronous_connection_failover_add_source('channel_rpl','10.100.20.206',
3306,'', 70);
mysql> SELECT asynchronous_connection_failover_add_source('channel_rpl','10.100.20.206',
3307,'', 80);

# 查看异步复制源信息
mysql>SELECT *
    FROM performance_schema.replication_asynchronous_connection_failover;
+--------------+---------------+------+-------------------+--------+--------------+
| CHANNEL_NAME | HOST          | PORT | NETWORK_NAMESPACE | WEIGHT | MANAGED_NAME |
+--------------+---------------+------+-------------------+--------+--------------+
| channel_rpl  | 10.100.20.206 | 3306 |                   |   70   |              |
| channel_rpl  | 10.100.20.206 | 3307 |                   |   80   |              |
+--------------+---------------+------+-------------------+--------+--------------+
2 rows in set (0.00 sec)
```

```
# 删除异步复制源信息
mysql> SELECT asynchronous_connection_failover_delete_source('channel_rpl', '10.100.20.206', 3307,'');
+--------------------------------------------------------------------------------------------+
| asynchronous_connection_failover_delete_source('channel_rpl', '10.100.20.206', 3307,'')    |
+--------------------------------------------------------------------------------------------+
| The UDF asynchronous_connection_failover_delete_source() executed successfully.            |
+--------------------------------------------------------------------------------------------+
1 row in set (0.01 sec)

# 在目标 MySQL 服务上配置复制源信息
mysql> CHANGE REPLICATION SOURCE TO SOURCE_HOST='10.100.20.206',
  SOURCE_USER='repl',
  SOURCE_PASSWORD='123456',
  SOURCE_PORT=3306,
  SOURCE_CONNECTION_AUTO_FAILOVER=1,
  SOURCE_AUTO_POSITION = 1,
  SOURCE_CONNECT_RETRY = 1,
  SOURCE_RETRY_COUNT = 6
  FOR CHANNEL 'channel_rpl';
```

在主从架构中，如果主库 M 突然宕机，则从库 S 可能会被提升为新的主库。然而，如果此时原主库 M 恢复，则复制源可能会重新指向 M，导致复制冲突。在切换过程中，如果从库 S 上的事务已被写入并被清除（purge），那么在复制源切换后，复制可能会失败。为了避免这种情况，在实际应用中需要确保对复制源的变化进行适当控制。例如，可以通过脚本监控来实现对异步复制源信息的动态更新，这样可能会使切换更加便捷。另外，采用双主架构可以有效地处理这种切换，确保不会出现脑裂场景。

（2）MGR 集群复制源

图 3-12 所示为 MGR 集群下异步复制源实现方式。目标端可以是单节点，也可以是另一个 MGR 集群。

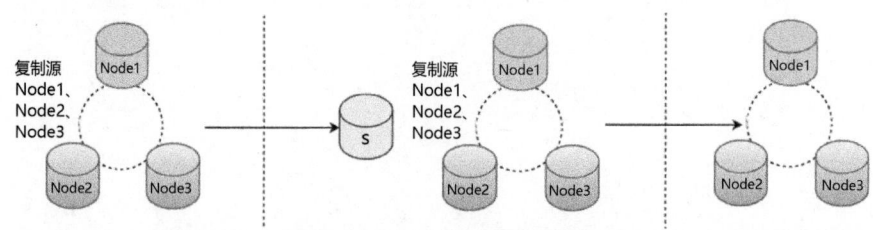

图 3-12　MySQL MGR 异步复制源实现架构

为了实现 MGR 的异步复制源功能，MySQL 提供了两个函数 asynchronous_connection_failover_add_managed 和 asynchronous_connection_failover_delete_managed 来管理 MGR 源列表，以便添加和删除 MGR 复制源服务器。添加异步复制源信息的命令行示例如下。

```
# 确认MGR集群名
mysql> SELECT @@group_replication_group_name;

# 查看MGR集群节点信息
mysql> SELECT * FROM performance_schema.replication_group_members;

# 添加MGR集群异步复制源
mysql> SELECT asynchronous_connection_failover_add_managed('channel_mgr',
'GroupReplication','d150c201-c02c-11eb-a374-38f3ab632406','10.132.20.206', 3306, '', 80, 60);

# 查看异步复制源信息
mysql> select * from
performance_schema.replication_asynchronous_connection_failover;
+--------------+---------------+------+------------------+--------+--------------------------------------+
| CHANNEL_NAME | HOST          | PORT | NETWORK_NAMESPACE | WEIGHT | MANAGED_NAME                         |
+--------------+---------------+------+------------------+--------+--------------------------------------+
| channel_mgr  | 10.100.20.206 | 3306 |                  | 60     | d150c201-c02c-11eb-a374-38f3ab632406 |
| channel_mgr  | 10.100.20.206 | 3307 |                  | 70     | d150c201-c02c-11eb-a374-38f3ab632406 |
| channel_mgr  | 10.100.20.206 | 3308 |                  | 80     | d150c201-c02c-11eb-a374-38f3ab632406 |
+--------------+---------------+------+------------------+--------+--------------------------------------+
3 rows in set (0.00 sec)

mysql> select * from
performance_schema.replication_asynchronous_connection_failover_managed;
+--------------+--------------------------------------+--------------+---------------------------------------------------+
| CHANNEL_NAME | MANAGED_NAME                         | MANAGED_TYPE | CONFIGURATION                                     |
+--------------+--------------------------------------+--------------+---------------------------------------------------+
| channel_mgr  | d150c201-c02c-11eb-a374-38f3ab632406 | GroupReplication | {"Primary_weight": 80, "Secondary_weight": 60} |
+--------------+--------------------------------------+--------------+---------------------------------------------------+
1 row in set (0.00 sec)

# 删除MGR异步复制源
mysql> SELECT asynchronous_connection_failover_delete_managed('channel_mgr',
'd150c201-c02c-11eb-a374-38f3ab632406');
```

如果使用MySQL Shell创建MGR集群，则会存在元数据同步问题。MGR集群的元数据保存在mysql_innodb_cluster_metadata库下，有变更时会同步，除此之外，mysql系统库内部用户也会同步变更信息。不同的MGR集群之间是不允许同步元数据的。

使用MySQL Shell搭建的MGR集群配置值时，需要使用多源复制过滤配置项。过滤复制信息的SQL语法如下所示。

```
CHANGE REPLICATION FILTER filter[, filter]
    [, ...] [FOR CHANNEL channel]
```

（3）双主之间复制

图 3-13 所示为两套双主架构之间的异步复制源同步方案，可以使用不同的异步复制源通道进行控制，然而交叉场景过于复杂，难以有效管理，因此建议放弃此方案。

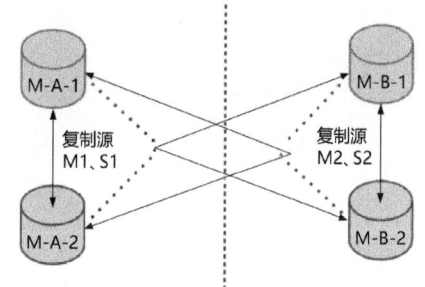

图 3-13　MySQL 中两套双主异步复制源相互同步实现架构

4. 注意事项

使用异步复制源时需要注意以下事项。

- 在通道的源列表中的所有源服务器上都必须存在相同的复制用户账户和密码。此账户用于连接到每个源服务节点。
- 复制用户账户必须具有 Performance Schema 表的 SELECT 权限，例如，GRANT SELECT ON performance_schema.* TO 'repl' 方式。
- 由于复制源使用的是异步机制，因此存在一定的延迟是很正常的。
- 当源端主节点发生故障时，在源列表中，具有更高优先级（权重）设置的另一个可用服务器将被优先考虑。

5. 总结

从一个集群（主从、双主、MGR 集群）到另一个主从集群的迁移可以被有效利用，但存在一些问题。例如，如上所述，当一个集群发生多次分裂后，同步的源是否仍然有效，目前无法有效判断，需要人为干预。

对于从一个 MGR 集群到另一个 MGR 集群的迁移，不建议使用，因为心跳问题可能导致脑裂等问题，这些新技术的稳定性和效果还需要进一步了解与熟悉。目前，还没有相关的实际案例。要做到无误差的心跳检测和各种场景下的正常切换，系统运行至少需要像主从集群一样稳定和可靠，这需要不断地进行尝试和优化。

3.3.3　MGR 集群容灾功能

MGR 高可用方案实现了副本冗余之间的最终一致性。在单机房场景下，MGR 集群的表现非常友好，但跨机房部署节点时却存在许多不可预测的因素。因此，在使用 MGR 集群的场景中，为了保证跨机房的数据同步，初期会采用主从复制的原理进行数据复制（异步复制源）。为此，在 MySQL 8.0.27 版本中，为 MGR 集群推出了 MySQL InnoDB ClusterSet 容灾方案。

1. InnoDB ClusterSet 介绍

InnoDB ClusterSet 容灾方案是指将一个主 MGR 集群和一个或多个在其他位置（如不同的数据中心）的副本连接起来。InnoDB ClusterSet 使用专用的 ClusterSet 复制通道自动管理从主集群到副本集群的复制。如果主集群由于数据中心的丢失或到它的网络连接的丢失而变得不可用，则可以使一个副本集群激活以恢复服务的可用性。复制通道还是属于异步复制，所以通常会有一定的复制延迟。

图 3-14 所示为 MySQL InnoDB ClusterSet 容灾架构。其中，主 MGR 集群把变更的数据通过 binlog 复制到其他机房 MGR 集群中回放，实现集群之间的数据同步，这需要 MySQL Shell 工具和 MySQL Router 中间件结合使用，对前端 App 基本无感知方式。

InnoDB ClusterSet 的特性如下。

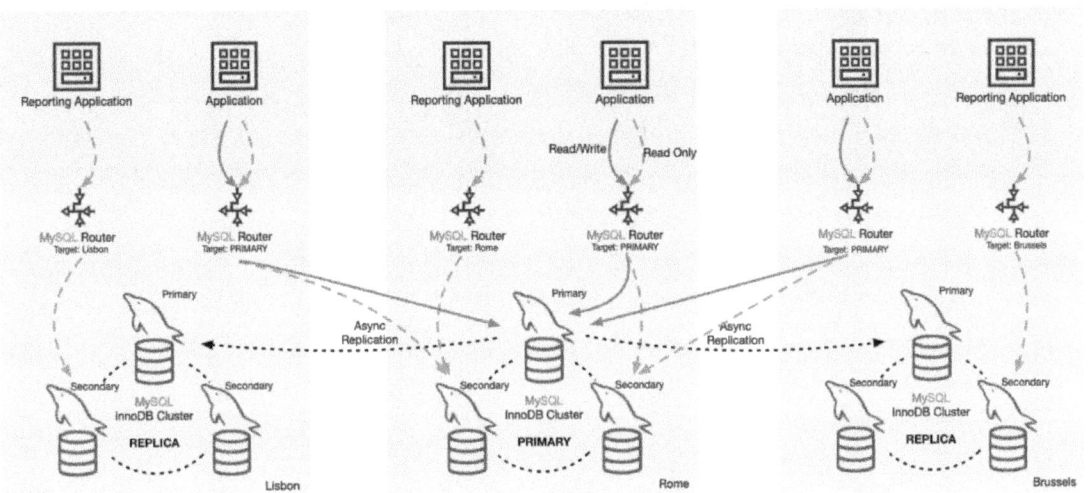

图 3-14　MySQL InnoDB ClusterSet 容灾架构

- InnoDB ClusterSet 将可用性置于一致性之上，以最大限度地提高灾难容忍度。正常的复制滞后或网络分区可能意味着在主集群出现问题时，部分或全部副本集群与主集群不完全一致。这意味着，在触发紧急故障切换时，未复制或不一致的事务会丢失，且只能手动恢复和协调（如果可访问）。因此，无法保证在发生紧急故障转移时数据会得到保留。
- InnoDB ClusterSet 作为解决方案，对写入性能产生显著影响，因为稳定和低延迟的网络对于 InnoDB 集群成员服务器相互通信以达成事务共识非常重要。
- InnoDB ClusterSet 不会自动将故障转移到副本集群。由于紧急故障切换可能导致丢失事务和数据一致性得不到保证，因此数据库管理员必须迅速做出执行紧急故障切换的决定。如果原始主集群保持联机状态，则应谨慎评估，并在确保数据完整性的前提下，考虑是否立即关闭它。
- InnoDB ClusterSet 仅支持异步复制，不能使用半同步复制。
- InnoDB ClusterSet 仅支持单主模式，不支持多主模式。
- InnoDB ClusterSet 的部署只能包含一个读写主集群，而所有复制副本集群都是只读的。不允许使用具有多个主集群的相互写入，因为这样的配置在集群出现故障时无法保证数据的一致性。
- InnoDB ClusterSet 不支持使用运行 MySQL 5.7 版本的实例。包含 MySQL 5.7 实例的 InnoDB 集群不能作为 InnoDB 集群集部署的一部分。

InnoDB ClusterSet 使用要求如下。

- 为了设置 InnoDB ClusterSet 部署，需要 MySQL 8.0.27 或更高版本、MySQL Shell 8.0.27 版本和 MySQL Router 8.0.27 或更高版本。
- InnoDB ClusterSet 版本必须为 2.1.0 版本或更高版本。当在集群上执行任何操作（如 dba.getCluster 命令）时，如果集群的元数据需要更新，则 AdminAPI 会发出警告。可以通过在 MySQL Shell 8.0.27 或更高版本中发出 dba.upgradeMetadata 命令，将元数据更新到适合 InnoDB ClusterSet 操作的版本。
- InnoDB 集群可以处于单主模式或多主模式，但 InnoDB ClusterSet 仅支持单主模式。这意味着在使用 InnoDB ClusterSet 时，集群必须处于单主模式。如果需要配置为单主模式，则可以在 MySQL Shell 中使用 cluster.switchToSinglePrimaryMode 命令进行设置。
- 必须在集群中的所有成员服务器上使用相同的值设置 group_replication_view_change_uid 参数，以便为视图更改事件提供替代 UUID。从 MySQL 8.0.27 版本开始，使用 dba.createCluster 命令创建的 InnoDB 集群将自动为这个参数生成并设置一个值。而在 MySQL 8.0.27 版本之前创建的 MGR 集群可能没有这个参数。在创建 InnoDB ClusterSet 时，系统会检查所有成员服务器上是否已正确设置此参数，如果没有设置或设置不正确，则系统将采取适当的措施（如警告或失败），以确保集群的一致性和可靠性。
- Cluster.rescan 命令可用于在 InnoDB 集群中的所有成员服务器上生成和设置 group_replication_view_change_uid 的值。在 MySQL Shell 8.0.29 版本之前，该命令会在扫描集群时自动执行此操作。但从 MySQL Shell 8.0.29 版本开始，必须重新启动集群才能实现更改。为了在扫描过程中自动生成和设置 group_replication_view_change_uid 的值，需要启用 updateViewChangeUuid 选项。在重新启动集群后，可以重试 InnoDB ClusterSet 的创建过程。
- 在 InnoDB 集群中，为了确保数据的一致性和集群的稳定性，任何成员服务器上都不允许有来自集群外部服务器的入站复制通道。但是，有两个例外情况：Group_Replication_applier 和 Group_Replication_recovery 通道，它们是由组复制自动创建的，因此是被允许的。这些例外通道用于集群内部的数据同步和恢复操作。
- 需要知道 MGR 集群服务器配置账户用户名和密码。此账户用于创建和配置 MGR 集群与 InnoDB ClusterSet 部署的成员服务器。每个成员服务器都需要这个账户信息，并且为了保持一致性，必须在集群中的所有成员服务器上使用相同的用户名和密码。建议使用强密码，并妥善保管，不要将其泄露给未经授权的人员。
- 在部署 InnoDB ClusterSet 之前，必须确保底层的 MGR 集群处于在线状态且健康状况良好。此外，还需要能够通过 MySQL Shell 访问到 MGR 集群的主要成员服务。这些都是成功创建 InnoDB ClusterSet 的必要条件。
- 在 InnoDB ClusterSet 环境中，可能有多个集群同时运行，每个集群可能负责处理不同类型的请求或数据。MySQL Router 可以帮助管理这些流量，确保客户端请求被正确地路由到相应的集群，从而实现负载均衡和高可用性。

MySQL 软件升级是维护数据库系统健康和安全的重要部分，尤其是在使用像 MGR 集群这样的复杂系统时。由于 MGR 集群不是由单一组件构成，因此在升级 InnoDB ClusterSet 中的服务器实例时需要谨慎，并且通常建议在业务访问量较低或无业务访问时进行，以避免升级过程中的任何潜在问题对用户造成影响。升级顺序如下。
- 升级 MySQL Router。
- 升级 MySQL Shell。
- 升级 MySQL 服务。

注意，升级后需要进行状态检查。

升级时用到的命令行如下所示。

```
<ClusterSet>.status()                    #检查状态
<ClusterSet>.status({extended:1})        #检查更详细的信息
<Cluster>.rescan()                       #重新扫描节点
dba.configureInstance()                  #函数会检查使实例能够用于 InnoDB 集群的所有设置
<ClusterSet>.listRouters()               #显示当前 Router 信息
```

2. InnoDB ClusterSet 部署

要部署一个 InnoDB ClusterSet,至少需要两个 MGR 集群。图 3-15 所示为部署 InnoDB ClusterSet 流程及其查询点。

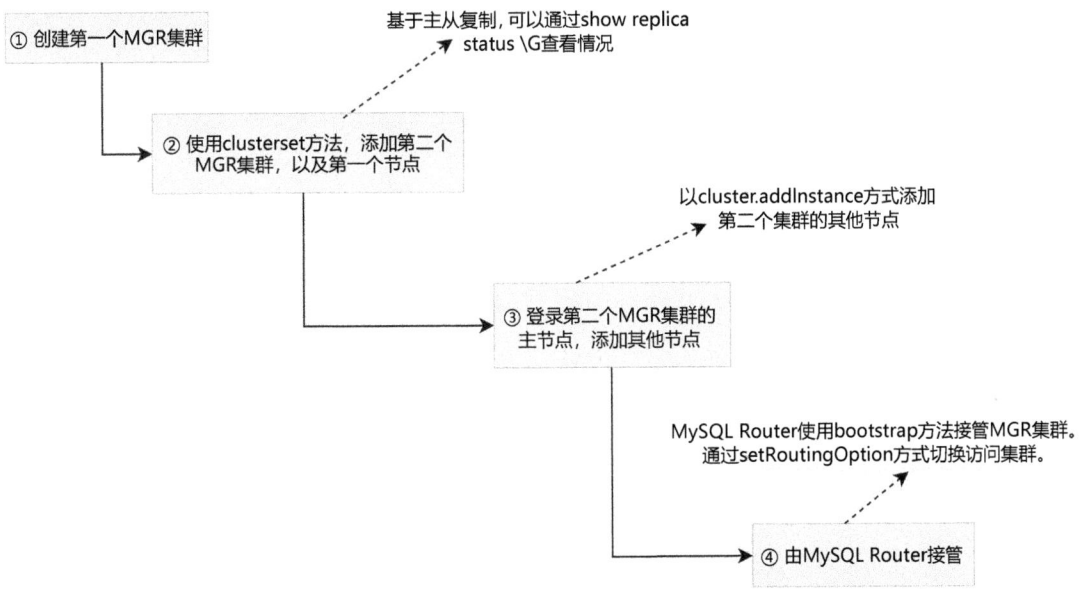

图 3-15　MySQL InnoDB ClusterSet 部署流程

具体部署 InnoDB ClusterSet 实例的命令行示例如下。

```
#1.获取原 cluster 信息
MySQL-JS > var cluster=dba.getCluster()

#2.创建 ClusterSet
MySQL-JS > myclusterset = cluster.createClusterSet('testclusterset')
A newClusterSet will be created based on the Cluster 'mgrCluster'.
* Validating Cluster 'mgrCluster' for ClusterSet compliance.
* Creating InnoDB ClusterSet 'testclusterset' on 'mgrCluster'...
* Updating metadata...
ClusterSet successfully created. Use ClusterSet.createReplicaCluster() to add
Replica Clusters to it.
<ClusterSet:testclusterset>

#3.查看 ClusterSet 状态
MySQL-JS> var myclusterset = dba.getClusterSet()
MySQL-JS> myclusterset.status()
{
    "clusters": {
        "mgrCluster": {
            "clusterRole": "PRIMARY",
            "globalStatus": "OK",
            "primary": "ss301:3380"
        }
```

```
    },
    "domainName": "testclusterset",
    "globalPrimaryInstance": "ss301:3380",
    "primaryCluster": "mgrCluster",
    "status": "HEALTHY",
    "statusText": "All Clusters available."
```

#4.添加第二个 MGR 集群的第一个 MySQL 节点
```
MySQL-JS > myclusterset.createReplicaCluster("192.168.244.132:3380",
"clustertwo", {recoveryProgress: 1, timeout: 10})
Setting up replica 'clustertwo' of cluster 'mgrCluster' at instance 'ss302:3380'.

A new InnoDB Cluster will be created on instance 'ss302:3380'.
...
Creating InnoDB Cluster 'clustertwo' on 'ss302:3380'..
Replica Cluster 'clustertwo' successfully created on ClusterSet 'testclusterset'.
<Cluster:clustertwo>
```

#5.查看创建的复制集群情况
```
MySQL-JS > myclusterset.status()
{
    "clusters": {
        "clustertwo": {
            "clusterRole": "REPLICA",
            "clusterSetReplicationStatus": "OK",
            "globalStatus": "OK"
        },
        "mgrCluster": {
            "clusterRole": "PRIMARY",
            "globalStatus": "OK",
            "primary": "ss301:3380"
        }
    },
    "domainName": "testclusterset",
    "globalPrimaryInstance": "ss301:3380",
    "primaryCluster": "mgrCluster",
    "status": "HEALTHY",
    "statusText": "All Clusters available."
```

集群新添加的节点的复制信息，可以通过查看主从状态的命令行确认。查看复制状态的命令行示例如下。

```
mysql> SHOW REPLICA STATUS \G;
*************************** 1. row ***************************
             Replica_IO_State: Waiting for source to send event
                  Source_Host: ss301
                  Source_User: mysql_innodb_cs_84
                  Source_Port: 3380
                Connect_Retry: 3
              Source_Log_File: mysql-bin.000004
```

```
          Read_Source_Log_Pos: 61961
             Relay_Log_File: relay-log-clusterset_replication.000002
              Relay_Log_Pos: 6129
       Relay_Source_Log_File: mysql-bin.000004
          Replica_IO_Running: Yes
         Replica_SQL_Running: Yes
...
              Auto_Position: 1
         Replicate_Rewrite_DB:
               Channel_Name: clusterset_replication
```

查看异步复制源 asynchronous_connection_failover 的配置信息。查看异步复制源信息的命令行示例如下。

```
mysql> select * from replication_asynchronous_connection_failover;
+----------------------+-------+------+-------------------+--------+--------------------------------------+
| CHANNEL_NAME         | HOST  | PORT | NETWORK_NAMESPACE | WEIGHT | MANAGED_NAME                         |
+----------------------+-------+------+-------------------+--------+--------------------------------------+
| clusterset_replication| ss301| 3380 |                   |   80   | 8080336f-ff45-11ed-9f9e-00505635a9dc |
| clusterset_replication| ss301| 3381 |                   |   60   | 8080336f-ff45-11ed-9f9e-00505635a9dc |
| clusterset_replication| ss301| 3382 |                   |   60   | 8080336f-ff45-11ed-9f9e-00505635a9dc |
+----------------------+-------+------+-------------------+--------+--------------------------------------+
```

从上述查询信息中可以看出，InnoDB ClusterSet 使用了 MySQL 主从复制技术和 asynchronous_connection_failover 异步复制源技术。

向第二个 MGR 集群 clustertwo 添加节点的命令行示例如下。

```
MySQL-JS > var cluster = dba.getCluster()
MySQL-JS > cluster.status()
{
    "clusterName": "clustertwo",
    "clusterRole": "REPLICA",
    "clusterSetReplicationStatus": "OK",
    "defaultReplicaSet": {
        "name": "default",
        "primary": "ss302:3380",
        "ssl": "DISABLED",
        "status": "OK_NO_TOLERANCE",
        "statusText": "Cluster is NOT tolerant to any failures.",
        "topology": {
            "ss302:3380": {
                "address": "ss302:3380",
                "memberRole": "PRIMARY",
                "mode": "R/O",
                "readReplicas": {},
                "replicationLagFromImmediateSource":"",
                "replicationLagFromOriginalSource":"",
                "role": "HA",
                "status": "ONLINE",
                "version": "8.0.33"
```

```
            }
        },
        "topologyMode": "Single-Primary"
    },
    "domainName": "testclusterset",
    "groupInformationSourceMember": "ss302:3380",
    "metadataServer": "ss301:3380"
}

MySQL-JS > cluster.addInstance('mgradmin@192.168.244.132:3381',{recoveryMethod:'clone'})

Validating instance configuration at 192.168.244.132:3381...

This instance reports its own address as ss302:3381

Instance configuration is suitable.
NOTE: Group Replication will communicate with other members using 'ss302:3381'. Use thelocal-
Address option to override.
...
* Waiting for server restart... ready
* ss302:3381 has restarted, waiting for clone to finish...
** Stage RESTART: Completed
* Clone process has finished: 212.23 MB transferred in about 1 second (~212.23 MB/s)
State recovery already finished for 'ss302:3381'
The instance 'ss302:3381' was successfully added to the cluster.
```

MGR 集群 clustertwo 的第二个节点登录服务查看复制状态。存在主从复制信息，但 Replica_IO_Running 和 Replica_SQL_Running 复制线程状态都是停止状态（NO）。查看复制状态的命令行示例如下。

```
mysql> show replica status \G;
*************************** 1. row ***************************
             Replica_IO_State:
                  Source_Host: ss301
                  Source_User: mysql_innodb_cs_84
                  Source_Port: 3380
                Connect_Retry: 3
              Source_Log_File: mysql-bin.000004
          Read_Source_Log_Pos: 56236
               Relay_Log_File: relay-log-clusterset_replication.000002
                Relay_Log_Pos: 4
        Relay_Source_Log_File: mysql-bin.000004
           Replica_IO_Running: No
          Replica_SQL_Running: No
...
                Auto_Position: 1
          Replicate_Rewrite_DB:
                 Channel_Name: clusterset_replication
```

查看两个 MGR 集群 mgrCluster 和 clustertwo 的状态与节点复制是否正常。查看 MGR 集群的命令行示例如下。

```
MySQL-JS > myclusterset=dba.getClusterSet()
<ClusterSet:testclusterset>
MySQL-JS > myclusterset.status({extended:1})
{
    "clusters": {
        "clustertwo": {
            "clusterRole": "REPLICA",
            "clusterSetReplication": {
                "applierStatus": "APPLIED_ALL",
                "applierThreadState": "Waiting for an event from Coordinator",
                "applierWorkerThreads": 4,
                "receiver": "ss302:3380",
                "receiverStatus": "ON",
                "receiverThreadState": "Waiting for source to send event",
                "replicationSsl": null,
                "source": "ss301:3380"
            },
            "clusterSetReplicationStatus": "OK",
            "globalStatus": "OK",
            "status": "OK",
            "statusText": "Cluster is ONLINE and can tolerate up to ONE failure.",
            "topology": {
                "ss302:3380": {
                    "address": "ss302:3380",
                    "memberRole": "PRIMARY",
                    "mode": "R/O",
                    "replicationLagFromImmediateSource":"",
                    "replicationLagFromOriginalSource":"",
                    "status": "ONLINE",
                    "version": "8.0.33"
                },
                "ss302:3381": {
                    "address": "ss302:3381",
                    "memberRole": "SECONDARY",
                    "mode": "R/O",
                    "replicationLagFromImmediateSource":"",
                    "replicationLagFromOriginalSource":"",
                    "status": "ONLINE",
                    "version": "8.0.33"
                },
                "ss302:3382": {
                    "address": "ss302:3382",
                    "memberRole": "SECONDARY",
                    "mode": "R/O",
                    "replicationLagFromImmediateSource":"",
                    "replicationLagFromOriginalSource":"",
                    "status": "ONLINE",
                    "version": "8.0.33"
                }
```

```
            },
            "transactionSet": "8080336f-ff45-11ed-9f9e-00505635a9dc:1-133:1000071,808037c5-
ff45-11ed-9f9e-00505635a9dc: 1-6, 86a473cf-ff42-11ed-b58a-00505635a9dc: 1-22, ce5f0588-ff64-11ed-
b1b2-0050563557d1:1-7",
            "transactionSetConsistencyStatus": "OK",
            "transactionSetErrantGtidSet":"",
            "transactionSetMissingGtidSet":""
        },
        "mgrCluster": {
            "clusterRole": "PRIMARY",
            "globalStatus": "OK",
            "primary": "ss301:3380",
            "status": "OK",
            "statusText": "Cluster is ONLINE and can tolerate up to ONE failure.",
            "topology": {
                "ss301:3380": {
                    "address": "ss301:3380",
                    "memberRole": "PRIMARY",
                    "mode": "R/W",
                    "status": "ONLINE",
                    "version": "8.0.33"
                },
                "ss301:3381": {
                    "address": "ss301:3381",
                    "memberRole": "SECONDARY",
                    "mode": "R/O",
                    "replicationLagFromImmediateSource":"",
                    "replicationLagFromOriginalSource":"",
                    "status": "ONLINE",
                    "version": "8.0.33"
                },
                "ss301:3382": {
                    "address": "ss301:3382",
                    "memberRole": "SECONDARY",
                    "mode": "R/O",
                    "replicationLagFromImmediateSource":"",
                    "replicationLagFromOriginalSource":"",
                    "status": "ONLINE",
                    "version": "8.0.33"
                }
            },
            "transactionSet": "8080336f-ff45-11ed-9f9e-00505635a9dc:1-133:1000071,808037c5-
ff45-11ed-9f9e-00505635a9dc: 1-6, 86a473cf-ff42-11ed-b58a-00505635a9dc: 1-22, ce5f0588-ff64-11ed-
b1b2-0050563557d1:1-5"
        }
    },
    "domainName": "testclusterset",
    "globalPrimaryInstance": "ss301:3380",
    "metadataServer": "ss301:3380",
    "primaryCluster": "mgrCluster",
```

```
        "status": "HEALTHY",
        "statusText": "All Clusters available."
}
```

若输出信息中的 status 状态值为 OK 或 ONLINE,则说明配置成功。到此,MGR 集群的 ClusterSet 配置完成。

3. MySQL Router 配置

在 InnoDB ClusterSet 中配置 MySQL Router 的过程与基础操作相似,但额外提供了对不同集群节点访问的 API 实现。MySQL Router 配置命令行如下所示。

```
shell $>mysqlrouter --bootstrap root@ip:3380 --directory /opt/mysqlrouter/mgr
--name='Rome1' --user=root
ClusterSet 'testclusterset' can be reached by connecting to:
## MySQL Classic protocol

- Read/Write Connections: localhost:6446
- Read/Only Connections:  localhost:6447

## MySQL X protocol
- Read/Write Connections: localhost:6448
- Read/Only Connections:  localhost:6449

MySQL-JS > myclusterset.listRouters()
{
    "domainName": "testclusterset",
    "routers": {
        "ss301::Rome1": {
            "hostname": "ss301",
            "lastCheckIn": null,
            "roPort": "6447",
            "roXPort": "6449",
            "rwPort": "6446",
            "rwXPort": "6448",
            "targetCluster": null,
            "version": "8.0.33"
        }
    }
}
```

配置 MySQL Router 访问第二个 MGR 集群实例。使用 setRoutingOption 方法切换节点的操作的命令行示例如下。

```
# 获取 MQR 集群信息,配置 MySQL Router
MySQL-JS > myclusterset.setRoutingOption('ss301::Rome1','target_cluster','clustertwo')
Routing option 'target_cluster' successfully updated in router 'ss301::Rome1'.

# 客户端链接验证
shell#> mysql -umgradmin -p123456 -h192.168.244.131 -P6447 -e "select @@hostname,@@port;"
+--------------------+-------------+
```

```
|@@hostname |@@port |
+-------------------+------------+
|ss302              | 3380       |
+-------------------+------------+
```

说明：在 InnoDB ClusterSet 配置中，只有一个 MySQL 实例支持写入操作。其他实例设置为只读模式（read_only）或超级只读模式（super_read_only）。setRoutingOption 方法用于动态更改 MySQL Router 的路由配置，但它一次只能应用于一个后端或路由。在多个 MySQL Router 进行切换时，有可能会导致访问不一致和写无法操作现象。

4. InnoDB ClusterSet 运维

当使用 InnoDB ClusterSet 跨机房部署 MGR 集群时，由于涉及多个组件和跨机房的复杂性，因此日常运维变得尤为重要。以下是 InnoDB ClusterSet 出现问题时的一些应对方式。

- 在进行紧急故障切换之后，需要采取额外的预防措施来确保数据的完整性和一致性。由于 InnoDB ClusterSet 的异步复制特性，因此在紧急故障切换过程中，副本集群可能与主集群存在事务集不同步的风险。这意味着在切换过程中，副本集群可能包含了一些尚未在主集群中提交的事务。必须将主集群与写流量或所有流量隔离开来。
- 当 InnoDB ClusterSet 中的复制集群（副本集群）的 GTID 集与主集群不一致时，意味着有一些事务在复制过程中丢失了，或者副本集群没有接收到主集群上的所有事务。集群日志会输出 OK_NOT_CONSISTENT 警告信息。在这种情况下，禁止访问副本集群，同时，可以选择跳过不一致的事务，或重新同步副本集群。
- 如果 InnoDB ClusterSet 中的一个集群因为某种原因无法修复或恢复，则使用 ClusterSet.removeCluster() 方法从 InnoDB ClusterSet 中移除它。之后可将它重新加入 InnoDB ClusterSet 中。
- 当对集群进行修复或维护时，可以使用 ClusterSet.request() 方法将其重新加入 InnoDB ClusterSet。此方法验证集群是否能够重新加入，更新和启动 InnoDB ClusterSet 复制通道，并从集群中删除任何无效状态。

5. 总结

InnoDB ClusterSet 优先考虑可用性而不是数据一致性，以便最大限度地提高容灾能力，数据一致性得到保证。因为基于主从异步复制，所以紧急故障转移会带来丢失事务的风险，并有一定概率造成 InnoDB ClusterSet 脑裂的情况。如果无法快速修复主集群以恢复可用性，则建议保留一个集群，关闭其他集群，然后在可能的情况下进行修复。

InnoDB ClusterSet 实现方式是两个集群之间的主从复制。虽然 MySQL Shell 的介入简化了配置过程，但整个过程仍比较复杂，运维 DBA 需要有一定的实操能力。

3.3.4 增强的多源复制功能

MySQL 多源复制用于将多个 MySQL 数据库中的数据变更信息复制到一个目标 MySQL 数据库中，以实现数据同步、合并或备份等功能。如果复制源服务节点没有将语句写入其二进制日志，则该语句不会被复制。

在 MySQL 8.0 版本之前，所有的复制过滤器都是全局的，因此它们会应用于所有的复制通道。这会导致在特定场景下存在限制，如 MGR 集群之间的同步。由于 mysql_innodb_cluster_metadata 是每个集群都有的，因此全局复制过滤器无法区分不同集群的需求，从而无法实现精准的表过滤。

另外，在多源复制的场景下，如果某些特定的库或表不需要从一个特定源复制，则全局过滤器也无法满足这种需求。

然而，MySQL 8.0 版本引入了 CHANNEL 通道复制过滤器，它允许为不同的复制通道配置独立的过滤规则。这种灵活性不仅解决了 MySQL 之前版本中全局过滤器的限制，还满足了多个 MGR 集群之间的同步需求。通过 CHANNEL 通道复制过滤器，可以精确地指定哪些表需要从特定源复制，从而提高了复制过程的灵活性和效率。

CHANNEL 通道复制过滤器使用"CHANNEL:数据库对象"方式进行配置。表 3-5 所示为指定复制通道的配置方式。

表 3-5 MySQL 指定复制通道配置方式

复制通道配置参数方式	说　明
replicate-do-db=channel:database_id	指定复制通道的被复制的数据库名
replicate-ignore-db=channel:database_id	指定复制通道的被忽略的数据库名
replicate-do-table=channel:table_id	指定复制通道的被复制的特定表名
replicate-ignore-table=channel:table_id	指定复制通道的被忽略的特定表名
replicate-rewrite-db=channel:db1-db2	指定复制通道的重写复制事件中的数据库名
replicate-wild-do-table=channel:table_pattern	使用通配符来指定复制通道的被复制的表名

下面是通过指定复制通道过滤的实例，相关命令行如下。

```
# 指定 channel_1 的数据库 db1 和 db2
mysql> CHANGE REPLICATION FILTER REPLICATE_DO_DB=(db1, db2) FOR CHANNEL 'channel_1';

# 指定 channel_1 的表 db1.t7，忽略表 db2.t7
mysql> CHANGE REPLICATION FILTER
    REPLICATE_WILD_DO_TABLE = ('db1.t7'),
    REPLICATE_WILD_IGNORE_TABLE = ('db2.t7') FOR CHANNEL 'channel_1';

# 忽略复制频道 channel_1 的 db2 库中的以 t1、t2 开头的表
mysql> CHANGE REPLICATION FILTER
    REPLICATE_WILD_DO_TABLE = ('db2.t1%','db2.t2%') FOR CHANNEL 'channel_1';

# 下面的语句将主数据库 db1 上的语句重写到从数据库 db2 上的语句
mysql> CHANGE REPLICATION FILTER REPLICATE_REWRITE_DB = ((db1, db2)) FOR CHANNEL 'channel_1';
mysql> CHANGE REPLICATION FILTER
    REPLICATE_REWRITE_DB = ((dbA,dbB), (dbC, dbD)) FOR CHANNEL 'channel_1';

# 清空 channel_1 的复制过滤信息
mysql> CHANGE REPLICATION FILTER REPLICATE_DO_DB=() FOR CHANNEL 'channel_1';
mysql> CHANGE REPLICATION FILTER REPLICATE_DO_TABLE=() FOR CHANNEL 'channel_1';
mysql> CHANGE REPLICATION FILTER REPLICATE_WILD_DO_TABLE=() FOR CHANNEL 'channel_1';
mysql> CHANGE REPLICATION FILTER Replicate_Ignore_DB=() FOR CHANNEL 'channel_1';
mysql> CHANGE REPLICATION FILTER Replicate_Wild_Ignore_Table=() FOR CHANNEL 'channel_1';
```

总之，MySQL 8.0 版本对多源复制的增强提供了更加灵活的控制方式，使得复制过程更加灵活和多样化。这为用户提供了更大的灵活性和便利性，可满足不同的业务需求和数据复制场景。

3.3.5 MySQL Shell 快速创建、纳管副本集和 MGR 集群功能

MySQL Shell 是 MySQL Server 的高级客户端和代码编辑器。除了和 MySQL 命令行客户端程序一样使用常规的 SQL 功能以外，MySQL Shell 还提供了 JavaScript 和 Python 的脚本功能，并包含多个 API，其中的 AdminAPI 用于操作 MGR 集群。图 3-16 所示为 MySQL Shell 实现的功能。

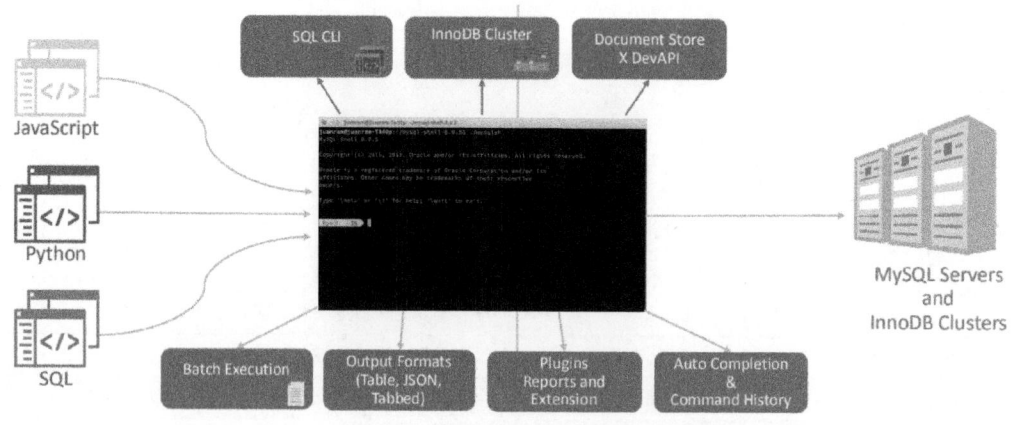

图 3-16 MySQL Shell 实现的功能

下面介绍如何使用 MySQL Shell 创建和纳管主从复制与 MGR 架构。对于 MySQL Shell 的安装，从官方下载对应环境中的 MySQL 版本的 MySQL Shell 的 TAR 包并解压缩后即可使用。

1. 副本集管理

在 MySQL Shell 中创建和纳管副本集（ReplicaSet，就是主从架构）时，需要检查参数配置的一致性、指定副本集的名称，以及添加从节点。在执行这些操作之前，需要在 MySQL 服务器上正确配置了复制所需的用户权限。

（1）副本集创建

下面是创建 ReplicaSet 的过程实例。检查副本集的命令行示例如下。

```
# 在创建副本集时,会自动进行检查
shell $>mysqlsh
MySQL Shell 8.0.34
MySQL-JS >\connect dbadmin@localhost:3380

# 检查当前配置是否满足创建副本集要求
MySQL-JS >dba.configureReplicaSetInstance()
Configuring local MySQL instance listening at port 3380 for use in an InnoDB ReplicaSet...
ERROR: The account 'dbadmin'@'%' is missing privileges required to manage an InnoDB ReplicaSet:
GRANT DELETE, INSERT, UPDATE ON mysql.* TO 'dbadmin'@'%' WITH GRANT OPTION;
For more information, see the online documentation.
Dba.configureReplicaSetInstance: The account 'dbadmin'@'%' is missing privileges required to manage an InnoDB ReplicaSet. (RuntimeError)

# 运行配置检查机制,如果出现错误,则需要按照要求更改
NOTE: Some configuration options need to be fixed:
+----------------------------------------+-----------------------+-----------------------+----------------------------------------+
```

```
|Variable                              |Current Value |Required Value |Note                       |
+--------------------------------------+--------------+---------------+---------------------------+
|replica_preserve_commit_order         |OFF           |ON             |Update the server variable |
+--------------------------------------+--------------+---------------+---------------------------+

+--------------------------------------+--------------+---------------+---------------------------+
|Variable                              |Current Value |Required Value |Note                       |
+--------------------------------------+--------------+---------------+---------------------------+
|binlog_transaction_dependency_tracking|COMMIT_ORDER  |WRITESET       |Update the server variable |
+--------------------------------------+--------------+---------------+---------------------------+
ERROR: local:3380: Instance must be configured and validated with dba.configureReplicaSetIn-
stance() before it can be used in areplicaset.
Dba.createReplicaSet: Instance check failed (MYSQLSH 51150)
```

按照上述提示要求修改配置和赋予权限。创建副本集的命令行示例如下。

```
MySQL-JS > var rs = dba.createReplicaSet("dbexample")
A new replicaset with instance 'schouse:3380' will be created.

* Checking MySQL instance at schouse:3380

This instance reports its own address as schouse:3380
Schouse01:3380: Instance configuration is suitable.

NOTE: TLS not available at 'schouse:3380', assuming replicationSslMode to be DISABLED
* Checking connectivity and SSL configuration...
* Updating metadata...
ReplicaSet object successfully created for schouse01:3380.
Use rs.addInstance() to add more asynchronously replicated instances to this replicaset and
rs.status() to check its status.
```

虽然已经成功创建 MySQL 副本集，但目前只有一个节点。通过命令行来查看副本集的状态和成员信息的示例如下。

```
MySQL-JS > var rs = dba.getReplicaSet()
You are connected to a member of replicaset 'dbexample'.
MySQL-JS > rs.status()
{
    "replicaSet": {
        "name": "dbexample",
        "primary": "schouse01:3380",
        "status": "AVAILABLE",
        "statusText": "All instances available.",
        "topology": {
            "schouse:3380": {
                "address": "schouse01:3380",
                "instanceRole": "PRIMARY",
                "mode": "R/W",
                "status": "ONLINE"
            }
        },
        "type": "ASYNC"
    }
}
```

说明：在上述输出信息中，需要确认 instanceRole 是主节点（PRIMARY），mode 是读写模式（R/W），status 是在线模式（ONLINE）。副本集指标显示正常。

在 MySQL 副本集中添加新节点，通常涉及几个步骤，包括克隆现有节点、配置新节点、启动复制，以及服务重新启动。副本集添加节点命令行示例如下。

```
MySQL-JS >rs.addInstance('root@ip:3381')
Adding instance to the replicaset...
* Performing validation checks
This instance reports its own address as schouse:3381
schouse:3381: Instance configuration is suitable.
* Checking async replication topology...
* Checking connectivity and SSL configuration...
* Checking transaction state of the instance...

Please select a recovery method [C]lone/[A]bort (default Abort): C
Please select a recovery method [C]lone/[A]bort (default Abort): C
* Updating topology
Waiting for clone process of the new member to complete. Press ^C to abort the operation.
* Waiting for clone to finish...
NOTE:schouse:3381 is being cloned from schouse01:3380
** Stage DROP DATA: Completed
** Clone Transfer
FILE COPY   ############################################=======  71%  In Progress
PAGE COPY   =================================================    0%   Not Started
REDO COPY   =================================================    0%   Not Started

NOTE:schouse:3381 is shutting down...
** Changing replication source of schouse02:3381 to schouse01:3380
** Waiting for new instance to synchronize with PRIMARY...
** Transactions replicated############################################################  100%
The instance 'schouse:3381' was added to the replicaset and is replicating from schouse:3380.
* Waiting for instance 'schouse:3381' to synchronize the Metadata updates with the PRIMARY...
** Transactions replicated  ############################################################  100%
```

再次查看副本集状态。第二个节点的 instanceRole 值为 SECONDARY，mode 值是 R/O（只读节点），status 值是 ONLINE（在线模式），SECONDARY 提示创建副本集成功。查看状态的命令行示例如下。

```
MySQL-JS > rs.status()
{
    "replicaSet": {
        "name": "dbexample",
        "primary": "schouse01:3380",
        "status": "AVAILABLE",
        "statusText": "All instances available.",
        "topology": {
            "schouse:3380": {
                "address": "schouse01:3380",
                "instanceRole": "PRIMARY",
                "mode": "R/W",
```

```
            "status": "ONLINE"
        },
        "schouse:3381": {
            "address": "schouse02:3381",
            "instanceRole": "SECONDARY",
            "mode": "R/O",
            "replication": {
                "applierStatus": "APPLIED_ALL",
                "applierThreadState": "Waiting for an event from Coordinator",
                "applierWorkerThreads": 4,
                "receiverStatus": "ON",
                "receiverThreadState": "Waiting for source to send event",
                "replicationLag": null,
                "replicationSsl": null
            },
            "status": "ONLINE"
        }
    },
    "type": "ASYNC"
  }
}
```

登录副本集节点的数据库，查看复制状态。查看复制状态的命令行示例如下。

```
mysql> SHOW REPLICA STATUS \G
*************************** 1. row ***************************
             Replica_IO_State: Waiting for source to send event
                  Source_Host: schouse
                  Source_User: mysql_innodb_rs_12912
                  Source_Port: 3380
                Connect_Retry: 60
              Source_Log_File: mysql-bin.000003
          Read_Source_Log_Pos: 51068
               Relay_Log_File: relay-log.000002
                Relay_Log_Pos: 5765
        Relay_Source_Log_File: mysql-bin.000003
           Replica_IO_Running: Yes
          Replica_SQL_Running: Yes
...
```

说明：在上述输出信息中，Replica_IO_Running 和 Replica_SQL_Running 的值都显示为 YES。副本集同步状态正常。副本集配置信息会保存在数据库的 mysql_innodb_cluster_metadata 系统库中。

（2）副本集纳管

在 MySQL Shell 中，如果想要纳管一个已经通过命令行方式配置好的副本集，那么，首先登录主节点，然后使用 createReplicaSet 命令，同时使用 adoptFromAR 选项。

下面为具体纳管的副本集的实例，相关命令行如下。

```
Shell JS > dba.createReplicaSet('testadopt',{'adoptFromAR':1})
A new replicaset with the topology visible from 'schouse:3380' will be created.
```

```
* Scanning replication topology...
** Scanning state of instance schouse01:3380
** Scanning state of instance schouse02:3381

* Discovering async replication topology starting with schouse:3380
Discovered topology:
-schouse01:3380: uuid=fbde18d9-b034-11ee-9b2b-00163e23e2cc read_only=no
-schouse02:3381: uuid=f356014b-9418-11ee-87e9-00163e23e2cc read_only=no
    - replicates from schouse:3380
    source="172.17.27.48:3380" channel= status=ON receiver=ON coordinator=ON applier0=ON
applier1=ON applier2=ON applier3=ON

* Checking configuration of discovered instances...

* Checking discovered replication topology...
schouse:3380 detected as the PRIMARY.
...
```

到此,成功纳管现有副本集,副本集名为"testadopt"。

(3)副本集管理注意事项

目前,使用 MySQL Shell 管理副本集的灵活性不如直接使用 SQL 命令行工具,并且还存在一些限制。MySQL Shell 管理副本集的限制如下。

- 目前无法提供双主配置和并行复制、多源复制相关的参数设置。
- 目前没有副本集直接可以使用的删除命令行。通常,需要首先通过 removeInstance 命令删除指定的节点,然后执行 STOP REPLICA 和 RESET REPLICA ALL 操作来停止与重置复制,最后,删除 mysql_innodb_cluster_metadata 系统库来完成副本集的删除过程。
- 从 MySQL 8.0.21 版本开始,为了从客户端流量中隐藏实例,引入了两个"内置"标签(_hidden 和 disconnect_existing_sessions_when_hidden),可以立即更改路由器的行为。通过_hidden 在特定实例上启用标签,可以指示 MySQL Router 将实例从客户端应用程序的候选目标列表中排除。考虑到 MySQL 路由器不断轮询元数据,其效果实际上是立竿见影的。
_hidden 设置的命令行示例如下。

```
#设置隐藏实例
MySQL-JS >cluster.setInstanceOption("ip:3380", "tag:_hidden", true)

# disconnect_existing_sessions_when_hidden 标记为隐藏实例的连接
MySQL-JS >replicaset.setInstanceOption("ip:3380",
"tag:_disconnect_existing_sessions_when_hidden", false).
```

(4)副本集管理命令集合

除了上述创建和纳管 MySQL 副本集命令以外,MySQL Shell 还提供了其他运维命令。表 3-6 所示为副本集支持的命令行。

表 3-6 副本集支持的命令行

命 令 行	说 明
rs=dba.createReplicaSet()	现有副本集创建和纳管
rs.addInstance()	添加节点

(续)

命 令 行	说 明
rs.disconnect()	断开 replicaset 对象使用的所有内部会话
rs.forcePrimaryInstance(instance, options)	在不可用副本集中执行故障转移
rs.getName()	获取副本集名
rs.rejoinInstance(instance[, options])	重新加入副本集中
rs.removeInstance(instance[, options])	删除节点
rs.setPrimaryInstance(instance, options)	指定主节点
rs.status([options])	查看副本集状态
rs.help([member])	查看帮助信息

说明：optisns 部分可以通过\help ReplicaSet.removeInstance 方式进行查看。

2. MGR 集群创建和纳管

（1）MGR 集群创建

通过 MySQL Shell 创建 MGR 集群的过程与创建传统的 MySQL 副本集类似，但也有一些特定的步骤和考虑因素。以下是创建包含 3 个节点的 MGR 集群的示例。

检查 3 个节点的配置是否满足 MGR 集群要求（如每张表都有主键、InnoDB 引擎等）的命令行示例如下。

```
shell $>mysqlsh
MySQL-JS > dba.checkInstanceConfiguration('dbadmin@192.168.1.1:3380')
The instance 'ens8:3380' is valid to be used in an InnoDB cluster.
{
        "status": "ok"
}
MySQL-JS > dba.checkInstanceConfiguration('dbadmin@192.168.1.2:3381')
MySQL-JS > dba.checkInstanceConfiguration('dbadmin@192.168.1.3:3382')

# 执行 dba.configureInstance()后再次检查实例配置
MySQL-JS > dba.configureInstance('dbadmin@192.168.1.1:3380')
Configuring localMySQL instance listening at port 3380 for use in an InnoDB cluster...

This instance reports its own address as192.168.1.1:3380
Clients and other cluster members will communicate with it through this address by default. If this is not correct, the report_host MySQL system variable should be changed.

applierWorkerThreads will be set to the default value of 4.
The instance'schouse:3380'is valid to be used in an InnoDB cluster.
The instance'schouse:3380'is already ready to be used in an InnoDB cluster.
MySQL-JS > dba.configureInstance('dbadmin@192.168.1.2:3381')
MySQL-JS > dba.configureInstance('dbadmin@192.168.1.3:3382')
```

在上述输出中，如果有提示信息，就需要按照要求进行更改。创建 MGR 集群的命令行示例如下。

```
# 连接主节点
MySQL-JS > \connect dbadmin@192.168.1.1:3380

# 创建集群
MySQL-JS > dba.createCluster('mgrCluster')
MySQL-JS > var cluster = dba.getCluster()

# 添加节点
MySQL-JS > cluster.addInstance('dbadmin@192.168.1.2:3381')
MySQL-JS > cluster.addInstance('dbadmin@192.168.1.3:3382')

# 查看集群状态
MySQL-JS > cluster.status()

# 删除 InnoDB Cluster
MySQL-JS > cluster.dissolve()

# 配置节点权重
MySQL-JS > var mycluster = dba.getCluster()
MySQL-JS > mycluster.addInstance('dbadmin@192.168.1.3:3382', {memberWeight:50})

# 指定一个新的主节点
MySQL-JS > cluster.setPrimaryInstance(192.168.1.2:3381')

# 使用 Cluster.switchToMultiPrimaryMode() 切换到多主模式
MySQL-JS >cluster.switchToMultiPrimaryMode()

# 使用 Cluster.switchToSinglePrimaryMode() 切换到单主模式
MySQL-JS > cluster.switchToSinglePrimaryMode('192.168.1.1:3380')
```

(2) MGR 集群纳管

在 MySQL Shell 中,如果想要纳管一个已经通过命令行方式配置好的 MGR 集群,那么首先登录主节点,然后使用 createCluster 命令,同时使用 adoptFromGR 选项。

下面为具体纳管的 MGR 集群的实例,相关命令行如下。

```
shell $>mysqlsh --uri dbadmin@192.168.1.1:3380
MySQL-JS > var cluster = dba.createCluster('testCluster', {adoptFromGR: true})
A new InnoDB cluster will be created based on the existing replication group on instance 'ip:3380'.
Creating InnoDB cluster 'testCluster' on 'ens8:3380'...

Adding Seed Instance...
Adding Instance 'ens8:3380'...
Adding Instance 'ens9:3381'...
Adding Instance 'ens10:3382'...
Resetting distributed recovery credentials across the cluster...
Cluster successfully created based on existing replication group.
MySQL-JS > cluster.status()
    {
```

```
            "clusterName": "testCluster",
            "defaultReplicaSet": {
                "name": "default",
                "primary": "ens8:3380",
                "ssl": "DISABLED",
                "status": "OK",
                "statusText": "Cluster is ONLINE and can tolerate up to ONE failure.",
                "topology": {
                    "ens8:3380": {
                        "address": "ens8:3380",
                        "mode": "R/W",
                        "readReplicas": {},
                        "replicationLag": null,
                        "role": "HA",
                        "status": "ONLINE",
                        "version": "8.0.34"
                    },
                    "ens8:3381": {
                        "address": "ens9:3381",
                        "mode": "R/O",
                        "readReplicas": {},
                        "replicationLag": null,
                        "role": "HA",
                        "status": "ONLINE",
                        "version": "8.0.34"
                    },
                    "ens8:3382": {
                        "address": "ens10:3382",
                        "mode": "R/O",
                        "readReplicas": {},
                        "replicationLag": null,
                        "role": "HA",
                        "status": "ONLINE",
                        "version": "8.0.34"
                    }
                },
                "topologyMode": "Single-Primary"
            },
            "groupInformationSourceMember": "ens8:3380"
}
```

(3) MGR 集群管理注意事项

目前,MySQL Shell 在管理 MGR 集群方面已经相当成熟,并且有很多实际的生产案例支持,但还需要注意以下内容。

- 当需要向 MGR 集群添加新节点或重新搭建节点时,MySQL Shell 使用 Clone 功能初始化数据。在这个过程中,Clone 操作可能会导致主节点的性能下降,甚至可能影响整个集群的稳定性和可用性。
- MySQL Shell 提供了一套丰富的 API 来管理 MySQL 和 MGR 集群,但在某些方面,这些 API

可能还不够完善。有些配置参数可能无法直接通过 MySQL Shell 的 API 进行设置，而需要通过传统的 my.cnf 配置文件或 SQL 命令行来进行配置。

在 MGR 中，每个节点都扮演着特定的角色，并且集群的健康和性能取决于这些节点的状态。MGR 集群和节点的状态是非常重要的。

MGR 集群的状态如下。

- OK：所有节点处于 ONLINE 大状态，有冗余节点。
- OK_PARTIAL：有节点不可用，但仍有冗余节点。
- OK_NO_TOLERANCE：有足够的处于 ONLINE 状态的节点，但没有冗余节点，例如，在有两个节点的集群中，若其中一个停止工作，集群就不可用了。
- NO_QUORUM：有节点处于 ONLINE 状态，但达不到法定节点数，此状态下集群无法写入，只能读取。
- UNKNOWN：不是 ONLINE 或 RECOVERING 状态，尝试连接其他实例查看状态。
- UNAVAILABLE：组内节点全处于 OFFLINE 状态，但实例在运行，可能是因为实例刚重启，还没加入集群。

MGR 集群的节点的状态如下。

- ONLINE：节点状态正常。
- OFFLINE：实例在运行，但没有加入任何集群。
- RECOVERING：实例已加入集群，正在同步数据。
- ERROR：同步数据发生异常。
- UNREACHABLE：与其他节点的通信中断，可能是网络问题，也可能是节点 crash。
- MISSING：节点已加入集群，但未启动 group replication 进程。

（4）MGR 集群管理命令行集合

除了上述创建和纳管 MGR 集群的命令行以外，MySQL Shell 还提供了其他运维命令。表 3-7 所示为 MGR 集群支持的命令行。

表 3-7 MGR 集群支持的命令行

命 令 行	说 明
dba.rebootClusterFromCompleteOutage()	重启所有 MGR 集群的节点
dba.dropMetadataSchema()	删除 MGR 集群管理库
dba.getCluster()	获取当前 MGR 集群名
cluster.checkInstanceState()	检查当前集群的状态
cluster.rejoinInstance()	脱离的 MGR 集群的节点重新加入节点
addcluster.dissolve()	删除 MGR 集群。指定变量 {force:true} 可进行强制删除
cluster.addInstance()	添加节点到 MGR 集群中
cluster.removeInstance()	在 MGR 集群中删除指定节点。指定变量 {force:true} 可进行强制删除
cluster.describe()	查看集群描述信息
cluster.setOption（option, value）	设置整个集群级别的参数
cluster.setInstanceOption（instance, option, value）	用于设置特定集群成员实例的参数

说明：option 部分可以通过 \help ReplicaSet.removeInstance 方式进行查看。

（5）MGR 集群常用的操作

在使用 MySQL Shell 管理 MGR 集群时，某些操作可能由于它们的复杂性和相互依赖性而容易混淆。这些操作包括重置集群、切换主节点、删除集群和重新加载集群。下面介绍这些操作的命令行示例和步骤。

重置 MGR 集群通常意味着删除所有关于组复制的配置和元数据，然后重新初始化集群。这通常是一个破坏性操作，因为它会删除所有组复制的状态，包括成员信息、全局事务 ID 等。以下是重置 MGR 集群的命令行示例。

```
#1.使用 MySQL Shell 登录主节点，删除管理库
MySQL-JS > dba.dropMetadataSchema()
#2.登录数据库，执行停止 MGR 集群命令
mysql> STOP GROUP_REPLICATION;
#3.清空日志，确保和从库中的表没有冲突
mysql> RESET MASTER;
mysql> RESET SLAVE;

#4.对于其他节点，使用 SQL 命令行进行操作
mysql> STOP GROUP_REPLICATION;
mysql> RESET MASTER;
mysql> RESET SLAVE;
```

将 MGR 集群从多主模式（Multi-Primary）更改为单主模式（Single-Primary）的命令行示例如下。

```
#1.使用 Cluster.switchToMultiPrimaryMode()切换到多主模式
MySQL-JS > cluster.switchToMultiPrimaryMode()

#2.指定一个新的主节点
MySQL-JS > cluster.setPrimaryInstance('192.168.244.129:3381')
```

删除和重建 MGR 集群是两个相对复杂的操作，因为涉及删除现有的集群配置并重新创建它。以下是删除和重建 MGR 集群的命令行示例。

```
#1.删除原来的集群
MySQL-JS > cluster.dissolve({force: true})

#2.登录 MGR 每个节点，修改 MGR 配置
mysql> set global group_replication_enforce_update_everywhere_checks=OFF;
mysql> set global group_replication_single_primary_mode=ON;

#3.重新创建集群
MySQL-JS > var cluster = dba.createCluster('mysqlCluster')
MySQL-JS > cluster.addInstance('dbadmin'@ip:3381')
```

当整个 MGR 集群由于某种原因（如硬件故障、网络分区、所有节点同时宕机等）使得所有节点都处于 OFFLINE 状态时，需要重新恢复集群的连通性和配置。以下是从完全 OFFLINE 状态重新引导集群的命令行示例。

```
#执行 rebootClusterFromCompleteOutage 命令，可恢复集群
MySQL-JS > dba.rebootClusterFromCompleteOutage('mlampCluster')
```

3. 总结

通过 MySQL Shell 创建与纳管副本集和 MGR 集群的过程通常比较快捷和方便，特别是在熟悉其命令和流程后。MySQL Shell 提供了许多内置的命令和功能，使得副本集和集群的配置、管理变得相对简单。其中 MySQL Shell 管理 MGR 集群的技术非常成熟。

3.3.6 MySQL Router+MGR 方式实现高可用

MySQL Router 是轻量级中间件，它提供了应用程序和任何后端 MySQL 服务之间的透明路由功能。通过有效地将数据库流量路由到适当的后端 MySQL 服务，MySQL Router 确保了高可用性和可伸缩性（无状态）。

1. MySQL Router 介绍

MySQL Router 是 MySQL Proxy（读写分离中间件）的替代方案，MySQL 官方不建议将 MySQL Proxy 用于生产环境，并且已经不提供 MySQL Proxy 的下载。在后续使用中，作者发现 MySQL Router 没有出色的表现，因为它除了读写分离以外，没有其他作用，也无法动态更改配置。MySQL Router 通常用于解决"MySQL 集群规模性迁移"问题，如跨机房部署、流量迁移和异构兼容，以及 MySQL 集群规模性宕机时快速切换等。

MySQL Router 中间件本身不会对请求"拆包"（unpackage），所以无法在 MySQL Router 中间件上实现诸如"SQL 审计""隔离""限流""分库分表"等功能。但是 Router 提供了插件（C 语言）机制，用户可以开发自己的插件来扩展 Router 的额外特性。

图 3-17 所示为 MySQL Router 架构，它可以有效地管理应用连接，同时还支持 MySQL 的读写分离。

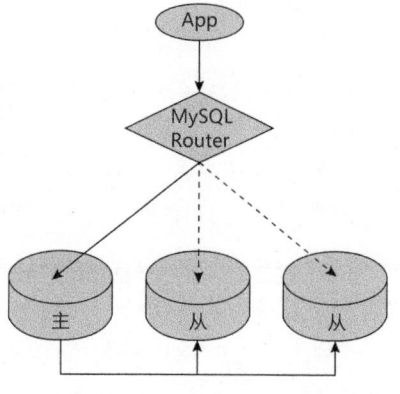

图 3-17 MySQL Router 架构

MySQL Router 的工作流程如下。

- 客户端连接：MySQL 客户端连接到 MySQL Router。
- 服务检查：MySQL Router 检查可用的后端 MySQL 服务列表。根据配置的策略（如负载均衡算法、故障转移机制等），MySQL Router 会选择一个合适的后端 MySQL 服务。
- 建立连接：MySQL Router 协助在应用程序和所选的后端 MySQL 服务之间建立一个 TCP 连接。
- 心跳检测：在连接建立后，MySQL Router 会进行心跳检测，以确保连接仍然有效并且后端 MySQL 服务处于活动状态。
- 故障检测与处理：如果 MySQL Router 检测到连接的 MySQL 服务出现故障（例如，连接丢失、服务无响应等），则它会发送故障信息给应用程序，并提示应用程序断开与 MySQL 服务的连接。
- 重试与重新路由：应用程序重试连接到 MySQL Router 时，MySQL Router 会再次根据配置的策略选择一个其他可用的后端 MySQL 服务，并重复上述流程。

简单理解，MySQL Router 作为一个中间件或代理，介入了应用程序和 MySQL 服务之间的直接连接。通过 MySQL Router，应用程序可以更加灵活地与后端 MySQL 服务进行交互，同时 MySQL Router 也提供了高可用性和负载均衡等特性。

MySQL Router 支持以下功能。

- Failover（故障切换）：MySQL Router 能够识别后端 MySQL 服务中的集群或副本集的配置，并在主服务器发生故障时自动将流量重定向到健康的从服务器。这降低了应用程序需要处理的复杂性，并确保了数据的持续访问。
- Load Balancing（负载均衡）：MySQL Router 通过其负载均衡功能，可以根据预定义的策略（如轮询、最小连接数等）将客户端连接分发到多个 MySQL 服务。这确保了单个服务器不会因为过多的负载而过载，从而提高了整体系统的稳定性和性能。
- Pluggable Architecture（可插拔体系结构）：MySQL 提供了创建自己的定制插件的能力，从而提供了无限的可能性。MySQL Router 是端口转发机制，不检查数据包或修改它们，从而提供最大的吞吐量。

对于 MySQL Router 部署建议，出于性能和安全考虑，MySQL Router 不应与应用程序部署在同一主机上。尽管官方文档指出，当前版本的简单重定向连接路由与直连数据库相比，性能损失仅为约 1%，但仍建议根据实际情况进行自行测试以确认性能表现。

对于 MySQL Router 的安装，从官方下载对应环境中的 MySQL 版本的 MySQL Router 的 TAR 包并解压缩后即可使用。

2. MySQL Router 配置副本集读写分离架构

使用 MySQL Router 配置副本集（主从复制），需要手动为主节点和从节点编写配置信息。在副本集架构中，为了实现读写分离，需要配置不同的端口来提供服务。从原先的 MySQL 的 3306 端口衍生出专门的读写端口（如 7001），以及纯读端口（如 7002）。同时，需要设置 routing_strategy 来确定负载方式，以及指定 destinations 来提供服务的节点信息。

使用 MySQL Router 配置主从读写信息的配置文件的内容如下所示。

```
shell $> vi /opt/mysqlrouter/config/mysqlrouter.conf
[DEFAULT]        # 基础配置信息
logging_folder          = /opt/mysqlrouter/log
plugin_folder           = /opt/idc/mysql-router/lib/mysqlrouter
config_folder           = /opt/mysqlrouter/config
runtime_folder          = /opt/mysqlrouter/run
data_folder             = /opt/mysqlrouter/data

client_connect_timeout  = 2
connect_timeout         = 2
read_timeout            = 30
max_connections         = 2048
# keyring_path = /var/lib/keyring-data
# master_key_path = /var/lib/keyring-key

[logger]         # 日志配置
level = DEBUG
timestamp_precision = second

[routing:primary]       # 副本集写入节点配置信息
# To be more transparent, use MySQL Server port 3306
bind_address            = 0.0.0.0
bind_port               = 7001
```

```
routing_strategy    = first-available
mode                = read-write
destinations        = 192.168.244.130:3410

[routing:secondary]     # 副本集读节点配置信息
# To be more transparent, use MySQL Server port 3306
bind_address        = 0.0.0.0
bind_port           = 7002
routing_strategy    = round-robin
mode                = read-only
destinations        = 192.168.244.144:4310,192.168.244.130:3410

# If no plugin is configured which starts a service, keepalive
# will make sure MySQL Router will not immediately exit. It is
# safe to remove once Router is configured.
[keepalive]         # 副本集心跳检测配置信息
interval            = 60
```

在上述 MySQL Router 的配置中，[routing:primary] 标签用于指定主服务（主节点）的配置信息；[routing:secondary] 标签用来指定在读写分离场景中，读操作应该被路由到的从服务。routing:secondary 可以配置多个，以支持多个从服务的负载均衡。

3. MySQL Router 配置 MGR 集群

在 MGR 集群中，MySQL Router 被用作代理，从而隐藏网络上的多个 MySQL 实例，并将数据请求映射到 MGR 集群中的一个实例。在生产环境中，推荐进行 MySQL Router 与 MGR 集群的集成部署。在与 MGR 集群的配合使用中，MySQL Router 的重要性得到了充分体现。

如图 3-18 所示，MySQL Router 作为一个端口转发层，位于应用与 MGR 集群之间，其功能类似于 LVS（Linux Virtual Server 的简称，采用 IP 负载均衡技术）。

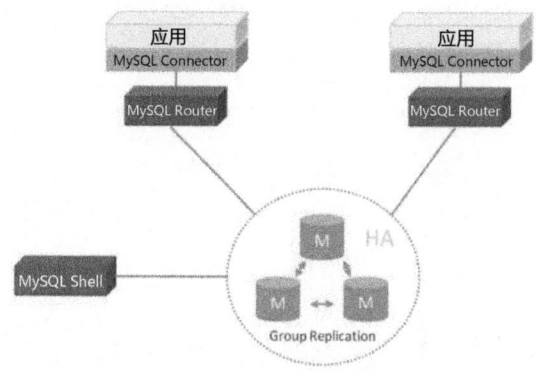

图 3-18　MySQL Router+MGR 部署架构

对于 MGR 集群，MySQL Router 的配置文件不需要像副本集那样手动编写。相反，可以使用 MySQL Router 的命令行工具，通过指定 --bootstrap 参数，以动态加载的方式自动生成配置文件。以下是一个生成配置文件的命令行示例。

```
shell $>mysqlrouter --bootstrap root@192.168.244.129:3380 --directory
/opt/mysqlrouter/mgr --user=root
```

图 3-19 所示为通过 --bootstrap 自动添加配置信息的过程，包括生成配置文件、创建账号确认、读写 IP 和端口配置，以及纯读 IP 和端口配置等步骤。其他未明确指出的配置将采用默认值。

```
[root@ss30 ~]# mysqlrouter --bootstrap root@192.168.244.129:3380 --directory /opt/mysqlrouter/mgr --user=root
Please enter MySQL password for root:
# Bootstrapping MySQL Router instance at '/opt/mysqlrouter/mgr'...

- Creating account(s) (only those that are needed, if any)
- Verifying account (using it to run SQL queries that would be run by Router)
- Storing account in keyring
- Adjusting permissions of generated files
- Creating configuration /opt/mysqlrouter/mgr/mysqlrouter.conf

# MySQL Router configured for the InnoDB Cluster 'mgrCluster'

After this MySQL Router has been started with the generated configuration

    $ mysqlrouter -c /opt/mysqlrouter/mgr/mysqlrouter.conf

the cluster 'mgrCluster' can be reached by connecting to:

## MySQL Classic protocol

- Read/Write Connections: localhost:6446
- Read/Only Connections:  localhost:6447

## MySQL X protocol

- Read/Write Connections: localhost:64460
- Read/Only Connections:  localhost:64470
```

图 3-19　MySQL Router 自动添加配置信息

下面为 MySQL Router 配置生成的文件列表，包含 .conf、.key、.log 和 sh 格式的脚本等。

```
shell $>tree ./mysqlrouter
├── config                              # 配置目录
│   ├── mysqlrouter.conf                # 配置文件
│   └── mysqlrouter.key                 # key 文件
├── data                                # 数据目录
├── log                                 # 日志目录
│   └── log.log                         # 日志文件
└── run                                 # 脚本
    ├── start.sh                        # 启动脚本
    └── stop.sh                         # 关闭脚本
```

由 MGR 集群的 MySQL Router 生成的具体配置文件 mysqlrouter.conf 的内容如下。

```
# File automatically generated during MySQL Router bootstrap
[DEFAULT] # 基础配置信息
user=root
logging_folder=/opt/mysqlrouter/mgr/log
runtime_folder=/opt/mysqlrouter/mgr/run
data_folder=/opt/mysqlrouter/mgr/data
keyring_path=/opt/mysqlrouter/mgr/data/keyring
master_key_path=/opt/mysqlrouter/mgr/mysqlrouter.key
connect_timeout=15
read_timeout=30
dynamic_state=/opt/mysqlrouter/mgr/data/state.json

[logger] # 日志级别
```

```
level = INFO

[metadata_cache:mgrCluster] # MGR 集群元数据库信息
cluster_type=gr
router_id=1
user=mysql_router1_zesa3g471bns
metadata_cluster=mgrCluster
ttl=0.5
use_gr_notifications=0

[routing:mgrCluster_rw] # MGR 集群读写节点配置信息
bind_address=0.0.0.0
bind_port=6446
destinations=metadata-cache://mgrCluster/?role=PRIMARY
routing_strategy=first-available
protocol=classic

[routing:mgrCluster_ro] # MGR 集群纯读节点配置信息
bind_address=0.0.0.0
bind_port=6447
destinations=metadata-cache://mgrCluster/?role=SECONDARY
routing_strategy=round-robin-with-fallback
protocol=classic

[routing:mgrCluster_x_rw] # MGR 集群读写 MySQL Shell API 配置信息
bind_address=0.0.0.0
bind_port=64460
destinations=metadata-cache://mgrCluster/?role=PRIMARY
routing_strategy=first-available
protocol=x

[routing:mgrCluster_x_ro] # MGR 集群纯读 MySQL Shell API 配置信息
bind_address=0.0.0.0
bind_port=64470
destinations=metadata-cache://mgrCluster/?role=SECONDARY
routing_strategy=round-robin-with-fallback
protocol=x
```

在 MGR 集群路由创建过程中，会在 MySQL 里自动创建账号并赋予权限。图 3-20 所示为账号信息。

如图 3-20 所示，在创建 MGR 集群的路由过程中，MySQL 会自动创建账号（mysql_innodb_cluster_129、mysql_innodb_cluster_1291、mysql_innodb_cluster_1292 和 msyql_router1_zesa3g471bns）并赋予相应的权限。

通过 MySQL Router 访问 MGR 集群，以验证其连接和配置是否正确。连接测试的命令行示例如下。

图 3-20　MySQL Router 在数据库中创建的账号信息

```
# 确认是否能读取数据库信息
shell $> mysql -h192.168.244.130 -P6447 -udbadmin -p
mysql> show databases;
+--------------------------------+
| Database                       |
+--------------------------------+
| information_schema             |
| mysql_innodb_cluster_metadata  |
| mysql                          |
| performance_schema             |
| sys                            |
+--------------------------------+
5 rows in set (0.00 sec)


# 查看是否连接的是主节点信息
shell $> mysql -h192.168.244.130 -P6447 -udbadmin -p123456 --protocol=TCP -N -r -B -e"select @@
hostname;SHOW VARIABLES LIKE 'server_uuid';"
+--------------+
| @@hostname   |
+--------------+
| schouse      |
+--------------+
1 row in set (0.00 sec)


+---------------+--------------------------------------+
| Variable_name | Value                                |
+---------------+--------------------------------------+
| server_uuid   | fbde18d9-b034-11ee-9b2b-00163e23e2cc |
+---------------+--------------------------------------+
1 row in set (0.01 sec)
```

4. MySQL Router 参数和元数据缓存

正确配置 MySQL Router 的参数以及保持元数据的准确性和实时性对于确保 MySQL Router 高效、安全工作至关重要。

(1) 配置参数

MySQL Router 的配置参数决定了其如何与后端 MySQL 服务进行交互，包括如何路由连接请求、如何处理身份验证、如何管理连接池等。这些配置参数直接影响 MySQL Router 的性能、安全性和可用性。表 3-8 所示为 MySQL Router 目前提供的 20 个基本配置参数。

表 3-8 MySQL Router 基本配置参数

参数	类型	说明
config_folder	String	配置文件的路径
connect_timeout	Integer	在与元数据服务器的连接尝试被认为超时之前的秒数
core-file	Boolean	控制是否生成 core 文件
event_source_name	String	仅限 Microsoft Windows 平台。定义 MySQL 路由器在 Microsoft Windows 上作为服务运行时使用的服务名称
keyring_path	String	密钥环文件的路径
logging_folder	String	日志的路径
master_key_path	String	主密钥路径
master-key-reader	String	读取主密钥的脚本
master-key-writer	String	写入主密钥的脚本
max_total_connections	Integer	允许客户端连接的最大总数
pid_file	String	运行 PID 文件的位置
plugin_folder	String	路由器插件的路径
runtime_folder	String	运行时文件的路径
sinks	String	用于接收配置的日志数据的日志记录方法
thread_stack_size	Integer	分配给每个线程堆栈的内存大小（KB）
unknown_config_option	String	如果遇到未知配置选项，则发送错误类型
user	String	MySQL Router 启动用户
read_timeout	String	读取超时
mode	String	读写或读读模式
routing_strategy	String	负载模式

下面抽取重要的参数进行说明。

- connect_timeout：超时时间，默认值是 30s。这个值应该根据网络环境和服务器性能进行调整。如果网络延迟较高或者服务器响应较慢，则可能需要增加这个值。
- max_total_connections：默认值是 512。这个值应该根据 MySQL 服务的性能和预期的并发连接数进行调整。如果 MySQL 和应用服务器性能都较高，并且需要处理大量并发连接，则可以适当增加这个值。
- thread_stack_size：默认值是 64KB。通常情况下，这个值不需要频繁调整。但是，如果应用程序需要处理大量并发请求，并且出现了堆栈溢出错误，则可以考虑增加这个值。

- read_timeout：默认值是 30s。这个值应该根据网络环境和服务器性能进行调整。如果网络延迟较高或者服务器响应较慢，则可能需要增加这个值。
- mode：支持两种模式（read-write、read-only）。根据应用程序的需求选择合适的模式。
- routing_strategy：负载模式。它决定了路由器如何为客户端的连接请求选择后端服务器。不同的路由策略会对性能、可用性和数据一致性产生不同的影响。表 3-9 所示为 MySQL Router 支持的负载模式。

表 3-9 MySQL Router 负载模式

负载模式	说 明
first-available	新连接选择列表路由到第一个可用的服务器。如果出现故障，则使用下一个可用的服务器。这个循环一直持续到所有服务器都不可用为止
next-available	新连接选择列表路由到第一个可用的服务器。如果出现故障，则使用下一个可用的服务器。这个循环一直持续到所有服务器都不可用为止。与 first-available 不同的是，如果一个服务器被标记为不可到达，那么它将被丢弃，并且再也不会使用
round-robin	每个新连接都以循环方式连接到下一个可用的服务器
round-robin-with-fallback	每个新连接都以循环方式连接到下一个可用的辅助服务器。如果辅助服务器不可用，则以循环方式使用主列表中的服务器

说明：PRIMARY、SECONDARY 和 PRIMARY_AND_SECONDARY 默认选择 round-robin 负载模式。

（2）元数据缓存

元数据是 MySQL Router 用来了解后端 MySQL 服务器状态、角色和地址等关键信息的数据。MySQL Router 使用这些元数据来做出路由决策，确保连接请求被正确地路由到相应的服务器。如果元数据不准确或过时，则可能会导致连接请求被路由到错误的服务器，从而影响应用程序的性能和可用性。表 3-10 所示为 MySQL Router 缓存信息设置参数。

表 3-10 MySQL Router 缓存信息设置参数

参　数	类　型	说　明
uth_cache_refresh_interval	Numeric	身份验证缓存刷新尝试的间隔时间
auth_cache_ttl	Numeric	如果不刷新，则缓存变为无效时间
bootstrap_server_addresses	String	带有元数据的 MySQL 服务器，以逗号分隔的列表
cluster_type	String	MySQL 高可用模式（主从架构、MGR 架构）
metadata_cluster	String	MGR 集群名
router_id	Integer	Router ID
ssl_ca	String	SSL CA 文件
ssl_capath	String	SSL CA 文件的路径
ssl_crlpath	String	SSL CRL 文件的路径
ssl_mode	String	SSL 模式
tls_version	String	TLS 版本信息
ttl	Integer	生存时间（秒）

(续)

参　　数	类　　型	说　　明
use_gr_notifications	Integer	组复制通知行为
user	String	访问 MySQL Server 元数据架构的 MySQL 用户
dynamic_state		元数据持久化文件路径

下面抽取重要的参数进行说明。

- bootstrap_server_addresses：该参数用于指定一个或多个初始的 MySQL 服务器地址，MySQL Router 会从这些服务器中获取集群的初始元数据。然而，这种方法的一个限制是，如果指定的引导服务器不可用或发生变化，则 MySQL Router 可能无法及时更新其元数据缓存。此参数在 MySQL 8.0.14 中被弃用，不再由引导进程生成。另外，添加了 dynamic_state 参数作为替代。
- dynamic_state：跟踪和存储活动的 MGR 集群的服务器地址，并在 MySQL Router 重启时加载到内存中。这个功能是由 Bootstrap 激活的。在 MGR 集群状态发生变化时（例如，一个新的服务器被添加到集群中，或者一个现有的服务器出现故障并被移除），MySQL Router 会更新这个文件，以反映最新的集群状态。

5. 常见问题

下面探讨 MySQL Router 的常见问题。

1）MySQL Router 安装在哪里？

官方建议，MySQL Router 可安装在应用程序服务器所在的同一主机上，以减少网络延迟并提高性能。但是，这不是必需的，因为路由器可以安装在任何主机上。然而，出于网络架构或安全性考虑，可以将其安装在单独的主机上。

2）可以运行多个 MySQL Router 实例吗？

可以运行多个 MySQL Router 实例以满足不同的需求。例如，可能希望为不同的应用程序或服务提供不同的路由规则，或者为了高可用性和故障转移而运行多个实例。但是，每个实例都会消耗少量的资源，因此需要根据系统的整体资源限制和性能需求来规划实例的数量。

3）MySQL Router 会检查数据包吗？

MySQL Router 主要负责连接管理和路由决策，而不对传输的数据包内容进行检查。

4）MySQL Router 影响性能吗？

引入 MySQL Router 会在一定程度上影响性能，因为它需要处理连接和路由逻辑。然而，这种影响通常很小，通常在 1% 左右，并且可以通过优化配置和硬件资源来最小化影响。

5）MySQL Router 的版本有 2.× 和 8.×，应该使用哪个版本？

推荐使用 8.× 版本。2.× 版本是为了与早期产品兼容。8.× 版本是 MySQL Router 的较新版本，包含许多新功能、性能改进和安全性更新。除此之外，MySQL Router 版本和 MySQL Server 版本要保持一致。

6）每个 MySQL Router 实例支持多少并发连接？

MySQL Router 的并发连接数的多少取决于多个因素，包括操作系统的限制、配置和硬件资源。在 MySQL Router 2.1.5 和 8.0.4 及更高版本中，由于使用了 poll() 函数，因此支持高达 5000 个并发连接。而在早期版本中，由于使用 select() 函数，因此只支持 500 个并发连接。这些数字都是理论上的最大值，实际并发连接数可能会受到其他因素的影响，如网络带宽、CPU 性能等。

6. 总结

总体来说，MySQL Router 从原理、安装、配置到部署使用都相对简单直观。它能够满足简单的高可用性应用场景的需求。然而，需要注意的是，MGR 集群的配置不能手动进行，因为相关的元数据存储在 mysql_innodb_cluster_metadata 数据库中。如果 MGR 集群出现故障，则需要准备一些应急措施来应对。

3.3.7 ProxySQL+MGR 方式实现高可用

MySQL 推出的全新高可用架构 MGR 集群，得到了推广和广泛的应用，这一过程也带动了原先较少使用的组件的普及。特别是 ProxySQL 这个读写分离的中间件，现在扮演着 MySQL Router 的角色，与 MGR 集群协同工作。

1. ProxySQL 介绍

ProxySQL 是一个用 C++语言开发的轻量级开源软件，其性能和功能可满足读写中间件所需的绝大多数需求。ProxySQL 的配置数据是基于 SQLite 进行存储的，目前已经更新至 2.6 版本。

（1）功能

ProxySQL 支持以下功能。

- 基本的读写分离，且方式有多种。
- 可定制基于用户、基于库、基于语句的规则对 SQL 语句进行路由。换句话说，规则很灵活。基于库和基于语句的规则，可以实现简单的分片。虽然不支持分表，但可以通过规则实现分表的效果。
- 支持动态加载配置，即一般可以在线修改配置，但有少部分参数还是需要重启来生效。
- 支持对 SQL 语句的路由，可以针对某个语句分配执行实例。
- 使用连接池方式连接后端数据库，而且是多路复用（Multiplexing），可提高资源利用率和传输速率。
- 可以对后端数据库的主机和用户设置最大连接数限制，以确保资源得到合理的分配和管理。
- 在延迟超过阈值、ping 延迟超过阈值、网络不通或宕机情况下，自动下线后端数据库。
- 强大的规则路由引擎：实现读写分离、查询重写、SQL 流量镜像等。
- 支持预处理语句（Prepared Statement），这可以提高 SQL 语句的执行效率和安全性。
- 支持查询缓存（Query Cache），从而加速对相同查询的响应速度。
- 支持负载均衡，与 Gelera Cluster（MariaDB 的集群解决方案）结合使用可实现自动故障转移（Failover）。
- 将所有配置保存写入 SQLite 库中。同时配置信息可以动态更新。
- 支持对查询的路由的灵活控制。

（2）ProxySQL 结构

图 3-21 所示为 ProxySQL 的结构。核心组件说明如下。

- Monitoring 负责监控后端数据库的健康状态，检测主从复制的延时，并能够在必要时临时下线不正常的数据库实例。
- User Auth 负责处理底层后端数据库的用户认证和进行凭证管理。
- Query Processor 用于匹配查询规则，并根据规则决定是否缓存查询、是否将查询加入黑名单、是否重新路由、是否重写查询和是否镜像查询到其他主机组。

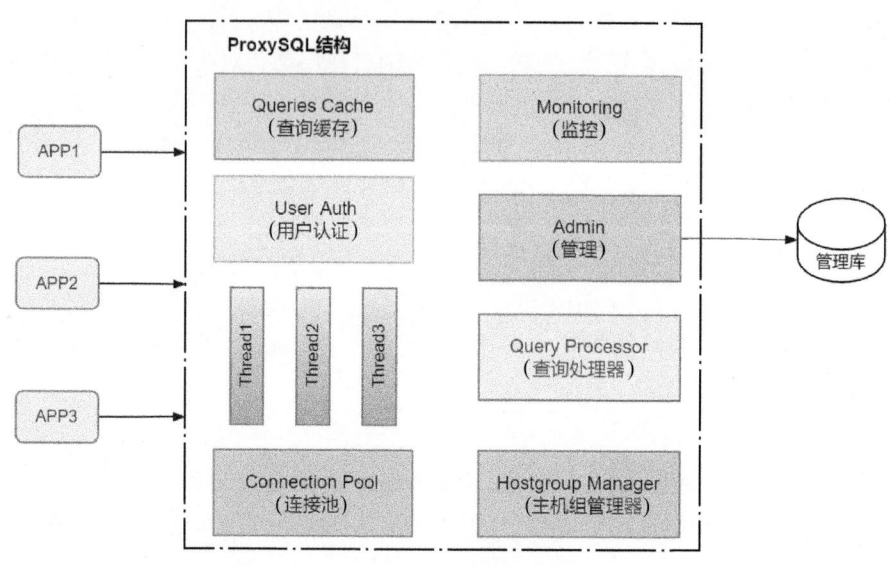

图 3-21　ProxySQL 结构

- Connection Pool 负责创建和管理与后端数据库的连接，这些连接可以被多个前端应用程序共享使用。
- Hostgroup Manager 负责管理和发送 SQL 请求到后端数据库，同时跟踪与管理 SQL 请求的状态和性能。

（3）系统库结构

ProxySQL 自身共有 5 个系统库，分别为 3 个保存在内存中的库和两个保存在磁盘中的 SQLite 默认库。通过 Proxy 的 6032 管理端口登入，就可以查看系统库。查看系统库配置的信息的命令行如下所示。

```
shell $> mysql -uadmin -p****** -h127.0.0.1 -P6032
mysql: [Warning] Using a password on the command line interface can be insecure.
Welcome to the MySQL monitor.  Commands end with ; or \g.
Your MySQL connection id is 2
Server version: 5.5.30 (ProxySQL Admin Module)
mysql> show databases;
+-----+---------------+-------------------------------------------+
|seq  |name           |file                                       |
+-----+---------------+-------------------------------------------+
|0    |main           |                                           |
|1    |disk           |/var/lib/proxysql/proxysql.db              |
|2    |stats          |                                           |
|3    |monitor        |                                           |
|4    |stats_history  |/var/lib/proxysql/proxysql_stats.db        |
+-----+---------------+-------------------------------------------+
5 rows in set (0.00 sec)
```

默认连接的就是 main 库，所有的配置更改都必须在这个库中进行，disk 存档库不会直接受到影响。具体每个库的作用如下。

- main 库是内存配置数据库，表里存放后端数据库实例、用户验证、路由规则等信息。以 runtime_开头的表中有 ProxySQL 当前运行的配置内容，不能通过 DML 语句修改，只能修改对应的不以 runtime_开头的（在内存里的）表，然后执行 LOAD 使其生效，最后执行 SAVE 使其保存到硬盘以供下次重启加载。
- disk 库保存配置信息。这个库的数据需要持久化到磁盘。SQLite 是一个进程内的库，实现了自给自足的、无服务器的、零配置的、事务性的 SQL 数据库引擎。
- stats 库是 ProxySQL 运行抓取的统计信息，包括到后端各命令的执行次数、流量、进程列表、查询种类汇总/执行时间等。
- monitor 库存储 monitor 模块收集的信息，主要是对后端数据库的健康/延迟检查。
- stats_history：统计信息历史库。

（4）核心配置表

ProxySQL 的 6 张核心配置表各自承担着不同的作用，它们共同协作以确保 ProxySQL 能够高效、安全地处理数据库请求。以下介绍这些核心配置表的作用。

mysql_servers 表是后端 MySQL 数据记录信息表。表 3-11 所示为具体表结构信息。

表 3-11　mysql_servers 表结构信息

字段	说明
hostgroup_id	包含此 mysqld 实例的主机组。注意，同一实例可以是多个主机组的一部分
hostname、port	mysqld 实例的 TCP 端点
gtid_port	ProxySQL Binlog Reader 侦听 GTID 跟踪的后端服务器端口
status	● ONLINE：后端服务器完全可以运行 ● SHUNNED：后端服务器暂时停止使用，因为在时间太短或复制延迟超过允许阈值的情况下连接错误太多 ● OFFLINE_SOFT：当服务器进入 OFFLINE_SOFT 模式时，不再接受新的传入连接，而现有连接将保持不变，直到它们变为非活动状态为止。换句话说，连接一直在使用，直到当前事务完成为止。这允许优雅地分离后端 ● OFFLINE_HARD：当服务器进入此模式时，现有连接被丢弃，而新的传入连接也不被接受。这相当于从主机组中删除服务器，或暂时将其从主机组中取出以进行维护工作
weight	服务器相对于其他权重的权重越大，从主机组中选择服务器的概率就越高
compression	如果该值大于 0，则与该服务器的新连接将使用压缩技术
max_connections	ProxySQL 将向此后端服务器打开的最大连接数。即使此服务器具有最高权重，但一旦达到此限制，就不会向其打开新连接。请确保后端配置了正确的 max_connections 值，以避免 ProxySQL 尝试超出该限制
max_replication_lag	ProxySQL 将定期监视复制延迟，如果超出此阈值，则它将暂时不提供从节点读服务，直到复制延迟小于此阈值，才能提供从节点读服务
max_latency_ms	定期监视 ping 时间。如果主机的 ping 时间大于 max_latency_ms，则它将从连接池中排除（尽管服务器还保持 ONLINE 状态）

mysql_users 表用于应用连接 ProxySQL 的用户配置信息。表 3-12 所示为具体表结构信息。

表 3-12 mysql_users 表信息

字 段	说 明
username、password	用于连接 mysqld 或 ProxySQL 实例的凭据
active	将在数据库中跟踪 active=0 的用户,但永远不会在内存数据结构中加载
default_hostgroup	此用户发送的查询没有匹配规则,则生成的流量将发送到指定的主机组
default_schema	默认情况下连接应更改的架构
transaction_persistent	如果一个事务内存在多条 SQL 语句,则它们都只会路由到一个主机组中
frontend	如果设置为 1,则此用户名和密码对用于对 ProxySQL 实例进行身份验证
backend	如果设置为 1,则此用户名和密码对用于针对任何主机组对 mysqld 服务器进行身份验证
max_connections	定义特定用户的最大允许前端连接数

mysql_replication_hostgroups 表示用于配置读写组信息。表 3-13 所示为具体表结构信息。

表 3-13 mysql_replication_hostgroups 表信息

字 段	说 明
writer_hostgroup	发送所有请求的主机组,MySQL 中 read_only=0 的节点将分配给该主机组
reader_hostgroup	发送读取请求的主机组,应该定义查询规则或单独的只读用户将流量路由到此主机组,将 read_only=1 的节点分配给该主机组
check_type	执行只读检查时检查的 MySQL 变量,默认情况下为 read_only(也可以使用 super_read_only)。对于 AWS Aurora,应使用 innodb_read_only
comment	用户自定义内容。可以是集群存储内容的描述、添加或禁用主机组的提醒或某些检查器脚本处理的 JSON

mysql_group_replication_hostgroups 表用于配置 MGR 集群信息。表 3-14 所示为具体表结构信息。

表 3-14 mysql_group_replication_hostgroups 表信息

字 段	说 明
writer_hostgroup	发送所有请求的主机组,MySQL 中 read_only=0 的节点将分配给该主机组
backup_writer_hostgroup	定义写入节点的信息(保证 MySQL 参数 read_only=0)
reader_hostgroup	发送读请求的主机组,将 read_only=1 的节点分配给该主机组
offline_hostgroup	当 ProxySQL 监视并确定节点为 OFFLINE 时,它将被放入 offline_hostgroup
active	ProxySQL 监视主机组并在适当的主机组之间移动节点
max_writers	writer_hostgroup 中应允许的最大节点数,超过此值的节点将放入 backup_writer_hostgroup 中
writer_is_also_reader	确定是否应将同一个节点分别添加到 reader_hostgroup 和 writer_hostgroup 中
max_transactions_behind	ProxySQL 应允许的写入其后面的最大事务数,以防止读取落后过多(这是通过查询 MySQL 中 sys.gr_member_routing_candidate_status 表的 transactions_behind 字段来确定的)
comment	用户自定义内容。可以是集群存储内容的描述、添加或禁用主机组的提醒或某些检查器脚本处理的 JSON

mysql_galera_hostgroups 表在 ProxySQL 2.×及更高版本中可用。它定义了用于配置 MySQL 高可用架构的 Galera Cluster/Percona XtraDB Cluster 的主机组。表 3-15 所示为具体表结构信息。

表 3-15　mysql_galera_hostgroups 表信息

字　段	说　明
writer_hostgroup	发送所有请求的主机组，MySQL 中 read_only=0 的节点将分配给该主机组
backup_writer_hostgroup	定义写入节点的信息（保证 MySQL 参数 read_only=0）
reader_hostgroup	发送读请求的主机组，将 read_only=1 的节点分配给该主机组
offline_hostgroup	当 ProxySQL 监视并确定节点为 OFFLINE 时，它将被放入 offline_hostgroup
active	ProxySQL 监视主机组并在适当的主机组之间移动节点
max_writers	writer_hostgroup 中应允许的最大节点数，超过此值的节点将放入 backup_writer_hostgroup 中
writer_is_also_reader	确定是否应将同一个节点分别添加到 reader_hostgroup 和 writer_hostgroup 中
max_transactions_behind	ProxySQL 应允许的写入其后面的最大事务数，以防止读取落后过多（这是通过查询 MySQL 中 sys.gr_member_routing_candidate_status 表的 transactions_behind 字段来确定的）
comment	用户自定义内容。可以是集群存储内容的描述、添加或禁用主机组的提醒或某些检查器脚本处理的 JSON

mysql_query_rules 表用于定义规则信息。表 3-16 所示为具体表结构信息。

表 3-16　mysql_query_rules 表信息

字　段	说　明
rule_id	规则的唯一 ID。规则以 rule_id 顺序处理
active	查询处理模块将仅考虑 active=1 的规则，并且只将活动规则加载到运行时
username	匹配用户名的过滤条件。如果为非 NULL，则仅当使用正确的用户名建立连接时，查询才会匹配
schemaname	匹配 schemaname 的过滤条件
flagIN、flagOUT 和 apply	• 分别用来定义规则的入口和出口，从而实现链式规则 • 入口值 flagIN 设置为 0，并且在开始时仅考虑 flagIN=0 的规则 • 当为特定查询找到匹配规则时，将评估 flagOUT，如果其为 NOT NULL，则将使用 flagOUT 中的指定标志标记查询
client_addr	匹配来自特定源的流量
proxy_addr	匹配特定本地 IP 地址上的传入流量
proxy_port	匹配特定本地端口上的传入流量
match_digest match_pattern negate_match_pattern re_modifiers replace_pattern	与查询摘要匹配的正则表达式
destination_hostgroup	将匹配的查询路由到此主机组
cache_ttl cache_empty_result	设置缓存清除机制
timeout	执行匹配或重写查询的最大超时（以毫秒为单位）

(续)

字段	说明
retries	执行查询期间检测到失败时需要重新执行查询的最大次数,如果未指定重试次数,则应用全局参数 query_retries_on_failure
delay	延迟执行查询的毫秒数
error_msg 或 OK_msg	查询返回给客户端指定的消息
multiplex	• 如果为 0,则禁用多路复用。 • 如果为 1,且没有任何其他条件阻止此操作(如用户变量或事务),则可以重新启用多路复用。 • 如果为 2,则不会仅针对当前查询禁用多路复用。
log	记录查询内容
apply	当设置为 1 时,在匹配和处理此规则后,将不再评估进一步的查询
gtid_from_hostgroup	定义哪个主机组应该用作 GTID 一致性读取的领导者(通常是复制主机组对中定义的 WRITER 主机组)

(5)常用的命令行

ProxySQL 提供了一套丰富的运维命令,这些命令可以通过其管理接口(MySQL 协议)执行,以便对 ProxySQL 进行配置、监控、维护等操作。表 3-17 中是 ProxySQL 中常用的运维命令及其说明。

表 3-17 ProxySQL 中常用的运维命令及其说明

命令	说明
LOAD MYSQL USERS TO RUNTIME LOAD MYSQL USERS FROM MEMORY	将修改后的配置信息,刷新到内存中
SAVE MYSQL USERS TO MEMORY SAVE MYSQL USERS FROM RUNTIME	将配置刷新到内存中,并应用于运行环境
LOAD MYSQL USERS TO MEMORY LOAD MYSQL USERS FROM DISK	将磁盘中持久化的配置刷新到内存中
SAVE MYSQL USERS TO DISK SAVE MYSQL USERS FROM MEMORY	将内存中的配置保存到磁盘中
LOAD MYSQL USERS FROM CONFIG	将配置文件中的配置加载到内存中

执行某些命令(特别是涉及配置更改的命令)时需要谨慎,因为错误的配置更改可能会导致服务中断或其他不可预见的问题。在生产环境中操作之前,建议先在测试环境中验证更改的效果。

2. ProxySQL 配置主从读写分离

在 MySQL 的主从架构中,ProxySQL 可以实现读写分离。图 3-22 所示为在主从架构下 ProxySQL 的读写分离的实现。在此实现中,需要确保写操作(如 INSERT、UPDATE、DELETE)只发送到主节点(Master),而读操作(如 SELECT)被分发到一个或多个从节点(Slave)。这种配置可以提高系统的性能和可扩展性,因为读操作通常比写操作更频繁。

实现读写分离,涉及多个步骤和配置,需要确保 ProxySQL 能够正确地识别和管理 MySQL 的主从架构。下面先介绍一个详细的配置主从架构读写分离示例,再探讨在主从架构下使用 ProxySQL 时经常遇到的两个问题。

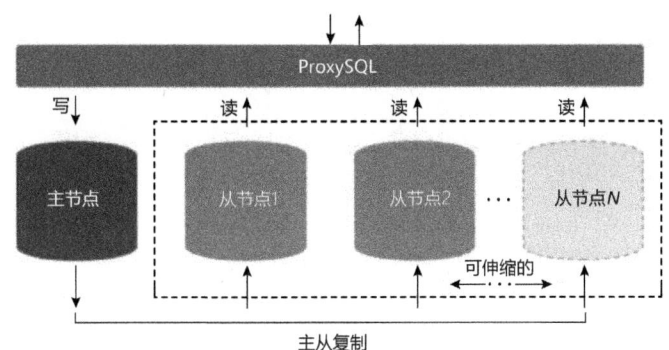

图 3-22　主从架构下 ProxySQL 的读写分离的实现

（1）配置主从架构读写分离

在 MySQL 中创建监控用户和访问用户。监控用户用于查看 MySQL 服务的状态。访问用户用于 MySQL 数据库的读写操作。创建用户的命令行示例如下。

```
# 登录 MySQL 数据库创建用户,并赋予权限
shell $> mysql -uroot -p***** -h127.0.0.1 -P3410
# 访问用户
mysql>GRANT ALL PRIVILEGES ON *.* TO 'dbadmin'@'%' identified by '******' WITH GRANT OPTION;
# 监控用户
mysql> create user monitor@'%' identified by '******';
Query OK, 0 rows affected (0.00 sec)
mysql> grant replication client on *.* to monitor@'%';
Query OK, 0 rows affected (0.01 sec)
```

在 ProxySQL 中添加 MySQL 节点信息。其中，需要确保 hostgroup_id、hostname 和 port 的组合是唯一的，以形成一个主键。配置 MySQL 节点的命令行示例如下。

```
# 登录 ProxySQL 管理数据库
shell $> mysql -uadmin -padmin -h127.0.0.1 -P6032
Your MySQL connection id is 3
Server version: 5.5.30 (ProxySQL Admin Module)

# 配置 MySQL 节点信息
mysql> use main
Database changed
mysql> INSERT INTO mysql_servers(hostgroup_id,hostname,port)
values(10,'192.168.244.130',3410);
mysql> INSERT INTO mysql_servers(hostgroup_id,hostname,port)
values(10,'192.168.244.130',3400);

# 查看配置信息和状态
mysql> select * from mysql_servers\G
*************************** 1. row ***************************
       hostgroup_id: 10
           hostname: 192.168.244.130
               port: 3410
          gtid_port: 0
```

```
              status: ONLINE
              weight: 1
         compression: 0
     max_connections: 1000
 max_replication_lag: 0
             use_ssl: 0
      max_latency_ms: 0
             comment:
1 row in set (0.00 sec)

# 加载到运行时,并保存到硬盘中
mysql> load mysql servers to runtime;
Query OK, 0 rows affected (0.00 sec)

mysql> save mysql servers to disk;
Query OK, 0 rows affected (0.03 sec)
```

在 ProxySQL 中,配置 ProxySQL 访问 MySQL 的用户信息。访问用户需要把访问的自身信息添加到 mysql_users 表中,包括用户名、密码、主信息。配置访问用户的命令行示例如下。

```
mysql> INSERT INTO mysql_users(username,password,default_hostgroup)
values('dbadmin','123456',10);

mysql> load mysql users to runtime;

mysql> save mysql users to disk;
Query OK, 0 rows affected (0.01 sec)
mysql> select * from mysql_users\G
*************************** 1. row ***************************
              username: dbadmin
              password: 123456
                active: 1   # active=1 表示用户生效,0 表示不生效
               use_ssl: 0
     default_hostgroup: 10
        default_schema: NULL
         schema_locked: 0
 transaction_persistent: 1  # 如果设置为 1,连接 ProxySQL 的会话后,如果在一个 hostgroup 上开启了事务,
                            # 那么后续的 SQL 语句都继续维持在这个 hostgroup 上,无论是否会匹配上其他路由规则,
                            # 直到事务结束。默认值是 0
          fast_forward: 0
               backend: 1
              frontend: 1
       max_connections: 10000    # 该用户允许的最大连接数
               comment:
1 row in set (0.00 sec)
```

注意:需要注意 active 和 transaction_persistent 两个列。只有当 active=1 时,用户才是有效的。对于 transaction_persistent 字段,当其值为 1 时,表示事务持久化。也就是说,当某个连接使用该用户开启了一个事务后,在事务提交或回滚之前,所有的语句都会被路由到同一个组中,以避免语句分散到不同的组。建议在创建用户之后将其设置为 1,以避免出现脏读、幻读等现象。

在 MySQL 主从复制的环境中，需要监控后端节点以确保 ProxySQL 可以正确地根据节点的 READ_ONLY 值来将它们分配到不同的主机组，从而实现读写分离。同时 ProxySQL 需要能够监控复制延迟，并根据预设的阈值来决定是否将读查询路由到某个从节点。如果延迟超过了设定阈值，那么 ProxySQL 应该避免将查询发送到那个节点。配置监控用户的命令行示例如下。

```
shell $> mysql -uadmin -padmin -h127.0.0.1 -P6032
mysql> set mysql-monitor_username='monitor';
mysql> set mysql-monitor_password='123456';
mysql> load mysql variables to runtime;
mysql> save mysql variables to disk;
```

配置完成之后，查看 ProxySQL 的运行状态和日志的命令行示例如下。

```
# MySQL 节点运行情况
mysql> select hostgroup_id,hostname,port,status,weight from mysql_servers;
+--------------+-----------------+------+--------+--------+
| hostgroup_id | hostname        | port | status | weight |
+--------------+-----------------+------+--------+--------+
| 10           | 192.168.244.130 | 3410 | ONLINE | 2      |
| 10           | 192.168.244.130 | 3400 | ONLINE | 1      |
+--------------+-----------------+------+--------+--------+

# 查看 read_only 日志
mysql> select * from mysql_server_read_only_log;

# 查看 replication_lag 的复制延迟日志
mysql> select * from mysql_server_replication_lag_log;

# 查看连接信息
mysql> select * from mysql_server_connect_log;

# 查看心跳连接信息
mysql> select * from mysql_server_ping_log;

# 查看执行过的 SQL 语句信息
mysql> select * from stats_mysql_query_digest;
```

想要验证 ProxySQL 是否实现了读写分离方案，可以使用命令行工具连接到 ProxySQL，并执行一些查询来检查读写请求是否被正确地路由到不同的后端 MySQL 节点。验证读写分离的命令行示例如下。

```
shell $> mysql -udbadmin -p****** -P6033 -h127.0.0.1 -e "select @@port"
+--------+
| @@port |
+--------+
| 3400   |
+--------+
shell $>mysql -udbadmin -p123456 -h127.0.0.1 -P6033 -e "start transaction;select @@port;commit;select @@port;"
+--------+
| @@port |
```

说明：SELECT 语句分发到从节点，指定事务的 SQL 语句分发到主节点执行。

（2）常见问题

在主从架构下使用 ProxySQL 时，会经常遇到以下两个问题。处理方式如下。

1）主从复制延迟。ProxySQL 会自动回避复制延迟较大的节点。如果节点 max_replication_lag 列设置为非零值，则 Monitor 模块会定期检查复制延迟。当复制延迟超过 30s 时，会自动回避这些延迟较大的节点。如果将 max_replication_lag 设置为 0，则代表不检查复制延迟。设置复制延迟示例如下。

```
# 查看节点延迟 max_replication_lag 信息
mysql> select hostgroup_id,hostname,port,max_replication_lag from mysql_servers;
+--------------+-----------------+------+---------------------+
| hostgroup_id | hostname        | port | max_replication_lag |
+--------------+-----------------+------+---------------------+
| 20           | 192.168.244.130 | 3400 | 0                   |
| 10           | 192.168.244.130 | 3410 | 0                   |
+--------------+-----------------+------+---------------------+
2 rows in set (0.00 sec)

# 设置复制延迟阈值为 30s
mysql> update mysql_servers set max_replication_lag=30 where hostgroup_id=10;
Query OK, 1 row affected (0.00 sec)

# 查看复制延迟是否设置成功
mysql> select hostgroup_id,hostname,port,max_replication_lag from mysql_servers;
+--------------+-----------------+------+---------------------+
| hostgroup_id | hostname        | port | max_replication_lag |
+--------------+-----------------+------+---------------------+
| 20           | 192.168.244.130 | 3400 | 0                   |
| 10           | 192.168.244.130 | 3410 | 30                  |
+--------------+-----------------+------+---------------------+
2 rows in set (0.00 sec)
```

说明：max_replication_lag 判断指标是通过主从复制延迟指标 Seconds_Behind_Master，该参数判断延迟的准确性不高，故个人建议将它作为参考功能。另外，max_replication_lag 仅适用于从节点。如果服务器未启用复制，则 Monitor 不会执行任何操作。

2）假死场景。在 ProxySQL 中，如果遇到 MySQL 主节点假死的情况，则可以采取一些措施来强制关闭与假死主库的连接，以防止应用继续访问无法响应的数据库服务。假死 MySQL 节点处理的命令行示例如下。

```
# 配置成 OFFLINE_HARD 状态 (当某后端 MySQL 设置为 OFFLINE_HARD 时, ProxySQL 将不会向它发送新的请求)
Admin> update runtime_mysql_servers set status="OFFLINE_HARD" where hostname=
'192.168.20.31' and port='3306';

# 或者删除节点
Admin> delete from mysql_servers where hostname='192.168.20.31' and port='3306';
Admin> load mysql servers to runtimeAdmin> save mysql serbers to disk;
```

3. ProxySQL 配置 MGR 集群

在将 ProxySQL 与 MGR 集群结合使用时,需要对 MySQL 和 ProxySQL 都进行相应的配置。

(1) MySQL 配置

在 MySQL 中创建监控用户和访问用户。监控用户用于查看 MySQL MGR 的状态,需要额外赋予其对 sys 库的读权限。访问用户用于 MySQL 数据库的读写操作。创建用户的命令行示例如下。

```
# 登录 MySQL 数据库创建用户,并赋予权限
shell $> mysql -uroot -p -S /opt/data8.0/mysql/mysql.sock
# 访问用户
mysql>GRANT ALL PRIVILEGES ON *.* TO 'mgradmin'@'%' identified by '******' WITH GRANT OPTION;

# 监控用户
mysql> CREATE USER 'monitor'@'%' IDENTIFIED BY '******';
Query OK, 0 rows affected (0.03 sec)

mysql> GRANT SELECT on sys.* to 'monitor'@'%';
Query OK, 0 rows affected (0.03 sec)
```

在 MySQL 的 sys 库中配置脚本(包含函数和视图)。GitHub 上提供了监控 MGR 集群函数和视图(脚本地址:https://proxysql.com/documentation/main-runtime/)。具体脚本内容如下。

```
mysql> USE sys;
# GTID 对比信息
mysql> DELIMITER $$
CREATE FUNCTION IFZERO(a INT, b INT)
RETURNS INT
DETERMINISTIC
RETURN IF(a = 0, b, a) $$


CREATE FUNCTION LOCATE2(needle TEXT(10000), haystack TEXT(10000), offset INT)
RETURNS INT
DETERMINISTIC
RETURN IFZERO(LOCATE(needle, haystack, offset), LENGTH(haystack) + 1) $$


CREATE FUNCTION GTID_NORMALIZE(g TEXT(10000))
RETURNS TEXT(10000)
DETERMINISTIC
RETURN GTID_SUBTRACT(g, '') $$

# 查看 MGR 集群 GTID 对比信息
CREATE FUNCTION GTID_COUNT(gtid_set TEXT(10000))
RETURNS INT
```

```
DETERMINISTIC
BEGIN
  DECLARE result BIGINT DEFAULT 0;
  DECLARE colon_pos INT;
  DECLARE next_dash_pos INT;
  DECLARE next_colon_pos INT;
  DECLARE next_comma_pos INT;
  SET gtid_set = GTID_NORMALIZE(gtid_set);
  SET colon_pos = LOCATE2(':',gtid_set, 1);
  WHILE colon_pos != LENGTH(gtid_set) + 1 DO
    SET next_dash_pos = LOCATE2('-',gtid_set, colon_pos + 1);
    SET next_colon_pos = LOCATE2(':',gtid_set, colon_pos + 1);
    SET next_comma_pos = LOCATE2(',',gtid_set, colon_pos + 1);
    IF next_dash_pos < next_colon_pos AND next_dash_pos < next_comma_pos THEN
      SET result = result +
        SUBSTR(gtid_set, next_dash_pos + 1,
               LEAST(next_colon_pos, next_comma_pos) - (next_dash_pos + 1)) -
        SUBSTR(gtid_set, colon_pos + 1, next_dash_pos - (colon_pos + 1)) + 1;
    ELSE
      SET result = result + 1;
    END IF
      SET colon_pos = next_colon_pos;
  END WHILE;
  RETURN result;
END $$

# 查看MGR集群Apply队列信息
CREATE FUNCTION gr_applier_queue_length()
RETURNS INT
DETERMINISTIC
BEGIN
  RETURN (SELECT sys.gtid_count( GTID_SUBTRACT( (SELECT
Received_transaction_set FROM performance_schema.replication_connection_status
WHERE Channel_name = 'group_replication_applier'), (SELECT
@@global.GTID_EXECUTED) )));
END $$

# 查看MGR集群主节点信息
CREATE FUNCTION gr_member_in_primary_partition()
RETURNS VARCHAR(3)
DETERMINISTIC
BEGIN
  RETURN (SELECT IF( MEMBER_STATE='ONLINE' AND ((SELECT COUNT(*) FROM
performance_schema.replication_group_members WHERE MEMBER_STATE != 'ONLINE') >=
((SELECT COUNT(*) FROM performance_schema.replication_group_members)/2) = 0),
'YES','NO') FROM performance_schema.replication_group_members JOIN
performance_schema.replication_group_member_stats USING(member_id));
END $$
```

```
# 查看 MGR 集群 READ_ONLY 属性信息
CREATE VIEW gr_member_routing_candidate_status AS SELECT
sys.gr_member_in_primary_partition() as viable_candidate,
IF( (SELECT (SELECT GROUP_CONCAT(variable_value) FROM
performance_schema.global_variables WHERE variable_name IN ('read_only',
'super_read_only')) != 'OFF,OFF'), 'YES', 'NO') as read_only,
sys.gr_applier_queue_length() as transactions_behind, Count_Transactions_in_queue as
'transactions_to_cert' from performance_schema.replication_group_member_stats; $$

DELIMITER;
```

配置完 MGR 集群监控脚本之后，通过其中配置的视图 gr_member_routing_candidate_status 查看 MGR 集群情况。查看 MGR 集群监控信息状态的命令行示例如下。

```
mysql> select * from gr_member_routing_candidate_status;
+------------------+-----------+--------------------+--------------------+
|viable_candidate  |read_only  |transactions_behind |transactions_to_cert|
+------------------+-----------+--------------------+--------------------+
|YES               |NO         |         0          |         0          |
+------------------+-----------+--------------------+--------------------+
1 row in set (0.00 sec)
```

表 3-18 所示为 ProxySQL 监控 MGR 集群的视图返回字段。

表 3-18　ProxySQL 监控 MGR 集群的视图返回字段

字段	说明
viable_candidate	可行的候选节点情况。YES 表示半数以上节点存活
read_only	是否将其他节点设置为只读属性
transactions_behind	事务是否存在延迟
transactions_to_cert	MGR 集群证书验证是否开启

（2）ProxySQL 配置

在 ProxySQL 中配置 MGR 集群时，需要定义后端服务，配置路由规则，并设置相应的主机组，以便在发生节点切换时，ProxySQL 能够自动更新并识别新的主节点。

下面是配置 ProxySQL 以管理 MGR 集群时定义指定主机组的 SQL 命令示例。这些主机组包括写组 50、备写组 60、读组 70 和离线组 80。

```
shell $> mysql -uadmin -p****** -h127.0.0.1 -P6032
Your MySQL connection id is 3
Server version: 5.5.30 (ProxySQL Admin Module)
mysql> use main
Database changed
# 配置写组
mysql>INSERT INTO mysql_servers(hostgroup_id,hostname,port)
values (50,'192.168.244.129',3380);

mysql>INSERT INTO mysql_servers(hostgroup_id,hostname,port)
values (50,'192.168.244.129',3381);
```

```
mysql>INSERT INTO mysql_servers(hostgroup_id,hostname,port)
values (50,'192.168.244.129',3382);

#配置写组的 MySQL 账号信息
mysql> INSERT INTO mysql_users(username,password,default_hostgroup)
values('mgradmin','*****',50);

#配置写组、备写组、读组和离线组
mysql> INSERT INTO mysql_group_replication_hostgroups
(writer_hostgroup,backup_writer_hostgroup,reader_hostgroup,offline_hostgroup,active,max
_writers,writer_is_also_reader,max_transactions_behind) values(50,60,70,80,1,1,0,100);

#配置持久化和刷新到内存中
mysql>load mysql servers to runtime;
save mysql servers to disk;
load mysql users to runtime;
save mysql users to disk;
load mysql variables to runtime;
save mysql variables to disk;
```

配置完成后，查看 MGR 集群节点的状态的命令行示例如下。

```
mysql> SELECT hostgroup_id, hostname,port, status FROM runtime_mysql_servers;
+--------------+-----------------+------+--------+
|hostgroup_id  |hostname         |port  |status  |
+--------------+-----------------+------+--------+
|50            |192.168.244.129  |3380  |ONLINE  |
|70            |192.168.244.129  |3381  |ONLINE  |
|70            |192.168.244.129  |3382  |ONLINE  |
+--------------+-----------------+------+--------+
```

通过 mysql_server_group_replication_log 表，查看是否可按照配置 MGR 集群正常访问。具体命令行示例如下。

```
mysql > SELECT time_start_us, hostname, port, viable_candidate, read_only, transactions_
behind,error FROM mysql_server_group_replication_log
    ORDER BY time_start_us DESC LIMIT 6;
+------------------+-----------------+-----+-----------------+----------+--------------------+-------+
|time_start_us     |hostname         |port |viable_candidate |read_only |transactions_behind |error  |
+------------------+-----------------+-----+-----------------+----------+--------------------+-------+
|1586094196050264  |192.168.244.129  |3382 |YES              |YES       |1                   |NULL   |
|1586094196049313  |192.168.244.129  |3381 |YES              |YES       |0                   |NULL   |
|1586094196048453  |192.168.244.129  |3380 |YES              |NO        |0                   |NULL   |
|1586094191055906  |192.168.244.129  |3382 |YES              |YES       |1                   |NULL   |
|1586094191055905  |192.168.244.129  |3381 |YES              |YES       |0                   |NULL   |
|1586094191055901  |192.168.244.129  |3380 |YES              |NO        |0                   |NULL   |
+------------------+-----------------+-----+-----------------+----------+--------------------+-------+
```

想要验证 ProxySQL 是否实现了 MGR 集群的读写分离方案，可以使用命令行工具连接到 ProxySQL，并执行一些查询来检查读写请求是否被正确地路由到不同的后端 MySQL 节点。验证 MGR

集群读写分离的命令行示例如下。

```
shell $>mysql -umgradmin -p123456 -h127.0.0.1 -P6033 -e "START TRANSACTION; SELECT
@@port; COMMIT; SELECT @@port;"
mysql: [Warning] Using a password on the command line interface can be insecure.
+--------+
| @@port |
+--------+
|  3380  |
+--------+
+--------+
| @@port |
+--------+
|  3381  |
+--------+
```

通过 stats_mysql_query_digest 表查看上述访问的节点信息，命令行示例如下。

```
mysql> SELECT hostgroup,digest_text FROM stats_mysql_query_digest
       ORDER BY digest_text DESC LIMIT 5;
+-----------+-----------------------------------+
| hostgroup | digest_text                       |
+-----------+-----------------------------------+
| 50        | start transaction                 |
| 50        | select @@version_comment limit ?  |
| 70        | select @@port                     |
| 50        | select @@port                     |
+-----------+-----------------------------------+
```

4. 总结

ProxySQL 作为一个数据库代理，提供了多种功能，包括读写分离、负载均衡、故障转移、查询缓存、查询重写等。尽管 ProxySQL 提供了丰富的功能和配置选项，但配置过程本身并不复杂。由于 ProxySQL 需要维护连接池并进行流量转发，因此需要一定的硬件资源来支持其运行。ProxySQL 的性能损耗主要来自于额外的网络跳转、查询解析和重写等过程。在大多数情况下，与直接连接数据库相比，引入 ProxySQL 可能会有 5%～18% 的性能损耗。

3.4 MySQL 8.0 新增功能

3.4.1 角色管理

在 MySQL 8.0 中，新增了角色功能，为数据库权限管理提供了更灵活和高效的方式。下面介绍 MySQL 8.0 中的角色功能及其实现方式。

1. 角色介绍

数据库里对应的权限都可以指定赋予。那么角色的作用是什么呢？角色是一个命名的权限集合。为了对许多拥有相似权限的用户进行分类管理，定义了角色的概念。与用户账户一样，角色也可以被授予和撤销特权。例如，当多个用户需要分配复杂又细致的权限时，角色的作用就体现出来了。通过将一组权限赋给一个角色，新用户只需要被分配这个角色，就能获得对应的权限。

在 MySQL 8.0 版本中，角色和用户的数据保存在同一张 mysql.user 系统表里。图 3-23 所示为保存在 mysql.user 系统表中的角色和用户的区别。两者的区别在于锁（account_locked）和过期（password_expired）字段值。角色默认是锁住的，密码永不过期。相反，用户默认是不锁的，密码没有过期设置。

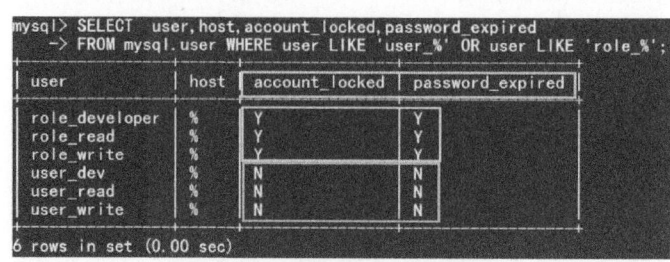

图 3-23　角色和用户区别字段

在 MySQL 8.0 版本中，角色的功能是通过用户实现的，另外，角色没有登录能力（角色没有密码，同时被锁住）。

如下示例给角色设置密码，解锁后可以正常登录 MySQL 服务。给角色设置密码，解锁的命令行示例如下。

```
mysql> ALTER USER 'role_developer'@'%' IDENTIFIED BY '123456';
mysql> ALTER USER 'role_developer'@'%' ACCOUNT UNLOCK;
```

说明：到这里可以大致理解，实际上角色和用户是相等的，只是通过关系绑在一起。

到这里对 MySQL 8.0 版本中的角色有了初步了解，下面介绍具体实现方式。

2. 角色实现方式

下面介绍角色实现示例。

可将角色与用户关联起来，即给用户赋予角色，具体命令行示例如下。

```
# 创建角色
mysql> DROP ROLE IF EXISTS 'role_developer'@'%','role_read'@'%',
       'role_write'@'%';
mysql> CREATE ROLE 'role_developer'@'%','role_read'@'%','role_write'@'%';

# 为角色赋予权限
mysql> GRANT ALL ON world.* TO 'role_developer';
mysql> GRANT SELECT ON world.* TO 'role_read';
mysql> GRANT INSERT, UPDATE, DELETE ON world.* TO 'role_write';

# 创建 3 个用户(读写、只读和开发)
mysql> DROP USER IF EXISTS 'user_dev'@'%','user_read'@'%','user_write'@'%';

mysql> CREATE USER 'user_dev'@'%' IDENTIFIED BY '123456';
mysql> CREATE USER 'user_read'@'%' IDENTIFIED BY '123456';
mysql> CREATE USER 'user_write'@'%' IDENTIFIED BY '123456';

# 查询用户和角色状态
mysql> SELECT user,host,account_locked,password_expired FROM mysql.user WHERE user LIKE
'user_%' OR user LIKE 'role_%';
```

```
+--------------------+--------+----------------+------------------+
| user               | host   | account_locked | password_expired |
+--------------------+--------+----------------+------------------+
| role_developer     | %      | Y              | Y                |
| role_read          | %      | Y              | Y                |
| role_write         | %      | Y              | Y                |
| user_dev           | %      | N              | N                |
| user_read          | %      | N              | N                |
| user_write         | %      | N              | N                |
+--------------------+--------+----------------+------------------+
6 rows in set (0.00 sec)

# 角色授予和撤销
## 授予
mysql> GRANT 'role_developer'@'%' TO 'user_dev'@'%';
mysql> GRANT 'role_developer'@'%' TO 'user_read'@'%' WITH ADMIN OPTION;
mysql> GRANT 'role_write'@'%' TO 'user_write'@'%' WITH admin OPTION;

## 撤销
mysql> REVOKE 'role_developer'@'%' FROM 'user_dev'@'%';
mysql> REVOKE 'role_developer'@'%' FROM 'user_read'@'%';
mysql> REVOKE 'role_write'@'%' FROM 'user_write'@'%';
```

在 MySQL 中创建角色和用户后,需要进行后续步骤,即激活角色功能,以便用户能够使用这些角色所赋予的权限进行操作。激活角色的命令行示例如下。

```
mysql> SET DEFAULT ROLE ALL TO 'user_dev'@'%';
mysql> SET DEFAULT ROLE ALL TO 'user_read'@'%';
mysql> SET DEFAULT ROLE ALL TO 'user_write'@'%';
```

3. 角色相关命令和配置参数

在 MySQL 8.0 版本中,为了提供对角色的更灵活和可维护的管理,引入了管理角色的一些新的命令和配置参数。

表 3-19 所示为角色相关的命令。

表 3-19 角色相关命令

命 令	说 明
CREATE ROLE、DROP ROLE	创建和删除角色
GRANT、REVOKE	是否激活角色
SHOW GRANTS	显示用户或角色所拥有的角色或者权限
SET DEFAULT ROLE	设置用户默认使用的角色
SET ROLE	改变当前会话的角色。可以指定下列子语句。 • NONE:无角色 • ALL EXCEPT 'role_write':除已命名的角色以外的所有角色 • ALL:所有角色
CURRENT_ROLE	显示当前会话的角色

(续)

命　令	说　明
WITH ADMIN OPTION	授予或撤销其他用户或角色的权限
ROLES_GRAPHML	以 SELECT 方式调用接口，返回 UTF 8 字符串 xml（graphml）格式的用户信息

表 3-20 所示为角色配置参数。

表 3-20　角色配置参数

参　　数	说　　明
mandatory_roles	允许定义用户登录时强制权限的角色
activate_all_roles_on_login	在用户登录时激活每个用户的对应的角色配置

给用户配置角色的命令行和配置参数方式如下。

```
#命令行方式
mysql> SET PERSIST mandatory_roles = 'role1,role2@%,r3@%.example.com';
mysql> SET PERSIST activate_all_roles_on_login = ON;

#配置参数方式
[msyqld]
mandatory_roles='role_developer'
activate_all_roles_on_login = ON
```

4. 总结

虽然角色功能在 MySQL 8.0 版本中已经被引入，但在实际应用中，其使用场景可能并不如预期那么广泛。这可能是因为已经习惯了传统的直接为用户分配权限的方式。了解如何创建和管理角色仍然是很有价值的（权限集中管理，简化权限分配，灵活的角色切换）。

3.4.2　直方图

直方图（Histogram）是一种统计报告图，通过一系列高度不等的纵向条纹或线段来展示数据的分布情况。在 MySQL 8.0 版本中，引入了直方图功能，为用户提供了一种可视化数据分布的新方法。这一功能的引入可能借鉴了其他数据库系统（如 Oracle 和 MariaDB）的相关实现。

1. 直方图介绍

直方图能近似获得一列的数据分布情况，通过将数据分到一系列桶（Bucket）中来实现。MySQL 会自动将这些数据划分到不同的桶，并自动决定创建哪种类型的直方图。

如图 3-24 所示，在源码实现中，MySQL 支持两种直方图类型：等宽直方图（singleton）、等高直方图（equi-height）。singleton 是指每个桶的宽度（即范围）是相等的。而 equi-height 意味着每个桶中的数据项数量相同，因此直方图的每个柱子的高度相等。

在图 3-25 所示两个直方图代码实现的对比中，equi-height 方式多了下限（Lower）和上限（Upper）指标。

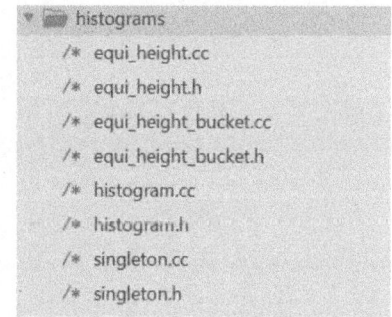

图 3-24　直方图实现源码目录

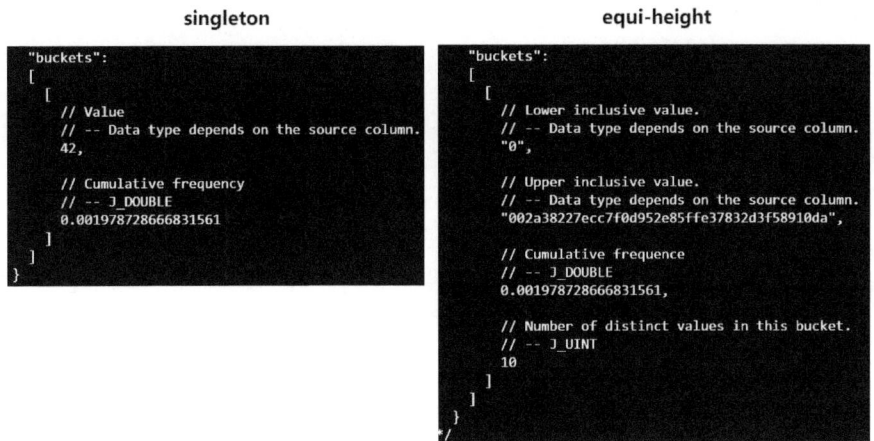

图 3-25　直方图实现源码对比

在 MySQL 8.0 版本中，直方图的使用是通过 ANALYZE TABLE…UPDATE HISTOGRAM 命令来实现的，该命令属于 ANALYZE TABLE 命令的一部分。ANALYZE TABLE 命令在 MySQL 中的主要作用是收集表和索引的统计信息，这些信息对于查询优化器来说非常重要，因为它可以帮助优化器制定更有效的查询计划。

2. 直方图实现方式

下面介绍直方图的使用示例。

下面对 employees 表中的 emp_no 字段更新统计信息（包括直方图）。对 emp_no 字段创建直方图的命令行示例如下。

```
# 对 emp_no 字段创建直方图
mysql> ANALYZE TABLE employees UPDATE HISTOGRAM ON emp_no WITH 32 BUCKETS;
+--------------------+-----------+----------+-------------------------------------------------+
|Table               |Op         |Msg_type  |Msg_text                                         |
+--------------------+-----------+----------+-------------------------------------------------+
|employees.employees |histogram  |status    |Histogram statistics created for column 'emp_no'.|
+--------------------+-----------+----------+-------------------------------------------------+
1 row in set (0.87 sec)
# 删除 emp_no 直方图
mysql> ANALYZE TABLE employees DROP HISTOGRAM ON emp_no ;
+--------------------+-----------+----------+-------------------------------------------------+
|Table               |Op         |Msg_type  |Msg_text                                         |
+--------------------+-----------+----------+-------------------------------------------------+
|employees.employees |histogram  |status    |Histogram statistics removed for column 'emp_no'.|
+--------------------+-----------+----------+-------------------------------------------------+
1 row in set (0.00 sec)
```

在 MySQL 8.0 版本中，统计信息，包括直方图，会存储在数据字典中。这些信息可以通过 information_schema 数据库中的 COLUMN_STATISTICS 视图来访问。直方图数据以灵活的 JSON 格式存储，这种格式便于扩展和解析。JSON 中包含桶的信息。桶是直方图的基本组成部分，每个桶都代表一个数据范围以及该范围内数据的出现频率或数量。值得注意的是，MySQL 默认会为直方图创建 100 个桶，但这个数量不是固定的，而是根据数据的分布和统计信息的需要来动态确定的。查看直方图信息的命令行示例如下。

```
mysql> SELECT * FROM INFORMATION_SCHEMA.COLUMN_STATISTICS \G;
*************************** 3. row ***************************
SCHEMA_NAME: employees
 TABLE_NAME: employees
COLUMN_NAME: emp_no
  HISTOGRAM: {"buckets": [[298001, 401193, 0.03130422847282816, 3193], [401194, 404386,
0.06260845694565632, 3193], [404387, 407579, 0.09391268541848449, 3193], [407580, 410772,
0.12521691389131265, 3193], [410773, 413965, 0.15652114236414083, 3193], [413966, 417158,
0.18782537083696899, 3193], [417159, 420351, 0.21912959930979717, 3193], [420352, 423544,
0.25043382778262533, 3193], [423545, 426737, 0.28173805625545350, 3193], [426738, 429930,
0.31304228472828166, 3193], [429931, 433123, 0.34434651320110980, 3193], [433124, 436316,
0.37565074167393797, 3193], [436317, 439509, 0.40695497014676610, 3193], [439510, 442702,
0.43825919861959434, 3193], [442703, 445895, 0.46956342709242250, 3193], [445896, 449088,
0.50086765556525060, 3193], [449089, 452281, 0.53217188403807880, 3193], [452282, 455474,
0.56347611251090700, 3193], [455475, 458667, 0.59478034098373510, 3193], [458668, 461860,
0.62608456945656330, 3193], [461861, 465053, 0.65738879792939140, 3193], [465054, 468246,
0.68869302640221960, 3193], [468247, 471439, 0.71999725487504780, 3193], [471440, 474632,
0.75130148334787590, 3193], [474633, 477825, 0.78260571182070420, 3193], [477826, 481018,
0.81390994029353230, 3193], [481019, 484211, 0.84521416876636050, 3193], [484212, 487404,
0.87651839723918870, 3193], [487405, 490597, 0.90782262571201680, 3193], [490598, 493790,
0.93912685418484500, 3193], [493791, 496983, 0.97043108265767310, 3193], [496984, 499999, 1.0,
3016]], "data-type": "int", "null-values": 0.0, "collation-id": 8, "last-updated": "2023-09-03 08:
46:28.838492", "sampling-rate": 1.0, "histogram-type": "equi-height", "number-of-buckets-speci-
fied": 32}

#查看采样率,如下所示
mysql> SELECT HISTOGRAM->>'$."sampling-rate"'
    FROM INFORMATION_SCHEMA.COLUMN_STATISTICS
    WHERE TABLE_NAME = "employees01" AND COLUMN_NAME = "emp_no";
+-----------------------------------------------+
| HISTOGRAM->>'$."sampling-rate"' |
+-----------------------------------------------+
| 1.0                                           |
+-----------------------------------------------+
1 row in set (0.01 sec)
```

说明：采样率为 1，意味着 emp_no 列的全部数据被读入内存以生成直方图统计信息。

3. 直方图其他事项

在参数方面，直方图参数 histogram_generation_max_mem_size 用于控制在创建和更新直方图时所需的内存大小，单位是 B。该参数的默认值是 20 000 000 B（约 19.1MB），最小值是 1 000 000B（约 976.6KB）。这是一个会话级（session）参数，意味着它仅影响当前的数据库会话。每次创建和更新直方图时，都会分配相应的内存，并在操作完成后释放，这有助于更有效地管理数据库资源。

直方图存在如下限制。

- 不支持加密表和临时表：由于加密表的数据是不可见的，因此不能为它们生成直方图。同样，临时表的生命周期较短，不需要持久化的直方图。

166

- 数据类型限制：直方图不支持几何类型（如空间数据）和 JSON 数据类型的列，因为这些数据类型具有特殊的结构和处理方式。
- 支持存储列和虚拟列：可以为存储列（实际存储在数据库中的列）和虚拟列（由表达式或函数计算得出的列）生成直方图。
- 不支持单列唯一索引：如果列被单列唯一索引覆盖，那么不能为该列生成直方图，因为索引本身已经提供了足够的信息来优化查询。
- 表类型和分析支持：直方图分析支持 InnoDB 和 MyISAM 表类型。同时，参数 innodb_read_only 必须关闭，因为直方图生成需要写操作。
- 缓存表统计信息：information_schema_stats_expiry 参数定义了缓存表统计信息在过期之前的时间段，这对于直方图的时效性管理很重要。
- 分区表支持：分析表支持分区表，这意味着可以为分区表的各个分区生成直方图，以优化针对这些分区的查询。
- 表锁和长时间运行的事务：分析表时，可能需要删除并重新创建表的直方图，这可能需要获取表的锁。如果有长时间运行的事务正在使用该表，则这些事务可能需要等待，直到分析完成为止。

直方图受 DDL 语句的如下影响。
- ALTER TABLE…CONVERT TO CHARACTER SET…命令行删除字符列的直方图，因为直方图会受到字符集更改的影响。但非字符列的直方图不受影响。
- 删除表：DDL 语句（如 DROP TABLE）会删除被删除表中所有列的直方图，因为表本身已不存在。
- 删除数据库：DROP DATABASE 语句会删除数据库中所有表的直方图，因为整个数据库都被删除了。
- 重命名表：重命名表不会直接删除直方图，而是将直方图与新的表名关联起来，这样直方图就可以继续为新表名提供优化信息。
- 修改列：如果 ALTER TABLE 语句涉及删除或修改列，那么将会删除该列的直方图，因为列的结构已经发生变化。
- 字符集转换：ALTER TABLE 语句中的字符集转换（如将字符列转换为不同的字符集）会删除这些字符列的直方图，因为字符集的变化可能影响数据的分布和直方图的准确性。非字符列的直方图不受此影响。

在性能方面，可在相同的实验环境下，对有无直方图的情况进行性能对比。通过使用特定的查询语句和数据集，会发现直方图的存在使查询速度提高了。这表明直方图在提升数据库性能方面具有重要作用。

在如下示例中，SQL 语句的执行计划中的 key（NULL→idx_dt_birth）、filtered（15.40→100.00）和 rows（299600→46146）都有了明显变化。具体命令行示例如下。

```
# 无直方图下执行计划
mysql> EXPLAIN SELECT * FROM employees WHERE birth_date>'1964-02-01'\G;
*************************** 1. row ***************************
           id: 1
  select_type: SIMPLE
        table: employees
   partitions: NULL
```

```
          type: ALL
 possible_keys: idx_dt_birth
           key: NULL
       key_len: NULL
           ref: NULL
          rows: 299600
      filtered: 15.40
         Extra: Using where
1 row in set, 1 warning (0.00 sec)
```

创建 birth_data 直方图

```
mysql> ANALYZE TABLE employees UPDATE HISTOGRAM ON birth_date WITH 35 BUCKETS;
+---------------------+-----------+----------+------------------------------------------------------------+
|Table                |Op         |Msg_type  |Msg_text                                                    |
+---------------------+-----------+----------+------------------------------------------------------------+
|employees.employees  |histogram  |status    |Histogram statistics created for column 'birth_date'.       |
+---------------------+-----------+----------+------------------------------------------------------------+
1 row in set (1.29 sec)
```

直方图下执行计划

```
mysql> EXPLAIN SELECT * FROM employees WHERE birth_date>'1964-02-01'\G
*************************** 1. row ***************************
            id: 1
   select_type: SIMPLE
         table: employees
    partitions: NULL
          type: range
 possible_keys: idx_dt_birth
           key: idx_dt_birth
       key_len: 3
           ref: NULL
          rows: 46146
      filtered: 100.00
         Extra: Using index condition
1 row in set, 1 warning (0.00 sec)
```

通过 OPTIMIZER_TRACE（优化器跟踪）功能，可跟踪实际执行情况。具体命令行示例如下。

```
mysql> SET OPTIMIZER_TRACE = "enabled=on";
Query OK, 0 rows affected (0.00 sec)
mysql> SET OPTIMIZER_TRACE_MAX_MEM_SIZE = 1000000;
Query OK, 0 rows affected (0.00 sec)

mysql> SELECT * FROM employees WHERE birth_date>'1964-02-01';
mysql> SELECT * FROM INFORMATION_SCHEMA.OPTIMIZER_TRACE\G
```

图 3-26 所示为优化器跟踪下 cost 的对比，左侧通过全表扫描方式扫描了 30345 行，右侧通过索引扫描方式扫描了 23749 行，最终会选择 23749 行扫描的执行计划。

图 3-26　OPTIMIZER_TRACE 功能

说明：在该示例当中，发现 MySQL 服务重新启动之后执行计划不参考直方图，进行的是全表扫描，需要官方优化。

4. 总结

在大数据场景下，根据性能对比，直方图的使用可能使查询性能提升 2～3 倍。然而，由于直方图需要额外的内存消耗，因此在使用前需要对数据库环境和数据量进行有效的评估。在 MySQL 8.0 版本中，当发现优化器选择的执行计划不是最优时，可以考虑使用直方图来改进。在目前版本中，直方图功能仅提供了基础支持，但预计未来会加入更多高级功能，类似于 Oracle 数据库的实现。

3.4.3　资源组

MySQL 是一个单进程、多线程的程序，其中包括多个后台线程（如 Master Thread、IO Thread、Purge Thread 等），以及用户线程。在 MySQL 8.0 版本之前，所有线程的优先级都是一样的，并且所有线程共享资源。

1. 资源组介绍

在 MySQL 8.0 中，引入了资源组（Resource Group）功能，用于管理复杂 SQL 语句和运行批任务。这些任务通常运行时间长且资源消耗大，可能会对其他线程的运行造成影响。通过资源组，可以为不同的线程指定和限制 CPU 资源的使用，从而减少对业务线程的影响。需要注意的是，目前版本中资源组主要控制的是 CPU 资源，并且资源分配的最小单位是逻辑 CPU 核心（vCPU）。

MySQL 中的资源组管理以下功能。

- 允许创建、更改和删除资源组，并允许将线程分配给资源组。优化器提示可以指定单个语句应该分配到哪个资源组。

- 资源组权限提供对哪些用户可以执行资源组操作的控制。
- information_schema 库中的 RESOURCE_GROUPS 表提供了关于资源组定义的信息。
- performance_schema 库中的 thread 表显示了每个线程的资源组分配情况。

在 MySQL 8.0 版本中，提供了丰富的 SQL 接口以及相关的权限和信息架构来支持资源组的管理。表 3-21 所示为资源组支持的命令行列表。

表 3-21 资源组命令行

资源组命令行	说　　明
CREATE RESOURCE GROUP	用于创建新的资源组
ALTER RESOURCE GROUP	用于修改现有资源组的属性，如 CPU 配额、线程优先级等
DROP RESOURCE GROUP	用于删除不再需要的资源组
优化器提示（如 /*+ RESOURCE_GROUP（group_name）*/）	允许将单个 SQL 语句的执行分配给特定的资源组
SET RESOURCE GROUP	允许将当前会话的线程分配给特定的资源组

2. 资源组实现方式

下面介绍资源组的创建、查看，指定线程，优化器提示用法，以及修改资源组配置等。

MySQL 默认会创建两个资源组（RESOURCE_GROUP_NAME 列），一个是 USR_default，另一个是 SYS_default。首先创建一个新的资源组。创建和查看资源组的命令行示例如下。

```
# 创建资源组,指定优先级为 10
mysql> CREATE RESOURCE GROUP Batch TYPE = USER VCPU = 0 THREAD_PRIORITY = 10;
# 查看资源组
mysql> SELECT * FROM information_schema.resource_groups;
+---------------------+---------------------+-----------------------+----------+----------------+
| RESOURCE_GROUP_NAME | RESOURCE_GROUP_TYPE | RESOURCE_GROUP_ENABLED | VCPU_IDS | THREAD_PRIORITY |
+---------------------+---------------------+-----------------------+----------+----------------+
| USR_default         | USER                |                     1 | 0x302D30 |              0 |
| SYS_default         | SYSTEM              |                     1 | 0x302D30 |              0 |
| Batch               | USER                |                     1 | 0x30     |             10 |
+---------------------+---------------------+-----------------------+----------+----------------+
```

在 MySQL 中，通过指定使用资源组，可以将创建的 Batch 资源组绑定到执行的线程上。这通常通过以下两种方式实现。

(1) 方式一

从 PERFORMANCE_SCHEMA.THREADS 表中查找需要绑定执行的线程 ID。注意，PERFORMANCE_SCHEMA.THREADS 表中的 THREAD_ID 和 SHOW PROCESSLIST 的 ID 不等同。查看需要绑定的线程的 THREAD_ID 的命令行示例如下。

```
mysql>SELECT THREAD_ID,NAME,TYPE<PROCESSLIST_ID,PROCESSLIST_USER,
        PROCESSLIST_HOST,RESOURCE_GROUP
      FROM performance_schema.threads where TYPE='FOREGROUND';
+---------------------+---------------------+-----------------------+----------+----------------+
| RESOURCE_GROUP_NAME | RESOURCE_GROUP_TYPE | RESOURCE_GROUP_ENABLED | VCPU_IDS | THREAD_PRIORITY |
+---------------------+---------------------+-----------------------+----------+----------------+
```

```
|    43 | thread/sql/event_scheduler     | 1 | NULL   | NULL      | SYS_default |
|    45 | thread/sql/compress_gtid_table | 1 | NULL   | NULL      | SYS_default |
|   302 | thread/sql/one_connection      | 1 | root   | localhost | USR_default |
+-------+--------------------------------+---+--------+-----------+-------------+
```

将线程与 Batch 资源组进行绑定的命令行示例如下。

```
mysql> SET RESOURCE GROUP Batch FOR 302;
```

查看线程对应的资源组绑定结果的命令行示例如下。

```
mysql> select THREAD_ID,NAME,TYPE<PROCESSLIST_ID, PROCESSLIST_USER,
       PROCESSLIST_HOST,RESOURCE_GROUP
FROM performance_schema.threads where TYPE='FOREGROUND';
+-----------+--------------------------------+---------------------+------------------+------------------+----------------+
| THREAD_ID | NAME   | TYPE<PROCESSLIST_ID           | PROCESSLIST_USER | PROCESSLIST_HOST | RESOURCE_GROUP |
+-----------+--------------------------------+---------------------+------------------+------------------+----------------+
|    43     | thread/sql/event_scheduler     | 1 | NULL | NULL    | SYS_default    |
|    45     | thread/sql/compress_gtid_table | 1 | NULL | NULL    | SYS_default    |
|   302     | thread/sql/one_connection      | 1 | root | localho | Batch          |
+-----------+--------------------------------+---------------------+------------------+------------------+----------------+
```

（2）方式二

可采用优化器提示（Optimizer Hints）方式指定 SQL 语句使用的资源组。优化器提示方式绑定资源组的命令行示例如下。

```
mysql> SELECT /*+ RESOURCE_GROUP(Batch) */ * FROM t2;
```

在 MySQL 中，修改资源组配置是为了优化数据库的性能，确保资源能够按照业务需求和优先级进行分配。

当 CPU 资源不够的时候，可以修改资源组的配置。具体命令行示例如下。

```
mysql> ALTER RESOURCE GROUP Batch VCPU = 10-20;
```

修改资源组优先级的命令行示例如下。

```
mysql> ALTER RESOURCE GROUP Batch THREAD_PRIORITY = 5;
```

禁止使用某些资源组的命令行示例如下。

```
mysql> ALTER RESOURCE GROUP Batch DISABLE FORCE;
```

3. 资源组其他事项

在使用 MySQL 的资源组时，需要注意下列一些关键点，以确保有效地利用和管理资源。

- CREATE RESOURCE GROUP、ALTER RESOURCE GROUP 和 DROP RESOURCE GROUP 等 SQL 语句并不会被记录到 binlog 中，不会被复制到从库。
- 如果 MySQL 安装了线程池（Thread pool）插件，则无法使用资源组特性。
- 在 macOS 平台上，不支持资源组功能。
- 在 FreeBSD 和 Solaris 平台上，资源组优先级（priority）可能无法指定，所有线程可能默认在优先级 0 运行。
- 在 Linux 平台上，只有开启 CAP_SYS_NICE 特性才能使用资源组。开启 CAP_SYS_NICE 的命令行示例如下。

```
# 检查mysqld进程是否开启CAP_SYS_NICE特性
shell $>getcap /home/mysql/program/mysql8.0/bin/mysqld

# 给mysqld进程开启CAP_SYS_NICE特性
shell $>setcap cap_sys_nice+ep /home/mysql/program/mysql8.0/bin/mysqld

# 检查是否开启成功
shell $>getcap /home/mysql/program/mysql8.0/bin/mysqld
   /home/mysql/program/mysql8.0/bin/mysqld = cap_sys_nice+ep
```

- 在Windows平台上,线程优先级只有5个等级,分别为THREAD_PRIORITY_HIGHEST (−20~−10)、THREAD_PRIORITY_ABOVE_NORMAL (−9~1)、THREAD_PRIORITY_NORMAL (0)、THREAD_PRIORITY_BELOW_NORMAL (1~10)和THREAD_PRIORITY_LOWEST (11~19)。

4. 总结

MySQL 8.0版本中增加的资源管理特性为数据库资源的使用提供了更加灵活和便捷的管理方式。相比之下,Oracle在10g版本中就已经推出资源组特性,并在12c版本中进一步增强了其灵活性和易用性。据了解,在Oracle 12c中,CPU、内存和I/O都可以被指定到不同的资源组,并且还能够对SQL级别进行限制。虽然MySQL的资源组目前还相对简单,只能控制CPU资源,但相信随着版本的迭代,资源组能够管理和控制的资源类型会越来越多。

3.4.4　优化器提示

优化器是关系数据库的核心组件,它决定了SQL执行计划的优劣。在生产数据库中,由于数据不断变化以及估计准确性等因素,优化器并不总是能够产生最优的执行计划。在MySQL 8.0版本中,提供了优化器提示的功能,从而指导优化器选择特定的执行计划。

1. 优化器提示介绍

优化器提示方式通过向优化器的关键决策点提供具体决策,从而避免选择错误的执行计划,并能够迅速地解决执行计划走错的问题。在MySQL中,优化器提示的使用场景主要集中在系统上线前的SQL优化,以及系统上线后由于数据变化导致的执行计划变更上。当然,绝大部分运行中的执行计划还是相当准确的,但在某些特定情况下,使用优化器提示可以对性能进行微调,进一步提高查询效率。

在MySQL中,优化器提示可以应用于不同的作用域级别,这些级别通常与查询的结构和执行计划相关。表3-22所示为优化器提示四个作用域级别及其说明。

表3-22　MySQL优化器提示四个作用域级别及其说明

作用域级别	说明	作用域级别	说明
Global	影响全局语句	Table-level	影响查询块中的特定表
Query block	影响语句中的特定查询块	Index-level	影响表中的特定索引

表3-23所示为优化器提示支持的策略及其作用域。按照功能,可以将策略分类为连接顺序、表级优先、索引选择、子查询、语句执行时间、变量设置、资源组、不同算法应用和命名查询块。

表 3-23 MySQL 优化器提示支持的策略及其作用域

策略名	说明	作用域
BKA、NO_BKA	BAK 算法连接处理	Query block, Table-level
BNL、NO_BNL	在 MySQL 8.0.20 之前：影响块嵌套循环连接处理；在 MySQL 8.0.20 及以上版本：仅影响哈希连接优化	Query block, Table-level
DERIVED_CONDITION_PUSHDOWN、NO_DERIVED_CONDITION_PUSHDOWN	使用或忽略物化派生表的派生条件下推优化	Query block, Table-level
GROUP_INDEX、NO_GROUP_INDEX	GROUP BY 操作中使用或忽略索引扫描的指定索引	Index-level
HASH_JOIN、NO_HASH_JOIN	哈希连接优化	Query block, Table-level
INDEX、NO_INDEX	作为 JOIN_INDEX、GROUP_INDEX 和 ORDER_INDEX 的组合，或者作为 NO_JOIN_INDEX、NO_GROUP_INDEX 和 NO_ORDER_INDEX 的组合（在 MySQL 8.0.20 中添加）	Index-level
INDEX_MERGE、NO_INDEX_MERGE	索引合并优化	Table-level, Index-level
JOIN_FIXED_ORDER	使用 FROM 子句中指定的表顺序作为连接顺序	Query block
JOIN_INDEX、NO_JOIN_INDEX	使用或忽略任何访问方法的指定索引（在 MySQL 8.0.20 中添加）	Index-level
JOIN_ORDER	将指定的表顺序作为连接顺序	Query block
JOIN_PREFIX	设定第一个表顺序	Query block
JOIN_SUFFIX	设定最后一个表的顺序	Query block
MAX_EXECUTION_TIME	限制语句执行时间	Global
MERGE、NO_MERGE	派生表/视图合并到外部查询块	Table-level
MRR、NO_MRR	多范围读取优化	Table-level, Index-level
NO_ICP	索引条件下推优化	Table-level, Index-level
NO_RANGE_OPTIMIZATION	范围优化	Table-level, Index-level
ORDER_INDEX、NO_ORDER_INDEX	使用或忽略指定的索引来排序行（在 MySQL 8.0.20 中添加）	Index-level
QB_NAME	查询块分配名称	Query block
RESOURCE_GROUP	资源组	Global
SEMIJOIN、NO_SEMIJOIN	半连接策略。从 MySQL 8.0.17 开始，这也适用于反连接	Query block
SKIP_SCAN、NO_SKIP_SCAN	跳过扫描优化	Table-level, Index-level
SET_VAR	SET_VAR 在语句执行期间设置变量	Global
SUBQUERY	物化、IN-to-EXISTS 子查询策略	Query block

在 MySQL 8.0 版本中，优化器提示是通过/ * + ... * /风格的注释语法来实现的。这些提示允许用户为查询优化器提供关于如何执行查询的建议，从而可能影响查询的执行计划。以下是 MySQL 8.0 版本中优化器提示的具体语法：

```
/*+ BKA(t1) */
/*+ BNL(t1, t2) */
/*+ NO_RANGE_OPTIMIZATION(t4 PRIMARY) */
/*+ SET_VAR(join_buffer_size = 8M) */
/*+ MAX_EXECUTION_TIME(1000) */
/*+ BNL(t1) BKA(t2) */
/*+ QB_NAME(qb1) */
```

优化器提示不是在所有情况下都可以使用。下面是优化器提示的使用限制和规则。
- 优化器提示可以在 SELECT、UPDATE、INSERT、REPLACE 和 DELETE 语句中使用。同时，EXPLAIN 执行计划语句也是支持的。
- 对于重复的提示，如/＊+MRR（idx1）MRR（idx1）＊/，只会使用第一个提示，并发出警告，告知用户存在重复的提示。
- 查询块名称是标识符，这意味着必须遵循 MySQL 的标识符命名规则。例如，不能以数字开头、不能包含 MySQL 关键字等。
- 提示名称、查询块名称和策略名称在 MySQL 中是不区分大小写的。这意味着/＊+INDEX（t t_idx）＊/和/＊+index（t t_idx）＊/是等效的。
- 使用 SHOW WARNINGS 语句可以显示执行 SQL 语句时产生的警告，包括优化器提示可能产生的冲突或错误。

2. 优化器提示实现方式

下面介绍 MySQL 优化器提示的 8 种示例，以及它们是如何被用于影响查询优化器的决策的。

（1）Index_Level 提示

在 MySQL 8.0 版本之前，开发者使用 USE INDEX、IGNORE INDEX 或 FORCE INDEX（后跟索引名）来进行优化器提示。然而，从 MySQL 8.0 版本开始，这些优化器索引提示的使用方法变得更加简单和灵活。优化器索引提示语法（前面是表名，之后跟着多个索引名）如下所示。

```
hint_name([@query_block_name]tbl_name [index_name [, index_name] ...])
hint_name(tbl_name@query_block_name [index_name [, index_name] ...])
```

多个优化器索引提示的命令行示例如下。

```
#多个不使用索引提示
mysql> SELECT /*+ NO_INDEX(t1 PRIMARY, i_a, i_b, i_c) */ *
       FROM t1 WHERE a = 1 AND b = 2 AND c = 3 AND d = 4;
#多个索引合并提示
mysql> SELECT /*+ INDEX_MERGE(t1 i_a, i_b) NO_ORDER_INDEX(t1 i_b) */ *
       FROM t1 WHERE a = 1 AND b = 2 AND c = 3 AND d = 4;
#会话级别设置 optimizer_switch 结合场景
mysql> SET optimizer_switch='index_merge_intersection=off';
mysql> SELECT /*+ INDEX_MERGE(t1 i_b) */ *
       FROM t1 WHERE b = 1 AND c = 2 AND d = 3;
#优化器索引提示和传统索引提示的结合场景
mysql> SELECT /*+ INDEX_MERGE(t1 i_a, i_b, i_c) */ *
       FROM t1 IGNORE INDEX(i_a) WHERE b = 1 AND c = 2 AND d = 3;
```

当优化器索引提示和传统索引提示（如 USE INDEX、IGNORE INDEX 和 FORCE INDEX）同时存在时，优化器会优先考虑优化器索引提示。这是因为优化器索引提示提供了更细粒度的控制和更先进的优化逻辑。为了确认这一行为，可对比两种索引提示下的执行计划。具体对比的命令行示例如下。

```
mysql>EXPLAIN FORMAT=TREE SELECT en_name
       FROM city USE INDEX(idx_name) WHERE en_name='Guba';
+-----------------------------------------------------------------+
| EXPLAIN                                                         |
+-----------------------------------------------------------------+
```

```
       -> Covering index lookup on city using
                    idx_name (en_name='Guba')(cost=0.35 rows=1) |
+------------------------------------------------------------------------------------------+1
row in set (0.01 sec)
mysql> EXPLAIN FORMAT=TREE SELECT /*+ NO_INDEX(city idx_name) */ en_name
       FROM city USE INDEX(idx_name) WHERE en_name='Guba';
+------------------------------------------------------------------------------------------+
|EXPLAIN                                                                                   |
+------------------------------------------------------------------------------------------+|
-> Filter: (city.en_name = 'Guba')   (cost=3.6 rows=1)
   -> Table scan on city   (cost=3.6 rows=26)                                              |
+------------------------------------------------------------------------------------------+1
row in set (0.00 sec)
```

通过对比上述两个 EXPLAIN 命令的输出结果，可以观察优化器选择/*...*/方式的索引提示的执行计划。

(2) SET_VAR 提示

在 MySQL 8.0 之前，可以通过 SET 语句临时设置会话级别的用户参数，这些设置在单个会话的持续时间内有效，直到会话结束或被显式更改为止。现在，MySQL 8.0 版本提供了更多的灵活性和便利性，通过 SET_VAR 方式，允许在单个语句或事务中临时设置参数，而不需要在整个会话中保持这些更改。

SQL 语句级别设置优化器参数提示的 SET_VAR 的命令行示例如下。

```
SELECT /*+ SET_VAR(sort_buffer_size = 16M) */ * FROM city ORDER BY en_name;
INSERT /*+ SET_VAR(foreign_key_checks=OFF) */ INTO city VALUES(1);
SELECT /*+ SET_VAR(optimizer_switch='use_invisible_indexes=off') */ 1;
SELECT /*+ SET_VAR(optimizer_switch='mrr_cost_based=off')
           SET_VAR(max_heap_table_size=1G) */ 1;
```

并非所有的会话参数都允许通过 SET_VAR 命令来动态设置。如果出现 WARNING 提示，则应该通过命令行（SHOW WARNINGS）来获取更多信息。SET_VAR 方式的命令行示例如下。

```
mysql> SELECT /*+ SET_VAR(collation_server = 'utf8mb4') */ 1;
+-----+
| 1 |
+-----+
| 1 |
+-----+
1 row in set, 1 warning (0.00 sec)
mysql> SHOW WARNINGS;
+-------------+------+----------------------------------------------------------------+
|Level   |Code |Message                                                              |
+-------------+------+----------------------------------------------------------------+
|Warning|3637 |Variable 'collation_server' cannot be set using SET_VAR hint.        |
+-------------+------+----------------------------------------------------------------+
1 row in set (0.00 sec)
```

除此之外，如果同一个语句中出现了几个具有相同参数名的提示，则优化器参数提示会选择第一个提示，忽略其他提示并发出警告。在复制场景中，会忽略复制语句中的 SET_VAR 提示，以避免潜在的安全问题。

(3) Resource_Group 提示

MySQL 8.0 版本之后，由于引入了 Resource_Group 特性，因此可以通过设置 SQL 语句的优化器资源组提示，从而有效利用系统资源。例如，如果有一个需要高优先级的读操作，则可以将其分配到一个专门为读操作设计的资源组，以确保它能够在需要时获得足够的系统资源。同样，对于写操作或批量处理任务，可以创建不同的资源组，以满足不同的性能要求。

设置 SQL 语句级别的资源组提示的命令行示例如下。

```
# 指定资源组，执行用户需要 SUPER、RESOURCE_GROUP_ADMIN 或 RESOURCE_GROUP_USER 权限
mysql> SELECT /*+ RESOURCE_GROUP(USR_default) */ * FROM t1;
```

(4) Subquery 提示

在 MySQL 8.0 中，子查询（Subquery）优化得到了显著的提升，特别是在处理 IN 和 EXISTS 子查询时。MySQL 优化器现在会更智能地处理这些子查询，并且可以通过不同的策略来执行，包括半连接（semi-join）和传统的子查询策略。

在子查询的半连接策略里包含 4 种实现方式：DUPSWEEDOUT、FIRSTMATCH、LOOSESCAN 和 MATERIALIZATION。通过优化器子查询提示，可以选择不同的策略进行优化。Subquery 提示命令行示例如下。

```
mysql> EXPLAIN FORMAT=TREE SELECT   /*+ SEMIJOIN(@subq1 MATERIALIZATION, DUPSWEEDOUT)
       */ *
    FROM city
    WHERE CountryCode IN (SELECT /*+ QB_NAME(subq1) */ Code FROM country);
+----------------------------------------------------------------------------------+
| EXPLAIN                                                                          |
+----------------------------------------------------------------------------------+
-> Inner hash join (city.CountryCode = `<subquery2>`.`Code`)  (cost=17.8 rows=13)
    -> Table scan on city  (cost=1.26 rows=26)
    -> Hash
        -> Table scan on <subquery2>  (cost=1.76..3.81 rows=5)
            -> Materialize with deduplication  (cost=1.25..1.25 rows=5)
                -> Table scan on country  (cost=0.75 rows=5)         |
+----------------------------------------------------------------------------------+
1 row in set (0.00 sec)

mysql> EXPLAIN FORMAT=TREE SELECT
          /*+ NO_SEMIJOIN(@subq1 MATERIALIZATION, DUPSWEEDOUT) */ *
          FROM city
          WHERE CountryCode IN (SELECT /*+ QB_NAME(subq1) */ Code FROM country);
+----------------------------------------------------------------------------------+
| EXPLAIN                                                                          |
+----------------------------------------------------------------------------------+
-> Hash semijoin (country.`Code` = city.CountryCode)  (cost=13.3 rows=26)
    -> Table scan on city  (cost=3.6 rows=26)
    -> Hash
        -> Table scan on country  (cost=0.0135 rows=5)               |
+----------------------------------------------------------------------------------+
1 row in set (0.00 sec)
```

(5) Naming_Query_Blocks 提示

QB_NAME（Naming_Query_Blocks）提示是一种命名查询块的优化器提示，它允许用户为查询中的特定部分指定一个名称，这有助于理解和调试复杂的 SQL 查询。当有一个长且复杂的查询，其中包含多个子查询、连接和联合时，使用 QB_NAME 可以帮助用户更清晰地了解哪些优化器提示应用于哪些部分。QB_NAME 提示的语法如下。

```
mysql> SELECT /*+ QB_NAME(qb1) MRR(@qb1 t1) BKA(@qb2) NO_MRR(@qb3t1 idx1, id2)
*/ ...
    FROM (SELECT /*+ QB_NAME(qb2) */ ...
    FROM (SELECT /*+ QB_NAME(qb3) */ ... FROM ...)) ...
mysql> SELECT /*+ BKA(@`my hint name`) */ ...
    FROM (SELECT /*+ QB_NAME(`my hint name`) */ ...) ...
```

(6) Execution_Time 提示

Execution_Time 提示可为特定的 SELECT 语句设置一个最大执行时间（MAX_EXECUTION_TIME）限制。当查询的执行时间超过设置的时间限制时，MySQL 将终止该查询。这个优化器执行时间提示可以帮助防止长时间运行的查询阻塞数据库或消耗过多的资源。查询设置的最大执行时间的单位是毫秒（ms）。

注意以下几点。

- Execution_Time 提示只对 SELECT 语句有效，并且不支持在存储过程或函数中使用。
- 这个提示仅对查询的执行时间设置了一个上限，并不保证查询一定会在这个时间限制内完成。实际执行时间取决于许多因素，包括数据库负载、资源可用性、索引使用等。
- 设置的最大执行时间是在语句开始执行时计算的，而不是在查询的某个特定部分。这意味着如果查询在达到时间限制之前已经执行了很长时间，它仍然可能会被终止。
- Execution_Time 提示只对当前的查询会话有效，并且只对设置了该提示的特定查询有效。它不会影响其他查询或会话。

Execution_Time 提示的命令行示例如下。

```
mysql> SELECT /*+ MAX_EXECUTION_TIME(1000) */ * FROM t1 INNER JOIN t2 WHERE ...
```

(7) Join_Order 提示

在 MySQL 中，优化器在选择如何连接表时，会基于统计信息、索引、查询条件等多种因素来做出决策。然而，有时可能想要影响优化器的决策，或者确保它以特定的顺序连接表，这时，可以使用 Join_Order 提示来指导优化器。

连接提示包含 JOIN_FIXED_ORDER、JOIN_ORDER、JOIN_PREFIX 和 JOIN_SUFFIX 等方式。Join_Order 提示命令行示例如下。

```
mysql> SELECT
   /*+ JOIN_PREFIX(t2, t5@subq2, t4@subq1)
       JOIN_ORDER(t4@subq1, t3)
       JOIN_SUFFIX(t1) */
   COUNT(*) FROM t1 JOIN t2 JOIN t3
           WHERE t1.id IN (SELECT /*+ QB_NAME(subq1) */ id FROM t4)
           AND t2.id IN (SELECT /*+ QB_NAME(subq2) */ id FROM t5);
```

(8) Table_Level 提示

在 MySQL 中，表级（Table_Level）提示允许影响查询优化器处理特定的表或表连接的方式。

这些提示可以帮助优化器选择更有效的连接策略或影响派生表、视图和公共表表达式（Common Table Expressions，CTEs）的处理方式。

表级提示包含 BKA、NO_BKA、BNL、NO_BNL、DERIVED_CONDITION_PUSHDOWN、NO_DERIVED_CONDITION_PUSHDOWN、HASH_JOIN、NO_HASH_JOIN、MERGE 和 NO_MERGE 等方式。表级提示语法示例如下。

```
mysql> SELECT /*+ NO_BKA(t1, t2) */ t1.* FROM t1 INNER JOIN t2 INNER JOIN t3;
mysql> SELECT /*+ NO_BNL() BKA(t1) */ t1.* FROM t1 INNER JOIN t2 INNER JOIN t3;
mysql> SELECT /*+ NO_MERGE(dt) */ * FROM (SELECT * FROM t1) AS dt;
```

说明：在优化器表级提示场景下，考虑的因素还是驱动表、多表合并等。

3. 总结

优化器提示是在 MySQL 8.0 版本中优化 SQL 的另一种深度技术手段。语句级别实现，多种组合策略选择，更深入地控制了优化器成本模型决策。在使用优化器提示之前，建议进行充分的测试，确保 SQL 语句提高了查询性能，并且不会影响其他查询或系统性能。

3.4.5 新增的优化器行为标志

在 MySQL 中，optimizer_switch 参数是一个用于控制查询优化器行为的机制。这个参数允许用户通过打开或关闭不同的行为标志来调整查询优化器的行为。每个行为标志都有一个 ON 或 OFF 值，用于控制查询优化器在生成执行计划时使用的各种启发式算法。

以下是在 MySQL 8.0 版本中新增的优化器行为标志。

- 哈希连接（hash_join）。
- 隐藏索引（use_invisible_indexes）。
- 排序索引（prefer_ordering_index）。
- 子查询转换为派生表（subquery_to_derived）。
- 派生条件回移（derived_condition_pushdown）。
- 跳过扫描（skip_scan）。
- 超图（hypergraph_optimizer）。

下面介绍新增的优化器行为标志（除了人们熟悉的 hash_join 以外）。

1. use_invisible_indexes

use_invisible_indexes 行为标志用来控制查询优化器是否不隐藏索引来构建查询的执行计划。隐藏索引是一种特殊类型的索引，在优化器选择索引时默认是不可见的。这允许保留索引以供将来的使用，同时避免它在当前的查询优化中被选中。这对于那些知道不是最优的索引，但又不想立即删除的情况特别有用。use_invisible_indexes 行为标志的命令行示例如下。

```
# 模拟表结构和数据
mysql> CREATE table `members` (
    id int unsigned NOT NULL AUTO_INCREMENT,
    first_name varchar(100) DEFAULT NULL,
    last_name varchar(100) DEFAULT NULL,
    age INT DEFAULT '0',
    create_time timestamp DEFAULT CURRENT_TIMESTAMP,
    update_time timestamp DEFAULT current_timestamp() ON UPDATE current_timestamp(),
```

```
    primary KEY (id),KEY idx_first_last(first_name(3),last_name(3)),KEY idx_age(age)
);
mysql> INSERT INTO `members`(id,first_name,last_name,age)
    VALUES(1,'AAAAA','BBBBB',12),(2,'AAAAA','CCCCC',20),(3,'DDDDD','EEEEE',30);

# 索引 age 隐藏
mysql> ALTER TABLE members ALTER INDEX idx_age INVISIBLE;
Query OK, 0 rows affected (0.01 sec)
Records: 0  Duplicates: 0  Warnings: 0

# 执行计划全表扫描
mysql> EXPLAIN FORMAT=TREE SELECT * FROM members WHERE age > 10;
+-------------------------------------------------------------------------+
| EXPLAIN                                                                 |
+-------------------------------------------------------------------------+
| -> Filter: (members.age > 10)  (cost=0.55 rows=1)
    -> Table scan on members   (cost=0.55 rows=3) |
+-------------------------------------------------------------------------+
1 row in set (0.00 sec)
# 优化器可用隐藏索引打开
mysql> SET SESSION optimizer_switch='use_invisible_indexes=on';
Query OK, 0 rows affected (0.00 sec)

# 执行计划使用 age 隐藏索引
mysql> EXPLAIN FORMAT=TREE SELECT * FROM members WHERE age > 10;
+----------------------------------------------------------------------------------+
| EXPLAIN                                                                          |
+----------------------------------------------------------------------------------+
| -> Index range scan on members using idx_age over (10 < age), with index condition: (members.
age > 10)   (cost=1.61 rows=3) |
+----------------------------------------------------------------------------------+
1 row in set (0.00 sec)
```

说明：当业务逻辑复杂、SQL 语句难以快速更改，或者对索引的性能影响不确定时，创建隐藏索引，并打开 use_invisible_indexes 标志，可以作为一种应急对策来尝试改善查询性能。

2. prefer_ordering_index

prefer_ordering_index 行为标志影响优化器为带有 LIMIT 子句的 ORDER BY 或 GROUP BY 查询选择执行计划的方式。当 prefer_ordering_index 设置为 ON 时，优化器会倾向于使用与 ORDER BY 或 GROUP BY 子句中列的顺序相匹配的索引（称为有序索引），即使其他可能的执行计划会更快。因此它可能并不总是完全正确的。从 MySQL 8.0.21 版本开始，对 optimizer_switch 参数的 prefer_ordering_index 行为标志默认禁用这种优化算法。prefer_ordering_index 行为标志的命令行示例如下。

```
# 排序优先开启
mysql>SET optimizer_switch = "prefer_ordering_index=on";
# 优先 ORDER BY 和 LIMIT 选择主键
mysql> EXPLAIN FORMAT=TREE SELECT *
    FROM members WHERE age> 15 order by `id` DESC LIMIT 1;
+-----------------------------------------------------------------------------------------+
```

```
| EXPLAIN                                                                              |
+--------------------------------------------------------------------------------------+
|-> Limit: 1 row(s)  (cost=0.161 rows=0.333)
    -> Filter: (members.age > 15)  (cost=0.161 rows=0.333)
        -> Index scan on members using PRIMARY (reverse)  (cost=0.161 rows=1) |
+--------------------------------------------------------------------------------------+
1 row in set (0.00 sec)

# 排序优先关闭
mysql> SET optimizer_switch = "prefer_ordering_index=off";
# 选择合理执行计划
mysql> EXPLAIN FORMAT=TREE SELECT *
       FROM members WHERE age> 15 order by `id` DESC LIMIT 1;
+--------------------------------------------------------------------------------------+
| EXPLAIN                                                                              |
+--------------------------------------------------------------------------------------+
|-> Limit: 1 row(s)  (cost=0.55 rows=1)
   -> Sort: members.id DESC, limit input to 1 row(s) per chunk  (cost=0.55 rows=3)
       -> Filter: (members.age > 15)  (cost=0.55 rows=3)
           -> Table scan on members  (cost=0.55 rows=3) |
+--------------------------------------------------------------------------------------+
1 row in set (0.00 sec)
```

说明：prefer_ordering_index 行为标志目前发现只能在范围查询中使用。

3. subquery_to_derived

从 MySQL 8.0.21 版本开始，优化器增加了 subquery_to_derived 行为标志（默认关闭）。优化器在许多情况下能够将 SELECT、WHERE、JOIN 或 HAVING 子句中的标量子查询转换为派生表上的左外连接（在某些情况下是内连接），这种转换有助于优化器更有效地执行查询，特别是在涉及复杂连接和过滤的情况下。其中：

- subquery 是包含在 SELECT 中的子查询（不在 FROM 子句中）。
- derived 是包含在 FROM 子句中的子查询。MySQL 会将结果存放在一个临时表中，也称为派生表（derived）。

在大多数情况下，启用此行为标志不会产生明显的性能改进（在许多情况下，甚至会使查询运行得更慢），但还要视情况而定，所以选择默认值即可。

subquery_to_derived 行为标志可以在 SQL 语句执行之后，通过 SHOW WARNINGS 信息查看改写情况。具体命令行示例如下。

```
mysql> SET SESSION optimizer_switch='subquery_to_derived=OFF';
    Query OK, 0 rows affected (0.00 sec)

mysql> EXPLAIN FORMAT=TREE SELECT * FROM t1 WHERE t1.a > (SELECT COUNT(a) FROM t2);
+--------------------------------------------------------------------------------------+
| EXPLAIN                                                                              |
+--------------------------------------------------------------------------------------+
|-> Filter: (t1.a > (select #2))  (cost=0.383 rows=1.33)
    -> Table scan on t1  (cost=0.383 rows=4)
    -> Select #2 (subquery in condition。run only once)
```

```
            -> Aggregate: count(t2.a)   (cost=0.65 rows=1)
                -> Table scan on t2   (cost=0.45 rows=2)              |
+------------------------------------------------------------------------+
1 row in set (0.00 sec)
mysql> SHOW WARNINGS;
+----------+------+----------------------------------------------------+
|Level  |Code  |Message                                                |
+----------+------+----------------------------------------------------+
|Note   |1003  |/* select#1 */ select `test`.`t1`.`a` AS `a`
      from `test`.`t1` where (`test`.`t1`.`a` > (/* select#2 */ select count(`test`.`t2`.`a`)
from `test`.`t2`)) |
+----------+------+----------------------------------------------------+
1 row in set (0.00 sec)

mysql> SET SESSION optimizer_switch='subquery_to_derived=ON';
mysql> EXPLAIN FORMAT=TREE SELECT * FROM t1  WHERE t1.a > (SELECT COUNT(a) FROM t2);
+------------------------------------------------------------------------+
|EXPLAIN                                                                 |
+------------------------------------------------------------------------+
   |-> Filter: (t1.a > derived_1_2.`COUNT(a)`)   (cost=3.91 rows=1.33)
      -> Inner hash join (no condition)   (cost=3.91 rows=1.33)
         -> Table scan on t1   (cost=0.383 rows=4)
         -> Hash
            -> Table scan on derived_1_2   (cost=3.26..3.26 rows=1)
               -> Materialize   (cost=0.75..0.75 rows=1)
                  -> Aggregate: count(t2.a)   (cost=0.65 rows=1)
                     -> Table scan on t2   (cost=0.45 rows=2) |
+------------------------------------------------------------------------+
1 row in set (0.00 sec)
mysql>  SHOW WARNINGS;
+----------+------+----------------------------------------------------+
|Level  |Code  |Message                                                |
+----------+------+----------------------------------------------------+
|Note   |1003  |/* select#1 */ select `test`.`t1`.`a` AS `a` from `test`.`t1` join (/* select
#2 */ select count(`test`.`t2`.`a`) AS `COUNT(a)` from `test`.`t2`) `derived_1_2` where (`test`.
`t1`.`a` > `derived_1_2`.`COUNT(a)`) |
+----------+------+----------------------------------------------------+
1 row in set (0.00 sec)
```

说明：从上述 WARNINGS 信息中可以看出，当 subquery_to_derived=ON 时，生成 derived_1_2 表以进行关联。

4. derived_condition_pushdown

MySQL 8.0.22 版本中引入了派生条件回移（Derived Condition Pushdown）的优化功能。derived_condition_pushdown 行为标志通过减少派生表处理的行数来提高查询执行的效率。

derived_condition_pushdown 行为标志优化的 SQL 语句如下。

```
mysql> SELECT * FROM (SELECT i, j FROM t1) AS dt
    WHERE i > constant;
# 通过派生条件回移优化后出现类似如下形式(将 WHERE 条件放到派生表里)
mysql> SELECT * FROM (SELECT i, j FROM t1 WHERE i > constant);
```

从上述语句变化中可以看出，将外部 WHERE 条件应用到派生表的处理过程中，通过减少需要处理的行数来加快查询的执行速度。

从 MySQL 8.0.29 版本开始，派生表条件下推优化可以用于 UNION 查询。不过，也会在一些情况下条件不能下推，如派生表是内部表、包含子查询的条件、使用了公共表表达式、存在 LIMIT 子句、涉及变量的赋值等场景。

5. skip_scan

从 MySQL 8.0.13 版本开始，当执行范围扫描且 WHERE 条件没有匹配索引前缀时，MySQL 支持一种新的跳跃范围扫描（Skip Scan Range）方式。优化器是否能使用这种新的扫描方式，取决于特定的行为标志 skip_scan（默认值是 ON）。

skip_scan 行为标志的命令行示例如下。

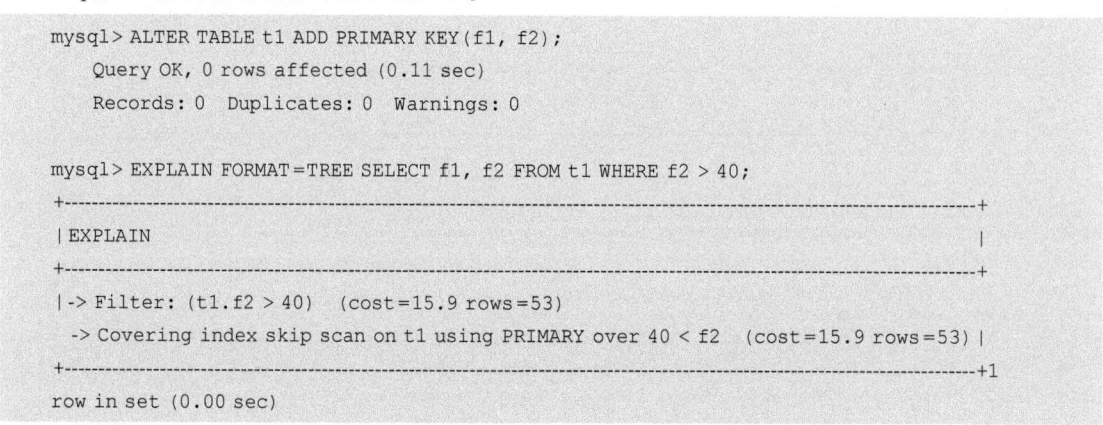

从上述执行计划中可以看出，虽然组合索引前缀是 f1，但还是使用索引。

6. hypergraph_optimizer

从 MySQL 8.0.23 版本开始，官方宣布优化器将支持基于 hypergraph（超图）的模型。然而，目前这个特性在实际应用中还不被支持，仅能在 Debug 模式下进行验证和测试。

在数学中，hypergraph 是一个更广义的图的概念，用于描述对象之间更复杂的关系。在数据算法中，hypergraph 可以用于表达数据点之间的多种关联方式，这些关联可能不是简单的点对点关系，而是可以涉及多个数据点的复杂集合。图 3-27 所示就是通过边界算法将相关的数据点、范围等标识出来。

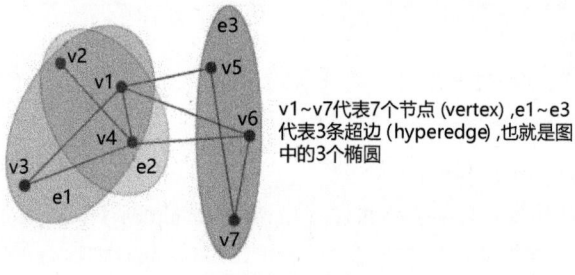

图 3-27 超图关系实现方式

查看优化器行为标志 hypergraph_optimizer 默认是 OFF。通过命令行更改 hypergraph_optimizer 的行为标志，提示不是 Debug 模式。查看和变更 hypergraph_optimizer 行为标志的命令行示例如下。

```
# 查看优化器行为标志
mysql>SHOW VARIABLES LIKE 'optimizer_switch' \G;
*************************** 1. row ***************************
Variable_name: optimizer_switch
        Value: index_merge=on,index_merge_union=on,index_merge_sort_union=on,index_
merge_intersection=on,engine_condition_pushdown=on,index_condition_pushdown=on,mrr=on,mrr_
cost_based=on,block_nested_loop=on,batched_key_access=off,materialization=on,semijoin=on,
loosescan=on,firstmatch=on,duplicateweedout=on,subquery_materialization_cost_based=on,use
_index_extensions=on,condition_fanout_filter=on,derived_merge=on,use_invisible_indexes=
off,skip_scan=on,hash_join=on,subquery_to_derived=off,prefer_ordering_index=on,hypergraph_
optimizer=off,derived_condition_pushdown=on
1 row in set (0.00 sec)
#
mysql> SET SESSION optimizer_switch='hypergraph_optimizer=ON';
ERROR 3999 (42000): The hypergraph optimizer does not yet support 'use in non-debug builds'
```

在 MySQL 中，多表关联的处理方法包括 Nest Loop Join 和 Hash Join 等算法。其中，Hash Join 在 MySQL 8.0 版本中得到了更多的优化。此外，目前还处在开发和测试阶段中的优化方法——hypergraph算法，特别适合利用 B+树索引的场景。

7. 总结

优化器行为的变更可能会导致 MySQL 不同版本之间 SQL 语句的性能表现不同。尽管大多数更改旨在提高性能，但在某些复杂查询的情况下，也可能出现性能下降的情况。因此，在升级 MySQL 版本时，需要特别关注优化器的变化。在此，作者对名为 hypergraph（超图）的优化方案特别感兴趣，因为它有可能通过结合 B+树索引来显著提高多表关联查询的性能。然而，需要注意的是，该方案的具体实现和性能提升程度还需要进一步验证和评估。

3.4.6 DDL 即时操作

从 MySQL 8.0.21 版本开始，引入了 DDL 即时（INSTANT）操作特性，这一功能是由腾讯游戏 DBA 团队贡献的。

MySQL 的 DDL 的 INSTANT 操作旨在优化数据定义语言（DDL）操作，使得某些 DDL 操作能够立即完成，而不需要长时间锁定表或元数据。这种即时操作能够减少 DDL 变更对正在运行的应用程序的影响，因为可以在不阻塞其他数据操纵语言（DML）语句的情况下完成。

INSTANT 操作支持如下 DDL 操作类型。

- 即时添加列。可以即时添加一个新列到表中，而不需要重建整个表，也不会锁定表。
- 添加或删除虚拟列。虚拟列不会实际存储数据，只存在于表的元数据中，用于在查询中计算和展示数据。因此，添加或删除虚拟列可以即时完成，不需要修改实际的数据行。
- 添加或删除列默认值。列默认值可以被即时添加或删除，不会改变表中已存在的数据，因此这个操作也是即时的，不会锁定表。
- 修改 ENUM 或 SET 列的定义。对于 ENUM 和 SET 类型的列，可以修改其定义，包括添加、删除或重新排序枚举值。这种修改也是即时的，因为它只改变列的元数据，而不会改变表中已经存在的数据。
- 更改索引类型。某些情况下，可以即时更改索引的类型。但是，并非所有索引类型之间的转换都支持即时操作，这取决于具体的索引类型和存储引擎。

- 重命名表。这个操作会立即更新元数据,使得新名称可用于访问表,而旧名称不再可用。需要注意的是,在重命名过程中,表的实际数据不会被移动或复制。

DDL 的 INSTANT 操作由 ALTER 命令的 ALGORITHM 算法参数指定。INSTANT 语法如下。

```
ALTER TABLE tbl_name
[alter_specification [, alter_specification] ...]
[partition_options]
ALGORITHM [=] {DEFAULT | INSTANT | INPLACE | COPY}
```

下面是 DDL 的 INSTANT 操作支持的命令行示例。

```
# 模拟表
mysql> CREATE TABLE `t_temp` (
  `id` bigint(20) NOT NULL AUTO_INCREMENT,
  `name` varchar(255) NOT NULL,
  `age` tinyint(4) NOT NULL,
  `create_time` datetime NOT NULL,
  `update_time` datetime NOT NULL,
  PRIMARY KEY (`id`),
  key (name)
);

# 支持添加一个字段,加到末尾
mysql> ALTER TABLE t_temp ADD addr varchar(10),ALGORITHM=INSTANT;
mysql> ALTER TABLE t_temp ADD COLUMN sex ENUM('M','F'), ALGORITHM = INSTANT;

# 支持初始化值
mysql> ALTER TABLE t_temp ALTER COLUMN age SET DEFAULT 10, ALGORITHM = INSTANT;

# 支持重命名表名
mysql> ALTER TABLE t_temp1 RENAME TO t_temp2, ALGORITHM = INSTANT;

# 支持变化长度
mysql> ALTER TABLE t_temp MODIFY COLUMN addr varchar(10),ALGORITHM=INSTANT;

# 支持 DROP 字段
mysql> ALTER TABLE t_temp DROP COLUMN addr, ALGORITHM = INSTANT;
```

下面是对索引的 DDL 的 INSTANT 操作的命令行示例。

```
# 支持主键操作
mysql> ALTER TABLE t_temp DROP PRIMARY KEY, ADD PRIMARY KEY(id),ALGORITHM = INSTANT;

# 支持对存在的索引更改
mysql> ALTER TABLE t_temp DROP INDEX idx_nm, ADD INDEX idx_nm(name),ALGORITHM = INSTANT;

# 支持索引长度变更
mysql> ALTER TABLE t_temp DROP INDEX idx_nm, ADD INDEX idx_nm(name(2)),ALGORITHM = INSTANT;

# 支持唯一索引
```

```
mysql> ALTER TABLE t_temp ADD UNIQUE INDEX index_nm3(name);
mysql> ALTER TABLE t_temp DROP INDEX index_nm3, ADD INDEX
index_nm3(name),ALGORITHM = INSTANT;

# 支持全文索引
mysql> ALTER TABLE t_temp ADD FULLTEXT index_nm4(name);
mysql> ALTER TABLE t_temp DROP INDEX index_nm4, ADD FULLTEXT index_nm4(name),ALGORITHM = IN-
STANT;

# 支持 InnoDB 行格式变更。但不能是 COMPRESSED 或行溢出
mysql> ALTER TABLE t_temp ROW_FORMAT = DYNAMIC;

# 不支持外键。在外键关联的时候,不允许删除
mysql> ALTER TABLE t_temp DROP INDEX index_nm5, ADD INDEX
index_nm5(name),ALGORITHM = INSTANT;

# 不支持无索引的列,添加索引失败
mysql> ALTER TABLE t_temp ADD INDEX addr(addr),ALGORITHM = INSTANT;
ERROR 1845 (0A000): ALGORITHM=INSTANT is not supported for this operation. Try ALGORITHM=
COPY/INPLACE.
```

下面是 DDL 的 INSTANT 操作限制。

- 只能顺序加列,新列只能被添加到表的末尾,不能在现有列的中间插入新列。这是因为 INSTANT 操作要求不修改或移动现有的数据行,而中间插入列会需要更改数据行的结构,因此不支持这种操作。
- 不支持临时表,因为临时表通常用于存储短暂的、会话特定的数据,并且临时表的生命周期与数据库连接相关联。由于这种短暂性和会话特定的性质,MySQL 可能认为在这些表上应用 INSTANT 操作没有太大的价值,因此不支持它。临时表只能使用复制方式执行 DDL 语句。
- 不支持那些在数据词典表空间中创建的表。这是因为数据词典表空间中的表有自己的文件,并且 MySQL 可能需要更复杂的操作来修改这些表的结构,这超出了 INSTANT 操作的范围。
- 数据字典中的表不能使用 INSTANT 算法。MySQL 的数据字典中存储了关于数据库对象的元数据。直接修改数据字典中的表通常需要更复杂的操作,因为这些更改可能会影响整个数据库的结构和完整性。

MySQL 8.0 版本中的 INSTANT 操作用于优化 DDL 操作,特别是添加新列的操作,通过避免移动或复制现有数据行来实现表的快速修改列的功能,从而方便了数据库日常元数据的更改,提高了性能。

3.4.7 增强密码机制

对于一款数据库来说,数据库系统访问控制是非常重要的。从 MySQL 8.0.4 版本开始,MySQL 默认身份验证插件从 mysql_native_password 改为 caching_sha2_password。相应地,libmysqlclient 也使用 caching_sha2_password 作为默认的身份验证机制。当然,在 MySQL 8.0 版本中,两种密码机制都是得到支持的。除此之外,在用户安全控制方面,添加了一些更加实用的功能。

1. 密码重用策略

MySQL 允许对重用以前的密码进行限制。重用限制可以基于密码更改次数和时间确定。可以全局地建立重用策略，并且可以将单个账户设置为遵循全局策略或使用特定的某个账户行为覆盖全局策略。

- 账户的密码历史记录由过去分配的密码组成。MySQL 可以限制从这个历史记录中选择新密码。例如，如果密码修改次数设置为 3，则新密码不能与最近 3 次的密码相同。
- 根据时间限制账户，即不能从历史记录中更新于指定天数的密码中选择新密码。例如，如果密码重复使用间隔设置为 60 天，则新密码不能是最近 60 天内所选择的密码。

表 3-24 所示为 MySQL 8.0 密码重用策略的控制参数。

表 3-24　MySQL 8.0 密码重用策略的控制参数

参数	说明
password_history	用于根据所需的最少密码更改次数控制以前密码的重用。也就是说，密码多次变更后，可以重新使用之前的密码
password_reuse_interval	根据经过的时间控制以前密码的重用

密码重用策略设置方式和命令行示例如下。

```
# 在 my.cnf 配置文件中修改
[mysqld]
password_history=6
password_reuse_interval=365

# 全局设置密码重用策略
mysql>SET PERSIST password_history = 6;
mysql>SET PERSIST password_reuse_interval = 365;

# 采取全局值密码重用策略
mysql> CREATE USER 'tuser'@'localhost'
  PASSWORD HISTORY DEFAULT PASSWORD REUSE INTERVAL DEFAULT;
mysql> ALTER USER 'tuser'@'localhost'
  PASSWORD HISTORY DEFAULT PASSWORD REUSE INTERVAL DEFAULT;

# 单个用户设置密码重用策略
mysql>CREATE USER 'tuser'@'localhost'
     PASSWORD HISTORY 5 PASSWORD REUSE INTERVAL 100 DAY;
mysql> ALTER USER 'tuser'@'localhost'
     PASSWORD HISTORY 10 PASSWORD REUSE INTERVAL 365 DAY;

# 同一个密码多次使用错误提示
mysql> ALTER USER 'tuser'@'localhost' IDENTIFIED BY '123456';
ERROR 3638 (HY000): Cannot use these credentials for 'tuser@localhost'
              because they contradict the password history policy
```

说明：从上述操作中可以得知，MySQL 8.0 版本会保留密码历史记录，以便实施密码重用的限制功能。

2. 验证旧密码策略

从 MySQL 8.0.13 版本开始，可以指定要替换的当前密码来验证更改账户密码。这使得 DBA 能够防止用户在没有证明当前密码的情况下更改密码。

- 如果账户设置为 PASSWORD REQUIRED CURRENT，则密码更改必须指定当前密码。
- 如果账户设置为 PASSWORD REQUIRED CURRENT OPTIONAL，则密码更改不需要指定当前密码。

表 3-25 所示为 MySQL 8.0 版本验证旧密码的策略。

表 3-25　MySQL 8.0 验证旧密码策略

账户设置	设置值	更改时是否需要当前密码
PASSWORD REQUIRE CURRENT	OFF	是
PASSWORD REQUIRE CURRENT	ON	是
PASSWORD REQUIRE CURRENT OPTIONAL	OFF	否
PASSWORD REQUIRE CURRENT OPTIONAL	ON	否
PASSWORD REQUIRE CURRENT DEFAULT	OFF	否
PASSWORD REQUIRE CURRENT DEFAULT	ON	是

验证旧密码设置方式和命令行示例如下。

```
# 在my.cnf配置文件中修改
[mysqld]
password_require_current=ON

# 以命令行方式持久化到配置文件中
mysql> SET PERSIST password_require_current = ON;

# 创建密码当前验证用户
mysql> DROP USER if EXISTS 'tuser'@'localhost';
mysql> CREATE USER 'tuser'@'localhost' PASSWORD REQUIRE CURRENT;
mysql> ALTER USER 'tuser'@'localhost' PASSWORD REQUIRE CURRENT;

# 更改密码,提示需要旧密码输入
mysql> SELECT USER();
+-----------------+
| USER()          |
+-----------------+
| tuser@localhost |
+-----------------+
1 row in set (0.00 sec)
mysql> ALTER USER USER() IDENTIFIED BY '1234567';
ERROR 3892 (HY000): Current password needs to be specified in the REPLACE clause in order to change it.

mysql> ALTER USER USER() IDENTIFIED BY '1234567' REPLACE '123456';
Query OK, 0 rows affected (0.00 sec)
```

上述操作中，使用 REPLACE 方式验证旧密码语句，在 binlog 记录里省略原密码，以避免向其写入明文密码。下面通过 mysqlbinlog 工具分析密码更改的 binlog 内容，密码更改会以加密的哈希值形式记录在 binlog 中，如下所示。

```
# at 4874
# 230302 13:06:55 server id 129  end_log_pos 5109 CRC32 0x4430efad  Query
    thread_id=16exec_time=0error_code=0Xid = 154
SET TIMESTAMP=1677733615.932292/*!*/;
ALTER USER 'tuser'@'localhost' IDENTIFIED WITH 'mysql_native_password' AS '*
6A7A490FB9DC8C33C2B025A91737077A7E9CC5E5'
/*!*/;
SET @@SESSION.GTID_NEXT='AUTOMATIC'/* added by mysqlbinlog *//*!*/;
DELIMITER;
# End of log file
```

3. 双重密码支持

从 MySQL 8.0.14 版本开始，允许用户账户具有双重密码，即用户设置主密码和辅助密码。这样做的好处是，在 DBA 变动时，如果没有管理密码的相关记录，就不需要使用 skip-grant-tables 重新启动服务，而是通过辅助密码进行密码重置。当然，在日常数据库维护中，只允许使用主密码。

双重密码设置有如下注意事项。

- 当分配新的主密码时，ALTER USER 和 SET PASSWORD 语句的 RETAIN CURRENT PASSWORD 子句会将账户当前密码保存为其辅助密码。
- ALTER USER 的 DISCARD OLD PASSWORD 子句丢弃账户辅助密码，只保留主密码。
- 管理员操作辅助密码，需要 APPLICATION_PASSWORD_ADMIN 权限。

辅助密码设置的命令行示例如下。

```
# 添加辅助密码
mysql>ALTER USER 'tuser'@'localhost'
    IDENTIFIED BY 'abc123' RETAIN CURRENT PASSWORD;

# 丢弃账户辅助密码
    mysql> ALTER USER 'tuser'@'localhost' DISCARD OLD PASSWORD;
```

辅助密码记录在 mysql.user 表的 User_attributes 属性上面。查看辅助密码记录信息的命令行示例如下。

```
mysql> SELECT * FROM mysql.user WHERE user='tester'\G
*************************** 1. row ***************************
              Host: localhost
              User: tuser
...
            plugin: mysql_native_password
authentication_string: *6691484EA6B50DDDE1926A220DA01FA9E575C18A
...
    User_attributes: {"additional_password":
"*6BB4837EB74329105EE4568DDA7DC67ED2CA2AD9"}
```

4. 随机密码生成

从 MySQL 8.0.18 版本开始，CREATE USER、ALTER USER 和 SET PASSWORD 语句可以为用户账户生成随机密码，而不需要数据库管理员指定的明文密码。这大大简化了密码管理过程，尤其是当需要创建大量用户账户时,无须为每个账户设置和记录复杂的明文密码。默认情况下,生成的随机密码长度为 20 个字符。这个长度可以通过 generated_random_password_length 系统变量进行控制。该参数的有效值范围为 5~255 个字符,允许数据库管理员根据需要调整密码的长度。

在通过 RANDOM 语句为用户生成随机密码时,新密码会被存储在 MySQL 的 mysql.user 系统表中。密码会根据账户的身份验证插件进行适当的哈希处理,以确保其在存储时是加密的。这样,即使密码在生成后是明文返回的(作为执行结果的一部分),存储在数据库中的也仍然是加密后的密码。

使用 RANDOM 方式创建用户随机密码的命令行示例如下。

```
mysql> CREATE USER 'tuser'@'localhost' IDENTIFIED BY RANDOM PASSWORD;
+--------+-----------+--------------------+------------+
| user   | host      | generated password | auth_factor|
+--------+-----------+--------------------+------------+
| tester | localhost | dcG0UQbjr<,lbk:x:%.:|          1 |
+--------+-----------+--------------------+------------+
1 row in set (0.00 sec)
```

需要注意的是,虽然随机生成的密码可以简化初始密码设置的过程,但在实际应用中,通常建议用户在收到随机生成的密码后,尽快登录并更改为自己的密码,以增强账户的安全性。

5. 登录失败跟踪

从 MySQL 8.0.19 版本开始,数据库管理员可以配置用户账户在连续多次登录失败后被暂时锁定。登录失败指的是客户端在尝试连接时未能提供正确的密码。这种配置不包括由于未知用户、网络问题或其他非密码错误原因导致的连接失败。在标准的 MySQL 配置中,每个用户账户只能有一个与之关联的密码,因此,只要提供正确的密码,用户就可以成功登录。

使用 CREATE USER 和 ALTER USER 语句中的 FAILED_LOGIN_ATTEMPTS 与 PASSWORD_LOCK_TIME 选项可以为每个账户都配置所需的登录失败次数和锁定时间。

- FAILED_LOGIN_ATTEMPTS:此选项指示是否跟踪指定错误密码的账户登录尝试,可指定连续错误密码导致临时账户锁定的次数。
- PASSWORD_LOCK_TIME:此选项指定在连续多次登录尝试提供错误密码后锁定账户的时间。其值可指定为账户保持锁定的天数,或者 UNBOUNDED(用于指定当账户进入临时锁定状态时,该状态的持续时间是无限的,直到账户解锁后才会结束)。

用户设置登录失败次数的命令行示例如下。

```
mysql> CREATE USER 'tuser'@'localhost' IDENTIFIED BY '******'
       FAILED_LOGIN_ATTEMPTS 3 PASSWORD_LOCK_TIME 7;
mysql> FLUSH PRIVILEGES;
```

为创建的用户验证登录 3 次失败的命令行示例如下。

```
# 第一次登录失败
shell# mysql -utuser-p -S /opt/data8.0/data/mysql.sock
Enter password:
ERROR 1045 (28000): Access denied for user 'tuser'@'localhost' (using password: YES)

# 第二次登录失败
shel# mysql -utuser-p -S /opt/data8.0/data/mysql.sock
Enter password:
```

```
ERROR 1045 (28000): Access denied for user 'tuser'@'localhost' (using pas sword: YES)

#第三次登录失败
shel# mysql -utuser-p -S /opt/data8.0/data/mysql.sock
Enter password:
ERROR 3955 (HY000): Access denied for user 'tuser'@'localhost'. Account is blocked for 3 day
(s) (3 day(s) remaining) due to 3 consecutive failed logins.
```

上述最终提示中显示，由于连续 3 次登录失败，因此账户被阻止登录 3 天。

重置用户登录失败的次数可分为以下两种情况。

1）对所有账号进行全局重置登录失败次数。

- MySQL 服务重新启动。
- 执行 FLUSH PRIVILEGES 命令行。
- 如果使用 skip-grant-tables 授权表启动服务器，则会导致不读取授权表，从而禁用失败的登录跟踪。

2）单独用户重置登录失败次数。

- 此用户登录成功。
- 锁定持续时间已过。在这种情况下，失败的登录计数在下次登录尝试时重置。
- 执行 ALTER USER 语句将 FAILED_LOGIN_ATTEMPTS 或 PASSWORD_LOCK_TIME（或两者）设置为任何值。它们将会被重置。
- 执行 ALTER USER ... UNLOCK 语句。

除此之外，对于 PROXY USER 用户代理，则是对代理用户跟踪，而不是跟踪被代理用户。

6. 总结

MySQL 8.0 版本在密码安全方面引入了多项实用功能，如更强的密码策略验证和连接控制，从而极大地增强了账号的安全性。在实际部署中，建议数据库管理员充分了解和利用这些功能，以提高数据库的安全性。

3.4.8 增强的 JSON 功能

目前很多应用环境中都能看到 JSON 的灵活应用。JSON 格式在数据处理的不同阶段都表现出极大的灵活性，能够清晰地表示和区分不同层次的数据。从 MySQL 5.7 版本开始，MySQL 支持 JSON 数据类型，但注意，在 MySQL 8.0 版本之前，JSON 类型数据并不是单独的数据类型，而是被存储为字符串。然而，在 MySQL 8.0 版本中，JSON 首次作为原生数据类型被支持，提供了自动验证的 JSON 文档以及优化的存储格式。

1. JSON 介绍

在 MySQL 8.0 版本中，JSON 数据类型提供以下功能。

- 自动验证存储在 JSON 列中的 JSON 文档。若为无效文档，则产生错误。
- 优化的存储格式。存储在 JSON 列中的 JSON 文档被转换为允许快速读取访问文档元素的内部格式（二进制格式）。
- 对文档元素的快速读取访问。当再次读取 JSON 文档时，不需要重新解析文本获取该值。通过键或数组索引直接查找子对象或嵌套值，而不需要读取文档中的所有值。
- 存储 JSON 文档所需的空间大致与 LONGBLOB 或 LONGTEXT 相同。

- 存储在 JSON 列中的任何 JSON 文档的大小都受限于 MAX_ALLOWED_PACKET 参数。
- JSON 类型数据可以支持将默认值设置为 NULL。

目前，MySQL 8.0 版本的 JSON 类型支持 32 个普通函数和两个空间函数，这些函数提供了丰富的功能，进一步增强了 MySQL 在处理 JSON 数据时的能力。表 3-26 所示为 MySQL 8.0 版本的 JSON 类型支持的普通函数。

表 3-26 MySQL 8.0 版本的 JSON 类型支持的普通函数

JSON 类型支持的普通函数	说　　明
->	返回值从 JSON 列进行 ison_extract 操作的简洁写法
->>	返回值从 JSON 列进行 json_unquote（column ->path）操作的简洁写法
JSON ARRAY	创建 JSON 数组
JSON ARRAY APPEND	JSON 数据中再追加数据
JSON ARRAY INSERT	插入 JSON 数组
JSON CONTAINS	是否包含特定对象的 JSON 文档路径
JSON CONTAINS PATH	JSON 数据中包含任何数据路径
JSON DEPTH	JSON 文档的最大深度
JSON EXTRACT	返回 JSON 数据，相当于 json_unquote（json extract()）函数
JSON INSERT	将数据插入到 JSON 文档
JSON KEYS	提取 JSON 中的键值
JSON LENGTH	在 JSON 文档中的元素数
JSON MERGE PATCH	合并的 JSON 文件，免去重复键的值
JSON MERGE PRESERVE	合并的 JSON 文件，保存重复键
JSON OBJECT	创建 JSON 对象
JSON OVERLAPS	比较两个 JSON 文档，如果有共同的键值对或数组元素，则返回 TRUE（1），否则返回 FALSE（0）
JSON PRETTY	以 JSON 格式输出文档
JSON QUOTE	引用 JSON 文档
JSON REMOVE	从 JSON 文件中删除数据
JSON REPLACE	对 JSON 文件的值进行替换
JSON SCHEMA VALID	根据 JSON 模式验证 JSON 文档。如果验证成功，则返回 TRUE 或 1。如果验证失败，则返回 FALSE 或 0
JSON SCHEMA VALIDATION REPORT	根据 JSON 模式验证 JSON 文档；以 JSON 格式返回验证结果报告，包括成功或失败（包括失败原因）
JSON SEARCH	在 JSON 数据中查询路径
JSON SET	将数据插入到 JSON 文档
JSON STORAGE FREE	部分更新后 JSON 列值的二进制表示中释放的空间
JSON STORAGE SIZE	用于存储 JSON 文档的二进制表示形式的空间
JSON TABLE	将 JSON 表达式中的数据作为关系表返回
JSON TYPE	返回 JSON 值的类型

(续)

JSON 类型支持的普通函数	说明
JSON_UNQUOTE	去除 JSON 字符串的引号,将值转换成 string 类型
JSON_VALID	判断是否为合法的 JSON 文档
JSON_VALUE	从 JSON 文档中提取由路径所指向的值:以 VARCHAR(512) 或指定类型返回该值
MEMBER OF	如果第一个操作数匹配作为第二个操作数传入的 JSON 数组中的任何元素,则返回 true(1),否则返回 false(0)

表 3-27 所示为 MySQL 8.0 版本的 JSON 类型支持的空间函数。

表 3-27 MySQL 8.0 版本的 JSON 类型支持的空间函数

JSON 类型支持的空间函数	说明
ST_AsGeoJSON	从几何图形生成 GeoJSON 对象。对象字符串具有连接字符集和排序
ST_GeomFromGeoJSON	从 GeoJSON 表示构造一个 PostGIS 几何对象

在 MySQL 8.0 版本中,JSON 数据类型的比较和排序遵循一套特定的规则。JSON 数据类型比较和排序的规则如下。

- JSON 值可以使用=、<、<=、>、>=、<>、!=和<=>操作符进行比较。
- JSON 值不支持 BETWEEN、IN( )、GREATEST( )和 LEAST( )比较操作符或函数。
- 对于列出的比较操作符和函数,一种变通方法是将 JSON 值转换为本地 MySQL 数值或字符串数据类型,以便具有一致的非 JSON 标量类型。也就是说,转换成需要的 MySQL 字段继续换算,这也算是一种折中方案。
- JSON 值的比较分为两个级别。第一级比较基于比较值的 JSON 类型。如果类型不同,则仅由类型优先级来决定比较结果。如果两个值具有相同的 JSON 类型,则使用特定类型的规则进行第二级比较:BLOB>BIT>OPAQUE>DATETIME>TIME>DATE>BOOLEAN>ARRAY>OBJECT>STRING>INTEGER 或 DOUBLE>NULL。

在 MySQL 8.0 版本中,JSON 类型支持与其他类型值之间的转换。使用 CAST 函数可将 JSON 类型的值转换为其他类型。JSON 和非 JSON 值之间的转换命令行示例如下。

```
mysql>SET @j5 = '{"id":123, "name":"kevin","age":20, "time":"2023-06-01 01:00:00"}'.
Query OK, 0 rows affected (0.00 sec)

mysql>SELECT CAST(JSON_EXTRACT(@j5,'$.age') AS UNSIGNED);
+---------------------------------------------+
| CAST(JSON_EXTRACT(@j5,'$.age') AS UNSIGNED) |
+---------------------------------------------+
|                                          20 |
+---------------------------------------------+
1 row in set (0.00 sec)
```

在 MySQL 8.0 版本中,对于 JSON 值的聚合,NULL 值和其他数据类型一样被忽略。除 MIN、MAX 和 GROUP_CONCAT 以外,非 NULL 值被转换为数字类型并聚合。对于数字标量的 JSON 值,可能会出现截断和精度损失(取决于值)。

2. JSON 数据类型使用索引的方式

在 MySQL 中，JSON 列本身不能直接创建索引，但可以通过从 JSON 列中提取标量值来创建索引，从而更有效地利用 MySQL 的优势。

JSON 数据类型所支持的索引如下。

- JSON 列，像其他二进制类型的列一样，不直接索引。相反，可以在生成的列上创建索引，从 JSON 列中提取标量值。
- MySQL 优化器还会在匹配 JSON 表达式的虚拟列上寻找兼容的索引。
- 从 MySQL 8.0.17 版本开始，InnoDB 存储引擎支持 JSON 数组上的多值索引。
- MySQL NDB Cluster 8.0 版本支持 JSON 列和 MySQL JSON 函数，包括在从 JSON 列生成的列上创建索引，作为无法索引 JSON 列的解决方案。每个 NDB 表最多支持 3 个 JSON 列。

（1）JSON 虚拟列索引

MySQL 中的 JSON 虚拟列索引是通过在 JSON 列上创建一个虚拟列（也称为生成列或存储列），并在该虚拟列上创建索引来实现的。JSON 虚拟列索引创建的列的语法如下。

```
col_name data_type [GENERATED ALWAYS] AS (expr)
    [VIRTUAL | STORED] [NOT NULL | NULL]
    [UNIQUE [KEY]] [[PRIMARY] KEY]
    [COMMENT 'string']
```

上述语法中 VIRTUAL 或 STORED 关键字表示列值是如何存储的，这对虚拟列的使用影响非常大。

- VIRTUAL：VIRTUAL 列的值在需要时被计算，通常在查询读取行数据之后、返回结果之前。VIRTUAL 列的值不会持久化到磁盘上，但会在查询执行时存储在内存中。VIRTUAL 列的计算没有固定的性能或资源限制，它依赖于实际的查询负载和服务器性能。
- STORED：STORED 列的值会在数据被插入或更新时计算，并持久化到磁盘上。STORED 列会占用额外的存储空间，并且可以像常规列一样建立索引以优化查询性能。
- 如果不明确指定列的类型（VIRTUAL 或 STORED），则默认会创建 VIRTUAL 列。

在 JSON 中创建虚拟列的命令行示例如下。

```
#创建表的同时支持虚拟列
mysql>  CREATE TABLE `jemp` (
        id BIGINT NOT NULL AUTO_INCREMENT PRIMARY KEY,
        c JSON,d JSON,
        g INT GENERATED ALWAYS AS  (c->"$.id") STORED,
        INDEX i (g)
    );
#插入数据
mysql> INSERT INTO jemp (c,d) VALUES
('{"id": "1", "name": "Fred"}','{"user":"Fred",  "user_id":1, "zipcode":[14471,14531]}'), ('{"id": "2", "name": "Wilma"}',
'{"user":"Wilma",  "user_id":2, "zipcode":[24472,24532]}' ),
('{"id": "3", "name": "Jack"}','{"user":"Jack", "user_id":3, "zipcode":[34473,34533]}'   ), ('{"id": "4", "name": "Betty"}',
'{"user":"Betty",  "user_id":4, "zipcode":[44474,44534]}' );
Query OK, 4 rows affected (0.02 sec)

#通过执行计划查看是否使用虚拟列索引
```

```
mysql> EXPLAIN FORMAT=TREE SELECT c->>"$.name" AS name FROM jemp WHERE g > 2;
+-----------------------------------------------------------------------------+
| EXPLAIN                                                                     |
+-----------------------------------------------------------------------------+
-> Index range scan on jemp using i over (2 < g), with index condition: (jemp.g > 2)  (cost=1.16 rows=2)
+-----------------------------------------------------------------------------+
row in set (0.00 sec)
# 通过 WARNINGS 信息查看改写的 SQL 语句
mysql> SHOW WARNINGS \G
*************************** 1. row ***************************
  Level: Note
   Code: 1003
Message: /* select#1 */ select json_unquote(json_extract(`db1`.`jemp`.`c`,'$.name')) AS `name` from `db1`.`jemp` where (`db1`.`jemp`.`g` > 2)1 row in set (0.00 sec)
```

(2) JSON 多值索引

JSON 多值索引，可以为 JSON 数组中的每个元素创建索引，从而加速 MEMBER OF、JSON_CONTAINS 和 JSON_OVERLAPS 等函数的查询。JSON 多值索引的命令行示例如下。

```
# 添加多值索引
mysql> ALTER TABLE jemp ADD INDEX zips( (CAST(d->'$.zipcode' AS UNSIGNED ARRAY)) );

# 使用 MEMBER OF 函数进行查询,查看执行计划
mysql> EXPLAIN FORMAT=TREE SELECT * FROM jemp
       WHERE 24472 MEMBER OF(d->'$.zipcode');
+-----------------------------------------------------------------------------+
| EXPLAIN                                                                     |
+-----------------------------------------------------------------------------+
-> Filter:json'24472' member of (cast(json_extract(d,_utf8mb4'$.zipcode') as unsigned array))  (cost=0.35 rows=1)
    -> Index lookup on jemp using zips (cast(json_extract(d,_utf8mb4'$.zipcode') as unsigned array)=json'24472')  (cost=0.35 rows=1) |
+-----------------------------------------------------------------------------+

# 使用 JSON_CONTAINS 函数进行查询,查看执行计划
mysql> EXPLAIN FORMAT=TREE SELECT * FROM jemp
       WHERE JSON_CONTAINS(d->'$.zipcode', CAST('[14471,14531]' AS JSON));
+-----------------------------------------------------------------------------+
| EXPLAIN                                                                     |
+-----------------------------------------------------------------------------+
-> Filter:json_contains(cast(json_extract(d,_utf8mb4'$.zipcode') as unsigned array),json'[14471, 14531]')  (cost=1.41 rows=2)
    -> Index range scan on jemp using zips over (14531 MEMBER OF (json_extract(d,_utf8mb4'$.zipcode'))) OR (14471 MEMBER OF (json_extract(d,_utf8mb4'$.zipcode')))  (cost=1.41 rows=2) |
+-----------------------------------------------------------------------------+
row in set (0.00 sec)
# 使用 JSON_OVERLAPS 函数进行查询,查看执行计划
    mysql> EXPLAIN FORMAT=TREE SELECT * FROM jemp
        WHERE JSON_OVERLAPS(d->'$.zipcode', CAST('[44474,94582]' AS JSON));
```

```
+----------------------------------------------------------------------------------+
| EXPLAIN                                                                          |
+----------------------------------------------------------------------------------+
-> Filter:json_overlaps(cast(json_extract(d,_utf8mb4'$.zipcode') as unsigned array),json
'[44474, 94582]')   (cost=1.41 rows=2)
    -> Index range scan on jemp using zips over (94582 MEMBER OF
(json_extract(d,_utf8mb4'$.zipcode'))) OR (44474 MEMBER OF
(json_extract(d,_utf8mb4'$.zipcode')))   (cost=1.41 rows=2) |
+----------------------------------------------------------------------------------+1
row in set (0.01 sec)
```

从上述示例可以看出，JSON 列数据的查询在 MySQL 中仍然基于 B 树索引方式，而 JSON 只是数据保存的一种方式。

3. 总结

MySQL 8.0 版本通过增强其 JSON 功能，带来了许多优势，使得它更加灵活，能够处理更复杂的数据结构。通过引入对 JSON 的更多原生支持和优化，MySQL 能够更快地传输和存储 JSON 数据，并提供更有效的查询能力。其中，虚拟列索引（也称为生成列或存储列）是一个重要的特性，它允许在 JSON 字段上创建索引，从而解决了传统 JSON 字段查询的性能问题。

3.4.9 增强的 EXPLAIN 功能

EXPLAIN 是 MySQL 中用于分析查询执行计划的命令。当在一个 SELECT、DELETE、INSERT、REPLACE 或 UPDATE 语句前加上 EXPLAIN 关键字时，MySQL 将返回优化器关于语句执行计划的信息。这些信息对于理解查询的性能瓶颈以及如何优化查询都非常有帮助。

在 MySQL 8.0 版本中，对 EXPLAIN 命令进行了进一步的加强和改进，提供了更多的格式和选项来帮助用户理解查询的执行计划。以下是增强的 EXPLAIN 命令的基本语法和选项。

```
EXPLAIN                     # 原用法
EXPLAIN FORMAT=TREE
EXPLAIN FORMAT=JSON
EXPLAIN ANALYZE
EXPLAIN [options] FOR CONNECTION
```

1. FORMAT=TREE 输出

在 MySQL 8.0 中，通过 EXPLAIN FORMAT=TREE 指定格式时，将获得一个树形结构的输出的执行计划。这个树形结构清楚地展示了查询的各个部分是如何相互关联的，以及在执行时的顺序。EXPLAIN FORMAT=TREE 的命令行示例如下。

```
mysql> EXPLAIN FORMAT=TREE SELECT * FROM employees
    WHERE first_name LIKE 'Ho%' LIMIT 5;
+----------------------------------------------------------------------------------+
| EXPLAIN                                                                          |
+----------------------------------------------------------------------------------+
-> Limit: 5 row(s)   (cost=2238 rows=5)
    -> Index range scan on employees using idx_fname over ('Ho' <= first_name <= 'Ho'), with in-
dex condition: (employees.first_name like 'Ho%')   (cost=2238 rows=1864)
                 |
+----------------------------------------------------------------------------------+
1 row in set (0.00 sec)
```

在 EXPLAIN 的基本语法中,通常执行计划的顺序遵循以下原则:在所有组中,ID 值越大,优先级越高,越先执行。如果 ID 相同,则可以认为是一组,从上往下顺序执行。但有些特殊情况下也会存在例外。因此,辨别顺序非常麻烦,需要一定的经验。但树方式层次明显,不会出现上述问题。

2. FORMAT=JSON 输出

在 MySQL 8.0 中,通过 EXPLAIN FORMAT=JSON 指定格式时,将获得一个 JSON 结构的输出的执行计划。JSON 格式的输出非常简洁,数据格式更加紧凑,易于阅读。但 JSON 格式的输出内容可能非常长,特别是当查询涉及多个表、索引和连接操作时。这可能会导致在控制台输出时屏幕滚动,使得阅读不太方便。

EXPLAIN FORMAT=JSON 的命令行示例如下。

```
mysql> EXPLAIN FORMAT=JSON SELECT * FROM employees
        WHERE first_name LIKE 'Ho%' LIMIT 5;
+----------------------------------------------------------------------------+
|EXPLAIN                                                                     |
+----------------------------------------------------------------------------+
{
  "query_block": {
    "select_id": 1,
    "cost_info": {
      "query_cost": "2237.81"
    },
    "table": {
      "table_name": "employees",
      "access_type": "range",
      "possible_keys": [
        "idx_fname"
      ],
      "key": "idx_fname",
      "used_key_parts": [
        "first_name"
      ],
      "key_length": "58",
      "rows_examined_per_scan": 1864,
      "rows_produced_per_join": 1864,
      "filtered": "100.00",
      "index_condition": "(`employees`.`employees`.`first_name` like 'Ho%')",
      "cost_info": {
        "read_cost": "2051.41",
        "eval_cost": "186.40",
        "prefix_cost": "2237.81",
        "data_read_per_join": "247K"
      },
      "used_columns": [
        "emp_no",
```

```
              "birth_date",
              "first_name",
              "last_name",
              "gender",
              "hire_date"
           ]
         }
       }
} |
+-------------------------------------------------------------------------------+
1 row in set, 1 warning (0.01 sec)
```

EXPLAIN 命令中的 TREE 和 JSON 格式提供了查询的预评估计划，也就是查询优化器在查询执行前估计的执行策略。这些输出中的成本信息（如 cost、read_cost、eval_cost、prefix_cost、data_read_per_join 等）是基于查询优化器的统计信息得出的估计值，用于指导查询优化器选择最佳的查询执行计划。

3. EXPLAIN ANALYZE

在 MySQL 8.0 中，EXPLAIN ANALYZE 的输出类似于 EXPLAIN FORMAT=TREE，但它提供了实际执行时的统计信息，而不仅仅是优化器的估计。这意味着可以看到查询在实际执行过程中的行为，而不仅仅是基于优化器的预测。EXPLAIN ANALYZE 可以用于 SELECT 语句，以及针对多个表的 UPDATE 和 DELETE 语句。

EXPLAIN ANALYZE 的命令行示例如下。

```
mysql> EXPLAIN ANALYZE SELECT * FROM employees
       WHERE first_name LIKE 'Ho%' LIMIT 5;
+-----------------------------------------------------------------------------------------+
| EXPLAIN                                                                                 |
+-----------------------------------------------------------------------------------------+
-> Limit: 5 row(s)  (cost=2238 rows=5) (actual time=1.51..5.96 rows=5 loops=1)
    -> Index range scan on employees using idx_fname over ('Ho' <= first_name <= 'Ho'), with
index condition: (employees.first_name like 'Ho%')  (cost=2238 rows=1864) (actual time=1.51..
5.95 rows=5 loops=1) |
+-----------------------------------------------------------------------------------------+1
row in set (0.01 sec)
```

从上述 EXPLAIN ANALYZE 的执行计划输出结果中可见，增加了 actual time 来表示查询的实际执行时间。为了确认查询是否实际执行，在无其他访问的情况下，观察执行 EXPLAIN ANALYZE 命令前后 Innodb_pages_read（InnoDB 存储引擎读取的页面数）的变化值。如图 3-28 所示，Innodb_pages_read 的值从 1094 增加到了 1188，这意味着在执行 EXPLAIN ANALYZE 命令之后，InnoDB 存储引擎从硬盘读取了数据页，进行了实际的磁盘 I/O 操作。

4. EXPLAIN FOR CONNECTION

EXPLAIN FOR CONNECTION 返回当前在给定连接中正在执行查询的解释信息。对于正在长时间运行或执行延迟的 SQL，可以通过这种方式获取当前的执行计划。connection_id 是连接标识符，可以通过执行 SHOW PROCESSLIST 或 SELECT CONNECTION_ID() 语句获得。

通过 EXPLAIN FOR CONNECTION 查看当前 SQL 语句执行计划的命令行示例如下。

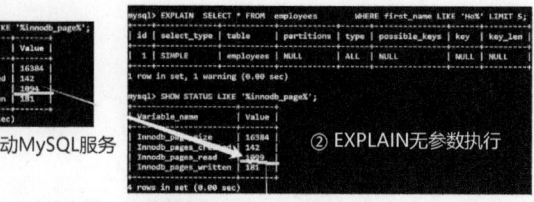
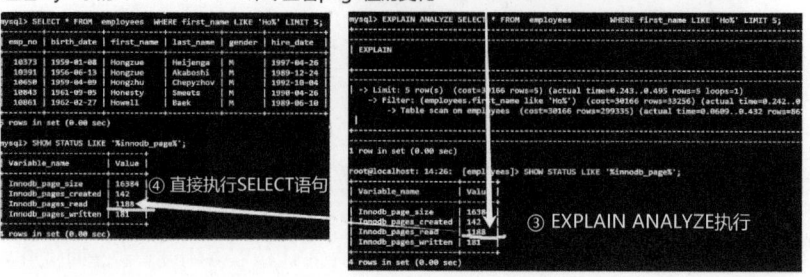

图 3-28 Innodb_pages_read 变化情况

```
#窗口1
mysql> SELECT CONNECTION_ID();
+-----------------+
| CONNECTION_ID() |
+-----------------+
|              14 |
+-----------------+
1 row in set (0.00 sec)
mysql> select * from employees;

#窗口2
mysql> EXPLAIN FOR CONNECTION 14;
*************************** 1. row ***************************
           id: 1
  select_type: SIMPLE
        table: employees
   partitions: NULL
         type: ALL
possible_keys: NULL
          key: NULL
      key_len: NULL
          ref: NULL
         rows: 299600
     filtered: 100.00
        Extra: NULL
1 row in set, 1 warning (0.00 sec)
```

说明：当第一次执行 EXPLAIN FOR CONNECTION 时，可以得到执行计划信息，但随后多次执行时，不再返回执行计划信息。这有可能是语句的执行阶段状态变更或功能不完善导致的。

5. 总结

在 MySQL 8.0 中，EXPLAIN 命令得到了扩展和增强，为数据库管理员和开发者提供了更多关于查询执行计划的详细信息。由于 EXPLAIN ANALYZE 命令具有实际执行查询的特性，因此应该谨慎使用。

3.4.10 GIPK 隐藏主键可视化功能

在 MySQL 中，主键的重要性不言而喻。主键不仅唯一标识表中的每一行数据，还是维持表组织和数据完整性的关键。特别是在复制过程中，主键使得系统能够准确地定位需要更新的行，从而避免了全表扫描，显著提高了性能和效率。

1. GIPK 介绍

从 MySQL 8.0.30 版本开始，支持 GIPK（Generated Invisible Primary Keys）隐藏主键可视化功能。GIPK 是指，当在创建表时没有明确指定主键时，InnoDB 存储引擎会自动为该表生成一个不可见（即用户无法直接查询或修改）的主键。这个主键用于唯一标识表中的每一行数据，以确保数据的完整性和一致性。

在 MySQL 中对 GIPK 的定义如下所示。

```
my_row_id BIGINT UNSIGNED NOT NULL AUTO_INCREMENT INVISIBLE PRIMARY KEY
```

这个定义创建了一个自增的主键，名为 my_row_id，类型为 BIGINT UNSIGNED。由于 BIGINT UNSIGNED 的范围非常广，因此在日常使用中不用担心这个值会用完。默认情况下，它对客户端是不可见的。

默认情况下，GIPK 功能不开启，需要把参数 sql_generate_invisible_primary_key 设置成 ON，才可以启动。GIPK 功能启用的命令行示例如下。

```
# 先查看 GIPK 是否开启
mysql> SHOW GLOBAL VARIABLES LIKE 'sql_generate_invisible_primary_key';
+------------------------------------+-------+
|Variable_name                       |Value  |
+------------------------------------+-------+
|sql_generate_invisible_primary_key  |OFF    |
+------------------------------------+-------+
1 row in set (0.01 sec)
# 开启 GIPK 功能
mysql> SET global sql_generate_invisible_primary_key=ON;
Query OK, 0 rows affected (0.00 sec)

# 创建无显式表
mysql> CREATE TABLE `pk_table` (
  `id` int NOT NULL,
  `name` varchar(25) NOT NULL
) ENGINE=InnoDB;

# 查看表结构
mysql> SHOW CREATE TABLE `pk_table`;
+----------+--------------------------------------------------------------+
|Table     |Create Table                                                  |
+----------+--------------------------------------------------------------+
```

```
|pk_table |    CREATE TABLE `pk_table` (
 `my_row_id` bigint unsigned NOT NULL AUTO_INCREMENT /*!80023 INVISIBLE */,
 `id` int NOT NULL,
 `name` varchar(25) COLLATE utf8mb4_bin NOT NULL,
 PRIMARY KEY (`my_row_id`)
) ENGINE=InnoDB DEFAULT CHARSET=utf8mb4 COLLATE=utf8mb4_bi;
# 插入数据
mysql> INSERT INTO `pk_table` (id,name) VALUES(1,'K');
Query OK, 1 row affected (0.01 sec)
```

从上述结果中可以看出，自动创建了 my_row_id 的隐藏主键。下面再次确认，GIPK 的创建 SQL 语句是否已写到 binlog 中。通过 mysqlbinlog 工具解析 binlog 文件。解析上述操作的 binlog 记录内容如下所示。

```
shell $> mysqlbinlog --no-defaults --base64-output=decode-rows -vv mysql-bin.000001
SET @@SESSION.GTID_NEXT='22228e8c-b0ee-11ec-a2d1-00163e23e2cc:19'/*!*/;
# at 276
#230829  8:52:29 server id 129   end_log_pos 566 CRC32 0x606635ee    Query
    thread_id=8  exec_time=0  error_code=0Xid = 16
...
CREATE TABLE `pk_table` (
 `my_row_id` bigint unsigned NOT NULL AUTO_INCREMENT /*!80023 INVISIBLE */,
 `id` int NOT NULL,
 `name` varchar(25) COLLATE utf8mb4_bin NOT NULL,
 PRIMARY KEY (`my_row_id`)
)
/*!*/;
BEGIN
/*!*/;
# at 719
#230829  8:52:32 server id 129   end_log_pos 789 CRC32 0xf9317c57   Rows_query
# INSERT INTO `pk_table` (id,name) VALUES(1,'K')
# at 789
#230829  8:52:32 server id 129   end_log_pos 849 CRC32 0x23abac8e   Table_map:
`db8`.`pk_table` mapped to number 95
# has_generated_invisible_primary_key=1
...
### INSERT INTO `test`.`pk_table`
### SET
###   @1=1 /* LONGINT meta=0 nullable=0 is_null=0 */
###   @2=1 /* INT meta=0 nullable=0 is_null=0 */
###   @3='K' /* VARSTRING(100) meta=100 nullable=0 is_null=0 */
```

解析的 binlog 内容记录了 my_row_id 的创建语句，并且数据插入语句中记录了 has_generated_invisible_primary_key 属性，这意味着在数据库复制的过程中，将与不可见主键相关的操作记录到了二进制日志中。

2. 影响和注意事项

下面是 MySQL 中使用 GIPK 不可见主键的影响和注意事项。

1）在 SELECT 查询时，可以控制 GIPK 是否可见。通过使用 ALTER TABLE ... SET invisible

来切换 GIPK 的可见性，命令行示例如下。

```
mysql> ALTER TABLE `pk_table` ALTER COLUMN my_row_id SET invisible;
Query OK, 0 rows affected (0.01 sec)
Records: 0  Duplicates: 0  Warnings: 0

mysql> SELECT * FROM pk_table;
+----------------+------+--------+
|my_row_id |id |name |
+----------------+------+--------+
|        1 | 1 |W     |
+----------------+------+--------+
1 row in set (0.00 sec)

mysql> ALTER TABLE `pk_table` ALTER COLUMN my_row_id SET invisible;
Query OK, 0 rows affected (0.01 sec)
Records: 0  Duplicates: 0  Warnings: 0

mysql> SELECT * FROM pk_table;
+------+--------+
|id |name |
+------+--------+
| 1 |W     |
+------+--------+
1 row in set (0.00 sec)
```

2）在 MySQL 中，每个表只能有一个自增列。因为 GIPK 使用 my_row_id 占据了自增属性，所以无法再创建自增属性字段。验证自增字段是否已创建的命令行示例如下。

```
mysql> SHOW GLOBAL VARIABLES LIKE 'sql_generate_invisible_primary_key';
+------------------------------------------+----------+
|Variable_name                             |Value |
+------------------------------------------+----------+
|sql_generate_invisible_primary_key |ON    |
+------------------------------------------+----------+
1 row in set (0.00 sec)

mysql> CREATE TABLE `pk_table1` (
       `id` int NOT NULL,
       `iid` bigint unsigned NOT NULL auto_increment,
       `name` varchar(25) NOT NULL,
        key(`iid`)
       );
ERROR 4109 (HY000): Failed to generate invisible primary key. Auto-increment column already exists.
```

3）my_row_id 已经成为 MySQL 关键字，不能随意使用。关键字错误提示示例如下。

```
mysql> CREATE TABLE `pk_table2` (
    my_row_id int NOT NULL,
      `name`varchar(25) NOT NULL
    );
ERROR 4108 (HY000): Failed to generate invisible primary key. Column 'my_row_id' already exists.
```

4）在启用 GIPK 时，无法删除 my_row_id 列，需要重新选择主键。删除 my_row_id 的命令行示例如下。

```
mysql> ALTER TABLE pk_table DROP PRIMARY KEY,ADD primary key (id);
ERROR 1235 (42000): This version of MySQL doesn't yet support 'existing primary key drop without adding a new primary key. In @@sql_generate_invisible_primary_key=ON mode table should have a primary key. Please add a new primary key to be able to drop existing primary key.'
```

5）在使用 mysqldump 逻辑导出 GIPK 表时，需要添加 --skip-generated-invisible-primary-key 选项，从输出中排除生成的不可见的主键。逻辑导出命令行示例如下。

```
shell $>mysqldump -uroot -p123456 --set-gtid-purged=OFF
--skip-generated-invisible-primary-key -B test  > pt_table.sql

shell $> cat pt_table2.sql
CREATE DATABASE /*!32312 IF NOT EXISTS*/ `test` /*!40100 DEFAULT CHARACTER SET utf8mb4 COLLATE utf8mb4_bin */ /*!80016 DEFAULT ENCRYPTION='N' */;
USE `teset`;
--
-- Table structure for table `pk_table`
--
DROP TABLE IF EXISTS `pk_table`;
/*!40101 SET @saved_cs_client     = @@character_set_client */;
/*!50503 SET character_set_client = utf8mb4 */;
CREATE TABLE `pk_table` (
  `id` int NOT NULL,
  `name`varchar(25) CHARACTER SET utf8mb4 COLLATE utf8mb4_bin NOT NULL
) ENGINE=InnoDB DEFAULT CHARSET=utf8mb4 COLLATE=utf8mb4_bin;
/*!40101 SET character_set_client = @saved_cs_client */;
```

说明：目前开源 MyDumper 逻辑导出工具还没有实现导出 GIPK 列的功能。

3. 总结

MySQL 的 GIPK 功能解决了无主键下的性能问题。但在使用它时，也需要考虑其限制和约束，如自增属性和列属性的限制等。因此，需要权衡其带来的好处和潜在的限制，确保它符合应用的需求和预期的行为。

3.4.11 参数修改持久化功能

在 MySQL 8.0 之前的版本中，通过 SET GLOBAL 命令修改的全局参数，只会影响 MySQL 服务当前的内存中的值，这些更改在服务重启后不会保持。为了使更改在服务重启后仍然有效，需要手动编辑 MySQL 的配置文件，并在其中添加或修改相应的参数。这种手动编辑配置文件的做法存在容易遗忘和出错的风险。

1. 持久化功能介绍

从 MySQL 8.0 版本开始，引入了 PERSIST 和 PERSIST_ONLY 命令，这些命令用于将参数更改持久化到配置文件中。具体来说，PERSIST 命令会使动态参数的新值立即生效并同时写入到自动生成的 mysqld-auto.cnf 文件中。而 PERSIST_ONLY 命令则用于将参数值写入 mysqld-auto.cnf 文件，这些更改在 MySQL 服务重新启动之前不会生效。

2. 持久化命令行使用方式

下面使用 MySQL 的持久化功能并查看已持久化的参数。

使用 PERSIST 命令将 max_connect_errors 参数的值持久化的命令行示例如下。

```
# 使用 PERSIST 设置值
mysql> SET PERSIST max_connect_errors=99999;
Query OK, 0 rows affected (0.00 sec)
# 查看参数是否生效
mysql> SELECT * FROM performance_schema.persisted_variables;
+--------------------+----------------+
| VARIABLE_NAME      | VARIABLE_VALUE |
+--------------------+----------------+
| max_connect_errors | 99999          |
+--------------------+----------------+
1 row in set (0.00 sec)
```

持久化更改的参数以 JSON 格式保存到 mysqld-auto.cnf 文件中，该文件位于 MySQL 的数据目录中。在 Linux 下，可以使用文本编辑器或命令行工具（如 cat、less、grep 等）来查看文件内容。查看 mysqld-auto.cnf 文件的内容的命令行如下。

```
shell $> cat mysqld-auto.cnf
{"Version": 2, "mysql_dynamic_variables": {"max_connect_errors": {"Value": "99999", "Metadata": {"Host": "localhost", "User": "root", "Timestamp": 1693221812638159}}}}
```

注意：如果需要从 mysqld-auto.cnf 文件中删除持久化的所有参数的设置，则可使用 RESET PERSIST 命令行。RESET PERSIST 命令会清空 mysqld-auto.cnf 和 performance_schema.persisted_variables 中的内容，但对于已经修改完的参数的值，不会产生任何影响。

使用 PERSIST_ONLY 命令将 max_connect_errors 参数的值持久化，但不生效的命令行示例如下。

```
# 权限报错
mysql> SET PERSIST_ONLY max_connect_errors=11111;
ERROR 3630 (42000): Access denied。you need SYSTEM_VARIABLES_ADMIN and PERSIST_RO_VARIABLES_ADMIN privileges for this operation
# 赋予权限
mysql> GRANT SYSTEM_VARIABLES_ADMIN, PERSIST_RO_VARIABLES_ADMIN ON *.* TO 'root'@'localhost';
Query OK, 0 rows affected (0.00 sec)
mysql> FLUSH PRIVILEGES;
Query OK, 0 rows affected (0.00 sec)

# 更改参数
mysql> SET PERSIST_ONLY max_connect_errors=11111;
Query OK, 0 rows affected (0.01 sec)
```

```
# 参数没有生效
mysql> SHOW VARIABLES LIKE 'max_connect_errors';
+---------------------+-------+
| Variable_name       | Value |
+---------------------+-------+
| max_connect_errors  | 100   |
+---------------------+-------+
1 row in set (0.00 sec)
# 查看 mysqld-auto.cnf 文件,该参数的值已经修改成 1111
shell $> cat mysqld-auto.cnf
{"Version": 2, "mysql_dynamic_variables": {"max_connect_errors": {"Value": "11111", "Metadata": {"Host": "localhost", "User": "root", "Timestamp": 1712482594246249}}}}
```

除此之外,在 MySQL 8.0 版本中,当使用密钥环组件时,为了增强安全性,敏感参数值在持久化到配置文件中时默认会被加密。如果密钥环组件不可用,而 persist_sensitive_variables_in_plaintext 参数被设置为 ON,则 MySQL 会允许以未加密的格式存储这些参数值。这样做虽然方便了管理,但可能会降低安全性,因为未加密的参数值可能会被任何能够访问配置文件的用户查看。

3. 总结

MySQL 8.0 版本中引入的参数持久化功能解决了命令行和底层配置文件之间的一致性问题。这样,即使数据库服务重启,更改后的参数设置也会保留下来,确保配置的一致性。尽管参数持久化功能减少了 DBA 疏忽的风险,但它并不能完全取代手动配置管理。因为配置文件又多了,所以需要额外关注和维护持久化配置文件。

3.4.12 克隆插件功能

MySQL 8.0.17 版本推出了一个重量级的插件——克隆插件(Clone Plugin)。该插件可以克隆数据库的一个副本。由于克隆功能是官方推出的,因此它可以作为全量备份的一种替代方案,从而有可能减少 DBA 对 Percona XtraBackup 等第三方备份工具的依赖。

下面将详细介绍克隆插件。

1. 克隆插件介绍

克隆插件允许用户将当前 MySQL 实例进行本地或者远程的克隆操作,该插件通过对存储在 InnoDB 中的 schema(数据库)、table(表)、tablespaces(表空间)和 data dictionary metadata(数据字典元数据)的数据进行在线物理快照来实现其功能。

克隆插件的主要用途如下。

- 快速备份和恢复:克隆插件提供了一个快速的方法来创建数据库实例的完整备份。与逻辑备份相比,物理备份通常更快,因为直接复制数据文件和索引,而不是逐行读取和写入数据。
- 高可用性解决方案:在 MGR 集群中,克隆插件用于快速添加新成员。DBA 可以使用该插件创建现有成员的副本,并将其部署为新节点,从而简化扩展和维护过程。
- 灾难恢复:克隆插件也可以用于灾难恢复场景。通过创建一个物理克隆,可以在不同的数据中心拥有数据库的即时副本。这样,在发生自然灾害或硬件故障时,可以迅速切换到备份实例。
- 开发和测试环境:开发人员可以克隆生产数据库以创建用于测试和开发的新实例。
- 迁移和升级:当需要迁移数据库或升级 MySQL 版本时,克隆插件可以用来创建一个当前实例的副本，然后在这个副本上进行迁移或升级工作，而不影响生产系统。

MySQL 8.0 版本中的克隆插件的加载和使用方式与其他插件类似。克隆插件的共享库 mysql_clone.so 文件位于 mysql/lib/plugin 目录下。克隆插件配置和加载的命令行示例如下。

```
# 配置my.cnf方式
[mysqld]
plugin-load-add=mysql_clone.so
clone=FORCE_PLUS_PERMANENT

# 启动加载
shell $>mysqld_safe --defaults-file=/etc/my.cnf --plugin-load-add=mysql_clone.so
--user=mysql &

# SQL 语句加载
mysql> INSTALL PLUGIN clone SONAME 'mysql_clone.so';
Query OK, 0 rows affected (0.01 sec)

# SQL 语句卸载
mysql> UNINSTALL PLUGIN clone;
Query OK, 0 rows affected (0.01 sec)

# 查看克隆插件是否加载成功
mysql> SELECT PLUGIN_NAME, PLUGIN_STATUS
       FROM INFORMATION_SCHEMA.PLUGINS WHERE PLUGIN_NAME = 'clone';
+-------------+---------------+
| PLUGIN_NAME | PLUGIN_STATUS |
+-------------+---------------+
| clone       | ACTIVE        |
+-------------+---------------+
1 row in set (0.00 sec)
```

MySQL 的克隆插件提供了一系列参数来配置和控制克隆操作的行为。这些参数在执行克隆操作的 MySQL 服务器实例上进行配置，并且很多情况下建议源数据库和目标数据库都进行相应的设置。这些参数都是动态参数，可以在运行时进行调整。表 3-28 所示为 12 个克隆参数。

表 3-28 MySQL 8.0 克隆参数

参　　数	默认值和值范围	配置对象	说　　明
Clone_autotune_concurrency	默认值：Yes	Recipient	动态生成用于远程克隆操作的其他线程，以优化数据传输速度。此设置仅适用于接收方 MySQL 服务器实例
Clone_max_concurrency	默认值：16，范围：1~128	Recipient	定义远程克隆操作的最大并发线程数。默认值是 16。更多的线程可以提高克隆性能，但也会减少允许同时进行的客户端连接的数量，这会影响现有客户端连接的性能
Clone_max_data_bandwidth	默认值：0，范围：0~1048576	Recipient	定义远程克隆操作每秒的最大数据传输速率，单位为 MB/s。此变量有助于管理克隆操作的性能影响
Clone_max_network_bandwidth	默认值：0，范围：0~1048576	Recipient	指定远程克隆操作每秒的最大近似网络传输速率（以 MB/s 为单位）
Clone_valid_donor_list	—	Recipient	定义远程克隆操作的有效主机地址

(续)

参　数	默认值和值范围	配置对象	说　明
Clone_enable_compression	默认值：OFF	Donor、Recipient	允许在远程克隆操作期间压缩网络层的数据。压缩以 CPU 为代价节省网络带宽。启用压缩可以提高数据传输速率
Clone_ddl_timeout	默认值：0，范围：0~2592000	Donor、Recipient	执行克隆操作时等待备份锁的时间（以秒为单位）。此设置同时应用于发送方和接收方 MySQL 服务器实例。值 0 表示克隆操作不需要备份锁。在这种情况下，如果同时尝试 DDL 操作，操作将失败并出现错误
Clone_buffer_size	默认值：4M，范围：1048576~268435456	Donor、Recipient	执行克隆操作时使用的内存大小
Clone_ssl_ca		Recipient	克隆所需的 CA 证书
Clone_ssl_key		Recipient	指定私钥文件的路径
Clone_ssl_cert		Recipient	客户端公钥证书文件的路径
Clone_block_ddl	默认值：OFF	Donor	在克隆操作期间对实施 MySQL 服务器实例启用独占备份锁定，从而阻止执行并发 DDL 操作

2. 本地克隆

本地克隆指的是从本地 MySQL 服务器实例复制数据到同一台服务器上的另一个目录的过程，如图 3-29 所示。

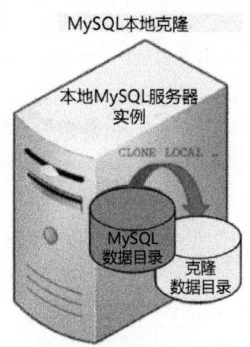

图 3-29　本地克隆

本地克隆操作前，先创建备份目录，并且赋予执行 BACKUP_ADMIN 权限。具体命令行示例如下。

```
# 创建克隆文件夹，赋予 mysql 用户该文件夹的所有权
shell $> mkdir -p /opt/backup/20230801
shell $> chown -R mysql:mysql /opt/backup/20230801

# 创建克隆数据库账号和赋予权限
mysql> CREATE USER 'clone_user'@'%' identified by '123456';
Query OK, 0 rows affected (0.01 sec)
mysql> GRANT BACKUP_ADMIN ON  *.*  TO 'clone_user'@'%';
Query OK, 0 rows affected (0.01 sec)
# 用克隆账号登录数据库，执行克隆命令
mysql> clone LOCAL DATA DIRECTORY = '/opt/backup/20230801/bak';
Query OK, 0 rows affected (0.58 sec)
```

本地克隆过程相对简单。图 3-30 所示为克隆数据库列表和克隆完成后的物理备份文件。需要注意的是，克隆过程通常不包括复制系统数据库，如 information_schema 和 performance_schema，因为这些数据库包含有关数据库服务器内部状态的信息，而非实际的应用程序数据。此外，如果 db2 是一个空的数据库，不包含任何表，那么它也不会被包括在克隆过程中。同时，某些日志文件，如 log-error、slow_query_log_file、log-bin 和 relay-log 等，也不会被克隆。

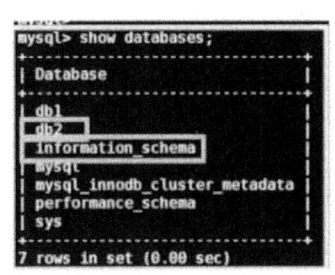

图 3-30 本地备份文件

3. 克隆恢复

下面基于本地克隆，文件（/opt/backup/20230801/bak）进行恢复，执行以下步骤。
1）停止数据库服务。
2）清空原数据目录，确保没有遗漏任何重要文件。
3）将备份的克隆文件或目录复制到数据目录下，确保包含所有必要的文件和子目录。
4）更改数据目录及其文件的所有权和权限，确保 MySQL 用户有适当的访问权限。
5）使用 touch 命令在数据目录下创建一个错误日志文件。
6）启动数据库服务，并检查错误日志文件以获取任何潜在问题或错误消息。
7）验证数据库是否正常运行和数据的一致性，确保恢复成功，之后搭建从节点。
克隆恢复的命令行示例如下。

```
# 克隆恢复步骤 1）~ 6）
shell $> cp -r /opt/backup/20230801/bak /opt/data8.1/mysql
shell $> touch /opt/data8.1/logs/mysql_err.log
shell $> chow -R mysql:mysql
shell $> mysqld_safe --defaults-file=/etc/clone.cnf --user=mysql  &

# 查看数据库 GTID 情况和搭建从节点（步骤 7）
mysql> show master status \G;
*************************** 1. row ***************************
             File: mysql-bin.000017
         Position: 1407
     Binlog_Do_DB:
 Binlog_Ignore_DB:
Executed_Gtid_Set: 22228e8c-b0ee-11ec-a2d1-00163e23e2cc:1-20085
1 row in set (0.00 sec)

# 直接创建从节点
mysql> CHANGE REPLICATION SOURCE TO
```

```
                SOURCE_HOST = '192.168.1.1',
                SOURCE_PORT = 3380,
                SOURCE_USER = 'repl',
                SOURCE_PASSWORD = '******',
                SOURCE_AUTO_POSITION=1;
```

说明：在 MySQL 克隆过程中，以下目录和文件对于克隆操作来说是必要的，并且用户不能手动修改。以下是对这些目录和文件的说明。

- #clone 目录：包含克隆操作使用的内部克隆文件。用于临时存储克隆数据和元数据。
- #ib_archive 目录：用于存储克隆操作期间的归档日志文件。
- *.#clone 文件：在远程克隆操作替换现有数据目录时创建的临时数据文件。

4. 远程克隆

MySQL 的远程克隆操作和本地克隆操作在克隆的数据方面没有区别。无论是从远程服务器克隆数据到本地，还是从本地服务器克隆数据到另一个本地目录，克隆过程的核心都是复制数据库的数据文件。

在远程克隆操作中，涉及两个 MySQL 服务器实例：一个是本地服务器实例，通常称为 Recipient（接收方），另一个是远程服务器实例，称为 Donor（发送方）。远程克隆操作的目标是将数据从 Donor 服务器传输到 Recipient 服务器。

图 3-31 所示为远程克隆方式。在远程克隆过程中，默认情况下，Recipient 服务器上的现有数据将被删除，并用从 Donor 服务器克隆过来的数据替换。这是为了确保数据的一致性和完整性。然而，可以选择将数据导入到 Recipient 服务器上的另一个目录，而不是默认的数据目录，这样可以避免删除现有数据。这通常涉及在导入数据时指定不同的目录路径。

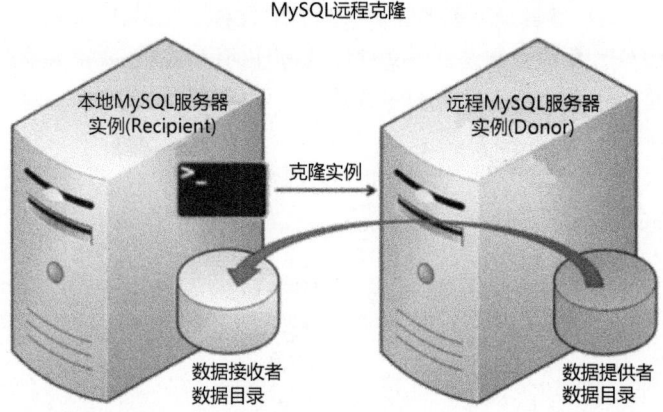

图 3-31　远程克隆

远程克隆有下列 13 个限制条件。

- 克隆插件必须在发送方（Donor）和接收方（Recipient）的 MySQL 服务器实例上都是活动的。
- 权限设置。以 Donor 身份登录的用户需要 BACKUP_ADMIN 特权来访问和传输来自克隆服务器的数据，并在操作期间阻塞 DDL。目标端 Recipient 用户需要 clone_ADMIN 特权来替换收件人数据，在操作期间阻塞 DDL，并自动重新启动服务器。clone_ADMIN 特权隐式地包括 BACKUP_ADMIN 和 SHUTDOWN 特权。

- Donor 和 Recipient 必须拥有相同的 MySQL 服务器版本。MySQL 8.0.17 版本及更高版本支持克隆插件。
- Recipient 必须有足够的磁盘空间来存放克隆数据。
- 由 SELECT FILE_NAME FROM INFORMATION_SCHEMA.FILES 语句识别出来的内容才会被克隆。
- 需要相同的 MySQL 服务器字符集和排序规则。
- innodb_page_size 和 innodb_data_file_path 参数设置必须相同。
- 在克隆加密或页面压缩的数据时，Donor 和 Recipient 必须具有相同的文件系统块大小。
- SSL 和密钥配置必须与主节点一样。
- Recipient 里必须设置 clone_valid_donor_list 源库的信息。
- 一次只允许一个克隆操作。
- 克隆插件以 1MB 大小的包和元数据的形式传输数据。因此，对于 Recipient 和 Donor 的 MySQL 服务器实例，需要的 max_allowed_packet 最小值是 2MB。
- Undo 表空间克隆期间，不能有变更。从 MySQL 8.0.18 版本开始，如果在克隆操作期间遇到重复的 Undo 表空间文件名，克隆就会失败。

下面是远程克隆的具体操作。

对于远程克隆，首先要在 Recipient 和 Donor 上加载克隆插件，创建克隆所需的账号和权限，之后配置 Donor 信息。远程克隆操作的命令行示例如下。

```
# 在 Recipient 和 Donor 上加载插件
mysql> INSTALL PLUGIN clone SONAME 'mysql_clone.so';
Query OK, 0 rows affected (0.02 sec)

# 在 Recipient 和 Donor 上查看插件
mysql> SELECT PLUGIN_NAME, PLUGIN_STATUS FROM INFORMATION_SCHEMA.PLUGINS WHERE PLUGIN_NAME = 'clone';
+-------------+---------------+
| PLUGIN_NAME | PLUGIN_STATUS |
+-------------+---------------+
| clone       | ACTIVE        |
+-------------+---------------+

# 在 Donor 上创建用户,并赋予其 BACKUP_ADMIN 和 clone_ADMIN 权限
mysql> CREATE USER 'clone_user'@'%' identified by '******';
Query OK, 0 rows affected (0.01 sec)
mysql> GRANT BACKUP_ADMIN,clone_ADMIN ON *.* TO 'clone_user'@'%';
Query OK, 0 rows affected (0.01 sec)

# 在 Recipient 上配置 Donor 的 IP 地址和端口信息
mysql> SET GLOBAL clone_valid_donor_list = "192.168.1.1:3380";
Query OK, 0 rows affected (0.00 sec)
mysql> SHOW VARIABLES LIKE 'clone_valid_donor_list';
+----------------------------+-------+
| Variable_name              | Value |
+----------------------------+-------+
```

```
|clone_valid_donor_list |192.168.1.1:3380    |
+-----------------------+------------------------+
# 在 Recipient 上执行克隆操作以覆盖本地数据
mysql> clone INSTANCE FROM 'clone_user'@'192.168.1.1':3380 IDENTIFIED BY '******';
Query OK, 0 rows affected (1.76 sec)

Restarting mysqld...
2023-08-01T15:41:51.659583Z mysqld_safe Number of processes running now: 0
2023-08-01T15:41:51.666882Z mysqld_safe mysqld restarted
```

如果不需要覆盖原数据库目录和重新启动数据库服务，则需要手动配置路径才能使用克隆数据。远程克隆的命令行示例如下。

```
mysql>clone INSTANCE FROM 'clone_user'@'192.168.1.3380':3380 IDENTIFIED BY
'******' DATA DIRECTORY='/opt/backup/20230801/bak02';
```

在远程克隆中，通过 performance_schema 提供的 clone_status、clone_progress 两张表来监控克隆的执行状态和进度。远程克隆监控的命令行示例如下。

```
mysql> SELECT * FROM performance_schema.clone_status \G
*************************** 1. row ***************************
             ID: 1
            PID: 0
          STATE: Completed
     BEGIN_TIME: 2023-08-0109:50:32.112
       END_TIME: 2023-08-0109:55:45.551
         SOURCE: 192.168.1.1:3380
    DESTINATION: LOCAL INSTANCE
       ERROR_NO: 0
  ERROR_MESSAGE:
    BINLOG_FILE: mysql-bin.000009
BINLOG_POSITION: 397
  GTID_EXECUTED: fbde18d9-b034-11ee-9b2b-00163e23e2cc:1-231
mysql> SELECT * FROM performance_schema.clone_progress;
+----+-----------+-----------+---------------------+---------------------+---------+----------+----------+---------+------------+---------------+
|ID | STAGE     | STATE     | BEGIN_TIME          | END_TIME            | THREADS | ESTIMATE | DATA     | NETWORK | DATA_SPEED | NETWORK_SPEED |
+----+-----------+-----------+---------------------+---------------------+---------+----------+----------+---------+------------+---------------+
| 1 | DROP DATA | Completed | 2023-08-0109:50:14  | 2023-08-0109:50:14  | 1       | 302.88 m | 0        | 0       | 0          | 0             |
| 2 | FILE COPY | Completed | 2023-08-0109:50:14  | 2023-08-0109:52:16  | 1       | 1.38 s   | 12004    | 0       | 0          | 0             |
| 3 | PAGE COPY | Completed | 2023-08-0109:52:16  | 2023-08-0109:53:16  | 1       | 104.43 ms| 356      | 0       | 0          | 0             |
| 4 | REDO COPY | Completed | 2023-08-0109:53:16  | 2023-08-0109:53:16  | 1       | 101.02 ms| 12       | 0       | 0          | 0             |
| 5 | FILE SYNC | Completed | 2023-08-0109:53:16  | 2023-08-0109:54:18  | 1       | 60.2 s   | 1039     | 0       | 0          | 0             |
| 6 | RESTART   | Completed | 2023-08-0109:54:18  | 2023-08-0109:55:44  | 0       | 34.02s   |          |         | 100%       |               |
| 7 | RECOVERY  | Completed | 2023-08-0109:55:44  | 2023-08-0109:55:45  | 0       | 1.19 s   |          |         | 100%       |               |
+----+-----------+-----------+---------------------+---------------------+---------+----------+----------+---------+------------+---------------+
```

表 3-29 所示为 clone_status 表信息。

表 3-29　MySQL 8.0 中 clone_status 表信息

克隆状态字段	说　　明
ID	当前 MySQL 服务器实例中的唯一克隆操作标识符
PID	执行克隆操作的会话的进程列表 ID
STATE	克隆操作的当前状态值包括未启动、正在进行、已完成和失败（Not Started、In Progress、Completed 和 Failed）
BEGIN_TIME	开始时间
END_TIME	结束时间
SOURCE	源端服务器信息
DESTINATION	目标端目录
ERROR_NO	失败报错编号
ERROR_MESSAGE	失败报错信息
BINLOG_FILE	复制数据的目标二进制日志文件的名称
BINLOG_POSITION	复制数据的目标二进制日志的位置
GTID_EXECUTED	最后一个克隆事务的 GTID 值

表 3-30 所示为 clone_progress 表信息。

表 3-30　MySQL 8.0 中 clone_progress 表信息

克隆进度字段	说　　明
ID	当前 MySQL 服务器实例中的唯一克隆操作标识符
STAGE	包括删除数据、文件复制、PAGE_COPY、REDO_COPY、FILE_SYNC、重新启动和恢复指标
STATE	克隆阶段的当前状态包括未启动（Not Started）、正在进行（In Progress）和已完成（Completed）
BEGIN_TIME	开始时间
END_TIME	结束时间
THREADS	阶段中使用的并发线程数
ESTIMATE	当前阶段的估计数据量，以字节为单位
DATA	当前状态下传输的数据量，以字节为单位
NETWORK	当前状态下传输的网络数据量，以字节为单位
DATA_SPEED	当前数据传输的实际速度，以 B/s 为单位。这个值可能与 clone_max_data_bandwidth 定义的请求最大数据传输速率不同
NETWORK_SPEED	当前网络传输的速度，以 B/s 为单位

5. 克隆注意事项

在生产环境中使用克隆功能时，需要注意以下事项。

- 对于网络负载，克隆操作可能会占据大量网络带宽，尽管可以通过参数设置进行调整，但仍需要一定的时间来优化。特别是在当前 MySQL 单库数据量达到 TB 级别的情况下，这一点尤为重要。

- 克隆操作会占用服务层的资源，在负载高的情况下，可能会让 MySQL 的线程阻塞。
- 目前克隆功能仅支持整个数据库的全备份，暂时不支持单库和单个表的克隆。
- 克隆功能会删除目标 MySQL 实例上的所有数据。因此，在使用克隆功能之前，请确保已经对目标 MySQL 实例上的数据进行了完整备份。
- MySQL 的错误日志中关于克隆操作的记录信息相对较少。错误日志记录的内容只有两行，参考意义基本没有。克隆错误日志记录信息如下所示。

```
[Warning][MY-013460][InnoDB]Clone removing all user data for provisioning:Started
[Warning][MY-013460][InnoDB]Clone removing all user data for provisioning:FInished
```

- 紧急情况下停止克隆操作。查询 performance_schema.clone_status 表以获取克隆的 PID 信息，之后使用 KILL 命令强制停止克隆操作。

6. 总结

以个人体验来说，目前克隆功能还不能完全替代 XtraBackup 备份的功能，两者各有优势和应用场景。在 MGR 集群环境中，初始化是通过克隆来实现的，但从上述提到的网络负载、线程占用、功能限制、大数据量传输的稳定性以及日志记录等方面来看，克隆功能还有待进一步完善。克隆功能的亮点在于它使用了官方企业版 mysqlbackup 的 Page Tracking 和 Redo Archiving 技术。而 XtraBackup 也是基于 Redo Archiving 功能进行工作的。对于这部分技术，值得深入了解和学习。

在使用 MySQL 8.0 版本时，建议同时掌握 XtraBackup 和克隆这两种物理备份恢复方法。

3.4.13 MySQL Shell 的逻辑备份恢复 API 功能

从 MySQL 8.0 版本开始提供的 MySQL Shell 功能，将 DBA 的工作提升到了新的水平。除了传统的 SQL 操作以外，MySQL Shell 还正式踏入了 Cloud 数据库服务和 Shell 操作数据库领域，其中 MGR 集群就是一个典型的代表。其中，在 MySQL 8.0.21 版本中新增的逻辑备份恢复 API，为 DBA 提供了更灵活和强大的逻辑备份恢复功能。

MySQL Shell 的逻辑备份恢复 API 通过支持多线程、控制速率、使用 zstd 压缩、块（chunk）并行导出和并行导入等特性，大大提高了备份和恢复的效率。这种并行备份恢复方式允许 DBA 在更短的时间内完成操作，减少了数据丢失的风险，并且能够更好地利用系统资源。此外，MySQL Shell 还支持将备份数据直接导出到第三方存储设备中，如 OCI 和 S3 等。这一特性使得备份数据的存储和管理更加灵活与方便，顺应了云计算和分布式存储的发展趋势。

MySQL Shell 的逻辑备份恢复提供 7 个 API，如下所示。

```
1. util.dumpInstance()        用于备份整个实例
2. util.dumpSchemas()         用于备份指定 schema
3. util.dumpTables()          用于备份表
4. util.loadDump()            用于恢复备份
5. util.importTable()         用于导入表
6. util.importJson()          用户 JSON 导入
7. util.export_table()        用于导出表
```

MySQL Shell 的逻辑备份恢复 API 为 DBA 提供了另一种并行备份恢复的选择，这相对于传统的工具（如 mysqldump、mysqlpump 和 MyDumper）来说，在处理大型数据库时效率可能更高。虽然传统的工具功能强大，但在备份和恢复大型数据库时可能效率有限。

1. 备份 API

MySQL Shell 的备份 API 功能非常强大，它允许用户在不同级别进行逻辑备份，包括整个实例、单独的数据库或特定的表。dump-schemas 选项使得用户可以指定要备份的 schema 级别，这对于只需要备份部分数据库的场景非常有用。

在执行备份过程中，MySQL Shell 会输出丰富的信息，包括线程、DDL（数据定义语言）、rows（行数据）、schema（模式结构）、tables（表结构）以及 Compressed（压缩信息）等。这些信息对于用户来说是非常有价值的，因为可以作为判断备份进度的依据，也可以帮助用户在出现问题时进行调试和排查。

备份 API 的命令行示例如下。

```
# 指定库导出
shell $>mysqlsh shadmin@198.168.1.1:3380 --util dump-schemas employees
--outputUrl=/tmp/dump

Please provide the password for 'shadmin@198.168.1.1:3380': ******
Save password for 'shadmin@198.168.1.1:3380'? [Y]es/[N]o/Ne[v]er (default No): N

Acquiring global read lock
Global read lock acquired
Initializing - done
1 schemas will be dumped and within them 11 tables, 2 views.
Gathering information - done
All transactions have been started
Locking instance for backup
Global read lock has been released
Writing global DDL files
NOTE: Could not select columns to be used as an index for table `employees`.`employees01`. Chunking has been disabled for this table, data will be dumped to a single file.
Running data dump using 4 threads.
NOTE: Progress information uses estimated values and may not be accurate.
Writing schema metadata - done
Writing DDL - done
Writing table metadata - done
Starting data dump100% (2.52M rows / ~2.51M rows), 744.64K rows/s, 28.66 MB/s uncompressed, 7.84 MB/s compressed
Dump duration: 00:00:03s
Total duration: 00:00:03s
Schemas dumped: 1
Tables dumped: 11
Uncompressed data size: 96.44 MB
Compressed data size: 27.11 MB
Compression ratio: 3.6
Rows written: 2518358
Bytes written: 27.11 MB
Average uncompressed throughput: 27.77 MB/s
Average compressed throughput: 7.81 MB/s
```

上述备份生成的文件结构如下所示。

MySQL Shell 的备份功能允许为不同的对象（如表、数据库或整个实例）生成单独的文件，这极大地简化了备份的管理和存储。每个对象都可以独立备份到一个文件中，在需要恢复特定的表或数据时，只需要恢复相应的文件，无须恢复整个数据库的备份。这种粒度级别的恢复能力大大提高了备份的灵活性和效率。

图 3-32 所示为两个以 .json 为扩展名的 API 备份文件内容，这些文件用于记录备份的汇总信息和一致性位置信息。其中，done.json 包含了当前数据库中所有表的名称和每张表的总行数，这是非常有用的信息，可以帮助管理员快速了解备份的概况。此外，.json 文件还可能包含 MySQL 的版本信息、导出时的 binlog 位置等元数据，这些都是在恢复过程中需要的重要信息。

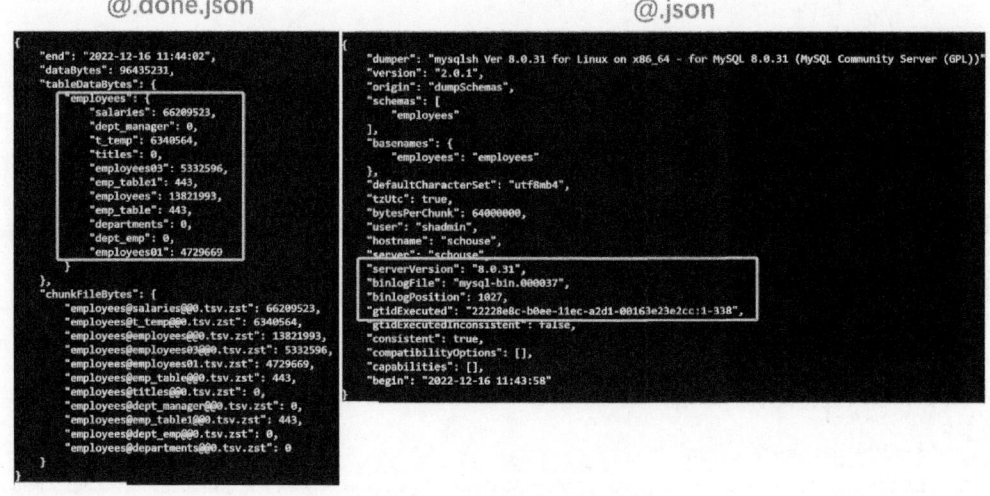

图 3-32　API 备份文件内容

表 3-31 所示为 MySQL Shell 备份 API 的核心控制参数。

表 3-31　MySQL Shell 备份 API 的核心控制参数

参　　数	说　　明
dryRun	显示 MySQL 数据库服务兼容性检查结果的信息
showProgress	显示（True）或隐藏（False）执行的进度信息
maxRate	备份期间每个线程每秒用于数据读取吞吐量的最大字节数
defaultCharacterSet	导出时使用的字符集
consistent	备份一致数据，使用 FLUSH TABLES WITH READ LOCK 语句，设置全局读锁，或者使用 LOCK TABLES 语句

(续)

参数	说明
threads	并行线程数。每个线程都有自己的 MySQL 实例连接。默认值为 4
compression	写入数据转储文件时使用的压缩算法：none、gzip、zstd。默认值：zstd

除此之外，MySQL Shell 备份 API 的 dump-instance（适用于整个 MySQL 实例的备份）和 dump-tables（适用于特定数据库表的备份）命令的使用相对直观与简单。两个备份 API 的命令行示例如下。

```
# 实例备份 util dump-instance <outputUrl> [<options>]
shell $> mysqlsh root@localhost --util dump-instance /tmp/dump
--excludeSchemas=employees

# 指定表备份 util dump-tables <schema><tables> --outputUrl=<str> [<options>]
shell $> mysqlsh shadmin@198.168.1.1:3380 -- util dump-tables employees
employees03 --outputUrl=/tmp/dump
```

2. 恢复 API

MySQL Shell 的恢复 API 提供的 loadDump 命令用于从之前创建的逻辑备份中恢复数据。loadDump 相对于传统的恢复方法具有下列一些明显的优点。

- 支持多线程并行操作，这意味着它可以同时从多个备份文件中恢复数据，从而大大提高恢复速度。这对于大型数据库或需要快速恢复的场景来说非常有用。
- 支持断点续传功能。
- 通过 progressFile 文件，loadDump 能够记录恢复操作的进度。如果在恢复过程中遇到中断（例如，由于系统崩溃或手动停止），则它可以从上次完成的地方继续执行，而不是重新开始。这大大减少了恢复操作所需的时间和复杂性。
- 支持延迟创建二级索引。在恢复数据时，支持延迟创建二级索引。这意味着在数据导入过程中不会立即创建索引，而是在所有数据导入完成后再创建。这可以显著提高数据导入的速度，特别是在有大量索引的数据库中。
- 使用 LOAD DATA LOCAL INFILE 命令导入数据。这意味着数据导入是在单个事务中完成的。这确保了数据的一致性，并减少了在恢复过程中可能出现的问题。
- 自动切割大文件。如果单个备份文件过大，则可以自动将其切割成较小的块，然后分别恢复。这样做可以避免产生过大的事务，从而减少恢复过程中的资源消耗和潜在的问题。

使用 loadDump 命令来恢复备份数据时，MySQL Shell 会自动读取备份文件夹中的元数据，并依据这些信息来重建表结构、索引以及加载数据。loadDump 命令恢复的示例如下。

```
MySQL-JS > util.loadDump("/tmp/dump/")
Loading DDL and Data from '/tmp/dump/' using 4 threads.
Opening dump...
Target is MySQL 8.0.31. Dump was produced from MySQL 8.0.31
NOTE: Load progress file detected. Load will be resumed from where it was left, assuming no external updates were made.
You may enable the 'resetProgress' option to discard progress for this MySQL instance and force it to be completely reloaded.
Scanning metadata - done
Executing common preamble SQL
Executing DDL - done
```

```
Executing view DDL - done
Starting data load
Executing common postamble SQL                    100% (257 bytes / 257 bytes),
0.00 B/s, 2 / 2 tables done
Recreating indexes - done
2 chunks (11 rows, 257 bytes) for 2 tables in 1 schemas were loaded in 0 sec (avg
throughput 257.00 B/s)
0 warnings were reported during the load.
```

loadDump 命令支持对单表结构创建和数据恢复操作。单表操作的命令行示例如下。

```
# 单独导入表结构
MySQL-JS > util.loadDump("/tmp/dump", { includeTables:["worldb.city"], loadDdl:
true,  loadData: false})
```

loadDump 命令输出执行日志以 .json 为扩展名的文件中 JSON 记录的信息,如下所示。

```
shell $》cat load-progress.22228e8c-b0ee-11ec-a2d1-00163e23e2cc.json
{"op":" SERVER-UUID","done": true,"timestamp": 1672108993998,"uuid":"22228e8c-b0ee-11ec-
a2d1-00163e23e2cc"}
{"op":"SCHEMA-DDL","done":false,"timestamp":1672109012555,"schema":"worldb"}
{"op":"TABLE-DDL","done":false,"timestamp":1672109012555,"schema":"worldb","table":"city"}
{"op":"TABLE-DDL","done":false,"timestamp":1672109012555,"schema":"worldb","table":"ct"}
{"op":"SCHEMA-DDL","done":true,"timestamp":1672109012578,"schema":"worldb"}
{"op":"TABLE-DATA","done":false,"timestamp":1672109012579,"schema":"worldb","table":
"city","chunk":0}
{"op":"TABLE-DATA","done":false,"timestamp":1672109012579,"schema":"worldb","table":
"ct","chunk":-1}
{"op":"TABLE-DATA","done":true,"timestamp":1672109012593,"schema":"worldb","table":
"city","chunk":0,"bytes":210,"raw_bytes":148,"rows":9}
```

3. 其他 API 功能

MySQL Shell 还提供了 exportTable 和 importTable 的 API,分别用于导出和导入单个表的数据。

(1) exportTable

MySQL Shell 的 exportTable 用于将数据库表的数据导出到文件中。这个功能通过执行 SQL 查询来获取数据,并将这些数据以用户指定的格式保存到文件中。与标准的 SQL SELECT ... INTO OUTFILE 语句类似,exportTable 支持多种导出文件格式,包括 default、csv、csv-unix 和 tsv。在使用 exportTable 时,用户还可以为导出的数据文件指定一系列的字段和行处理选项,包括指定行分隔符(TerminatedBy)、字段分隔符(fieldsTerminatedBy)、字段引用符(fieldsEnclosedBy)、字符型字段引用符(fieldsOptionallyEnclosed)和转义字符(fieldsScapedBy)等,以便按照用户的需求定制输出文件的格式。

表 3-32 所示为 MySQL Shell 的 exportTable API 控制字符符号列表。

表 3-32 exportTable API 控制字符符号

方言	行分隔符	字段分隔符	字段引用符	字符型字段引用符	转义字符
default	[LF]	[TAB]	[empty]	false	\
csv	[CR][LF]	,	"	true	\
csv-unix	[LF]	,	"	false	\
tsv	[CR][LF]	[TAB]	"	true	\

MySQL Shell 的 exportTable 的命令行示例如下。

```
MySQL-JS > util.exportTable("test.employees", "/tmp/employees.txt")
Initializing - done
Gathering information - done
Running data dump using 1 thread.
NOTE: Progress information uses estimated values and may not be accurate.
Starting data dump100% (300.02K rows / ~299.47K rows), 0.00 rows/s, 0.00 B/s
Dump duration: 00:00:00s
Total duration: 00:00:00s
Data size: 13.82 MB
Rows written: 300024
Bytes written: 13.82 MB
Average throughput: 13.82 MB/s

The dump can be loaded using:
util.importTable("/tmp/employees.txt", {
    "characterSet": "utf8mb4",
    "schema": "test",
    "table": "employees"
})
```

（2）importTable

MySQL Shell 的 importTable 的功能是将数据导入到 MySQL 数据库表中。这个功能允许用户从外部文件读取数据，并将这些数据导入到指定的数据库表中。importTable 支持多种文件格式，如 CSV、TSV 等，用户可以根据文件的实际格式进行选择。它还允许用户指定一些导入选项，如字段分隔符、字段引用符、行分隔符等，以便正确地解析文件中的数据。用户还可以指定导入模式，如是否替换现有数据、是否忽略错误等。

importTable 功能与 MySQL 的 LOAD DATA LOCAL INFILE 语句在功能上非常相似，都旨在高效地处理大量数据的导入。因此目标 MySQL 服务上的 LOCAL_INFILE 参数必须设置为 ON。MySQL Shell 的 importTable 的命令行示例如下。

```
MySQL-JS > \sql SET GLOBAL local_infile = 1;

MySQL-JS > util.importTable("/tmp/employees.txt",
{"characterSet": "utf8mb4","schema": "test","table": "employees",
"threads":4, "showProgress": true})

Importing from file '/tmp/employees.txt' to table `test`.`employees` in MySQL Server at 127.0.0.1:3380 using 1 thread
[Worker000] employees.txt: Records: 300024  Deleted: 0  Skipped: 0  Warnings: 0 100% (13.82 MB / 13.82 MB), 2.46 MB/s
File '/tmp/employees.txt' (13.82 MB) was imported in 6.0851 sec at 2.27 MB/s
    Total rows affected in test.employees: Records: 300024  Deleted: 0  Skipped: 0  Warnings: 0
```

表 3-33 所示为 importTable API 控制字符符号列表。

表 3-33 importTable API 控制字符符号

方言	行分隔符	字段分隔符	字段引用符	字符型字段引用符	转义字符
default	[LF]	[TAB]	[empty]	false	\
csv	[CR][LF]	,	"	true	\
csv-unix	[LF]	,	"	false	\
tsv	[CR][LF]	[TAB]	"	true	\

4. 总结

在 MySQL 8.0.27 版本之后，对 MySQL Shell 逻辑备份恢复 API 进行了重要的改进和稳定化，这为用户提供了更加高效和可靠的备份恢复解决方案。对于寻求高效备份恢复解决方案的用户来说，MySQL Shell 的逻辑备份恢复 API 是一个值得考虑的选项。

表 3-34 所示为 MySQL 的逻辑备份常用工具对比总结。

表 3-34 MySQL 逻辑备份常用工具对比

	mysqldump	mysqlpump （高版本弃用）	MyDumper	MySQL Shell 备份恢复 API
线程	单线程	多线程	多线程	多线程
压缩	支持	支持	支持	支持
远程	支持	支持	支持	支持
备份速度	慢	快	快	快
恢复速度	慢	慢	快	快
分割	不支持	不支持	支持	支持
第三方存储	不支持	不支持	不支持	支持

第 4 章
MySQL 8.0 的升级与迁移

升级 MySQL 数据库版本是为了获得更好的性能、功能或安全性。迁移是为了将数据从一个硬件设备或云平台迁移到另一个，或者将数据从一个版本的 MySQL 数据库迁移到另一个版本。迁移过程中需要保证数据的安全性和完整性，并通过迁移提高数据库的性能和可用性。

4.1 MySQL 8.0 的版本升级

对于应用软件来说，在使用一段时间后，不可避免地要进行升级，MySQL 数据库也是如此。把 MySQL 数据库升级到 8.0 版本主要出于以下考虑。
- 提升安全性（修复安全漏洞）。新版本会解决目前存在的安全问题。
- 提升性能和稳定性，如提供新的事务调度算法、并行复制和异步复制等。
- 提供新的高可用架构 MGR，实现全面的支持。
- 提供新的功能：Hash join、窗口函数、DDL 即时、JSON 支持、MGR 复制、直方图等。
- 进行运维规划：原始环境中版本太多，可统一管理版本。
- 修复版本 bug。低版本中很多 bug 无法修复，需要升级到 MySQL 8.0 版本。
- MySQL 的低版本维护周期已结束（大版本生命周期为 8 年）。

对于 MySQL 数据库升级，其间包含很多细致的工作，如方案确认、版本确认、功能确认、测试、准备、备份、验证、高可用切换等。前期需要投入很多精力进行准备，这样才能做到一步到位。

4.1.1 MySQL 数据库升级的方法

MySQL 版本升级，对于不同的场景，可以分为小版本升级和大版本升级。

小版本升级是指在同一个大版本序列内，选择最后位最大数值的软件版本升级。例如，在 MySQL 5.7 版本中，从 MySQL 5.7.21 升级到 MySQL 5.7.34 最新版本。小版本升级是为了修复 bug，以及添加一些功能点。具体特点如下。
- 不改变数据文件，升级速度快。
- 不可以跨操作系统、大版本进行升级。
- 升级方式采用就地升级（In-Place Upgrade）。

大版本升级是指从一个大序列到另一个更大序列的软件版本升级。例如，从 MySQL 5.7.×升级到 MySQL 8.0.×版本。MySQL 8.0 版本的体系架构有很大的变化，包括底层结构的变化、优化器算法的变更。大版本升级，需要做诸多兼容性测试和验证。具体特点如下。

- 适合不同操作系统中的 MySQL 版本升级，升级速度慢。
- 逻辑上可跨机器或操作系统进行升级。
- 搭建高配置从库，安装新版本 MySQL，配置从库同步。
- 使用第三方工具把数据逻辑导出，之后导入到新的版本环境中。

图 4-1 所示为 MySQL 大版本升级的三种常用方式。

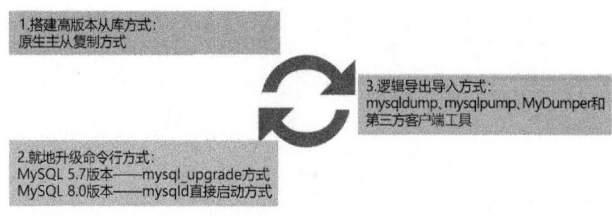

图 4-1 MySQL 常用升级方式

1. 搭建高版本从库方式

搭建一个全新的高版本从库，需要准备新的硬件资源和新版本从库软件。

MySQL 支持从一个版本系列复制到更高版本系列。可以从 MySQL 5.6 的源复制到运行 MySQL 5.7 的副本，从运行 MySQL 5.7 的源复制到运行 MySQL 8.0 的副本，以此类推。但是，在 MySQL 复制过程中，如果源端（主服务器）使用的某些特性或语句在新版本的从端（副本或复制服务器）上不再被支持，就可能导致复制中断或数据不一致。这通常发生在从一个较旧的 MySQL 版本迁移到一个新的大版本时。例如，MySQL 8.0 版本不再支持超过 64 个字符的外键名称、新版本 binlog 不兼容旧版本 binlog 等。

2. 就地升级命令行方式

就地升级（In-Place Upgrade）命令行方式是指先安装新版本的 MySQL 软件，之后使用 MySQL 安装包自带的 mysql_upgrade 执行脚本来完成升级。优先从库升级，之后切换主库。具体步骤如下。

- 新版本的安装包替代老版本的安装包。
- 以 mysql_upgrade 命令方式升级（MySQL 8.0 使用 mysqld --upgrade 参数）。
- 在线升级，主节点切换后确保原连接的会话已释放。

3. 逻辑导出和导入方式

图 4-2 所示为采用 mysqldump、mysqlpump、MyDumper 和 Navicate 等逻辑处理工具，实现数据的逻辑导出和导入。具体步骤如下。

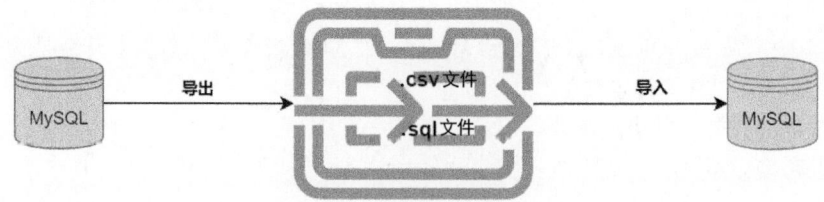

图 4-2 逻辑导出和导入方式

- 先导入库、表结构。
- 之后导入常规数据。
- 导入用户表信息。
- 导入除库、表对象之外的对象（存储过程、触发器、视图和事件）。
- 特殊类型处理，如 Blob、Text、Bit 字段的导入。

4.1.2 MySQL 8.0 升级的注意事项

1. MySQL 升级前涉及的工作内容

MySQL 升级前需要做的工作如下。
- 通过测试环境验证新版本数据库是否完全兼容已有应用。
- 确定新版本数据库不兼容的命令是否影响高可用复制。
- 确定 my.cnf 配置文件需要删除、废弃的参数和需要新配置的参数。
- 在保证系统稳定性的情况下选择平稳的过渡方式。例如，先升级一个从库，观察一段时间，之后升级所有从库。
- 获取准确的升级窗口时间，保证在最少停机时间内完成升级。
- 升级完之后，对新数据库进行验证，如版本是否正确、运维工具是否兼容、行数和表的数量是否与升级前一致等。
- 升级当中碰到未知问题的应对方式。
- 考虑回退方案。
- 升级前对数据库进行完全备份。

2. 深入了解 MySQL 8.0 特性

对于 MySQL 8.0 的新特性，需要进行深入了解，包括：
- 新旧功能的变化。升级前需要基本了解 MySQL 8.0 的一些功能，包括 Added in（添加）功能、Features Deprecated（弃用）功能、Features Removed（移除）功能等，如图 4-3 所示。详细介绍参见官网地址：https://dev.mysql.com/doc/refman/8.0/en/mysql-nutshell.html。
- 关键字是否兼容。有些关键字在命令行中是无法直接使用的。例如，直接使用关键字 added in 就会提示无法识别错误。详细介绍参见官网地址：https://dev.mysql.com/doc/refman/8.0/en/keywords.html。
- SQL 语句是否兼容。例如，在 MySQL 5.6 中可以执行 "SELECT id,count(*)FROM GROUPBY name" 语句，但在 MySQL 5.7 和 MySQL 8.0 中，如果 SQL_MODE 不一样，会导致得到不一样的结果集。
- 数据文件存储格式是否支持直接升级。新版本存储文件格式无法完全兼容。
- 现有版本中的数据库对象是否兼容新版本。例如，自定义函数、不规范的 SQL 语句，容易在新版本中无法执行。
- 默认密码策略的变更。MySQL 8.0 默认密码策略由 mysql_native_password 变更为 caching_sha2_password，旧的数据库驱动软件无法连接新版本数据库。

3. 需要注意的技术关键点

表 4-1 所示为 MySQL 8.0 升级时需要注意的技术关键点。

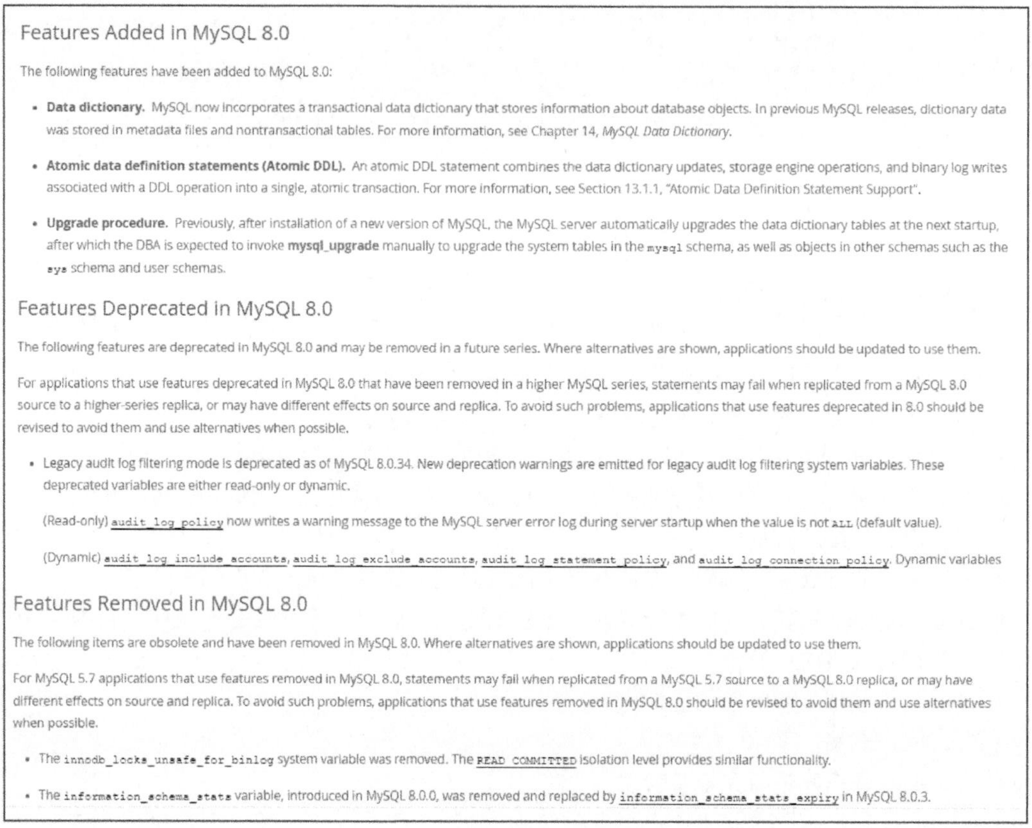

Features Added in MySQL 8.0

The following features have been added to MySQL 8.0:

- **Data dictionary.** MySQL now incorporates a transactional data dictionary that stores information about database objects. In previous MySQL releases, dictionary data was stored in metadata files and nontransactional tables. For more information, see Chapter 14, *MySQL Data Dictionary*.
- **Atomic data definition statements (Atomic DDL).** An atomic DDL statement combines the data dictionary updates, storage engine operations, and binary log writes associated with a DDL operation into a single, atomic transaction. For more information, see Section 13.1.1, "Atomic Data Definition Statement Support".
- **Upgrade procedure.** Previously, after installation of a new version of MySQL, the MySQL server automatically upgrades the data dictionary tables at the next startup, after which the DBA is expected to invoke `mysql_upgrade` manually to upgrade the system tables in the `mysql` schema, as well as objects in other schemas such as the `sys` schema and user schemas.

Features Deprecated in MySQL 8.0

The following features are deprecated in MySQL 8.0 and may be removed in a future series. Where alternatives are shown, applications should be updated to use them.

For applications that use features deprecated in MySQL 8.0 that have been removed in a higher MySQL series, statements may fail when replicated from a MySQL 8.0 source to a higher-series replica, or may have different effects on source and replica. To avoid such problems, applications that use features deprecated in 8.0 should be revised to avoid them and use alternatives when possible.

- Legacy audit log filtering mode is deprecated as of MySQL 8.0.34. New deprecation warnings are emitted for legacy audit log filtering system variables. These deprecated variables are either read-only or dynamic.

  (Read-only) `audit_log_policy` now writes a warning message to the MySQL server error log during server startup when the value is not `ALL` (default value).

  (Dynamic) `audit_log_include_accounts`, `audit_log_exclude_accounts`, `audit_log_statement_policy`, and `audit_log_connection_policy`. Dynamic variables

Features Removed in MySQL 8.0

The following items are obsolete and have been removed in MySQL 8.0. Where alternatives are shown, applications should be updated to use them.

For MySQL 5.7 applications that use features removed in MySQL 8.0, statements may fail when replicated from a MySQL 5.7 source to a MySQL 8.0 replica, or may have different effects on source and replica. To avoid such problems, applications that use features removed in MySQL 8.0 should be revised to avoid them and use alternatives when possible.

- The `innodb_locks_unsafe_for_binlog` system variable was removed. The `READ COMMITTED` isolation level provides similar functionality.
- The `information_schema_stats` variable, introduced in MySQL 8.0.0, was removed and replaced by `information_schema_stats_expiry` in MySQL 8.0.3.

图 4-3　MySQL 8.0 功能信息

表 4-1　MySQL 8.0 升级时需要注意的技术关键点

序号	技术关键点	说　　明
1	SQL_MODE 设置	SQL_MODE 控制 SQL 语句的规范。不同版本之间 SQL_MODE 的设置不一样
2	客户端 JDBC 驱动	不同的驱动支持的 MySQL 版本不同，设置参数也存在差异
3	密码过期，SSL 开启	default_password_lifetime＝0 表示密码不过期。skip_ssl 设置是否开启 SSL 认证
4	MySQL 8.0 密码策略	default_authentication_plugin 设置默认身份验证插件
5	分区表	是否支持分区表的升级，分区表的定义是否符合 MySQL 的要求
6	Percona Toolkit 和 XtraBackup	开源运维工具的支持
7	参数 innodb_file_format	底层数据文件格式保存参数
8	Text、Blob、Bit 等特殊类型字段	大字段是否支持升级
9	datetime、timestamp	时间范围是否超出设置值
10	文件路径	basedir、binlog、relay 路径
11	版本兼容性检查是否开启	如 show_compatibility_56 参数
12	innodb_fast_shutdown 参数	当关闭服务时，数据是否刷新到数据文件中

(续)

序号	技术关键点	说明
13	自定义函数，一些不规范的 SQL 语句	
14	Optimizer_switch 的执行计划变更	
15	Features Deprecate 弃用参数	
16	以 MySQL 高版本启动一次之后，无法用旧版本启动 MySQL 服务	
17	新版本软件安装方式	MySQL 支持 RPM 包、TAR 包和源码包安装方式。不同的安装方式会影响现有使用 MySQL 情况
18	字符集不一致	若采用的字符集不一致，则会导致保存数据格式变化、关联查询索引失效等问题
19	升级参数的选择	mysql_upgrade 或 mysqld 参数对数据更新存在不同的影响

下面抽取几个技术关键点进行说明。

（1）升级之后客户端 JDBC 驱动兼容问题

MySQL 自带多种驱动程序，这些驱动程序可以帮助开发者更加高效地操作数据库。JDBC 驱动程序、ODBC 驱动程序、.NET 驱动程序和 C++驱动程序分别适用于不同的开发语言与应用场景。不同的版本，驱动程序不一样。图 4-4 所示为 JDBC 对应 MySQL 不同的版本支持信息。

Table 2.1 Summary of Connector/J Versions

Connector/J version	JDBC version	MySQL Server version	JRE Required	JDK Required for Compilation	Status
8.0	4.2 [1]	5.6, 5.7, 8.0	JRE 8 or higher	JDK 8.0 or higher [3]	General availability. Recommended version.
5.1	3.0, 4.0, 4.1, 4.2	5.6 [2], 5.7 [2], 8.0 [2]	JRE 5 or higher [2]	JDK 5.0 AND JDK 8.0 or higher [3]	General availability

图 4-4 MySQL JDBC 版本兼容

（2）innodb_fast_shutdown 参数设置

innodb_fast_shutdown 参数用于设置 MySQL InnoDB 引擎的关闭模式。

- 设置 0：完成所有的 full purge 和 merge insert buffer 操作。
- 设置 1：默认，不需要完成上述操作，但会刷新缓冲池中的脏页。
- 设置 2：不完成上述两个操作，而是将日志写入日志文件，下次启动时，会执行恢复操作 recovery。
- 没有正常地关闭数据库。例如，使用 kill 命令或 innodb_fast_shutdown＝2 时，需要进行恢复操作。

（3）RPM 安装和 TAR 安装的区别

RPM 安装路径是固定的，所以在一个系统上只能安装一个 MySQL 版本。TAR 安装方式可以灵活地选择安装目录，可以安装多个不同的 MySQL 版本。需要选择合理的软件安装方式。

（4）字符集不一致

字符集的排序顺序和大小写敏感设置需要保证一致。若字符集不一致，则数据排列顺序有所变化，返回的结果集会不一样。

（5）mysql_upgrade 参数使用注意事项

该参数表示可以选择性升级对应内容，如下所示。

- -s，--upgrade-system-tables：只升级系统表，不尝试升级数据。

- -f, --force：遇到警告错误，强制执行。

4. 升级之后的检查项

当升级到 MySQL 8.0 版本之后，升级任务并没有结束，而是需要进行检查工作。下面是涉及的具体检查事项：

- 升级完成之后，需要重启数据库，确认是否能正常启动。
- 检查升级之后的版本信息是否一致。
- 升级完成之后，检查命令行系统时间是否正常获取。
- 检查 Blob、Text 和 Bit 字段。版本升级时字段容易截断，丢失数据。
- 检查存储过程、触发器和事件和功能是否正常。
- 检查高可用节点切换是否正常。
- 检查业务用户是否能正常访问数据库服务。

5. 升级中错误处理

MySQL 的 Error 日志检查对于确认是否升级成功而言必不可少，下面列举了一些升级中的常见错误。

（1）文件格式 innodb-file-format-check 的警告信息

文件格式 innodb-file-format-check 值 Barracuda 无效。该警告信息如下所示。

```
[Warning] /usr/local/mysql-5.7/bin/mysqld: ignoring option '-innodb-file-format-check' due to invalid value 'Barracuda'
```

说明：检查 innoDB 文件格式，因为值"Barracuda"无效。innodb-file-format-check 在 MySQL 5.7中不存在，需要注释掉配置参数。

（2）复制账号密码不安全的警告信息

MySQL 复制用户名和密码信息存储在 master 信息库中，这种方式是不安全的。具体警告信息如下所示。

```
[Warning] Storing MySQL user name or password information in the master info repository is not secure and is therefore not recommended. Please consider using the USER and PASSWORD connection options for START SLAVE; see the 'START SLAVE Syntax' in the MySQL Manual for more information.
```

说明：主要是由 MASTER_HEARTBEAT_PERIOD 参数导致的。

解决方法1：复制进程主节点心跳包时间间隔参数调大。命令行：CHANGE MASTER TO MASTER_HEARTBEAT_PERIOD = 30。

解决方法2：设置 slave_net_timeout = 3600（秒）。

（3）无效的服务参数的错误提示信息

当升级提示完成时，若访问数据库服务不正常，则会提示如下错误信息。

```
shell $>mysql -uroot -p ******
Error occurred: Cannot setup server variables
```

说明：无法设置服务器参数，如果在执行 MySQL 8.0 登录过程中提示此类错误，则说明错误是由权限导致的，在 my.cnf 文件中加入 skip-grant-tables 跳过就行了，执行完成后再去掉它。

（4）时间字段的错误提示信息

timestamp 字段默认值问题的错误提示如下所示。

```
Error Code: 1048.Column 'create_time' cannot be null
```

说明：列"create_time"不能为 null，即在 explicit_defaults_for_timestamp=on 的情况下，不允许出现 null 值。

在 MySQL 里，timestamp 的范围是'1970-01-01 00:00:01'~'2038-01-19 03:14:07'。早期的版本没有这方面的强行检查，所以需要人为干涉。

（5）数据字典的错误提示信息

MySQL 8.0 版本升级过程中数据字典的错误提示信息如下所示。

```
[ERROR][MY-011091][Server] Data dictionary upgrade prohibited by the
command line option '--no_dd_upgrade'.
[ERROR][MY-010020][Server] Data Dictionary initialization failed.
```

说明：禁止升级数据字典，数据字典初始化失败。表示数据字典功能已被禁用，也就是 MySQL 数据字典安装或者更新失败了。一旦数据字典操作失败，MySQL 就会无法正常运行。需要新的数据文件，重新升级。

6. 总结

MySQL 版本升级不是一件简单的事情，需要花费一定的精力和编写计划，还需要协调好各方面，提前准备工作必不可少。这样才可以考虑全面，并且做到无论碰到什么问题，都能应付。

4.1.3 MySQL 5.7 升级至 MySQL 8.0 的步骤

为了平稳地升级到 MySQL 8.0 版本，官方提供了便利的功能。在 MySQL 8.0 中，mysql_upgrade 客户端已被弃用。升级客户端时执行的操作现在由服务完成。使用新版本的 MySQL 8.0 软件，指定数据目录就可以完成升级。启动过程中，会自行升级完成数据字典和用户表。升级之后也不需要重新启动服务。

图 4-5 所示为 MySQL 5.7 升级到 MySQL 8.0 时官方推荐的升级方式。首先使用 MySQL Shell 进行升级前的兼容性检查，在最终确认没问题之后，用新版本软件进行启动。

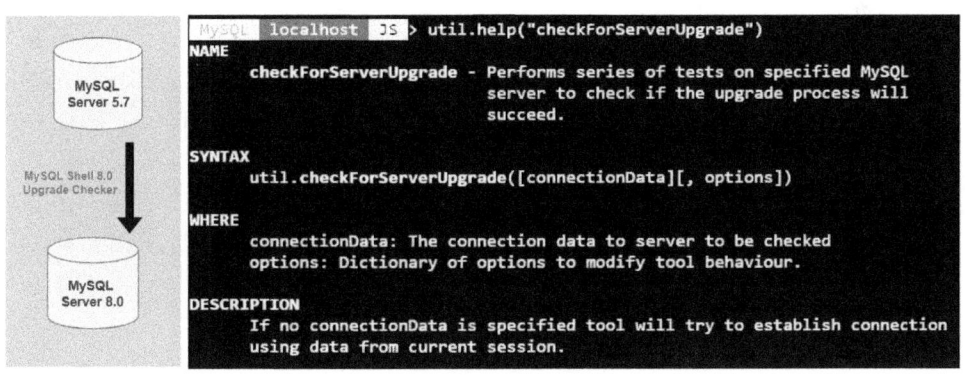

图 4-5　MySQL 8.0 官方推荐升级方式

图 4-6 所示为官方提供的 MySQL Shell 进行升级前的兼容性检查的步骤，提示存在的错误、警告和通知等内容。

图 4-7 所示为按照提示的要求需要进行更改。错误提示信息必须修复，否则升级会失败。

```
[root@schouse idc]# mysqlsh -- util checkForServerUpgrade  dbadmin:123456@172.17.27.48:3306 --target-version=8.0.29 --output-format=JSON
--config-path=/etc/my5.7.cnf
{
    "serverAddress": "172.17.27.48:3306",
    "serverVersion": "5.7.37-log - MySQL Community Server (GPL)",
    "targetVersion": "8.0.29",
    "errorCount": 1,
    "warningCount": 14,
    "noticeCount": 2,
    "summary": "1 errors were found. Please correct these issues before upgrading to avoid compatibility issues.",
    "checksPerformed": [
        {
            "id": "oldTemporalCheck",
            "title": "Usage of old temporal type",
            "status": "OK",
            "detectedProblems": []
        },
        {
            "id": "reservedKeywordsCheck",
            "title": "Usage of db objects with names conflicting with new reserved keywords",
            "status": "OK",
            "detectedProblems": []
        },
        {
            "id": "utf8mb3Check",
            "title": "Usage of utf8mb3 charset",
            "status": "OK",
            "detectedProblems": []
15) Removed system variables
  Error: Following system variables that were detected as being used will be
    removed. Please update your system to not rely on them before the upgrade.
  More information:
    https://dev.mysql.com/doc/refman/8.0/en/added-deprecated-removed.html#optvars-removed

  innodb_file_format - is set and will be removed

16) System variables with new default values
  Warning: Following system variables that are not defined in your
    configuration file will have new default values. Please review if you rely on
    their current values and if so define them before performing upgrade.
  More information:
    https://mysqlserverteam.com/new-defaults-in-mysql-8-0/

  back_log - default value will change
  event_scheduler - default value will change from OFF to ON
  innodb_autoinc_lock_mode - default value will change from 1 (consecutive) to
    2 (interleaved)
  innodb_flush_neighbors - default value will change from 1 (enable) to 0
    (disable)
  innodb_max_dirty_pages_pct - default value will change from 75 (%)  90 (%)
  innodb_max_dirty_pages_pct_lwm - default value will change from 0 (%) to 10

Errors:   1
Warnings: 14
Notices:  2
```

图 4-6　MySQL Shell 升级前的兼容性检查

```
15) Removed system variables
  Error: Following system variables that were detected as being used will be
    removed. Please update your system to not rely on them before the upgrade.
  More information:
    https://dev.mysql.com/doc/refman/8.0/en/added-deprecated-removed.html#optvars-removed

  innodb_file_format - is set and will be removed
  innodb_support_xa - is set and will be removed
  log_warnings - is set and will be removed, consider using log_error_verbosity
    instead
  query_cache_size - is set and will be removed
  query_cache_type - is set and will be removed
```

图 4-7　MySQL Shell 提示信息

1. MySQL 8.0 升级操作

MySQL 8.0 升级具体步骤如下。

（1）登录 MySQL 5.7 正常关闭服务

将 innodb_fast_shutdown 的参数设置为 0（默认值为 1，表示在 InnoDB 引擎关闭时，不需要清除所有数据和日志，以及合并缓存里的插入数据，只需要刷脏页。在设置为默认值后，当用 MySQL 8.0 第一次启动服务时，会出现无法兼容问题而报错），之后安全关闭。操作命令行如下所示。

```
# 关闭 InnoDB 参数前的确认
mysql> show variables like 'innodb_fast_shutdown';
+----------------------+-------+
|Variable_name         |Value  |
+----------------------+-------+
|innodb_fast_shutdown  |1      |
+----------------------+-------+
1 row in set (0.00 sec)

# 确保数据已刷到硬盘上,将该参数值更改成 0
mysql> set global innodb_fast_shutdown=0;
Query OK, 0 rows affected (0.01 sec)

mysql> shutdown;
Query OK, 0 rows affected (0.00 sec)
```

（2）使用 MySQL 8.0 直接启动服务

MySQL 服务启动之后，查看 MySQL 错误日志。命令行如下所示。

```
# 启动 MySQL 服务
shell $> /opt/mysql8.0.34/bin/mysqld_safe --defaults-file=/etc/my3400.cnf
--user=mysql&
[1] 15400
shell $> 2023-04-25T13:07:16.591560Z mysqld_safe Logging to
'/opt/data3400/logs/error.log'.
2023-04-25T13:07:16.636879Z mysqld_safe Starting mysqld daemon with databases from
/opt/data3400/mysql

# 打开另一个窗口来查看错误日志
shell $> tail -f /opt/data3400/logs/mysql_error.log
...
```

2. MySQL 8.0 升级之后的基本验证

登录 MySQL 服务，检查是否成功升级为 MySQL 8.0 版本。具体命令行如下所示。

```
# 登录服务确认
shell $>mysql -uroot -p -S /opt/data3400/mysql/mysql.sock
Enter password:
Welcome to the MySQL monitor.  Commands end with ; or \g.
Your MySQL connection id is 10
Server version: 8.0.34 MySQL Community Server - GPL
Copyright (c) 2000, 2020, Oracle and/or its affiliates.All rights reserved.
Oracle is a registered trademark of Oracle Corporation and/or its
affiliates.Other names may be trademarks of their respective
owners.
Type 'help;' or '\h' for help.Type '\c' to clear the current input statement.

# 查看版本信息
mysql> select version();
```

```
+-----------+
| version() |
+-----------+
| 8.0.34    |
+-----------+
1 row in set (0.01 sec)

# 基础的数据字典确认是否存在 MyISAM 引擎表
mysql> SELECT table_schema,table_name,engine FROM information_schema.tables
WHERE engine='MyISAM'
        AND TABLE_SCHEMA  IN ('INFORMATION_SCHEMA','performance_schema',
        'sys','mysql');

# 表数量确认
mysql> SELECT TABLE_SCHEMA, ENGINE,COUNT(*)
    FROM INFORMATION_SCHEMA.TABLES
    WHERE TABLE_SCHEMA NOT IN ('INFORMATION_SCHEMA','PERFORMANCE_SCHEMA','SYS','MYSQL')
                AND TABLE_TYPE='BASE TABLE'
                GROUP BY TABLE_SCHEMA, ENGINE;

# 查看用户信息
mysql> SELECT user,host,plugin,password_expired,password_lifetime,account_locked from mysql.user;
```

说明：MySQL 5.7 版本和 MySQL 8.0 版本的最大区别就是前者系统表里无 MyISAM 引擎表。在高可用场景下，可以先执行从节点，确认没问题后再升级主节点。剩下的就是检查业务是否能正常访问和使用数据库。

3. 总结

MySQL 8.0 版本的升级步骤还是比较简单的。操作前一定要做好备份。对于使用 MySQL 8.0 版本启动过一次的数据库，无法使用 MySQL 5.7 版本再次启动。如果升级中出现问题，则需要删除升级过程中的所有文件，从头开始操作。图 4-8 所示为升级到 MySQL 8.0 的事项。

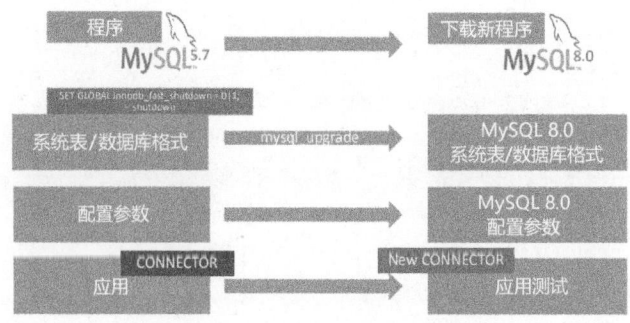

图 4-8　升级到 MySQL 8.0 的事项

4.2 MySQL 8.0 的迁移

一直以来，应用系统中，诸多企业以商业数据库为主（如 Oracle、DB2、SQL Server）。但随着业务的移动化和互联网化，IT 应用架构向微服务化转变。在技术自主可控和降低 IT 成本的总体要求下，一些重要的业务系统开始使用 MySQL 数据库。除此之外，原先使用 MySQL 的企业，因为一些 IT 资源或策略的变更，需要从现有 IDC 机房迁移到另一个 IDC 机房，这样的场景下，就需要进行数据库迁移，如 Oracle 迁移到 MySQL、SQL Server 迁移到 MySQL、云 MySQL 迁移到本地 MySQL 等。

对于这种重大变更，存在着较大的迁移风险，如何合理选择迁移方案至关重要。

4.2.1 MySQL 数据库迁移方案设计

数据库迁移时要保证应用系统运行稳定性，减少因其他因素所造成的损失，并尽量减少停机时间，减小对业务中断的影响。因此，数据库迁移，应当从如下方面考虑：
- 系统最长允许的停机窗口时间。
- 系统的数据量。
- 迁移后业务程序的兼容性及性能稳定性。
- 迁移当中出现异常的处理。尤其是在业务已经在新平台运行以后出现异常时实施应急预案。
- 与迁移系统的关联系统的兼容性及数据库的性能稳定性。
- 迁移前后的数据一致性校验。

1. 迁移方案设计

常用的 MySQL 迁移方案有如下 3 种。
- 通过 MySQL 自身的复制功能实现迁移。需要测试效率和时间，如网络、速度等。
- 通过第三方开源软件进行迁移。需要测试迁移工具的效率和时间，如网络、速度等。
- 通过脚本+触发器等方式实现实时同步数据操作。需要编写比较多的脚本，进行严密的测试。

根据现状选择合适的迁移方案。数据量方面，可以选择全量迁移或增量迁移。全量迁移是将原数据库中的所有数据迁移到目标数据库中。增量迁移是只迁移原数据库中新增或修改的数据。

2. 实施流程

为了能够完整地实现 MySQL 数据库迁移，同时确保满足集中整合后系统的连续可用、高效运行需求，需要整理迁移的每个实施环节。图 4-9 所示为 MySQL 8.0 数据库升级迁移的每个实施环节。

表 4-2 所示为对每个实施环节的说明。

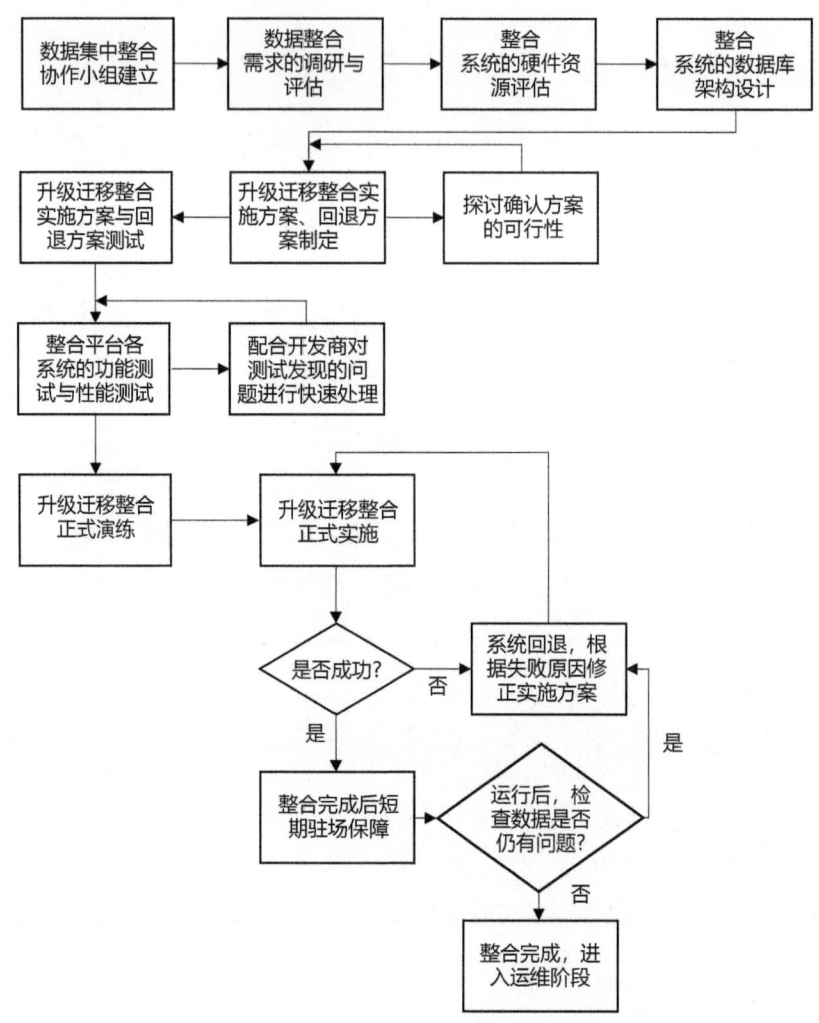

图 4-9　MySQL 8.0 迁移实施环节

表 4-2　实施环节说明

阶　　段	步骤说明
建立协助小组	建立协助小组，明确小组的组织架构和成员，确定各方职责权益，梳理流程和沟通方式，为整个实施推进管理提供必要的组织结构
需求的调研与评估	对需要进行 MySQL 数据库迁移的系统进行业务类型、操作特性以及数据等不同层面的调研，评估其数据集中的可行性以及可能存在的隐患等，通过评估最终确认需要集中的系统，并为后续确定 MySQL 数据库迁移方案提供所需信息
硬件资源评估	根据需要集中的多套系统的峰值性能信息叠加，推测集中之后的硬件资源需求量，为系统的硬件配置选型提供参照数据
数据库架构设计	根据需要 MySQL 数据库迁移的多套系统的硬件资源需求量以及系统的可靠性等需求，规划设计系统的数据库架构和参数，以满足 MySQL 数据库迁移后的业务需求
方案的制定与选型	根据需要 MySQL 数据库迁移的各系统数据量及停机时间窗口，选择合适的 MySQL 数据库迁移方法，并对每个需要 MySQL 数据库迁移的业务系统制定实施方案和回退方案

（续）

阶 段	步 骤 说 明
实施方案与回退方案测试	在测试环境下测试迁移方案和回退方案，确定方案的可行性和时间窗口
功能测试与性能测试	在完成各业务系统迁移的测试环境下测试各业务系统的运行情况，确定是否存在功能、性能、稳定性等方面的影响，并提出相应的解决方案。主要分为以下几种测试。 （1）迁移测试 ● 测试实施方案与实施步骤。 ● 测试相应的回退方案。 ● 确定步骤的可行性、难度、时间和风险。 （2）功能测试 ● 业务应用改变。 ● 数据库改变。 ● 主机名、IP 地址改变。 （3）性能测试 ● 硬件性能测试。 ● 压力测试。 ● 重点测试数据库改变所带来的性能变化
实施的正式演练	在有条件的情况下，严格按照整合实施方案流程进行升级迁移整合实施，修正实施方案中的所有问题，通过演练使各方熟悉协作配合过程
方案的正式实施	在用户给定的停机时间窗口实施各业务系统的 MySQL 数据库迁移，确保各系统数据迁移到目标平台上
实施回退	在正式实施时，若因出现故障造成失败，或者业务运行在目标平台后发现数据存在问题，则可参照回退方案将系统回退到原系统库中，以确保业务可用

3. 风险控制

数据迁移并非只是简单地将业务数据移动到新的数据库中，在进行迁移时，必须考虑系统中各业务数据的各种需求、特性等。因此，需要对系统进行深度调研，对迁移需求进行梳理，减少不合理系统迁移需求带来的风险，同时针对各种迁移可能面临的问题提出相应的应对策略，从而形成真正可行的数据迁移方案，确保迁移后各业务系统仍然能够稳定、高效运行。

数据迁移关键点如下。

（1）性能

根据业务情况预估充足的硬件资源，以及进行充分且全面的业务性能测试和各种优化，都是确保性能良好的关键。

因此，想要在新环境数据库中保证原有业务系统仍然能够高效运行，必须在迁移实施前进行充分的性能测试。需要对迁移系统平台进行功能、性能、压力、稳定性、完整性、系统扩展性等多维度测试，借助相关测试工具和方法，通过技术手段，实现模拟业务场景的实效测试，并对测试中的各种性能瓶颈进行优化处理，确保迁移系统上线运行时的高效性。

（2）数据完整性

在不同 MySQL 版本中，字符集、函数、复制、优化器等技术点均可能存在差异。数据迁移时，不合理的迁移方法、不正确的环境配置，都可能导致系统数据的紊乱，尤其是涉及多字节数据（如中文）时，很容易在数据迁移时形成乱码，因此必须对系统的各库字符集进行调研，同时测试异构字符集的数据是否可以完整地迁移到新系统库，确保数据的完整性。

（3）网络的互通性

各个业务库的网络地址均不相同，网段也不完全一样，迁移库采取的新IP地址是否能够被所有的终端应用和外围系统所访问，这是在迁移前首先需要确定的，这也意味着迁移时可能需要相关业务系统上的关联配合调整。

（4）应用对主机和系统的依赖性

在过去较早的应用中，有些应用将数据库服务器的IP地址或主机名写入程序中，或者应用需要访问数据库平台上的目录或文件。在这样的情况下，一旦在迁移时变更了数据库IP地址或主机名，或者新平台的目录结构、目录权限等发生变化，应用中的部分功能就会无法正常运行。对于此类应用系统，需要提前发现并规划相应的方案。

（5）系统回退需求

在系统迁移后，业务系统将在新的架构上运行，但由于个别系统特殊的业务特性，或者部分系统在切换到新系统运行后出现功能或性能方面的问题，由此导致系统不能正常支撑个别业务系统运行，需要进行回退，将业务切换回原有系统，同时需要保证在新系统运行时段内的数据不能丢失，能够带回原有系统。

表 4-3 所示为回退方案。

表 4-3 回退方案

风险点	造成后果	规避方法	回退策略
迁移实施中遇到不可解决的问题，如升级时遇到数据库 bug	迁移失败	最大程度重视迁移方案的测试及演练，形成标准化迁移实施文档	即时回退
迁移工作超时	迁移失败或者业务运营延误	最大程度重视迁移方案的测试及演练，形成标准化迁移实施文档。在测试过程中评估迁移时间，精确到分钟	即时回退

4. 总结

每种迁移方案的光环下都隐藏着专业工程师的汗水，每个数据迁移项目或任务都是一次未知的挑战，只有凭借丰富的经验和不断的演练，才能保证迁移成功。

4.2.2　MySQL 8.0 数据库迁移工具及注意事项

通过数据库迁移方法论，对现有生产环境中的 MySQL 数据库进行详细调研及梳理，就可以进行迁移了。

1. 迁移方法

抛开业务逻辑的因素，根据不同的版本、平台和停机时间需求，有不同的可选路径可决定迁移方法和工具。表 4-4 所示为常用的 MySQL 迁移方法。

表 4-4 迁移方法

类别	迁移方法	优点	缺点
异构	SQL LOAD	操作简单、速度快、选择数据范围灵活	需要自定义开发批量操作、无法支持 CLOB 等特殊字段
	OGG	商用软件，拥有广泛的数据库平台支持、灵活的复制架构、基于日志的实时数据同步、稳定性高等特点	对维护技能有一定的要求，费用高
	ETL 软件	使用简单方便，可定时同步	批量处理大量表时需要定制化配置
	MySQL 移植工具	安装简单，可自动创建表	不可定制、技术支持较弱
	定制迁移工具	可高度定制，保证最佳性能和最短停机时间	需要编写大量的代码和验证测试
同构	MySQL 基础复制	基于本身的复制机制，稳定且保证最短停机时间	异步实现

说明：不同的数据库版本、不同的组件安装、不同的应用开发特征都会增加迁移计划的复杂性和差异性。

下列第三方迁移工具还是比较不错的，可以访问其官方地址，具体了解一下。注意，有些工具在超过一定的数据量时会收费。同构数据库可以直接采取 MySQL 的复制功能，异构数据库按照需求选择即可。对于一些有开发能力的 DBA，可以尝试自己写 Python 脚本。

- OGG（GoldenGate）：同时支持 Oracle、SQL Server 迁移到 MySQL 上。
- SQLyog。
- Navicat Premium。
- Mss2sql。
- DB2DB。

2. 迁移中的注意事项

在 MySQL 迁移中会存在一些细节上的问题，下面探讨常见的 6 种。

（1）字符集一致

MySQL 字符集有两个维度，一个是狭义的字符集，另一个是校对规则。狭义的字符集（Character sets）是指一系列符号以及符号对应的编码的集合，如英文字母可以用 ASCII 编码、中文可以用 GBK 或者 UTF-8 编码。校对规则（Collations）则是指一种比较字符的规则。这种比较规则决定了 MySQL 如何进行排序以及如何对字符比较大小。例如，Oracle 字符集 AL32UTF8 转换成对应的 MySQL 支持的字符集是 utf8mb4。MySQL 服务的字符集和校对规则配置与客户端保持一致是非常重要的，这样可以避免在数据插入、查询、更新和排序时出现字符编码不一致的问题。字符集和校对规则的不一致可能会导致数据损坏、乱码或查询结果不正确。

（2）数据类型合理

在选择数据库的数据类型时，需要综合考虑多个因素，包括数据的可移植性、查询性能、存储效率，以及不同数据库管理系统（如 MySQL、Oracle、PostgreSQL 等 DBMS）之间的兼容性和特定限制。每种 DBMS（数据库管理系统）都有其独特的数据类型，而且即使是相似的数据类型，在不同的系统中也可能有不同的限制和特性。例如，Oracle 中的 CLOB（Character Large Object）和 MySQL 中的 LONGTEXT 都是用来存储大量文本数据的数据类型。虽然它们在功能上是相似的，但在具体的限制和性能上可能会有所不同。Oracle 的 CLOB 类型最大长度可以达到 4GB，而 MySQL

的 LONGTEXT 类型虽然也可以存储大约 4GB 的数据，但是，将接近这个限制的数据量加载到 MySQL 服务中可能会导致数据库的性能问题，如 MySQL 数据库服务变慢，甚至可能导致内存不足。

（3）设置主键

有些源表没有设置主键，但对于 MySQL 来说主键的作用非常大，特别是在后期维护和复制环节中，所以必须设置主键。

（4）迁移时间和数据量评估

在迁移过程中，评估所需时间和数据量是至关重要的，尤其是对于需要在线不间断提供服务的业务系统。基于这些评估，可以确定是采用全量迁移还是增量迁移的方式。

（5）数据库对象的迁移

除了表结构和数据以外，数据库中的其他对象（如视图、存储过程、函数、触发器以及索引等）都是迁移过程中需要关注的重要部分。由于不同的 DBMS 有其特有的 SQL 语法和特性，因此在迁移过程中，这些对象通常需要被重新编写或调整以适应新的数据库系统。在 MySQL 高负载场景中，视图、存储过程、函数和触发器的性能确实可能影响整体性能，因为这些数据库对象可能会增加查询的复杂性，消耗更多的 CPU 资源。因此，在迁移过程中，除了确保这些对象的正确性以外，还需要进行性能测试和优化，以确保它们在 MySQL 中能够高效运行。

（6）校验数据

当数据迁移完成后，确保数据迁移的正确性、完整性和无遗漏是数据迁移过程中至关重要的环节。在面临数据类型不同和数据量偏大的挑战时，都无法完全保证数据迁移过程中不会出错。因此，在数据迁移后，仍然需要保持对数据的持续监控和验证，确保数据的准确性和完整性。一般场景下采用如下验证方式校验数据。

- 查询对应数据总行数来判断数据是否存在问题。
- 用 create_time 或 update_time 时间字段验证数据。
- 抽取部分数据进行验证。
- 迁移过程中的日志信息，如警告、错误、进度等信息。

3. 利用参数提升迁移效率

在迁移数据过程中，可以修改 MySQL 基础参数，获取最大的性能，如可以关闭自适应哈希功能、Doublewrite、Redo 日志，以及调整缓存区大小、各种日志文件的大小和底层 I/O 刷新机制。下面介绍数据迁移过程中可以调优的常用参数。

（1）innodb_buffer_pool_size 参数

该参数是 InnoDB 加速优化首要参数。这个参数主要缓存 InnoDB 表的索引、数据、插入数据时的缓冲池。缓冲池越大越好。

（2）innodb_buffer_pool_instances 参数

该参数是 InnoDB 缓冲池拆分成的区域数量。对于数 GB 规模缓冲池的系统，通过减少不同线程读写缓冲页面的争用，将缓冲池拆分为不同实例，有助于改善并发性。

（3）innodb_flush_log_at_trx_commit 参数

该参数用于指定 Redo 刷新机制。如果 innodb_flush_log_at_trx_commit 设置为 0，那么日志缓冲区将每秒一次地写入 Redo 文件中，并且 Redo 文件的刷新磁盘操作同时进行。在该模式下，提交事务时，不会主动触发写入磁盘的操作，迁移性能最好。

注意：由于进程调度策略问题，因此这个"每秒执行一次刷新磁盘操作"并不是保证 100%

的"每秒"。

（4）sync_binlog 参数

该参数用于指定刷新 binlog 机制。sync_binlog 的设置为 0，和操作系统刷其他文件的机制一样，MySQL 不会同步到磁盘中，而是依赖操作系统来刷新 binary log。

注意：如果启用了 autocommit，那么每一个插入语句就会有一次写操作。否则，每个事务对应一个写操作。

（5）max_allowed_packet 参数

该参数用于指定 MySQL 服务器端和客户端在一次传送数据包的过程中最大允许的数据包大小。它可加大网络带宽的使用。如果该参数值设置过小，在导出或导入大容量数据，特别是 CLOB 数据时，可能会出现异常："Packets larger than max_allowed_packet are not allowed"。

（6）innodb_log_file_size

该参数用于指定 InnoDB 日志文件大小，设置得太大，会影响 MySQL 崩溃恢复的时间，太小则会增加 I/O 负担，所以应调整为合适的大小。在数据导入时，先把这个值调大一点，避免无谓的缓冲池的 flush 操作，但也不能设置得太大，因为会明显增加 InnoDB 的日志写入操作，而且会造成操作系统需要更多的磁盘缓存开销。

（7）innodb_log_buffer_size

该参数用于指定 InnoDB 用于将日志文件写入磁盘时的缓冲区大小（字节数）。为了实现较高写入吞吐率，可增大该参数的默认值。一个大的日志缓冲区让一个大的事务运行，不需要在事务提交前写日志到磁盘，因此，如果事务有很多更新（update）、插入（insert）和删除（delete）的记录，只要日志缓冲区足够大，就能节省磁盘 I/O 开销。

4. 总结

在 MySQL 8.0 迁移过程中，没有哪一种方法或工具是最好的，只有最合适的。数据的检验非常重要，有时迁移过程很顺利，校验时却发生错误，这个时候必须重来。重复迁移是很正常的。每次迁移都可能需要很长时间，往往会有错误发生，要做好再迁移的准备。迁移过程中的日志记录非常重要，一旦出现故障，可以从问题点开始继续迁移。

第 5 章
MySQL 8.0 性能优化

MySQL 性能优化是数据库管理员（DBA）和数据库开发人员的必备技能。MySQL 优化，一是，找出系统瓶颈，提高 MySQL 数据库整体的性能；二是，需要合理的结构设计和参数调整，提高用户操作响应的速度；三是，尽可能节省系统资源，以便系统可以提供承载更大负荷的服务。

5.1 MySQL 8.0 性能优化概述

对于保存核心数据的数据库，如果使用不当，则在性能上会存在诸多问题，而且，随着时间的推移，性能问题会越来越严重，因此解决性能问题迫在眉睫。本章将介绍关于 MySQL 8.0 性能优化的内容。

5.1.1 性能优化的作用与方法

数据库性能优化是 DBA 必须掌握的技能之一，它可以有效地提升数据库的响应速度和并发性能，提高系统的可靠性和稳定性，以满足不断增长的数据需求和业务需求。

1. 数据库性能优化的作用

数据库性能优化的作用主要体现在以下 4 个方面。

（1）提高数据访问速度

随着数据量的增加，数据库的性能可能会变得越来越差，影响业务运营和用户体验。通过数据库性能优化，可以提高数据的访问速度和效率，缩短查询和操作时间，使数据库能够更快地处理数据和查询请求，提高数据的并发处理能力和效率。

（2）降低硬件资源使用率

通过数据库性能优化，可以优化执行计划，避免不必要的资源浪费和性能瓶颈，同时减少数据库的负载和压力，提高数据库的响应时间和效率。

（3）降低数据库维护成本

数据库的不良设计，可能导致维护成本高和维护难度大。通过数据库性能优化，可以优化数据库结构和查询语句，减少维护工作量，降低维护成本。

（4）提高用户体验

通过数据库性能优化，可以提高用户的体验水平和满意度，提高应用程序的性能和响应速度，增强用户的忠诚度。

2. 数据库性能优化的方法

无论是关系数据库 MySQL 还是 NoSQL，每个数据库软件都有固有的特性，同时这些数据库软件都运行在硬件资源上，因此需要从多个维度考虑性能问题。所以，对于数据库性能优化，不能只着眼于数据库软件，而是要从整体出发。

图 5-1 所示为 MySQL 性能问题排查思路，涉及从底层软件运行环境到业务操作。

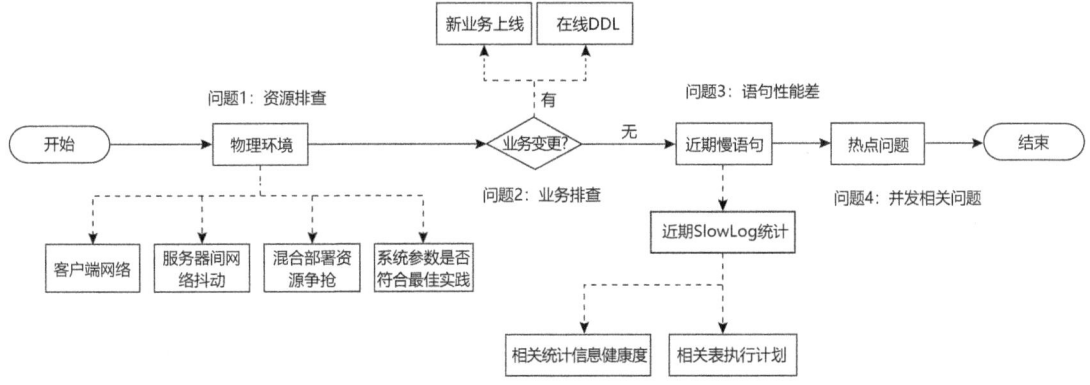

图 5-1　MySQL 性能问题排查思路

图 5-2 所示为影响 MySQL 性能的因素。自顶向下分为 4 层，分别是硬件、存储系统、存储结构、具体实现。

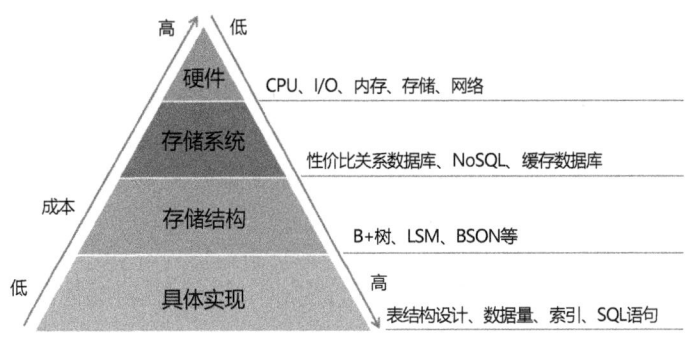

图 5-2　影响 MySQL 性能的因素

MySQL 性能优化方法可以按照 4 层结构分为 5 大类。
- 资源配置优化。
- 选择合适的存储引擎。
- 数据量优化。
- 表结构优化。
- SQL 语句优化。

层与层之间是紧密联系的，每层的上层是该层的载体。因此，越往顶层，越能决定性能的上限，优化的成本会越高，性价比越低。以底层的具体实现为例，MySQL 索引的优化成本应该是最低的，可以说，在加了索引后，无论是 CPU 消耗还是响应时间，都会降低。但加索引方式是有局限性的，因为诸多索引的加入，会引发未知的问题（如 MySQL 内部维护索引，导致插入性能下降）。当"具体实现"层没有可以优化的空间时，就需要往上一层（"存储结构"层）进行优化（如分库分表、压缩数据量等）。

如果在"存储结构"层的优化没有效果，就得继续往上一层进行优化。下面具体介绍上述5类优化方法。

（1）资源配置优化

对于数据库软件而言，资源就是根基。当业务规模逐渐扩大时，所需的数据库资源会逐渐增加，高负载下的数据库服务缓慢，如果不进行优化，那么资源就有可能被浪费。因此需要考虑如何有效地利用这些资源，提高数据库服务效率。

- 数据库配置：首先需要对数据库进行合理的配置。可以将所有的数据库资源分为多个数据库组，并给每个组分配一定的处理能力。
- 请求分配：根据请求的类型、资源组的负载情况，选择一个可以处理当前请求的资源组进行处理。
- 负载均衡：当资源组的负载达到一定程度时，自动进行负载均衡，把一部分业务请求分配到其他资源组中进行处理。
- 监控和调整：对数据库资源的利用情况进行监控和调整，确保数据库资源的稳定和安全。

（2）选择合适的存储引擎

MySQL支持多种存储引擎，每种存储引擎都有其各自的功能、优点和缺点。存储引擎负责处理数据库表中数据的存储、检索和管理任务。为MySQL表选择合适的存储引擎是至关重要的，这会直接影响应用程序的性能、可靠性和可扩展性。推荐选择InnoDB存储引擎，因为InnoDB存储引擎支持事务、行级锁、MVCC（多版本并发控制）、数据恢复。同时，InnoDB也被设计用来最有效地利用内存和CPU资源，以提供高性能的存储解决方案。

（3）数据量优化

无论是哪种存储方式，数据量越小，自然查询性能就越高，随着数据量增大，资源的消耗（CPU、磁盘读写繁忙）和耗时也会越来越高。对于数据量调整，有以下两种优化方法。

1）数据归档。数据归档的目的是解决数据存储的问题。通过合理的数据归档策略和过程，可以最大限度地提高性能。

- 数据归档可以按照不同的维度来进行。例如，按照时间、重要性、类型等进行分类和归档。这样可以确保只有最常用且重要的数据在主表上，而不常用且重要性较低的数据则被归档到其他历史表中。
- 数据归档的过程需要对数据进行整理、清理和校验，以确保被归档的数据的完整性和准确性。
- 数据归档可以节省存储空间，使系统更高效地运行。数据归档可以提供备份，以防数据的丢失和损坏，从而增加数据的可靠性和安全性。此外，数据归档还可以用于合规性要求和审计目的，满足相关法规和标准。

2）分库分表。分库分表是在海量数据下，由于单库、单表数据量过大，导致数据库性能持续下降，而产生的技术方案。

- 分库：从单个数据库拆分成多个数据库的过程，将数据散落在多个数据库中。
- 分表：从单张表拆分成多张表的过程，将数据散落在多张表内。

分库有明显的优缺点。分库的优点是提升性能、增加可用性。其缺点是，当需要聚合的时候，需要多次合并计算才能得到结果，或者需要将数据导入到其他NoSQL平台进行汇总，这增加了操作的复杂性。

（4）表结构优化

在 MySQL 数据库表结构设计中，应该保证表字段的正确性、保证表关系清晰明确、选取合适的字段类型、保证数据的完整性、尽量可扩展、将频繁查询的字段设置索引，尽量避免使用存储过程、触发器和视图。这些原则都是经实战总结而来的，可以帮助用户设计出高效、稳定且可扩展的 MySQL 数据库表结构。然而，实际应用中还需要根据具体的业务需求和技术环境进行灵活调整与优化。

（5）SQL 语句优化

在 MySQL 数据库后期的维护过程中，绝大多数问题都出在 SQL 语句上面。在 MySQL 数据库实际应用中，SQL 语句的性能和正确性常常是导致问题的关键所在。为了确保数据库的稳定运行和高效性能，编写正确、规范、高效的 SQL 语句是至关重要的。

- 因为 MySQL 的 InnoDB 索引组织表特性，所以，对于数据操作，SQL 语句已索引立足点进行操作。
- 尽量避免编写过于复杂的 SQL 语句，因为它们可能难以理解和维护，同时也可能导致性能问题。例如，复杂 SQL 语句中生成的中间表索引失效。
- SQL 语句的书写格式应与官方标准相适应。
- SQL 语句中避免使用关键字，因为会与 MySQL 系统中的定义冲突。
- SQL 语句中严格区分大小写。虽然绝大多数情况下 MySQL 在 Linux 平台上对大小写不敏感，但总有一些场景需要，所以要养成好的习惯。

5.1.2 性能分析需要收集的 11 类信息

在 MySQL 数据库的日常运维中，当遇到性能问题时，通常需要收集一系列信息来进行分析，以找到问题的根源。

接下来介绍分析数据库性能问题时，通常需要获取的一些关键信息。

1. 基本信息

MySQL 数据库基本信息包含参数、统计计数、复制状态，这些信息可方便了解数据库的运行状态、参数是否合理、复制是否正常，以应对数据库的故障和进行性能问题的排查。

收集 MySQL 数据库基础信息的命令行如下所示。

```
mysql> \s;
mysql> SHOW GLOBAL STATUS;
mysql> SHOW GLOBAL VARIABLES;
mysql> SHOW SLAVE STATUS \G;
```

2. 数据库连接信息

通过查看 MySQL 数据库的连接信息，可以了解哪些线程正在运行、这些线程正在执行的具体 SQL 语句是什么，以及这些线程当前的运行状态等关键信息。这些信息对于诊断数据库性能问题、识别长时间运行的查询，以及理解数据库的并发活动模式等都至关重要。实际上，可以通过 MySQL 的 PROCESSLIST 命令来获取这些连接信息的具体输出。查询数据库连接信息的命令行如下所示。

```
mysql>select USER,HOST,DB,COMMAND, TIME,STATE, INFO from
information_schema.processlist where COMMAND<>'Sleep' limit 10;
```

说明：需要特别关注 STATE（状态值）和 TIME（执行时间）。

3. 错误日志

MySQL 的错误日志记录了服务启动、运行或停止的事件信息。它可以帮助追踪、调试和解决问题。查看错误日志路径和记录级别的命令行如下所示。

```
mysql> SHOW VARIABLES WHERE Variable_name in('log_error_verbosity','log_error');
+-------------------------------+------------------------------------+
|Variable_name                  |Value                               |
+-------------------------------+------------------------------------+
|log_error                      |/opt/data8.0/logs/mysql_err.log     |
|log_error_verbosity            |3                                   |
+-------------------------------+------------------------------------+
```

说明：log_error_verbosity 参数决定了哪些消息会被写入错误日志。其值可以是 0（只记录错误）、1（记录错误和警告）、2（记录错误、警告和注意信息）、3（记录所有消息）。

4. 慢查询日志

MySQL 的慢查询日志记录了执行时间超过预设阈值的 SQL 语句。这些语句通常是引发性能瓶颈问题的候选者，因此分析慢查询日志是优化数据库性能的重要步骤。查看慢查询日志相关配置的命令行如下所示。

```
mysql> SHOW VARIABLES LIKE '%slow%';
+-----------------------------+-----------------------------+
|Variable_name                |Value                        |
+-----------------------------+-----------------------------+
|log_slow_admin_statements    |ON                           |
|log_slow_extra               |OFF                          |
|log_slow_slave_statements    |OFF                          |
|slow_launch_time             |2                            |
|slow_query_log               |ON                           |
|slow_query_log_file          |/opt/data8.0/logs/slow.log   |
+-----------------------------+-----------------------------+
```

MySQL 的慢查询日志是以易读的文本格式记录的，从而使其分析更为便捷。常用的工具，如 mysqldumpslow 和 pt-query-digest，为解析和分析慢查询日志提供了便捷方式，可帮助数据库管理员和开发人员识别性能瓶颈与优化 SQL 语句。

5. 锁信息

在 MySQL 中，尤其是在 InnoDB 存储引擎中，对锁的使用和监控是性能调优与避免死锁的关键。通过将 information_schema 和 performance_schema 中的表关联 SQL 语句，可以获取关于锁、阻塞线程、用户、IP 地址和端口等的信息。这些信息对于诊断性能瓶颈、识别死锁原因以及优化数据库性能都非常有帮助。获取锁相关的阻塞线程的 SQL 查询示例如下所示。

```
mysql> SELECT locked_table,
       locked_index,
       locked_type,
       blocking_pid,
       concat(T2.USER,'@',T2.HOST) AS "blocking(user@ip:port)",
       blocking_lock_mode,
       blocking_trx_rows_modified,
       waiting_pid,
       concat(T3.USER,'@',T3.HOST) AS "waiting(user@ip:port)",
       waiting_lock_mode,
       waiting_trx_rows_modified,
```

```
        wait_age_secs,
        waiting_query
FROM sys.x$innodb_lock_waits T1
LEFT JOIN INFORMATION_SCHEMA.processlist T2 ON T1.blocking_pid=T2.ID
LEFT JOIN INFORMATION_SCHEMA.processlist T3 ON T3.ID=T1.waiting_pid;
```

上述语句中，INFORMATION_SCHEMA.processlist 表提供了当前正在运行的所有线程的信息；sys.x$innodb_lock_waits 表提供了哪些事务正在等待锁以及它们等待了多长时间的信息，这有助于识别潜在的锁争用和性能瓶颈。除此之外，"information_schema.INNODB_TRX，performance_schema"下的 data_locks、data_lock_waits 和 metadata_locks 表提供了锁与锁等待的详细信息。

6. InnoDB 状态信息

查看 InnoDB 存储引擎的内部状态信息，包括锁的信息、事务状态、缓冲池状态等，是为了更好地了解数据库的性能瓶颈、优化数据库配置或进行故障排除。获取 InnoDB 内部状态的命令行如下所示。

```
mysql> SHOW ENGINE INNODB STATUS \G;
mysql> SHOW ENGINE INNODB MUTEX;
```

说明：MySQL 中的 latch 与 lock 都称为锁，在 InnoDB 中，lock 针对的是事务，latch 针对的是线程。latch 又可以分为 mutex 和 rw_lock。latch 的目的是保证并发的线程操作临界资源的正确性。mutex 量指的是一种用于保护一些临界资源的使用的信号量。当有线程需要使用这些临界资源时，会请求获得 mutex 量，请求成功的线程进入临界区，而请求失败的线程只能等待这个 mutex 被释放。

7. binlog 文件

在 MySQL 中，二进制日志（binlog）是一个非常重要的日志文件，它记录了所有对数据库执行更改的语句，包括数据定义语言（DDL），如 CREATE、ALTER、DROP 等语句，以及数据操纵语言（DML），如 INSERT、UPDATE、DELETE 等语句。这些日志以事件的形式存储，并且每一条更新数据的语句的执行时间也会被记录下来。此外，binlog 还包含一些其他的元数据，如事务的 ID、执行时间、提交时间、服务器唯一 ID 等。binlog 文件会存储在服务器的数据目录下，并且可以通过 mysqlbinlog 工具来分析这些日志文件，查看具体的操作信息。

8. 监控信息

监控信息是数据库性能分析中的一个重要环节，通过分析监控信息，可发现数据库的性能问题，优化数据库性能，并确保系统的稳定运行。获取所需的信息如下。

- OS 信息：查看 CPU TOP、iostat 状态、网络流量等。
- MySQL 信息：包括连接数、active 线程数、锁等待、临时表使用情况、TPS、QPS、网络输入与输出等信息。

9. MGR 状态信息

如果高可用方案采取 MGR（MySQL Group Replication，组复制是 MySQL 提供的一种高可用和负载均衡的复制解决方案）集群，就需要额外获取 MGR 集群运行状态信息。查看 MGR 运行情况的命令行如下所示。

```
# MGR 集群 GTID 信息
mysql>SHOW MASTER STATUS \G;

# 集群节点状态信息
```

```
mysql>SELECT * FROM performance_schema.replication_group_member_stats\G;
```

```
# MGR 成员间的角色和状态信息
mysql> SELECT a.member_id, a.member_host, a.member_state, a.member_role,
    b.channel_name, b.count_transactions_in_queue,
    b.count_transactions_remote_in_applier_queue
FROM performance_schema.replication_group_members a,
    performance_schema.replication_group_member_stats b
WHERE a.member_id=b.member_id order by a.member_role;
```

10. pstack 堆栈信息

pstack 是一个在 Linux 系统上常用的命令行工具，它用于生成指定进程的堆栈跟踪。通过堆栈跟踪，可以看到进程在某一时刻的调用栈，从而了解哪些函数正在被执行，以及这些函数是如何被调用的。这对于调试程序、分析性能问题或诊断崩溃原因非常有用。输出堆栈信息的命令行如下所示。

```
shell $>pstack $mysqld_pid>/tmp/pstack.info
```

说明：在日常运维中，不能使用 pstack 命令，有可能会导致 MySQL 服务挂起（hang）或变得不可用。但在出现 MySQL 服务挂起或不可用时，使用 pstack 命令来抓取堆栈信息是一个很好的诊断手段。

11. 网络包信息

当网络出现异常信息时，可以使用 tcpdump 命令进行网络抓包，捕获并分析网络传输的数据包。数据包里包含 SQL 语句、参数、源端 IP 地址信息、包分发状态等信息。输出网络传输包信息的命令行如下所示。

```
shell $>tcpdump -i ens33 tcp port 3306 and host 192.168.244.130 -w ./mysqltcp.pcap
```

说明：导出的 pcap 包需要使用 Wireshark 工具查看。无法对加密网络包进行分析。

5.1.3 导致性能突发事件的十大原因

性能突发事件是指在特定时刻和地点突然发生的、对数据库系统性能产生较大影响或危害性的不可预测事件。这些事件对业务运营和数据完整性构成严重威胁。性能突发事件并非凭空出现，它们往往由多种因素引发，包括难以控制的客观因素、对数据库的理解盲区，以及忽视细微的性能问题而逐渐积累等。这些因素可能导致服务中断、数据丢失或损坏，甚至可能影响整个企业的运营。以下是作者总结的导致数据库性能突发事件的十大原因。

1. 低级程序错误

低级程序错误通常指的是由于编程者没有遵循官方提供的数据库语法规则、逻辑，或未按照标准方式使用 API 等导致的错误。这些错误有可能会导致程序无法正常运行。

2. 数据库结构设计随意、混乱

数据库结构没有采取合理、科学的方式设计。例如：
- 没有主键的表。主键是表中唯一标识每条记录的字段或字段组合。没有主键的表会导致数据不完整和重复记录的问题。在复制环节中，主键还有助于快速定位变更的数据。
- 索引不合理（如过多索引、不必要的索引、索引列选择不当等）会导致维护开销增加，同时可能降低查询性能。
- 表设计没有考虑后续相关处理。例如未考虑自增键（常用于生成唯一 ID），在需要唯一标识记录的场景下，可能会导致问题。又如，TEXT、JSON 等字段的使用未做合理设计，可能会导致查询性能下降或数据存储不经济。
- 字段设计不合理（如数据类型选择不当、字段名过长、字段数量过多等）会浪费存储空间，降低查询性能，可能会增加额外的 I/O 开销。
- 不稳定的功能设计（如频繁更改表结构、使用不稳定的数据类型等）可能导致系统不稳定和难以维护。

3. SQL 写法随意，引发全表扫描

如果 SQL 语句优化不够，则可能会引发全表扫描（Full Table Scan），这是一种非常耗时的操作，因为它会检查表中的每一行数据，从而导致性能突发事件。

4. 传输大数据量的耗时 SQL 语句

在使用存储过程、触发器，或查询返回大量数据时，如果并发量较大，则可能会导致资源耗尽，进而影响系统性能和稳定性。

5. 系统设计没有对海量数据应用场景进行考虑

当系统设计没有针对海量数据应用场景进行充分考虑时，可能会导致资源瓶颈，如 CPU 资源耗尽，尤其是在执行大批量导入、数据清理任务和复杂的报表统计任务时。这些操作通常需要大量的计算资源，如果系统设计不当，很容易引发性能问题。

6. 不同业务逻辑在处理过程中的相互影响

当不同业务逻辑使用同一张表进行数据的修改时，可能导致死锁问题。死锁是指两个或多个事务在争夺资源时，因互相等待对方释放资源而造成的一种互相僵持的状态。在数据库系统中，死锁通常发生在多个事务试图以不同的顺序锁定资源时。

7. 系统架构问题导致数据不一致

MySQL 双主架构中，数据库系统包含两个主节点，这两个主节点都可以处理读写操作，而不仅仅是单一的主节点处理所有写操作。变更数据通过复制进行同步。如果复制过程出现延迟，则可能会导致数据不一致，进而引发读取错误、插入重复数据问题，最终造成复制冲突，破坏高可用架构，并导致业务处理逻辑混乱。

8. 访问量突然暴增

如果系统突然面临访问量暴增的情况，尤其是在没有提前准备或预期的情况下，则可能会导致数据库负载升高、性能下降，甚至数据库服务中断。

9. 高危命令的执行

高危命令的执行是数据库管理中的一个重要风险点，特别是当这些命令被不恰当地使用或由未经授权的用户执行时。例如，LOCK TABLE、FLUSH TABLES、在线 DDL、底层 RM 等命令可能导致数据库性能下降、数据丢失或系统崩溃。

10. 硬件故障或老化

硬件故障或老化是常见的导致系统性能下降的原因，特别是在长时间运行的系统中。随着硬件组件的老化，它们的性能可能会逐渐降低，甚至发生故障，这会直接影响数据库系统的整体性能和利用率。

5.1.4 性能监控指标

数据库性能监控是数据库管理中至关重要的一个环节。通过收集和分析数据库的各种运行指标，管理员和开发者能够深入了解数据库的健康状况、性能瓶颈以及潜在的错误和风险。这有助于预防或快速应对可能出现的问题，确保数据库系统的稳定运行和高效服务。

表 5-1 所示为 MySQL 性能的监控指标列表，通过这个列表可以了解数据库的运行状态，并据此进行性能优化。

表 5-1 MySQL 监控指标

类别	监控项名称	监控频率	警告阈值	严重阈值	告警值单位
SQL 语句执行情况	每秒执行事务数（TPS，单位：次/s）	1~5min	8000	10000	次/s
	每秒执行操作数（QPS，单位：次/s）	1~5min	30000	100000	次/s
	每秒 SELECT 语句执行次数（单位：次/s）	1~5min	30000	50000	次/s
	每秒 INSERT 语句执行次数（单位：次/s）	1~5min	5000	10000	次/s
	每秒 UPDATE 语句执行次数（单位：次/s）	1~5min	5000	10000	次/s
	每秒 DELETE 语句执行次数（单位：次/s）	1~5min	5000	10000	次/s
	每秒 COMMIT 语句执行次数（单位：次/s）	1~5min	5000	10000	次/s
	每秒 ROLLBACK 语句执行次数（单位：次/s）	1~5min	1	10	次/s
连接相关指标	最大连接数（max_connections，单位：个）	1~3min	2000	3000	次/s
	已创建的线程数（Threads_created，单位：个）	1~3min	—	—	—
	缓存的线程数（Threads_cached，单位：个）	1~3min	—	—	—
	运行的线程数（Threads_running，单位：个）	1~3min	1500	2000	次/s
	当前打开（连接）的连接数（Threads_connected，单位：个）	1~3min	2000	97	次/s
	已用连接百分比（%）	1~3min	80%	100%	次/s
流量状态（千兆网络）	内网出流量（bytes_sent，单位：KB/秒）	1~3min	512MB	800MB	KB/s
	内网入流量（bytes_received，单位：KB/秒）	1~3min	512MB	800MB	KB/s
InnoDB 文件读写次数	InnoDB 平均每秒从文件中读取的次数	1~3min	—	—	—
	InnoDB 平均每秒从文件中写入的次数	1~3min	—	—	—
	InnoDB 平均每秒进行 fsync() 操作的次数	1~3min	—	—	—
	InnoDB 平均每秒读取的数据量，单位为 KB	1~3min	—	—	—
	InnoDB 平均每秒写入的数据量，单位为 KB	1~3min	—	—	—
InnoDB 缓冲池状态	InnoDB 空页数	1~3min	—	—	—
	InnoDB 总页数	1~3min	—	—	—
	InnoDB 物理读请求数	1~3min	—	—	—
	InnoDB 逻辑读请求数	1~3min	—	—	—
	InnoDB 缓冲池的写次数	1~3min	—	—	—
	InnoDB 缓冲池中脏页的数目	1~3min	—	—	—
	InnoDB 缓冲池中刷新页请求的数目	1~3min	—	—	—

(续)

类别	监控项名称	监控频率	警告阈值	严重阈值	告警值单位
InnoDB 缓冲池状态	InnoDB 缓冲池的读命中率	1~3min	—	—	—
	InnoDB 缓冲池的利用率	1~3min	—	—	—
InnoDB 日志	平均每秒向日志文件完成的 fsync( ) 写数量	1~3min	—	—	—
	平均每秒写入日志文件的字节数	1~3min	—	—	—
	平均每秒向日志文件的物理写次数	1~3min	—	—	—
	平均每秒日志写请求数	1~3min	—	—	—
InnoDB 行	InnoDB 行删除量（单位：次/s）	1~3min	—	—	—
	InnoDB 行插入量（单位：次/s）	1~3min	—	—	—
	InnoDB 行读取量（单位：次/s）	1~3min	—	—	—
	InnoDB 行更新量（单位：次/s）	1~3min	—	—	—
	InnoDB 等待行锁次数	1~3min	—	—	—
	InnoDB 行锁定花费的总时间，单位为 ms	1~3min	—	—	—
	InnoDB 行锁定的平均时间，单位为 ms	1~3min	—	—	—
MyISAM 读写次数	MyISAM 平均每秒从缓冲池中读取的次数	1~3min	—	—	—
	MyISAM 平均每秒向缓冲池中写入的次数	1~3min	—	—	—
	MyISAM 平均每秒从硬盘上读取的次数	1~3min	—	—	—
	MyISAM 平均每秒向硬盘上写入的次数	1~3min	—	—	—
MyISAM 缓冲池	MyISAM 平均每秒 key buffer 利用率	1~3min	—	—	—
	MyISAM 平均每秒 key buffer 读命中率	1~3min	—	—	—
	MyISAM 平均每秒 key buffer 写命中率	1~3min	—	—	—
临时表使用情况	使用临时文件次数（单位：次/s）	1~3min	—	—	—
	硬盘中创建的临时表的数量	1~3min	—	—	—
	内存中创建的临时表的数量	1~3min	—	—	—
	临时表利用率=Created_tmp_disk_tables/Created_tmp_tables	1~3min	80%	95%	次/s
块使用情况	键缓存内未使用的块数量（单位：个）	1~3min	—	—	—
	键缓存内使用的块数量（单位：个）	1~3min	—	—	—
	键缓存读取数据块次数（单位：次/s）	1~3min	—	—	—
	硬盘读取数据块次数（单位：次/s）	1~3min	—	—	—
	数据块写入键缓存次数（单位：次/s）	1~3min	—	—	—
	数据块写入磁盘次数（单位：次/s）	1~3min	—	—	—
打开表或文件使用情况	已经打开表的数量	1~3min	—	—	—
	打开文件总数	1~3min	—	—	—
	当前 InnoDB 打开表的数量（innodb_num_open_files，单位：个）	1~3min	—	—	—

(续)

类别	监控项名称	监控频率	警告阈值	严重阈值	告警值单位
复制信息	复制进程中 I/O 线程运行状态（slave_io_running，参数取值有 Yes、No 和 Connecting）	1min	Connecting、No	Connecting、No	次/s
	复制进程中 SQL 线程运行状态（slave_sql_running，参数取值有 Yes 和 No）	1min	Connecting、No	Connecting、No	次/s
	主从差距时间（seconds_behind_master，单位：s）	1min	3s	5s	次/s
数据量信息	数据大小	1h	—	—	—
	索引大小	1h	—	—	—
	磁盘空间	5min	—	—	—
	磁盘使用率	5min	80%	90%	次/s
性能指标	慢查询数	1~3min	2	10	次/s
	全表扫描数	1~3min	—	—	—
	死锁	1~3min	1	5	次/s

通过采集和分析监控指标数据，可以在数据库出现问题时及时识别和排除故障，从而避免问题升级为灾难性级别。对于 MySQL 数据库，有多种开源工具可用于监控，包括 Zabbix、Prometheus 和 Percona Monitoring and Management（PMM）等。虽然这些工具监控的指标类似，但在数据的展示、用户界面、集成和告警机制等方面可能有所不同。

5.2 MySQL 8.0 性能优化的关键点

5.2.1 数据库配置优化

MySQL 提供了大量的配置参数来调整和优化数据库服务的行为。这些参数对于数据库管理员（DBA）来说非常重要，因为它们会直接影响 MySQL 的性能、安全性、资源使用以及许多其他方面。

在 MySQL 8.0 中，虽然总共有 650 多个配置参数，但 DBA 通常只需要关注其中的一小部分，这些参数通常与 DBA 的具体需求、硬件环境以及工作负载紧密相关。以下是一些常用的 MySQL 配置参数，共分为 7 类，这些参数在大多数情况下都需要进行优化和调整。

- 缓存设置相关参数。
- I/O 刷新策略设置相关参数。
- 网络设置相关参数。
- 线程设置相关参数。
- 日志设置参数。
- 连接用户数设置相关参数。
- 其他设置参数。

1. 缓存设置相关参数

表 5-2 所示为 MySQL 缓存相关调优参数列表。

表 5-2　MySQL 缓存相关调优参数

参　　数	说　　明	备　　注
innodb_buffer_pool_size	建议配置成实际能使用物理内存的 50%~60%	
read_buffer_size	读缓冲区：1~4MB	初期为 2MB，最大可以设置为 8MB
sort_buffer_size	排序缓冲区：1~4MB	初期为 2MB，最大可以设置为 8MB
join_buffer_size	连接缓冲区：1~4MB。read_buffer_size、sort_buffer_size 和 join_buffer_size 这 3 个值对齐	初期为 2MB，最大可以设置为 8MB
read_rnd_buffer_size	2~8MB，是 read_buffer_size 的两倍	初期为 4MB，最大可以设置为 16MB
tmp_table_size、max_heap_table_size	临时表缓冲区，32~128MB 既可	
thread_stack	连接临时缓冲区，一直保留在内存里，建议设置为 24~64MB	
innodb_log_buffer_size	Redo log 刷新缓冲区：4~16MB	
binlog_cache_size	binlog 刷新缓冲区：1~32MB	

2. I/O 刷新策略设置相关参数

表 5-3 所示为 MySQL I/O 相关刷新策略参数列表。

表 5-3　MySQL I/O 相关刷新策略参数

参　　数	说　　明	备　　注
innodb_lru_scan_depth	InnoDB 缓冲池中页的 LRU（Least Recently Used）算法的扫描深度，默认值为 1024，建议改成 512	
innodb_io_capacity 和 innodb_io_capacity_max	单位是页，只有在频繁写操作的时候才有意义，根据系统可以支持多大的 IOPS 进行调整。在 SSD 硬盘中，最高可达 5000，建议设置为 3000~5000	SAS：200~1000；SSD：2000~5000；PCI-E：10000~50000
innodb_adaptive_hash_index	自适应哈希索引，系统自动控制	在出现 flush 耗时比较长的日志的时候，建议关闭
innodb_flush_log_at_trx_commit 和 sync_binlog	Redo 日志和 binlog 落盘参数，建议保持为 1，但从库延迟非常大的时候，可以临时调整为 0	
innodb_redo_log_capacity	Redo 日志文件占用的磁盘空间，采取默认值即可	

3. 网络设置相关参数

表 5-4 所示为 MySQL 网络相关调优参数列表。

表 5-4　MySQL 网络相关调优参数

参　　数	说　　明	备　　注
max_allowed_packet	网络传输包限制大小，可以考虑设置为 128MB、512MB 或 1GB	千兆以太网：512MB；万兆以太网：1GB
slave_max_allowed_packet	Slave SQL 线程可以处理的最大数据包大小由参数 slave_max_allowed_packet 控制。这是限制 binlog 事件大小	千兆以太网：512MB；万兆以太网：1GB

(续)

参数	说明	备注
slave_pending_jobs_size_max	当前需要执行事件所需的内存大小，需要大于主库 max_allowed_packet 的大小	
net_read_timeout 和 slave_net_timeout	复制进程设置的网络超时的两个参数，维持默认值，很少修改	

4. 线程设置相关参数

表 5-5 所示为 MySQL 线程相关调优参数列表。

表 5-5 MySQL 线程相关调优参数

参数	说明	备注
innodb_purge_threads	与 purge 线程有关	设置的数值建议小于等于物理 CPU 总核数
innodb_read_io_threads	与 read 线程有关，基本异步	设置的数值建议小于等于物理 CPU 总核数
innodb_write_io_threads	与 write 线程有关，基本异步	设置的数值建议小于等于物理 CPU 总核数
innodb_page_cleaners	与 page 线程有关，基本异步	建议采用默认值。如果 CPU 核数大于等于 16，则可以设置为 8
table_open_cache_instances	表缓存打开的实例数	设置的数值等于物理 CPU 总核数的 2 倍
innodb_buffer_pool_instances	InnoDB 缓冲池划分的区域数，每个内存保证最少分配 1GB	设置的数值建议小于等于物理 CPU 总核数

5. 日志设置参数

表 5-6 所示为 MySQL 日志相关调优参数列表。

表 5-6 MySQL 日志相关调优参数

参数	说明
log_error_verbosity	错误日志记录。必须设置为 3
slow_query_log	慢查询日志开启
long_query_time	慢查询日志记录时间间隔为 1~3s
log_bin	MySQL 的 binlog 开启
binlog_format	binlog 格式。必须为 ROW 模式
expire_logs_days	binlog 在硬盘中保留天数。建议为 7 天
binlog_expire_logs_seconds	binlog 在硬盘中保留时间（单位为秒），如 604800（7 天）
innodb_redo_log_capacity	Redo 日志文件占用的磁盘空间，默认值为 1GB
innodb_print_all_deadlocks	输出死锁信息到错误日志中

6. 连接用户数设置相关参数

表 5-7 所示为 MySQL 连接用户数相关调优参数列表。

表 5-7 MySQL 连接用户数相关调优参数

参数	说明	备注
max_connections	同时允许的最大客户端连接数	建议最多 3000
max_connect_errors	定义了一个客户端在连接 MySQL 服务时允许的最大连接错误次数。如果某个客户端在错误次数上达到了 max_connect_errors 设置的值，则 MySQL 服务将无条件地阻止该客户端的进一步连接尝试，直到这个计数器被重置为止	建议设置为 1000000
skip_ssl	使用非加密的连接数据库服务	默认开启，但建议关闭
admin_address、admin_port	管理员账号链接地址、端口	建议设置默认值
mysqlx_port	X 协议提供一个非常类似 MongoDB 的 Shell 服务	建议关闭
default_authentication_plugin	建议采用原生密码方式，即设置 mysql_native_password	按照实际需求设置
skip_name_resolve	是否进行主机名 DNS 解析	建议禁止

7. 其他设置参数

表 5-8 所示为 MySQL 其他可调优参数列表。

表 5-8 MySQL 其他可调优参数

参数	说明	备注
table_open_cache	所有线程打开的表的数量。增加这个值会增加 mysqld 进程需要的文件描述符的数量。可以通过检查 Opened_tables 状态变量来检查是否需要增加表缓存。3000 值	
table_definition_cache	控制总的.frm、ibd 文件的数量，建议设置为 3000	
open_files_limit	MySQL 服务能够打开的最大文件描述符数量，建议设置为 65535	建议与/etc/security/limits.conf 配置信息一致
innodb_open_files	InnoDB 操作文件句柄的数量，建议设置为 65535	建议与/etc/security/limits.conf 配置信息一致
innodb_thread_concurrency	InnoDB 存储引擎可以同时执行的线程数目，有助于控制并发操作	该参数的设置范围为 64~128。如果 CPU 性能更高，则可以将该参数值向上调整。或者，可以将该参数设置为 0，交由操作系统来控制
innodb_deadlock_detect	发起死锁检测，主动回滚一个事务，让其他事务继续执行	建议开启，但会影响部分性能
innodb_lock_wait_timeout	等待锁的事件超时时间	考虑业务设置，要是无业务设置，建议设置为 3~10s
skip_slave_start	重新启动是否开启复制进程	建议关闭
lower_case_table_names	大小写敏感参数	在必须初始化的时候，设置为 1

5.2.2 库、表、字段和索引的设计优化

对于数据库来说，库、表、字段和索引的合理设计至关重要。不恰当的设计会导致性能问题、数据冗余、维护困难，以及带来可扩展性方面的挑战。在开发初期，由于时间压力和快速开发的需求，设计上的考虑可能不够全面，这可能导致在开发过程中进行重复工作，以及为后续数据库运维带来困扰。

以下是一些关于如何合理设计 MySQL 数据库的 InnoDB 引擎的库、表、字段和索引的建议。

1. 全局设计的原则

全局设计的原则要求 MySQL 数据库的库、表、字段和索引遵循以下原则。

- 库名、表名、字段名必须使用小写字母并采用下画线分隔。
- 库名、表名、字段名禁止超过 32 个字符，所有表都需要添加注释，禁止拼音与英文混用。
- 禁止使用 MySQL 的保留字。
- 临时库、临时表名必须以 tmp 为前缀并以日期为后缀。
- 备份库、备份表名必须以 bak 为前缀并以日期为后缀。
- MySQL 列长度有一个硬限制，即单个行的总大小（包括所有的列）不能超过 65535B（即约 64KB）。这个限制是由 MySQL 存储引擎的内部实现方式所决定的。BLOB 或 TEXT 列的内容不超过其最大大小限制，例如，BLOB 列的最大大小为 65535B，而 TEXT 列的最大大小为 65535B 的约 4 倍，即约 262144B。
- MySQL 数据库单实例表数目必须小于 3000 个，单表数据量建议控制在 5000 万行以内。
- 数据库单个表中的列的数量应该小于 30 个，以避免行溢出和性能下降的问题。
- 存储引擎选择 InnoDB 引擎。它支持 ACID 事务、行级锁定以及外键约束，提供了高性能和数据完整性。
- 表必须有主键，如自增主键，禁止使用 VARCHAR 类型字段作为主键语句设计。
- 表设计时应确保每个表都有一个主键。在没有合适的主键时，推荐使用自增的整数类型作为主键，而不是 VARCHAR 类型。

➢ 自增主键是一种特殊类型的主键，通常用于整数类型（如 INT 或 BIGINT）的字段。自增主键的值会自动递增，数据写入可以提高插入性能，同时可以避免页频繁分裂，减少表碎片，提升空间和内存的使用效率。

➢ VARCHAR 类型字段用于存储可变长度的字符串。虽然 VARCHAR 类型字段可以作为主键使用，但通常不推荐这样做，因为 VARCHAR 类型的主键可能导致性能问题，在索引时也会占用更多的存储空间，这些都可能会降低查询效率。

- 禁止使用外键。外键会导致表与表之间耦合。当执行 UPDATE 或 DELETE 操作时，数据库需要确保相关的外键约束不会被违反，会进行额外的检查。在某些高并发和大数据量的业务场景中，外键可能会带来性能问题，甚至会造成死锁。

2. 字段设计的原则

在设计 MySQL 数据库时，字段设计应遵循以下原则。

- 在数据库设计中，合理选择字段类型是非常重要的，因为它直接影响存储效率、性能、以及存储空间的使用。不同的字段类型有不同的存储范围和存储空间需求。

- 尽量避免使用 NULL 值。NULL 值在 MySQL 中有特殊的含义，它表示缺失或未知的数据，而不是表示零或空字符串。NULL 值的处理在数据库内部需要额外的逻辑和存储空间，而且可能使查询优化变得更为复杂。
- 不同表中相同内容的字段采用统一的名称和类型是非常重要的，这有助于保持数据库的一致性和可维护性。
- 手机号码必须使用 20 位 VARCHAR 类型字段存储，这主要考虑到国际化和未来可能的扩展。
- 尽量不要为 VARCHAR 字段预留过多的空间。VARCHAR 是一个变长字段，它只会占用必要的空间来存储实际的数据。然而，如果为 VARCHAR 类型字段预留了过大的空间，则可能会浪费存储空间、降低 I/O 性能和占用更多的内存空间。
- 禁止使用 TEXT、BLOB 等大字段类型。大字段会浪费更多的磁盘和内存空间，大量非必要的大字段查询会刷掉热数据，导致内存命中率急剧降低，影响数据库性能。
- 禁止使用 ENUM，可使用 TINYINT 代替，因为增加新的 ENUM 值要进行 DDL 操作，ENUM 的内部实际存储的就是整数。
- 不保存大字段数据，特别是在处理图片、视频、音频或其他大型二进制文件时。将这些大文件存储在文件系统中，并在数据库中仅保存这些文件的 URI（统一资源标识符）或路径。

3. 索引设计的原则

在设计 MySQL 数据库时，索引设计应遵循以下原则。
- 单表索引建议控制在 5 个以内，单索引字段数不允许超过 5 个。过多的索引会对表的 INSERT、UPDATE 和 DELETE 操作带来很大的性能影响，从而影响数据库的稳定性。
- 禁止在更新频繁、区分度不高的字段上建立索引。当表中的某些列频繁更新时，相应的索引也需要频繁地进行维护，这会导致维护成本增加，并且索引的碎片率也会增加，降低了索引的效率。区分度不高的索引无法实现有效的数据过滤，可能需要进行全表扫描，这会导致查询性能的下降。
- 建立组合索引，必须把区分度高的字段放在前面。数据库查询优化器通常从左到右使用索引的字段。因此，将区分度高的字段放在前面意味着查询优化器可以更早地使用索引来过滤掉不需要的行，从而减少需要检查的行数。这有助于提高查询效率，特别是对于那些涉及大量数据的查询。
- 推荐使用自增列作为主键，这在 MySQL 数据库中具有多方面的优势。首先，自增主键能够加快数据插入的速度，因为 InnoDB 存储引擎会按照主键的顺序来存储数据。其次，使用自增主键能够降低页分裂（page split）的频率，这种分裂发生在一行数据的大小超过当前页的空间大小时，可能会导致性能下降。最后，自增主键通常可以预测，这对于某些应用程序来说非常有用，如需要生成连续唯一标识符的场景。
- 避免冗余或重复索引。冗余或重复索引不仅浪费存储空间，还可能降低数据库的性能。重复的索引意味着数据库系统需要维护更多的索引结构，这增加了数据插入、更新和删除操作的开销，因为每次数据变更时，所有相关的索引都需要更新。
- 其他。对于低基数列，如性别，应避免建立索引，因为执行计划很大可能使用全表扫描代替。

5.2.3　SQL 语句优化

MySQL 的 InnoDB 存储引擎是一个索引组织表，这就意味着每张表都有一个主键，如果没有显式创建，则底层会自动创建一个 6B 的主键，主键不仅包含索引的键值，还包含记录其他列的信息。所以，对于数据的访问和操作，最快的就是通过主键进行。以下总结了 SQL 语句优化的若干条原则。

- 禁止使用 SELECT *，只获取必要的字段。读取不需要的列会增加 CPU、I/O、网络的消耗，不能有效地利用覆盖索引。使用 SELECT * 容易在增加或者删除字段后出现程序 bug。
- 对于语句的 SELECT、UPDATE 和 DELETE 操作，WHERE 条件必须是索引字段。如果 WHERE 子句中的条件不是索引字段，那么 InnoDB 将不得不执行全表扫描来查找满足条件的行，这会显著降低查询性能，特别是在大型表中。
- 禁止使用 INSERT INTO t_××× VALUES（×××）。使用 INSERT INTO t_××× VALUES（×××），必须显式指定插入的列属性，因为不指定值容易在增加或者删除字段后出现程序 bug，插入语句失败。
- 禁止使用属性隐式转换。列类型隐式转换，禁止在 WHERE 条件的列上使用函数或者表达式。
- 禁止负向查询，以及 % 开头的模糊查询。负向查询条件：NOT、! =、<>、! <、! >、NOT IN、NOT LIKE 等，会导致全表扫描。以 % 开头的模糊查询，会导致全表扫描。
- 禁止使用跨库查询，禁止大表使用 JOIN 查询，禁止大表使用子查询。这些操作会产生临时表，消耗较多内存与 CPU，极大影响数据库性能，建议将子查询转换成关联查询。
- 应用程序必须捕获 SQL 异常，并有相应的处理。在数据库交互中，可能会遇到各种各样的错误，如连接失败、查询错误、事务冲突等。如果没有适当的异常处理，这些错误可能会导致应用程序崩溃或产生不可预测的行为。
- 禁止核心业务使用复杂 SQL 语句查询。复杂的 SQL 语句查询可能会导致数据库性能下降，尤其是在处理大量数据时。这可能会影响核心业务的响应时间和吞吐量。复杂的查询通常更难以理解和维护。将复杂的 SQL 语句查询拆分成单表简单查询可以提高查询的性能和可读性。
- SQL 中避免出现 NOW()、RAND()、SYSDATE()、CURRENT_USER() 等不确定结果的函数。这些语句在复制场景下，可能引起主从数据不一致。
- 批量数据的插入和删除操作。对于表批量插入操作，以单词 1000 行/次方式提交事务。对于表删除数据，必须使用主键索引，删除建议为 1000 行/次。（TRUNCATE 操作有概率阻塞其他线程，建议使用 DELETE 操作。）
- 避免 IN 子语句。当使用 IN 或 NOT IN 子句来查询包含大量数据的表时，查询性能可能会受影响，尤其是当 IN 或 NOT IN 列表中包含大量值时。这是因为对于列表中的每个值，数据库都需要单独查询表，这会增加 I/O 操作和 CPU 计算的次数。
- 避免不必要的排序。当处理大量数据时，不必要的排序会消耗大量的计算资源。如果排序操作无法在内存中完成，而是需要使用磁盘进行临时存储，那么性能会大大降低。
- 安全性。无论是使用 SELECT，还是使用 "破坏力极大" 的 UPDATE 和 DELETE 语句，一定要检查 WHERE 条件判断的完整性，不要在运行时出现数据的大量丢失。如果不确定，最好先用 SELECT 语句带上相同条件来验证一下结果集，检验条件是否正确。

5.3 MySQL 8.0 性能优化实践

5.3.1 SQL 语句执行性能的指标——QRTi

MySQL Enterprise Monitor（EM）是 MySQL 企业版提供的一个强大的监控工具，它能够帮助数据库管理员和开发者实时监控与分析 MySQL 服务器的性能。官网提供一个月的企业版试用期，读者可自行下载研究。

EM 工具通过 QRTi（查询响应时间）来判断 SQL 语句性能是否存在问题。QRTi 指标通常有 3 个阈值范围：Optimum（最佳）、Acceptable（可接受）和 Unacceptable（不可接受）。这些阈值是根据不同的业务需求、数据库配置和硬件性能设定的，旨在帮助数据库管理员快速识别哪些查询的响应时间在理想范围内、哪些可能需要关注，以及哪些需要立即采取行动进行优化。表 5-9 所示为 EM 工具对数据库 QRTi 指标的定义。

表 5-9 EM 工具 QRTi 指标

类　型	默　认　值	赋　值	描　述	颜　色
Optimum	100ms	1.00（100%）	最佳时间	绿色
Acceptable	100~400ms	0.50（50%）	可接受时间	黄色
Unacceptable	超过 400ms	0.00（0%）	不可接受时间	红色

QRTi 值计算例子如下。

如果执行 100 次摘要/规范查询，其中 60 次在 100ms（最佳时间框架）内完成，30 次在 100~400ms（可接受时间框架）之间完成，其余 10 次超过 400ms（不可接受时间框架），那么 QRTi 得分如下所示：

```
((60 +(30 / 2) +(10 * 0)) / 100) = 0.75
```

说明：MySQL 官网对于查询响应时间的基本要求为在 0.5s 以内。这个得分表明 75% 的查询在最佳或可接受的时间框架内完成。虽然这个得分不是最高的，但它也表明大多数查询的性能是可接受的。

图 5-3 所示为 EM 工具中 QRTi 指标内容，分成两部分，第一部分使用不同的颜色区分 SQL 语句的执行时间，最上层的蓝颜色表示最佳级别，往下的黄颜色表示可接受级别，再往下的红颜色表示不可接受级别；第二部分包括具体 SQL 语句的延迟曲线图、总消耗时间、总执行时间和扫描行数等信息。

图 5-4 所示为 EM 工具中具体语句统计信息。显示语句执行所消耗的时间（Excution Time）、读取行数（Rows）、报错告警信息（Error, Warning）、临时表操作数（Temp）、连接扫秒方式（Join, Scan）等信息。

通过启用 general_log 来查看 EM 的触发动作，图 5-5 所示为当执行 EM 查询分析（Query Analyzer）时，通过定时抽取数值，并进行相减来得到 SQL 执行时的情况。

通过分析上述 SQL 语句记录，可以了解 MySQL 的性能指标主要是基于 performance_schema 库来获取的。其中，events_statements_summary_by_digest 用于收集当前和最近的语句事件数据，并在这些汇总表中对这些信息进行聚合。

图 5-3　EM 工具 QRTi 指标

Query Analyzer > 详细　　　　　　　　　　　　　　　　Statement Digest: ...668ebb 显示 Copy

From 2021-6-4 11:17:24 to 2021-6-4 11:47:24.

Normalized SQL Statement

```
SELECT `extract_schema_from_file_name` ( `performance_schema` . `file_summary_by_
instance` . `FILE_NAME` ) AS `table_schema` ,
       `extract_table_from_file_name` ( `performance_schema` . `file_summary_by_i
nstance` . `FILE_NAME` ) AS `table_name` ,
       SUM ( `performance_schema` . `file_summary_by_instance` . `COUNT_READ` ) A
S `count_read` ,
       SUM ( `performance_schema` . `file_summary_by_instance` . `SUM_NUMBER_OF_B
YTES_READ` ) AS `sum_number_of_bytes_read` ,
       SUM ( `performance_schema` . `file_summary_by_instance` . `SUM_TIMER_READ`
) AS `sum_timer_read` ,
       SUM ( `performance_schema` . `file_summary_by_instance` . `COUNT_WRITE` )
AS `count_write` ,
       SUM ( `performance_schema` . `file_summary_by_instance` . `SUM_NUMBER_OF_B
YTES_WRITE` ) AS `sum_number_of_bytes_write` ,
       SUM ( `performance_schema` . `file_summary_by_instance` . `SUM_TIMER_WRITE
` ) AS `sum_timer_write` ,
       SUM ( `performance_schema` . `file_summary_by_instance` . `COUNT_MISC` ) A
S `count_misc` ,
       SUM ( `performance_schema` .
```

统计

Execution Count: 2
Total Execution Time: 435ms
Min Execution Time: 187ms
Max Execution Time: 248ms
Avg Execution Time: 218ms
Total Lock Time: 2ms
Total Rows: 200
Total Rows Examined: 200
Avg Rows: 100
Total Errors: 0
Total Warnings: 0
Temp Tables: 6
Temp Disk Tables: 0
Temp Tables Avg: 3
Temp Tables Disk Pct: 0
Select Full Join: 0
Select Full Range Join: 0
Select Range: 0
Select Range Check: 0
Select Scan: 6
Sort Merge Passes: 0
Sort Range: 0
Sort Rows: 1,304
Sort Scan: 2
Database: mysql
First Seen: 2021-6-4 11:37:13

图 5-4　EM 工具中具体语句统计信息

```
2021-06-04T01:06:28.222093Z 63961 Query SELECT digest AS `digest`,
            schema_name AS `schema`,
            count_star AS `execCount`,
            IFNULL(sum_timer_wait * 0.000001, 0) AS `execTimeTotal`,
            IFNULL(min_timer_wait * 0.000001, 0) AS `execTimeMin`,
            IFNULL(max_timer_wait * 0.000001, 0) AS `execTimeMax`,
            IFNULL(sum_lock_time * 0.000001, 0) AS `lockTimeTotal`,
            sum_errors AS `errorCount`,
            sum_warnings AS `warningCount`,
            sum_rows_affected + sum_rows_sent AS `rowsTotal`,
            sum_rows_examined AS `rowsExaminedTotal`,
            sum_created_tmp_disk_tables AS `createdTmpDiskTables`,
            sum_created_tmp_tables AS `createdTmpTables`,
            sum_select_full_join AS `selectFullJoin`,
            sum_select_full_range_join AS `selectFullRangeJoin`,
            sum_select_range AS `selectRange`,
            sum_select_range_check AS `selectRangeCheck`,
            sum_select_scan AS `selectScan`,
            sum_sort_merge_passes AS `sortMergePasses`,
            sum_sort_range AS `sortRange`,
            sum_sort_rows AS `sortRows`,
            sum_sort_scan AS `sortScan`,
            sum_no_index_used AS `noIndexUsedCount`,
            sum_no_good_index_used AS `noGoodIndexUsedCount`,
            digest_text AS `normalizedText`,
            unix_timestamp(first_seen)*1000 AS `firstSeen`,
            unix_timestamp(last_seen)*1000 AS `lastSeen`
       FROM performance_schema.events_statements_summary_by_digest
      WHERE digest IS NOT NULL
        -- even though i want a (from,to) set based on last_seen, it's only at 1s
        -- precision, so if i elide a seconds worth of samples, it could miss some.
        -- note, then this requires the delta'izing algorithm to elide exec counts
        -- that haven't changed
        AND UNIX_TIMESTAMP(last_seen)*1000 >= 1622768728140
```

图 5-5 EM 工具抓取记录

5.3.2 通过 events_statements_summary_by_digest 表发现问题 SQL 语句

1. events_statements_summary_by_digest 表的统计信息

events_statements_summary_by_digest 表提供了对 SQL 语句摘要的聚合统计信息，而不是单个语句实例的详细信息。它对于识别性能瓶颈和优化查询非常有用。首先，查看这张表汇总表记录的详细信息。查询命令行如下所示。

```
mysql>SELECT * FROM performance_schema.events_statements_summary_by_digest limit 1\G;
*************************** 1.row ***************************
              SCHEMA_NAME: NULL
                   DIGEST: 44e35cee979ba420eb49a8471f852bbe15b403c89742704817dfbaace0d99dbb
              DIGEST_TEXT: SELECT @@`version_comment` LIMIT ?
               COUNT_STAR: 8
           SUM_TIMER_WAIT: 1566500000
           MIN_TIMER_WAIT: 87700000
           AVG_TIMER_WAIT: 195800000
           MAX_TIMER_WAIT: 489900000
            SUM_LOCK_TIME: 0
               SUM_ERRORS: 0
             SUM_WARNINGS: 0
        SUM_ROWS_AFFECTED: 0
            SUM_ROWS_SENT: 8
        SUM_ROWS_EXAMINED: 8
SUM_CREATED_TMP_DISK_TABLES: 0
     SUM_CREATED_TMP_TABLES: 0
```

```
                SUM_SELECT_FULL_JOIN: 0
          SUM_SELECT_FULL_RANGE_JOIN: 0
                    SUM_SELECT_RANGE: 0
              SUM_SELECT_RANGE_CHECK: 0
                     SUM_SELECT_SCAN: 0
               SUM_SORT_MERGE_PASSES: 0
                      SUM_SORT_RANGE: 0
                       SUM_SORT_ROWS: 0
                       SUM_SORT_SCAN: 0
                   SUM_NO_INDEX_USED: 0
              SUM_NO_GOOD_INDEX_USED: 0
                          FIRST_SEEN: 2021-06-03 11:04:10.553945
                           LAST_SEEN: 2021-06-03 14:34:15.950372
                        QUANTILE_95: 501187233
                        QUANTILE_99: 501187233
                       QUANTILE_999: 501187233
                   QUERY_SAMPLE_TEXT: select @@version_comment limit 1
                   QUERY_SAMPLE_SEEN: 2021-06-03 14:34:15.950372
             QUERY_SAMPLE_TIMER_WAIT: 4899000001 row in set (0.00 sec)
```

表 5-10 所示为 events_statements_summary_by_digest 的显示字段及其说明。

表 5-10　events_statements_summary_by_digest 的显示字段及其说明

字　　段	说　　明
SCHEMA_NAME	库名
DIGEST	对 SQL 使用 MD5 而产生的 32 位字符串
DIGEST_TEXT	将语句中的值部分用问号代替,用于 SQL 语句归类
COUNT_STAR	事件计数
SUM_TIMER_WAIT	总的等待时间
MIN_TIMER_WAIT	最小等待时间
AVG_TIMER_WAIT	平均等待时间
MAX_TIMER_WAIT	最大等待时间
SUM_LOCK_TIME	锁时间总时长
SUM_ERRORS	错误的总次数
SUM_WARNINGS	警告的总次数
SUM_ROWS_AFFECTED	影响的总行数
SUM_ROWS_SENT	返回客户端的总行数
SUM_ROWS_EXAMINED	扫描行的总数目
SUM_CREATED_TMP_DISK_TABLES	创建磁盘临时表的总次数
SUM_CREATED_TMP_TABLES	创建临时表的总次数
SUM_SELECT_FULL_JOIN	全表连接扫描的总次数
SUM_SELECT_FULL_RANGE_JOIN	范围全扫描的总次数
SUM_SELECT_RANGE	范围方式扫描的总次数

(续)

字　　段	说　　明
SUM_SELECT_SCAN	全表扫描的总次数
SUM_SORT_MERGE_PASSES	合并排序的总次数
SUM_SORT_RANGE	范围排序的总次数
SUM_SORT_ROWS	排序记录的总次数
SUM_SORT_SCAN	表排序扫描总次数
SUM_NO_INDEX_USED	没有使用索引的总次数
FIRST_SEEN	第一次执行时间
LAST_SEEN	最后一次执行时间
QUANTILE_95	95%的语句延迟低于这个值（单位为皮秒）
QUANTILE_99	99%的语句延迟低于这个值（单位为皮秒）
QUANTILE_999	99.9%的语句延迟低于这个值（单位为皮秒）
QUERY_SAMPLE_TEXT	生成的摘要值的样本 SQL 语句
QUERY_SAMPLE_SEEN	样本语句的最近一次执行的时间
QUERY_SAMPLE_TIMER_WAIT	样本语句的等待时间

说明：events_statements_summary_by_digest 表中的每一行代表一个唯一的 SQL 语句摘要（digest），摘要是通过规范化 SQL 文本（去除空格、换行符和注释等）生成的。这样，对于相似的 SQL 语句，即使它们的文本表示略有不同，也会被归类到同一个摘要中。为了使用 events_statements_summary_by_digest 表，需要确保 performance_schema 已经启用，并且已经配置了适当的仪器（instrument）来收集语句事件数据。此外，由于 events_statements_summary_by_digest 表中的数据是聚合统计信息，因此不支持 DELETE 操作，但可以使用 TRUNCATE 命令来清空表中的所有数据。

2. events_statements_summary_by_digest 表的作用

通过 events_statements_summary_by_digest 表中的 SQL 语句摘要统计信息，可以发现问题 SQL 语句，从而实现性能瓶颈识别、查询优化、错误监控。

（1）性能瓶颈识别

通过分析 events_statements_summary_by_digest 表中的数据，可以识别出哪些 SQL 语句的执行次数最多、平均执行时间最长或者错误次数最多，从而确定性能瓶颈所在，为优化查询提供依据。

（2）查询优化

通过查看 events_statements_summary_by_digest 表中的 DIGEST_TEXT、COUNT_STAR、AVG_TIMER_WAIT 等列，可以了解哪些查询需要优化。例如，如果发现某个查询的执行时间特别长，就可以考虑对该查询进行优化，如重写查询、添加索引或调整数据库配置等。

（3）错误监控

虽然 events_statements_summary_by_digest 表不记录具体的错误类型，但它会记录发生错误的语句摘要统计信息。通过监控这些信息，可以及时发现并处理 SQL 语句执行过程中的错误。

在实际应用中，应该定期分析该表中的数据，并根据分析结果进行相应的优化操作。

3. events_statements_summary_by_digest 表参数设定

events_statements_summary_by_digest 表记录信息可通过参数加以设定。

(1) 是否启用 SQL 语句统计

通过 setup_consumers 表确认是否激活,默认是激活。查询语句如下所示。

```
mysql> SELECT * FROM performance_schema.setup_consumers
    WHERE name = 'statements_digest';
+-------------------+---------+
| NAME              | ENABLED |
+-------------------+---------+
| statements_digest | YES     |
+-------------------+---------+
1 row in set (0.00 sec)
```

说明:MySQL 在默认安装环境下,会自动记录所有的 SQL 语句。提供了查看历史 SQL 语句性能的排查方式。

(2) SQL 语句长度限制参数

记录的 SQL 语句的长度不能超过设置的参数,否则截断。查看语句长度参数的命令行如下所示。

```
mysql> SHOW VARIABLES LIKE '%max_digest_leng%';
+------------------------------------+-------+
| Variable_name                      | Value |
+------------------------------------+-------+
| max_digest_length                  | 1024  |
| performance_schema_max_digest_length | 1024  |
+------------------------------------+-------+
2 rows in set, 1 warning (0.01 sec)
```

说明:单位为 B。

第一个参数 max_digest_length 是会话级别的。

第二个参数 performance_schema_max_digest_length 是语句级别的,只对 performance_schema 起作用。如果 performance_schema_max_digest_length 小于 max_digest_length,则相对于原始语句的复制将被截断。

(3) 表记录总行数参数

performance_schema_digests_size 参数限制 events_statements_summary_by_digest 表记录的最大行数。默认值是 10000,如果超过这个最大值,新的 SQL 语句就无法插入这张表中。查看最大记录行数的命令行如下所示。

```
mysql> SHOW VARIABLES LIKE '%performance_schema_digests_size%';
+---------------------------------+-------+
| Variable_name                   | Value |
+---------------------------------+-------+
| performance_schema_digests_size | 10000 |
+---------------------------------+-------+
1 row in set, 1 warning (0.00 sec)
```

5.3.3 使用 statement_analysis 视图分析 SQL 语句

statement_analysis 是 MySQL 中 sys 库下的一个视图,它主要用于分析和展示 SQL 语句的执行统计信息。其数据来源主要是 events_statements_summary_by_digest 表。statement_analysis 视图通过汇总和规范化这些数据,为数据库管理员和开发者提供了一个方便查看与分析 SQL 语句性能的界面。通过这个视图，可以查看哪些 SQL 语句执行得比较频繁、哪些语句占用了较多的时间，以及哪些语句可能存在性能问题等。

查看执行过的 SQL 语句的命令行如下所示。

```
mysql> SELECT * FROM sys.statement_analysis order by total_latencydesc limit 1 \G;
*************************** 1.row ***************************
            query: SELECT * FROM `performance_schema`.`clone`
               db: employees
        full_scan: 
       exec_count: 1
        err_count: 1
       warn_count: 0
    total_latency: 906.02 us
      max_latency: 906.02 us
      avg_latency: 906.02 us
     lock_latency: 5.00 us
      CPU_latency: 0 ps
        rows_sent: 0
    rows_sent_avg: 0
    rows_examined: 0
rows_examined_avg: 0
    rows_affected: 0
rows_affected_avg: 0
       tmp_tables: 0
  tmp_disk_tables: 0
      rows_sorted: 0
sort_merge_passes: 0
max_controlled_memory: 16.09KiB
 max_total_memory: 144.59KiB
           digest: 3df3df5b2250262d8420af4e72a8e42d6f386ae9ca4ad381e1f2c224660ba6a6
       first_seen: 2023-09-03 17:25:04.522271
        last_seen: 2023-09-03 17:25:04.522271
1 row in set (0.03 sec)
```

表 5-11 所示为 sys.statement_analysis 表的字段及其说明。

表 5-11　sys.statement_analysis 表的字段及其说明

字　　段	说　　明
query	SQL 语句
db	SQL 语句执行的数据库名
full_scan	全表扫描的次数，无全表扫描时显示为空
exec_count	总执行次数
err_count	总执行错误次数
warn_count	总警告次数
total_latency	总执行时间
max_latency	最大执行时间

(续)

字 段	说 明
avg_latency	平均执行时间
lock_latency	锁等待总时间
CPU_latency	CPU 耗时总时间
rows_sent	总返回行数
rows_sent_avg	平均返回行数
rows_examined	总扫描行数
rows_examined_avg	平均扫描行数
tmp_tables	临时表创建总数
tmp_disk_tables	临时表落盘总次数
rows_sorted	排序次数
sort_merge_passes	合并排序次数
max_controlled_memory	最大消耗内存
max_total_memory	总消耗内存
digest	语句哈希后的摘要
first_seen	第一次语句执行时间
last_seen	最后一次语句执行时间

总之，statement_analysis 视图是 MySQL 数据库中一个非常重要的性能分析工具，它可以帮助数据库管理员和开发者更好地优化 SQL 语句的性能。

5.3.4 通过分析 sys 库的存储过程排查性能问题

在 MySQL 8.0 中，除了查看慢查询语句、分析 binlog 等常规手段以外，sys 库提供了一些有用的存储过程来帮助数据库管理员和开发者分析排查数据库性能问题。这些存储过程是专门为性能诊断、监控和故障排除而设计的。和传统的慢查询日志、SHOW 命令以及 EXPLAIN 分析相比，这些工具提供了更高级和更易于理解的性能洞察。

下面介绍 sys 库下性能排查的 3 个存储过程。

1. diagnostics：诊断当前服务运行状态

diagnostics 存储过程将创建当前 MySQL 服务整体运行报告。该报告每 30s 进行一次迭代，使用当前的 performance_schema 最多运行 30s。因为内容比较多，所以可以采用 tee 命令将结果标准输出到文件中。需要开启 secure_file_priv 参数。创建的报告会输出到.out 文件中。调用 diagnostics 存储过程的命令行如下所示。

```
mysql> tee diag.out;
mysql> CALL sys.diagnostics(30, 30, 'current');
mysql> notee;
```

报告输出的指标非常多，也非常全面。表 5-12 所示为 diagnostics 存储过程输出指标及其说明。

表 5-12 diagnostics 存储过程输出指标及其说明

输出指标	说明
GLOBAL VARIABLES	MySQL 服务配置的全局变量
SHOW MASTER STATUS	binlog 文件的当前位置和名称,以及 GTID 信息
SHOW ENGINE INNODB STATUS	InnoDB 存储引擎的状态信息,包括锁等待、缓冲池状态、事件日志状态等
SELECT * FROM sys.processlist	MySQL 服务器上运行的所有进程的列表
Performance schema 的 setup_instruments	用于配置各种性能事件的采集器
SHOW ENGINE PERFORMANCE_SCHEMA STATUS	查看 Performance Schema 自身库状态信息
sys.memory_by_host_by_current_bytes	按照主机和当前内存使用量对数据库中的连接进行分组,这有助于识别哪些主机占用了较多的内存
Overall host_summary	按照主机分组的数据库活动的摘要信息,这有助于识别哪些主机对数据库的负载较大
host_summary_by_file_io_type	按照主机和文件 I/O 类型(读/写)对数据库活动进行分组,这有助于识别哪些主机执行了较多的文件 I/O 操作
host_summary_by_statement_latency	按照主机和查询延迟对数据库活动进行分组,这有助于识别哪些主机上的查询较慢
Overall io_by_thread_by_latency	按照线程和延迟分组的 I/O 活动的摘要信息,这有助于识别哪些线程执行了较慢或较多的 I/O 操作
Overall io_global_by_file_by_bytes	按照文件和字节数分组的全局 I/O 活动的摘要信息,这有助于识别哪些文件进行了较多的读/写操作
Overall schema_index_statistics	数据库中所有索引的统计信息。它有助于识别哪些索引需要优化或重建
Overall schema_table_statistics	数据库中所有表的统计信息。它有助于识别哪些表很大或需要优化

通过了解上述指标,可以了解数据库服务器的性能、健康状况、资源使用情况以及是否存在潜在的问题或瓶颈。执行 diagnostics 存储过程,输出诊断报告的命令行如下所示。

```
mysql > CALL sys.diagnostics(30, 30, 'current');
+-----------------------------+
| summary                     |
+-----------------------------+
| Disabled 1 thread           |
+-----------------------------+
1 row in set (0.02 sec)


+-----------------------------+-------------------------------------+
| Name                        | Value                               |
+-----------------------------+-------------------------------------+
| Hostname                    | schouse                             |
| Port                        | 3380                                |
| Socket                      | /opt/data8.0/data/mysql.sock        |
| Datadir                     | /opt/data8.0/data/                  |
```

```
| Server UUID              | 22228e8c-b0ee-11ec-a2d1-00163e23e2cc |
|--------------------------|--------------------------------------|
| MySQL Version            | 8.0.34                               |
| Sys Schema Version       | 2.1.2                                |
| Version Comment          | MySQL Community Server - GPL         |
| Version Compile OS       | Linux                                |
| Version Compile Machine  | x86_64                               |
|--------------------------|--------------------------------------|
| UTC Time                 | 2023-08-28 06:22:54                  |
| Local Time               | 2023-08-28 14:22:54                  |
| Time Zone                | +08:00                               |
| System Time Zone         | CST                                  |
| Time Zone Offset         | 08:00:00                             |
+---------------------------+-------------------------------------+
17 rows in set (0.02 sec)

========================
     Configuration
========================
+-------------------------------------------------------+
                                                      +1
row in set (0.01 sec)


+--------------------------------------+
| The following output is:             |
+--------------------------------------+
| GLOBAL VARIABLES                     |
+--------------------------------------+

+-----------------------------------------------+---------------------------------+
| Variable_name                                 | Variable_value                  |
+-----------------------------------------------+---------------------------------+
| activate_all_roles_on_login                   | OFF                             |
| admin_address                                 |                                 |
| admin_port                                    | 33090                           |
| admin_ssl_crl                                 |                                 |
| admin_tls_version                             | TLSv1.2,TLSv1.3                 |
| back_log                                      | 1000                            |
| basedir                                       | /opt/idc/mysql8.0.28/           |
| big_tables                                    | OFF                             |
| bind_address                                  | 198.168.1.1                     |
| binlog_cache_size                             | 32768                           |
...
+--------------------------------------+
| The following output is:             |
+--------------------------------------+
| SHOW MASTER STATUS                   |
+--------------------------------------+
1 row in set (15.07 sec)
```

```
+------------------+--------+-------------+-----------------+--------------------------------------+
| File             |Position|Binlog_Do_DB |Binlog_Ignore_DB |Executed_Gtid_Set                     |
+------------------+--------+-------------+-----------------+--------------------------------------+
|mysql-bin.000025  | 56904  |             |                 |4af6a158-aa6d-11eb-82f2-00163e23e2cc:1-511629|
+------------------+--------+-------------+-----------------+--------------------------------------+
1 row in set (15.07 sec)

+---------------------------------+
|The following output is:         |
+---------------------------------+
|SHOW ENGINE INNODB STATUS        |
+---------------------------------+

=====================================
2022-03-22 10:16:48 139752869254912 INNODB MONITOR OUTPUT
=====================================
Per second averages calculated from the last 1 seconds-----------------
BACKGROUND THREAD----------------
srv_master_thread loops: 386 srv_active, 0 srv_shutdown, 773146 srv_idle
srv_master_thread log flush and writes: 0
...
+-----------------------------------------+
|The following output is:                 |
+-----------------------------------------+
|SELECT * FROM sys.processlist            |
+-----------------------------------------+


+--------------------------------------------------+
|The following output is:                          |
+--------------------------------------------------+
|Performance Schema Setup - Instruments            |
+--------------------------------------------------+

+--------------------------+------------+------------+
|InstrumentClass           |EnabledPct  |TimedPct    |
+--------------------------+------------+------------+
|error                     |   100.00   |    0.00    |
|idle                      |   100.00   |  100.00    |
|memory/archive            |   100.00   |    0.00    |
|memory/blackhole          |   100.00   |    0.00    |
|memory/client             |   100.00   |    0.00    |
|memory/component_sys_vars |   100.00   |    0.00    |
...
+-----------------------------------------+
|The following output is:                 |
+-----------------------------------------+
|Delta io_by_thread_by_latency            |
+-----------------------------------------+
1 row in set (0.50 sec)
```

```
+------------------------------+--------+---------------+-------------+-------------+-------------+-----------+---------------+
|user                          |total   |total_latency  |min_latency  |avg_latency  |max_latency  |thread_id  |processlist_id |
+------------------------------+--------+---------------+-------------+-------------+-------------+-----------+---------------+
|log_flusher_thread            | 54600  |1.15 min       |108.69 us    |1.26 ms      |88.44 ms     |    16     |     NULL      |
|log_writer_thread             |108516  |6.99 s         |994.15 ns    |64.45 us     |87.38 ms     |    18     |     NULL      |
|page_flush_coordinator_thread |10293   |1.99 s         |2.94 us      |193.37 us    |8.96 ms      |    13     |     NULL      |
|io_write_thread               | 1107   |852.79 ms      |52.75 us     |770.36 us    |8.72 ms      |    12     |     NULL      |
...
```

对于 diagnostics 存储过程输出的性能检测报告，虽然完整，但因为涉及大量的指标和数据，目前没有一个规范化的格式或展示方式，所以解读报告变得非常费劲。

2. ps_trace_statement_digest：追踪语句执行状态

ps_trace_statement_digest 存储过程用于追踪 SQL 语句执行状态统计信息。通过 events_statements_summary_by_digest 表中 DIGEST 列的 MD5 值，在轮询时间内计算运行指标（Explain+Time+Event latency 方式）并分析性能。调用 ps_trace_statement_digest 存储过程的命令行如下所示。

```
# 下面语句中的参数值'891ec6860f98ba46d89dd20b0c03652c'、10、0.1、TRUE 和 TRUE 分别代表语句 MD5、分析
需要的时间、分析间隔时间、是否截断 truncate 表、是否自动开启需要的 consumers 消费表
mysql> CALL sys.ps_trace_statement_digest('891ec6860f98ba46d89dd20b0c03652c', 10, 0.1,
TRUE, TRUE);
+--------------------------------+
| SUMMARY STATISTICS             |
+--------------------------------+
| SUMMARY STATISTICS             |
+--------------------------------+
1 row in set (9.11 sec)

+------------+-----------+----------+-----------+---------------+------------+------------+
|executions  |exec_time  |lock_time |rows_sent  |rows_examined  |tmp_tables  |full_scans  |
+------------+-----------+----------+-----------+---------------+------------+------------+
|    21      |4.11 ms    |2.00 ms   |    0      |      21       |     0      |     0      |
+------------+-----------+----------+-----------+---------------+------------+------------+
1 row in set (9.11 sec)

+----------------------------------------+--------+-----------+
|event_name                              |count   |latency    |
+----------------------------------------+--------+-----------+
|stage/sql/statistics                    |  16    |546.92 us  |
|stage/sql/freeing items                 |  18    |520.11 us  |
|stage/sql/init                          |  51    |466.80 us  |
...
|stage/sql/cleaning up                   |  18    |11.92 us   |
|stage/sql/executing                     |  16    |6.95 us    |
+----------------------------------------+--------+-----------+
17 rows in set (9.12 sec)

+----------------------------------------+
| LONGEST RUNNING STATEMENT              |
+----------------------------------------+
| LONGEST RUNNING STATEMENT              |
```

```
+--------------------------------+
1 row in set (9.16 sec)
+----------+----------+----------+-----------+---------------+------------+----------+
|thread_id |exec_time |lock_time |rows_sent  |rows_examined  |tmp_tables  |full_scan |
+----------+----------+----------+-----------+---------------+------------+----------+
|   166646 |618.43 us |1.00 ms   |         0 |             1 |          0 |        0 |
+----------+----------+----------+-----------+---------------+------------+----------+
1 row in set (9.16 sec)
# Truncated for clarity...
+---------------------------------------------------------------+
|sql_text                                                       |
+---------------------------------------------------------------+
|selecthibeventhe0_.id as id1382_, hibeventhe0_.createdTime ... |
+---------------------------------------------------------------+
1 row in set (9.17 sec)

+----------------------------------+-----------+
|event_name                        |latency    |
+----------------------------------+-----------+
|stage/sql/init                    |8.61 us    |
|stage/sql/init                    |331.07 ns  |
...
|stage/sql/freeing items           |30.46 us   |
|stage/sql/cleaning up             |662.13 ns  |
+----------------------------------+-----------+
18 rows in set (9.23 sec)

+----+-----------+-----------+------+-------------+---------+-------+-----------+------+-------+
|id  |select_type|table      |type  |possible_keys|key      |key_len|ref        |rows  |Extra  |
+----+-----------+-----------+------+-------------+---------+-------+-----------+------+-------+
|  1 |SIMPLE     |hibeventhe0_|const|fixedTime    |fixedTime|775    |const,const|    1 |NULL   |
+----+-----------+-----------+------+-------------+---------+-------+-----------+------+-------+
1 row in set (9.27 sec)
Query OK, 0 rows affected (9.28 sec)
```

追踪语句执行状态报告可以分为以下 3 个部分。

- 第一部分包含语句执行的总次数（executions）、总时间（exec_time）、锁定时间（lock_time）和返回行数（rows_sent）等信息。从这些信息中可以分析出语句的执行效率。
- 第二部分包含执行语句的事件（event）的延迟（latency）信息。延迟大说明执行性能不好。
- 第三部分显示执行计划。可以了解 SQL 语句是否使用了合适的索引、是否有可能发生全表扫描或额外的文件排序等问题。

3. statement_performance_analyzer：分析语句执行性能

statement_performance_analyzer 存储过程用于收集 events_statements_summary_by_digest 表中的数据，然后对比两个不同时间点的快照，以生成性能变化的增量报告。

statement_performance_analyzer 存储过程在执行时需要两个变量作为输入：计算方式（用于确定如何计算性能差异或指标的方式）和 SET 方式（用于指定如何设置或选择数据集的方式）。两个变量的说明如下。

表 5-13 所示为计算方式的变量说明列表。

表 5-13 计算方式的变量说明

计算方式	说明
snapshot	存储快照。默认情况下，对性能模式 events_statements_summary_by_digest 表的当前内容做一个快照，通过设置 in_table，可以覆盖它以复制指定表的内容。快照存储在 sys 架构 tmp_digests 临时表中
overall	根据 in_table 指定的表的内容生成一个分析器。对于整体分析，in_table 可以利用 NOW( ) 来使用一个新的快照。这将覆盖现有的快照。如果 in_table 为 NULL 且不存在快照，则创建一个新的快照
delta	生成 delta 分析。这个增量是在 in_table 指定的引用表和快照之间计算的，所以快照必须存在。此操作使用 sys 架构 tmp_digests_delta 临时表
create_tmp	创建一个适合存储快照以供以后使用的临时表（如用于计算增量）
create_table	创建一个适合存储快照以供以后使用的普通表（如用于计算增量）
save	将快照保存到 in_table 指定的表中。表必须存在并且具有正确的结构。如果不存在快照，则创建新的快照
cleanup	删除用于快照和增量的临时表

表 5-14 所示为生成分析表的视图说明。

表 5-14 生成分析表的视图说明

视图参数名	参考视图
with_runtimes_in_95th_percentile	使用 statements_with_runtimes_in_95th_percentile 视图
analysis	使用 statement_analysis 视图
with_errors_or_warnings	使用 statements_with_errors_or_warnings 视图
with_full_table_scans	使用 statements_with_full_table_scans 视图
with_sorting	使用 statements_with_sorting 视图
with_temp_tables	使用 statements_with_temp_tables 视图
custom	使用自定义视图

下面通过实例介绍使用不同变量所进行的性能分析。

执行 events_statements_summary_by_digest 存储过程，输出所有语句中的 95% SQL 语句分析报告。报告输出命令行如下所示。

```
# 创建临时表
mysql> CALL sys.statement_performance_analyzer('create_tmp','mydb.tmp_digests_ini',
NULL);
Query OK, 0 rows affected (0.08 sec)

# 创建第一个 digest 表的快照
mysql> CALL sys.statement_performance_analyzer('snapshot', NULL, NULL);
Query OK, 0 rows affected (0.02 sec)

# 保存到表中
```

```
mysql> CALL sys.statement_performance_analyzer('save', 'mydb.tmp_digests_ini', NULL);
Query OK, 0 rows affected (0.00 sec)
```

等待 60 秒
```
mysql> DO SLEEP(60);
Query OK, 0 rows affected (1 min 0.00 sec)
```

创建第二个快照
```
mysql> CALL sys.statement_performance_analyzer('snapshot', NULL, NULL);
Query OK, 0 rows affected (0.02 sec)
```

生成分析表,使用 with_runtimes_in_95th_percentile 视图
```
mysql> CALL sys.statement_performance_analyzer('overall', NULL, 'with_runtimes_in_95th_percentile');
+--------------------------------------------------------------+
| Next Output                                                  |
+--------------------------------------------------------------+
| Queries with Runtime in 95th Percentile                      |
+--------------------------------------------------------------+
...
```

生成 delta 分析
```
mysql> CALL sys.statement_performance_analyzer('delta', 'mydb.tmp_digests_ini', 'with_runtimes_in_95th_percentile');
+--------------------------------------------------------------+
| Next Output                                                  |
+--------------------------------------------------------------+
| Queries with Runtime in 95th Percentile                      |
+--------------------------------------------------------------+
1 row in set (0.03 sec)
*************************** 1.row ***************************
              query: SELECT * FROM employees where id=?
                 db: user
          full_scan:
         exec_count: 32
          err_count: 1
         warn_count: 0
      total_latency: 35.00 s
        max_latency: 60.00 s
        avg_latency: 80.00 s
          rows_sent: 31
      rows_sent_avg: 50
      rows_examined: 10
  rows_examined_avg: 3
         first_seen: 2024-01-14 10:49:34.606095
          last_seen: 2024-03-14 10:49:34.606095
             digest: 7f306ac5372b766bc1eaf8ac104e2d686e501cd1ac25cd07d2ad20dc94bff135
1 row in set (0.01 sec)
...
```

通过查看上述性能分析报告中的基础信息,如全表扫描、执行次数、错误次数、警告次数、延迟时间和返回行数等,可以对 SQL 语句的执行效率进行初步评估,从而发现可能存在的问题并进行相应的优化。

说明:使用 SET 方式变量,可以获取不同类型的 SQL 语句性能报表,同时 sys.statement_performance_analyzer.limit 参数可以提供 TOP N 的 SQL 语句输出。

5.3.5 通过监控 InnoDB 存储引擎进行性能优化

MySQL 的体系架构划分为服务层和存储引擎层。服务层包含查询解析、优化、缓存、内置函数等核心功能,而存储引擎层则负责数据的存储和检索。想要更深入地了解 InnoDB 存储引擎层的运行情况,就需要通过命令行来监控 InnoDB 存储引擎层的状态变化值。这些命令提供了关于 InnoDB 内部操作的详细信息,包括锁定、事务、缓冲池状态等,从而有助于进行性能分析和优化。

下面介绍 InnoDB 状态指标的获取方式及说明,并探讨如何从这些指标中发现问题。

1. InnoDB 状态信息

SHOW ENGINE INNODB STATUS 是 MySQL 中用于获取 InnoDB 存储引擎内部状态和诊断信息的 SQL 命令。这个命令的输出提供了关于 InnoDB 存储引擎的许多重要信息,包括锁定状态、事务状态、缓冲池的使用情况、I/O 操作、死锁等。这些信息可以帮助用户监控和调优 InnoDB 存储引擎的性能。表 5-15 所示为该命令的输出的 10 类信息。

表 5-15　InnoDB 状态信息

信　　息	说　　明
Header	输出当前日期以及距离上次操作执行的间隔时间
BACKGROUND THREAD	显示 InnoDB 后台线程运行状态
SEMAPHORES	显示 InnoDB 存储引擎使用的信号量和互斥量的状态
TRANSACTIONS	输出当前正在进行的事务以及它们的状态,可以查看每个事务的 ID、状态、锁定等信息
LATEST DETECTED DEADLOCK	输出最近一次死锁信息。只有产生过死锁,才会有显示
FILE	输出 InnoDB 存储引擎正在进行的文件 I/O 操作,可以查看每个文件的名称、读写次数、字节数等信息
INSERT BUFFER AND ADAPTIVE HASH INDEX	输出 InnoDB 存储引擎使用的插入缓冲区和自适应哈希索引的状态
LOG	输出 InnoDB 存储引擎正在进行的日志操作,可以查看日志缓冲区、日志文件、日志写入次数等信息
BUFFER POOL AND MEMORY	输出 InnoDB 存储引擎缓冲池和内存使用的详细信息。如果设置了多个缓冲池实例,那么显示每个缓冲池信息
ROW OPERATIONS	输出 InnoDB 存储引擎中正在进行的行操作的详细信息

SHOW ENGINE INNODB STATUS 命令的输出提供了大量的关于 InnoDB 存储引擎内部状态的信息,这些信息反映了当前某个模块的运行情况,可从这些数值上发现问题,并进行优化。

具体说明如下。

(1)BACKGROUND THREAD(后台线程)信息

在 BACKGROUND THREAD 输出信息中,可以了解到不同类型的后台线程(如 master thread、

io_thread、insert buffer thread 等）的状态，以及它们的活跃程度。后台线程输出信息如下。

```
----------------
BACKGROUND THREAD
----------------
srv_master_thread loops: 194 srv_active, 0 srv_shutdown, 14709 srv_idle
srv_master_thread log flush and writes: 3627
```

表 5-16 所示为 BACKGROUND THREAD 各项指标及其说明。

表 5-16 BACKGROUND THREAD 各项指标及其说明

指 标	说 明
srv_master_thread loops	主线程的循环次数。主线程在每次循环过程中都会休眠（sleep），休眠的时间为 1s。在每次循环的过程中都会选择 active、shutdown、idle 中的一种状态执行。因为主线程在不停循环，所以其值是随时间递增的
srv_active	主线程选择 active 状态执行。active 数量的增加与数据表、数据库更新操作有关，与查询无关，如插入数据、更新数据、修改表等
srv_shutdown	这个指标的值一直为 0，因为它的值只有在 MySQL 服务关闭的时候才会增加
srv_idle	这个指标的值在主线程空闲的时候增加，即没有任何数据库改动操作时
log_flush_and_write	主线程会在后台定期刷新日志，日志刷新时间由参数 innodb_flush_log_at_timeout 控制

在 BACKGROUND THREAD 输出信息中，active 和 idle 都是描述线程状态的关键词。
- active：表示线程当前正在执行某种任务，处于活跃状态。
- idle：表示线程当前没有执行任务，处于空闲状态。

通过对比 active 和 idle 的值，可以对 MySQL 服务器的负载情况有一个大致的了解。
- 如果 active 值很高而 idle 值很低，就意味着 InnoDB 的后台线程非常忙碌，可能在进行大量的磁盘 I/O 操作、日志刷新或数据页刷新等。这通常表明服务器负载较高，需要关注性能瓶颈或资源使用情况。
- 如果 idle 值很高而 active 值很低，就表示 InnoDB 的后台线程大部分时间都处于空闲状态。这可能是因为服务器的负载较轻，或者数据库操作相对较少。然而，如果这种情况持续存在，也可能表明服务器资源利用不足，可以进一步优化或增加负载。

（2）SEMAPHORES（信号量）信息

在 InnoDB 的 SEMAPHORES 输出信息中，提供了关于 InnoDB 存储引擎内部信号量的详细信息。信号量是数据库用来控制并发访问共享资源的机制，可确保多个线程在访问同一资源时不会发生冲突。以下是一个 SEMAPHORES 信息输出示例。

```
----------
SEMAPHORES
----------
OS WAIT ARRAY INFO: reservation count 1947
OS WAIT ARRAY INFO: signal count 1870
RW-shared spins 0, rounds 0, OS waits 0
RW-excl spins 0, rounds 0, OS waits 0
RW-sx spins 0, rounds 0, OS waits 0
Spin rounds per wait:26.00 RW-shared, 5.00 RW-excl, 1.00 RW-sx
```

表 5-17 所示为 SEMAPHORES 各项指标及其说明。

表 5-17　SEMAPHORES 各项指标及其说明

指标	说明
OS WAIT ARRAY INFO	操作系统等待数组的信息，该数组是一个插槽数组，表示 InnoDB 使用了多少次操作系统的等待。 保留统计（reservation count）：显示了 InnoDB 分配插槽的频度。 信号计数（signal count）：衡量的是线程通过数组得到信号的频度
RW-shared spins	显示读写的共享锁的计数器的数值
RW-excl spins	显示读写的排他锁的计数器
RW-sx spins	显示共享排他锁的计数器的数值

在 SEMAPHORES 输出信息中，可以找到不同类型的信号量以及它们当前的状态，如等待（wait）和信号（signal）的数量，以及不同类型的信号量（如 RW-SHARED、RW-EXCL、BINLOG 和 LOG 等）。这些信息可以帮助诊断并发问题，确定哪些类型的资源是瓶颈。

（3）TRANSACTIONS（事务）信息

TRANSACTIONS 是 InnoDB 事务的统计信息，显示当前正在运行的所有事务的详细信息，包括事务 ID、开始时间、锁定的资源、等待的锁等。事务输出信息如下所示。

```
------------
TRANSACTIONS
------------
Trx id counter 87984263
Purge done for trx's n:o < 87984251 undo n:o < 0 state: running but idle
History list length 4
LIST OF TRANSACTIONS FOR EACH SESSION:
---TRANSACTION 421653598543936, not started
0 lock struct(s), heap size 1136, 0 row lock(s)
```

表 5-18 所示为 TRANSACTIONS 各项指标及其说明。

表 5-18　TRANSACTIONS 各项指标及其说明

指标	说明
Trx id counter	这个 ID 是一个系统变量，随着每次新事务的产生而增加
Purge done	正在进行清除操作的事务 ID
History list length	记录了撤销空间内未清除的事务的个数

在 TRANSACTIONS 输出信息中，可以通过查看哪些事务持有锁并等待其他事务释放资源，确定是否存在锁争用或死锁。此外，还可以分析事务的执行时间，以确定是否有长时间运行的事务阻塞了其他事务。

说明： 显示的事务信息是 SQL 语句级别的。也就是说，可能会看到正在执行或已经执行的部分 SQL 语句，而不是完整的事务。

（4）LATEST DETECTED DEADLOCK（死锁）信息

LATEST DETECTED DEADLOCK 的输出信息是最近检测到的死锁信息。它提供关于死锁的详细信息，包括导致死锁的事务及其持有的锁和等待的锁等。死锁是指两个或多个事务在执行过程中，因争夺资源而造成的一种相互等待的现象。死锁输出信息如下所示。

```
------------------------
LATEST DETECTED DEADLOCK
------------------------
2023-07-17 15:05:19 0x7f780c9db700
*** (1) TRANSACTION:
TRANSACTION 67715809, ACTIVE 1 sec starting index read
mysql tables in use 3, locked 1
LOCK WAIT 3 lock struct(s), heap size 1136, 8 row lock(s), undo log entries 1

*** (2) WAITING FOR THIS LOCK TO BE GRANTED:
RECORD LOCKS space id 104 page no 1827 n bits 128 index PRIMARY of table `test`.`id` trx id
67715810 lock_mode X locks rec but not gap waiting
Record lock, heap no 51 PHYSICAL RECORD: n_fields 24; compact format; info bits 0
```

MySQL 对死锁的处理机制是，当 InnoDB 存储引擎检测到死锁时，它会选择一个事务作为"牺牲品"，然后回滚该事务，以便其他事务可以继续执行。

上述输出信息中的指标介绍如下。

- mysql tables in use 3：使用 3 张表。
- TRANSACTION 67715809：67715809 为事务编号。
- LOCK WAIT 3 lock struct(s)：事务的锁链表的长度为 3。
- heap size 1136：为事务分配的锁堆内存大小。
- 8 row lock(s)：锁住 8 行数据。
- undo log entries 1：已记录 Undo 日志。
- PRIMARY of table `test`.`id`：表明 test 表的 id 为主键。
- lock_mode X locks rec but not gap waiting：不是范围锁，而是排他锁。

通过分析这些信息，可以确定导致死锁的原因，并采取相应的措施来避免类似的死锁情况再次发生，这可能包括重新设计表结构、优化查询语句、调整事务逻辑或更改并发控制参数等。

(5) FILE I/O（文件的输入/输出）信息

FILE I/O 信息是关于 InnoDB 对文件系统的读写操作的信息。FILE I/O 信息如下所示。

```
--------
FILE I/O
--------
I/O thread 0 state: waiting for completed aio requests ((null))
I/O thread 1 state: waiting for completed aio requests (insert buffer thread)
I/O thread 2 state: waiting for completed aio requests (read thread)
I/O thread 3 state: waiting for completed aio requests (read thread)
I/O thread 4 state: waiting for completed aio requests (read thread)
I/O thread 5 state: waiting for completed aio requests (read thread)
I/O thread 6 state: waiting for completed aio requests (write thread)
I/O thread 7 state: waiting for completed aio requests (write thread)
I/O thread 8 state: waiting for completed aio requests (write thread)
Pending normal aio reads: [0, 0, 0, 0], aio writes: [0, 0, 0, 0],
 ibuf aio reads:
Pending flushes (fsync) log: 0; buffer pool: 0
```

```
3308 OS file reads, 151405 OS file writes, 71962 OSfsyncs

0.00 reads/s, 0avg bytes/read, 0.00 writes/s, 0.00 fsyncs/s
```

表5-19所示为FILE I/O各项指标及其说明。

表5-19　FILE I/O各项指标及其说明

指标	说明
thread	I/O线程的状态（插入缓冲池线程、日志线程和读取线程）
Pending normal aio reads	待处理的异步I/O读写请求数量
Pending flushes (fsync) log	日志和缓冲池的挂起刷新（fsync）数量
reads/s、writes/s和fsyncs/s	从系统启动到当前时间点的OS文件的每秒读、写和fsync的次数

这些信息（InnoDB的I/O操作的各种统计信息）对于评估MySQL数据库的文件系统性能、监控磁盘活动以及诊断与I/O相关的性能问题非常有帮助。需要注意的是，这些统计信息是从InnoDB启动时开始累积的，所以它们会随着时间而增加。

(6) LOG（日志）信息

LOG输出信息提供了关于InnoDB的Redo日志和Undo日志的检查点信息。日志输出信息如下所示。

```
LOG
...
Log sequence number 21666084204
Log flushed up to   21666084204
Pages flushed up to 21619373558
Last checkpoint at  21619373558
0 pending log flushes, 0 pending chkp writes
65821807 log i/o's done, 3.02 log i/o's/second
----------------------
```

表5-20所示为LOG各项指标及其说明。

表5-20　LOG各项指标及其说明

指标	说明
Log sequence number	表示Redo日志缓冲区的LSN
Log flushed up to	表示Redo日志文件中的LSN
Pages flushed up to	表示缓冲池最旧脏页的LSN
Last checkpoint at	表示上一次检查点所在位置的LSN

在日志输出信息中，检查点（checkpoint）对于确保数据的持久性至关重要，它标记了某个时刻，在这个时刻之前的所有修改都已经确保被写入到磁盘上的数据文件中。通过这些信息，可以判断InnoDB是否需要执行更多的FLUSH操作来将缓冲区中的数据写入磁盘。例如，如果Checkpoint age的值很大，就可能意味着InnoDB需要尽快执行检查点，以减少未刷盘数据的风险。

此外，检查点也是一个重要的一致性检查指标。检查点的位置与Redo日志文件的当前写入位置之间的差距也很重要。如果差距过大，则可能意味着InnoDB需要更频繁地执行检查点来减少未刷盘的数据量，这可能会对性能产生影响。

(7) INSERT BUFFER AND ADAPTIVE HASH INDEX（插入缓冲区和自适应哈希索引）信息

INSERT BUFFER AND ADAPTIVE HASH INDEX 输出信息是 InnoDB 存储引擎的插入缓冲区（Insert Buffer）和自适应哈希索引（Adaptive Hash Index）的状态与统计信息。插入缓冲区和自适应哈希索引的输出信息如下所示。

```
-------------------------------------
INSERT BUFFER AND ADAPTIVE HASH INDEX
-------------------------------------
Ibuf: size 1, free list len 0, seg size 2, 0 merges
merged operations:
 insert 0, delete mark 0, delete 0
discarded operations:
 insert 0, delete mark 0, delete 0
Hash table size 2365241, node heap has 50 buffer(s)
Hash table size 2365241, node heap has 22 buffer(s)
Hash table size 2365241, node heap has 273 buffer(s)
Hash table size 2365241, node heap has 49 buffer(s)
Hash table size 2365241, node heap has 112 buffer(s)
Hash table size 2365241, node heap has 12 buffer(s)
Hash table size 2365241, node heap has 221 buffer(s)
Hash table size 2365241, node heap has 117 buffer(s)
6968.28 hash searches/s, 169.32 non-hash searches/s
---
```

表 5-21 所示为 INSERT BUFFER AND ADAPTIVE HASH INDEX 各项指标及其说明。

表 5-21 INSERT BUFFER AND ADAPTIVE HASH INDEX 各项指标及其说明

类型	指标	说明
插入缓冲区	size	插入缓冲区的大小（以页为单位）
	free list len	插入缓冲区空闲列表的长度
	seg size	插入缓冲区的分段大小
	merges	指示插入缓冲区合并的次数
	merged operations	插入缓冲区合并操作的统计信息，包括插入、合并记录和合并操作的数量
	discarded operations	由于某些原因而被丢弃的操作数量
自适应哈希索引	Hash table size	自适应哈希索引的大小（以字节为单位）
	node heap has	自适应哈希索引节点堆中的缓冲区数量

自适应哈希索引是 InnoDB 用于优化某些查询性能的一种机制。它会监视某些索引的搜索模式，并基于这些模式创建哈希索引以加速查找。插入缓冲区则用于优化非聚集索引的插入操作，通过将变更缓存起来并在后台合并，以减少磁盘 I/O。如果看到有关插入缓冲区或自适应哈希索引的性能问题或错误消息，就需要调整 InnoDB 的配置参数或调查查询模式以优化它们的性能。

(8) BUFFER POOL AND MEMORY（缓冲池和内存）信息

BUFFER POOL AND MEMORY 输出信息提供了关于 InnoDB 存储引擎的缓冲池和内存使用的详细信息。缓冲池和内存输出信息如下所示。

```
----------------------
BUFFER POOL AND MEMORY
----------------------
Total large memory allocated 8795455488
Dictionary memory allocated 688406
Buffer pool size    524224
Free buffers        499307
Database pages      24061
Old database pages  8718
Modified db pages   1532
Pending reads       0
Pending writes: LRU 0, flush list 0, single page 0
Pages made young 31, not young 0
0.00 youngs/s, 0.00 non-youngs/s
Pages read 268, created 23793, written 361754
0.00 reads/s, 0.00 creates/s, 0.00 writes/s
Buffer pool hit rate 1000 / 1000, young-making rate 0 / 1000 not 0 / 1000
Pages read ahead 0.00/s, evicted without access 0.00/s, Random read ahead 0.00/s
LRU len: 24061, unzip_LRU len: 0
I/O sum[0]:cur[0], unzip sum[0]:cur[0]
----------------------
INDIVIDUAL BUFFER POOL INFO
----------------------
---BUFFER POOL 0
Buffer pool size    65528
Free buffers        62209
Database pages      3213
Old database pages  1166
Modified db pages   258
Pending reads       0
Pending writes: LRU 0, flush list 0, single page 0
Pages made young 10, not young 0
0.00 youngs/s, 0.00 non-youngs/s
Pages read 40, created 3173, written 61429
0.00 reads/s, 0.00 creates/s, 0.00 writes/s
Buffer pool hit rate 1000 / 1000, young-making rate 0 / 1000 not 0 / 1000
Pages read ahead 0.00/s, evicted without access 0.00/s, Random read ahead 0.00/s
LRU len: 3213, unzip_LRU len: 0
I/O sum[0]:cur[0], unzip sum[0]:cur[0]
```

表 5-22 所示为 BUFFER POOL AND MEMORY 各项指标及其说明。

表 5-22　BUFFER POOL AND MEMORY 各项指标及其说明

指标	说明
Total large memory allocated	表示缓冲池向操作系统申请的连续内存空间大小
Dictionary memory allocated	表示 InnoDB 数据字典分配的内存大小
Buffer pool size	表示缓冲池的总大小，以字节为单位
Buffer Pool Instances	表示缓冲池实例的数量。InnoDB 可以配置多个缓冲池实例以在多核系统上实现更好的并发性能

(续)

指　标	说　明
Free buffers	表示缓冲池中当前空闲的缓冲区页数
Database pages	表示当前在缓冲池中缓存的数据库页数
Old database pages	表示 LRU 链表中 old 区域的节点的数据页数
Modified db pages	表示脏页数量
Pending reads	表示正在等待从磁盘上加载到缓冲池中的页数
Pending writes LRU	表示从 LRU 链表刷新到磁盘中的页数
Pending writes flush list	表示即将从 flush 链表刷新到磁盘中的页面数量
Pending writes single page	表示即将以单个页面的形式刷新到磁盘中的页面数量
Pages made young	表示 LRU 链表中曾经从 old 区域移动到 young 区域头部的节点数量
not young	表示 LRU 链表由于不符合时间间隔的限制而不能将节点移动到 young 区域的数量
youngs/s	表示每秒从 old 区域移动到 young 区域头部的节点数量
non-youngs/s	表示每秒由于不满足时间限制而不能从 old 区域移动到 young 区域头部的节点数量
Pages read、created、written	表示读取、创建和写入了多少页。后面的数值是读取、创建、写入的速率
LRU len	表示 LRU 链表中节点的数量

缓冲池是 InnoDB 用于缓存数据和索引的地方，它是 InnoDB 性能调优中的一个关键组件，可以帮助了解缓冲池的使用情况，以及是否存在性能瓶颈。例如，如果缓冲池命中率很低，那么可能需要增加缓冲池的大小以提高性能。Pending writes LRU 持续增长，可能是大事务或 I/O 性能达到瓶颈，无法满足工作负载的需求。

（9）ROW OPERATIONS（行操作）信息

ROW OPERATIONS 输出信息提供了关于 InnoDB 存储引擎中当前正在进行的行级操作（INSERT、UPDATE、DELETE 和 READ）的详细信息。行操作输出信息如下所示。

```
--------------
ROW OPERATIONS
--------------
0 queries inside InnoDB, 0 queries in queue
10 read views open inside InnoDB
Process ID=104404, Main thread ID=140169563342592, state: sleeping
Number of rows inserted 12124635687, updated 44153088, deleted 6372, read 46090409718
554.31 inserts/s, 2.02 updates/s, 0.00 deletes/s, 8177.21 reads/s
----------------------------
```

表 5-23 所示为 ROW OPERATIONS 各项指标及其说明。

表 5-23　ROW OPERATIONS 各项指标及其说明

指　标	说　明
queries in queue	表示 InnoDB 的行级锁等待队列中查询等待数量
read views open inside InnoDB	表示当前有多少这样的读视图是活跃的
inserts/supdates/sdeletes/sreads	表示当前数据库每秒插入、更新、删除、读的行数

ROW OPERATIONS 提供关于数据库工作负载的有价值的信息。例如，如果 queries inside InnoDB 数字增长很快，而 queries in queue 始终为 0，则可能意味着数据库正在处理大量的行级操作，行级锁争用导致查询延迟。若 read views open inside InnoDB 数字很高，则可能意味着系统中有许多长时间运行的事务，这可能会影响性能。

2. InnoDB 互斥锁信息

使用 SHOW ENGINE INNODB MUTEX 命令可以查看 InnoDB 存储引擎中互斥锁（mutex）和读写锁（rw-lock）的统计信息。这些统计信息对于诊断性能问题和了解锁争用情况非常有用。查看互斥锁的命令行如下所示。

```
# 设置 latch 参数值以显示更详细的互斥锁统计信息
mysql> SET GLOBAL innodb_monitor_enable='latch';
mysql> SHOW ENGINE INNODB MUTEX;
+------------+----------------------------+------------------------------------+
| Type       | Name                       | Status                             |
+------------+----------------------------+------------------------------------+
| InnoDB     | SRV_SYS                    | spins=3511,waits=100,calls=107     |
| InnoDB     | TRX                        | spins=359,waits=10,calls=11        |
| InnoDB     | LOG_FILES                  | spins=214,waits=7,calls=7          |
| InnoDB     | LOG_WRITER                 | spins=150,waits=5,calls=5          |
| InnoDB     | BUF_POOL_FLUSH_STATE       | spins=102,waits=3,calls=3          |
| InnoDB     | SRV_SYS_TASKS              | spins=60,waits=2,calls=2           |
| InnoDB     | LOG_FLUSHER                | spins=34,waits=1,calls=1           |
| InnoDB     | sum rwlock: buf0buf.cc:794 | waits=17                           |
+------------+----------------------------+------------------------------------+
8 rows in set (0.01 sec)
```

表 5-24 所示为 MUTEX 输出字段及其说明。

表 5-24　MUTEX 输出字段及其说明

字　　段	说　　明
Type	显示 InnoDB
Name	互斥锁类型的名称，对于读写锁，该字段生成读写锁的源文件，以及创建该文件中的行号
Status	spins：表示旋转次数； waits：表示互斥锁等待的次数； calls：指示请求互斥对象的次数

互斥锁是 InnoDB 存储引擎用来保护其内部数据结构和资源不被并发访问所破坏的一种机制。互斥锁确保了在任何时候只有一个线程可以访问特定的资源或执行特定的任务。通过输出的互斥锁的名称、等待的线程数、锁定的时间等，可以诊断性能问题和了解锁争用情况，以及了解哪些互斥锁是性能瓶颈。

3. InnoDB 打开的表的信息

SHOW OPEN TABLES 命令用于列出当前在表缓存中打开的非临时表，以便了解哪些表当前处于打开状态，这有助于诊断性能问题或了解锁争用情况。查看 InnoDB 打开表的命令行如下所示。

```
mysql> SHOW OPEN TABLES where In_use >0;
+----------+--------+--------+-------------+
|Database  |Table   |In_use  |Name_locked  |
+----------+--------+--------+-------------+
|test      |seckill |   71   |      0      |
+----------+--------+--------+-------------+
1 row in set (0.00 sec)
```

表 5-25 所示为 InnoDB 打开表输出的字段及其说明。

表 5-25　InnoDB 打开表输出的字段及其说明

字　　段	说　　明
Database	打开表的数据库名称
Table	打开的表名称
In_use	表示有多少线程正在使用某张表
Name_locked	指示表名是否被锁，这一般发生在以 DROP 或 RENAME 命令操作这张表时。所以，SHOW OPEN TABLES 命令不能帮助解答以下常见的问题，如当前某张表是否有死锁、谁拥有表上的这个锁等

通过查看 In_use 和 Name_locked 的值，可以更好地了解数据库表的使用情况和锁定状态，以便采取相应的优化措施。如果一个表长时间保持打开状态，那么它可能会消耗系统资源，并可能会妨碍其他查询或事务正常执行。

5.3.6　问题 SQL 语句优化命令行

MySQL 数据库运行中的绝大多数性能问题都发生在 SQL 语句上面。对 SQL 语句的性能问题进行筛选并进行相应的优化是一项持续且重要的工作。下面介绍优化问题 SQL 语句的 3 个命令行。

1. EXPLAIN 命令行

EXPLAIN 命令行是 MySQL 中一个非常有用的命令，它允许数据库管理员和开发人员在不实际执行 SQL 语句的情况下查看语句的执行计划。EXPLAIN 命令行适用于 SELECT、DELETE、INSERT、REPLACE 和 UPDATE 语句，当 EXPLAIN 命令行与它们一起使用时，MySQL 会显示优化器关于该语句执行计划的信息。也就是说，MySQL 解释了它将如何处理这条语句，包括有关表如何连接和以何种顺序连接的信息，EXPLAIN 会展示预执行计划信息。虽然不是真正的执行情况的反馈，但也是 MySQL 内部算法推演出来的预执行计划，还是比较准确的。

特别是能将在 B+树中是否有效地使用了索引和正确地选择了索引等情况反馈出来，让数据库管理员和开发人员能更好地设计表结构和编写 SQL 语句。

在 EXPLAIN 命令行的输出结果集中，通过 key 字段提示信息，就能看出是否使用了索引。EXPLAIN 命令行如下所示。

```
mysql> EXPLAIN SELECT * FROM  employees WHERE first_name LIKE '%Ho%' LIMIT 5\G
*************************** 1.row ***************************
           id: 1
  select_type: SIMPLE
        table: employees
   partitions: NULL
         type: ALL
```

```
        possible_keys: NULL
                  key: NULL
              key_len: NULL
                  ref: NULL
                 rows: 299512
             filtered: 11.11
                Extra: Using where
1 row in set, 1 warning (0.00 sec)
```

说明：具体用法可以参考 3.4.9 节中增强的 EXPLAIN 功能。

通过对 EXPLAIN 命令行的不断实践和学习，数据库管理员和开发人员可以逐步提高他们的查询优化技能，从而能更有效地管理数据库和提高应用程序的性能。

2. PROFILE 命令行

PROFILE 命令行是 MySQL 中用来分析当前会话中语句执行时资源消耗情况的命令，可以显示当前会话过程中执行语句的资源使用情况。PROFILE 命令行语法如下所示。

```
SHOW PROFILE [type [, type] ...] [FOR QUERY n] [LIMIT row_count [OFFSET offset]] type: { ALL | BLOCK IO | CONTEXT SWITCHES | CPU | IPC | MEMORY | PAGE FAULTS | SOURCE | SWAPS }
```

说明：对应 type 说明如下。
- ALL：显示所有性能信息。
- BLOCK IO：显示块 I/O 操作的次数。
- CONTEXT SWITCHES：显示上下文切换次数，不管是主动还是被动。
- CPU：显示用户 CPU 时间、系统 CPU 时间。
- IPC：显示发送和接收的消息数量。
- MEMORY：当前没有实现。
- PAGE FAULTS：显示页错误数量。
- SOURCE：显示源码中的函数名称与位置。
- SWAPS：显示 SWAP 的次数。

下面是 PROFILE 使用方式。首先需要开启 PROFILE 功能，然后实际执行 SQL 语句，同时把资源使用情况记录到系统内存表中。通过命令行查看资源消耗情况。具体 PROFILE 使用命令行如下所示。

```
# 查看 PROFILE 功能是否开启
mysql> SHOW VARIABLES LIKE '%profiling%';
+------------------------+-------+
| Variable_name          | Value |
+------------------------+-------+
| have_profiling         | YES   |
| profiling              | OFF   |
| profiling_history_size | 15    |
+------------------------+-------+

# 开启 PROFILE 功能
mysql> SET profiling = 1;
Query OK, 0 rows affected, 1 warning (0.00 sec)
```

```
# 执行语句
mysql> SELECT * FROM employees WHERE first_name like 'Ho%' limit 5;
+--------+------------+------------+-----------+--------+------------+
|emp_no  |birth_date  |first_name  |last_name  |gender  |hire_date   |
+--------+------------+------------+-----------+--------+------------+
|  11105 |1956-12-17  |Holgard     |McAlpine   |M       |1990-01-13  |
|  11484 |1959-04-08  |Holgard     |Siepmann   |M       |1995-01-22  |
|  15162 |1960-05-01  |Holgard     |Koblitz    |F       |1990-08-07  |
|  16372 |1961-02-12  |Holgard     |Terkki     |M       |1991-06-08  |
|  16374 |1963-05-23  |Holgard     |Nergos     |F       |1987-05-21  |
+--------+------------+------------+-----------+--------+------------+
5 rows in set (0.00 sec)

# 查看 PROFILE 功能记录语句编号
mysql> SHOW PROFILES;
+----------+------------+-----------------------------------------------------------------+
|Query_ID  |Duration    |Query                                                            |
+----------+------------+-----------------------------------------------------------------+
|1         |0.00568950  |SELECT * FROM employees WHERE first_name like 'Ho%' limit 5      |
+----------+------------+-----------------------------------------------------------------+
1 row in set, 1 warning (0.00 sec)

# 查看执行语句的资源使用情况
mysql> SHOW PROFILE FOR QUERY 1;
+------------------------------+----------+
|Status                        |Duration  |
+------------------------------+----------+
|starting                      |0.004549  |
|Executing hook on transaction |0.000030  |
|starting                      |0.000016  |
|checking permissions          |0.000319  |
|Opening tables                |0.000055  |
|init                          |0.000008  |
|System lock                   |0.000009  |
|optimizing                    |0.000011  |
|statistics                    |0.000071  |
|preparing                     |0.000205  |
|executing                     |0.000357  |
|Sending data                  |0.000190  |
|Sorting result                |0.000010  |
|end                           |0.000007  |
|query end                     |0.000005  |
|waiting for handler commit    |0.000011  |
|closing tables                |0.000008  |
|freeing items                 |0.000022  |
|cleaning up                   |0.000010  |
+------------------------------+----------+
17 rows in set, 1 warning (0.00 sec)
```

表 5-26 所示为执行的 SQL 语句每个过程的意义说明列表。

表 5-26　PROFILE 过程意义说明

状　态	说　明
starting	开始执行
Executing hook on transaction	开启对应事务
checking permissions	检查权限
Opening tables	打开表
init	初始化
System lock	系统锁
optimizing	优化
statistics	统计
preparing	准备
executing	执行
Sending data	发送数据
Sorting result	排序
end	结束
query end	查询结束
waiting for handler commit	当前事务正在等待其数据修改被提交到存储引擎
closing tables	关闭表或去除临时表
freeing items	释放资源
cleaning up	清理

PROFILE 命令行也可以仅显示 CPU 指标信息。仅显示 CPU 指标信息的 PROFILE 命令行如下所示。

```
mysql> SHOW PROFILE CPU FOR QUERY 1;
+------------------------------------+----------+----------+------------+
| Status                             | Duration | CPU_user | CPU_system |
+------------------------------------+----------+----------+------------+
| starting                           | 0.000132 | 0.000083 |   0.000047 |
| Executing hook on transaction      | 0.000019 | 0.000012 |   0.000007 |
| starting                           | 0.000013 | 0.000008 |   0.000005 |
| checking permissions               | 0.000009 | 0.000006 |   0.000003 |
| Opening tables                     | 0.000047 | 0.000030 |   0.000017 |
| init                               | 0.000012 | 0.000007 |   0.000004 |
| System lock                        | 0.000013 | 0.000018 |   0.000005 |
| optimizing                         | 0.000015 | 0.000009 |   0.000006 |
| statistics                         | 0.000052 | 0.000014 |   0.000108 |
| preparing                          | 0.000036 | 0.000017 |   0.000009 |
| executing                          | 0.000390 | 0.000077 |   0.000033 |
| end                                | 0.000008 | 0.000005 |   0.000303 |
| query end                          | 0.000007 | 0.000004 |   0.000002 |
| waiting for handler commit         | 0.000015 | 0.000010 |   0.000006 |
```

```
| closing tables                      | 0.000012 | 0.000007 | 0.000004 |
| freeing items                       | 0.000017 | 0.000011 | 0.000006 |
| logging slow query                  | 0.000059 | 0.000037 | 0.000022 |
| cleaning up                         | 0.000014 | 0.000009 | 0.000005 |
+-------------------------------------+----------+----------+----------+
18 rows in set, 1 warning (0.00 sec)
```

说明：当因资源（I/O、CPU、网络）缺乏而影响 SQL 语句执行效率时，可以通过 PROFILE 功能获取相关信息，特别是关于 I/O、CPU、网络等方面的问题，从而有效地定位性能问题。

3. OPTIMIZER_TRACE 命令行

OPTIMIZER_TRACE 命令行的执行过程是优化器跟踪实际执行的过程，可帮助理解 MySQL 优化器所采取的决策和行动。显示优化器跟踪参数的命令行如下所示。

```
mysql> SHOW VARIABLES LIKE '%optimizer_trace%';
+------------------------------+-------------------------------------------------------------------------+
| Variable_name                | Value                                                                   |
+------------------------------+-------------------------------------------------------------------------+
| optimizer_trace              | enabled=off,one_line=off                                                |
| optimizer_trace_features     | greedy_search=on,range_optimizer=on,dynamic_range=on,repeated_subselect=on |
| optimizer_trace_limit        | 1                                                                       |
| optimizer_trace_max_mem_size | 1048576                                                                 |
| optimizer_trace_offset       | -1                                                                      |
+------------------------------+-------------------------------------------------------------------------+
5 rows in set (0.00 sec)
```

优化器跟踪参数说明如下。

- optimizer_trace：可通过设置 enabled 来启用或禁用 OPTIMIZER_TRACE 功能。one_line 决定了跟踪信息的存储方式，为 on 时表示使用单行存储，否则以 JSON 树的标准展示形式存储。
- optimizer_trace_features：该参数中存储了跟踪信息中可控的输出项，可以通过调整该参数显示更多的信息。可调整的指标有 greedy_search、range_optimizer、dynamic_range 和 repeated_subselect。
- optimizer_trace_limit：控制 optimizer_trace 展示多少条结果，默认值为 1。
- optimizer_trace_max_mem_size：optimizer_trace 记录的内存的大小，如果跟踪信息超过这个大小，则信息将会被截断。
- optimizer_trace_offset：optimizer_trace 的偏移量。和 LIMIT 一样，optimizer_trace_offset 从 0 开始计算。默认值-1 表示没有限制。

下面是 OPTIMIZER_TRACE 命令行使用方式。通过 4 个阶段命令就可以输出优化器实际执行信息。具体 OPTIMIZER_TRACE 使用的命令行如下所示。

```
1.SET OPTIMIZER_TRACE="enabled=on";
2.执行 SQL 语句
3.SELECT * FROM INFORMATION_SCHEMA.OPTIMIZER_TRACE limit 30 \G;
4.禁用 OPTIMIZER_TRACE 功能
5.SET OPTIMIZER_TRACE="enabled=off";
```

图 5-6 所示为跟踪过程中每个阶段的说明。

```
{
  "steps":[
    {
      "join_preparation":{…}  准备阶段
    },
    {
      "join_optimization":{
        "select#":1,
        "steps":[
          {
            "condition_processing":{…}  WHERE 条件优化处理
          },
          {
            "substitute_generated_columns":{…}  替换虚拟列
          },
          {
            "table_dependencies":[1]  表之间依赖关系处理
          },
          {
            "ref_optimizer_key_uses":[0]  ref使用的索引处理
          },
          {
            "rows_estimation":[1]  估计表行数和扫描代价
          },
          {
            "considered_execution_plans":[1]  对比执行计划，选择最优执行计划
          },
          {
            "attaching_conditions_to_tables":{…}  改造原有的WHERE条件
          },
          {
            "finalizing_table_conditions":[1]  最终定义的条件
          },
          {
            "refine_plan":[1]  改善之后的执行计划
          }
        ]
      }
    },
    {
      "join_execution":{…}  执行阶段
    }
  ]
}
```

图 5-6　MySQL 中 OPTIMIZER_TRACE 解析树

说明：跟踪过程分为 3 个阶段，其中优化阶段包含 9 个步骤，每个步骤都有基本说明。特别是在主从架构（所有条件相等）下碰到 SQL 语句执行效率不一样时，可通 OPTIMIZER_TRACE 方式查找原因。

注意，上述 SQL 语句优化命令行中，EXPLAIN、PROFILE 和 OPTIMIZER_TRACE 命令需要实际执行才能获得分析指标。因此，在使用这些命令时，需要合理使用以确保有效地评估和优化 SQL 查询性能。

5.3.7　定位导致 CPU 使用率高的问题

在实际中，经常会出现 MySQL 服务器单个或多个 CPU 使用率非常高的情况。在这种情况下，进行资源使用监控分析是找出性能问题的关键步骤。这种分析可以帮助了解服务器的运行状况，确定资源消耗的主要方面，从而找到性能瓶颈和优化点。

图 5-7 所示为 Linux 平台下 CPU 存在的 5 种状态，每种状态都代表了 CPU 的不同使用情况。
- 用户空间（us）：为用户模式。这是 CPU 执行用户应用程序代码的时间。当应用程序在运行时，它运行在用户空间，执行的是用户态指令。
- 系统空间（sys）：也称为内核空间。这是 CPU 执行操作系统内核代码的时间。当系统需要执行一些管理任务，如处理系统调用、管理硬件资源或执行内核态的驱动程序代码时，会切换到系统空间。

- 空闲（Idle）：这是 CPU 没有执行任务，等待新的任务到来的时间。
- 等待 I/O（wa）：CPU 等待 I/O 操作（如磁盘读写）完成的时间。当进程需要等待外部设备完成数据传输时，它会进入等待状态，此时 CPU 会空闲下来以等待 I/O 操作完成。
- 软硬中断（hi & si）：软硬中断是内核为了处理异步事件而中断进程执行的一种方式。这些事件包括网络接收、定时器等。其中软中断在内核空间执行，它的处理更加温和，因为它不会打断正在执行的代码。

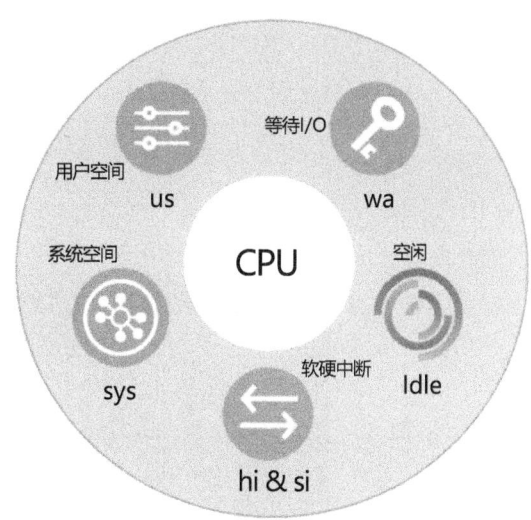

图 5-7 CPU 存在的 5 种状态

图 5-8 所示为 Linux 平台下使用 top 命令查看到的 CPU 使用情况。

图 5-8 Linux 平台下 CPU 使用情况

说明：从图 5-8 中的结果来看，sy 和 hi&si 基本由系统自动控制，干涉部分不是太多。us、wa 和 Idle 有一定的优化空间，可有效地使用资源。

表 5-27 所示为 top 命令行返回的 CPU 使用情况的各项指标及其说明。

表 5-27 top 命令行返回的 CPU 使用情况的各项指标及其说明

指标	说明
us	用户空间占用 CPU 百分比
sy	内核空间占用 CPU 百分比
ni	用户进程空间内改变过优先级的进程占用 CPU 百分比
id	空闲 CPU 百分比
wa	等待输入输出的 CPU 时间百分比
hi	硬件中断服务占用的 CPU 时间百分比
si	表示 CPU 处理软中断所花费的时间。软中断是由软件程序（如网络收发、定时调度等）发出的中断信号，特点是延迟执行
st	st 不是 top 命令标准输出的一部分。表示被窃取（steal）的时间，即虚拟机中的 CPU 等待宿主机的 CPU 资源的时间

通过上述介绍，已经对 CPU 的相关基础知识有了了解。接下来，将探讨 MySQL 在 CPU 使用方面的表现。在以往的 MySQL 案例中，因为使用上的一些问题，经常会导致高 CPU 使用率。这些问题包括连接数增加、执行效率低的查询 SQL、哈希连接或多表合并连接、存储过程、读写 I/O 慢以及参数设置不合理等。下面将详细介绍在哪些情况下 MySQL 会导致 CPU 使用率上升。

1. 通过 SHOW PROCESSLIST 命令行定位导致 CPU 使用率上升的 SQL 语句

导致 MySQL 服务器 CPU 使用率上升的 SQL 语句通常是那些复杂、低效或者未经优化的查询。这些查询可能因为各种原因（如缺少索引、使用不当的函数、全表扫描等）而执行缓慢，从而占用过多的 CPU 资源。

首先通过查看图 5-9 中展示的 SHOW PROCESSLIST 命令行输出结果，可以从"State"列中获取到关于 MySQL 服务器当前运行状态的反馈信息。

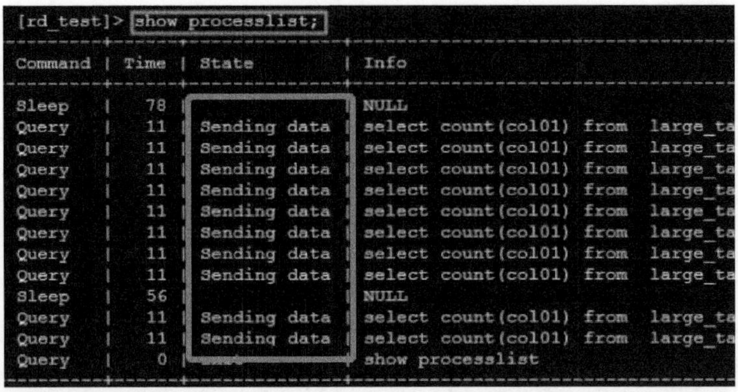

图 5-9 MySQL 服务器当前运行状态的反馈信息

如果 SQL 语句执行时间过长，并且其状态为以下值之一，那么这些语句很有可能存在性能问题。

- Sending data
- Copying to tmp table
- Copying to tmp table on disk
- Sorting result
- Creating sort index

下面具体介绍每个状态以及什么场景下会出现类似问题。

（1）Sending data

"Sending data"这个状态的名称很具有误导性，因为它不仅仅涉及发送数据，还包括了"收集 + 发送"数据的操作。当这个状态持续时间过长时，通常出现在以下场景中。

- 在执行查询语句时，查询条件没有使用索引字段。
- 如果在 SQL 语句中没有直接使用主键进行查询，而是使用了非主键索引，那么可能需要进行回表操作来获取完整的数据行，然后再将这些数据返回给客户端。
- 当查询返回的行数过多时，会导致频繁的输入/输出（I/O）交互。

（2）Copying to tmp table 和 Copying to tmp table on disk

这两个状态指的是生成临时表时占用的内存空间，当内存空间不足时，临时表会落到磁盘上。临时表使用过多通常体现在多表 JOIN 操作时缓存设置不合理。除此之外，ALTER 语句在某些情况下（如算法方式为 copy）也会产生大量的临时表。

（3）Sorting result

这个状态所指非临时表的结果集需要进行大量的排序操作，而基本上这些 SQL 语句的排序字段上没有索引。当这种情况大量出现时，就会导致 CPU 使用率飙升。针对这种情况，提出以下优化建议。

- 添加索引以提高查询速度；使用组合索引，特别是经常同时出现在查询条件中的多个列上；尽量坚持使用两张表以内的 JOIN 方式，以减少查询的复杂性并提高执行效率。这样，查询的执行成本将会大幅度减少。
- 隐式转换导致索引失效。隐式转换（Implicit conversion）是指当操作符或函数期望特定类型的参数，但接收到的是另一种类型时，会发生隐式转换。虽然 MySQL 会自动尝试进行类型转换，但这种隐式转换可能会导致性能下降。因此，建议始终明确指定数据类型，避免不必要的隐式转换，以确保代码的正确性和效率。
- 避免在 SQL 查询中对用于索引的列调用函数，因为这会导致索引失效，进而降低查询性能。

（4）Creating sort index

这个状态所指语句生成中间临时表并在这些表上创建临时索引，这与 Sorting result 情况类似，都是在进行排序操作。当这种情况大量出现时，就会导致 CPU 使用率飙升。针对这种情况，提出以下优化建议。

- 增加 sort_buffer_size 参数的值，MySQL 可以分配更多的内存用于排序操作，这可能会减少创建临时表和索引的需求。
- 尝试重写或修改查询以减少需要排序的数据量。例如，通过添加合适的索引、使用更精确的 WHERE 子句或改变查询逻辑来减少结果集的大小。

除此之外，在紧急情况下，若无法更改上述内容，则可以通过参数控制并发度（innodb_thread_concurrency）、执行时间（max_execution_time）。

2. 定位导致 CPU 上升的 I/O 操作

I/O 操作可能导致 CPU 使用率上升，这是因为 I/O 操作通常是异步的，需要 CPU 进行管理和调度。当 I/O 请求被发起时，CPU 可能会等待 I/O 完成，这个等待过程可能导致 CPU 使用率上升。图 5-10 所示为 CPU 在处理 I/O 请求时的状态。

表 5-28 所示为 I/O 请求状态及其说明。

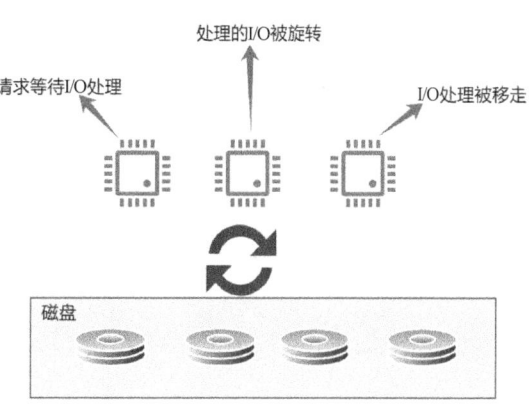

图 5-10 CPU 处理 I/O 请求时的状态

表 5-28 I/O 请求状态及其说明

请求状态	说明
请求等待 I/O 处理	当一个进程或线程需要执行 I/O 操作时，它会向系统发出请求。这个请求会被放入一个等待队列中，等待被操作系统处理
处理的 I/O 被旋转	这里的"旋转"指的是 CPU 在等待 I/O 操作完成时的自旋等待状态。也就是说，CPU 会循环检查 I/O 操作是否完成，而不是进入深度睡眠或执行其他任务
I/O 处理被移走	一旦 I/O 操作完成，相关的处理信息（可能是指完成通知、数据缓冲区等）会从等待队列中移除，并且 CPU 可以继续执行其他任务，或者如果 I/O 操作是原任务的一部分，那么任务可以继续执行

MySQL 的 I/O 操作可能导致 CPU 使用率上升，这通常是由于以下几个原因造成的。

- 磁盘 I/O 等待：当 MySQL 执行大量读写操作时，如果磁盘性能不足或存在其他 I/O 争用，那么数据库进程可能需要等待磁盘操作完成。在等待期间，CPU 可能会处于空闲状态，但由于等待时间较长，因此会给人一种 CPU 使用率上升的错觉。
- 频繁的 I/O 操作：如果 MySQL 需要频繁地读取或写入数据，那么会导致大量的 CPU 时间用于处理这些 I/O 请求。特别是在处理大量小文件或小块数据时，这种情况尤为明显。
- I/O 调度不当：操作系统中的 I/O 调度算法可能会影响 MySQL 的性能。如果调度算法选择不当，则可能会导致 I/O 操作不够高效，从而增加 CPU 的负担。
- 不合理的查询和索引：复杂的查询或缺乏适当索引的表可能导致 MySQL 执行大量的磁盘 I/O 操作，从而增加 CPU 的负载。
- 若 MySQL 的 I/O 相关参数（innodb_flush_log_at_trx_commit、sync_binlog、innodb_io_capacity 和 sync_relay_log）设置不合理，则可能会导致 I/O 堆积并在一次性刷新时引起 CPU 使用率的上升。

3. 通过 CPU 的 PID 定位导致 CPU 使用率上升的 SQL 语句

在 MySQL 系统中，可以通过 CPU 的 PID 来定位到正在执行的 SQL 语句。这涉及使用操作系统的工具（如 top 或 htop）来查看 CPU 使用率最高的进程，然后结合 MySQL 的命令来获取该进程正在执行的 SQL 语句。如下面的示例所示。

步骤 1：在 Linux 下使用 ps 命令结合 grep 来查找 MySQL（通常是 mysqld）进程的 PID。再使用 ps 命令来查看 MySQL（mysqld）进程的 PID，并且确定该进程当前正在哪个 CPU 核心上执行。查找 mysqld 进程的命令行如下。

```
shell $>ps -efaux | grep mysqld
mysql    3247 3367 0 Mar14 ?    00:07:43 mysqld
--defaults-file=/etc/my.cnf--basedir=/opt/idc/mysql8.0
--datadir=/opt/data8.0/data--port=3380

shell $> ps -o pid,psr,comm -p 3247
PID PSR COMMAND
3247  3  mysql
```

上述输出表示 PID 为 3247 的进程（名为 mysql）目前在编号为 3 的 CPU 上运行着。如果该进程没有被固定，则 PSR 列的值可能会随着内核对该进程的调度而改变，从而显示该进程当前正在哪个 CPU 上运行。

步骤 2：在 Linux 下使用 top 命令指定 PID 进入显示界面，然后按<F>键，接着使用上下键选择 P=Last Used CPU，并按空格键，出现"＊"即可，最后按<Esc>键退出，这时 top 命令行界面上显示"P=Last Used CPU"列，这一列会显示每个进程最近运行在哪个 CPU 核心上。top 命令行操作如下。

```
shell $> top -p 3247
top - 08:44:16 up 4 days, 17:42,  2 users,  load average: 0.03, 0.04, 0.00
Tasks:   1 total,   0 running,   1 sleeping,   0 stopped,   0 zombie
%CPU0  :  0.0 us,  1.3 sy,  0.0 ni, 98.7 id,  0.0 wa,  0.0 hi,  0.0 si,  0.0 st
MiB Mem :   1826.7 total,     97.4 free,    850.1 used,    879.1 buff/cache
MiB Swap:      0.0 total,      0.0 free,      0.0 used.    819.2 avail Mem
  PID USER      PR  NI    VIRT    RES    SHR S  %CPU  %MEM     TIME+ COMMAND P
172591 mysql    20   0 2397188 611976  42068 S   1.2  32.7  11:16.65 mysql   3
```

步骤 3：再使用 top 命令以线程模式（top -H -p <mysqld 进程 ID>）查看 MySQL 进程的线程并找出导致 CPU 使用率飙升的线程。查看线程的 top 命令行如下。

```
shell $> #top -H -p 1821
top - 21:04:24 up 10 min,  1 user,  load average: 0.00, 0.02, 0.04
Threads:  31 total,   0 running,  31 sleeping,   0 stopped,   0 zombie
%CPU(s): 90.0 us,  0.2 sy,  0.0 ni, 99.8 id,  0.0 wa,  0.0 hi,  0.0 si,  0.0 st
KiB Mem:   1867048 total,   390212 free,   1227176 used,   249660 buff/cache
KiB Swap:  2097148 total,  2097148 free,         0 used.   454824 avail Mem


  PID USER      PR  NI    VIRT    RES    SHR S %CPU %MEM    TIME+ COMMAND
 1942 mysql     20   0 12.327g 1.082g  12004 S  90 60.8   0:00.26 mysqld
 1821 mysql     20   0 12.327g 1.082g  12004 S  90 60.8   0:13.02 mysqld
 1924 mysql     20   0 12.327g 1.082g  12004 S  90 60.8   0:00.00 mysqld
```

步骤 4：根据上面输出的结果，已经知道哪个 MySQL 线程的 CPU 使用率高，通过下面的 SQL 语句，可最终定位问题 SQL 语句。查找问题 SQL 语句如下所示。

```
mysql> SELECT a.THREAD_OS_ID,b.id,b.user,b.host,
b.db,b.command,b.time,b.state,b.info
FROM performance_schema.threads a,information_schema.processlist b
WHERE b.id = a.processlist_id and a.THREAD_OS_ID=<具体 PID>;
```

在找到 CPU 使用率高的问题 SQL 语句之后，下一步是进行 SQL 优化。

第 6 章
MySQL 8.0 的运维管理

数据库运维管理是指对数据库的运维工作进行管理和控制，以确保数据库的稳定运行和高效性能。本章将介绍 MySQL 8.0 的运维管理工作、关键点和运维中的实践操作。

6.1 MySQL 8.0 运维管理概述

6.1.1 数据库运维管理的作用

数据库运维管理是指对数据库系统进行全面、系统的管理、监控和维护的过程，其作用主要有以下四个方面。

1. 提高数据库的稳定性

数据库是企业科学量化管理的基础，同时也是企业发展的承载和依托。在运行阶段，必须加强对数据库的规范管理，并定期进行检查、维护和监控，帮助及时有效地处理发现的问题，以确保系统能够持续、安全、可靠运行，从而减少系统故障和数据丢失的风险。

2. 提高数据库的性能

通过加强监控告警机制和实施优化措施，提升数据库和系统的性能，迅速响应并有效解决故障，减少系统停机时间，从而确保业务的连续性和可用性。

3. 提高数据库的可靠性

根据企业的具体业务需求和要求，综合考虑业务需求、可用性、性能和扩展性，提供高可靠、高效且安全的架构方案。在确保设备硬件稳定运行的同时，及时消除潜在隐患。对于存在安全隐患的部分，会提出相应的备份和应对策略，以保证运行期间的正常工作。

4. 助力企业业务增长和创新

通过优化数据库配置、调整存储结构以及合理规划容量，降低数据库运维成本，最大限度地利用资源，从而帮助企业更加专注于核心业务，推动业务增长并提升其创新能力。

6.1.2 数据库运维管理的主要工作

在日常工作中，数据库管理员需要进行多种任务，以保证数据库系统的稳定性、安全性和高效性。以下是数据库运维工作中的主要任务。

1. 数据库安装部署

在数据库的安装和部署过程中，需要根据业务需求合理分配资源，并安装部署适合的 MySQL 数据库版本。这个过程包括选择合适的 MySQL 版本、进行参数设置、配置安全选项等。

2. 数据库监控告警

数据库管理员需要实时监控数据库的各项性能指标，如 CPU 利用率、内存利用率、磁盘容量、TPS、QPS 和锁等信息。当异常发生时，通过设定的一些阈值触发告警。通过这些监控指标，可以分析潜在的性能问题，并及时进行优化，以提高数据库的响应速度和吞吐量，从而保证数据库的稳定性。

3. 数据库备份恢复

当高可用架构全部失效时，数据库的备份文件是数据恢复的最后手段。在 MySQL 备份体系中，支持多种备份手段，如逻辑备份、物理备份、热备和冷备等，同时都有对应的成熟工具支持。对于备份策略，可以通过全量备份、增量备份以及 binlog 恢复到指定的时间点。在日常工作中，应该包括常规性测试备份文件的可用性和数据恢复操作，以应对可能的数据丢失风险。

4. 数据库故障排查

在数据库运行过程中，故障排查是必要的。数据库可能会遇到各种问题，如查询性能下降、死锁、主从延迟、数据库停滞（或挂起）和严重崩溃等情况。数据库管理员应及时响应这些问题，第一时间恢复业务运行，并在事后分析原因，采取措施防止类似问题再次发生。

5. 数据库安全管理

由于数据库中存储着重要的企业数据，因此数据库安全管理至关重要。需要利用 MySQL 软件提供的安全机制，如防火墙设置、SSL 加密等，来进行安全配置。例如，更改数据库服务端口（默认端口 3306 容易被攻击者利用）、严格管理用户的访问权限、限制底层文件访问，以及加密存储的数据。此外，还需要定期检查和修复安全漏洞，及时更新数据库软件版本，以应对新出现的安全威胁。

6. 数据库日常操作

数据库日常操作包括表结构的更改、数据的导入和导出，以及高可用管理等。图 6-1 展示了这些操作的流程，涉及 DML 操作、DDL 操作和高可用管理等。这些操作需要经过层层审核，最终才能应用到生产环境中。

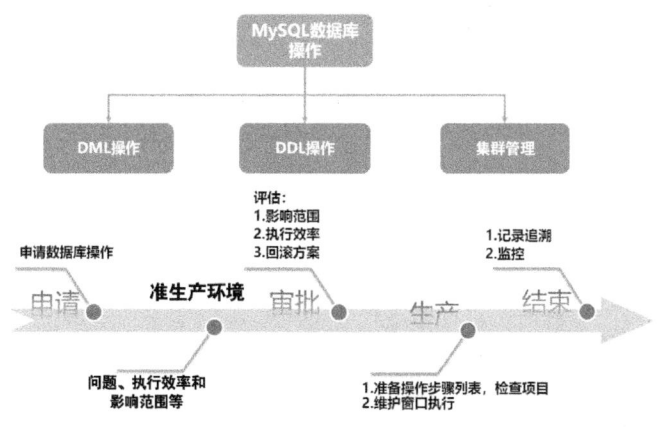

图 6-1 MySQL 日常操作

7. 数据库文档和技能培训

数据库管理员需要编写操作手册和维护文档,以便记录数据库的配置和管理信息。此外,数据库管理员应定期进行培训,以掌握数据库的规范,从而提高数据库的应用水平。

6.2 MySQL 8.0 运维管理的关键点

6.2.1 高频使用的运维管理操作

在 MySQL 数据库的运维工作中,数据库管理员经常需要借助各种命令来完成各种任务。为了帮助更好地掌握这些常用命令,以下整理了 18 个高频使用的运维管理命令,并提供了其使用示例。

1. 获取连接用户信息

通过查看特定用户连接的会话级别参数设置及状态变量,观察其会话连接行为,从而辅助定位和解决连接类问题。

【例 6-1】检查进程 ID 为 19 的用户连接的字符集设置。同时,也可以选择不指定 PROCESSLIST_ID 条件,以便查看数据库中所有用户的连接信息。相应的 SQL 语句如下所示。

```
mysql> SELECT T1.VARIABLE_NAME,
       T1.VARIABLE_VALUE,
       T2.PROCESSLIST_ID,
       concat(T2.PROCESSLIST_USER,"@",T2.PROCESSLIST_HOST),
       T2.PROCESSLIST_DB,
       T2.PROCESSLIST_COMMAND
FROM PERFORMANCE_SCHEMA.VARIABLES_BY_THREAD T1,
    PERFORMANCE_SCHEMA.THREADS T2
WHERE T1.THREAD_ID = T2.THREAD_ID
  AND T1.VARIABLE_NAME LIKE 'character%'
  AND PROCESSLIST_ID ='19';
*************************** 1.row ***************************
       VARIABLE_NAME: character_set_client
       VARIABLE_VALUE: gbk
       PROCESSLIST_ID: 19
concat(T2.PROCESSLIST_USER,"@",T2.PROCESSLIST_HOST): tester@localhost
       PROCESSLIST_DB: test
       PROCESSLIST_COMMAND: Query
*************************** 2.row ***************************
       VARIABLE_NAME: character_set_connection
       VARIABLE_VALUE: utf8mb4
       PROCESSLIST_ID: 19
concat(T2.PROCESSLIST_USER,"@",T2.PROCESSLIST_HOST): tester@localhost
       PROCESSLIST_DB: db1
       PROCESSLIST_COMMAND: Query
...
6 rows in set (0.01 sec)
```

【例 6-2】检查进程 ID 为 20 的用户连接是否关闭了 sql_log_bin 参数。相应的 SQL 语句如下所示。

```
mysql> SELECT T1.VARIABLE_NAME,
       T1.VARIABLE_VALUE,
       T2.PROCESSLIST_ID,
       concat(T2.PROCESSLIST_USER,"@",T2.PROCESSLIST_HOST) AS 'User@Host',
       T2.PROCESSLIST_DB,
       T2.PROCESSLIST_COMMAND
FROM PERFORMANCE_SCHEMA.VARIABLES_BY_THREAD T1,
     PERFORMANCE_SCHEMA.THREADS T2
WHERE T1.THREAD_ID = T2.THREAD_ID
  AND T1.VARIABLE_NAME LIKE 'sql_log_bin';
    Query OK, 0 rows affected ( 0.00 sec)
```

【例 6-3】 检查进程 ID 为 24 的用户连接的网络流量变化。相应的 SQL 语句如下所示。

```
mysql> SELECT T1.VARIABLE_NAME,
       T1.VARIABLE_VALUE,
       T2.PROCESSLIST_ID,
       concat(T2.PROCESSLIST_USER,"@",T2.PROCESSLIST_HOST) AS 'User@Host',
       T2.PROCESSLIST_DB,
       T2.PROCESSLIST_COMMAND
FROM PERFORMANCE_SCHEMA.STATUS_BY_THREAD T1,
     PERFORMANCE_SCHEMA.THREADS T2
WHERE T1.THREAD_ID = T2.THREAD_ID
  AND T2.PROCESSLIST_USER = 'root'
  AND PROCESSLIST_ID= 24
  AND VARIABLE_NAME LIKE 'Byte%';
+----------------+----------------+----------------+----------------+----------------+--------------------+
|VARIABLE_NAME   |VARIABLE_VALUE  |PROCESSLIST_ID  |User@Host       |PROCESSLIST_DB  |PROCESSLIST_COMMAND |
+----------------+----------------+----------------+----------------+----------------+--------------------+
|Bytes_received  |224             |24              |root@127.0.0.1  |NULL            |Sleep               |
|Bytes_sent      |182             |24              |root@127.0.0.1  |NULL            |Sleep               |
+----------------+----------------+----------------+----------------+----------------+--------------------+
2 rows in set (0.00 sec)
```

2. 查找长时间执行 SQL 语句的用户连接

长时间执行 SQL 语句的用户连接，因为占有资源不释放，可能会导致数据库性能下降、堵塞情况发生，所以需要尽快处理。查找执行 SQL 语句时间超过 5s 的用户连接的命令行示例如下。

```
mysql> SELECT trx_mysql_thread_id AS PROCESSLIST_ID,
       NOW(),
       TRX_STARTED,
       TO_SECONDS(now())-TO_SECONDS(trx_started) AS TRX_LAST_TIME,
       USER,
       HOST,
       DB,
       TRX_QUERY
FROM INFORMATION_SCHEMA.INNODB_TRX trx
INNER JOIN INFORMATION_SCHEMA.processlist pcl
ON trx.trx_mysql_thread_id=pcl.id
WHERE trx_mysql_thread_id != connection_id()
```

```
      AND TO_SECONDS(now())-TO_SECONDS(trx_started) >= 5;
+----------------+--------+-------------------+-------------------+--------+--------+------+---------------+
| PROCESSLIST_ID | NOW()  | TRX_STARTED       | TRX_LAST_TIME     | User   | Host   | DB   | TRX_QUERY     |
+----------------+--------+-------------------+-------------------+--------+--------+------+---------------+
| 24 | 2023-08-16 02:49:52 | 2023-08-16 02:41:15 | 517 | root | 127.0.0.1:58682 | db | NULL |
select name,age from table... |
+----------------+--------+-------------------+-------------------+--------+--------+------+---------------+
1 row in set (0.01 sec)
```

上述输出中显示了进程 ID、执行开始时间、用户名、IP 地址、数据库名和执行的 SQL 语句。

3. 跟踪元数据锁

MySQL 的元数据锁主要是指 MDL（MetaData Lock），用于保护元数据的一致性，防止在读写事务并发时导致的数据不一致问题。元数据库锁会堵塞，影响表的所有操作，最终造成连接数达到上限。为了追踪和定位与元数据锁相关的阻塞问题，可以开启元数据锁跟踪功能。然而，请注意，开启跟踪功能会消耗额外的内存和性能资源，因此在日常运维中不建议频繁使用。开启元数据锁跟踪的命令行示例如下。

```
# 临时开启,动态生效 UPDATE
mysql> UPDATE performance_schema.setup_consumersSET ENABLED = 'YES'
       WHERE NAME ='global_instrumentation';
mysql> UPDATE performance_schema.setup_instrumentsSET ENABLED = 'YES'
       WHERE NAME ='wait/lock/metadata/sql/mdl';

# 在配置文件中添加,重启后生效
[mysqld]
performance-schema-instrument = wait/lock/metadata/sql/mdl=ON
```

【例 6-4】为了确保 DDL 语句可以顺利执行，需要终止当前持有 MDL 锁的会话。

当 DDL 语句被阻塞时，通常是因为存在未提交或未释放的事务占用了所需资源。为了解决这个问题，可以尝试查找并结束那些运行时间超过 DDL 语句的事务。相应的 SQL 语句如下所示。

```
# 查找事务运行时间大于等于 DDL 等待时间的线程
mysql> SELECT trx_mysql_thread_id AS PROCESSLIST_ID,
         NOW(),
         TRX_STARTED,
         TO_SECONDS(now())-TO_SECONDS(trx_started) AS TRX_LAST_TIME,
         USER,
         HOST,
         DB,
         TRX_QUERY
    FROM INFORMATION_SCHEMA.INNODB_TRX trx
    JOIN INFORMATION_SCHEMA.processlist pcl ON trx.trx_mysql_thread_id=pcl.id
    WHERE trx_mysql_thread_id != connection_id()
      AND TO_SECONDS(now())-TO_SECONDS(trx_started) >=
        (SELECT MAX(Time)
         FROM INFORMATION_SCHEMA.processlist
         WHERE STATE='Waiting for table metadata lock'
           AND INFO LIKE 'alter%table%' OR INFO LIKE 'truncate%table%');
```

```
+---------------+---------------------+---------------------+--------------+------+-----------+------+-----------+
| PROCESSLIST_ID| NOW()               | TRX_STARTED         | TRX_LAST_TIME| User | Host      | DB   | TRX_QUERY |
+---------------+---------------------+---------------------+--------------+------+-----------+------+-----------+
| 253           | 2023-02-24 01:42:11 | 2023-02-24 01:41:24 |           47 | root | localhost | NULL | NULL      |
+---------------+---------------------+---------------------+--------------+------+-----------+------+-----------+
1 row in set (0.00 sec)

# KILL 掉长事务,释放持有的 MDL 资源
mysql> kill 253;
```

说明：在执行基于运行时间的会话清理操作时，由于 MySQL 元数据信息记录有限，因此可能无法准确区分正在执行重要业务逻辑的事务和可以安全结束的事务，从而增加了误 KILL 掉"无辜"长事务的风险。这种"误杀"是无法完全避免的，因此在执行相关操作时需要特别谨慎。

当 KILL 掉阻塞源后，可能会存在 DDL 语句与被阻塞的 SQL 语句同时加锁的情况。在这种情况下，由于加锁冲突，某些事务可能无法正常进行，因此，这类受影响的事务也需要进行 KILL 操作。定位 SQL 语句如下所示。

```
# 查找事务开始时间等于 DDL 语句事务开始时间的线程
mysql> SELECT trx_mysql_thread_id AS PROCESSLIST_ID,
       NOW(),
       TRX_STARTED,
       TO_SECONDS(now())-TO_SECONDS(trx_started) AS TRX_LAST_TIME,
       USER,
       HOST,
       DB,
       TRX_QUERY
    FROM INFORMATION_SCHEMA.INNODB_TRX trx JOIN INFORMATION_SCHEMA.processlist pcl
    ON trx.trx_mysql_thread_id=pcl.id
    WHERE trx_mysql_thread_id != connection_id()
      AND trx_started =
        (SELECT MIN(trx_started)
         FROM INFORMATION_SCHEMA.INNODB_TRX
         GROUP BY trx_started HAVING count(trx_started)>=2)
      AND TRX_QUERY NOT LIKE 'alter%table%'
      OR TRX_QUERY IS NULL;
+---------------+---------------------+---------------------+--------------+------+-----------+------+-----------+
| PROCESSLIST_ID| NOW()               | TRX_STARTED         | TRX_LAST_TIME| User | Host      | DB   | TRX_QUERY |
+---------------+---------------------+---------------------+--------------+------+-----------+------+-----------+
| 255           | 2023-02-24 01:42:44 | 2023-02-24 01:42:33 |           11 | root | localhost | NULL | NULL      |
+---------------+---------------------+---------------------+--------------+------+-----------+------+-----------+
1 row in set (0.00 sec)
# KILL 掉阻塞源
mysql> kill 255;
```

【例 6-5】 为了保证业务不被阻塞，KILL 掉阻塞的 DDL 语句的用户连接，以确保业务的顺畅进行。定位 SQL 语句如下所示。

```
# 查找DDL语句所在用户连接
mysql> SELECT *
    FROM INFORMATION_SCHEMA.PROCESSLIST
    WHERE INFO LIKE 'ALTER%TABLE%';

# 如果processlist表中没有记录,则可以查看events_statements_current表
mysql> SELECT p.*,c.SQL_TEXT
    FROM performance_schema.events_statements_current AS c INNER JOIN performance_schema.
threads AS t
    ON c.THREAD_ID=t.THREAD_ID
    INNER JOIN INFORMATION_SCHEMA.processlist AS p
    ON p.id=t.PROCESSLIST_ID
    WHERE c.SQL_TEXT LIKE 'alter%table%'
    OR c.SQL_TEXT LIKE 'create%index%'
    OR c.SQL_TEXT LIKE 'drop%table%'
    OR c.SQL_TEXT LIKE 'truncate%table%'
    AND p.STATE='Waiting for table metadata lock';
+-----+------+-----------+------+---------+------+---------------------------------+---------------------------------------+
| ID  | USER | HOST      | DB   | COMMAND | TIME | STATE                           | INFO                                  |
+-----+------+-----------+------+---------+------+---------------------------------+---------------------------------------+
| 254 | root | localhost | NULL | Query   |  730 | Waiting for table metadata lock | alter table db.t1 add index (id)      |
+-----+------+-----------+------+---------+------+---------------------------------+---------------------------------------+1
row in set (0.00 sec)
# KILL掉DDL语句所在用户连接
mysql> kill 254;
```

4. 获取锁等待语句

在 MySQL 中,为了避免死锁,当不同线程之间发生锁抢占时,需要选择其中一个事务进行回滚,以确保业务能够继续运行。同时,查看锁等待相关的阻塞线程和被阻塞线程的详细信息至关重要,包括涉及的用户、IP 地址和端口信息,以便更好地理解和解决死锁问题。定位的 SQL 语句如下所示。

```
mysql> SELECT locked_table,
        locked_index,
        locked_type,
        blocking_pid,
        concat(T2.USER,'@',T2.HOST) AS "blocking(user@ip:port)",
        blocking_lock_mode,
        blocking_trx_rows_modified,
        waiting_pid,
        concat(T3.USER,'@',T3.HOST) AS "waiting(user@ip:port)",
        waiting_lock_mode,
        waiting_trx_rows_modified,
        wait_age_secs,
        waiting_query
FROM sys.x$innodb_lock_waits T1
LEFT JOIN INFORMATION_SCHEMA.processlist T2
    ON T1.blocking_pid=T2.ID
```

```
        LEFT JOIN INFORMATION_SCHEMA.processlist T3
            ON T3.ID=T1.waiting_pid;
*************************** 1.row ***************************
           locked_table: `db`.`t1`
           locked_index: PRIMARY
            locked_type: RECORD
           blocking_pid: 228
  blocking(user@ip:port): dba@127.0.0.1:56724
     blocking_lock_mode: X
 blocking_trx_rows_modified: 1
            waiting_pid: 231
   waiting(user@ip:port): root@127.0.0.1:50852
      waiting_lock_mode: S
  waiting_trx_rows_modified: 0
          wait_age_secs: 1
          waiting_query: insert into db.t1(id) values(2)
row in set, 3 warnings (0.00 sec)
```

若不关心阻塞相关的用户、IP 地址和端口信息，则可直接查询 innodb_lock_waits 表记录信息。查询阻塞 SQL 语句如下所示。

```
mysql> select * from sys.x$innodb_lock_waits\G
*************************** 1.row ***************************
              wait_started: 2023-02-23 02:14:22
                  wait_age: 00:00:32
             wait_age_secs: 32
              locked_table: `db`.`t1`
              locked_index: PRIMARY
               locked_type: RECORD
            waiting_trx_id: 7204404
       waiting_trx_started: 2023-02-23 02:14:18
           waiting_trx_age: 00:00:36
   waiting_trx_rows_locked: 1
 waiting_trx_rows_modified: 0
               waiting_pid: 213
             waiting_query: delete from db.t1 where id=200
           waiting_lock_id: 7204404:1994:3:4
         waiting_lock_mode: X
           blocking_trx_id: 7204394
              blocking_pid: 207
            blocking_query: select * from sys.x$innodb_lock_waits
          blocking_lock_id: 7204394:1994:3:4
        blocking_lock_mode: X
       blocking_trx_started: 2023-02-23 02:10:06
          blocking_trx_age: 00:04:48
  blocking_trx_rows_locked: 1
 blocking_trx_rows_modified: 1
     sql_kill_blocking_query: KILL QUERY 207
sql_kill_blocking_connection: KILL 2071 row in set, 3 warnings (0.00 sec)
```

说明： 除此之外，可以通过影响锁等待超时参数（innodb_lock_wait_timeout），避免类似问题。

5. 获取全局读锁语句

全局读锁操作会阻塞所有对这张表的写操作（部分情况下也会阻塞读操作），需要尽快处理掉相关事务。在 MySQL 8.0 中，可以通过查看 PERFORMANCE_SCHEMA.METADATA_LOCKS 表来识别全局读锁。当 OBJECT_TYPE 列为 TABLE 且 LOCK_STATUS 列为 GRANTED 时，表示已经成功加上全局读锁（FTWRL）。

【例 6-6】 KILL 掉持有全局读锁的会话，以恢复业务语句的正常运行。定位 SQL 语句如下所示。

```
mysql> SELECT processlist_id,
            mdl.OBJECT_TYPE,
            OBJECT_SCHEMA,
            OBJECT_NAME,
            LOCK_TYPE,
            LOCK_DURATION,
            LOCK_STATUS
    FROM performance_schema.metadata_locksmdl INNER JOIN performance_schema.threads thd ON mdl.owner_thread_id = thd.thread_id
    AND processlist_id <> connection_id()
    AND LOCK_DURATION='EXPLICIT';
+----------------+-------------+---------------+-------------+-----------+---------------+-------------+
|processlist_id  |OBJECT_TYPE  |OBJECT_SCHEMA  |OBJECT_NAME  |LOCK_TYPE  |LOCK_DURATION  |LOCK_STATUS  |
+----------------+-------------+---------------+-------------+-----------+---------------+-------------+
|           231  |GLOBAL       |NULL           |NULL         |SHARED     |EXPLICIT       |GRANTED      |
|           231  |COMMIT       |NULL           |NULL         |SHARED     |EXPLICIT       |GRANTED      |
+----------------+-------------+---------------+-------------+-----------+---------------+-------------+
2 rows in set (0.00 sec)

# 终止 FTWRL 执行的用户线程 ID
mysql> kill 231;
```

【例 6-7】 终止执行时间大于 FTWRL 执行时间的用户线程 ID，确保 FTWRL 执行成功。定位 SQL 语句如下所示。

```
mysql> SELECT T2.THREAD_ID,
            T1.ID AS PROCESSLIST_ID,
            T1.User,
            T1.Host,
            T1.db,
            T1.Time,
            T1.State,
            T1.Info,
            T3.TRX_STARTED,
            TO_SECONDS(now())-TO_SECONDS(trx_started) AS TRX_LAST_TIME
    FROM INFORMATION_SCHEMA.processlist T1 LEFT JOIN PERFORMANCE_SCHEMA.THREADS T2
    ON T1.ID=T2.PROCESSLIST_ID LEFT JOIN INFORMATION_SCHEMA.INNODB_TRX T3
    ON T1.id=T3.trx_mysql_thread_id
    WHERE T1.TIME >=
```

```
            (SELECT MAX(Time)
            FROM INFORMATION_SCHEMA.processlist
            WHERE INFO LIKE 'flush%table%with%read%lock')
        AND Info IS NOT NULL;
*************************** 1.row ***************************
             THREAD_ID: 284
         PROCESSLIST_ID: 246
                  User: dba
                  Host: localhost
                    db: NULL
                  Time: 364
                 State: User sleep
                  Info: select * from db.t1 where sleep(1000000000)
           TRX_STARTED: 2023-02-23 14:57:23
         TRX_LAST_TIME: 364
1 rows in set (0.00 sec)
```

6. 统计内存使用情况

在 MySQL 的使用场景中，内存泄漏是一个常见问题。为了有效地识别和解决这些问题，可以使用 MySQL 的 performance_schema 来获取系统表的内存统计信息。然而，需要注意的是，performance_schema 默认情况下只提供部分的内存统计数据，要获取更全面的内存使用情况，可能需要手动启用更多的内存统计表。全局内存统计打开命令行示例如下所示。

```
# 动态开启,开启后开始统计
mysql> update performance_schema.setup_instruments set
enabled = 'yes' where name like 'memory%';


# 在配置文件中添加,重启后生效
[mysqld]
performance-schema-instrument='memory/%=COUNTED'
```

下面检查数据库实例的内存消耗分布情况。在 sys 库中，有多个与内存相关的视图，这些视图可以帮助用户定位和分析内存溢出问题。统计当前各内存模块信息的 SQL 语句如下所示。

```
mysql> SELECT event_name,
       current_alloc FROM sys.memory_global_by_current_bytes
WHERE event_name LIKE 'memory%innodb%';
+----------------------------------+-----------------+
| event_name                       | current_alloc   |
+----------------------------------+-----------------+
| memory/innodb/buf_buf_pool       | 134.31 MiB      |
| memory/innodb/log0log            | 32.01 MiB       |
| memory/innodb/mem0mem            | 15.71 MiB       |
| memory/innodb/lock0lock          | 12.21 MiB       |
| memory/innodb/os0event           | 8.37 MiB        |
| memory/innodb/hash0hash          | 4.74 MiB        |
| ...                              |                 |
+----------------------------------+-----------------+
42 rows in set (0.01 sec)
```

7. 统计分区表信息

在 MySQL 的使用场景中，因为分区表存在分区锁，所以，如果进行不当的操作（如不指定分区键的操作），则可能会锁住所有分区的数据。为了避免这种情况，需要检查分区表的设计是否合理，如分区键的选择和分区数量的设置是否基于业务需求。统计分区表信息的 SQL 语句如下所示。

```
mysql> SELECT TABLE_SCHEMA,
        TABLE_NAME,
        count(PARTITION_NAME) AS PARTITION_COUNT,
        sum(TABLE_ROWS) AS TABLE_TOTAL_ROWS,
        CONCAT(ROUND(SUM(DATA_LENGTH) / (1024 * 1024), 2),'M') DATA_LENGTH,
        CONCAT(ROUND(SUM(INDEX_LENGTH) / (1024 * 1024), 2),'M') INDEX_LENGTH,
        CONCAT(ROUND(ROUND(SUM(DATA_LENGTH + INDEX_LENGTH)) / (1024 * 1024),2),'M') TOTAL_SIZE
    FROM INFORMATION_SCHEMA.PARTITIONS
    WHERE TABLE_NAME NOT IN ('sys','mysql','INFORMATION_SCHEMA',
                             'performance_schema')
      AND PARTITION_NAME IS NOT NULL
    GROUP BY TABLE_SCHEMA, TABLE_NAME
    ORDER BY sum(DATA_LENGTH + INDEX_LENGTH) DESC ;
```

TABLE_SCHEMA	TABLE_NAME	PARTITION_COUNT	TABLE_TOTAL_ROWS	DATA_LENGTH	INDEX_LENGTH	TOTAL_SIZE
db	t1	365	0	5.70M	17.11M	22.81M
db	t2	391	0	6.11M	0.00M	6.11M
db	t3	4	32556	2.28M	0.69M	2.97M
db	t4	26	0	0.41M	2.44M	2.84M
db	t5	4	0	0.06M	0.00M	0.06M
db	t6	4	0	0.06M	0.00M	0.06M

6 rows in set (1.04 sec)

统计库名为 db、表名为 e 的分区表的 SQL 语句如下所示。

```
mysql> SELECT TABLE_SCHEMA,
        TABLE_NAME,
        PARTITION_NAME,
        PARTITION_EXPRESSION,
        PARTITION_METHOD,
        PARTITION_DESCRIPTION,
        TABLE_ROWS,
        CONCAT(ROUND(DATA_LENGTH / (1024 * 1024), 2),'M') DATA_LENGTH,
        CONCAT(ROUND(INDEX_LENGTH / (1024 * 1024), 2),'M') INDEX_LENGTH,
        CONCAT(ROUND(ROUND(DATA_LENGTH + INDEX_LENGTH) / (1024 * 1024),2),'M') TOTAL_SIZE
    FROM INFORMATION_SCHEMA.PARTITIONS
    WHERE TABLE_SCHEMA NOT IN ('sys',
                               'mysql',
                               'INFORMATION_SCHEMA',
                               'performance_schema')
      AND PARTITION_NAME IS NOT NULL
```

```
  AND TABLE_SCHEMA='db'
  AND TABLE_NAME='e';
+--------------+------------+----------------+--------------------+------------------+
| TABLE_SCHEMA | TABLE_NAME | PARTITION_NAME | PARTITION_EXPRESSION | PARTITION_METHOD |
| PARTITION_DESCRIPTION | TABLE_ROWS | DATA_LENGTH | INDEX_LENGTH | TOTAL_SIZE |
+--------------+------------+----------------+--------------------+------------------+
| db           | e          | p0  | id      | RANGE  | 50       | 4096  | 0.20M | 0.09M | 0.30M |
| db           | e          | p1  | id      | RANGE  | 100      | 6144  | 0.28M | 0.13M | 0.41M |
| db           | e          | p2  | id      | RANGE  | 150      | 6144  | 0.28M | 0.13M | 0.41M |
| db           | e          | p3  | id      | RANGE  | MAXVALUE | 16172 | 1.52M | 0.34M | 1.86M |
+--------------+------------+----------------+--------------------+------------------+
4 rows in set (0.00 sec)
```

8. 统计长时间未更新的表

长时间没有更新操作的表可能是存储历史数据的表或者已经不再需要维护的表，为了优化数据库性能和管理，应该将这些表归档处理。可以检查 INFORMATION_SCHEMA.TABLES 系统表，查找那些 UPDATE_TIME 列为 NULL 的表（表示自服务启动以来一直未更新过数据）。在确认这些表是长期不更新后，可以采取适当的归档措施。

查看未更新表的 SQL 语句如下所示。

```
mysql> SELECT TABLE_SCHEMA,
              TABLE_NAME,
              UPDATE_TIME
       FROM INFORMATION_SCHEMA.TABLES
       WHERE TABLE_SCHEMA NOT IN ('SYS','MYSQL','INFORMATION_SCHEMA',
                                  'PERFORMANCE_SCHEMA')
         AND TABLE_TYPE='BASE TABLE'
ORDER BY UPDATE_TIME ;
+--------------+------------+---------------------+
| TABLE_SCHEMA | TABLE_NAME | UPDATE_TIME         |
+--------------+------------+---------------------+
| db           | t1         | NULL                |
| db           | t2         | NULL                |
| db           | t3         | NULL                |
| db           | t4         | 2020-02-16 07:45:29 |
| db           | t5         | 2020-02-16 16:52:01 |
+--------------+------------+---------------------+
22 rows in set, 1 warning (0.01 sec)
```

9. 统计数据库基础信息

通过收集数据库的基础信息，包括库大小、表大小、字符集信息等，评估当前的设置是否合理、数据库容量是否足够应对未来需求，以及各数据库和表使用的字符集是否一致。统计实例中各数据库大小的 SQL 语句如下所示。

```
mysql> SELECT TABLE_SCHEMA,
              round(SUM(data_length+index_length)/1024/1024,2) AS TOTAL_MB,
              round(SUM(data_length)/1024/1024,2) AS DATA_MB,
              round(SUM(index_length)/1024/1024,2) AS INDEX_MB,
```

```
          COUNT(*) AS TABLES
    FROM INFORMATION_SCHEMA.tables
    WHERE TABLE_SCHEMA NOT IN ('sys','mysql','INFORMATION_SCHEMA',
                    'performance_schema')
    GROUP BY TABLE_SCHEMA
    ORDER BY 2 DESC;
+--------------+----------+---------+----------+--------+
| TABLE_SCHEMA | TOTAL_MB | DATA_MB | INDEX_MB | TABLES |
+--------------+----------+---------+----------+--------+
| cloud        |   229.84 |  223.02 |     6.83 |     41 |
| db           |    66.42 |   30.56 |    35.86 |     31 |
| dks          |    14.41 |    9.70 |     4.70 |    621 |
| test         |     0.06 |    0.06 |     0.00 |      4 |
| db2          |     0.03 |    0.03 |     0.00 |      2 |
+--------------+----------+---------+----------+--------+
5 rows in set, 1 warning (0.91 sec)
```

统计某库下各表占有物理空间大小的 SQL 语句如下所示。

```
mysql> SELECT TABLE_SCHEMA, TABLE_NAME TABLE_NAME,
CONCAT(ROUND(data_length / (1024 * 1024), 2),'M') data_length,
CONCAT(ROUND(index_length / (1024 * 1024), 2),'M') index_length,
CONCAT(ROUND(ROUND(data_length + index_length) / (1024 * 1024),2),'M') total_size,Engine
    FROM INFORMATION_SCHEMA.TABLES
    WHERE TABLE_SCHEMA NOT IN ('INFORMATION_SCHEMA',
                    'performance_schema','sys','mysql')
        AND TABLE_SCHEMA='db'
    ORDER BY (data_length + index_length) DESC LIMIT 10;
+--------------+------------+-------------+--------------+------------+--------+
| TABLE_SCHEMA | table_name | data_length | index_length | total_size | engine |
+--------------+------------+-------------+--------------+------------+--------+
| db           | t1         | 5.70M       | 22.81M       | 28.52M     | InnoDB |
| db           | t2         | 15.19M      | 9.59M        | 24.78M     | InnoDB |
| db           | t3         | 6.11M       | 0.00M        | 6.11M      | InnoDB |
...
+--------------+------------+-------------+--------------+------------+--------+
10 rows in set, 1 warning (0.01 sec)
```

统计某库下表的字符集基本信息的 SQL 语句如下所示。

```
mysql> SELECT TABLE_SCHEMA,
        TABLE_NAME, table_collation, engine, table_rows
    FROM INFORMATION_SCHEMA.tables
    WHERE TABLE_SCHEMA NOT IN ('INFORMATION_SCHEMA','sys','mysql',
                    'performance_schema')
      AND TABLE_TYPE='BASE TABLE'
      AND TABLE_SCHEMA='db'
    ORDER BY table_rows DESC ;
+--------------+------------+-------------------+--------+------------+
```

```
|TABLE_SCHEMA         |table_name          |table_collation         |engine  |table_rows |
+---------------------+--------------------+------------------------+--------+-----------+
|db                   |t1                  |utf8_general_ci         |InnoDB  |    159432 |
|db                   |t2                  |utf8mb4_general_ci      |InnoDB  |     32556 |
|db                   |t3                  |utf8mb4_general_ci      |InnoDB  |      2032 |
...
+---------------------+--------------------+------------------------+--------+-----------+
25 rows in set, 1 warning (0.01 sec)
```

10. 统计存储引擎使用情况

在 MySQL 中，不同的存储引擎具有不同的机制。例如，MyISAM 引擎不支持事务，而 InnoDB 引擎则支持。然而，这并不意味着在所有需要事务支持的场合都必须使用 InnoDB，因为还有其他存储引擎和因素需要考虑。统计存储引擎分布情况的 SQL 语句如下所示。

```
mysql> SELECT TABLE_SCHEMA, ENGINE,COUNT(*)
    FROM INFORMATION_SCHEMA.TABLES
    WHERE TABLE_SCHEMA NOT IN ('INFORMATION_SCHEMA','PERFORMANCE_SCHEMA',
                               'SYS','MYSQL')
    AND TABLE_TYPE='BASE TABLE'
    GROUP BY TABLE_SCHEMA,ENGINE;
```

统计非 InnoDB 存储引擎表的 SQL 语句如下所示。

```
mysql> SELECT TABLE_SCHEMA,TABLE_NAME,TABLE_COLLATION,ENGINE,TABLE_ROWS
    FROM INFORMATION_SCHEMA.TABLES
    WHERE TABLE_SCHEMA NOT IN ('INFORMATION_SCHEMA','SYS','MYSQL',
                               'PERFORMANCE_SCHEMA')
      AND TABLE_TYPE='BASE TABLE'
      AND ENGINE NOT IN ('INNODB')
    ORDER BY TABLE_ROWS DESC ;
```

11. 统计主键和唯一键使用情况

MySQL 中主键和唯一键的重要性不言而喻。InnoDB 存储引擎使用索引组织表，对数据操作和复制都起着重要作用。因此，对于性能而言，设置合适的主键字段至关重要。

统计无主键、唯一键和索引的表情况的 SQL 语句如下所示。

```
mysql> SELECT T1.TABLE_SCHEMA, T1.TABLE_NAME
    FROM INFORMATION_SCHEMA.COLUMNS T1 JOIN INFORMATION_SCHEMA.TABLES T2
      ON T1.TABLE_SCHEMA=T2.TABLE_SCHEMA
        AND T1.TABLE_NAME=T2.TABLE_NAME
    WHERE T1.TABLE_SCHEMA NOT IN ('SYS','MYSQL',
                                  'INFORMATION_SCHEMA','PERFORMANCE_SCHEMA')
      AND T2.TABLE_TYPE='BASE TABLE'
      AND T1.TABLE_SCHEMA='db'
    GROUP BY T1.TABLE_SCHEMA, T1.TABLE_NAME
    HAVING MAX(COLUMN_KEY) ='';
```

下面查看无主键、唯一键和仅有二级索引表。该类型表由于无高效索引，因此从库回放 SQL 时容易导致复制延迟。统计无主键、唯一键的表情况的 SQL 语句如下所示。

```
mysql> SELECT T1.TABLE_SCHEMA,T1.TABLE_NAME
    FROM INFORMATION_SCHEMA.COLUMNS T1 JOIN INFORMATION_SCHEMA.TABLES T2
    ON T1.TABLE_SCHEMA=T2.TABLE_SCHEMA
    AND T1.TABLE_NAME=T2.TABLE_NAME
    WHERE T1.TABLE_SCHEMA NOT IN ('SYS','MYSQL',
                                  'INFORMATION_SCHEMA','PERFORMANCE_SCHEMA')
      AND T2.TABLE_TYPE='BASE TABLE'
      AND T1.COLUMN_KEY != ''
    GROUP BY T1.TABLE_SCHEMA,T1.TABLE_NAME
    HAVING group_concat(COLUMN_KEY) NOT REGEXP 'PRI|UNI';
```

统计无主键和有唯一键的表情况的 SQL 语句如下所示。

```
mysql> SELECT T1.TABLE_SCHEMA,
       T1.TABLE_NAME
    FROM INFORMATION_SCHEMA.COLUMNS T1 JOIN INFORMATION_SCHEMA.TABLES T2 ON T1.TABLE_SCHEMA=T2.
TABLE_SCHEMA AND T1.TABLE_NAME=T2.TABLE_NAME
    WHERE T1.TABLE_SCHEMA NOT IN ('SYS','MYSQL','INFORMATION_SCHEMA',
                                  'PERFORMANCE_SCHEMA')
    AND T2.TABLE_TYPE='BASE TABLE'
    GROUP BY T1.TABLE_SCHEMA, T1.TABLE_NAME
    HAVING group_concat(COLUMN_KEY) NOT REGEXP 'PRI|UNI';
```

12. 展示数据库实时负载的 Shell 脚本

通过官方提供的 mysqladmin 管理工具，实时展示 MySQL 负载情况。Shell 脚本内容如下所示。

```
shell $> mysqladmin  extended-status  -i1|awk 'BEGIN{local_switch=0}
    $2 ~ /Queries $/            {q=$4-lq;lq=$4;}
    $2 ~ /com_commit $/         {c=$4-lc;lc=$4;}
    $2 ~ /Com_rollback $/       {r=$4-lr;lr=$4;}
    $2 ~ /Com_select $/         {s=$4-ls;ls=$4;}
    $2 ~ /Com_update $/         {u=$4-lu;lu=$4;}
    $2 ~ /Com_insert $/         {i=$4-li;li=$4;}
    $2 ~ /Com_delete $/         {d=$4-ld;ld=$4;}
    $2 ~ /Innodb_rows_read $/        {irr=$4-lirr;lirr=$4;}
    $2 ~ /Innodb_rows_deleted $/     {ird=$4-lird;lird=$4;}
    $2 ~ /Innodb_rows_inserted $/    {iri=$4-liri;liri=$4;}
    $2 ~ /Innodb_rows_updated $/     {iru=$4-liru;liru=$4;}
    $2 ~ /Innodb_buffer_pool_read_requests $/    {ibprr=$4-libprr;libprr=$4;}
    $2 ~ /Innodb_buffer_pool_reads $/     {ibpr=$4-libpr;libpr=$4;}
    $2 ~ /Threads_connected $/  {tc=$4;}
    $2 ~ /Threads_running $/    {tr=$4;
        if(local_switch==0)
            {local_switch=1; count=16}
        else {
            if(count>15) {
                count=0;
                print
"-------------------------------------------------------------------------------------------------";
                print "Time----|  QPS |Commit Rollback TPS |select insert update delete |
read inserted updated deleted |logical physical |Tcon Trun";
```

```
                    print
"-------------------------------------------------------------------------------- ";
                }else{
                    count+=1;
                    printf "%s |%-5d| \n ",strftime("%H:%M:%S"), lq
                    printf "%s |%-5d|%-6d %-7d %-5d|%-7d %-7d %-5d %-6d|%-7d %-7d %-7d %-7d|
%-6d  %-9d|%-4d %-2d \n",strftime("%H:%M:%S"),q,c,r,c+r,s,u,i,d,irr,ird,iri,iru,ibprr,ib-
pr,tc,tr;
                }
            }
}'
```

13. 通过分析 binlog 文件统计 DDL 和 DML 执行次数

通过统计 binlog 文件中 DDL（数据定义语言）和 DML（数据操纵语言）的执行次数来检查是否存在过大的事务或 DDL 执行语句堵塞了其他表的查询和变更操作。

统计指定 binlog 文件中 DML 执行次数的命令行示例如下所示。

```
shell $> mysqlbinlog --no-defaults --base64-output=decode-rows -v -v
mysql-bin.000007 |awk '/###/{if($0~/UPDATE|INSERT|DELETE/)count[ $2"
" $NF]++}END{for(i in count)print i,"\t",count[i]}'|column -t|sort -k3nr
```

统计指定 binlog 文件中 DDL 执行次数的命令行示例如下所示。

```
shell $> mysqlbinlog --no-defaults --base64-output=decode-rows -v -v
mysql-bin.000007 |awk 'BEGIN{IGNORECASE=1} {if($0~/alter/)count[ $1" " $2" " $3"
" $NF]++}END{for(i in count)print i,"\t",count[i]}'|column -t|sort -k3nr
```

14. 通过分析 binlog 文件查看并行复制是否合理

通过分析 binlog 中的分组的并行指标，分析并行复制是否存在问题（如并行的事务数量大于 5000 个或没有进行并行）。分析 binlog 中的并行指标的命令行示例如下所示。

```
shell $> mysqlbinlog mysql-bin.000004 --start-position=20087624 |grep -o
'last_committed.*'  |sed 's/=// /g'|awk '{print $2"\t"$4}'|awk '{count++;print
$0;} END{print "total count is ",count}'
```

15. 查找设计不合理的字符集表

选择不合理的字符集会导致乱码等问题。在 MySQL 8.0 版本中，utf8mb4 字符集可以用来表示 Unicode 标准中的任何字符。所以建议统一使用 utf8mb4 字符集。查找非 uft8mb4 字符集的 SQL 语句示例如下所示。

```
mysql> SELECT TABLE_SCHEMA,TABLE_NAME,TABLE_COLLATION
       FROM information_schema.TABLES
    WHERE TABLE_COLLATION not like 'utf8%'
       and table_schema not in ('information_schema','mysql','performance_schema','sys');
```

如果数据库、表或字段的字符集设置与服务器或查询时的字符集不一致，则可能会导致数据存储或检索的问题，这可能间接地影响索引的使用效果。在 MySQL 中，字符集可以在服务器、数据库、表和字段等多个级别进行设置，因此确保字符集的一致性是很重要的。

确认字符集是否一致的方式如下。

步骤 1：查看系统字符集。

```
mysql> SHOW GLOBAL VARIABLES LIKE 'collation%';
```

步骤 2：查询与系统字符集不一样的数据库的 SQL 语句。

```
mysql> SELECT b.SCHEMA_NAME, b.DEFAULT_CHARACTER_SET_NAME,
b.DEFAULT_COLLATION_NAME
    FROM information_schema.SCHEMATA b
    WHERE b.SCHEMA_NAME not in
('information_schema','mysql','performance_schema','sys') and
b.DEFAULT_COLLATION_NAME<>@@collation_server;
```

步骤 3：查询与系统字符集不一样的表和字段的 SQL 语句。

```
mysql> SELECT distinct tschema,tname,tcoll
    FROM (
      select a.TABLE_SCHEMA astschema, a.TABLE_NAME as tname,a.TABLE_COLLATION as tcoll
    FROM information_schema.TABLES a
    WHERE a.TABLE_SCHEMA not in ('information_schema','mysql','performance_schema','sys')
    and a.TABLE_COLLATION<>@@collation_server
    UNION
    SELECT a.TABLE_SCHEMA astschema, TABLE_NAME as tname, a.COLLATION_NAME as tcoll
      FROM information_schema.COLUMNS a
      WHERE a.TABLE_SCHEMA not in ('information_schema','mysql','performance_schema','sys')
    and a.COLLATION_NAME<>@@collation_server ) as aa ;
```

16. 排查 CPU 使用率过高问题

在系统资源中，CPU 对 MySQL 的性能和响应能力有直接影响，而编写质量差的 SQL 语句有可能导致 CPU 使用率飙升，尤其是在处理大量数据时。当单核 CPU 的使用率达到 100%时，意味着该核心上的所有处理能力都被占用，这可能导致系统响应缓慢，甚至出现超时或崩溃。

想要查看 CPU 使用率飙升的 MySQL 线程，可使用 top -H -p <mysqld 进程 ID>命令行，示例如下所示。

```
#top -H -p 1821
top - 21:04:24 up 10 min,  1 user,  load average: 0.00, 0.02, 0.04
Threads:  31 total,   0 running,  31 sleeping,   0 stopped,   0 zombie
%Cpu(s): 90.0 us,  0.2 sy,  0.0 ni, 99.8 id,  0.0 wa,  0.0 hi,  0.0 si,  0.0 st
KiB Mem:  1867048 total,   390212 free,  1227176 used,   249660 buff/cache
KiB Swap: 2097148 total,  2097148 free,        0 used.   454824 avail Mem

PID USER      PR  NI    VIRT    RES    SHR S %CPU %MEM     TIME+ COMMAND
1942 mysql    20   0 12.327g 1.082g  12004 S   90 60.8   0:00.26 mysqld
1821 mysql    20   0 12.327g 1.082g  12004 S   90 60.8   0:13.02 mysqld
1924 mysql    20   0 12.327g 1.082g  12004 S   90 60.8   0:00.00 mysqld
```

根据 CPU 使用率最高的 PID，定位问题 SQL 语句，如下所示。

```
mysql>SELECT a.THREAD_OS_ID,b.id,b.user,b.host,b.db,
            b.command,b.time,b.state,b.info
  FROM performance_schema.threads a,information_schema.processlist b
  WHERE b.id = a.processlist_id
        and a.THREAD_OS_ID=<具体 PID>;
```

17. 批量终止应用会话

在某些场景中，可能需要终止数据库连接或特定的线程，尤其是在出现性能问题、故障转移（如主从切换）或需要中断长时间运行的查询时。

批量终止（KILL）执行时间大于 10s，不处在 SLEEP 状态的连接的 SQL 语句如下所示。

```
mysql> SELECT concat('KILL ',id,';')
    FROM information_schema.processlist
    WHERE time>10  and db is not null and command!='sleep'
        into outfile '/tmp/a.txt';
Query OK,  rows affected (0.00 sec)
mysql> source  /tmp/a.txt;
Query OK, 0 rows affected (0.00 sec)
```

通过使用 mysqladmin 命令行，批量终止指定用户运行的连接。例如，终止用户为 tester 的所有会话的命令行示例如下所示。

```
Shell $> mysqladmin -uroot -p processlist|awk -F "|"'{if($3 == "tester")print $2}'|xargs -n 1 mysqladmin -uroot -p
```

18. 重新启动 MySQL 服务

RESTART 命令是从 MySQL 客户端会话中重新启动 MySQL 服务的命令。一般重启服务的应用场景如下。

- MySQL 服务出现阻塞，无法处理请求。
- 对静态系统变量进行配置更改，这些更改只能在服务重新启动后生效。
- 在日常运维窗口中进行硬件升级、迁移等操作。
- 在复杂的环境中，尽管不确定确切原因，但重启服务可能是解决问题的有效尝试。
- 需要清空缓存、释放内存资源。

关闭 MySQL 服务的传统方式有 4 种：注册为系统服务、使用 mysqladmin 命令行工具、在客户端执行 SHUTDOWN 命令和在 Linux 下执行 KILL 命令来强制结束进程。

重新启动服务的命令行示例如下所示。

```
mysql> RESTART;
Query OK, 0 rows affected (0.00 sec)
```

MySQL 的 Error 日志中会输出重启动服务信息，如下所示。

```
[System] [MY-011086] [Server] Received RESTART from user root.  Restarting mysqld (Version: 8.0.34).
```

之后可通过 Linux 操作系统来观察 MySQL 的线程 ID 情况，以确认是否已重新启动服务。如图 6-2 所示，ps 命令会跟踪 MySQL 的 PID 的变化。

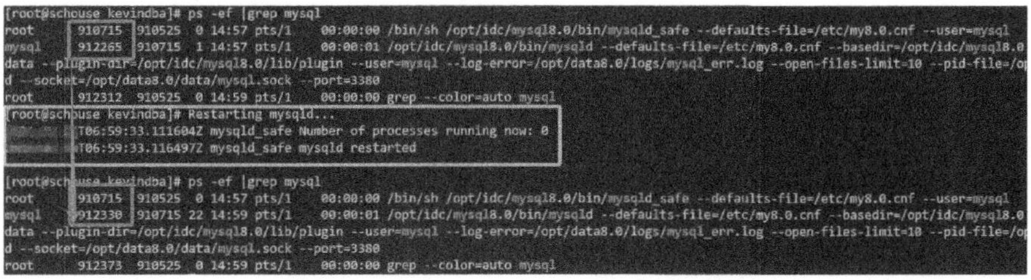

图 6-2　MySQL 重启服务时的 PID 跟踪

说明：mysqld_safe 的守护 PID 没有变化，但 mysqld 的 PID 变了。除此之外，执行 RESTART 命令需要 shutdown 权限。

6.2.2 运维管理中的高危操作

在实际工作中，经常遇到一些临时的应急需求，可能导致 DBA 在没有充分准备的情况下直接在生产环境上执行操作。此外，有时也会存在一些未经深思熟虑的命令行操作在生产环境中被执行的情况，这可能会引发生产数据库事故。

那么 MySQL DBA 运维中哪些动作是危险性操作？下面逐个介绍。

1. SHUTDOWN/RESTART 命令（MySQL 8.0 版本）

SHUTDOWN 与 RESTART 命令分别用于关闭和重新启动 MySQL 服务，但它们都是危险性操作，因此需要格外注意以下场景。

- 在平时运维中，避免随意执行这些命令，以免在生产环境中无意识执行而导致服务中断。
- 当大事务（如 LOAD DATA 或大表 DELETE 操作）正在运行时，执行关闭或重新启动 MySQL 服务的命令可能会导致事务大量回滚或数据库无法正常关闭。

2. KILL 命令

对于 MySQL 的 KILL 命令，它不是立即终止进程，而是触发底层的一致性回滚动作。特别是在大事务中，KILL 命令可能导致回滚过程非常耗时，可能长达 30 分钟~3 小时，甚至更长时间。因此，在以下场景中执行 KILL 命令需要格外谨慎。

- 如果 MySQL 进程正在执行大事务（例如，包含大量数据修改的操作），使用 KILL 命令终止该进程将触发回滚操作。这可能导致数据库负载显著增加，并影响其他事务的执行。
- 当服务器负载过高，CPU、I/O 等资源紧张时，使用 KILL 命令可能导致系统资源进一步紧张，因为回滚操作通常需要大量的系统资源来完成。

3. FLUSH BINARY LOGS 命令

该命令用于截断旧的 binlog 日志，并生成新的 binlog 日志。同时，这也会触发 expire_logs_days 参数来清除过期的 binlog。然而，在高可用架构下，执行这样的操作需要特别谨慎。以下是需要禁止执行此操作的场景。

- 存在从库尚未同步的 binlog：如果主库的 binlog 中包含了尚未同步到从库 relay log 中的事件，那么截断 binlog 操作可能会导致从库无法继续同步。这可能导致数据不一致。
- 正在执行大事务：如果数据库正在执行大事务，那么截断 binlog 操作可能会中断事务的正常执行。在某些情况下，这可能导致事务回滚，进而增加数据库的负载并可能导致性能下降。

说明：binlog 记录的是数据变更操作，如被清除，在出现故障时无法进行恢复。

4. SET sql_log_bin=0 操作

该命令行用于设置会话（SESSION）级别不记录 binlog 日志。这通常用于一些特殊情况，比如当从库需要额外创建索引时，不希望这些操作被记录并分发到其他从库。虽然这种操作是为了优化和管理数据库，但执行后可能会带来潜在的风险（比如，数据不一致、数据无法恢复和审计无法查到）。

5. RESET MASTER 和 RESET SLAVE 命令

RESET 命令是在 MySQL 数据库管理中用于重置主服务器和从服务器的复制状态与二进制日志的命令。RESET 命令行的影响如下。

- RESET MASTER 重置全局事务标识符（GTID）。之前所有的 GTID 都不再有效，复制点信息会丢失。
- RESET MASTER/SLAVE 清除 binlog 日志。之前所有的复制日志都会丢失。
- RESET SLAVE 重置从服务器上的复制信息，包括复制位置、复制用户和密码等。

6. TRUNCATE TABLE 命令

TRUNCATE TABLE 命令用于快速删除表中的所有数据。执行过程中，它会锁定表（元表锁），但这个锁定时间通常非常短暂。然而，使用 TRUNCATE 命令存在以下风险。

- 不会记录详细数据到 binlog 日志：由于 TRUNCATE 操作不会记录详细的数据变更到 binlog，因此在需要数据恢复时，可能无法找到具体的操作数据。
- 自增序列不重置：TRUNCATE TABLE 命令不会重置表的自增序列。
- 潜在的 bug 风险：尽管不常见，但 TRUNCATE 命令在某些情况下可能会触发数据库的 bug，这可能导致数据库服务无法提供。

7. DROP DATABASE 命令

DROP DATABASE 命令用于删除整个数据库及其包含的所有表和底层物理文件。在执行此命令之前，务必做好充分的备份，因为一旦数据库被删除，所有数据和结构都将永久丢失。DROP 命令存在如下风险。

- binlog 日志不记录详细数据内容。因此，在需要恢复数据时，无法通过 binlog 提供被删除的具体内容。
- 底层数据和结构文件物理删除。如果后续需要恢复数据，则将无法从物理文件中获取任何信息。

8. FLUSH TABLES 命令

该命令用于刷新表或锁定表以进行只读操作，有阻塞数据库其他操作的风险。FLUSH 操作内容如下。

- 关闭所有打开的表。
- 强制关闭所有正在使用的表。
- 刷新查询缓存和预准备语句缓存。
- 不会刷新脏块。
- 上全局 COMMIT 锁。

9. LOCK TABLES READ/WRITE 命令

该命令用于锁定表，这会导致其他用户无法对该表进行写入操作，但不会阻塞读取操作。

10. ALTER DATABASE dbname READ ONLY 命令（MySQL 8.0 版本）

该命令用于将数据库设置为只读属性，即不允许写入。

11. slave_skip_errors 参数设置

该参数用于复制过程，允许从服务器在遇到指定错误代码时跳过这些错误。然而，滥用此参数可能会导致主从数据库之间的数据不一致。

12. PURGE BINARY LOGS 命令

该命令用于清除不再需要的 binlog 文件。在复制状态下，需要确保定期清除无用的 binlog 文件，如果清除错误，则会导致复制中断。

13. DML 操作的 WHERE 条件中没有使用索引

如果在 DML 操作的 WHERE 条件中没有有效利用索引，则会执行全表扫描。特别是在处理大量数据时，这种情况对性能的影响会更加显著，可能导致查询速度显著下降，甚至影响整个数据库的性能。

14. 在线 DDL 操作

在线结构变更操作，无论是官方的 Online DDL 还是第三方工具（如 pt-osc 和 gh-ost），都存在一定的问题。因此，特别是在高负载的情况下，应谨慎执行。相关问题点如下。

- Online DDL 可能导致元数据锁。
- pt-osc 在处理大量表时可能意外停止，导致触发器无法删除。
- gh-ost 会写入 binlog 日志，当数据量大时，延迟会逐渐增加，可能导致操作无法继续。

15. ANALYZE TABLE 或 OPTIMIZE TABLE 命令锁表

锁表，缓存移除，对表数据无法进行修改，严重的话会影响业务。

执行 ANALYZE TABLE 或 OPTIMIZE TABLE 命令时，可能会对表进行锁定或重写，从而阻止对表数据的并发修改。这可能导致业务处理暂时受阻，甚至可能导致数据库连接数增加。

16. 未开启 binlog 的 ROW 模式的操作

在没有定期备份、没有配置高可用节点（如复制集中的其他节点），以及没有开启 ROW 模式的 binlog 的情况下，如果 MySQL 数据库发生故障，那么通过事务性操作进行的数据更改可能无法完全回逆。这可能导致数据丢失或不一致。

17. 执行 rm 命令前要三思

rm 命令会物理删除文件，即使某些进程仍在使用这些文件。虽然某些进程可能通过 lsof 命令被标记为可删除，但这并不意味着文件的内容可以全部恢复。一旦文件被物理删除，恢复它们就变得非常困难或不可能。

18. 未使用 mysqld_safe 守护进程启动服务

如果不使用 mysqld_safe 守护进程来启动 MySQL 服务，那么服务将没有守护进程，这可能导致 MySQL 服务在异常关闭后无法自行启动。

19. 在生产环境中执行测试命令

在生产环境中执行 MySQL 测试命令时，由于未经验证的命令直接被执行，因此会导致数据库负载加大，进而影响业务的正常运行。

20. 在使用客户端工具时选择了错误的数据库实例

在使用客户端工具（特别是基于 IP 连接管理环境的工具）时，如果不慎选择了错误的数据库实例，则可能会导致操作失误。因此，连接数据库环境时，务必注意以下操作要点。

- 特别是使用开发工具（如 Navicat、SQLyog）时，需要严格控制权限。
- 在执行数据操作命令前，必须核对实例信息，确保操作正确无误。

21. MySQL 使用的硬盘空间不足时的操作

如果 MySQL 使用的硬盘空间不足，则可能会导致在线服务中的 MySQL 事务丢失或数据库文件损坏。因此，在进行诸如备份、迁移数据文件等操作之前，务必确认硬盘空间是否充足。

22. 从库允许写操作

在 MySQL 高可用数据库架构中，如果从库（通常用于读取操作以提高性能）被错误地允许写入数据，那么会导致主从库之间的数据不一致，从而使整个高可用架构失去其预期的效果。

23. 打开多个窗口操作数据库

如果同时在多个窗口操作重要数据库，就容易误将命令发送到所有 MySQL 数据库节点上，导致不必要的风险。

24. 未加密的情况，敏感数据从线上同步到线下

如果敏感字段不进行加密处理，则线上数据同步到线下时可能导致敏感数据泄露。尽管数据本身可能无法做到完全透明，但可以通过特定的函数算法来替换或转换敏感字段，从而降低数据泄露的风险。

25. 操作系统表

对 MySQL 的 system 表以及 information_schema 库下的表进行删除、更改、创建和清空操作，都有可能导致数据库无法启动。

26. 更改整体影响参数

更改整体影响参数时需要谨慎，因为有些参数对数据库的全局性能有重要影响。以下是一些影响较大的参数。

- lower_case_table_names：这个参数决定了表名的大小写敏感性。它只能在数据库初始化时设置，之后不建议更改。如果需要更改，则建议采用逻辑导出导入的方式，以避免潜在的问题。
- innodb_buffer_pool_size：这个参数可以在线调整，用于设置 InnoDB 存储引擎的缓冲池大小。但在高负载情况下调整此参数可能导致持续的性能问题或系统阻塞。

27. 疲倦时操作线上环境

当人感到疲倦时，反应速度、判断力和注意力都会有所下降，这可能导致在操作线上环境时发生误操作。对于任何复杂操作，都应该事先制定详细的操作步骤，引入第二个人来协助确认和检查。疲倦时操作线上环境是非常危险的，应该采取一系列措施来确保操作的正确性和安全性。

6.2.3 运维管理中常用的官方工具

MySQL 是普及广泛的数据库，其开源社区非常活跃。谈到官方运维工具，广大读者最熟悉的是逻辑备份工具 mysqldump。除了 mysqldump 以外，MySQL 还提供了其他实用的运维工具。接下来，将详细介绍这些工具。

对于 MySQL 8.0 版本，官方提供了 31 个工具，如图 6-3 所示，可以大致分为四大类：服务类工具、运维类工具、配置类工具和其他实用工具。

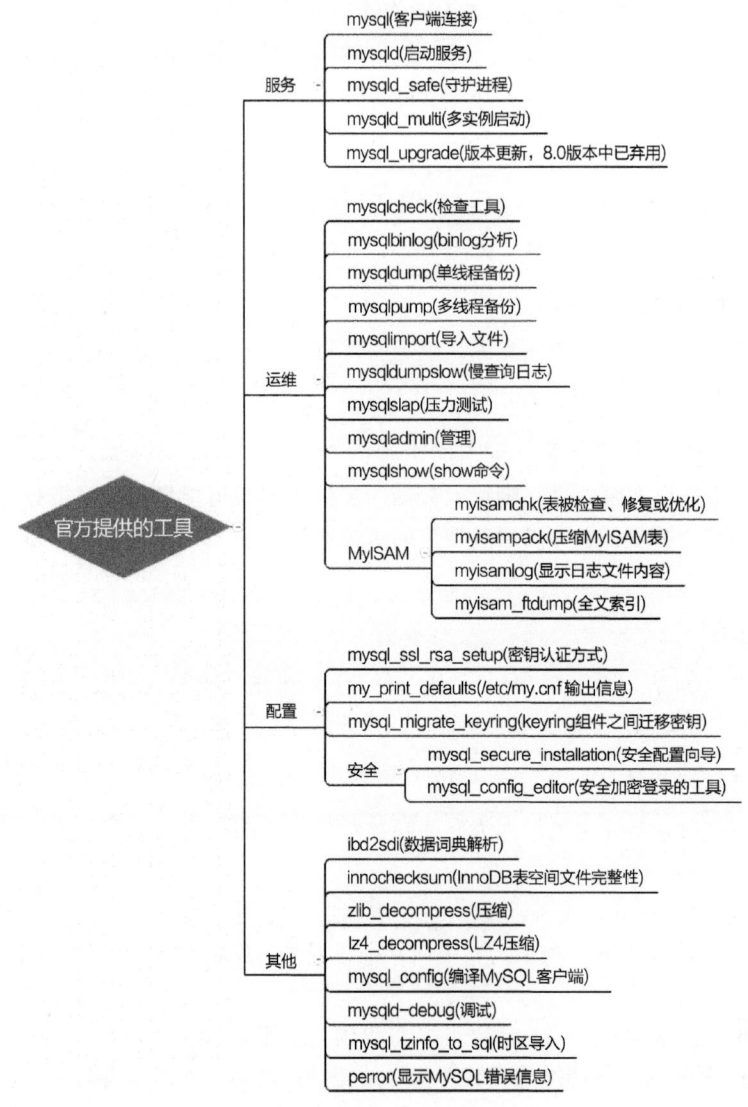

图 6-3　MySQL 8.0 官方工具分类

MySQL 运维类工具是指用于维护和管理 MySQL 数据库的工具。运维类工具包括备份工具、监控工具、性能优化工具等，用于确保数据库的稳定性和高效运行。下面将介绍几种常用的 MySQL 运维类工具。

（1）mysqladmin

mysqladmin 是一个用于执行管理操作的客户端工具。可以使用它来检查服务器的配置和当前状态，以及创建和删除数据库等。其优点在于，无须登录到 MySQL 操作界面，即可直接执行命令。表 6-1 展示了 mysqladmin 支持的参数。

表 6-1 mysqladmin 支持的参数

参数	说明	参数	说明
password	修改密码	extended-status	等同于 show global status 命令行
status	查看状态，相当于 show status 命令行	processlist	查看执行的 SQL 语句信息
variables	相当于 show variables 命令行	flush-tables	刷新所有表
shutdown	关闭 MySQL 服务	flush-threads	刷新线程缓存
create	创建数据库	kill	终止服务器线程
debug	服务器的错误日志包含与事件调度程序相关的输出	ping	检查服务器是否可用
		reload	重新加载授权表
Drop	删除数据库	refresh	刷新所有表并关闭或打开日志文件
flush-logs	切割日志	start-slave	开启复制进程
flush-privileges	重新加载授权表（与 reload 类似）	stop-slave	停止复制进程
flush-status	初始化状态值		

下面是一个使用 mysqladmin 命令来统计 MySQL 服务器状态的示例。通过指定 sleep 时间获取服务器状态变化值。mysqladmin 的命令行示例如下所示。

```
# 使用 mysqladmin extended-status 命令获得的 MySQL 的性能指标
# 进行差值计算;加上参数 --relative(-r)
# --sleep(-i) 就可以指定刷新的频率
shell> mysqladmin -uroot -p****** extended-status --relative --sleep=1
```

（2）mysqlslap

mysqlslap 是 MySQL 自带的一个简单的性能压力测试工具，用于模拟多客户端对 MySQL 服务器的负载情况。这个工具主要用于测试数据库在高并发环境下的性能表现，如查询性能、写入性能等。虽然 mysqlslap 提供的功能相对单一，但它是轻量级的，并且非常适合于基本的性能测试和模拟数据负载。使用 mysqlslap 创建表、插入和查询的命令行示例如下所示。

```
# concurrency 表示并行线程数,iterations 表示重复次数
shell>mysqlslap --delimiter=";" --create="CREATE TABLE a (b int);INSERT INTO a VALUES (23)"
--query="SELECT * FROM a" --concurrency=50 --iterations=200
```

（3）mysqlbinlog

mysqlbinlog 是 MySQL 提供的一个实用工具，用于处理二进制日志文件（binlog）。二进制日志记录了数据库更改的所有信息，如表的创建、更改、删除，以及数据的插入、更新和删除等操作。这对于复制操作、数据恢复和审计等任务都非常有用。

使用 mysqlbinlog，可以以可读的形式查看二进制日志中的事件，也可以将其内容提取出来以进行进一步处理。

（4）mysqlcheck

mysqlcheck 是 MySQL 提供的一个用于检查、修复、优化和分析表的命令行工具。该工具会在执行时给表加上读锁，这意味着在表被检查、修复或优化期间，其他客户端只能读取数据，不能写入数据。因此，mysqlcheck 通常在数据库负载较低的时候执行，以避免对用户产生太大影响。

（5）mysqldumpslow

mysqldumpslow 工具用于解析慢查询日志文件并总结其中的内容。通过 mysqldumpslow，可以

更容易地识别和优化那些执行时间较长的查询。mysqldumpslow 无法通过指定时间范围来过滤慢查询，但可以使用多种选项来定制它的输出，以满足需求。

（6）mysqlshow

mysqlshow 工具提供了一个简单的接口来查看数据库、表，以及表的列和索引的摘要信息。这个工具通常用于快速检查数据库的结构，而不需要登录到 MySQL 命令行客户端。在实际环境中基本不使用。

（7）mysqlimport

mysqlimport 工具允许用户从文本文件导入数据到 MySQL 数据库中。这个工具特别适用于导入大量数据到 MySQL 表中，并且它提供了一种比使用 LOAD DATA INFILE SQL 语句更简便的命令行接口。

（8）mysqldump 和 mysqlpump

mysqldump 和 mysqlpump 都是 MySQL 提供的用于执行逻辑备份的工具。它们通过生成一组 SQL 语句来记录数据库的结构（即对象定义，如表结构、索引、触发器等）以及数据，从而允许用户通过执行这些 SQL 语句来重新创建数据库及其内容。这些工具通常用于备份数据库，或者将数据库从一个服务器迁移到另一个服务器。

对于 mysqlpump，比较常用的是用户导出功能，如下所示。

```shell
shell> mysqlpump --exclude-databases=% --users
```

（9）MyISAM 支持的工具

针对 MyISAM 引擎，MySQL 提供了一组专用的工具来维护和管理 MyISAM 表。在 MySQL 中，InnoDB 存储引擎通常是默认的。了解如何使用这些工具进行维护和管理仍然是有必要的。下面介绍针对 MyISAM 引擎的工具。

- myisamchk：用于检查和修复 MyISAM 表。
- myisampack：用于压缩 MyISAM 表。
- myisamlog：显示 MyISAM 的日志文件内容。
- myisam_ftdump：用于管理 MyISAM 的全文索引。

6.2.4 运维管理中常用的周边工具

上一节介绍了 MySQL 运维管理中常用的官方工具，本节将介绍 MySQL 运维管理中常用的周边工具。这些工具在实际环境中得到了广泛应用，绝大部分都是开源性质的。如图 6-4 所示常用的 MySQL 周边工具。

掌握了这些 MySQL 周边工具（原理和使用技巧）后，MySQL 的日常运维工作会变得更加简单、高效。下面介绍 MySQL 生产环境中常用的八类 MySQL 周边工具。

1. 在线 DDL 变更工具

在线 DDL 变更中，pt-osc 和 gh-ost 都是常用的在线变更 DDL 的工具。

（1）pt-osc

pt-osc 是 Percona 公司开发的一个非常好用的 DDL 工具，全称为 pt online schema-change，是 Percona Toolkit 工具集中的一个组件，用于在不中断 MySQL 服务的情况下执行 MySQL 数据库的 DDL 更改。这个工具能够在不锁定表的情况下进行表结构变更，这对于需要保持数据库可用性的生产环境来说是非常有用的。

图 6-5 所示为 pt-osc 实现原理，即使用 MySQL 触发器实现在线 DDL 操作。

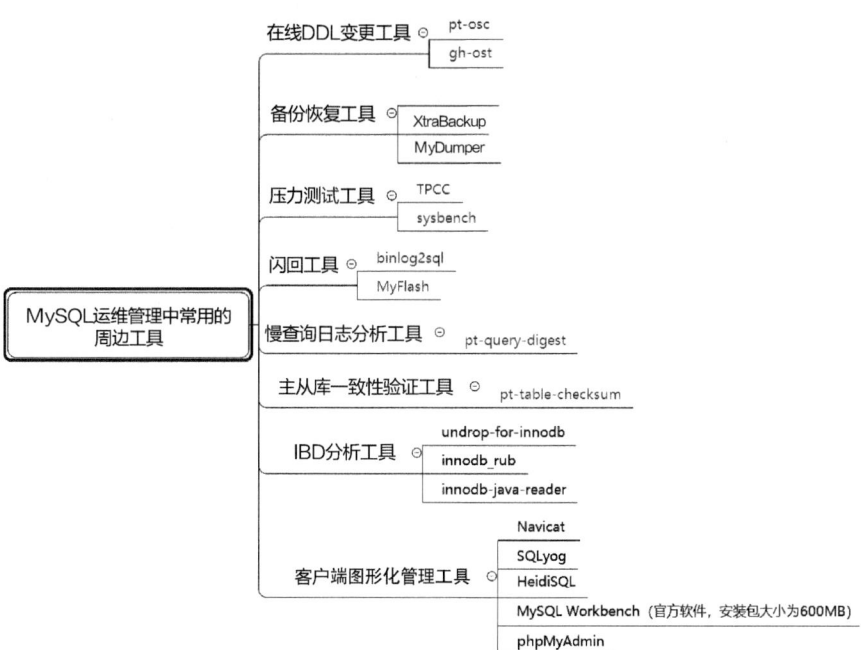

图 6-4　MySQL 开源社区工具

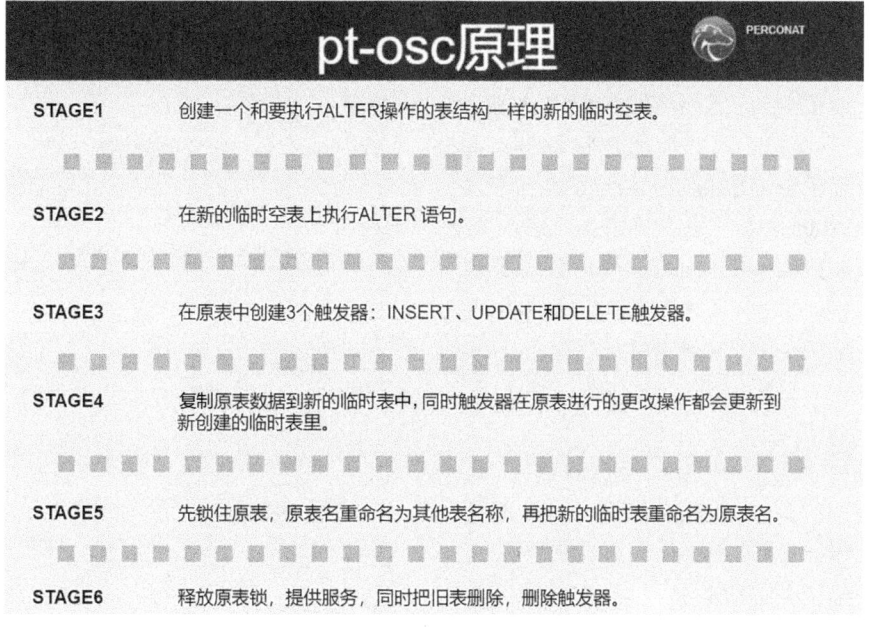

图 6-5　pt-osc 实现原理

（2）gh-ost

gh-ost 是一个由 GitHub 发布的开源工具，专门用于 MySQL 数据库的在线 DDL 变更操作。与 pt-osc 类似，gh-ost 也允许在不中断服务的情况下更改数据库表结构。图 6-6 所示为 gh-ost 实现原理，它不依赖触发器，通过模拟从库，在 binlog 中获取增量变更，再异步应用到新表中。

说明：pt-osc 和 gh-ost 工具在最后阶段都会执行 RENAME 表名的动作。

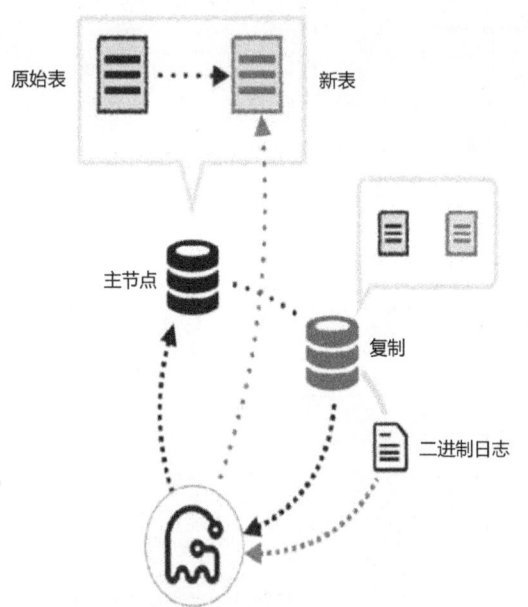

图 6-6 gh-ost 实现原理

2. 备份恢复工具

（1）XtraBackup

XtraBackup 是由 Percona 提供的对 InnoDB 做数据备份的工具，它支持在线热备份（备份时不影响数据读写），并且支持增量备份。它是商业备份工具 InnoDB 的 Hotbackup 的一个很好的替代品，也是目前开源物理备份工具的唯一选择。图 6-7 展示了 XtraBackup 的备份实现流程。它在备份过程中不断扫描和记录数据变更，实现了在线热备份和增量备份，是 InnoDB 数据库备份的一个强大工具。

（2）MyDumper

MyDumper 是一个开源的逻辑备份工具，由 C 语言编写，并由 MySQL 和 Facebook 等公司的开发与维护。MyDumper 工具包中还包含 myloader 工具，后者用于导入 SQL 语句。图 6-8 所示为 MyDumper 备份流程。对备份实例加读锁，阻塞写操作以建立一致性数据备份快照点，记录备份点 binlog 信息。创建工作线程，初始化备份任务队列，并向队列中推送非 InnoDB 表和 InnoDB 表的备份任务。

3. 压力测试工具

MySQL 压力测试工具可以模拟用户连接时

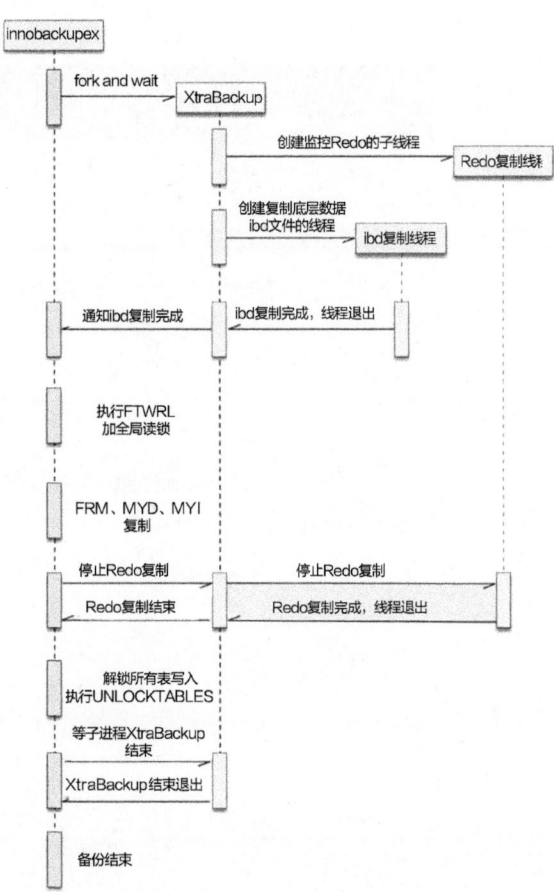

图 6-7 XtraBackup 备份实现流程

数据库在高并发、高负载情况下的性能表现，包括事务处理能力、响应时间、吞吐量等指标。通过模拟这些真实场景中的操作，可以了解数据库的性能瓶颈，从而进行优化以提高数据库的稳定性和性能。常用的压力测试工具有两个：TPCC 和 sysbench。

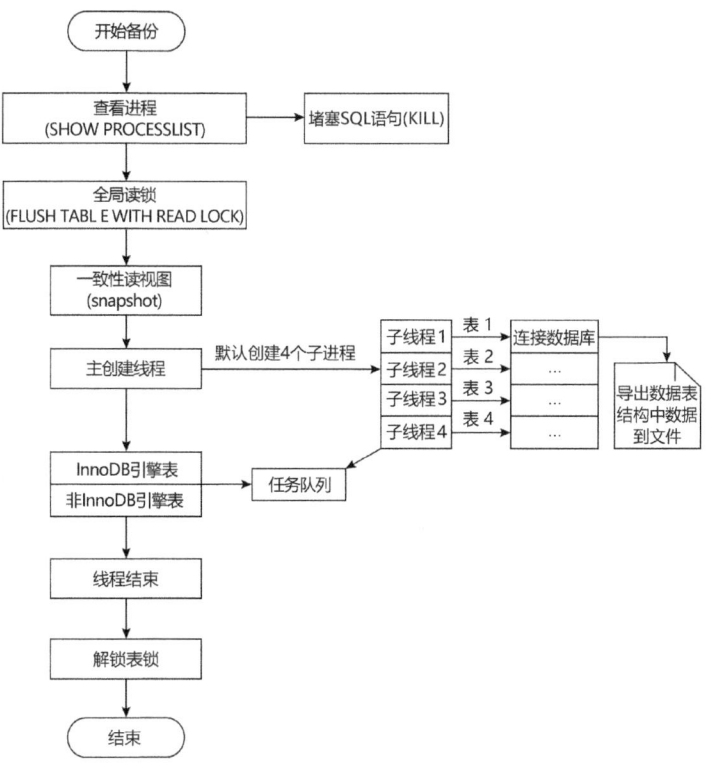

图 6-8　MyDumper 备份流程

（1）TPCC

MySQL 压力测试工具 TPCC 是一种用于模拟 OLTP（联机事务处理）负载的性能测试工具。它模拟了一个具有代表性的 OLTP 场景：在线订单处理系统。在这个场景中，假设有一个大型商品批发商，它拥有多个位于不同区域的仓库，每个仓库都有一定数量的商品库存，每个地区又有一定数量的客户。它可以模拟多个用户并发执行数据库事务，包括查询、更新、插入和删除等操作，以测试数据库在高并发情况下的性能表现。

图 6-9 所示为 TPCC 业务模型主要组成部分和它们之间的关系。
- TPCC 基准由 9 个表组成。这些表通过 10 个外键关系相连。
- 基础数据基于数据库初始加载期间生成的仓库数量（W）按基数缩放。
- 一个仓库对应 10 个地区、3 万个客户、30 万订单，每个订单又分为 New-Order 和 Order-line 两部分。New-Order 用来标示订单是否为新订单，在完成订单的发货操作后，其中的记录将被删除。Order-Line 是订单中进行批发的商品订购流水，每个订单有 5~10 个交易商品。
- Stock 用来记录 Warehouse 的库存情况，Item 记录 Warehouse 中存储的商品。
- 箭头方向可表明数据来源。

TPCC 模拟了在线订单系统的日常操作，通过调整并发用户数、事务混合比例等参数，可以模拟不同的负载情况，并对数据库的性能进行全面评估。

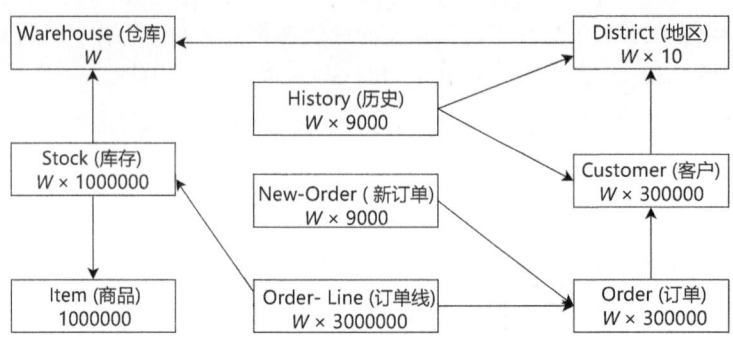

图 6-9　TPCC 业务模型

（2）sysbench

sysbench 是一个开源的、模块化的、跨平台的多线程性能测试工具，它可以用来模拟多种类型的负载（CPU、内存、磁盘 I/O），包括数据库负载。对于数据库性能测试，sysbench 可以模拟多种类型的 SQL 操作，如 INSERT、SELECT、UPDATE 和 DELETE，并且可以指定并发用户数、事务类型分布以及其他相关参数来生成特定的负载。

图 6-10 所示为 sysbench 的通用测试流程，包括准备阶段、预热阶段、测试阶段，最终生成的测试报告将提供详细的性能数据，帮助了解系统的性能表现和潜在瓶颈。

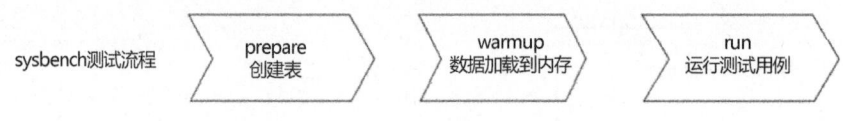

图 6-10　sysbench 的通用测试流程

说明： 上述两个测试工具的主要区别在于它们所模拟的业务流程和性能测试的侧重点不同。TPCC 测试注重于模拟真实的在线事务处理业务流程，而 sysbench 则更注重于测试单语句或单一操作的执行性能。

4. 闪回工具

MySQL 数据库闪回工具是指一种用于恢复数据库误操作的工具，它可以通过分析二进制日志（binlog）来回滚由于误删、误更新等操作导致的数据变更。这种工具通常要求 binlog 的格式为 ROW 模式，并且参数 binlog_row_image 设置为 FULL 模式，同时确保 binlog 文件被保留。恢复操作通过分析 binlog 中的 SQL 语句并以逆序（即时间顺序的倒序）回放来实现。具体来说，对于 DELETE 操作，通过 INSERT 操作来回滚；对于 INSERT 操作，通过 DELETE 操作来回滚；对于 UPDATE 操作，则通过相应的 UPDATE 操作来回滚。

下面介绍比较流行的开源的闪回工具。

（1）binlog2sql

binlog2sql 用于从 MySQL 的 binlog 中提取 SQL 语句。它通过分析数据库的运行状态和 BINLOG_DUMP 协议来获取 binlog 的内容。使用条件如下。

- 数据库服务必须处于在线状态。
- 开发语言为 Python。
- 仅支持 MySQL。

(2) MyFlash

MyFlash 直接从 MySQL 的 binlog 中提取标准的 SQL 语句，使得其效率更高。使用条件如下。
- 仅支持 MySQL。
- 仅回滚 DML（增、删、改）。
- 使用 C 语言开发，配合 mysqlbinlog 以及 sed、awk 等工具使用。
- 离线解析。

每个工具都有其独特的优点和限制。在选择时，需要根据具体的需求和场景来决定使用哪个工具。对于相同的 binlog 分析任务，MyFlash 的性能最好，其次是 binlog2sql。

5. 慢查询日志分析工具

pt-query-digest 是 Percona Toolkit 中的一个工具，它专门用于分析 MySQL 的查询日志。与官方提供的 mysqldumpslow 工具相比，pt-query-digest 具有更多优势，因为它不仅能够从普通日志和慢查询日志中分析 SQL，还能从二进制日志、SHOW PROCESSLIST 命令输出以及 Tcpdump 捕获的数据中进行分析。图 6-11 所示为 pt-query-digest 分析慢查询日志的显示内容，非常整洁，包含语句的 TOP 排名（Rank 列）、执行次数、总时间、平均时间、最大时间和具体 SQL 语句（"V/" 列）。

```
# Profile
# Rank Query ID                         Response time  Calls R/Call   V/
# ==== ================================ ============== ===== ======== ==
#    1 0x2A60E950668C2A9E6E3FE8E7DE07DC85 466.6741 77.5%     8 58.3343 91.30 SELECT test.t?
#    2 0xC54312D080D0904F754521EF548E3A43  91.6045 15.2%     3 30.5348 55.75 SELECT t?
#    3 0x59A74D08D407B5EDF9A57DD5A41825CA  30.0039  5.0%     3 10.0013  0.00 SELECT
# MISC 0xMISC                              14.0511  2.3%    52  0.2702  0.0 <23 ITEMS>

# Query 1: 0.00 QPS, 0.00x concurrency, ID 0x2A60E950668C2A9E6E3FE8E7DE07DC85 at byte 16890
# This item is included in the report because it matches --limit.
# Scores: V/M = 91.30
# Time range: 2021-01-06T09:56:31 to 2021-01-21T15:50:31
# Attribute    pct   total     min     max     avg     95%  stddev  median
# ============ === ======= ======= ======= ======= ======= ======= =======
# Count         12       8
# Exec time     77    467s      6s    206s     58s    202s     73s     32s
# Lock time     23     5ms   118us     2ms   675us     2ms   670us   570us
# Rows sent     18     213       1     206   26.62  202.40   66.61    0.99
# Rows examine   8     220       2     206   27.50  202.40   66.29    1.96
# Bytes sent    34   8.04k   8.04k   8.04k   8.04k   8.04k       0   8.04k
# Query size    18     394      31      60   49.25   59.77   13.60   59.77
# Bytes receiv   0
```

图 6-11　pt-query-digest 分析显示

6. 主从库一致性验证工具

pt-table-checksum 也是 Percona Toolkit 的组件，用于验证 MySQL 主从复制环境中的数据一致性。其原理是，首先，在主库上执行基于 statement 的 SQL 语句来生成主库数据块的 checksum；然后将相同的 SQL 语句传递到从库执行，并在从库上计算相同数据块的 checksum；最后，通过比较主从库上相同数据块的 checksum 值来判断数据是否一致。在配置 pt-table-checksum 时，如果选择 recursion-method 方式为 host，则会按照主机进行递归检查。图 6-12 所示为 pt-table-checksum 的输出报告。在该报告中，需要特别关注 DIFF_ROWS 的值。如果 DIFF_ROWS 大于 0，则说明主从库之间存在数据不一致的情况。

```
[root@ss30 tmp]# pt-table-checksum --no-check-binlog-format h=192.168.244.130,u=root,p=123456   --nocheck-plan --databases=db9 --recursion-me
thod=dsn=D=percona,t=dsns
Checking if all tables can be checksummed ...
Starting checksum ...
            TS ERRORS  DIFFS     ROWS  DIFF_ROWS  CHUNKS  SKIPPED     TIME TABLE
01-19T21:45:20      0      0        7          0       1        0    0.036 db9.course
01-19T21:45:20      0      0      401          0       1        0    0.042 db9.middle_user
01-19T21:45:20      0      0       17          0       1        0    0.031 db9.mysql_slow_query_review
01-19T21:45:20      0      0        5          0       1        0    0.038 db9.mysql_slow_query_review_history
01-19T21:45:20      0      0        0          0       1        0    0.032 db9.ss
01-19T21:45:20      0      1        3          0       1        0    0.034 db9.t1
01-19T21:45:20      0      0        1          0       1        0    0.035 db9.t2
01-19T21:45:20      0      0        1          0       1        0    0.031 db9.t3
01-19T21:45:20      0      0     1000          0       1        0    0.033 db9.t_temp
01-19T21:45:21      0      0       21          0       1        0    0.033 db9.test_optimizer
01-19T21:45:21      0      0        0          0       1        0    0.032 db9.tester001
01-19T21:45:21      0      0     1000          0       1        0    0.033 db9.tuser
01-19T21:45:21      0      0        3          0       1        0    0.032 db9.users
```

图 6-12　pt-table-checksum 关于主从库一致性验证报告

7. IBD 分析工具

分析 MySQL 数据文件（IBD 文件）是在数据库"死"机且无法启动，同时没有高可用性架构和备份的情况下，尝试恢复表结构和数据的最后手段之一。IBD 分析工具介绍如下。

- undrop-for-innodb：这个工具用于分析 IBD 文件以恢复误删除的数据。如果确保在误删除后没有新的数据写入到 IBD 文件中，那么数据恢复的可能性会更高。需要注意的是，中文解析可能存在问题，因此在处理包含中文数据的表时需要格外注意。
- innodb_rub：这个工具主要用于查看 InnoDB 数据库表的各种存储细节，并解析 InnoDB 文件。它对于学习数据库底层的存储结构非常有用。
- innodb-java-reader：这是一个用 Java 编写的工具，可以直接访问 MySQL 的 InnoDB 存储引擎文件。它提供了检查页面、通过主键和辅助键查找记录的功能，还可以生成页面热图，这有助于分析数据的分布和访问模式。

此外，还可以结合使用官方工具 innochecksum 来验证 IBD 文件的完整性，并为开发类似的恢复工具提供参考。

8. 客户端图形化管理工具

客户端图形化管理工具为数据库管理员和程序开发者提供了极大的便利，使得他们可以更加直观、高效地执行数据库操作、SQL 开发和管理任务。这些工具通常都提供了数据导入导出、表结构设计、查询执行、性能监控等功能，而且很多还提供了直观的可视化界面，大大降低了使用门槛。

下面介绍常用的客户端图形化管理工具。

- Navicat 支持多种主流的数据库系统，包括 MySQL、Oracle、PostgreSQL 等，方便用户在不同数据库之间进行切换和管理。它提供了数据建模、数据传输、同步、备份等多种高级功能。
- SQLyog 是一个专门为 MySQL 设计的图形化管理工具，轻巧且易于使用。它支持快速的数据导入导出，包括二进制数据转化为图片的功能。
- HeidiSQL 是用 Delphi 开发的简洁的图形化数据库管理工具，支持 MySQL、SQL Server、PostgreSQL、SQLite 等多种数据库。
- MySQL Workbench：作为 MySQL 官方提供的重量级管理工具，它集成了数据建模、SQL 开发、数据库管理等功能，特别适合进行数据库设计和建模，支持 ER 图设计等。
- phpMyAdmin：phpMyAdmin 是一个以 PHP 为基础，以 Web-Base 方式架构在网站主机上的 MySQL 数据库管理工具，让管理者可用 Web 接口管理 MySQL 数据库。

6.3 MySQL 8.0 运维实践

6.3.1 binlog 文件查看和解析

MySQL 中的 binlog（二进制日志）是服务层日志，是实现数据复制功能的基础。在日常的数据库维护中，binlog 文件的管理至关重要，同时，其内容也经常被用于诊断问题。为了查看和解析 binlog 文件，官方提供了两种常用的方法：一种是直接在 MySQL 客户端命令行界面中执行相关命令，另一种是使用名为 mysqlbinlog 的独立工具。下面将详细介绍这两种查看和解析 binlog 的方法。

1. 使用客户端命令行查看 binlog 内容和解析 binlog 文件

对于 MySQL 的 binlog，客户端命令行提供了 SHOW 和 PURGE 等命令来帮助用户查看与管理 binlog 文件。

使用 SHOW BINARY LOGS 命令行显示 binlog 文件的列表信息（包含文件名、是否加密等），如下所示。

```
mysql> SHOW BINARY LOGS;
+------------------+-----------+-----------+
| Log_name         | File_size | Encrypted |
+------------------+-----------+-----------+
| mysql-bin.000086 |      1073 | No        |
| mysql-bin.000087 |       197 | No        |
+------------------+-----------+-----------+
2 rows in set (0.00 sec)
```

使用 SHOW EVENTS 命令行显示 binlog 中的事件。这个命令通常用于调试和恢复，因为它允许查看在特定时间或位置发生的数据库更改信息。同时它也可以解析 relaylog 文件（因为在复制过程中，从库从主库获取 binlog 事件，并将这些事件写入 relaylog 文件中，所以 binlog 等价于 relaylog）。查看 binlog 和 relaylog 的相关事件的语法如下所示。

```
# binlog event SHOW BINLOG EVENTS
    [IN 'log_name']
    [FROM pos]
    [LIMIT [offset,] row_count]
# Relay event SHOW RELAYLOG EVENTS
    [IN 'log_name']
    [FROM pos]
    [LIMIT [offset,] row_count]
    [channel_option]
channel_option:
    FOR CHANNEL channel
```

使用 SHOW EVENTS 命令行查看 binlog 记录的变更内容，需要指定 binlog 或 relaylog 文件，同时再组合其他 SQL 写法（IN+FROM+LIMIT）等方式进行查询。SHOW EVENTS 命令行示例如下所示。

```
mysql> SHOW BINLOG EVENTS IN 'mysql-bin.000087';
mysql> SHOW BINLOG EVENTS IN 'mysql-bin.000087' FROM 1008;
mysql> SHOW BINLOG EVENTS IN 'mysql-bin.000087' FROM 1008 limit 2,5;
mysql> SHOW RELAYLOG EVENTS IN 'relay-log.000004' FROM 1062 LIMIT 5,2;
```

图 6-13 所示为使用 SHOW BINLOG EVENTS 解析 binlog 文件内容的窗口输出信息。解析内容包含文件名、开始位置、时间、服务 ID、结束位置和具体时间内容（Info 列）。SHOW BINLOG EVENTS 命令在输出大量事件时可能会导致屏幕滚动非常快，难以阅读和解析。此外，当事件非常频繁时，输出规模可能会非常庞大，难以管理和分析。因此，实际运维工作中很少使用它。

Log_name	Pos	Event_type	Server_id	End_log_pos	Info
mysql-bin.000087	4	Format_desc	1302	123	Server ver: 5.7.29-log, Binlog ver: 4
mysql-bin.000087	123	Previous_gtids	1302	194	39d0a7f2-702c-11ea-92a0-000c29b9a76d:1-48
mysql-bin.000087	194	Gtid	1302	259	SET @@SESSION.GTID_NEXT= '39d0a7f2-702c-11ea-92a0-000c29b9a76d:49'
mysql-bin.000087	259	Query	1302	389	use `db7`; create table t1(id int ,name varchar(20) ,primary key(id))
mysql-bin.000087	389	Gtid	1302	454	SET @@SESSION.GTID_NEXT= '39d0a7f2-702c-11ea-92a0-000c29b9a76d:50'
mysql-bin.000087	454	Query	1302	525	BEGIN
mysql-bin.000087	525	Rows_query	1302	590	# insert into t1(id ,name) values(1,'CCCC')
mysql-bin.000087	590	Table_map	1302	637	table_id: 176 (db7.t1)
mysql-bin.000087	637	Write_rows	1302	682	table_id: 176 flags: STMT_END_F
mysql-bin.000087	682	Xid	1302	713	COMMIT /* xid=1061 */
mysql-bin.000087	713	Gtid	1302	778	SET @@SESSION.GTID_NEXT= '39d0a7f2-702c-11ea-92a0-000c29b9a76d:51'
mysql-bin.000087	778	Query	1302	849	BEGIN
mysql-bin.000087	849	Rows_query	1302	915	# insert into t1(id ,name) values(2,'BBBBB')
mysql-bin.000087	915	Table_map	1302	962	table_id: 176 (db7.t1)
mysql-bin.000087	962	Write_rows	1302	1008	table_id: 176 flags: STMT_END_F
mysql-bin.000087	1008	Xid	1302	1039	COMMIT /* xid=1062 */
mysql-bin.000087	1039	Gtid	1302	1104	SET @@SESSION.GTID_NEXT= '39d0a7f2-702c-11ea-92a0-000c29b9a76d:52'
mysql-bin.000087	1104	Query	1302	1175	BEGIN
mysql-bin.000087	1175	Rows_query	1302	1240	# insert into t1(id ,name) values(3,'AAAA')
mysql-bin.000087	1240	Table_map	1302	1287	table_id: 176 (db7.t1)
mysql-bin.000087	1287	Write_rows	1302	1332	table_id: 176 flags: STMT_END_F
mysql-bin.000087	1332	Xid	1302	1363	COMMIT /* xid=1064 */

图 6-13　使用 SHOW BINLOG EVENTS 解析 binlog 文件内容

在 MySQL 中，PURGE BINARY LOGS 命令主要用于清理和释放不再需要的 binlog 文件所占用的磁盘空间，特别是在 binlog 文件积累过多并导致磁盘空间不足时。PURGE BINARY LOGS 语法和使用示例如下所示。

```
#语法：
PURGE { BINARY } LOGS {
    TO 'log_name'
   | BEFORE datetime_expr
}
#查看 binlog 文件
mysql> PURGE BINARY LOGS TO 'mysql-bin.000080';
+----------------------------+----------------+
| Log_name                   | File_size |
+----------------------------+----------------+
| mysql-bin.000080  |        194  |
| mysql-bin.000081  |        194  |
| mysql-bin.000082  |        194  |
| mysql-bin.000083  |        217  |
| mysql-bin.000084  |        408  |
| mysql-bin.000085  |        517  |
| mysql-bin.000086  |      7474  |
+----------------------------+----------------+
#清除某日的某个时间点之前的文件
mysql> PURGE BINARY LOGS BEFORE '2023-08-01 10:00:00';
Query OK, 0 rows affected, 1 warning (0.01 sec)
mysql> SHOW BINARY LOGS;
+----------------------------+----------------+
| Log_name                   | File_size |
+----------------------------+----------------+
```

```
| mysql-bin.000086 |     7474 |
+----------------------------+----------------+
1 row in set (0.00 sec)
```

说明：使用 PURGE BINARY LOGS 命令时需要谨慎，因为清除 binlog 文件将永久删除它，并且无法恢复。

2. 使用 mysqlbinlog 工具查看和解析 binlog 内容

mysqlbinlog 工具可以将 binlog 文件的内容转换为更易读的文本格式，这对于进行数据库审计、故障排查以及数据恢复等任务非常有用。由于不同 MySQL 版本可能采用不同的 binlog 格式，因此使用与 MySQL 服务器版本相匹配的 mysqlbinlog 客户端是必要的，以确保正确解析 binlog 文件的内容。

使用 mysqlbinlog 工具解析 binlog 文件的命令行示例如下。

```
shell $> mysqlbinlog mysql-bin.000001
# The proper term is pseudo_replica_mode, but we use this compatibility alias
# to make the statement usable on server versions 8.0.24 and older.
/*!50530 SET @@SESSION.PSEUDO_SLAVE_MODE=1*/;
/*!50003 SET @OLD_COMPLETION_TYPE=@@COMPLETION_TYPE,COMPLETION_TYPE=0*/;
...
BEGIN
/*!*/;
# at 311
# at 362
# 230915 16:34:44 server id 129  end_log_pos 409 CRC32 0xa4581e74  Table_map: `book`.`t` mapped to number 213
# has_generated_invisible_primary_key=0
# at 409
# 230915 16:34:44 server id 129  end_log_pos 449 CRC32 0x8b478fc6  Write_rows: table id 213 flags: STMT_END_F

BINLOG '
JBcEZROBAAAALwAAAJkBAAAAANUAAAAAAAEABGJvb2sAAXQAAQMAAQMAAQEBAHQeWKQ=
JBcEZR6BAAAAKAAAAMEBAAAAANUAAAAAAAEAAgAB/wABAAAAxo9Hiw==
'/*!*/;
# at 449
# 230915 16:34:44 server id 129  end_log_pos 480 CRC32 0x7cd4e85d  Xid = 427
COMMIT/*!*/;
SET @@SESSION.GTID_NEXT= 'AUTOMATIC' /* added by mysqlbinlog *//*!*/;
DELIMITER ;
# End of log file
/*!50003 SET COMPLETION_TYPE=@OLD_COMPLETION_TYPE*/;
/*!50530 SET @@SESSION.PSEUDO_SLAVE_MODE=0*/;
```

说明：上述使用 mysqlbinlog 工具解析出的内容中，除了具体时间、位置和表等信息以外，为了确保数据安全，实际更改的数据会以乱码形式显示。

通过使用特定的参数--base64-output=decode-row，可以利用 mysqlbinlog 工具来解析并查看 binlog 文件中记录的所有已执行的 SQL 语句内容，这对于分析具体操作内容和排查数据库性能问题都非常有帮助。使用 mysqlbinlog 工具显示 SQL 语句的命令行示例如下。

```
shell $> mysqlbinlog --no-defaults --base64-output=decode-row -v -v mysql-bin.000001
# The proper term is pseudo_replica_mode, but we use this compatibility alias
# to make the statement usable on server versions 8.0.24 and older.
...
BEGIN
/*!*/;
# at 311
#230915 16:34:44 server id 129    end_log_pos 362 CRC32 0x9d61fabc    Rows_query
# insert into t(c1) values(1)
# at 362
#230915 16:34:44 server id 129    end_log_pos 409 CRC32 0xa4581e74    Table_map: `book`.`t`
mapped to number 213
# has_generated_invisible_primary_key=0
# at 409
#230915 16:34:44 server id 129    end_log_pos 449 CRC32 0x8b478fc6    Write_rows: table id 213
flags: STMT_END_F
### INSERT INTO `book`.`t`
### SET
###   @1=1 /* INT meta=0 nullable=1 is_null=0 */
# at 449
#230915 16:34:44 server id 129    end_log_pos 480 CRC32 0x7cd4e85d    Xid = 427
COMMIT/*!*/;
SET @@SESSION.GTID_NEXT= 'AUTOMATIC' /* added by mysqlbinlog */ /*!*/;
DELIMITER ;
# End of log file
/*!50003 SET COMPLETION_TYPE=@OLD_COMPLETION_TYPE*/;
/*!50530 SET @@SESSION.PSEUDO_SLAVE_MODE=0*/;
```

说明：在 binlog 解析内容中，完整地显示了执行过的 SQL 语句和相关数据。

在使用 mysqlbinlog 工具时，通过指定参数-d 或--database=name，可以过滤并仅显示与特定数据库相关的操作。这对于分析特定数据库的活动和诊断问题都非常有用。例如，如果想查看名为 book 的数据库中的所有操作，则可以使用以下命令行。

```
shell $> mysqlbinlog --no-defaults -d book --base64-output=decode-row -v -v mysql-bin.000001
#230915 17:37:25 server id 129    end_log_pos 1095 CRC32 0x2130efbd    Write_rows: table id 213
flags: STMT_END_F
### INSERT INTO `book`.`t`
### SET
###   @1=3 /* INT meta=0 nullable=1 is_null=0 */
# at 1095
BEGIN
/*!*/;
# at 1493
#230915 17:53:35 server id 129    end_log_pos 1555 CRC32 0x10581639    Rows_query
# insert into t(id,name) values(1,'CUI')
# at 1555
# at 1607
# at 1651
#230915 17:53:35 server id 129    end_log_pos 1682 CRC32 0x76f5e2a2    Xid = 441
COMMIT/*!*/;
```

注意：mysqlbinlog 的参数-d，只能用于在本地系统上解析 binlog 文件，无法解析远程文件。

在 MySQL 数据库恢复中，回放 binlog（即使用 mysqlbinlog 工具将 binlog 事件应用到 MySQL 服务器）时，为了避免混淆或错误，或者提升回放 binlog 的速度，通过指定参数-D 或 --disable-log-bin，可以禁止恢复过程产生新的 binlog 记录。使用 mysqlbinlog 工具禁止写入新 binlog 记录的命令行示例如下。

```
shell $> mysqlbinlog --no-defaults -D --base64-output=decode-row -v -v  mysql-bin.000001
# The proper term is pseudo_replica_mode, but we use this compatibility alias
# to make the statement usable on server versions 8.0.24 and older.
/*!50530 SET @@SESSION.PSEUDO_SLAVE_MODE=1*/;
/*!32316 SET @OLD_SQL_LOG_BIN=@@SQL_LOG_BIN,SQL_LOG_BIN=0*/;
/*!50003 SET @OLD_COMPLETION_TYPE=@@COMPLETION_TYPE,COMPLETION_TYPE=0*/;
...
```

说明：mysqlbinlog 的参数-d 的作用就是解析信息中，额外添加 SQL_LOG_BIN=0 语句，不记录 binlog 日志，特别是恢复过程中 I/O 性能特别差的时候建议使用。

在使用 mysqlbinlog 工具时，通过指定参数-r 或>方式重定向输出到另一个文件中。使用 mysqlbinlog 工具重定向文件的命令行示例如下。

```
shell $> mysqlbinlog --no-defaults --base64-output=decode-row -v -v
mysql-bin.000001 > /opt/bin001.sql

shell $> mysqlbinlog --no-defaults --base64-output=decode-row -v -v mysql-bin.000001
-r /opt/bin002.sql
```

在使用 mysqlbinlog 工具时，通过指定参数--server-id，输出对应数据库服务器 ID 的，binlog 文件更改的内容。使用 mysqlbinlog 工具只显示指定服务器 ID 的更改内容的命令行示例如下。

```
shell $> mysqlbinlog --no-defaults --base64-output=decode-row --server-id=1301
-v -v mysql-bin.000001
```

注意：在 MySQL 复制中，每个参与复制的服务器（无论是主服务器还是从服务器）都需要有一个唯一的服务器 ID 来标识自己。当多个主服务器的 binlog 事件被复制到一个从服务器时，从服务器可以使用这些服务器 ID 来区分不同主服务器的 binlog 事件。

在使用 mysqlbinlog 工具时，通过指定参数--start-datetime（开始时间）和--stop-datetime（结束时间）或者--start-position（开始位置）和--stop-position（结束位置），可以查看在特定时间范围内或特定位置范围内的 binlog 事件。使用 mysqlbinlog 工具指定范围的命令行示例如下。

```
# 指定位置方式
shell $> mysqlbinlog --no-defaults --base64-output=decode-row -v -v
--start-position=123 --stop-position=290 mysql-bin.000001

shell $> mysqlbinlog --no-defaults --base64-output=decode-row -vv -j 123
--stop-position=290 mysql-bin.000001

# 指定时间方式
shell $> mysqlbinlog --no-defaults --base64-output=decode-row -vv
--start-datetime='2023-05-02 22:36:46' --stop-datetime='2023-05-02 23:25:46' mysql-bin.000001
```

使用 mysqlbinlog 工具时，在某些情况下（GTID 已存在，又想回放 binlog 内容），不希望输出

GTID信息，通过指定参数--skip-gtids，在输出中不显示任何GTID。使用mysqlbinlog工具不输出GTID信息的命令行示例如下。

```
shell $> mysqlbinlog --no-defaults --base64-output=decode-row -vv --skip-gtids mysql-bin.000001
# original_commit_timestamp=1694766884251264 (2023-09-15 16:34:44.251264 CST)
# immediate_commit_timestamp=1694766884251264 (2023-09-15 16:34:44.251264 CST)
/*!80001 SET @@session.original_commit_timestamp=1694766884251264*//*!*/;
/*!80014 SET @@session.original_server_version=80034*//*!*/;
/*!80014 SET @@session.immediate_server_version=80034*//*!*/;
SET @@SESSION.GTID_NEXT= '22228e8c-b0ee-11ec-a2d1-00163e23e2cc:1'/*!*/;# 不显示
```

说明：解析语句中无GTID信息。同时无GTID信息的日志，以便于在任何服务器上重复回放日志（如重复插入同样的数据场景）。

使用mysqlbinlog工具时，在某些情况下，希望指定或跳过GTID信息，可通过指定参数--include-gtids和--exclude-gtids实现。

- --include-gtids：指定需要回滚的GTID，支持GTID的单个和范围两种形式。
- --exclude-gtids：指定不需要回滚的GTID，用法同--include-gtids。

使用mysqlbinlog工具只输出GTID为1的事件的命令行示例如下。

```
shell $> mysqlbinlog --no-defaults --base64-output=decode-row -vv mysql-bin.000001
--include-gtids='22228e8c-b0ee-11ec-a2d1-00163e23e2cc:1'
# The proper term is pseudo_replica_mode, but we use this compatibility alias
# to make the statement usable on server versions 8.0.24 and older.
...
SET @@SESSION.GTID_NEXT= '22228e8c-b0ee-11ec-a2d1-00163e23e2cc:1'/*!*/;
...
### INSERT INTO `book`.`t`
### SET
###   @1=1 /* INT meta=0 nullable=1 is_null=0 */
# at 449
# 230915 16:34:44 server id 129  end_log_pos 480 CRC32 0x7cd4e85d  Xid = 427
COMMIT/*!*/;
SET @@SESSION.GTID_NEXT= 'AUTOMATIC' /* added by mysqlbinlog */ /*!*/;
DELIMITER ;
# End of log file
/*!50003 SET COMPLETION_TYPE=@OLD_COMPLETION_TYPE*/;
/*!50530 SET @@SESSION.PSEUDO_SLAVE_MODE=0*/;
```

使用mysqlbinlog工具排除GTID为1和2的事件的命令行示例如下。

```
shell $> mysqlbinlog --no-defaults --base64-output=decode-row -vv mysql-bin.000001
--exclude-gtids='22228e8c-b0ee-11ec-a2d1-00163e23e2cc:1-2'
# The proper term is pseudo_replica_mode, but we use this compatibility alias
# to make the statement usable on server versions 8.0.24 and older.
...
# at 126
# 230915 16:34:35 server id 129  end_log_pos 157 CRC32 0xcc3ad5b3
    Previous-GTIDs
# [empty]
# at 157
```

```
# at 236
# at 311
# at 362
# at 409
# at 449
# at 480
# at 559
# at 634
# at 685
# at 732
# at 772
# at 803
# 230915 17:37:25 server id 129   end_log_pos 882 CRC32 0x6bbc6116     GTID
   last_committed=2 sequence_number=3  rbr_only=yes
   original_committed_timestamp=1694770645114121
   immediate_commit_timestamp=1694770645114121  transaction_length=323
/*!50718 SET TRANSACTION ISOLATION LEVEL READ COMMITTED*//*!*/;
# original_commit_timestamp=1694770645114121 (2023-09-15 17:37:25.114121 CST)
# immediate_commit_timestamp=1694770645114121 (2023-09-15 17:37:25.114121 CST)
/*!80001 SET @@session.original_commit_timestamp=1694770645114121*//*!*/;
/*!80014 SET @@session.original_server_version=80034*//*!*/;
/*!80014 SET @@session.immediate_server_version=80034*//*!*/;
SET @@SESSION.GTID_NEXT= '22228e8c-b0ee-11ec-a2d1-00163e23e2cc:3'/*!*/;
...
```

使用 mysqlbinlog 工具时，通过指定参数--rewrite-db=from_name->to_name，可以重写数据库名。binlog 日志中出现的所有数据库对应的名称都会被重写为指定的新的库名。使用 mysqlbinlog 工具重写指定库名的命令行的示例如下。

```
shell $> mysqlbinlog --no-defaults --base64-output=decode-row -v -v -f --rewrite-db=
book->test mysql-bin.000001

# 部分输出内容:book 改成 test
# 230915 17:37:25 server id 129   end_log_pos 1095 CRC32 0x2130efbd    Write_rows:
table id 213 flags: STMT_END_F
### INSERT INTO `test`.`t`
### SET
###   @1=3 /* INT meta=0 nullable=1 is_null=0 */
```

说明： mysqlbinlog 工具重写指定库名的参数在多源复制场景中经常会用到，使用 binlog 方式把数据恢复到另一个数据库时，也会使用该参数。

使用 mysqlbinlog 工具时，通过指定参数--force-read 或-f，可以跳过无法识别的 binlog 事件，并继续读取其他事件。如果没有这个选项，当遇到无法识别的事件时，mysqlbinlog 会停止。

使用 mysqlbinlog 工具跳过无法识别事件的命令行示例如下。

```
shell $> mysqlbinlog --no-defaults --base64-output=decode-row -v -v -f
mysql-bin.000001
```

说明： 偶尔会碰上数据库故障导致 binlog 写入不完整的情况，这时，可以设置参数跳过日志。mysqlbinlog 工具可以从远程服务器读取 binlog 文件。需要指定下列参数。

- -R 选项指示 mysqlbinlog 命令从远程服务器读取日志文件。与--read-from-remote-server 作用相同。
- -h 指定远程服务器的 IP 地址。
- -p 将提示输入密码。默认情况下,它将使用 root 作为用户名。也可以使用 -u 选项指定用户名。
- -P 指定 MySQL 服务端口。

使用 mysqlbinlog 工具从远程服务器解析 binlog 的命令行示例如下。

```
shell $> mysqlbinlog --no-defaults -R -h 192.168.244.130 -uroot -p ****** -P 3410
--base64-output=decode-row -vv mysql-bin.000001
```

6.3.2 利用 setup_actors 命令进行资源使用统计

数据库中 SQL 语句的资源使用统计至关重要,因为它不仅是极具挑战性的部分,也是有效定位问题的方式。此外,这些资源使用统计信息还能协助 DBA 评估业务对数据的负载、确定硬件资源是否满足现有需求,以及检查参数配置是否合理。

尽管 MySQL 目前还没有像 Oracle 的 AWR 报告那样全面且细致的资源使用报告机制,但它提供了 PROFILE 命令行工具,这在统计资源使用方面提供了很大的帮助。然而,随着 MySQL 8.0 版本的广泛使用,官方已表示 PROFILE 语句将被弃用,并推荐使用 performance_schema.setup_actors 作为替代。这个新工具能够更精细地监控与统计数据库的性能和资源使用情况。

表 6-2 所示为 MySQL 中已被弃用的 PROFILE 功能显示的当前会话过程中执行的语句的资源使用统计指标类型列表。通过这些指标,可以很好地了解 SQL 语句消耗的资源情况。

表 6-2　PROFILE 统计指标

指 标	说 明
ALL	显示所有性能信息
BLOCK IO	显示块 I/O 操作的次数
CONTEXT SWITCHES	显示上下文切换次数,不管是主动还是被动
CPU	显示用户 CPU 时间、系统 CPU 时间
IPC	显示发送和接收的消息数量
MEMORY	当前没有实现
PAGE FAULTS	显示页错误数量
SOURCE	显示源码中的函数名称与位置
SWAPS	显示 SWAP 的次数

下面将介绍与 setup_actors 相关的统计项。

官方介绍指出,performance_schema 下的 setup_actors 表可用于限制按主机、用户或账户收集历史事件,以减少运行时开销和历史表中收集的数据量。这意味着通过用户级别的设置,可以自动对每条 SQL 语句统计资源使用情况。这种方式相较于原先的 PROFILE 功能更为全面。默认情况下,setup_actors 被配置为允许对所有前台线程进行监视,并收集历史事件的部分信息。如果需要收集所有信息,就需要开启收集器(instruments)指标。

通过正确配置 setup_actors,可以限制收集数据的范围,从而减少对系统性能的影响,并只收集真正关心的用户或主机的数据。

1. 用户指标设置

使用 setup_actors 表指定收集哪些操作用户的信息。用户指标设置的命令行示例如下。

```
# 查看收集用户信息
mysql> SELECT * FROM performance_schema.setup_actors;
+------+------+------+---------+---------+
| HOST | USER | ROLE | ENABLED | HISTORY |
+------+------+------+---------+---------+
| %    | %    | %    | YES     | YES     |
+------+------+------+---------+---------+
1 row in set (0.00 sec)

# 关闭所有用户收集信息
mysql> UPDATE performance_schema.setup_actors
    SET ENABLED = 'NO', HISTORY = 'NO'
    WHERE HOST = '%' AND USER = '%';
# 仅对 root@local 用户收集信息
mysql> INSERT INTO performance_schema.setup_actors
    (HOST,USER,ROLE,ENABLED,HISTORY)
    VALUES('localhost','root','%','YES','YES');
# 确认配置是否正常
mysql> SELECT * FROM performance_schema.setup_actors;
+-----------+------+------+---------+---------+
| HOST      | USER | ROLE | ENABLED | HISTORY |
+-----------+------+------+---------+---------+
| %         | %    | %    | NO      | NO      |
| localhost | root | %    | YES     | YES     |
+-----------+------+------+---------+---------+
2 rows in set (0.00 sec)
```

2. 监控指标设置

在监控指标设置方面,目前 instruments 提供了 1264 个指标。这些指标还在不断完善中。目前,默认已开启包括等待(wait)、执行阶段(stage)的 SQL 语句、内存(memory)等在内的 792 个指标的监控,涵盖了日常使用的 processlist、innodb status 等信息。查看 setup_instruments 表里配置的可用的指标的命令行示例如下。

```
mysql> SELECT NAME,ENABLED,TIMED FRO performance_schema.setup_instruments
    WHERE ENABLED='YES' limit 10;
+--------------------------------+---------+-------+
| NAME                           | ENABLED | TIMED |
+--------------------------------+---------+-------+
| wait/io/file/sql/binlog        | YES     | YES   |
| wait/io/file/sql/binlog_cache  | YES     | YES   |
| wait/io/file/sql/binlog_index  | YES     | YES   |
| statement/sql/kill             | YES     | YES   |
| statement/sql/analyze          | YES     | YES   |
| memory/sql/THD::sp_cache       | YES     | YES   |
| memory/sql/test_quick_select   | YES     | YES   |
| error                          | YES     | YES   |
```

```
...
+--------------------------------------------------+------------+----------+
792 rows in set (0.00 sec)
```

想要确保所有监控指标均被启用，可通过更新 setup_instruments 表来实现。由于某些指标可能已经默认启用，且当前没有详细的对应关系，因此将选择开启所有指标。setup_instruments 表指标开启的命令行示例如下。

```
# 开启对应类型监控指标
mysql> UPDATE performance_schema.setup_instruments
    SET ENABLED = 'YES', TIMED = 'YES'
    WHERE NAME LIKE '%statement/%';

mysql> UPDATE performance_schema.setup_instruments
    SET ENABLED = 'YES', TIMED = 'YES'
    WHERE NAME LIKE '%stage/%';
mysql> UPDATE performance_schema.setup_consumers
    SET ENABLED = 'YES'
    WHERE NAME LIKE '%events_statements_%';

mysql> UPDATE performance_schema.setup_consumers
    SET ENABLED = 'YES'
    WHERE NAME LIKE '%events_stages_%';
```

3. 查看资源使用统计信息

设置完统计用户和指标，通过 performance_schema 下的 events_statements_history_long 表查看对应的 SQL 语句资源使用情况。查看资源使用统计信息的命令行示例如下。

```
# 查询执行过的 SQL 事件 ID
mysql> SELECT EVENT_ID, TRUNCATE(TIMER_WAIT/1000000000000,6) as Duration, SQL_TEXT
       FROM performance_schema.events_statements_history_long
WHERE SQL_TEXT like '%t1%';
+----------+----------+-------------------------------------------------------+
| EVENT_ID | Duration | SQL_TEXT                                              |
+----------+----------+-------------------------------------------------------+
|    48    |  0.0065  | select * from db1.t1                                  |
|    84    |  0.0004  | select * from db1.t1                                  |
|    41    |  0.0082  | SELECT /*!40001 SQL_NO_CACHE */ * FROM `test1`        |
+----------+----------+-------------------------------------------------------+
# 再通过事件 ID 进行查询,目前只有耗时
mysql> SELECT event_name AS Stage, TRUNCATE(TIMER_WAIT/1000000000000,6) AS Duration
       FROM performance_schema.events_stages_history_long
WHERE NESTING_EVENT_ID=41;
+-------------------------------------+----------+
| Stage                               | Duration |
+-------------------------------------+----------+
| stage/sql/starting                  |  0.0001  |
| stage/sql/checking permissions      |  0.0000  |
| stage/sql/Opening tables            |  0.0000  |
```

```
| stage/sql/init              | 0.0000 |
| stage/sql/System lock       | 0.0000 |
| stage/sql/optimizing        | 0.0000 |
| stage/sql/statistics        | 0.0000 |
| stage/sql/preparing         | 0.0000 |
+-----------------------------+--------+
15 rows in set (0.00 sec)
```

说明：events_stages_history_long 表包含 N 个最近的跨所有线程全局结束的阶段事件记录。这些活动事件在结束后才会被添加到表中。当表已满时，新添加的行会替换最老的行，不论这一行是由哪个线程生成的。此外，events_stages_history_long 表仅支持使用 TRUNCATE TABLE 命令来删除所有行。

表 6-3 所示为 events_stages_history_long 表中字段及其说明。

表 6-3 events_stages_history_long 表中字段及其说明

字 段	说 明
THREAD_ID、EVENT_ID	开始时线程 ID 和事件 ID
END_EVENT_ID	该字段在事件开始时设置为 NULL，并在事件结束时更新为线程当前事件号
EVENT_NAME	产生事件的仪器的名称。这是 setup_instruments 表中的 NAME 值
SOURCE	源文件的名称，其中包含生成事件的经过检测的代码，以及发生检测的文件中的行号
TIMER_START、TIMER_END 和 TIMER_WAIT	TIMER_START 与 TIMER_END 值分别表示事件计时开始和结束的时间。TIMER_WAIT 是事件经过的时间（持续时间）
WORK_COMPLETED、WORK_ESTIMATED	WORK_COMPLETED 表示该阶段已经完成了多少工作单元，WORK_ESTIMATED 表示该阶段预计有多少工作单元
NESTING_EVENT_ID	嵌套该事件的事件的 EVENT_ID 值。事件的嵌套事件通常是语句事件
NESTING_EVENT_TYPE	嵌套事件类型。取值为 TRANSACTION、STATEMENT、STAGE 或 WAIT

除此之外，events_stages_history_long 表的记录行数受到参数 performance_schema_events_stages_history_long_size 的限制，默认只能记录 1000 条 SQL 语句，并且建议将这个限制增加到 10000 条。同时，这个参数是一个静态变量，需要修改配置文件，重新启动服务来生效。查看该参数的命令行示例如下。

```
mysql> SHOW VARIABLES LIKE '%events_stages_history_long%';
+-------------------------------------------------+-------+
| Variable_name                                   | Value |
+-------------------------------------------------+-------+
| performance_schema_events_stages_history_long_size | 10000 |
+-------------------------------------------------+-------+
```

6.3.3 数据库备份和恢复实践

MySQL 中的数据备份和恢复是数据库管理中的重要组成部分。它们对于确保数据的完整性、可用性和安全性起着至关重要的作用。

1. 数据库备份和恢复介绍

备份是指创建数据库、表或数据的副本，以便在数据丢失或损坏时能够恢复。恢复则是将备份的数据重新应用到数据库中以恢复其状态的过程。

数据库备份和恢复使用场景如下。

- 误删除操作：例如，误删除了一个数据库或表，在这种情况下，可以从备份中恢复这些对象。
- 数据库迁移或升级过程中出现问题：在数据库迁移或升级过程中，如果因发生错误而导致数据库无法启动，则可以使用备份进行恢复。
- 历史数据需求：如果需要回溯到过去某个时间点的数据，则备份可以提供这样的能力。
- 硬件故障：如果数据库服务器硬件出现故障，那么可以从历史备份中恢复数据，并在新硬件上重建数据库。

数据库备份的主要目的是在发生灾难性事件（如硬件故障、软件错误、自然灾害等）时，能够迅速、准确地恢复数据和业务，从而最小化数据丢失和业务中断的影响。备份的具体建议如下。

- 建设统一的备份服务器。部署一个集中式备份服务器，负责收集、存储和管理所有数据库实例的备份数据。
- 备份异常处理机制。设立备份任务监控和告警系统，实时监测备份任务的执行状态，并在备份失败或异常时及时通知管理员。备份失败后的自动恢复或重试机制，确保下一次备份能够正常进行。
- 逻辑、物理和镜像结合备份方式。根据不同的业务需求和场景，选择合适的备份方式。例如，对于关键业务数据，可以采用逻辑备份和物理备份相结合的方式，既保留可读的备份文件，又保留可直接恢复的物理文件。

如果数据库备份没有恢复的能力，那么它就失去了意义。因此，进行数据恢复操作是至关重要的。对于数据恢复，提出以下建议。

- 建设一个统一的恢复验证服务器，用于定期验证备份的有效性。
- 通过定期的恢复演练，可以确保备份数据在实际需要时能够成功恢复。
- 由于不可能对所有备份都进行实际还原的校验，因此建议采用文件 MD5 对比方式进行每日的基础校验，以确保备份数据的完整性。

2. 数据库备份工具的选择

在 MySQL 的官方和开源生态系统中，有一些常用的备份工具，它们可以根据备份的方式（逻辑或物理）进行分类（见图 6-14）。常用的工具如下。

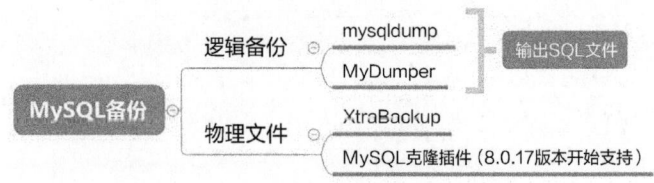

图 6-14　MySQL 常用的备份工具列表

- 逻辑导出：mysqldump、MyDumper。
- 物理导出：XtraBackup、MySQL 克隆插件。

mysqldump 是 MySQL 官方自带的逻辑备份工具。其工作原理是连接到 MySQL 服务，使用 SQL 语句导出数据库的结构和数据。其特点是简单易用，但可能对于大型数据库来说，其备份速度较慢，因为它依赖 SQL 语句的生成和执行。

MyDumper 是一个多线程的高性能逻辑备份工具。其工作原理类似于 mysqldump，但它使用多线程技术来加速备份过程。其适用于大型数据库的逻辑备份，能够快速完成备份任务。

XtraBackup 是一个由 Percona 公司开发的开源物理备份工具。其工作原理是直接复制数据库文件（通常是数据文件、索引文件和日志文件），而不是导出 SQL 语句。其通常比逻辑备份更快，因为它避免了 SQL 语句的生成和解析。但是，物理备份通常需要在相同或兼容的硬件和操作系统上恢复。

MySQL 克隆插件是 MySQL 8.0 版本中提供的物理全备份插件。其工作原理是直接复制数据库的物理文件，而不是导出 SQL 语句。它支持本地和远程克隆操作，可以用于快速创建数据库的副本。

上述提到的备份工具都有其适用场景和特点。下面将根据适用场景和特点进行简要概述。

- 数据量少于 10GB，使用 mysqldump、MyDumper 进行逻辑备份，否则（大于等于 10GB）建议使用 XtraBackup。
- 对于单表备份、表结构导出，使用逻辑备份工具 mysqldump 和 MyDumper。
- mysqldump 是单线程的，MyDumper 是多线程的。在逻辑导出场景中，单线程意味着在处理大型数据库时可能会比较慢。相比之下，使用多线程来加速备份过程，从而提高性能。
- 基于备份效率和性能表现选择备份工具的顺序是 XtraBackup、MyDumper、mysqldump。
- MySQL 8.0 版本的克隆功能是一个非常出色的功能，它需要脚本进行配合，并且在某些方面比 XtraBackup 更加出色。然而，由于它是以插件的形式存在的，因此会在 MySQL 服务中增加一定的负载。目前，业界对于这一功能的使用经验较少，因此还需要进一步完善和优化。

3. 数据库备份策略

对于不同的数据量，可以采取不同的备份策略，如结合全量备份（全备）和增量备份的方式。binlog 也需要定期进行备份。以下是常用备份策略。

如图 6-15 所示，对于少量的数据（数据量小于 300GB），采取全量+binlog 备份策略。全量备份每天进行一次，而 binlog 备份每小时进行一次。备份数据保留 3 个月，超过 3 个月的备份数据可以进行归档或删除，以释放存储空间。

业务名	数据库	周期	备份时间	备份	本地备份	本地保留	远程IP	远程备份目录	异机异地保留
运营管理数据库	Mysql	每天	4:01	全备	/dbbackup/YYYYMMDD/full/	1	127.0.0.1	/dbbackup/YYYYMMD/full/	3个月
	Mysql	每小时	整点:01	日志	/dbbackup/YYYYMMDD/log/	12	127.0.0.1	/dbbackup/YYYYMMD/log/	3个月

图 6-15 MySQL 少量数据的备份策略

如图 6-16 所示，对于大量的数据（数据量大于 300GB），采取全量+增量+binlog 备份策略。全量备份每周进行一次，在全量备份之后，增量备份每天进行一次（全量备份当日不进行增量备份），而 binlog 备份每小时进行一次。备份数据保留 3 个月，超过 3 个月的备份数据可以进行归档或删除，以释放存储空间。

业务名	数据库	周期	备份时间	备份	本地备份	本地保留	远程IP	远程备份目录	异机异地保留
业务数据库	Mysql	星期日	1:31	全备	/dbbackup/YYYYMMDD/full/	1	127.0.0.1	/dbbackup/YYYYMMDD/full/	3个月
	Mysql	星期一	3:01	增量	/dbbackup/YYYYMMDD/increm/	1	127.0.0.1	/dbbackup/YYYYMMDD/increm/	3个月
	Mysql	星期二	3:01	增量	/dbbackup/YYYYMMDD/increm/	1	127.0.0.1	/dbbackup/YYYYMMDD/increm/	3个月
	Mysql	星期三	3:01	增量	/dbbackup/YYYYMMDD/increm/	1	127.0.0.1	/dbbackup/YYYYMMDD/increm/	3个月
	Mysql	星期四	3:01	增量	/dbbackup/YYYYMMDD/increm/	1	127.0.0.1	/dbbackup/YYYYMMDD/increm/	3个月
	Mysql	星期五	3:01	增量	/dbbackup/YYYYMMDD/increm/	1	127.0.0.1	/dbbackup/YYYYMMDD/increm/	3个月
	Mysql	星期六	3:01	增量	/dbbackup/YYYYMMDD/increm/	1	127.0.0.1	/dbbackup/YYYYMMDD/increm/	3个月
	Mysql	每小时	整点:01	日志	/dbbackup/YYYYMMDD/log/	12	127.0.0.1	/dbbackup/YYYYMMDD/log/	3个月

图 6-16　MySQL 大量数据的备份策略

4. 数据库备份最佳实践

MySQL 的备份最佳实践涉及多个方面，包括备份频率、备份类型、备份存储和恢复策略。下面是结合图 6-17 给出的 MySQL 备份最佳实践的建议。

- 使用 XtraBackup 进行物理备份。
- 使用 MyDumper 进行逻辑备份，支持并行逻辑备份恢复。
- 备份文件可以存储在本地或者 NFS 上。
- 使用闪回工具解析 binlog 文件以进行闪回恢复。
- 对于重要系统，可以利用延迟从库进行备份和恢复。
- 在条件允许的情况下，每天进行系统级别的镜像备份，并随后进行去重处理。

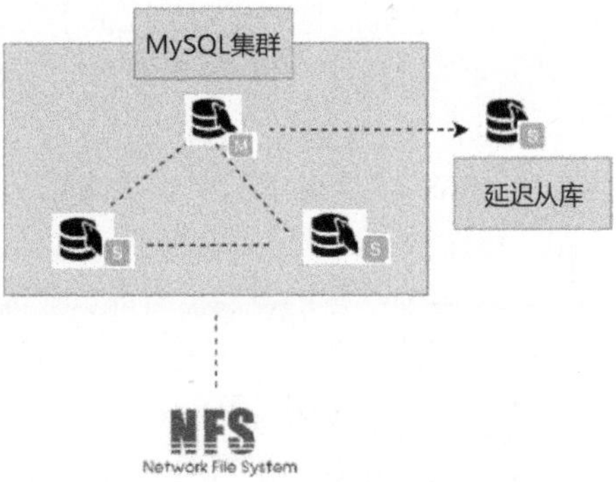

图 6-17　MySQL 备份最佳实践

5. 备份恢复验证

备份恢复验证则是对备份数据进行检查和测试，确保在需要时能够成功恢复数据，并且恢复后的数据库在功能和数据完整性上都没有问题。

备份恢复验证内容如下。

1）备份的文件大小是否存在很大差异。
2）备份文件里是否存在一致性验证内容，全局事务编号和 binlog 文件具体位置。
3）备份日志是否最终输出完成的语句。
4）在验证环境中，备份文件恢复后，以下检查项有异常。

- 与生产环境对比统计数据库对象的数量和行数。
- 运行 CHECK TABLE 命令来检查表的完整性和健康状态。
- 对比数据。通过比较两个数据库的特定表的数据来实现，或者使用更复杂的工具来比较整个数据库的结构和数据。
- 对于应用程序使用的数据库，恢复后应该在测试环境中运行应用程序的测试用例，以确保恢复后的数据库能正常工作，并且所有功能都没有问题。

通过这些步骤，可以确保 MySQL 数据库的备份恢复过程得到了充分的验证，从而提高了数据的安全性和可恢复性。

6. 数据库备份过程中常见问题

MySQL 备份涉及多个方面，其中出现的一些问题可能导致备份失败或数据不一致。常见的备份问题如下。

- Redo 日志过小，导致每秒检查失败。
- 在备份期间，执行 DDL 语句，导致备份失败。
- 备份过程中的网络不稳定，导致备份失败。
- 磁盘空间不足，导致备份失败。
- 不合理的参数配置，导致备份失败或数据不一致。
- 硬件设备的故障（如磁盘故障、内存故障等）导致备份失败或数据损坏。
- 备份软件本身存在缺陷或兼容性问题，导致备份失败或数据不一致。

下面介绍备份过程中的常见问题。

【例 6-8】 XtraBackup 物理备份的 GTID 不一致，在 MySQL 5.7 版本和 MySQL 8.0 版本中都会存在这种情况。

在备份过程中发现 xtrabackup_binlog_info 文件中记录的 GTID 与备份文件中记录的 GTID 不一致，这可能会导致复制问题或数据不一致。

如图 6-18 所示的备份示例显示，已执行过的 338fa273 的开始 GTID 为 1-9969，但 xtrabackup_binlog_info 文件中记录的 338fa273 的开始 GTID 却是 1-473，同时备份文件中也显示 GTID 为 1-473。

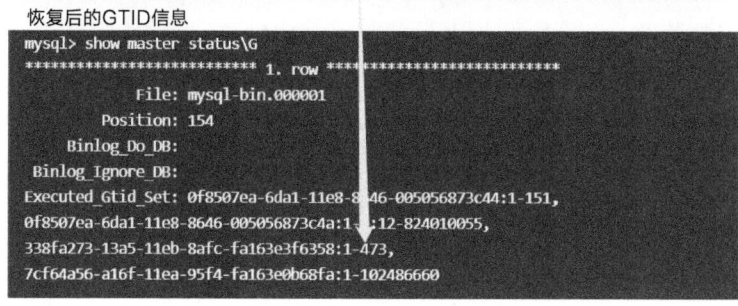

图 6-18　MySQL 备份的 GTID 不一致问题

分析问题：xtrabackup_binlog_info 文件中的 GTID 信息可以通过全局系统变量 gtid_executed 或 mysql.gtid_executed 表来获取。在 MySQL 5.7 和 MySQL 8.0 版本中，由于不同的机制或配置，备份和恢复时可能会出现 GTID 不一致的情况。

解决方案：在执行备份前，可以执行 FLUSH LOGS 命令来确保新的二进制日志文件已被创建。之后，通过 SHOW MASTER STATUS 查看当前二进制日志的位置，并与 mysql.gtid_executed 表中记录的 GTID 进行比较，以验证它们是否一致。

【例 6-9】 慢 SQL 语句导致备份失败。

在备份过程中，如果遇到了执行缓慢的 SQL 语句，则可能会导致备份进程被阻塞，从而引发备份失败。XtraBackup 提供了一些参数来控制和处理这种情况。

- 阻塞语句进行 KILL 操作。XtraBackup 开始执行 FLUSH TABLES WITH READ LOCK（FTWRL）时，如果某些 SQL 查询语句执行得特别慢，备份进程会被这些查询阻塞。为了解决这个问题，XtraBackup 提供了 kill-long-query-type（指定哪些类型的慢查询应该被终止）和 kill-long-queries-timeout（设置一个超时时间，超过这个时间的查询将被视为慢查询并被终止）参数。
- 等待长时间执行的 SQL 语句完成，之后进行备份。在等待过程中，可以使用参数 ftwrl-wait-query-type（获取 FTWRL 之前需要等待的查询类型）、ftwrl-wait-timeout（获取 FTWRL 之前等待的超时时间）和 ftwrl-wait-threshold（在执行 FTWRL 之前，等待正在运行 SQL 的时间）来控制 XtraBackup 的行为。
- no-lock 设置：表示关闭 FTWRL 的表锁。这样做可能会增加备份期间数据不一致的风险，因为其他事务可能仍在修改数据。

【例 6-10】DDL 语句导致备份失败。

在数据库备份过程中，若遇到 DDL 语句，如 ALTER TABLE、CREATE INDEX 等，有时会导致备份失败。这是因为 DDL 语句会修改表结构，而备份工具（如 XtraBackup）在备份过程中通常期望数据库保持静态或只读状态。DDL 操作会违反这一假设，导致备份失败。备份失败的提示信息如下所示。

```
InnoDB: Last flushed lsn: 3375345258517 load_index lsn 3379255303757
InnoDB: An optimized (without redo logging) DDLoperation has been performed.All modified pa-
ges may not have been flushed to the disk yet.
PXB will not be able take a consistent backup.Retry the backup operation.
```

分析问题：XtraBackup 在备份 InnoDB 数据时有两种线程：Redo 拷贝线程和 IBD 数据拷贝线程。XtraBackup 进程开始执行后，首先启动 Redo 拷贝线程，从最新的 checkpoint 点开始顺序拷贝 redo.log，随后启动 IBD 数据拷贝线程，进行 IBD 数据的拷贝。如果刷新 Redo 丢失，备份就会失败。当刷新大量数据或 Redo 刷新跟不上时，可能导致刷新 Redo 丢失，最终备份会失败。由提示信息可知，DDL 操作上次刷新的 lsn：3375345258517 已经丢失。

解决方案：避免备份期间执行 DDL 操作。

【例 6-11】大量事务持续刷新日志，同时 I/O 性能表现不佳，导致备份失败。

当 InnoDB 产生日志的速度远高于 XtraBackup 复制的速度时，会出现 InnoDB 日志被截断的情况，导致备份失败。备份失败的提示信息如下所示。

```
XtraBackup: error: log block numbers mismatch:
XtraBackup: error: expected log block no.201901064, but got no.208192508 from the log file.
XtraBackup: error: it looks like InnoDB log has wrapped around before XtraBackup could process
 all records due to either log copying being too slow, or log files being too small.
XtraBackup: Error: XtraBackup_copy_logfile()failed.
```

分析问题：经检查发现，预期的 Redo 日志块编号为 201901064，但从日志文件中实际读取的编号却为 208192508。这种不匹配表明在日志复制过程中出现了问题。可能的原因包括 Redo 日志复制速度过慢，导致备份工具未能及时捕获所有必要的日志条目，或者 Redo 日志文件大小配置不当，导致日志文件在日志条目生成速度超过其处理速度时被截断。最终导致备份失败。

解决方案：为了解决 InnoDB 日志复制速度不足和 I/O 性能问题，可以考虑增加 Redo 日志文件的大小和数量，以提高日志处理能力。此外，也可以选择在负载低峰时段进行备份，以减少备份过程中数据库的压力。

【例 6-12】备份空间不足,导致备份失败。

备份空间不足,即使备份进程已经开始,也可能因为无法存储所有的备份数据而导致备份失败。备份空间不足的提示信息如下所示。

```
XtraBackup: Error writing file '/tmp/xbtempevLQbf' (Errcode: 28 - No space left on device)
XtraBackup: Error: write to logfile failed
XtraBackup: Error: XtraBackup_copy_logfile() failed.
```

分析问题:查看备份日志发现,在尝试写入文件"/tmp/xbtempevLQbf"时发生了错误,因为设备上没有剩余空间,因此备份无法继续进行。

解决方案:清理不必要的文件,增加备份存储空间。

【例 6-13】备份软件的处理机制不太友好,导致备份失败。

备份软件的处理机制不够友好时,可能会引发备份失败。这可能是由于备份软件在处理大量事务或高并发场景时性能不足,或者软件本身的缺陷、配置错误、资源限制等原因导致的。备份日志里已经报出错误,但 XtraBackup 线程一直存在。提示信息如下所示。

```
failed to execute query SET SESSION lock_wait_timeout=31536000,MySQL server has gone away.
```

分析问题:查看备份日志发现,执行查询 SET SESSION lock_wait_timeout = 31536000 失败,MySQL 服务器已退出。gone away 错误的常见原因如下。

- MySQL 连接超时(受参数 wait_timeout 和 interactive_timeout 的控制)。
- MySQL 连接被 KILL。
- MySQL 实例重启。

解决方案:通过优化配置、升级版本、调整备份策略或使用其他备份工具来解决这个问题。

【例 6-14】使用 NFS 挂载时,备份的底层物理文件不一致问题。

使用 NFS(Network File System)进行 MySQL 备份时,特别是在使用像 XtraBackup 这样的工具时,可能会遇到数据一致性方面的问题,这主要是 NFS 客户端缓存机制造成的。

NFS 客户端会缓存从 NFS 服务器读取的数据,以提高性能。当数据库服务器从 NFS 挂载点读取数据时,它可能会从本地缓存中读取,而不是从 NFS 服务器中读取。同样,当数据库服务器向 NFS 写入数据时,这些数据可能首先被写入本地缓存,而不是直接发送到 NFS 服务器。这会导致物理备份文件可能包含了尚未刷新到 NFS 服务器的数据,而这些数据可能仍然只存在于数据库服务器的本地缓存中。

解决方案:禁用 NFS 客户端缓存或使用同步备份(确保在备份过程中,数据库服务器的所有写操作都同步到 NFS 服务器)。具体说明可以参考 Percona 官方说明。

6.3.4 数据库热数据加载设置

MySQL 服务在关闭和重新启动时耗时特别长,导致长时间等待。特别是在 innodb_buffer_pool_size 设置值较大时,这个现象会更加明显。这是由于 InnoDB 缓冲池在内存中的数据量较大,导致在关闭时需要更长的时间来将这些数据写回磁盘,而在启动时则需要更长的时间来从磁盘加载这些数据。

下面介绍 MySQL 服务在启动和关闭时数据落盘与加载的设定,从而评估和选择合适的热数据加载选项。

1. 热数据加载参数

在 MySQL 5.6 版本中，InnoDB 存储引擎引入了一些优化，其中包括缓存池持久化功能，它旨在减少 MySQL 服务在重启时加载 InnoDB 缓冲池的时间。通过缓冲池持久化，InnoDB 能够在数据库服务器关闭时，将缓冲池中的热数据（经常被访问的数据）写入磁盘，并在服务器重启时快速加载这些数据。缓冲池持久化和加载参数的查看的命令行示例如下。

```
mysql>SHOW VARIABLES WHERE variable_name LIKE 'innodb_buffer_pool_dump%' OR variable_name LIKE 'innodb_buffer_pool_load%';
+-------------------------------------------+-------+
| Variable_name                             | Value |
+-------------------------------------------+-------+
| innodb_buffer_pool_dump_at_shutdown       | ON    |
| innodb_buffer_pool_dump_now               | OFF   |
| innodb_buffer_pool_dump_pct               | 25    |
| innodb_buffer_pool_load_abort             | OFF   |
| innodb_buffer_pool_load_at_startup        | ON    |
| innodb_buffer_pool_load_now               | OFF   |
+-------------------------------------------+-------+
6 rows in set (0.01 sec)
```

表 6-4 所示为缓冲池持久化和加载参数及其说明。

表 6-4　缓冲池持久化和加载参数及其说明

参　　数	说　　明
innodb_buffer_pool_dump_at_shutdown	在关闭时把热数据持久化到本地磁盘
innodb_buffer_pool_dump_now	采用手动方式把热数据持久化到本地磁盘
innodb_buffer_pool_dump_pct	指定每个缓冲池最近使用的页读取和转储的百分比，范围是 1~100，默认值是 25。例如，有 4 个缓冲池，每个缓冲池有 100 个页，并且 innodb_buffer_pool_dump_pct 设置为 25，则持久化每个缓冲池中最近使用的 25 个页
innodb_buffer_pool_load_abort	是否要中止缓冲池加载操作，默认是关闭
innodb_buffer_pool_load_at_startup	在启动时把热数据加载到内存
innodb_buffer_pool_load_now	采用手动方式把热数据加载到内存

MySQL 缓冲池持久化和加载的配置参数，决定了 MySQL 服务启动和关闭有多快，也决定了持久化多少数据和索引。然而，当数据库服务器重启时，不加载持久化的数据，就需要从磁盘重新加载数据和索引。

2. 保存缓冲区文件

在关闭 MySQL 服务时，InnoDB 存储引擎会将内存中的热数据持久化到磁盘上。这些数据通常保存在 ib_buffer_pool 文件中。这个文件位于由 innodb_data_home_dir 指定的数据目录下。查看缓冲池持久化物理文件如下所示。

```
shell $> ll
total 278536
-rw-r--r-- 1 mysql mysql        4335 Sep 19  2023 ib_buffer_pool
```

如果 MySQL 服务因为"死"机或通过 pkill -9 命令强制终止,那么热数据不会被持久化,因为这种情况下没有机会进行有序的关闭和清理过程。

3. 热数据加载调整

在 MySQL 服务启动后,InnoDB 存储引擎会自动尝试将热数据加载到缓冲池中以提高性能。可以使用 SHOW STATUS 命令行查看缓冲池持久化文件的加载情况。查看缓冲池持久化文件加载情况的命令行示例如下。

```
mysql> SHOW STATUS LIKE 'Innodb_buffer_pool_load_status';
+--------------------------------+-------------------------------------------------+
|Variable_name                   |Value                                            |
+--------------------------------+-------------------------------------------------+
|Innodb_buffer_pool_load_status  |Buffer pool(s) load completed at 200511 22:1102  |
+--------------------------------+-------------------------------------------------+
1 row in set (0.00 sec)
```

想要禁止 MySQL 服务在启动时将持久化文件加载到缓冲池,可以将 innodb_buffer_pool_load_at_startup 参数设置为 0。这将告诉 MySQL 在服务启动时忽略 ib_buffer_pool 文件(如果存在的话),即不加载其内容到缓冲池中。禁止加载缓冲池持久化文件的方式和命令行示例如下。

```
# 配置 my.cnf 文件后启动 MySQL 服务
[mysqld]
innodb_buffer_pool_dump_at_shutdown=1
innodb_buffer_pool_load_at_startup=0

# 查看参数
mysql> SHOW variables WHERE variable_name like 'innodb_buffer_pool_dump%'
    or variable_name like 'innodb_buffer_pool_load%';
+---------------------------------------+-------+
|Variable_name                          |Value  |
+---------------------------------------+-------+
|innodb_buffer_pool_dump_at_shutdown    |ON     |
|innodb_buffer_pool_dump_now            |OFF    |
|innodb_buffer_pool_dump_pct            |25     |
|innodb_buffer_pool_load_abort          |OFF    |
|innodb_buffer_pool_load_at_startup     |OFF    |
|innodb_buffer_pool_load_now            |OFF    |
+---------------------------------------+-------+
6 rows in set (0.01 sec)

# 查看加载情况
SHOW STATUS LIKE 'Innodb_buffer_pool_load_status';
+--------------------------------+-----------------------------------+
|Variable_name                   |Value                              |
+--------------------------------+-----------------------------------+
|Innodb_buffer_pool_load_status  |Loading of buffer pool not started |
+--------------------------------+-----------------------------------+
1 row in set (0.00 sec)
```

说明:Value 列中的 Loading of buffer pool not started 提示没有进行加载。

如果在 MySQL 运行时将持久化文件动态地加载到缓冲池中，那么手动加载的命令行示例如下。

```
mysql> SET GLOBAL innodb_buffer_pool_load_now=1;

mysql> SHOW STATUS LIKE 'Innodb_buffer_pool_load_status';
+--------------------------------+-----------------------------------------------+
| Variable_name                  | Value                                         |
+--------------------------------+-----------------------------------------------+
| Innodb_buffer_pool_load_status | Buffer pool(s) load completed at 230828 11:41:55 |
+--------------------------------+-----------------------------------------------+
1 row in set (0.00 sec)
```

当 MySQL 服务重新启动时，通过配置设置，动态加载持久化文件到缓冲池，提高 MySQL 服务重启后的性能。

对于内存分配足够的 MySQL 环境，热数据的好处非常多，不仅不需要从 I/O 读取数据，还可以提升查询性能。反之，当 MySQL 服务关闭后重新启动时，需要花费比较长时间和占用过大的缓存空间，占据的磁盘空间也很大。

6.3.5 Query Rewrite 插件的使用

从 MySQL 5.7 版本开始，支持语句重写（Query Rewrite）插件，它可以在执行语句之前，在 MySQL 服务层检查并修改接收到的 SQL 语句。MySQL 8.0 版本提供了对多种 SQL（SELECT、INSERT、REPLACE、UPDATE 和 DELETE）语句的重写支持。无论是独立语句还是 PREPARED（准备）语句，都可能被重写。但需要注意的是，视图或存储过程中的语句不会被重写。

下面介绍 Query Rewrite 实现方式和使用方式。

1. 插件工作原理

Query Rewrite 插件通过解析 SQL 语句并遍历其解析树，基于内存中的重写规则缓存对语句进行重写，并将修改后的语句返回给 MySQL 服务层进行处理。虽然在实际应用中很少使用 Query Rewrite 插件，但它仍然具有一定的辅助作用，能够解决一些特殊场景下的需求，如错误的 SQL 语句导致全库查询、使用了第三方软件但 SQL 执行错误且无法直接修改应用等。

Query Rewrite 插件使用名为 query_rewrite 的数据库。这个数据库中包含一个名为 rewrite_rules 的表，它为插件提供了持久化存储，用于决定是否重写语句的规则。用户通过修改存储在这个表中的规则集来控制插件的行为。插件通过设置规则表的信息来与服务层交互。

名为 query_rewrite 的数据库中还包含一个名为 flush_rewrite_rules 的存储过程，该存储过程调用 load_rewrite_rules 函数来执行加载操作，将规则表的内容加载到插件中。

通过查看 Query Rewrite 插件的配置参数和运行状态值，可以有效地了解插件的工作状态。从 MySQL 8.0.31 版本开始，该插件还引入了一个名为 SKIP_QUERY_REWRITE 的特权，用以保护特定用户的查询不被重写。

2. 插件启用

Query Rewrite 插件的启用方式是通过安装位于软件 share 目录下的 install_rewriter.sql 脚本来进行初始化部署。这个脚本包含 5 个主要部分：重写配置库、重写配置表、加载重写插件、创建重写接口函数和创建重写语句处理的存储过程。install_rewriter.sql 脚本内容如下所示。

```sql
shell#> cat mysql8.0/share/install_rewriter.sql
/* Copyright (c) 2015, 2022, Oracle and/or its affiliates.

   This program is free software; you can redistribute it and/or modify
   it under the terms of the GNU General Public License, version 2.0,
as published by the Free Software Foundation.
   ...
   You should have received a copy of the GNU General Public License
   along with this program; if not, write to the Free Software
   Foundation, Inc., 51 Franklin St, Fifth Floor, Boston, MA 02110-1301  USA */

# 重写配置库
CREATE DATABASE IF NOT EXISTS query_rewrite;

# 重写配置表
CREATE TABLE IF NOT EXISTS query_rewrite.rewrite_rules (
  id INT NOT NULL AUTO_INCREMENT PRIMARY KEY,
  pattern VARCHAR(5000) CHARACTER SET utf8mb4 COLLATE utf8mb4_bin NOT NULL,
  pattern_database VARCHAR(20) CHARACTER SET utf8mb4 COLLATE utf8mb4_bin,
  replacement VARCHAR(5000) CHARACTER SET utf8mb4 COLLATE utf8mb4_bin NOT NULL,
  enabled ENUM('YES', 'NO') CHARACTER SET utf8mb4 COLLATE utf8mb4_bin NOT NULL
    DEFAULT 'YES',
  message VARCHAR(1000) CHARACTER SET utf8mb4 COLLATE utf8mb4_bin,
  pattern_digest VARCHAR(64),
  normalized_pattern VARCHAR(100)
) DEFAULT CHARSET = utf8mb4 ENGINE = INNODB;

# 加载重写插件
INSTALL PLUGIN rewriter SONAME 'rewriter.so';

# 创建重写接口函数
CREATE FUNCTION load_rewrite_rules RETURNS STRING
SONAME 'rewriter.so';

# 创建重写语句处理的存储过程
DELIMITER //

CREATE PROCEDURE query_rewrite.flush_rewrite_rules()
BEGIN
  DECLARE message_text VARCHAR(100);
  COMMIT;
  SELECT load_rewrite_rules() INTO message_text;
  IF NOT message_text IS NULL THEN
    SIGNAL SQLSTATE '45000' SET MESSAGE_TEXT = message_text;
  END IF;
END //

DELIMITER ;
```

install_rewriter.sql 脚本的应用的命令行示例如下。

```
# 直接导入提供的 SQL 语句
shell $> mysql -uroot -p < /opt/idc/mysql8.0/share/install_rewriter.sql
```

Query Rewrite 插件的启用方式如下所示。

```
# 方法一：在 my.cnf 文件中自动配置插件加载
[mysqld]
rewriter_enabled=ON


# 方法二：以命令行方式动态加载
mysql> SET GLOBAL rewriter_enabled = ON; SET GLOBAL rewriter_enabled = OFF;
```

查看 Query Rewrite 插件是否正常启用的命令行示例如下。

```
# 查看插件状态、SQL 导入情况等
mysql> SHOW GLOBAL VARIABLES LIKE 'rewriter_enabled';
+------------------+-------+
| Variable_name    | Value |
+------------------+-------+
| rewriter_enabled | ON    |
+------------------+-------+
1 row in set (0.01 sec)

# 查看是否创建了 query_rewrite 库
mysql> SHOW DATABASES;
+--------------------+
| Database           |
+--------------------+
| information_schema |
| mysql              |
| performance_schema |
| query_rewrite      |
| sys                |
+--------------------+
7 rows in set (0.01 sec)

# 查看 query_rewrite 库下的对象
mysql> show create table query_rewrite.rewrite_rules;
+---------------+--------------------------------------------------------------------------------+
| Table         | Create Table                                                                   |
+---------------+--------------------------------------------------------------------------------+
| rewrite_rules | CREATE TABLE `rewrite_rules` (
  `id` int NOT NULL AUTO_INCREMENT,
  `pattern` varchar(5000) CHARACTER SET utf8mb4 COLLATE utf8mb4_bin NOT NULL,
  `pattern_database` varchar(20) CHARACTER SET utf8mb4 COLLATE utf8mb4_bin DEFAULT NULL,
  `replacement` varchar(5000) CHARACTER SET utf8mb4 COLLATE utf8mb4_bin NOT NULL,
  `enabled` enum('YES','NO') CHARACTER SET utf8mb4 COLLATE utf8mb4_bin NOT NULL DEFAULT 'YES',
  `message` varchar(1000) CHARACTER SET utf8mb4 COLLATE utf8mb4_bin DEFAULT NULL,
  `pattern_digest` varchar(64) DEFAULT NULL,
  `normalized_pattern` varchar(100) DEFAULT NULL,
```

```
    PRIMARY KEY (`id`)
) ENGINE=InnoDB AUTO_INCREMENT=2 DEFAULT CHARSET=utf8mb4
COLLATE=utf8mb4_0900_ai_ci
+---------------------+----------------------------------------------------------------+
1 row in set (0.00 sec)
```

3. 插件使用方式

Query Rewrite 插件通过设置规则来替换原始的 SQL 语句。在这个过程中，插件使用 "?" 作为占位符来替换相应的变量。同时，它还提供了 Warnings 提示机制，以便用户可以查看替换过程中产生的任何信息或警告。

下面提供几个插件实践示例，展示如何使用 Query Rewrite 插件来重写 SQL 语句。

【例 6-15】通过重写 SELECT 语句来进行加法计算。加法计算的重写命令行示例如下。

```
#1.设置规则
mysql> INSERT INTO query_rewrite.rewrite_rules (pattern, replacement)
VALUES('SELECT ? ', 'SELECT ? + 1');
Query OK, 1 row affected (0.00 sec)

#2.规则加载
mysql> CALL query_rewrite.flush_rewrite_rules();
Query OK, 1 row affected (0.00 sec)

#3.检查是否发生重写
mysql> SELECT 10;
+--------+
| 10 + 1 |
+--------+
|     11 |
+--------+
1 row in set, 1 warning (0.00 sec)

mysql> SHOW WARNINGS \G
*************************** 1.row ***************************
  Level: Note
   Code: 1105
Message: Query 'SELECT 10' rewritten to 'SELECT 10 + 1' by a query rewrite plugin1
row in set (0.00 sec)
```

【例 6-16】通过重写 INSERT 语句来替换 VALUES 值。值替换的重写命令行示例如下。

```
#1.创建测试表
mysql>CREATE TABLE `t1` (
  `id` int NOT NULL,
  `name` varchar(20) COLLATE utf8mb4_bin DEFAULT NULL,
  `age` int DEFAULT '0',
`create_time` datetime DEFAULT CURRENT_TIMESTAMP
) ENGINE=InnoDB;
Query OK, 0 rows affected (0.03 sec)

#2.创建 INSERT 替换规则:新数据 age 都是 50
```

```
mysql> INSERT INTO query_rewrite.rewrite_rules (pattern,
pattern_database,replacement) VALUES(' INSERT INTO t1(id,name,age)
VALUES(?,?,?)','db2','INSERT INTO t1(id,name,age) VALUES(?,?,50)');
Query OK, 1 row affected (0.00 sec)
```

#3.规则加载
```
mysql> CALL query_rewrite.flush_rewrite_rules();
Query OK, 1 row affected (0.00 sec)
```

#4.插入数据
```
mysql> INSERT INTO t1(id,name,age) values(1,"C",100);
Query OK, 1 row affected, 1 warning (0.00 sec)

mysql>SHOW Warnings;
+-------+------+------------------------------------------------------------------------------+
|Level  |Code  |Message                                                                       |
+-------+------+------------------------------------------------------------------------------+
|Note   |1105  |Query 'insert into t1(id,name,age) values(1,"C",100)'
        rewritten to 'INSERT INTO t1(id,name,age) VALUES(1,'C',50)' by a query rewrite plugin |
+-------+------+------------------------------------------------------------------------------+
1 row in set (0.00 sec)
```

#5.验证
```
mysql>select * from t1;
+-----+------+------+---------------------+
|id   |name  |age   |create_time          |
+-----+------+------+---------------------+
|   1 |C     |  50  |2022-10-20 12:04:37  |
+-----+------+------+---------------------+
1 row in set (0.00 sec)
```

【例 6-17】通过重写 INSERT 语句来将其替换成 DELETE 语句。语句替换的重写命令行示例如下。

#1.INSERT 语句替换成 DELETE 语句规则
```
mysql> INSERT INTO query_rewrite.rewrite_rules (pattern,
pattern_database,replacement)   VALUES(' INSERT INTO t1(id,name,age)
VALUES(?,?,?)','db2','DELETE FROM t1 WHERE id=?');
Query OK, 1 row affected (0.00 sec)
```

#2.加载规则
```
mysql> CALL query_rewrite.flush_rewrite_rules();
Query OK, 1 row affected (0.01 sec)
```

#3.查看初始化数据
```
mysql> select * from t1;
+-----+------+------+---------------------+
|id   |name  |age   |create_time          |
+-----+------+------+---------------------+
|   1 |C     |  50  |2022-10-20 12:04:37  |
+-----+------+------+---------------------+
1 row in set (0.00 sec)
```

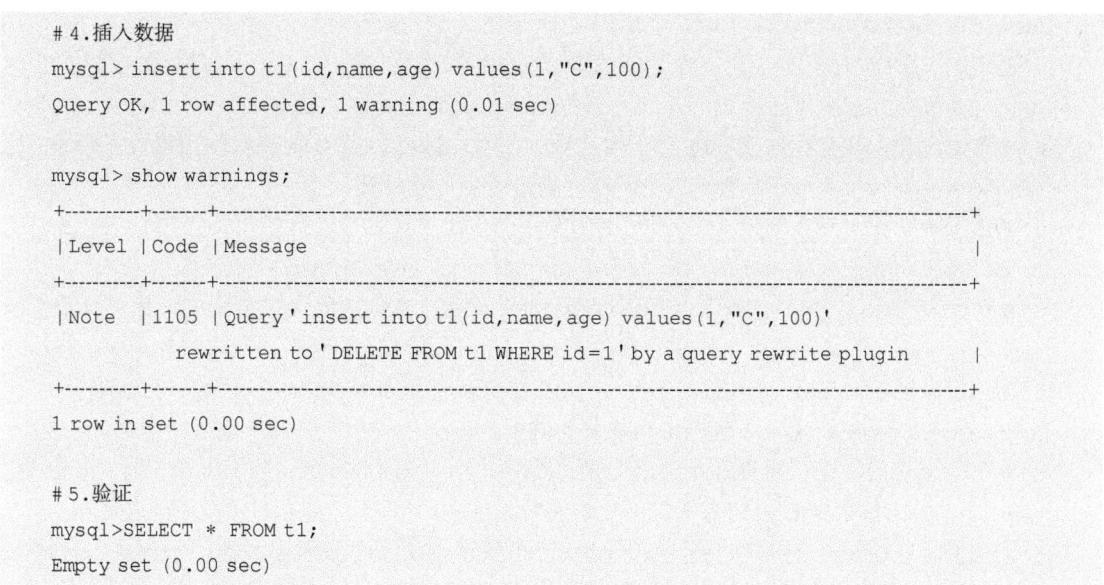

4. 插件参数和状态值

Query Rewrite 插件支持多个参数和状态值的查询，使得用户能够根据不同的场景和需求，灵活地调整查询条件和逻辑。

表 6-5 所示为重写插件参数的说明列表。

表 6-5　重写插件参数说明

参　数	说　明
rewriter_enabled	是否启用 Query Rewrite 插件
rewriter_enabled_for_threads_without_privilege_checks	特权检查的情况下执行的复制线程应用重写

表 6-6 所示为重写插件状态值的说明列表。

表 6-6　重写插件状态值说明

状态值	说　明
Rewriter_number_loaded_rules	从 rewrite_rules 表中成功加载到内存供 Query Rewrite 插件使用的重写规则的数量
Rewriter_number_reloads	rewrite_rules 表被加载到 Query Rewrite 插件使用的内存缓存中的次数
Rewriter_number_rewritten_queries	自 Query Rewrite 插件加载以来，由它重写的查询数
Rewriter_reload_error	rewrite_rules 表最近一次被加载到 Query Rewrite 插件使用的内存缓存中时是否发生了错误

5. 插件使用场景

使用 Query Rewrite 插件能快速改写 SQL 语句，从而消除慢查询，恢复数据库服务性能，该插件是一个值得考虑的选择。当然，触发器也可以实现某些功能，但它们无法对 SELECT 语句进行控制。由于 Query Rewrite 插件基于语法树进行解析和替换，因此它对数据库整体性能的影响非常小，几乎可以忽略不计。

Query Rewrite 插件可以在以下场景中使用。
- 复杂的 JOIN 操作：当 SQL 语句包含复杂的 JOIN 操作时，Query Rewrite 插件可以识别并优化这些操作，减少不必要的表扫描和数据处理，从而提高查询效率。
- 子查询优化：在存在性能问题的子查询中，Query Rewrite 插件能够识别出潜在的优化点，如将子查询转换为 JOIN 操作或使用临时表来减少重复计算。
- 函数和表达式重写：如果 SQL 语句中的某些函数或表达式导致性能下降，则 Query Rewrite 插件可以将它们替换为更高效的表达式或函数调用，以提高查询性能。
- 避免全表扫描：在某些情况下，如果没有合适的索引，查询可能会进行全表扫描，导致性能下降。Query Rewrite 插件可以识别这些查询，并尝试通过改写 SQL 语句来避免全表扫描。
- 简化复杂的 SQL 语句：复杂的 SQL 语句往往难以优化和维护。Query Rewrite 插件可以将这些复杂的语句简化为等效但更容易优化的形式。
- 索引未覆盖的查询：当查询无法充分利用索引时，Query Rewrite 插件可以识别这些查询，并尝试通过改写 SQL 语句来使索引更加有效。
- 避免使用不推荐的 SQL 特性：某些 SQL 特性可能在某些情况下导致性能问题。Query Rewrite 插件可以识别这些特性，并尝试使用更高效的替代方案。

需要注意的是，虽然 Query Rewrite 插件能够自动对 SQL 语句进行改写以优化性能，但它并不总是能够提供最佳的解决方案。因此，在使用 Query Rewrite 插件时，仍然需要数据库管理员或开发人员仔细审查改写后的 SQL 语句，确保它们仍然满足业务需求和性能要求。

此外，Query Rewrite 插件的使用也需要考虑其对数据库整体性能的影响。虽然它对单个查询的性能影响可能很小，但如果大量的查询都被频繁地改写，可能会对数据库服务器造成额外的负载。因此，在使用 Query Rewrite 插件时，应该谨慎评估其对整个系统的性能影响。

6.3.6 控制 InnoDB 的并发线程

在有限的硬件资源下，如何有效地控制 MySQL 的并发线程，以及 MySQL 数据库到底能支持多少并发线程，这些都是数据库管理员需要了解的重要问题。InnoDB 使用操作系统线程来处理来自用户的事务请求（事务在提交或回滚之前可能会向 InnoDB 发出多个请求）。

1. InnoDB 并发线程工作原理

在 MySQL 运行环境中，大多数工作负载都能良好运行，但这并不意味着并发线程的数量没有限制。实际上，MySQL 提供了多种机制来限制和控制并发执行的线程数量。可以通过设置参数，限制并发线程的数量。

如图 6-19 所示，当 MySQL 服务端接收到客户端的请求时，它会尝试分配一个线程来处理该请求。如果当前线程数量达到预设的最大限制，则新请求将被放入一个等待队列中，并且线程会休眠几微秒以减少 CPU 的占用。当有线程完成其任务并释放时，等待队列中的线程将被唤醒并分配空闲的线程来处理。每个线程在处理请求时都会持有一个票据（锁或资源），以确保在同一时间只有一个线程可以执行特定的事务。当所有可用票据都被使用时，新的请求必须等待，直到有线程释放其票据为止。

2. InnoDB 控制并发线程参数

在 MySQL 中，提供了对并发线程数的限制功能，但需要注意的是，这种限制通常针对的是 InnoDB 存储引擎的并发操作，而不是整个 MySQL 服务器的并发连接数。InnoDB 存储引擎会通过一些参数来控制其内部的并发行为。

第 6 章 MySQL 8.0 的运维管理

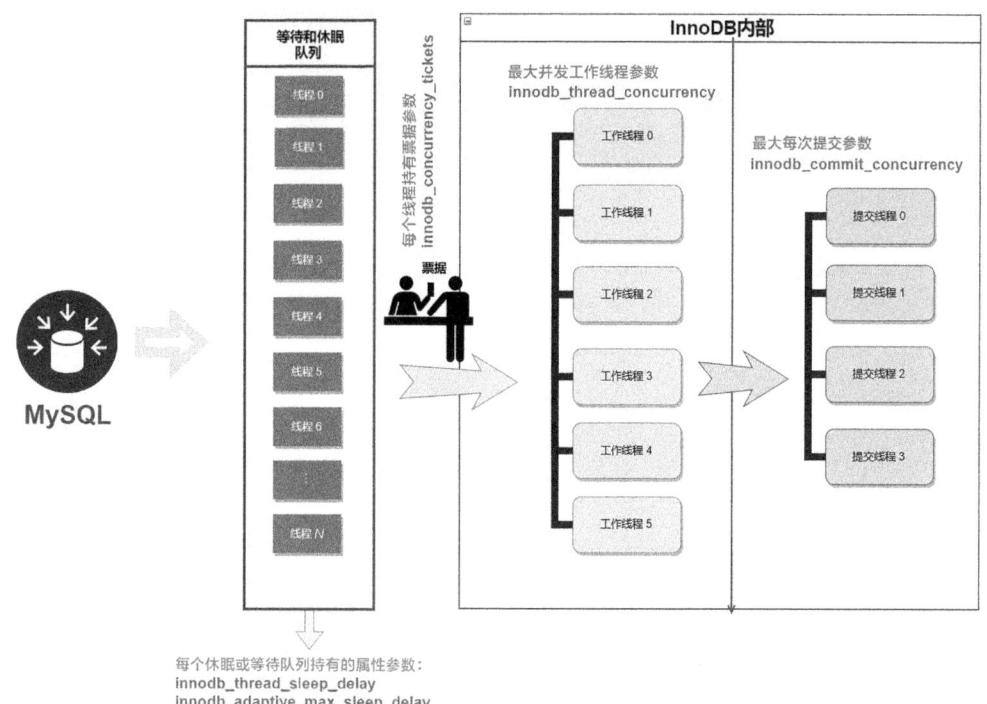

图 6-19 MySQL 并行限制

（1）控制并发线程数参数 innodb_thread_concurrency

innodb_thread_concurrency 参数用于控制并发线程的数量。它并不是用来限制 InnoDB 使用的虚拟 CPU 数量，而是用来控制 InnoDB 内部线程并发的数量。该参数的设置范围为 0~1000。当设置为默认值 0 时，意味着具有无限并发性（没有并发性检查）。禁用线程并发性检查允许 InnoDB 根据需要创建线程。此外，禁用线程并发性检查还会影响 InnoDB 的内部操作，包括禁用某些查询和显示 InnoDB 状态输出的行操作部分的队列计数器。

innodb_thread_concurrency 参数的正确设置对于确保 MySQL 性能至关重要，因为它可以根据工作负载、计算环境和 MySQL 版本来动态调整。为了找到最佳性能设置，DBA 需要测试一系列的值。由于 innodb_thread_concurrency 是一个动态参数，因此可以在实时测试系统中轻松地试验不同的设置。如果发现某个设置的表现不佳，DBA 可以迅速将其重置为默认值 0，以避免对性能造成进一步的影响。

innodb_thread_concurrency 参数取值建议如下。

- 当并发用户线程数小于 64 时，将 innodb_thread_concurrency 设置为 0 是一个好的起点。这意味着 InnoDB 存储引擎不会使用任何线程并发控制，而是依赖于操作系统的线程调度。
- 对于持续高负载或偶尔出现峰值的工作负载，可以从 innodb_thread_concurrency = 128 开始尝试，并逐步降低其值（如 96、80、64 等），直到找到最佳性能点为止。这种方法是通过试验和错误来找到最优设置的典型过程。
- 如果不想让 InnoDB 为用户线程提供超过一定数量的并发处理能力（例如，20 个线程），则可以将 innodb_thread_concurrency 设置为这个数（或者更低，取决于性能结果）。
- 如果目标是将 MySQL 服务与其他应用程序隔离开来，那么可以考虑将 mysqld 进程专门绑定到特定的虚拟 CPU 或核心上。

345

如果 MySQL 实例与其他应用程序共享同一台服务器的 CPU 资源，或者 MySQL 的工作负载和并发用户数量持续增加，那么 DBA 可能需要考虑调整 innodb_thread_concurrency 参数来优化性能。当然，innodb_thread_concurrency 值设置过高可能会导致线程间的竞争增加，从而导致性能下降。

（2）控制并发线程休眠参数 innodb_thread_sleep_delay

innodb_thread_sleep_delay 是 InnoDB 存储引擎的一个参数，它定义了当一个线程因为并发限制（由 innodb_thread_concurrency 控制）而被放入休眠队列之前，该线程应该休眠的时间长度，单位是微秒。默认值是 10000 微秒（即 10 毫秒）。如果将该参数设置为 0，那么线程将不会休眠，而是立即进入休眠队列。

此外，innodb_adaptive_max_sleep_delay 参数可以设置为 innodb_thread_sleep_delay 的最大值。当 InnoDB 根据当前的线程调度活动自动调整 innodb_thread_sleep_delay 时，它不会超过 innodb_adaptive_max_sleep_delay 所设置的值。这样，InnoDB 可以动态地调整线程的休眠延迟，以适应系统的工作负载变化。

（3）控制并发线程票据参数 innodb_concurrency_tickets

innodb_concurrency_tickets 参数用于控制每个线程在被允许进入 InnoDB 存储引擎之前可以获得的"票据"数量。这些"票据"可以看作线程在 InnoDB 内部执行操作的"许可证"。当线程的数量达到 innodb_thread_concurrency 的限制时，新到达的线程将被放入队列中等待。当一个线程完成其任务并释放资源时，它可能会释放一个或多个"票据"。这时，队列中的一个线程将获得这些"票据"并可以继续执行。这个过程会一直持续到该线程的"票据"用完。

innodb_concurrency_tickets 参数的设置对于平衡大小事务之间的并发性能非常重要。如果设置的值较小，那么小事务将与大事务公平竞争，因为它们只需要处理几行数据就可以用完它们的"票据"。然而，这可能导致大事务需要多次通过队列循环才能完成，从而增加了完成任务所需的时间。相反，如果设置的值较大，那么大事务将更少地受到队列的影响，因为它们可以在用完"票据"之前处理更多的行数据。这可以减少大事务等待队列末端位置的时间，但它们也可能让较小的事务等待更长的时间，从而造成"饿死"现象。

因此，调整 innodb_concurrency_tickets 参数的值时需要在大型和小型事务之间找到一个平衡点，以确保系统的整体性能最佳。这通常需要根据具体的工作负载、数据库大小、硬件配置以及应用程序的特性来进行调整，并进行性能测试以验证设置的效果。

（4）控制并发线程休眠参数 innodb_adaptive_max_sleep_delay

innodb_adaptive_max_sleep_delay 参数允许 InnoDB 存储引擎根据当前工作负载自动调整 innodb_thread_sleep_delay 的值。当设置为非零值时，InnoDB 可以动态地增加 innodb_thread_sleep_delay 的值，直到达到 innodb_adaptive_max_sleep_delay 所指定的最大值为止。这个最大值是以微秒为单位的（默认值为 150000 微秒）。这个参数对于拥有超过 16 个 InnoDB 线程的繁忙系统特别有用，特别是在同时处理数百或数千个连接的 MySQL 系统中。

innodb_thread_sleep_delay 和 innodb_adaptive_max_sleep_delay 这两个参数主要用于控制 InnoDB 线程在并发情况下的行为，特别是当线程数量达到 innodb_thread_concurrency 的限制时。innodb_adaptive_max_sleep_delay 的默认值（150000 微秒，即 0.15 秒）提供了一个上限，确保即使在非常繁忙的情况下，线程也不会过于频繁地尝试进入 InnoDB，从而避免过多的上下文切换。

对于大多数系统来说，使用默认值通常是合适的，但在某些特定的、高并发的场景下，可能需要根据实际情况调整该参数。

(5)控制并发提交参数 innodb_commit_concurrency

innodb_commit_concurrency 是 InnoDB 存储引擎的一个参数,用于控制同时提交事务的线程数。当多个事务试图同时提交时,这个参数有助于避免资源争用和潜在的性能问题。

- 默认值:0,这意味着没有并发提交的限制,允许任意数量的事务同时提交。
- 最大值:1000,即在任何时候最多允许 1000 个事务同时提交。
- 动态调整限制:innodb_commit_concurrency 的值不能在运行时从 0 动态地更改为非零值,反之亦然。但是,可以在非零值之间进行动态调整。

因此,对于需要限制同时提交事务数量的高并发环境,可以根据实际需求和性能监测结果来适当调整这个参数的值。

3. 控制并发线程建议

对于 MySQL 数据库,限制处理能力通常会涉及多个层面和策略(包括硬件资源管理、查询优化、索引策略等),而不仅仅是并行线程的数量。参数 innodb_thread_concurrency 的设置范围为 0~1000,当超过这个范围时,MySQL 内部处理上会存在瓶颈。但实际场景下,对该参数采用默认值是推荐的做法,这意味着 MySQL 将依赖其内部机制来管理并发线程。

在 MySQL 中,通常在如下场景中限制并发线程。

- 多个实例部署在同一服务器上:如果多个 MySQL 实例都部署在同一台服务器上,并且某个实例的性能占比较高,导致其他实例得不到足够的资源,那么这时可以考虑限制该高性能实例的并发线程数,以确保其他实例也能获得必要的资源。
- 系统资源成为瓶颈:当数据库面临大量的并发读写操作,且服务器 CPU、内存或磁盘 I/O 成为瓶颈时,限制并发线程可以帮助减少资源竞争,从而提高整体性能。通过限制线程数,可以减少同时进行的操作数量,降低资源消耗。
- 资源受限的环境:在虚拟环境或共享硬件资源的环境中,资源分配通常是有限的。限制并发线程可以防止数据库消耗过多的资源,从而影响其他应用程序或服务的性能。
- 性能调优:在进行性能调优时,如果发现调整并发参数能够改善 I/O 性能和 CPU 处理能力,那么可以根据实际情况调整这些参数。这可能涉及增加或减少并发线程的数量,以达到最佳的性能。
- 高并发 OLTP 系统:在高并发的 OLTP 系统中,锁竞争可能会成为性能瓶颈。限制并发线程可以减少锁竞争,因为较少的线程同时访问数据,从而降低了锁等待时间。
- 从库延迟控制:在主从复制架构中,从库的延迟是一个需要关注的问题。如果从库不允许延迟过高,那么可以通过限制主库的并发线程数来减缓写入的速度,从而降低从库的复制延迟。这通常是在复制延迟成为问题时采取的一种控制手段。

6.3.7 备份中全局读锁 FTWRL 对数据库的影响

在 MySQL 中,存在 FLUSH TABLES WITH READ LOCK(FTWRL)命令。这个命令通常在执行热备份(如使用 XtraBackup 和 mysqldump)时使用。由于 FTWRL 命令的特殊性,它会阻止对数据库表的所有写操作,直到命令被释放为止,因此,在执行此命令期间,数据库可能会变得不可用。由于存在这种锁定机制,因此执行 FTWRL 时需要格外小心,以避免对生产环境造成不必要的影响。

使用 XtraBackup 和 mysqldump 工具备份时,输出 FTWRL 的命令行示例如下。

```
# XtraBackup
shell $ > XtraBackup --defaults-file=/etc/my.cnf --no-server-version-check --backup -uroot -p
--socket=/tmp/mysql.sock --target-dir=/tmp/back
    XtraBackup version 8.0.29-22 based on MySQL server 8.0.29 Linux (x86_64) (revision id:
c31e7ddcce3)
    230819 11:26:05  version_check Connecting to MySQL server with DSN 'dbi:mysql:;mysql_read_
default_group=XtraBackup;port=3381;mysql_socket=/opt/data8.1/data/mysql.sock' as 'root'
(using password: YES).
    ...
    /tmp/back/undo_0012022-08-19T11:26:08.834591+08:00 0 [Note] [MY-011825] [XtraBackup] Execu-
ting FLUSH TABLES WITH READ LOCK...

# mysqldump 备份一致性命令
    shell $>mysqldump -uroot -p -S /tmp/mysql.sock --set-gtid-purged=OFF  --all-databases --master-
data=2 --single-transaction >/tmp/full.sql
    2023-08-19T16:33:11.448512+08:00  120 Query  /*!80000 SET SESSION information_schema_stats
_expiry=0 */
    2023-08-19T16:33:11.448706+08:00  120 Query  SET SESSION NET_READ_TIMEOUT = 86400, SESSION
NET_WRITE_TIMEOUT=86400
    2023-08-19T16:33:11.448866+08:00  120 Query  FLUSH /*!40101 LOCAL */ TABLES
    2023-08-19T16:33:11.451661+08:00  120 Query  FLUSH TABLES WITH READ LOCK
    ...
```

说明：在使用 mysqldump 备份时，指定参数 --master-data、--lock-all-tables 会引发 FLUSH TABLES 和 FLUSH TABLES WITH READ LOCK 的命令行输出。

1. FTWRL 机制

FTWRL 是 MySQL 中的一种机制，主要用于在热备份时保证数据的一致性。执行 FTWRL 操作时，MySQL 会做以下内部操作。

- 关闭所有打开的表，清理表缓存（close_cached_tables）。
- 使用全局读锁（lock_global_read_lock）。
- 上全局 COMMIT 锁（make_global_read_lock_block_commit），锁定所有数据库的所有表。

同时，执行 FTWRL 操作需要 FLUSH_TABLES 或 RELOAD 权限。尽管 FTWRL 与数据库的只读模式（READ_ONLY）有关，但它们并不是同一个概念。当数据库的 READ_ONLY 参数设置为 ON 时，它会阻止所有写操作，而 FTWRL 是一个更为具体的命令，用于在备份期间锁定所有表。

从相关描述中不难理解，操作表缓存时，必须获取与元表相关的锁。在 MySQL 中，每个表在内存中都有一个表缓存，这些缓存对象通过哈希链表进行维护。在执行 FTWRL 期间，如果有大的查询导致关闭表等待，那么所有访问该表的查询和更新操作都将被阻塞，直到 FTWRL 锁被释放为止。

总的来说，尽管 FTWRL 是确保数据一致性的有效手段，但由于其对数据库写操作的阻塞影响，因此使用时需要格外小心，以避免对生产环境造成不必要的干扰。

2. FTWRL 卡住现象

下面介绍 FTWRL 命令行导致数据库出现卡住情况的两个案例。

【例 6-18】FTWRL 触发导致无法提交事务的命令行示例如下。

```
# 会话 1:执行 FTWRL 命令
mysql> begin:
mysql> FLUSH TABLES WITH READ LOCK ;
Query OK, 0 rows affected (0.00 sec)
mysql> select connection_id();
+-----------------+
|connection_id() |
+-----------------+
|              32 |
+-----------------+
1 row in set (0.00 sec)

# 会话 2:执行 INSERT 语句,语句在等待中
mysql> INSERT INTO t(c1) VALUES(4);

# 会话 3:查看运行情况
mysql> SHOW PROCESSLIST;
+----+------+-----------+------+---------+------+----------------------------+---------------------------+
| Id | User | Host      | db   | Command | Time | State                      | Info                      |
+----+------+-----------+------+---------+------+----------------------------+---------------------------+
| 9  | root | localhost | NULL | Sleep   |   13 |                            | NULL                      |
| 10 | root | localhost | book | Query   |    9 | Waiting for global read lock | INSERT INTO t(c1) VALUES(4) |
| 11 | root | localhost | NULL | Query   |    0 | init                       | SHOW PROCESSLIST          |
+----+------+-----------+------+---------+------+----------------------------+---------------------------+
3 rows in set (0.00 sec)
```

说明：在上述 State 列中，Waiting for global read lock 状态是指等待全局读锁。这种情况在执行了 FLUSH TABLES WITH READ LOCK 命令或设置了全局 READ_ONLY 参数时出现。

【例 6-19】在 MySQL 备份场景下，如果有一个长时间运行的 SQL 语句或批量操作正在执行，则可能导致备份进程无法进行，并且会阻塞其他数据库操作。如图 6-20 所示，长时间运行的事务 1 语句占用了系统资源，导致备份进程无法获取所需的锁，因此备份、事务 2 和事务 3 被阻塞。

图 6-21 所示为 MySQL 数据库执行 FTWRL 时的运行状态。在 State 列中可以看到 "Waiting for table flush" 状态。线程正在执行 FLUSH TABLES，等待所有线程关闭它们的表，或者线程得到一个通知：表的底层结构已经改变，它需要重新打开表来获得新的结构。但是，要重新打开表，必须等到其他所有线程都关闭了这个表。当线程使用 FLUSH TABLES 或下列语句之一：FLUSH TABLES tbl_name、ALTER table、RENAME table、REPAIR table、ANALYZE table 和 OPTIMIZE table 时，会产生此通知。

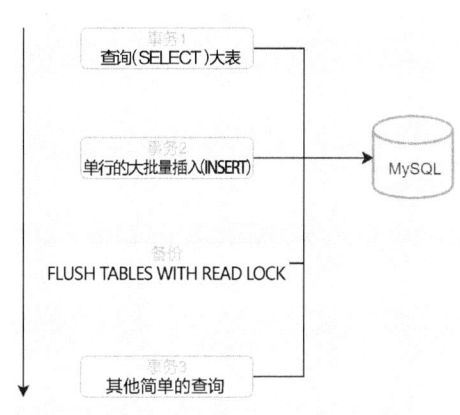

图 6-20　MySQL 备份 FTWRL 场景

3. FTWRL 对不同引擎的影响

MySQL 8.0 对 XtraBackup 和其他备份解决方案在处理 InnoDB 与 MyISAM 存储引擎时的 FTWRL

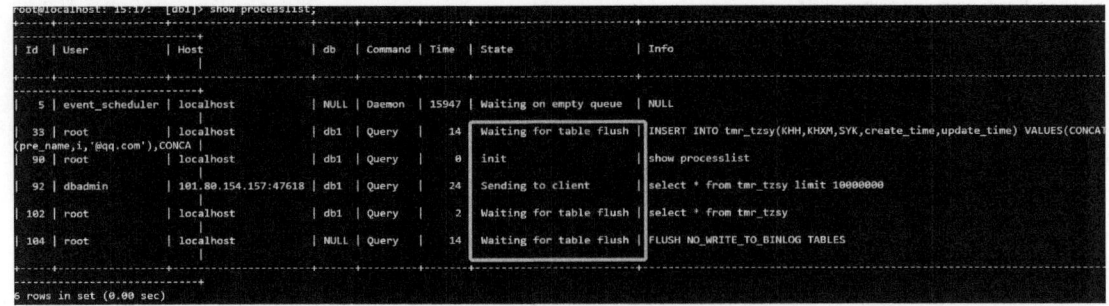

图 6-21　MySQL 状态显示情况

行为进行了优化。图 6-22 所示为不同存储引擎下 FTWRL 行为的出现情况。在存在 MyISAM 引擎表的情况下，FTWRL 行为仍然会出现；而在仅存在 InnoDB 引擎表时，FTWRL 行为则不会出现。

图 6-22　MySQL 8.0 版本不同存储引擎触发 FTWRL 机制

说明：在备份执行命令行中使用了 FLUSH NO_WRITE_TO_BINLOG BINARY LOGS 命令。该命令用于关闭当前的 binlog 文件并开启一个新的日志文件。

在 MySQL 5.6 和 5.7 版本中，由于系统表通常使用 MyISAM 存储引擎，因此这些版本中的 FTWRL 锁是避免不了的，因为当对 MyISAM 表执行某些操作时，MySQL 需要获取 FTWRL 锁。

在 MySQL 8.0 版本中，由于系统表中 MyISAM 存储引擎已被移除，因此保证系统表不会使用 MyISAM 引擎，从而不会上 FTWRL 锁。但这并不意味着业务表也不能使用 MyISAM 引擎。MySQL 8.0 仍然支持 MyISAM 引擎，用户可以选择在业务表中使用它。

统计引擎表类型和数量的 SQL 语句如下所示。

```
# 在 MySQL 5.7 版本环境中,mysql 系统库中存在 MyISAM 引擎表
mysql> SELECT TABLE_SCHEMA, ENGINE, COUNT(*)
    FROM INFORMATION_SCHEMA.TABLES
    WHERE TABLE_TYPE='BASE TABLE'
    GROUP BY TABLE_SCHEMA,ENGINE;
+--------------------+--------------------+----------+
| TABLE_SCHEMA       | ENGINE             | COUNT(*) |
+--------------------+--------------------+----------+
| mysql              | CSV                |        2 |
```

```
| mysql                 | InnoDB             |       19 |
| mysql                 | MyISAM             |       10 |
| performance_schema    | PERFORMANCE_SCHEMA |       87 |
| sys                   | InnoDB             |        1 |
+-----------------------+--------------------+----------+
4 rows in set (0.87 sec)

# 在 MySQL 8.0 版本环境中,系统库中不存在 MyISAM 引擎表
mysql> SELECT TABLE_SCHEMA, ENGINE, COUNT(*)
    FROM INFORMATION_SCHEMA.TABLES
    WHERE TABLE_TYPE='BASE TABLE'
    GROUP BY TABLE_SCHEMA,ENGINE;
+-----------------------+--------------------+----------+
| TABLE_SCHEMA          | ENGINE             | COUNT(*) |
+-----------------------+--------------------+----------+
| mysql                 | InnoDB             |       36 |
| mysql                 | CSV                |        2 |
| performance_schema    | PERFORMANCE_SCHEMA |      113 |
| sys                   | InnoDB             |        1 |
+-----------------------+--------------------+----------+
4 rows in set (0.00 sec)
```

4. FTWRL 被替代成 LOCK INSTANCE FOR BACKUP

在 MySQL 8.0 中,为了解决 FTWRL 锁对系统性能的影响问题,引入了 LOCK INSTANCE FOR BACKUP 机制。该锁允许在数据库进行在线备份时,仍然可以进行 DML 操作,如 INSERT、UPDATE、DELETE 等,从而减少了备份对生产环境的影响。同时,LOCK INSTANCE FOR BACKUP 也防止了那些可能导致备份快照不一致的操作。

LOCK INSTANCE FOR BACKUP 有以下作用和使用限制。

- 防止文件被创建、重命名或删除。
- 阻塞 REPAIR TABLE、TRUNCATE TABLE、OPTIMIZE TABLE,以及账户管理语句。
- 阻止对 InnoDB 重做日志中没有记录的修改 InnoDB 文件的操作。
- 阻止发出 PURGE BINARY LOGS 命令。
- 允许:只影响用户创建的临时表的 DDL 操作。实际上,当持有备份锁时,属于用户创建的临时表的文件可以被创建、重命名或删除。
- 允许:创建二进制日志文件。
- 使用 UNLOCK INSTANCE 释放当前会话持有的备份锁。如果会话终止,则会话持有的备份锁会被释放。
- 执行需要 BACKUP_ADMIN 权限,在从早期版本到 MySQL 8.0 的本地升级时,BACKUP_ADMIN 权限会自动授予具有 RELOAD 权限的用户。

LOCK INSTANCE FOR BACKUP 和 FTWRL 可组合使用。组合命令行示例如下。

```
#用法1
LOCK INSTANCE FOR BACKUP;
FLUSH TABLES tbl_name [, tbl_name] ...WITH READ LOCK;
UNLOCK TABLES;
UNLOCK INSTANCE;
```

```
# 用法 2
FLUSH TABLES tbl_name [, tbl_name] ...WITH READ LOCK;
LOCK INSTANCE FOR BACKUP;
UNLOCK INSTANCE;
UNLOCK TABLES;
```

MySQL 官方文档并没有推荐或支持组合使用这两种锁。事实上，由于 LOCK INSTANCE FOR BACKUP 和 FTWRL 的目的与效果不同，将它们组合使用可能会导致不可预见的行为或冲突，因此，在大多数情况下，使用 LOCK INSTANCE FOR BACKUP 应该足以满足在线备份的需求，同时允许在备份期间进行 DML 操作。如果需要更严格的 FTWRL 锁来保证数据一致性，那么将限制 DML 操作，并可能对系统性能产生更大的影响。

除此之外，MySQL 参数 lock_wait_timeout 定义 LOCK INSTANCE FOR BACKUP 语句在放弃之前，需要等待获取锁的时间。可以合理设置这个参数时间。

5. 针对 FTWRL 问题的应对方案

对于由 FTWRL 命令导致的 Waiting for global read lock 和 Waiting for table flush 现象，最佳的应对方式可能是优化备份策略，极端情况下可以终止（KILL）执行 FTWRL 命令的进程，释放资源。

下面列出避免使用 FTWRL 命令的方法。

- 无论是 MySQL 5.7 版本还是 MySQL 8.0 版本，均采取 InnoDB 引擎。
- 备份脚本在从库上执行即可，这样备份不会影响到主库的性能。
- 备份选择负载低的时间段进行，并建议暂停长时间运行的 SQL 语句。
- 采用 MySQL 8.0 版本，对该版本的 LOCK INSTANCE FOR BACKUP 提供了更细粒度的控制，允许在备份期间进行更多的操作。

除了上述提到的方法，在使用 XtraBackup 工具备份时，可以通过提供的参数来进行优化。表 6-7 所示为 MySQL 和 XtraBackup 控制锁机制参数列表。

表 6-7 MySQL 和 XtraBackup 控制锁机制参数

类　型	参　数	说　明
XtraBackup	ftwrl-wait-timeout	在执行 FTWRL 之前，如果被活跃会话阻塞了，就等待其执行完成；如果超时时间到后活跃会话还没执行完，则备份失败退出
XtraBackup	ftwrl-wait-threshold	在执行 FTWRL 之前，如果有超过该设置时间的活跃会话，则 FTWRL 将会等待，直到超过 ftwrl-wait-timeout 时备份失败退出
XtraBackup	kill-long-query	当 kill-long-queries-timeout 设置非零值时，限制 innobackup 可以终止的语句类型（SELECT 或者 ALL）
XtraBackup	kill-long-queries-timeout	在执行 FTWRL 过程中，不会立刻终止阻塞 FTWRL 执行的活跃会话，而是等待设置的时间，默认值是 0，即不会终止任何会话
MySQL	innodb_lock_wait_timeout	锁等待的时间

6.3.8 如何快速删除大量数据

在实际 MySQL 生产环境中，由于业务数据量的增长预测不准确，随着时间的推移，数据逐渐累积，最终达到像 Oracle、DB2 等那样的大数据量级别。大数据量可能给 MySQL 的性能、备份恢复时间等方面带来严峻挑战。为了保证现有 MySQL 数据库服务维持现有性能，需要在进行数据归档的同时执行删除数据操作。

下面介绍 MySQL 中的数据删除方式，以及如何安全地删除大量数据，并探讨在删除过程中可以优化的 MySQL 参数。

1. 数据删除方式

在 MySQL 中，删除数据主要有三种命令：DELETE、DROP 和 TRUNCATE。

- DELETE 操作属于 DML 命令，它会逐行删除数据。这种操作消耗时间较长，并且会产生大量的 binlog，可能导致主从延迟。
- DROP 操作是 DDL 命令，它会一次性删除表结构、数据和底层数据文件。如果数据文件过大，那么 DROP 操作会产生大量的 I/O，导致磁盘 I/O 开销增加和 CPU 负载过高，会对数据库整体服务产生影响。因此，当还需要表结构时，DROP 操作可能不是最佳选择。
- TRUNCATE 操作也是 DDL 命令，它会一次性清除表中的数据，但表结构保持不变。需要注意的是，TRUNCATE 操作不会重置自增值。尽管 TRUNCATE 操作速度快，且对表的锁定时间短，但在某些版本中可能存在不完善之处，可能导致数据库挂起（hang up）。

除了上述各自的特点以外，这 3 种删除操作还可能对 MySQL 数据库服务和操作系统产生影响。例如，大量的 binlog 产生可能导致主从延迟，而大量的 I/O 操作则可能导致磁盘 I/O 开销飙升和 CPU 负载过高，从而直接影响数据库的整体性能和服务。

2. 安全删除方式介绍

下面将介绍 4 种安全删除方式，它们都是基于经验总结的大量数据删除方法。采用这种方式，可以最大限度地减少对现有业务的影响。

（1）使用 DELETE 命令行批次处理方式

在 MySQL 中使用 DELETE 命令删除大量数据时，为了提高性能和减少锁定的时间，可以采用批次处理的方式，即每次删除一定数量的行（1000~5000 行），而不是一次性删除所有行。同时，为了确保删除操作的效率，应该确保在 DELETE 语句中使用适当的索引来定位要删除的行。

（2）使用 RENAME TABLE 命令行变更表名，之后创建新表以替代原表

在 MySQL 中，RENAME TABLE 命令用于重命名表名。执行此命令时，需要确保业务在重命名期间停止访问，因为重命名是即时生效的，会导致表被锁定。或者，先停止从库的复制进程，在从库上执行重命名操作，并在完成从库重命名后，进行复制切换，确保主库上的写操作也指向新的表名。重命名表在 MySQL 负载低时再进行删除操作。

（3）Linux 硬链接删除方式

在 Linux 操作系统中，硬链接是一种特殊的文件链接方式。当多个文件名同时指向同一个 inode 时，删除其中任何一个文件名都非常迅速，因为这只是删除了一个指向文件数据块的指针，而文件本身的物理数据块并没有被删除。这种特性可以有效地避免产生高 I/O，从而保护数据库性能。当系统负载较低时，可以删除这些硬链接文件，以释放它们所占用的磁盘空间。

Linux 下硬链接创建的命令行示例如下。

```
# 硬链接方式,没有 -s 参数
shell $> ln test.idb  test_bak.idb
```

（4）使用 TRUNCATE 命令行

在 MySQL 中，TRUNCATE 命令行不是删表再建表的操作，而是删除表中的所有数据。它会将

表截断为一个空表，表的结构、索引等定义保持不变。需要注意的是，TRUNCATE 命令行在执行后无法回滚事务，也不能使用 WHERE 子句来指定删除特定行。因此在使用 TRUNCATE TABLE 命令行时，需要格外小心，确保不会误删数据。还需要考虑使用 TRUNCATE 命令行时可能带来的"hang 住"的风险。

3. 删除大量数据优化参数

删除大量数据在 MySQL 运维当中也是高危操作，需要谨慎执行，选择合理的方案。在删除大量数据时，应该避开业务繁忙时期，在负载较低的情况下执行。为了提升效率，可以考虑将 MySQL 的双 1 参数（binlog 刷新参数 sync_binlog 和 Redo 日志刷新参数 innodb_flush_log_at_trx_commit）的值从 1 改为 0，以便由操作系统接管数据落盘的操作。

双 1 参数更改的命令行示例如下。

```
mysql> set global sync_binlog=0;
Query OK, 0 rows affected (0.00 sec)
mysql> set global innodb_flush_log_at_trx_commit=0;
Query OK, 0 rows affected (0.00 sec)
```

说明：在删除数据任务结束之后，需要把双 1 参数更改成原先的值，因为若发生系统崩溃和故障，则可能导致 binlog 和 Redo 日志没有持久化到磁盘，造成数据丢失。

第 7 章
MySQL 8.0 故障分析

MySQL 数据库在运行过程中,难免会碰到各类故障。当遇到数据库故障时,进行分析并提出相关解决办法,是数据库管理员必不可少的工作,从而防止数据库再次发生类似故障。

7.1 MySQL 8.0 故障分析概述

7.1.1 MySQL 8.0 的故障类型

MySQL 数据库在运行过程中可能会遇到多种故障,以下是一些常见的故障类型。

1. 运行中的错误

运行中的错误是指在 MySQL 运行时出现的异常错误。比较常见的介绍如下。

- 连接通信错误。当客户端尝试连接到 MySQL 服务时,可能会出现连接失败、连接丢失或通信中断的问题。原因可能是网络不稳定、防火墙设置问题、MySQL 服务配置问题(如 max_connections 达到上限)或客户端连接问题(如错误的连接参数)等。
- 内存溢出。MySQL 进程使用的内存超过其分配的限制,导致系统资源耗尽。原因可能是查询优化不足、数据表设计不合理、不恰当的缓冲池设置或大量并发连接等。
- 磁盘 I/O 错误。MySQL 在进行数据读写操作时可能遇到磁盘输入/输出错误。原因可能是硬盘故障、文件系统错误、磁盘空间不足或 MySQL 配置中的 I/O 参数不当等。
- 服务器过载。MySQL 服务器由于处理过多的请求或执行资源密集型任务而变得响应缓慢或停止响应。原因可能是高并发连接、复杂的查询、资源不足(如 CPU、内存、磁盘 I/O 等),或者不恰当的系统或 MySQL 配置等。

2. 配置错误

配置错误是指在 MySQL 配置方面的问题。比较常见的介绍如下。

- 数据库参数配置错误,导致 MySQL 服务无法提供服务。例如,用户连接爆满、内存分配不足、I/O 刷新慢等问题。
- 服务器资源配置不足,导致资源负载非常高。例如,磁盘空间不足、物理内存不足、虚拟机公用资源不足等问题。
- 网络带宽配置不足,导致数据传输速度受限。例如,受限制带宽、共享网络资源等问题。

3. 数据损坏

数据损坏是指 MySQL 数据库中存储的数据出现错误或无法访问的情况。造成这种损坏的原因如下。

- 内存错误。如果内存出现故障或不稳定,则可能会导致数据库操作失败,进而造成数据损坏。
- 硬件故障。如果硬盘驱动器故障或电源故障,则可能影响数据库的稳定性和完整性,导致数据损坏。
- 数据库文件损坏或丢失。文件系统错误、不当的操作(数据库 KILL 操作)等都有可能导致数据损坏或丢失。
- 软件错误或缺陷。MySQL 系统本身、相关的驱动程序、操作系统依赖包,以及数据库内部应用程序中的错误或缺陷都有可能导致数据损坏。

4. 运维故障

运维故障是指因日常操作不当、管理不严谨或执行错误的任务而引发数据丢失或服务中断等。比较常见的介绍如下。

- DDL 命令执行不当。例如,DDL 命令锁定表影响其他用户的并发操作,导致性能下降或超时错误;大型 DDL 操作消耗大量系统资源(CPU、内存、磁盘空间等),导致系统性能下降或崩溃。
- 高可用复制破坏。例如,其他操作(备份、复制大文件等)导致网络延迟或中断,主从节点之间的数据同步失败;复制错误、手动干预或其他原因,导致主从节点之间的数据不一致,破坏高可用性。
- 备份不合理。备份文件过大可能会占用过多存储空间,甚至导致磁盘空间不足。备份时未考虑数据库性能,可能导致备份过程中数据库性能下降或"死"机。

5. 安全问题

安全问题是指数据库存在一些对外暴露的隐患和管理漏洞。比较常见的介绍如下。

- 数据泄露。例如,敏感、机密或受保护的数据在没有授权的情况下被访问、披露或丢失。
- 系统漏洞。例如 SQL 注入漏洞、权限问题漏洞和密码问题漏洞等。
- 未授权访问。例如未经授权的用户或系统能够访问到数据库资源,包括数据、表结构等。

7.1.2 MySQL 8.0 故障分析方法

在 MySQL 数据库运维中,当突发故障时,需要有一套系统化、阶梯式的排查流程来快速定位问题并采取相应的应对措施。图 7-1 所示为一套典型的阶梯式故障排查流程。

图 7-1 MySQL 典型的阶梯式故障排查流程

下面介绍故障排查流程中每个阶梯需要排查的内容。

1. 观察现象

通过观察现象，可以推断出可能的原因和影响范围，从而采取相应的措施来恢复服务。以下是针对几个现象进行的故障分析。

- 对于数据库的部分请求可用，部分请求报错。这可能是由于数据库中的某些数据表、视图或存储过程存在问题，导致部分查询或操作失败。此外，某些请求可能耗尽了服务器资源（如 CPU、内存或数据库连接），导致资源瓶颈，新的连接请求被拒绝。
- 原有业务可以继续使用，但新的访问无法使用。新上线的功能或更新可能包含 bug，导致新的访问请求失败。这可能是由于代码错误、数据库结构不匹配或数据迁移问题导致的。
- 整体不可用。这可能是由于硬件故障（如硬盘故障、内存故障和网络堵塞等）或数据库服务本身的故障（如配置错误、软件 bug 等）导致的。

2. 监控指标

通过监控的指标，了解当前数据库服务情况。举例如下。

- CPU 使用率达到 99%。CPU 使用率高通常意味着数据库正在执行大量的计算工作，可能是因为复杂的查询、索引缺失、数据表结构不合理或应用程序逻辑导致的。
- 硬盘空间满。硬盘空间满可能是由于日志文件、临时文件、数据文件等不断增长而未得到及时清理或归档。
- MySQL 连接数满。连接数满通常意味着数据库连接资源耗尽，可能是因为并发连接过多或连接未得到及时释放。
- MySQL 主从复制延迟高。复制延迟高可能是由于网络问题、从库硬件性能不足或大量的写操作在主库上执行导致从库追赶不上。

3. 分析日志

运行日志是所有软件的基本配置机制，用于记录系统运行情况和异常事件等信息。举例如下。

- 系统日志通常记录了操作系统层面的各种事件，包括内核消息、系统启动和关闭消息、安全事件、磁盘错误、应用程序崩溃 OOM（Out of Memory，内存溢出）等。
- MySQL 错误日志记录了 MySQL 服务器运行过程中的错误、警告和其他重要消息。这些日志对于诊断数据库问题非常有帮助，因为它们包含了关于数据库操作失败、查询执行错误、连接问题等的信息。

4. DBA 排查问题 SQL 语句

对于 DBA 来说，具备一套常用且已经验证过的 SQL 查询语句，用于快速排查和监控数据库性能及状态，是日常工作中不可或缺的技能。通过 MySQL 的系统表，如 information_schema、performance_schema 和 sys，可以获取大量关于数据库运行状况的信息。在紧急情况下，能够快速定位问题至关重要，因此，事先准备好这些查询语句并熟悉它们的输出是非常必要的。

5. 分析原理

数据库的原理能够帮助用户深入了解数据库体系架构、存储和运行机制，进而提高对数据库的编程和数据操作技能，应对故障也会得心应手。数据库原理包括体系架构、MVCC、主从复制、组提交、并行复制和 MGR 等。

原理学得通透，再加上丰富的实践经验与领域知识，就能在 DBA 道路上走得更远。

6. 源码分析和调试

MySQL 的源码结构庞大且复杂，涵盖了众多的技术领域。深入理解 MySQL 的源码对于 DBA 来说是一项非常有价值的工作，它不仅能够帮助 DBA 更深入地了解 MySQL 的内部实现原理（MySQL 的架构设计、存储引擎、查询优化器、事务处理等），还能够提高在故障排查、性能优化等方面的能力。

源码分析和调试需要较高的技术水平，包括熟练掌握 C/C++ 语言、数据库原理、操作系统原理等。同时，还需要具备一定的算法和数据结构知识，以及对多线程、并发控制等复杂概念的理解。

通过研究 MySQL 的源码，DBA 可以不断提高自己的技术水平，以便更好地理解和应对数据库相关的各种挑战。同时，这也是对 MySQL 社区发展的一种贡献，通过参与开源项目的开发和维护，推动数据库技术的不断进步。

7.2 MySQL 8.0 故障分析关键点

7.2.1 日志信息

在数据库故障分析中，日志信息扮演着非常重要的角色。通过仔细分析这些日志，DBA 和开发人员可以深入了解系统的运行状态、性能瓶颈以及潜在的问题。日志中记录了各种事件、错误和警告，为故障排查提供了宝贵的线索。

对于 MySQL 故障分析，下面以 Linux 平台为例，说明如何进行日志分析。

1. 操作系统日志

操作系统日志通常包含有关系统事件、错误、警告和其他与操作系统运行相关的信息。MySQL 作为一个运行在操作系统上的服务，其性能和稳定性与它和操作系统之间的交互密切相关。通过操作系统日志，可以了解到操作系统层面可能对 MySQL 服务产生影响的各种问题。例如，如果 MySQL 服务依赖于特定的系统资源（如内存、CPU、磁盘空间），那么当这些资源出现瓶颈或故障时，操作系统日志中可能会有相应的记录，帮助 DBA 识别和解决潜在的问题。

【例 7-1】MySQL 进程无故被 KILL 的情况会被记录在 /var/log/messages 日志中。记录的日志信息如下所示。

```
Out of memory: Kill process 17406 (mysqld) score 9 or sacrifice child
Killed process 17406, UID 27, (mysqld) total-vm:73034440kB, anon-rss:31852248kB, file-rss: 80kB,shmem-rssokb
```

说明：上述信息是指 mysqld 进程号为 17406 的进程被操作系统终止处理。total-vm：当时进程使用的总的虚拟内容大小。anon-rss：当时即匿名内存。file-rss：当时映射到设备和文件上的内存页。

关于 MySQL 的 OOM 故障，当系统内存不足时，Linux 内核会启动 OOM killer 机制。OOM killer 会选择一个占用内存最多的进程并终止它，以释放内存资源，防止系统因内存耗尽而崩溃。当 MySQL 运行在内存不足的环境中时，如果服务器的物理内存不足，或者 MySQL 的 InnoDB 缓存设置得过大，使得 MySQL 进程占用了大量内存，那么 MySQL 进程就可能被 OOM killer 选中并终止，导致服务中断。针对这种情况，增加内存资源或优化 MySQL 的内存使用参数。

2. 系统内核 dmesg 日志

dmesg 日志信息会显示 Linux 系统在启动和运行过程中产生的内核消息。这些消息通常涉及底层硬件、驱动程序、内存管理、网络堆栈等方面的信息。MySQL 服务通过系统调用与内核进行交互，所以 MySQL 的内核调用信息通常不会直接显示在 dmesg 的输出中，然而，在某些情况下，如果 MySQL 的操作触发了内核的错误或警告消息，这些消息就可能会出现在 dmesg 日志中。如果 MySQL 在尝试访问某个硬件设备时遇到了问题，或者它的内存分配请求失败了，那么这些异常情况就可能会通过内核消息反映出来。

【例 7-2】mysqld 服务进程突然崩溃（crash），在内核 dmesg 日志中的记录如下所示。

```
shel $> dmesg -T
mysgld[203563]   segfault at 0 ip null sp 00007fof9dd9b170   error 14 mysqld[400000+107600]
mysgld[8535] segfault at 0 ip null sp 00007ffcab34e670 error 14 in
mysqld.libread.so[7f038f8e1000+107600]
```

说明：上述信息内容介绍如下。
- at 0：变量内容。
- ip：指令地址。
- sp 00007ffcab34e670：栈顶地址。
- error 14：错误类型。
- mysqld.libread.so：崩溃的服务。
- [400000+107600]：库在内存映射中的基地址。

内核崩溃会导致 MySQL 服务崩溃。这种情况通常涉及多个层面的问题，可能关联到 MySQL 代码、操作系统底层内核以及数据文件等多个方面。针对这种情况，需要综合考虑多个方面，采取适当的应对措施。具体来说，可以通过升级内核和 MySQL 版本来修复已知问题，优化系统资源以减轻系统压力，以及检查和修复文件系统来确保数据文件的完整性与一致性。通过这些措施，可以提高系统的稳定性和可靠性，从而减少类似故障的发生。

3. MySQL 错误日志

MySQL 错误日志记录了 MySQL 服务器的启动和关闭过程中发生的任何错误或警告，以及 MySQL 服务运行过程中发生的错误、警告、异常和其他关键事件的信息。通过 MySQL 的错误日志，可以快速定位问题并解决之。

【例 7-3】MySQL 进程之间互斥信号量不足时记录的错误日志如下所示。

```
[ERROR][FATAL]InnoDB: Semaphore wait has lasted > 600 seconds.We intentionally crash the
server because it appears to be hung.
```

说明：MySQL 后台线程 srv_error_monitor_thread 发现存在阻塞超过 600s 的 latch 锁，如果仍没有释放，将最终导致崩溃。

【例 7-4】MySQL 数据库用户连接达到了 max_connections 参数的限制，导致新的连接请求无法被接受。记录的错误信息如下。

```
[ERROR] 1040 (HY000): Too many connections.
```

说明：这个错误通常发生在有大量并发连接请求的情况下，即超过了 MySQL 服务器配置的最大连接数。可以通过命令行方式将最大连接数调高。

【例 7-5】MySQL 服务的 binlog 文件占满了磁盘空间，导致无法继续写入新的数据。记录的错误信息如下。

```
[ERROR] Disk is full writing '/data/mysql/binlog/mysql-bin.000014' (Errcode: 16044192
- No space left on device).Waiting for someone to free space...
[ERROR] Retry in 60 secs.Message reprinted in 600 secs
```

说明：当磁盘空间不足时，MySQL 无法创建新的 binlog 文件或写入现有文件，这会导致数据库写入操作失败。此时，需要手动删除旧的 binlog 文件以释放磁盘空间。

4. MySQL 的 binlog 日志

MySQL 的 binlog 是服务层记录数据更改的日志文件，同时也是主从复制的基础。MySQL 的 binlog 日志内容可以协助排查以下问题。

- 主从复制问题：在主从复制架构中，MySQL 从节点使用 binlog 来同步数据。如果主从复制出现问题，则可以通过检查 binlog 来确定哪些事件在从节点上未正确执行，从而解决复制延迟或数据不一致等问题。
- 性能优化问题：binlog 中记录了语句发生时间、执行时长等信息，可以用于分析数据库性能瓶颈。通过查看 binlog 中的执行时长和频率，可以找出性能较差的查询语句并进行优化。

【例 7-6】通过 mysqlbinlog 命令行分析 binlog 中记录的操作内容。具体命令行示例如下。

```
# 具体语句查看
shll#> mysqlbinlog --no-defaults --base64-output=decode-rows -v -v
--start-position=214813221mysql-bin.000001
BEGIN
/*!*/。
# at 3004450
# 230903 19:11:55 server id 129 end_log_pos 3004524 CRC32 0x99503d86 Rows_query
# update bug set num=num-1
# at 3004524
# 230903 19:11:55 server id 129 end_log_pos 3004578 CRC32 0xe99d27b3 Table_map: `test`.`buy`
mapped to number 318
```

说明：通过分析 binlog，可以看到具体的 SQL 语句，以确定是否有可以优化的地方。还可以分析是否是大事务导致的长时间持有锁，影响数据库的并发性能。

【例 7-7】通过 mysqlbinlog 命令行分析 binlog 日志，查看对某个表执行的频率。具体命令行示例如下。

```
# 执行语句效率统计
shll#> mysqlbinlog --no-defaults --base64-output=decode-rows -v -v mysql-bin.000001 |
awk '/###/{if($0~/UPDATE|INSERT|DELETE/)count[$2""$NF]++}END{for(i in count)print
i,"\t",count[i]}'| column -t | sort -k3nr

UPDATE test.buy 5121158378
INSERT test.log 7812251
DELETE test.job 55320694
```

说明：通过 mysqlbinlog 命令行工具分析 binlog 日志，并查看对特定表执行的频率，可以了解哪些表操作频繁，导致锁竞争，进而识别潜在的性能瓶颈。

7.2.2 监控指标

数据库监控是指对数据库系统进行实时监控和分析，以全面了解其运行情况。这种监控对于及时发现和解决数据库系统的性能、安全和可靠性问题至关重要。特别是在发生故障时，这些监控数据会为 DBA 提供有力的分析和判断依据，从而加快问题的定位和修复速度。

下面介绍 MySQL 监控的指标和分析问题点。

1. MySQL 基础监控指标

基础监控指标主要包含以下几个方面，用于分析资源使用情况、MySQL 服务状态和负载情况。

- 服务器资源使用情况。包括 CPU 使用率、I/O 性能（读写速度、IOPS 等）、内存使用情况、磁盘容量使用大小和剩余大小。这些指标有助于综合分析服务器的整体性能和资源利用情况，帮助识别是否存在性能瓶颈或资源问题。
- MySQL 服务状态。包括 MySQL 服务是否存活（心跳检测等）。这个指标可确认 MySQL 服务正在运行并且可以从外部访问。
- MySQL 的数据库使用情况。包括数据文件大小、索引大小、表数量和碎片率。这些指标有助于分析数据库内部是否存在大型操作或过多的碎片，导致影响数据库的性能。
- MySQL 的性能指标。包括 QPS（每秒查询数）和 TPS（每秒事务数）指标。这些指标有助于分析当时 MySQL 的负载情况，帮助判断是否存在过高的负载，从而导致 MySQL 服务无法承载。
- MySQL 的 InnoDB 内存使用情况。包括缓冲池和日志缓冲区的大小和使用情况。这些指标有助于分析内存瓶颈或内存不足的问题。
- MySQL 的用户连接情况。包括已连接的用户数和活动的用户连接数。这些指标有助于确认是否存在连接数已满的问题。

2. MySQL 性能监控指标

性能监控指标可以帮助分析系统的负载情况和性能瓶颈。

- 锁信息（死锁、行锁和表锁）。有助于分析并发问题、死锁原因。
- 慢查询的次数或语句内容的监控。有助于分析 SQL 语句执行效率低的原因，如是否缺少索引、是否使用了低效的查询方式等。
- 缓存命中率、缓存使用情况。有助于分析缓存是否配置不足和存在瓶颈。
- 临时表使用情况。有助于分析 SQL 语句性能问题。例如，通过减少不必要的 JOIN 操作、优化 WHERE 子句等方式减少临时表的使用。
- 连接、排序次数。有助于分析 SQL 语句多表连接、排序操作中的性能瓶颈。
- 全表扫描次数。有助于分析全表扫描的 SQL 语句，可以优化索引设计，避免不必要的全表扫描。
- InnoDB I/O 操作信息。有助于分析磁盘 I/O 的瓶颈，调整 InnoDB 的 I/O 相关参数。
- 当前运行线程的次数。有助于分析某些线程占用资源过多或执行效率低下的问题。
- 回滚语句的统计。有助于分析哪些事务因为错误而回滚，从而优化事务逻辑，减少回滚操作。
- Error 日志错误次数。有助于分析错误次数和具体错误信息，减少因错误导致的 MySQL 故障的发生。

3. MySQL 复制指标

MySQL 高可用是基于逻辑的复制方式。高可用节点的可用性非常重要。监控复制指标主要包含以下内容。

- 监控主库 binlog 和从库 relaylog 的文件数与大小。当文件大小超过设置阈值时，说明存在大事务（如长时间运行的事务、锁争用等），这可能导致延迟或阻塞发生。通过对比主库和从库的文件数与大小，可以判断是否存在文件丢失或传输延迟问题。
- 复制延迟时间。当主节点上有数据更新时，这些更新需要被同步到从节点上，复制延迟时间就是衡量这一过程所需时间的指标。复制延迟时间的增加表明数据从主节点同步到从节点的过程遇到了问题，其中包含大事务（涉及大量的数据更改）、资源方面存在瓶颈、锁竞争等问题。
- 复制进程状态。在 MySQL 的复制过程中，有两个关键的线程：接收日志线程（I/O 线程）和回放线程（SQL 线程）。这两个线程的状态对于确保复制的正常运行至关重要。如果其中任何一个线程的状态不是"Yes"，就需要进一步检查该线程的错误日志或状态信息，以确定问题的原因。
- 复制错误累计数。这一指标记录了从节点在复制数据时发生的错误次数。较少的复制错误数通常意味着 MySQL 的可靠性较高。当复制错误累计数较高时，表明存在一些问题，如网络不稳定、磁盘故障、配置错误和 MySQL 自身的 bug 等。

7.2.3 诊断工具

MySQL 数据库的基本问题涉及存储性能、操作系统兼容性、应用程序应用、数据库设置 4 个方面。故障分析并没有想象的那么难，当诊断工具应用于排查时，可以产生意想不到的效果。

1. Linux 平台中常用的性能分析工具

以 Linux 平台为例，介绍常用的性能分析工具和分析点，通过认真学习，读者就可以诊断绝大多数性能问题了。

（1）top（系统进程监控）

top 命令是 Linux 下常用的性能分析工具，能够实时显示系统中各个进程的资源占用状况，其功能类似于 Windows 的任务管理器。top 命令会定期显示并更新所有正在运行的进程列表，包含 CPU 的使用、内存的使用、交换内存、缓存大小、缓冲区大小等信息。此外，它还可以显示进程控制、用户和其他重要命令。查看进程信息的 top 命令行示例如下。

```
shell $>top
top - 10:33:42 up 18:55,  2 users,  load average: 0.00, 0.00, 0.00
Tasks: 100 total,   1 running,  99 sleeping,   0 stopped,   0 zombie
%Cpu0  :  1.0 us,  0.7 sy,  0.0 ni, 97.7 id,  0.0 wa,  0.3 hi,  0.3 si,  0.0 st
MiB Mem :   1826.7 total,    352.2 free,    635.6 used,    838.9 buff/cache
MiB Swap:      0.0 total,      0.0 free,      0.0 used.   1043.5 avail Mem

    PID USER      PR  NI    VIRT    RES    SHR S  %CPU  %MEM     TIME+ COMMAND
   2645 mysql     20   0 2320152 518044  40292 S   0.3  27.7   4:21.69 mysqld
    768 root      20   0  221564  38072  36412 S   0.0   2.0   0:00.85 sssd_nss
...
```

说明：关注 us（使用率）和 PID 对应 %CPU 指标即可。举例说明如下。

- %Cpu0 1.0 us：整个操作系统 CPU 使用率。
- 2645 mysql 0.3：MySQL 进程的 CPU 使用率。

当内存和 CPU 使用率过高时，可以通过 top 命令突出显示那些正在运行的进程。通过查看命令行输出，可以迅速了解系统的运行状态和性能瓶颈。

（2）vmstat（虚拟内存统计）

vmstat 命令在 Linux 平台上用于显示虚拟内存、内核线程、磁盘、系统进程、I/O 模块、中断、CPU 活跃状态等信息。查看整体系统指标的命令行示例如下。

```
shell $>vmstat
procs -----------memory---------- ---swap-- -----io---- -system-- ------cpu-----
 r  b   swpd   free   buff  cache   si   so    bi    bo   in   cs us sy id wa st
 5  0      0 360132   2104 856984    0    0    12     6  590  136  1  1 98  0  0
```

说明：

- r：可运行进程数。
- b：阻塞等待 I/O 完成的进程数。
- memory：内存，默认单位为 KB。
- swpd：使用的交换内存量。
- free：空闲内存量。
- buff：用作缓冲区的内存量。
- cache：活动内存的数量。
- swap：交换，单位为 kbit/s。
- si：从磁盘交换到内存的交换页数量。
- so：从内存交换到磁盘的交换页数据。
- io：输入输出，单位为块/s。
- bi：从块设备接收到的字节数。
- bo：发送到块设备的字节数。
- system：系统。
- in：每秒的中断次数。
- cs：每秒上下文切换的次数。

通过上述这些信息，可以诊断以下问题。

- 内存问题：帮助用户了解系统的内存使用情况和虚拟内存情况，从而及时发现和解决内存泄漏、内存不足等问题。
- 磁盘 I/O 问题：帮助用户了解系统的磁盘 I/O 性能状况，从而及时发现和解决磁盘读写瓶颈、I/O 等待时间过长等问题。
- CPU 问题：帮助用户了解系统的 CPU 使用情况，从而及时发现和解决 CPU 过载、性能瓶颈等问题。

（3）lsof（打开文件列表）

lsof 命令在 Linux 或 UNIX 平台用于显示打开的文件和进程。打开的文件主要包括磁盘文件、网络套接字、管道、设备和进程。通过这个命令，可以很容易地看出哪些文件正在使用。查看 MySQL 进程打开的文件的命令行示例如下。

```
shell $>lsof |grep mysqld
mysqld_sa 1099   root    cwd    DIR 253,1     4096  50642560 /home/kevindba
mysqld_sa 1099   root    rtd    DIR 253,1      244       128 /
```

```
mysqld_sa 1099  root    txt  REG 253,1   1219248  33615736 /usr/bin/bash
mysqld    2645  mysql   cwd  DIR 253,1       4096  51169506 /opt/data8.0/data
mysqld    2645  mysql   DEL  REG 0,18       28213 /[aio]
mysqld    2645  mysql   0r   CHR   1,3        0t0     8887 /dev/null
...
```

这个命令会列出所有打开的文件，并通过 grep 过滤出包含"mysql"的行。这将会显示 MySQL 进程打开的所有文件，包括配置文件、日志文件、数据文件等。

通过上述这些信息，可以诊断以下问题。

- 配置问题：如果配置文件被修改并且 MySQL 无法正常启动或运行，则可以在 lsof 输出中查看配置文件是否被正确加载。
- 资源瓶颈：通过检查打开的文件数量和类型，可以了解 MySQL 是否受到文件描述符限制或其他资源瓶颈的影响。
- 数据完整性问题：如果数据文件被意外删除或损坏，则 lsof 可以帮助确定哪些文件正在被 MySQL 使用，从而判断数据完整性是否受到影响。

（4）tcpdump（网络数据包分析器）

tcpdump 是一种使用广泛的抓取网络数据包的工具。它支持针对网络层、协议、主机、网络或端口的过滤并对其进行详细的检查，这对于网络调试、性能分析、故障排查和安全审计等任务非常有用。查看和抓取的命令行示例如下。

```
shell $>tcpdump -i eth0
tcpdump: verbose output suppressed, use -v or -vv for full protocol decode
listening on eth0, link-type EN10MB (Ethernet), capture size 262144 bytes
10:40:02.316213 IP iZuf6178v14ipc59jbbpfnZ.ssh > 301.237.105.96.55931: Flags [P.], seq
4249759042:4249759254, ack 3741082870, win 274, length 212
10:40:02.316818 IP iZuf6178v14ipc59jbbpfnZ.50822 > 118.100.2.136.domain: 13477+ PTR?
101.105.237.116.in-addr.arpa.(45)
10:40:02.324111 IP 116.437.205.96.55931 >iZuf6178v14ipc59jbbpfnZ.ssh: Flags [.], ack
212, win 1027, length 0
...
```

说明：抓取的网络数据包不支持可读性，需要另外添加 tcpdump 参数解析，也可以保存成扩展名为 pcap 的文件，使用 Wireshark 等软件进行查看。

使用 tcpdump 抓取 MySQL 数据包并分析，可以找出 MySQL 的性能瓶颈、网络问题或安全漏洞等。例如，可以分析数据包的传输延迟、丢包情况、重传情况等，从而判断网络连接的稳定性和性能；还可以分析数据包的内容，查看是否有异常或恶意的 SQL 语句、注入攻击等问题。

（5）netstat（网络统计）

netstat 命令在 Linux 上用于查看网络连接、路由表、接口统计信息等。通过 netstat 命令，可以查看当前系统所有的网络连接情况，包括占用的端口等。它对于监控网络性能和解决网络等相关问题都非常实用。查看 mysqld 进程相关的网络连接情况的命令行示例如下。

```
shell $> netstat -a | grep mysql
tcp6       0      0 [::]:mysqlx              [::]:*                    LISTEN
unix  2    [ ACC ]    STREAM     LISTENING     10443249 /opt/data8.0/data/mysql.sock
unix  2    [ ACC ]    STREAM     LISTENING     10443247 /tmp/mysqlx.sock
```

说明：需要关注 State（状态）值。

- LISTENING：监听中，服务端需要打开一个 socket 进行监听，侦听来自远方 TCP 端口的连接请求。
- ESTABLISHED：已连接，代表一个打开的连接，双方可以进行或已经在进行数据交互。

使用 netstat 查看 MySQL 网络信息，可以诊断 MySQL 相关的以下问题。

- MySQL 服务是否正在运行。使用 netstat 可以查看 MySQL 服务端口是否处于监听状态。如果 MySQL 服务正在运行并且配置正确，就应该能够看到该端口处于 LISTENING 状态。
- MySQL 连接问题。如果 MySQL 服务正在运行但客户端无法连接，就可以使用 netstat 检查是否有任何连接正在尝试连接到 MySQL 端口。此外，还可以查看是否有防火墙规则阻止了连接。
- 客户端连接情况。通过 netstat，可以查看 MySQL 服务器的网络连接数量、状态以及数据传输量。这有助于识别是否有过多的连接、长时间打开的连接或数据传输瓶颈。

（6）iostat（输入/输出统计）

iostat 用于监视系统输入/输出设备负载情况。它可以报告 CPU、各个分区、逻辑磁盘、物理磁盘、网络文件系统（如 NFS）及异步 I/O 的统计信息。使用 iostat 查看 I/O 性能的命令行示例如下。

```
shell $>iostat -x -d 2
Linux 4.18.0.e18_2.x86_6409/06/2023 _x86_64_(1 CPU)
Device r/s   w/s rkB/s   wkB/s  rrqm/s  wrqm/s  %rrqm  %wrqm r_await w_await aqu-sz rareq-sz wareq-sz svctm  %util
vda     0.16  0.40 11.76   5.76   0.00    0.02   0.22   4.85   0.98    1.02    0.00   72.90
14.40   0.58  0.03
```

说明：对于上述指标，关注以下几个指标即可。

- Device：/dev 目录下的磁盘（或分区）名称。
- r/s：每秒从磁盘读取数据的大小，单位为 KB/s。
- w/s：每秒写入磁盘的数据的大小，单位为 KB/s。
- r_await、w_await：读、写平均等待时间。
- %util：查看设备使用率。

MySQL 数据库的性能往往受到底层存储系统的影响，特别是当涉及大量的数据读写操作时。使用 iostat，可以收集关于磁盘 I/O 的关键指标，这些指标可以帮助识别是否存在磁盘性能问题。

2. MySQL 分析工具

MySQL 故障排查时常用的两个工具：官方自带工具 mysqlbinlog 和慢查询日志分析工具 pt-query-digest（在 6.2.3 节和 6.2.4 节中有详细的介绍）。

（1）使用 mysqlbinlog 分析 binlog 文件

mysqlbinlog 是 MySQL 官方提供的一个实用工具，它用于处理和分析 binlog 文件。binlog 日志记录了所有更改数据库数据的语句，这些语句是以"事件"的形式记录的。通过使用 mysqlbinlog 工具，DBA 可以读取和解析这些日志，从而帮助排查数据库故障。

当数据库出现问题时，如数据不一致、丢失或意外的更改，DBA 可以使用 mysqlbinlog 来查看二进制日志中记录的操作，以便找出导致问题的原因。通过查看这些操作，DBA 可以理解在特定时间点发生了什么，以及为什么会发生。此外，mysqlbinlog 还可以与其他工具（如 grep）结合使用，以过滤和搜索特定的日志条目，从而快速定位问题。

总之，mysqlbinlog 是一个强大的工具，它可以帮助 DBA 在排查数据库故障时获取关键信息，并了解导致问题的具体原因。

（2）使用 pt-query-digest 分析慢查询日志

pt-query-digest 是 Percona Toolkit 工具集中较为常用的工具，常用于分析慢查询日志（slow log）。它也可以分析 MySQL 数据库的 binlog、通用日志，同时还可以使用利用 show processlist 或 tcpdump 抓取的 MySQL 协议数据来进行分析。

通过 pt-query-digest 分析 MySQL 的慢查询日志，可找出执行时间较长的查询语句，统计各个查询语句的执行次数，发现可能导致性能下降的问题，如锁争用、不合适的索引等，从而进行优化。

7.2.4 SQL 语句

SQL（Structured Query Language，结构化查询语言）是数据库和应用系统之间的交互语言，主要用于查询、插入、更新和删除数据库中的数据。数据库系统定义了严格的 SQL 语法规则，但这并不意味着所有符合规则的 SQL 语句都能成功执行或不会产生错误。不当或低效的 SQL 语句可能会降低数据库的性能，甚至导致故障问题发生。

1. 确认 SQL 语句是否高效的方法

MySQL 提供通过查看执行计划和资源消耗来评估 SQL 语句效率的方式，具体有以下 3 种。

- 使用 EXPLAIN 输出查询计划来进行分析。
- PROFILE 语句可以显示当前会话过程中执行的语句的资源使用情况，从而帮助用户有效地了解资源的使用情况。
- OPTIMIZER_TRACE 可以跟踪实际执行的过程，帮助理解 MySQL 优化器所采取的决策措施。

2. SQL 语句需要结合 MySQL 的 InnoDB 引擎特性使用

MySQL 的 InnoDB 引擎表是索引组织表，意味着表中的数据实际上是按照主键的顺序存储的。所有的操作都是以主键为基础的。因此，在编写 SQL 语句时，充分利用 InnoDB 的这种特性是非常重要的。需要注意如下几点。

- MySQL 主键的重要性不言而喻。在数据更改中，主键用于唯一标识每一行数据；在多表 JOIN 操作中，主键与索引协同工作，以便更快地定位数据；在复制过程中，基于主键的值来识别和复制数据行，从而优化复制性能。
- 在 SQL 查询中，如果 WHERE 条件中没有使用索引字段，那么 MySQL 数据库将不得不进行全表扫描，这种情况下查询性能将会很差。为了提高查询性能，应该尽量在 WHERE 子句中使用索引字段。
- 在 SQL 查询中，当使用多表关联（JOIN）生成一个中间表时，中间表的字段索引会失效。索引失效意味着后面的关联查询使用全表扫描。
- 在 SQL 查询中，当查询需要返回表中大部分数据（大致 30%）时，即使查询条件使用了索引，也会出现索引失效的情况，因为有些场景下，返回大部分数据比全表扫描慢，使用索引可能并不总是有效的。因此，查询返回的数据量也要有效控制。

3. 参数配置是否合理

MySQL 的参数配置对于数据库的性能和稳定性至关重要。不同的参数配置会影响内存使用、磁盘 I/O、网络带宽、并发处理等多个方面。

- MySQL 的内存参数设置不合理可能会显著影响表之间的交互以及整个数据库系统的资源使用率。涉及的参数有 join buffer、sort buffer、read buffer 和 tmp_table 等。

- MySQL 的网络交互和并发限制设置不当可能导致服务无法有效处理更多的请求，这可能会引发一系列问题，包括请求堆积、CPU 资源和网络资源的过度使用，最终可能导致这些资源耗尽。涉及的参数有 max_connections、max_allowed_packet 和 innodb_thread_concurrency 等。

7.3 MySQL 8.0 典型故障分析实践

7.3.1 导致服务器 OOM 的故障分析

在 MySQL 运行环境中，MySQL 进程的内存占用率上升到 90% 及以上会导致服务器发生 OOM 故障，这是比较常见的现象。MySQL 内存使用率过高的原因有很多，普遍情况是由于使用不当造成的。除此之外，MySQL 本身可能存在的缺陷也会导致内存占用过高。下面将具体介绍排查服务器 OOM 故障的思路和方法。

1. 内存参数设置检查

MySQL 的内存分配参数设置不合理时，可能在启动时占用过多的内存，最终导致系统内存不足。为了解决这个问题，需要根据系统的实际情况调整内存参数。这些参数包括缓冲池大小、连接内存、临时表内存等。

MySQL 的内存主要分为全局内存和线程级内存。

- 全局内存：这些内存在 MySQL 启动时一次性分配，并且不会在运行时释放。常见的全局内存参数包括 innodb_buffer_pool_size、key_buffer_size 和 innodb_log_buffer_size。需要根据系统的可用内存、工作负载等因素来合理配置这些参数。
- 线程级内存：每个连接到 MySQL 服务的线程都需要分配自己的内存。这些内存包括线程栈（thread_stack，默认大小为 256KB）以及其他操作所需的内存（如排序、连接、临时表等）。线程级内存只有在需要时才会被使用，并在操作完成后释放。

在 MySQL 中，统计正在使用的内存信息的命令行示例如下。

```
mysql> SELECT @@key_buffer_size, @@innodb_buffer_pool_size,
    @@innodb_log_buffer_size,@@tmp_table_size,
    @@read_buffer_size,@@sort_buffer_size,
    @@join_buffer_size,    @@read_rnd_buffer_size,
    @@binlog_cache_size,    @@thread_stack,
(SELECT COUNT(host) FROM  information_schema.processlist where command<>'Sleep')as run_connet \G
*************************** 1.row ***************************
        @@key_buffer_size: 8388608
@@innodb_buffer_pool_size: 1073741824
 @@innodb_log_buffer_size: 16777216
         @@tmp_table_size: 67108864
       @@read_buffer_size: 131072
       @@sort_buffer_size: 2097152
       @@join_buffer_size: 2097152
   @@read_rnd_buffer_size: 4194304
      @@binlog_cache_size: 32768
           @@thread_stack: 524288
              run_connet: 1
1 row in set (0.00 sec)
```

说明：query_cache_size 参数在 MySQL 8.0 版本中已经废弃，所以不会计算。返回的 run_connet 列统计正在运行中的线程数。

2. 统计存储过程、函数、触发器和视图的使用情况

MySQL 中的存储过程、函数、触发器和视图可以带来很大的便利，通过封装和抽象化数据库逻辑，使得应用代码更加简洁和易于维护。然而，不当使用这些数据库对象有可能导致内存溢出的问题。读者应注意如下情况。

- 递归调用：如果存储过程和函数中的递归调用没有正确的终止条件或递归层次过深，则可能会导致栈溢出或消耗大量内存。
- 大型结果集：如果一个存储过程或函数返回了非常大的结果集，而没有有效地处理这些数据，则可能会导致内存占用迅速增加。
- 复杂的查询逻辑：复杂的 SQL 查询，特别是涉及多表连接、子查询、临时表等，可能会消耗大量内存。
- 视图嵌套：一个视图可以嵌套其他视图，如果嵌套层次过多或视图定义复杂，则可能会增加查询处理的复杂性和内存消耗。
- 数据类型选择不当：使用不合适的数据类型，比如将大量的文本数据存储在不适当的列类型中，会导致内存使用上升。

下面统计 MySQL 中这些对象的使用情况。

（1）存储过程和函数统计

MySQL 中存储过程和函数处理机制是先编译 SQL 语句再执行。在高并发或复杂逻辑下，可能导致内存和 CPU 的使用急剧增加。统计存储过程和函数的命令行示例如下。

```
# MySQL 5.7 版本
mysql> SELECT db,type,count(*)
       FROM mysql.proc
       WHERE db not in ('mysql','information_schema','performance_schema','sys')
       GROUP BY db, type;

# MySQL 8.0 版本
mysql> SELECT Routine_schema, Routine_type
       FROM information_schema.Routines
       WHERE Routine_schema not in
       ('mysql','information_schema','performance_schema','sys')
       GROUP BY Routine_schema, Routine_type;
```

（2）触发器统计

触发器是一种在数据库中自动执行操作的机制，当特定事件（如 INSERT、UPDATE 或 DELETE 操作）发生在指定的表上时，它会触发并执行相应的操作。虽然触发器本身不直接占用大量内存，但如果触发器执行的操作复杂或频繁触发，则可能会导致资源争用、锁竞争和性能下降。在极端情况下，如果多个资源密集型触发器同时执行，或者因触发器设计不当导致内存等资源泄漏，则可能会对数据库服务器的内存和性能造成严重影响。统计触发器使用情况的命令行示例如下。

```
mysql> SELECT TRIGGER_SCHEMA, count(*)
FROM information_schema.triggers
       WHERE TRIGGER_SCHEMA not in
```

```
        ('mysql','information_schema','performance_schema','sys')
    GROUP BY TRIGGER_SCHEMA;
```

(3)视图统计

视图是一种虚拟表,是多个表的数据的逻辑显示。视图并不存储数据。如果视图创建不合理,则会大量生成临时表,大量消耗内存,容易变成数据库系统内存上升的潜在隐患。统计视图使用情况的命令行示例如下。

```
mysql> SELECT TABLE_SCHEMA, COUNT(TABLE_NAME)
    FROM information_schema.VIEWS
    WHERE TABLE_SCHEMA not in
        ('mysql','information_schema','performance_schema','sys')
    GROUP BY TABLE_SCHEMA;
```

在 MySQL 中,内存泄漏通常是由应用程序或库中的编程错误引起的,尽管 MySQL 本身或其组件在罕见情况下也可能出现此类问题。不恰当的配置或查询设计可能导致 MySQL 使用过多的内存,进而引发操作系统的 OOM 现象。通过检查 MySQL 的配置参数和查询设计,评估是否存在潜在的内存泄漏问题。接下来,将通过分析系统统计数据和实际内存使用情况,来进一步诊断和解决潜在的性能问题。

3. 系统内存使用情况

MySQL 系统库提供了一些内存统计指标,这些指标可以通过查询 MySQL 的系统表来获取。了解这些指标可以帮助评估 MySQL 服务器的内存使用情况和性能表现,以便进行必要的优化和调整。

(1)总内存使用统计

了解 MySQL 当前占用的总内存,可以帮助判断 MySQL 服务的内存使用与实际分配的内存是否存在不一致的情况。统计当前占用的总内存的命令行示例如下。

```
mysql> SELECT SUM(CAST(replace(current_alloc,'MiB','')  as DECIMAL(10, 2)))
    FROM sys.memory_global_by_current_bytes
    WHERE current_alloc like '%MiB%';
```

(2)对事件统计内存使用情况

了解 MySQL 当前事件使用的内存,可以帮助判断特定事件所消耗的内存,以及在事件完成后是否存在使用内存不释放或内存泄漏的情况。统计 MySQL 不同事件使用内存的情况的命令行示例如下。

```
# 总的事件内存值
mysql> SELECT event_name,
    SUM(CAST(replace(current_alloc,'MiB','')  as DECIMAL(10, 2)))
    FROM sys.memory_global_by_current_bytes
    WHERE current_alloc like '%MiB%' GROUP BY event_name
      ORDER BY SUM(CAST(replace(current_alloc,'MiB','')  as DECIMAL(10, 2))) DESC;
# 不同事件使用内存合计值
mysql> SELECT event_name,
      sys.format_bytes(CURRENT_NUMBER_OF_BYTES_USED)
    FROM performance_schema.memory_summary_global_by_event_name
    ORDER BY CURRENT_NUMBER_OF_BYTES_USED DESC
    LIMIT 10;
```

(3) 对账号统计内存使用情况

通过统计 MySQL 访问账号的内存使用情况，可以发现哪些账号在使用过多的内存，并且可以结合其他内存使用信息来判断账号是否存在内存泄漏或内存不释放的问题。统计 MySQL 访问账号占用内存情况的命令行示例如下。

```
mysql>SELECT user,event_name,current_number_of_bytes_used/1024/1024 as
    MB_CURRENTLY_USED
    FROM performance_schema.memory_summary_by_account_by_event_name
    WHERE host<>"localhost"
    ORDER BY current_number_of_bytes_used
    DESC LIMIT 10;
```

说明：对账号进行内存统计是有必要的，因为许多环境都集成了第三方插件和模拟从库，这些插件和模拟从库很容易引发内存不释放的问题。

(4) 对线程对应事件统计内存使用情况

通过统计 MySQL 线程对应事件的内存使用情况，可以了解某个线程在处理 SQL 语句时是否存在内存泄漏或不释放内存的问题。统计 MySQL 线程占用内存情况的命令行示例如下。

```
mysql> SELECT thread_id,event_name,
    sys.format_bytes(CURRENT_NUMBER_OF_BYTES_USED)
    FROM performance_schema.memory_summary_by_thread_by_event_name
    ORDER BY CURRENT_NUMBER_OF_BYTES_USED DESC
    LIMIT 10;
+-----------+----------------------------------+------------------------------------------------+
| thread_id | event_name                       | sys.format_bytes (CURRENT_NUMBER_OF_BYTES_USED) |
+-----------+----------------------------------+------------------------------------------------+
|         1 | memory/innodb/memory             | 400.70 KiB                                     |
|         1 | memory/mysqld_openssl/openssl_malloc | 337.76 KiB                                 |
|         1 | memory/sql/dd::objects           | 329.98 KiB                                     |
...
+-----------+----------------------------------+------------------------------------------------+
3 rows in set (0.01 sec)
```

(5) 对 SQL 语句统计内存使用情况

大多数 SQL 语句在查询完成后会自动释放分配的内存。通过统计 SQL 语句的内存使用情况，可以了解某个 SQL 语句在处理时是否占用了大量内存，以及在查询完成后是否自动释放了这些内存。统计 MySQL 里执行过的语句的内存使用情况的命令行示例如下。

```
mysql> SELECT m.thread_idtid,
    m.user,
    esc.DIGEST_TEXT,
    m.current_allocated,
    m.total_allocated
    FROM sys.memory_by_thread_by_current_bytes m,
    performance_schema.events_statements_current esc
    WHERE m.`thread_id` = esc.THREAD_ID \G
*************************** 1.row ***************************
tid: 49
user: root@localhost
DIGEST_TEXT: SELECT `m`.`thread_id` `tid`, `m`.`user`, `esc`.`DIGEST_TEXT`,
```

```
`m`.`current_allocated`, `m`.`total_allocated` FROM `sys`.
`memory_by_thread_by_current_bytes` `m`, `performance_schema`.
`events_statements_current` `esc` WHERE `m`.`thread_id` = `esc`.`THREAD_ID`
current_allocated: 8.72 MiB
total_allocated: 137.96 MiB
1 row in set (0.05 sec)
```

（6）开启所有内存统计指标

MySQL 的内存统计默认已经启用基础监控，这对于大多数场景来说已经足够。然而，如果需要更深入地了解内存使用情况，就需要启用更强大的内存监控功能。额外的监控功能可以通过调整 Performance Schema 的配置来实现。启用更强大的内存监控功能可能会对 MySQL 的性能产生一定的影响，因此在生产环境中需要谨慎操作。此外，更改配置后通常需要重新启动 MySQL 服务，这样才能使更改生效。开启所有内存统计指标的命令行示例如下。

```
#配置开启所有内存统计指标
[mysqld]
performance-schema-instrument ='memory/%=COUNTED'

#开启所有内存统计指标命令行
mysql> UPDATE performance_schema.setup_instruments SET ENABLED = 'YES' WHERE NAME LIKE 'memory/%';

#关闭所有内存统计指标命令行
mysql> UPDATE performance_schema.setup_instruments SET ENABLED = 'NO' WHERE NAME LIKE 'memory/%';
```

MySQL 提供的更详细的内存监控系统表列表如下所示。

```
mysql> select * from sys.x$memory_by_host_by_current_bytes;
mysql> select * from sys.x$memory_by_thread_by_current_bytes;
mysql> select * from sys.x$memory_by_user_by_current_bytes;
mysql> select * from sys.x$memory_global_by_current_bytes;
mysql> select * from sys.x$memory_global_total;
mysql> select * from performance_schema.memory_summary_by_account_by_event_name;
mysql> select * from performance_schema.memory_summary_by_host_by_event_name;
mysql> select * from performance_schema.memory_summary_by_thread_by_event_name;
mysql> select * from performance_schema.memory_summary_by_user_by_event_name;
mysql> select * from performance_schema.memory_summary_global_by_event_name;
```

说明：从表名就可以了解到包含哪些信息，如主机、线程、用户、当前占用内存、占用的总内存。找到对应问题事件或线程后，可以进一步排查导致内存占用率高的指标。

4. 通过系统工具查看内存使用情况

下面介绍 Linux 下常用的内存使用统计命令行工具。首先使用 top、free、ps 在系统级别确定是否存在内存使用过多的问题。如果存在，则可以从 top 的输出中确定哪一个 MySQL 进程占用了大量内存。除此之外，pmap 工具能够帮助确定进程是否有内存泄漏问题。

（1）top 命令

top 命令是 Linux 下常用的性能分析工具，能够实时显示系统中各个进程的资源占用状况。通过〈Shift + M〉快捷键，可以查看按内存使用量排序的进程列表，从而了解实际内存使用情况。

在 top 命令的输出中，需要关注对应服务进程的 RES（物理内存使用量）指标，它表示进程当前使用的、未被交换出的物理内存大小。使用 top 查看内存资源使用的命令行示例如下。

```
shell $>top
top - 12:57:37 up 192 days,  2:12,  2 users,  load average: 0.00, 0.00, 0.00
Tasks: 101 total,   3 running,  98 sleeping,   0 stopped,   0 zombie
%Cpu(s):  1.7 us,  0.6 sy,  0.0 ni, 97.8 id,  0.0 wa,  0.0 hi,  0.0 si,  0.0 st
MiB Mem :   1826.3 total,     99.6 free,    673.5 used,   1053.2 buff/cache
MiB Swap:      0.0 total,      0.0 free,      0.0 used.    996.7 avail Mem

    PID USER      PR  NI    VIRT    RES    SHR S  %CPU  %MEM     TIME+ COMMAND
 627215 mysql     20   0 1107512 504228  39956 S   0.6  27.0  60:06.41 mysqld
    556 root      20   0  176576  63020  51064 S   0.0   3.4   4:43.95 systemd-journal
    922 root      20   0 1054152  53912  51940 S   0.0   2.9  14:07.49 rsyslogd
    820 root      20   0  425544  17116   2312 S   0.0   0.9  29:06.76 tuned
 670250 root      20   0   63436  15104   9492 S   0.0   0.8   0:00.06 mysql
```

（2）free 命令

free 命令是 Linux 系统中用于显示内存使用情况的工具。它提供了系统总内存、已用内存、空闲内存、交换空间（swap）等信息，可帮助用户了解系统的内存使用状况。使用 free 查看内存资源使用的命令行示例如下。

```
shell $>free
              total        used        free      shared  buff/cache   available
Mem:        1870124      688256      103356        2004     1078512     1022084
Swap:             0           0           0
```

free 输出指标的说明如下。
- total：表示总的物理内存大小。
- used：显示已经使用的物理内存和交换空间。
- buff/cache：显示被缓冲区和缓存使用的物理内存大小。
- available：显示还可以被应用程序使用的物理内存大小。
- Swap 行：交换空间的使用情况。

通过运行 free 命令并观察其输出，可以了解系统的内存使用情况是如何随时间变化的，从而采取相应的措施来优化或解决内存相关的问题。

（3）ps 命令

ps 命令是 Linux 系统中用于查看系统中运行的进程信息的工具。ps 命令可以显示当前活动的进程以及它们的详细信息，如 PID（进程 ID）、PPID（父进程 ID）、CPU 使用率、内存使用情况、进程优先级、进程状态等。查看 MySQL 进程使用内存情况的命令行示例如下。

```
shell $> ps eo user,pid,vsz,rss $(pgrep -f 'mysqld')
USER       PID     VSZ     RSS
root       215945   12960    2356
mysql      217246 1291540  241824
root       221056   12960    2428
mysql      374243 1336924  408752
```

说明：ps 输出指标的说明如下。
- USER：进程对应的用户。

- PID：进程 ID。
- VSZ：进程内存空间的大小。
- RSS：全称为 Resident Set Size，表示常驻内存的大小。

（4）pmap 命令

pmap 命令是 Linux 系统的一个调试和运维工具，用于查看进程的内存映射信息。通过 pmap 命令，可以查看 MySQL 进程的内存映射信息，包括每个内存段的起始地址、大小、访问权限和映射文件等，从而了解 MySQL 进程的内存分配和使用情况。

下面通过 pmap 统计记录 MySQL 进程（PID 为 22837）的内存使用情况，并且每两秒记录一次，使用 while 循环命令。具体命令行示例如下。

```
shell $> while true; do pmap -d 22837 | tail -1; sleep 2; done
# MySQL 客户端连接
mapped: 1333484K    writeable/private: 729992K    shared: 18588K
mapped: 1333484K    writeable/private: 729992K    shared: 18588K
# 执行 use db 命令
mapped: 1333484K    writeable/private: 730252K    shared: 18588K
mapped: 1333484K    writeable/private: 730252K    shared: 18588K
# 执行 show table 命令
mapped: 1333484K    writeable/private: 730744K    shared: 18588K
mapped: 1333484K    writeable/private: 730744K    shared: 18588K
```

通过 pmap 命令统计 MySQL 在特定 SQL 语句前后（如新的连接、使用数据库和显示表）捕获其内存映射信息时的内存使用情况。从这些捕获信息中，观察内存使用量的增长趋势，以及是否有异常增长的情况。

说明：pmap 输出指标的说明如下。
- mapped：占用内存的文件的大小。
- writeable/private：表示进程所占用的私有地址空间大小，也就是该进程实际使用的内存大小。
- shared：表示一个进程和其他进程共享的内存大小。

5. 总结

对于 MySQL 内存使用高导致的 OOM 问题，可以按照以下几点进行排查。
- 检查参数设置是否合理。
- 需要通过 ps 库进行排查。
- 利用 Linux 工具分析内存使用情况。
- 在官方 Bug 里查找内存泄漏（memory leak）相关问题，确定是否存在修复 Bug 的新的 MySQL 版本。

通过以上排查，若没有找到原因，则可以更换服务器或主从切换观察，也可以进行版本升级（代价不小）。如果可能，提供一个实际环境，进行逐步调试，抓取内存变化，以确定导致内存泄漏的具体原因。

7.3.2 导致 Got an error reading communication packet 提示的故障分析

在 MySQL 的错误日志中，经常会遇到与通信故障相关的告警。若告警信息是 "Got an error reading communication packet"，则意味着客户端在尝试从 MySQL 服务读取数据时遇到了问题。下面将分析这个告警出现的原因,并提供解决方案。

1. MySQL 指标分析

首先,当通信故障错误出现时,MySQL 的状态值 Aborted_clients 和 Aborted_connects 的计数会增加。这两个状态变量描述了由于客户端没有正确关闭连接而导致中断的连接数以及那些尝试登录 MySQL 但失败的连接数量。下面将详细地分析这两个状态变量以及可能导致它们的计数增加的原因。

(1) Aborted_connects

Aborted_connects 是 MySQL 的一个状态变量,当客户端尝试连接但未能成功时,这个值会递增。以下是导致 Aborted_connects 计数增加的一些原因。

- 客户端没有权限访问 MySQL 服务。
- 客户端输入的 MySQL 数据库密码错误。
- 连接数据包不包含正确信息,这可能是客户端或网络的错误导致的。
- 超过连接时间限制。主要由参数 connect_timeout 控制,其默认值是 10s,通常情况下,除非网络环境非常不稳定,否则不应该出现超时。

下面是访问密码错误时,Aborted_connects 计数递增的示例。

首先使用 FLUSH 命令重置 MySQL 的 Abort 状态的统计指标,然后使用错误的密码登录 MySQL 服务,最后查看 Abort 状态计数。具体操作如下所示。

```
mysql>FLUSH STATUS;
Query OK, 0 rows affected (0.00 sec)
mysql> show status like 'Abort%';
+-----------------------------+---------+
|Variable_name                |Value    |
+-----------------------------+---------+
|Aborted_clients              |0        |
|Aborted_connects             |0        |
+-----------------------------+---------+
2 rows in set (0.00 sec)

shell $>mysql -utester -p***** -h192.168.1.1 -P3380
mysql: [Warning] Using a password on the command line interface can be insecure.
ERROR 1045 (28000): Access denied for user 'tester'@'192.168.1.1' (using password: YES)

# 查看 Aborted_connects 状态计数,发现增加了 1
mysql>SHOW GLOBAL STATUS LIKE 'Aborted_connects';
+-----------------------------+---------+
|Variable_name                |Value    |
+-----------------------------+---------+
|Aborted_connects             |1        |
+-----------------------------+---------+
1 rows in set (0.01 sec)
```

(2) Aborted_clients

Aborted_clients 也是 MySQL 的一个状态变量,它表示客户端连接被服务器异常终止的次数。当客户端成功连接到 MySQL 服务器,但由于某种原因连接被不正常地断开或终止时,这个值就会增加。以下是导致 Aborted_clients 计数增加的原因。

- 客户端程序在用户操作后未正常关闭连接,而是直接退出或被操作系统终止,没有发送适当的关闭信号给 MySQL 服务器。
- 如果客户端在一段时间内没有发送任何请求到服务器，并且超过了 wait_timeout（非交互式连接）或 interactive_timeout（交互式连接）的设置值，那么服务器会自动关闭该连接。
- 客户端程序在数据传输过程中可能因为内存不足、程序崩溃或其他原因突然终止，导致连接被中断。

下面是程序未调用关闭连接函数导致的通信数据包出错和 Aborted_clients 计数递增的示例。

使用 Python 脚本模拟客户端程序在退出之前未调用 mysql_close( ) 关闭 MySQL 连接的场景。脚本如下所示。

```
##Python 代码##
!/usr/bin/python
#-*- coding: UTF-8 -*-
import pymysql

# 打开数据库连接
db = pymysql.connect(host='192.168.1.1, port=3380, database='sbtest',
                     charset='utf8',user='dbadmin', password='******')

# 使用 cursor()方法创建一个游标对象 cursor
cursor = db.cursor()

# 使用 execute()方法执行 SQL 查询
cursor.execute("SELECT * FROM t1;")

# 使用 fetchone()方法获取单条数据
data = cursor.fetchone()
print("Database version: %s " % str(data))
# 关闭数据库连接
db.close()
```

查看 MySQL 的错误日志和 Aborted_clients 计数情况。具体命令行如下所示。

```
# 错误日志记录提示信息
shell $> tail -5 /opt/data8.0/logs/mysql_err.log
 [Note] Aborted connection 32 to db: 'sbtest' user:'dbadmin' host:'192.168.1.1'Got an error reading communication packets
 [Note] Aborted connection 32 to db: 'sbtest' user:'dbadmin' host:'192.168.1.1'Got an error reading communication packets
 [Note] Aborted connection 32 to db: 'sbtest' user:'dbadmin' host:'192.168.1.1'Got an error reading communication packets

# 查看 Aborted_clients 状态计数
mysql> SHOW GLOBAL STATUS LIKE 'Aborted_clients';
+-----------------------------+---------+
|Variable_name                |Value    |
+-----------------------------+---------+
|Aborted_clients              |1        |
+-----------------------------+---------+
1 rows in set (0.01 sec)
```

（3）Aborted_connects 和 Aborted_clients 同时增加的情况

下面为 Aborted_connects 和 Aborted_clients 这两个值同时增加的原因。

- 如果 max_allow_packet 参数设置得太小，则客户端尝试发送的数据包可能会超过服务器允许的最大值，导致连接尝试失败。
- 如果操作系统中的网络驱动程序存在缺陷或错误，则它可能会干扰网络连接的建立，导致 Aborted_connects 计数增加。
- 如果 MySQL 使用的线程池存在缺陷，则可能在处理连接请求时发生读取错误，导致连接尝试失败。
- 如果服务器或客户端的 TCP/IP 配置不正确（例如，IP 地址、端口号、网络掩码等），那么连接请求可能会失败。
- 网络硬件故障，包括以太网卡、集线器、交换机或连接电缆的故障，都可能导致连接请求失败，从而增加 Aborted_connects 的计数。

下面是 MySQL 的 max_allowed_packet 参数设置过小，导致通信数据包出错和 Aborted_clients 计数递增的示例。

首先设置 MySQL 传送的网络包大小为 1KB。设置网络包大小的命令行如下所示。

```
mysql>SET GLOBAL max_allowed_packet=1024;
Query OK, 0 rows affected, 1 warning (0.00 sec)
```

在使用 Python 脚本并发执行多个插入任务时，会发现网络带宽上升，通信数据包出错。并发插入脚本如下所示。

```
##Python 脚本##
#!/usr/bin/env python
#-*- coding: utf8 -*-
import pymysql

# 创建连接
db = pymysql.connect(host='127.0.0.1', port=3306, user='dbadmin', passwd='******',
db='sbtest', charset='utf8')

# 创建游标
for num in rang(1,1000):
    coursor = db.cursor()
    sql=""" INSERT INTO t1(id,name,addr) VALIES(20,
      'AAAAAAAAAAAAAAAAAAAAAAAAAAAAAAAAAAAAAAAAAAAAAAAAAAAAAAAAAAAA',
      'BBBBBBBBBBBBBBBBBBBBBBBBBBBBBBBBBBBBBBBBBBBBBBBBBBBBBBBBBBBB')"""
try:
  # 执行 SQL 语句
  ursor.execute(sql)
  # 提交到数据库
  db.commit()
except:
  # 如果出现错误,则回滚
  print(num,"rollback")
  db.rollback()
```

图 7-2 所示为监控网络包传输情况。当网络接收包 recv 和网络发送包 send 值都大于 1KB 时，会在 MySQL 错误里出现 "Got an error reading communication packet" 信息。

```
---total-cpu-usage---- -dsk/total- -net/total- ---paging-- ---system--
usr sys idl wai hiq siq| read  writ| recv  send|  in   out | int   csw
  1   9  87   3   0   0|    0   11M|1289k  119|   0     0 | 10k   11k
  1   8  87   3   0   0|    0 8918k|1158k  107|   0     0 |9353  9895
  1   9  87   3   0   0|    0   13M|1403k  130|   0     0 | 11k   12k
  1   9  87   3   0   0|    0   14M|1428k  132|   0     0 | 11k   12k
  1   9  87   3   0   0|    0   14M|1275k  118|   0     0 | 10k   11k
  1   8  89   2   0   0|    0   11M|1157k  107|   0     0 |9235   10k
  0   1  99   0   0   0|    0  5280k|  18k 2123|   0     0 | 784   540
  0   1  99   0   0   0|    0  5280k|  18k 2123|   0     0 | 784   540
```

图 7-2 通过 dstat 命令查看传输的网络包

2. 案例分析

上面介绍了通信数据包出错的各种场景以及 MySQL 内部使用 Abort 状态计数来判断问题的示例。下面将提供在实际生产环境中，因 timeout 设置不合理，导致通信数据包出错的案例。

（1）错误信息记录

在制作运维月报过程中，发现标签数据库的 MySQL 连接异常率较高。查看错误日志后，发现大量 "Abort Connection（Got an error reading communication packets）" 错误。这些错误在日志中的分布较为均匀，且没有明显的时间规律。

（2）分析过程

首先，了解业务访问 MySQL 数据库的实现流程。如图 7-3 所示，应用部署在 5 台服务器上，这些服务器先通过 VIP 访问 HAProxy（VIP 是浮动 IP，一般通过 Keepalived 实现，只能绑定在一个 HAProxy 节点上，实现 HAProxy 1 和 HAProxy 2 的高可用方案），再通过 HAProxy 实现负载均衡，以便更均匀地分配对 MySQL 数据库的访问请求。

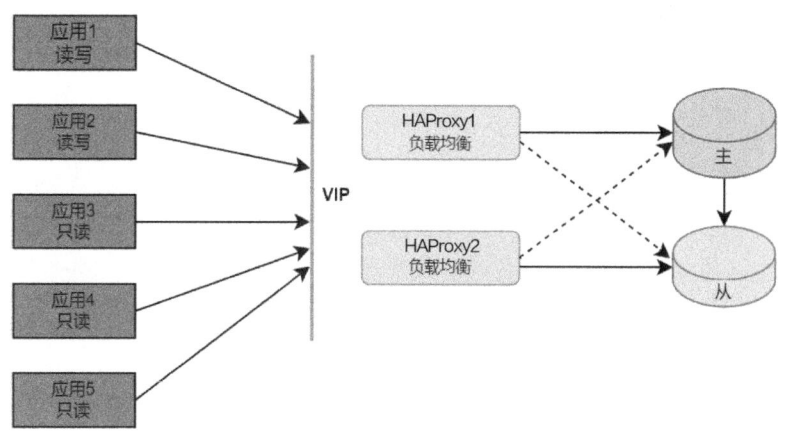

图 7-3 业务访问 MySQL 流程

MySQL 错误日志中输出的 "Got an error reading" 问题表明，虽然客户端已经成功连接到数据库，但连接异常终止了。问题可能出自 Tomcat、HAProxy 和 MySQL 三者中的某一个。初步判定这个问题是由这三者之间的超时处理机制引起的。

下面为三者的超时参数设置以及解释。

Tomcat 中的超时参数设置如下所示。

```
# Tomcat 超时时间设置
hummer.datasource.hikari.idle-timeout=60000    # MySQL 连接:idle 状态的最大时长
hummer.datasource.hikari,max-lifetime-1800000   # MySQL 连接:生命周期
hummer.datasource.hikari.connection-timeout=30000  # MySQL 连接:超时时间
```

HAProxy 中的超时参数设置如下所示。

```
defaults
...
    timeout queue 1m     # 一个请求在队列里的超时时间
    timeout connect 10s  # 连接超时
    timeout client 1m    # 客户端超时
    timeout server 1m    # 服务器端超时
    timeout check 10s    # 检测超时
    maxconn 65535        # 每个进程可用的最大连接数
```

MySQL 中的超时参数设置如下所示。

```
mysql>SHOW VARIABLES LIEK '%_timeout%';
+-------------------------------+---------+
|Variable_name                  |Value    |
+-------------------------------+---------+
|connect_timeout                |10       |  # 连接过程中握手的超时时间
|wait_timeout                   |600      |  # 不活跃的连接超时时间
|interactive_timeout            |600      |  # 交互方式不活跃的连接超时时间
|net_read_timeout               |30       |  # 在网络条件不好的情况下起作用
|net_write_timeout              |60       |  # 在网络条件不好的情况下起作用
...
+-------------------------------+---------+
26 rows in set (0.00 sec)
```

下面使用排除法。

步骤 1：先调整 MySQL 的参数，将 wait_timeout 和 interactive_timeout 均设置为 8h，并在 MySQL 的错误日志里观察到 "Got an error reading..." 的错误信息分布较为均匀，而不是集中式大量出现。这表明 MySQL 服务本身可能不是问题的根源，因为如果是 MySQL 的问题，那么调整这些超时参数可能会使错误更加集中或者没有变化。

步骤 2：当直接将 Tomcat 连接到 MySQL 数据库，而绕过 HAProxy 时，MySQL 错误日志中的 "Got an error reading..." 的错误信息不再出现。这强烈表明问题可能出在 HAProxy 上。之后，注释了 HAProxy 的 timeout client 和 timeout server 参数，问题就不再出现了。这进一步确定了问题是由于 HAProxy 的超时设置引起的。

HAProxy 官方提供的 timeout client 和 timeout server 参数解释如下：在 HTTP 模式下，客户端与服务器之间的交互涉及多个阶段，包括请求发送和响应接收。在这些过程中，不活动超时（inactivity timeout）是一个关键配置项，它定义了客户端或服务器在特定时间内无数据交换时，连接被视为不活动的时限。也就是说，对于第一阶段，最好设置 "timeout http request"，以便更好地保护 HAProxy 免受类似 Slowloris 的攻击。参数 timeout client 是特定于前端的，但可以在 "默认值" 部分一次性指定。这实际上是最简单的解决方案之一，不要忘记它。未指定超时将导致无限超时，不建议这样做。这样的用法可以令人接受并正常工作，但是在启动期间会报告一个警告，因为如果系统的超时也没有配置，则可能导致系统中过期会话的累积。

（3）解决方案

通过上述分析，提供以下解决方案。

- 在 Tomcat 中添加心跳查询操作。通过发送测试 SQL 语句，能够定期检测连接的健康状态，确保连接在空闲时不会因超时而被断开。这种方法能够从根本上解决因超时导致的连接中断但需要修改所有通过 HAProxy 连接到 MySQL 的应用端的配置，这涉及大量的工作量和潜在的风险。
- 在 HAProxy 配置文件中注释掉参数 timeout client，不需要改动应用端的配置。这样可以避免 HAProxy 因为超时而断开连接。虽然注释掉 timeout client 可以避免 HAProxy 断开连接，但这样会让连接的健康状态完全依赖于 TCP 连接的机制，无法有效检测连接是否仍然有效，可能会导致无效的连接长时间存在，影响系统的稳定性和性能。
- 忽略 MySQL 错误日志中的 Abort Connection 记录。无须修改任何配置，不会对现有系统产生任何影响。可以避免大量的 Abort Connection 记录对错误日志的干扰。但忽略这些记录意味着无法得知有多少连接因为超时而被中断，这可能会掩盖潜在的问题。

3. 总结

当遇到"Got an error reading..."这类问题时，首先查看 MySQL 的错误日志记录，以了解错误详情。然后，观察 MySQL 中的 Status 的 Aborted_clients 和 Aborted_connects 值是否波动较大。同时，检查相关应用的日志以获取更多线索。结合这些信息，可以考虑优化 MySQL 的配置参数，如调整连接超时设置，以解决或缓解问题。

7.3.3 导致服务器信号量不足的故障分析

MySQL 的处理机制包括利用主线程来协调多个工作线程，这些线程共同处理来自客户端的请求。每个线程都可以访问共享资源，如数据库文件和内存缓存。为了避免不同线程之间的竞争和冲突，MySQL 不仅使用自身的锁机制，还利用信号量和其他同步机制来管理线程对共享资源的访问，从而确保系统的稳定性和数据的安全性。

在使用 MySQL 中，经常会碰到的信号量（Semaphore）错误信息如下。

```
[ERROR] [FATAL]InnoDB: Semaphore wait has lasted > 600 seconds.We intentionally crash the server because it appears to be hung.
[ERROR] [FATAL]InnoDB: Warning: a long semaphore wait:
813 --Thread 139957495039744 has waited atbtr0cur.cc line 545 for 241.00 seconds the semaphore:
814 X-lock (wait_ex) on RW-latch at 0x7f4a60043da8 created in file dict0dict.cc line 2341
```

说明：上述信息的大致意思是，InnoDB 引擎在信号量等待已持续超过 600s，被挂起。最终，导致 MySQL 服务崩溃。

1. MySQL 中的信号量应用

操作系统的信号量应用于 MySQL 服务的以下方面。

- 并发控制：在 MySQL 中，信号量可以用于控制多个线程之间的并发访问，以避免不同线程之间的竞争和冲突。
- 内存管理：在 MySQL 中，信号量可以用于管理内存资源，以确保系统的稳定性和可靠性。
- 文件系统：在 MySQL 中，信号量可以用于控制文件系统的访问，以避免不同线程之间的竞争和冲突。
- 网络连接：在 MySQL 中，信号量可以用于控制网络连接的访问，以确保系统的稳定性和可靠性。

MySQL 内部机制中,存在两种类型的信号量:互斥信号量和内部同步。

- 互斥信号量:它用于保护共享资源,防止不同线程之间的竞争和冲突。互斥信号量通常用于保护共享缓存、文件系统、网络连接等资源。
- 内部同步:它用于控制线程的同步和通信,是一种高级的线程同步机制。通常用于控制线程的等待和唤醒,以实现线程之间的协作。

表 7-1 所示为目前官方提供的 MySQL 的 InnoDB 引擎信号量及其说明。

表 7-1 MySQL 的 InnoDB 引擎信号量及其说明

信号量	说明
Mutex Rounds	InnoDB 内部同步数组的信号量/互斥锁旋转次数
Mutex Spin Waits	InnoDB 信号量/互斥锁等待内部同步数组的数量
Os Reservation Count	InnoDB 信号量/互斥锁等待被添加到内部同步数组的次数
Os Signal Count	InnoDB 线程使用内部同步数组发出信号的次数
Rw exl Os Waits	InnoDB 等待给 Os 的独占(写)信号量的个数
Rw excel Rounds	InnoDB 同步数组中独占(写)信号量旋转的次数
Rw excel Spins	InnoDB 同步数组中独占(写)信号量 spin 等待的次数
Rw Shared Os Waits	InnoDB 向 Os 提供的共享(读)信号量等待数
Rw Shared Rounds	InnoDB 同步数组内共享(读)信号量的旋转次数
Rw Shared Spins	InnoDB 同步数组内共享(读)信号量 spin 等待的数量
Spins Per Wait Mutex	InnoDB 信号量/互斥锁旋转数与等待内部同步数组的互斥锁旋转数的比例
spin Per Wait Rw Excl	InnoDB 独占(写)信号量/互斥锁自旋轮数与内部同步数组内自旋等待数的比例
spin Per Wait Rw Shared	InnoDB 共享(读)信号量/互斥锁自旋轮数与内部同步数组内自旋等待数的比例

2. MySQL 中记录信号量信息的地方

通过以下信息,了解一下 MySQL 里信号量的指标。信号量信息记录在 INNODB STATUS 状态中。查看信号量信息的命令行示例如下。

```
mysql> SHOW ENGINE INNODB STATUS \G;
----------
SEMAPHORES
----------
OS WAIT ARRAY INFO: reservation count 68581015, signal count 218437328
--Thread 140653057947392 has waited at btr0pcur.c line 437 for 0.00 seconds the semaphore:
S-lock on RW-latch at 0x7ff536c7d3c0 created in file buf0buf.c line 916
a writer (thread id 140653057947392) has reserved it in mode exclusive
Mutex spin waits 1157217380, rounds 1783981614, OS waits 10610359
RW-shared spins 103830012, rounds 1982690277, OS waits 52051891
RW-excl spins 43730722, rounds 602114981, OS waits 3495769
```

说明:5.3.5 节详细介绍了 SEMAPHORES 指标信息的内容。

如果有高并发的工作负载,那么 SEMAPHORES 记录了信号量信息,它包含了两种数据形式:事件计数器以及可选的当前等待线程的列表。

(1) OS WAIT ARRAY INFO：reservation count 68581015，signal count 218437328

这行给出了关于操作系统等待数组的信息，它是一个插槽数组，InnoDB 在数组里为信号量保留了一些插槽，操作系统用这些信号量给线程发送信号，使线程可以继续运行，以完成它们等着做的事情。这一行还会显示 InnoDB 使用了多少次操作系统的等待，保留统计（reservation count）显示了 InnoDB 分配插槽的频度，信号计数（signal count）衡量的是线程通过数组得到信号的频度，从而反映了操作系统等待的频度。

(2) --Thread 140653057947392 has waited at btr0pcur.c line 437 for 0.00 seconds the semaphore

这部分显示的是当前正在等待互斥量的 InnoDB 线程，在这里可以看到有一个线程正在等待，每一个线程都是以--Thread <数字> has waited 开始的。这一段内容在正常情况下应该是空的（即查看的时候没有这部分内容），除非服务器有高并发的工作负载，促使 InnoDB 采取让操作系统等待的措施。

(3) 计数器信息

日志里输出的信息解释如下。

- "Mutex spin waits 1157217380，rounds 1783981614，OS waits 10610359" 行显示的是与互斥量相关的几个计数器。
- "RW-shared spins 103830012，rounds 1982690277，OS waits 52051891" 行显示读写的共享锁的计数器。
- "RW-excl spins 43730722，rounds 602114981，OS waits 3495769" 显示读写的排他锁的计数器。

(4) MUTEX 竞争信息

空转等待的成本相对较低，但是要不停地检查一个资源能否被锁定，这种方式会消耗 CPU 周期，因为当处理器在等待 I/O 时，一般都有一些空闲的 CPU 周期可用，即使没有空闲的 CPU 周期，那么空等也要比其他方式更加经济。然而，当另一个线程能做一些事情的时候，空转等待也还会独占 CPU。空转等待的替代方案就是让操作系统做上下文切换，这样，当一个线程在等待时，另一个线程就可以被运行，然后，通过等待数组里的信号量发出信号，唤醒那个"沉睡"的线程。通过信号量来发送信号是比较有效的，但是上下文切换成本很高，每秒钟几千次的切换会带来大量的系统开销。通过 INNODB MUTEX 命令行，可了解等待信息。查看 MUTEX 竞争信息的命令行示例如下。

```
mysql> show engine innodb mutex;
+----------+---------------------------+---------+
|Type      |Name                       |Status   |
+----------+---------------------------+---------+
|InnoDB    |rwlock: log0log.cc:846     |waits=3  |
+----------+---------------------------+---------+
1 row in set (0.01 sec)
```

说明： 5.3.5 节详细介绍了 MUTEX 指标信息的内容。

3. Redo 日志刷新机制影响信号量

在 InnoDB 存储引擎中，当数据被修改时，相应的修改操作会先被写入 Redo 日志，然后再异步地刷新到磁盘上的数据文件中。这个过程涉及信号量的使用，以确保对 Redo 日志和数据文件的并发访问是安全的。表 7-2 所示为 MySQL Redo 日志信号量参数列表。

表 7-2 MySQL Redo 日志信号量参数

Redo 日志信号量参数	说明
innodb_log_wait_for_flush_spin_hwm	定义最大平均日志刷新时间，超过此时间，用户线程在等待刷新重做时不再自旋。默认值是 400μs
innodb_log_spin_cpu_abs_lwm	定义在等待刷新重做时用户线程不再旋转的最小 CPU 使用量。取值为 CPU 内核占用率之和。例如，默认值 80 为单个 CPU 核心的 80%。在具有多核处理器的系统上，值 150 表示一个 CPU 核心的使用率为 100%，另一个 CPU 核心的使用率为 50%
innodb_log_spin_cpu_pct_hwm	定义在等待刷新重做时用户线程不再旋转的最大 CPU 使用量。该值表示为所有 CPU 核心综合总处理能力的百分比。默认值为 50%

在 Redo 日志的刷新过程中，上述参数使用信号量来协调不同线程之间的访问。

4. 信号量设置调优

针对 MySQL 中的信号量不足问题，可以从自旋锁轮询参数、自适应哈希索引和 OS 信号量参数 3 个方面进行优化。然而，具体的优化措施应根据实际情况进行评估和测试，以确保数据库的稳定性和性能。

（1）自旋锁轮询参数

InnoDB 的互斥锁和 rw 锁通常会保留很短的时间间隔。在多核系统中，线程在休眠前一段时间内连续检查其是否可以获得互斥锁或 rw 锁可能会更有效。如果互斥锁或 rw 锁在此期间可用，则线程可以立即在同一时间片中继续执行。但是，多线程过于频繁地轮询共享对象（如互斥锁或 rw 锁）可能会导致"缓存乒乓"，导致处理器使彼此缓存的部分无效。InnoDB 通过强制轮询之间的随机延迟同步轮询活动来最小化这个问题。随机延迟被实现为自旋等待循环。

1）innodb_spin_wait_delay 参数。innodb_spin_wait_delay 是指自旋锁的轮询之间的最大延迟。这种机制的底层实现取决于硬件和操作系统的组合，因此延迟并不对应于固定的时间间隔。自旋等待循环的持续时间由循环中出现的 PAUSE 指令的数量决定。这个数字是通过随机选择一个从 0 到 innodb_spin_wait_delay（但不包括其本身）值的整数，并将该值乘以 50 来生成的。（乘数 50 在 MySQL 8.0.16 之前是硬编码的，之后是可配置的。）当 innodb_spin_wait_delay 设置为 6 时，从以下范围中随机选择一个整数，如下所示。

```
{0, 1, 2, 3, 4, 5}
# 选择的整数乘以 50,得到 6 个可能的 PAUSE 指令值之一
{0,50,100,150,200,250}
```

对于这组值，250 是在自旋等待循环中可以出现的 PAUSE 指令的最大数目。将 innodb_spin_wait_delay 设置为 5 会得到一个包含 5 个可能值 {0,50,100,150,200} 的集合，其中 200 是 PAUSE 指令的最大数目，以此类推。通过这种方式，innodb_spin_wait_delay 设置控制自旋锁轮询之间的最大延迟。

在所有处理器内核共享快速缓存的系统中，可以通过设置 innodb_spin_wait_delay = 0 来减少最大延迟或完全禁用繁忙循环。

2）innodb_spin_wait_pause_multiplier 参数。innodb_spin_wait_pause_multiplier 是 MySQL 8.0.16 中引入的参数，它提供了一种解释 PAUSE 指令持续时间差异的方法。定义一个乘数，用于确定线程等待获取互斥锁或 rw 锁时发生的自旋等待循环中 PAUSE 指令的数量。例如，假设 innodb_spin_wait_delay 设置为 6，将 innodb_spin_wait_pause_multiplier 的值从 50（默认值和之前的硬编码值）

减少到 5，会生成一组较小的 PAUSE 指令值，如下所示。

```
# innodb_spin_wait_delay
{0, 1, 2, 3, 4, 5}

# PAUSE 指令值
{0,5,10,15,20,25}
```

增加或减少 PAUSE 指令值的能力允许针对不同的处理器架构微调 InnoDB。例如，较小的 PAUSE 指令值适用于具有相对较长的 PAUSE 指令的处理器体系结构。

3）innodb_sync_spin_loops 参数。innodb_sync_spin_loops 是指线程在被挂起之前等待 InnoDB 互斥锁释放的次数。InnoDB 有一个多阶段等待的策略，首先，它会试着对锁进行空转等待，如果经历了一个预设的空转等待周期（设置 innodb_sync_spin_loops 配置变量命令）之后还没有成功，就会退到成本更高、更复杂的等待数组中。

（2）自适应哈希索引

自适应哈希索引（Adaptive Hash Index）是 InnoDB 存储引擎的一种优化机制，如图 7-4 所示，它会监控对表上二级索引的查找操作，当发现某个二级索引被频繁访问，即成为热数据时，InnoDB 会为这个二级索引的热门值建立哈希索引，以加快查询速度。

自适应哈希索引能够显著提高查询性能。然而，如果查询语句中涉及回表操作，即需要通过二级索引的值去查找主键索引以获取完整数据行，那么直接使用主键进行查询可能会更加高效。因此，在 MySQL 的使用场景中，合理利用主键进行查询是非常重要的。

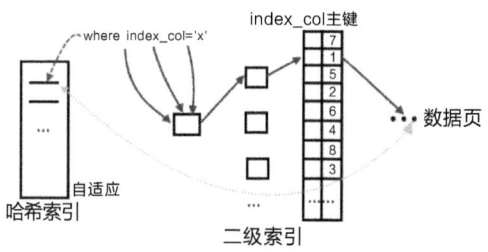

图 7-4 MySQL 自适应哈希索引

自适应哈希索引的特性如下。

- 自适应哈希索引存储在 InnoDB 的缓冲池中。缓冲池是 InnoDB 用于缓存数据和索引的地方，因此创建哈希索引会增加缓冲池的使用量。如果缓冲池的大小有限，则创建大量的哈希索引可能会影响其他数据和索引的缓存效率。
- 自适应哈希索引主要用于等值查询的加速。当查询能够直接通过哈希索引找到对应的值时，查询速度会显著提高。但是，对于范围查询（如 WHERE index_col BETWEEN 'a' AND 'z'）或者其他非等值查询（如 WHERE index_col > 'xxx'），哈希索引无法提供有效的帮助。
- 由于哈希索引是非聚集的，并且不存储记录的物理顺序，因此它们不能用于排序操作。
- 自适应哈希索引的创建和管理完全由 MySQL 的 InnoDB 存储引擎自动完成。用户不能直接创建或删除哈希索引，也不能干预其维护过程。

在某些工作负载下，哈希索引查找的加速收益可能远远超过监视索引查找和维护哈希索引结构的额外开销。然而，在很高的工作负载（如多个并发连接）中，对自适应哈希索引的访问有时会成为资源争用的来源。对于使用 LIKE 操作符和 % 通配符的查询，自适应哈希索引可能不会带来明显的性能提升。

在 MySQL 5.6 版本中，考虑到很难预测自适应哈希索引是否适合特定的系统和工作负载，所以更适合禁用它。在 MySQL 5.7 和 MySQL 8.0 中继续沿用了对自适应哈希索引的分区策略。每个索引都绑定到一个特定的分区,并由单独的锁存器保护。分区的数量由 innodb_adaptive_hash_index_parts 参数控制,默认是 8 个分区,最大可以设置为 512 个分区。早期版本中,自适应哈希索引特性由一个锁存器保护,这在高工作负载下可能成为资源争用的瓶颈。可以通过使用 SHOW ENGINE INNODB STATUS 输出的"INSERT BUFFER AND ADAPTIVE HASH INDEX"来监视自适应哈希索引的使用情况和争用状态。具体命令行示例如下。

```
mysql> SHOW ENGINE INNODB STATUS \G
-------------------------------------
INSERT BUFFER AND ADAPTIVE HASH INDEX
-------------------------------------
Ibuf: size 1, free list len 2235, seg size 2237, 0 merges
merged operations:
 insert 0, delete mark 0, delete 0
discarded operations:
 insert 0, delete mark 0, delete 0
Hash table size 276707, node heap has 0 buffer(s)
Hash table size 276707, node heap has 0 buffer(s)
Hash table size 276707, node heap has 0 buffer(s)
Hash table size 276707, node heap has 3 buffer(s)
Hash table size 276707, node heap has 0 buffer(s)
Hash table size 276707, node heap has 0 buffer(s)
Hash table size 276707, node heap has 0 buffer(s)
Hash table size 276707, node heap has 1 buffer(s)
0.00 hash searches/s, 0.00 non-hash searches/s
```

说明:默认采取 8 个分区,每个区包含 276707 个哈希值,这个内存开销算在 InnoDB 缓冲池的大小上。

结合官方说明、已知的 bug 以及社区经验,建议在某些特定情况下关闭自适应哈希索引功能。关闭自适应哈希索引命令行示例如下。

```
mysql> SHOW GLOBAL VARIABLES LIKE 'innodb_adaptive_hash_index';
+----------------------------+-------+
|Variable_name               |Value  |
+----------------------------+-------+
|innodb_adaptive_hash_index  |ON     |
+----------------------------+-------+
1 row in set (0.01 sec)

mysql> SET GLOBAL innodb_adaptive_hash_index=OFF;
Query OK, 0 rows affected (0.00 sec)

mysql> SHOW GLOBAL VARIABLES LIKE 'innodb_adaptive_hash_index';
+----------------------------+-------+
|Variable_name               |Value  |
+----------------------------+-------+
|innodb_adaptive_hash_index  |OFF    |
+----------------------------+-------+
1 row in set (0.01 sec)
```

在初期部署 MySQL 时,考虑关闭 innodb_adaptive_hash_index 特性。对于轻量级的 MySQL 数据

库，关闭此特性可能带来的性能提升难以准确评估，并且可能存在一些隐患。然而，这一决策应当基于针对具体负载的性能测试结果。如官方文档所述，如果存在多个线程在等待 RW-LATCH 的情况下产生竞争，可以考虑增加自适应哈希索引分区的数量，或者完全禁用自适应哈希索引特性。

至于 hash join 和 innodb_adaptive_hash_index 之间的关联，它们都是基于哈希的技术，但服务于不同的目的。hash join 是一种查询优化技术，用于连接两个或多个表，特别是在没有合适索引的情况下。而 innodb_adaptive_hash_index 则是一个内存中的数据结构，用于加速对 InnoDB 表中数据的访问。

（3）OS 信号量参数

在 Linux 系统中，信号量是进程间同步互斥的一种方式，它用于协调不同进程或线程之间对共享资源的访问。每个进程通信需要的信号灯或者 IPC 标志，以及每个连接的 PROCESS 都要分配一个信号量。kernel.sem 是 Linux 系统用来控制信号量的参数。查看 kernel.sem 配置值的命令行示例如下。

```
# cat /proc/sys/kernel/sem
250     32000     32     128
```

说明：
- 第一列，表示每个信号集中的最大信号量数目。
- 第二列，表示系统范围内的最大信号量总数目。
- 第三列，表示每个信号发生时的最大系统操作数目。
- 第四列，表示系统范围内的最大信号集总数目。

可以通过/proc/sys/kernel/sem 文件或 sysctl 命令调整参数。调整参数的命令行示例如下。

```
# sysctl -w kernel.sem="1000 64000 100 256"
```

kernel.sem 是信号量结构体，内核中提供了一组系统函数用于访问它。可使用 ipcs 命令查看当前系统的信号量列表，如下所示。

```
# ipcs -a
------ Shared Memory Segments --------
key          shmid       owner     perms     bytes      nattch     status
0x7401833d   2654208     root      600       4          0
0x00000000   3145729     root      600       4194304    9          dest
0x7401833c   2621442     root      600       4          0
0xd201012b   3080195     root      600       1720       2
```

说明： 可以看到当前系统所有的信号量以及进程的信息。
- key：共享内存的唯一标识。
- shmid：共享内存的编号。
- owner：创建的用户。
- perms：权限。
- bytes：创建的大小。
- nattch：连接到共享内存的进程数。

如果 OS 信号量的设置过大，则可能会导致系统资源的过度消耗，包括 CPU 和内存。特别是，当大量的信号量被创建和销毁时，可能会增加系统调用的次数，从而导致 CPU 的 sys 使用率上升。因此，合理设置 OS 信号量的数量是非常重要的。

7.3.4 Undo 日志无法清理导致阻塞数据库的故障分析

在 MySQL 的运行环境中，虽然从节点的资源使用率非常低且没有大事务，但会发现主从复制延迟逐渐增大。通过查看 InnoDB 状态和错误日志，发现如下信息。

```
mysql> SHOW ENGINE INNODB STATUS \G
*************************** 1.row ***************************
  Type:InnoDB
  Name:
Status:
=====================================
2023-07-30 16:57:46 0x7f3cc41f1700 INNODB MONITOR OUTPUT
=====================================
Per second averages calculated from the last 7 seconds----------------
BACKGROUND THREAD----------------
srv_master_thread loops: 77133688 srv_active, 0 srv_shutdown, 213108 srv_idle
srv_master_thread log flush and writes: 77346796
.................
TRANSACTIONS-----------
Trx id counter 65610880373Purge done for trx's n:o < 65610880373 undo n:o < 0 state: running but idle
History list length 2962710

# 错误日志
[Warning]InnoDB: Purge reached the head of the history list,
but its length is still reported as 2962026!
  Make a detailed bug report, and submit it to http://bugs.mysql.com
```

从上述信息中可以总结出，InnoDB 的 Purge 进程未能及时清理历史列表中的 Undo 日志，导致历史列表长度累积到 2962710。由于 Undo 日志不断积累而未被清理，因此最终导致了 MySQL 服务崩溃。在 MySQL 中，Undo 日志是用于支持事务回滚和多版本并发控制（MVCC）的核心日志，如果不及时清理，则占用的空间会不断增加，可能耗尽系统资源并导致数据库服务崩溃。

1. Purge 机制

当 SQL 语句删除一行时，InnoDB 不会立即从物理存储中删除该行，而是将其标记为删除。只有当这些被标记为删除的行不再需要用于多版本并发控制或回滚时，Purge 进程才会物理地删除这些行及其索引记录。

在 MySQL 的 SHOW ENGINE INNODB STATUS 命令的输出中，TRANSACTIONS 部分提供了关于当前活动事务和已完成但尚未被 Purge 进程清除的事务的信息。查看 Undo 日志 Purge 信息的命令行示例如下。

```
mysql> SHOW ENGINE INNODB STATUS;
...
------------
TRANSACTIONS
------------
Trx id counter 0 290328385 Purge done for trx's n:o < 0 290315608 undo n:o < 0 17
History list length 20
```

在上述 TRANSACTIONS 部分信息中，可找到一个名为 HISTORY LIST LENGTH 的值，这个值表示当前在 Purge 队列中等待清除的 Undo 日志条目的数量。如果 HISTORY LIST LENGTH 的值持续很高，并且 Purge 进程无法跟上事务的生成速度，那么通常意味着 Purge 进程可能受到了某种限制或延迟，如磁盘 I/O 性能问题、系统资源不足（如 CPU、内存），或者锁争用等。

Purge 进程是按照定期计划运行的，它负责解析和处理 Undo 日志页。这些 Undo 日志页位于历史列表中，该列表由 InnoDB 事务系统维护，包含了已提交事务的 Undo 日志页面。清除操作将处理完毕的 Undo 日志页从历史列表中释放，从而回收占用的空间。Purge 操作也有规则可寻，主要受以下 3 个参数影响。

- innodb_purge_threads 参数。该参数是指参与清除的线程数。默认值是 4。清除线程会在后台执行，清除工作会自动重新分配到可用的线程上。过多的活动清除线程可能会导致与用户线程的争用。在普遍情况下，采用默认值即可，但一定要比 CPU 物理核数小，同时，若 innodb_buffer_pool_instances 参数值设置得小，那么核数够的情况下不建议超过 16 个线程。
- innodb_purge_batch_size 参数。该参数用来控制每次从 Undo 日志历史列表中清除和解析的 Undo 日志页的数量。这个参数决定了 Purge 线程每次迭代时处理的 Undo 日志页的批量大小。默认值是 300，这意味着每次清除操作会处理 300 个 Undo 日志页。当然，在 CPU 资源足够，I/O 性能好的情况下，这个值适当减小也是很好的。
- innodb_max_purge_lag 参数。该参数是所需的最大清除延迟时间。当清除延迟超过 innodb_max_purge_lag 这个阈值时，INSERT、UPDATE 和 DELETE 操作将被延迟，以允许清除操作有时间赶上。默认值为 0，这意味着没有最大清除延迟。对于有问题的工作负载，一个典型的 innodb_max_purge_lag 可以设置为 1000000，假设事务很小，只有 100B 大小，并且允许有 100MB 的未清除的表行。

2. 分析问题

通过上面的介绍，已了解 Purge 机制和相关参数设置。Undo 日志没有被清理的原因是长时间运行的事务导致历史列表的长度增加。特别是在 InnoDB 的多版本并发控制时，为了保证数据的一致性和隔离性，这些长时间运行的事务会保留更多的 Undo 信息。以下是 3 种可能导致 Undo 日志历史列表长度增加的情况。

- 大量并发 DML 操作。当使用像 MyDumper 或 mysqldump 这样的工具，并且指定了 single transaction 选项时，这些工具会启动一个长时间运行的事务来确保数据的一致性。在这种情况下，由于事务持续的时间较长，因此它可能需要在 Undo 日志中保留更多的数据版本，直到事务提交为止。这会增加历史列表的长度。
- 未使用自动提交后的 SELECT 查询。如果一个事务被开始（例如，通过调用 START TRANSACTION），但没有显式地提交（COMMIT）或回滚（ROLLBACK），那么该事务将保持打开状态。在这个事务中执行的任何查询都会创建一个数据版本，该版本必须在 Undo 日志中保留，直到事务被提交或回滚为止。这样同样会导致历史列表的增长。
- 异步 Purge 操作。Purge 操作是异步执行的，意味着它不会立即清除不再需要的 Undo 日志条目。这有助于减少 Purge 操作对数据库性能的影响。但是，如果 Purge 操作因为各种原因（如 I/O 延迟、CPU 负载等）而滞后，那么 Undo 日志中的条目可能会在历史列表中保留更长的时间，从而导致其长度增加。

至此，结合上述分析，当排查这些内容（大量并发 DML 操作、未使用自动提交后的 SELECT

查询和异步 Purge 操作）时，发现存在未提交的事务，导致 Undo 日志无法清理。回滚这个事务，Undo 日志就会正常清理。

3. 总结

为了避免 Undo 日志对数据库性能产生负面影响，需要进行一些优化和调整。以下是给出的一些建议。

- innodb_buffer_pool_instances 和 innodb_purge_threads 的参数值小于物理 CPU 核数，以确保数据库系统能够有效地利用硬件资源。
- 使用 READ-COMMITTED 隔离级别可以减少锁竞争，提高并发性能。
- 大数据量下的逻辑备份工具 mysqldump 和 MyDumper 需要谨慎使用。应选择在系统负载较低的时间段进行备份，以减少对数据库性能的影响。可考虑使用物理备份工具 XtraBackup。
- 为了防止出现过大的 Purge 延迟，可以通过设置 innodb_max_purge_lag_delay 值来限制延迟。

7.3.5 导致服务器 CPU 的 sys 使用率过高的故障分析

MySQL 服务在运行一段时间后，随着并发量的增加，服务器的 CPU 的 sys 使用率也会上升，最终导致 QPS（每秒查询数）和 TPS（每秒事务数）下降，甚至出现服务性能瓶颈或阻塞的情况。

下面分析导致服务器的 CPU 使用率过高的问题。

1. 分析问题

首先，需要了解 CPU 的 sys（系统态）。其次，需要了解哪些情况下 MySQL 可能会影响 CPU 的 sys 使用率。基于这两点，可以进行性能排查。

- CPU 的 sys（或称为%sy）是操作系统内核态运行时间所占用的 CPU 百分比。当系统执行系统调用、管理硬件资源、处理中断或执行内核代码时，会处于内核态运行。如果 sys 的值较高，则可能表示系统正在执行大量的内核操作，如磁盘 I/O、网络处理、内存分配、进程调度等。高 sys 值问题可能是由于硬件资源不足、系统配置不当或软件问题引起的。
- MySQL 里的操作会导致 CPU 的 sys 使用率高，这可能有多种原因，主要包括调用系统时间函数、复杂的查询导致上下文切换以及内存资源不足时的内存整理。

2. 收集监控信息

- 使用 top 命令行查看资源使用情况。CPU 的使用率日常维持在 10%～20%，最高能达到 50%。使用 iostat 命令行查看 I/O 使用情况。硬盘的使用率为 20%。
- MySQL 当前运行线程数。查看线程运行状态的命令行示例如下。

```
mysql> SHOW GLOBAL STATUS LIKE 'Threads%';
+-------------------+-------+
| Variable_name     | Value |
+-------------------+-------+
| Threads_cached    | 33    |
| Threads_connected | 1649  |
| Threads_created   | 16392 |
| Threads_running   | 3     |
+-------------------+-------+
4 rows in set (0.00 sec)
```

- MySQL 的并发线程数。查看是否对并发线程数设置了限制，0 表示无限制。查看并发线程数的命令行示例如下。

```
mysql> SHOW VARIABLES LIKE 'innodb_thread_concurrency';
+---------------------------+-------+
| Variable_name             | Value |
+---------------------------+-------+
| innodb_thread_concurrency | 0     |
+---------------------------+-------+
1 row in set (0.00 sec)
```

- MySQL 错误日志和操作系统日志（/var/log/message）无异常信息。

从上述 MySQL 的指标来看，除了 CPU 的 sys 高以外，整体运行在正常范围内。使用 Linux 的 perf 指定 mysqld 进程分析内核调用情况。图 7-5 所示为使用 perf top 分析的信息。在发现 CPU 的 sys 利用率达到 79.15% 时，后台运行"__reset_isolation_suitable"调用会导致 CPU 的使用率突然达到 98%。进一步对内核进程上下文进行相关切换（finish_task_switch）以查看 CPU 使用率，发现运行"mov rdi, rax"时 CPU 的 sys 的使用率达到约 95%。

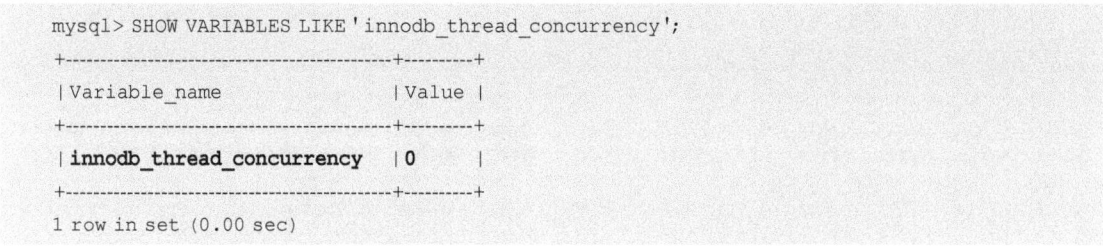

图 7-5　MySQL 调用系统内核

说明：上述 perf 指标的说明如下。

- mov rdi, rax：内存下的汇编指令。这条指令的功能是将 rax 寄存器中的值移动到 rdi 寄存器中。
- __reset_isolation_suitable：内存碎片管理接口。

经过上述内核调用分析，可以发现，问题都指向内核内存管理，尤其是与内存碎片管理有关。这基本上可以确定，CPU 的 sys 使用率高是内存碎片管理问题导致的。

3. CPU 与内存管理原理

CPU 是指中央处理器，其功能主要包括处理指令、执行操作、控制时间、处理数据。图 7-6 所示为 CPU 和内存之间的架构，分为 UMA 和 NUMA 两种。

- UMA 全称为 Uniform Memory Access，即一致性内存访问。多个 CPU 通过同一根总线来访问内存。无论多个 CPU 是访问内存的不同内存单元还是相同的内存单元，同一时刻，只有一个 CPU 能够访问内存。CPU 之间通过总线串行的方式访问内存，所以会出现访问瓶颈。

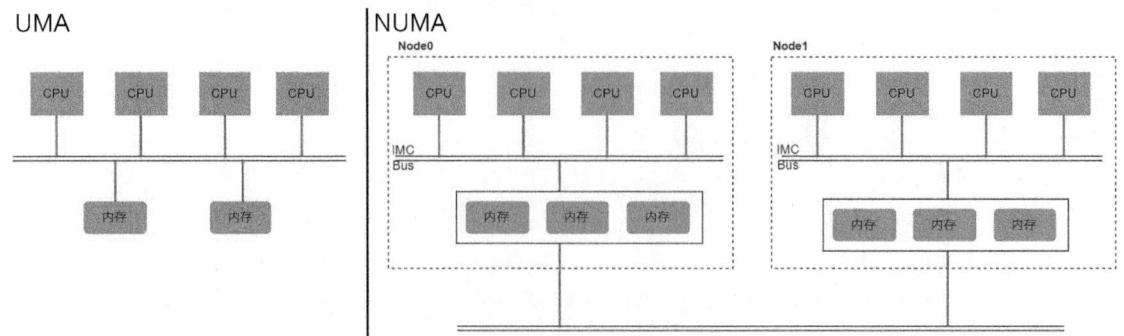

图 7-6　CPU 和内存之间的架构

- NUMA 全称为 Non-Uniform Memory Access，即非一致性内存访问。每个 CPU 都分配了一块内存，多个 CPU 可以同时并行访问各自的内存，读写内存的效率较高。

在 MySQL 环境中，更常见的是使用 UMA 架构，因为在 NUMA 架构下，虽然内存充裕，但仍然会触发 OOM，这是由内存碎片、不适当的内存分配策略，或者操作系统的内存管理策略不够高效导致的。

内存管理是操作系统内核中一个既复杂又有趣的领域。图 7-7 所示是 Linux 内核中管理物理内存的架构。物理内存的三级结构包括节点（Node）、区域（Zone）和页（Page）。

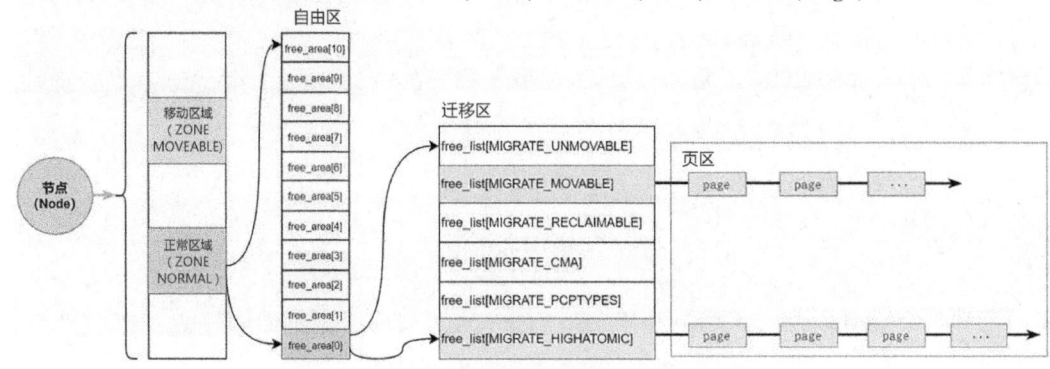

图 7-7　Linux 内核中管理物理内存的架构

对于 3 个区的说明如下。
- ZONE_NORMAL：正常使用的物理内存区域，大部分申请的内存都使用该区域。
- ZONE_HIGHMEM：只会出现在 32 位系统内，这是由于在 32 位系统中，物理内存最多能够直接映射内核中 896MB 内存。
- ZONE_MOVEABLE：可移动或回收区域，所管理的物理内存来自于 ZONE_NORMAL 或者 ZONE_HIGHMEM，主要是为了防止内存碎片和支持热插拔功能。内核在 ZONE_NORMAL 或者 ZONE_HIGHMEM 中根据配置划分出一片物理内存，称为 ZONE_MOVEABLE。

ZONE_MOVEABLE 有以下两个重要作用。
- 可以有效防止内存碎片化。

- 支持内存热插拔，尤其是在虚拟化场景，当不需要那么多物理内存时，可以释放给系统其他程序使用，当需要申请新的物理内存时，重新分配。

4. 定位问题

在找到问题点和了解了内存的一些基础理论后，如何确认是内存碎片导致的问题呢？Linux 内核提供了 pagetypeinfo、slabinfo 和 buddyinfo 等信息来帮助分析内存碎片情况。通过查看 buddyinfo，可以分析系统内存碎片的状况。查看 buddyinfo 信息的命令行示例如下。

```
shell $> cat /proc/buddyinfo |awk -v ps="`getconf PAGESIZE`" -v date="`date`" -v host="`hostname`" \
'BEGIN{printf("\n Fragmentation Report \n Low is Order 1-4, High is order 5-9, Normal is order
10-11 \n%s \t%s \n\n",host,date)} { \
L= ps * ( ($5 * 1) + ($6 * 2) + ($7 * 4) + ($8 * 8) ); \
H= ps * ( ($9 * 16) + ($10 * 32) + ($11 * 64) + ($12 * 128) + ($13 * 256)  ); \
N= ps * ( ($14 * 512) + ($15 * 1024) ); \
T=L+H+N; \
printf("%s \tTotal: %8dM \tLow: %02.2f%% \tHigh: %02.2f%% \tNormal: %02.2f%% \n", \
$1"" $2"" $3"" $4,T/1024/1024,(L/T)*100,(H/T)*100,(N/T)*100);}'

Fragmentation Report
Low is Order 1-4, High is order 5-9, Normal is order 10-11
schouseWed Sep  6 13:04:16 CST 2023

Node 0, zone DMA     Total:       15M Low: 0.00%High: 9.66%Normal: 92.19%
Node 0, zone DMA32   Total:     1012M Low: 4.28%High: 6.31%Normal: 89.34%
Node 0, zone DMA32   Total:    20480M Low: 99.80%High: 02.20%Normal: 0.00%
```

说明：关键指标说明如下。

- Normal：正常使用的物理内存区域，大部分申请的内存都使用该区域。
- Low：低水位，代表内存已经开始吃紧，需要启动回收页内核线程 kswapd 来回收内存。
- High：高水位，代表内存还是足够的。

从 buddyinfo 信息中发现，DMA 的 Low 区域是 0，已经没有可调用的内存，需要启动内存碎片整理。查看 zoneinfo 的信息的命令行如下。

```
# 单位均为页的数量,一个页默认大小是 4KB
shell $> cat /proc/zoneinfo
Node 0, zone DMA
  per-node stats
      nr_inactive_anon 448
      nr_active_anon 171214
      nr_inactive_file 121291
      nr_active_file 95855
      nr_unevictable 0
      nr_slab_reclaimable 13073
      ...
  pages free        2166      #当前可用的页面数
        min          94       #最低水位
        low         117       #低水位
        high        140       #高水位
        spanned    4095
```

```
            present      3998
            managed      3840
        protection: (0, 1781, 1781, 1781, 1781)
    nr_free_pages 2166
...
# 查看 page 的 min 大小设置
shell $ >cat /etc/sysctl.conf
vm.min_free_kbytes=10240
...
```

说明：关键指标说明如下。

- low：当空闲页的数量达到 page 的 low 所标定的数量的时候，kswapd 线程将被唤醒，并开始释放回收页。
- min：当空闲页的数量达到 page 的 min 所标定的数量的时候，分配页的动作和 kswapd 线程同步运行。
- high：当空闲页的数量达到 page 的 high 所标定的数量的时候，kswapd 线程将重新休眠，通常这个数值是 page_min 的 3 倍。

从 zoneinfo 信息中可以发现，page 的 low 已经大于 page 的 min 设置值（vm.min_free_kbytes）的 2 倍以上，kswapd 线程将被唤醒，并开始释放回收内存页。

到此，经过确认，服务器存在大量内存碎片，内存低水位达到了 0%，这导致时刻都在发生内存整理操作。由于频繁的内存整理，因此 CPU 的 sys 使用率上升到了 80%，严重影响了服务器的性能。

5. 解决方案

对于 Linux 下的内存碎片问题，可以通过以下三种方式来解决。

（1）内核内存参数更改

sysctl 是用于在 Linux 内核运行时动态地读取和设置内核参数的接口。通过 sysctl，可以查看和修改 /proc/sys 目录中的内核参数。这些参数涉及多个方面，包括 TCP/IP 网络堆栈、虚拟内存管理、文件系统和其他内核功能。

使用 sysctl 查看内存管理的参数的命令行如下。

```
shell#>sysctl -a | grep "vm\."
...
vm.extfrag_threshold = 500
vm.min_free_kbytes = 45056
vm.watermark_scale_factor = 10
...
```

通过调整上述 3 个内存指标，可以避免 CPU 的 sys 使用率高的问题。

1）内存压缩调整。

当向系统申请一大段连续内存时，如果找不到符合条件的连续页，就会触发内存压缩（memory compaction）。extfrag_threshold：参数用来控制出现内存压缩的概率。它是一个 0~1000 的整数。如果出现内存不够用的情况，Linux 会为当前系统的内存碎片情况打分，如果超过了 extfrag_threshold 这个值，kswapd 将触发内存压缩。这个值接近 1000 时，说明系统在内存碎片的处理上倾向于把旧的页换出，以符合申请的需要；而接近 0 时，表示系统在内存碎片的处理上倾向于做内存压缩。

通过修改 extfrag_threshold 内核参数来避免内存碎片处理。修改 extfrag_threshold 的命令行如下。

```
sysctl -w vm.extfrag_threshold = 1000
```

2）调整内存 min 水位线。

内存 min 水位线是 Linux 内核用于判断内存是否充足的一个阈值，当系统的空闲内存量低于该值时，内核会开始回收内存，以避免系统因内存不足而发生 OOM 错误。在多数情况下，建议将内存 min 水位线设置为总内存的 1%～3%。推荐设置为总内存的 2%，当内存资源紧张时，提前进行异步回收。设置内存 min 水位线的命令行如下。

```
sysctl -w vm.min_free_kbytes = memtotal_kbytes * 2%
```

说明：memtotal_kbytes * 2% 表示当前实例总内存的 2% 对应的内存大小。

3）调整内存 min 水位线和 low 水位线之间的差值。

内存 min 水位线和内存 low 水位线是 Linux 内核用于内存管理的两个阈值，用于触发不同的内存回收机制。可以通过内核的 watermark_scale_factor 调整内存 min 水位线和 low 水位线之间的差值，以应对业务突发申请内存的情况。默认情况下，watermark_scale_factor 的值为总内存的 0.1%，最小值为 min 水位线的 1/2。设置方式如下所示。

```
sysctl -w vm.watermark_scale_factor = value
```

说明：value 为手动设置的内存 min 水位线和 low 水位线之间的差值。

（2）定期整理内存

定期重新启动服务和服务器或定期进行内存整理操作，也可以实现内存碎片的整理。

1）定期进行内存整理。

可以在业务空闲时段，主动触发异步内存整理。主动触发内存整理的命令行如下。

```
echo 1 > /proc/sys/vm/compact_memory
```

2）定期手动释放缓存。

在以上措施均不能有效应对内存碎片化问题时，可以在业务空闲时段执行释放缓存（drop cache）的操作，然后内存会重新分配。释放缓存是避免内存碎片化问题的有效措施，但在释放缓存时，会出现短时间的系统性能抖动（因此这个方案不可取，只能应急）。释放内存缓存的命令行如下。

```
echo 3 > /proc/sys/vm/drop_caches
```

（3）升级内核和软件

除此之外，操作系统内核和软件升级也都是避免内存碎片问题的有效手段。随着技术的不断进步，新的内核版本和软件版本通常会包含性能改进、错误修复、安全增强和新功能等。这些更新可能会提高内存管理效率，减少内存碎片，提升系统整体性能。

第 8 章 MySQL 8.0 安全管理

随着互联网技术的不断发展和普及，以及数字化转型和云迁移的持续演进，MySQL 这一开源轻量级数据库得到了广泛的应用。MySQL 8.0 版本中引入了一系列新的安全功能和改进，包括更严格的新密码策略、增强的加密功能，以及改进的密码插件，以帮助保护 MySQL 的数据库安全。

8.1 MySQL 8.0 的安全管理概述

MySQL 数据库在使用中涉及众多敏感和重要的信息，如个人身份信息、银行账号信息以及公司商业机密等。因此，实施有效的安全管理技术对 MySQL 数据库至关重要。图 8-1 所示为 MySQL 数据库的安全技术框架，通过五个方面的安全点（访问控制、软件防护、数据脱敏、安全审计和数据加密）设置，以确保数据库的安全性。

图 8-1 MySQL 数据库安全技术框架

- 访问控制：涉及数据库的访问和操作流程的安全设计，如实施最小权限原则，确保只有必要的人员能够访问和操作数据库。
- 软件防护：通常包括使用最新版本的 MySQL 以获取最新的安全更新和补丁，以及配置 MySQL 的安全参数，如禁用不必要的账户、限制访问权限等。
- 数据脱敏：对于敏感数据（如个人身份信息、银行账号等），实施脱敏处理，即在不泄露实际数据的情况下，使用模拟数据进行测试和开发。
- 安全审计：定期对数据库的操作进行审计，以发现和防止任何潜在的安全威胁或违规行为。
- 数据加密：包括强密码策略的使用、定期更换密码，以及加密敏感数据的存储和传输。

8.1.1 MySQL 安全管理的作用

数据库安全管理是对数据库中的数据、程序和系统的安全进行管理与保护的过程。对于 MySQL 数据库而言，其安全管理的作用主要体现在以下 3 个方面。

1. 防止被黑客攻击

MySQL 数据库作为应用系统中的核心数据存储组件，承载着大量的关键数据和信息，因此成为黑客攻击的主要目标。通过实施有效的安全管理措施，可以大大降低数据库被黑客攻击的风险，保护系统内的敏感信息和保证数据安全。如果黑客成功入侵 MySQL 数据库，那么可能获取到敏感信息并掌握系统控制权，给整个系统带来严重的安全威胁。

2. 数据防泄露

许多应用程序将用户的隐私信息，如基础注册信息等，存储在 MySQL 数据库中。因此，数据安全变得尤为重要。数据库管理员必须确保用户信息得到充分保护，严格限制对敏感数据的访问，并采取措施避免数据泄露。随着数据安全上升到国家战略层面，《中华人民共和国数据安全法》和《中华人民共和国个人信息保护法》等法律法规的颁布实施，加强内部数据安全管理的同时，也需要满足法律法规的监管要求。

3. 灾难恢复

数据库是整个系统的关键组件，其安全性和稳定性对整个系统的运行至关重要。若数据库受到破坏或损坏，将可能对整个系统造成灾难性影响。因此，MySQL 数据库安全管理技术也涉及灾难恢复和备份策略的制定与实施，确保在数据库出现问题时能够迅速恢复数据，保障系统的持续运行。

8.1.2 MySQL 权限管理的作用

在 MySQL 数据库管理中，经常面临开发人员执行高危命令的风险，如删除（DROP）数据库和表，这可能导致数据丢失等严重后果。尽管可以通过备份和 binlog 来恢复数据，但这会给业务带来诸多不便。因此，数据库管理中的权限管理显得尤为重要。权限管理是通过合理的配置和管理，为每个用户分配适当的数据库权限，确保只能访问和操作其所需的数据对象。这包括用户的认证、授权和审计等方面，旨在保护数据库的机密性、完整性和可用性。

其作用主要体现在以下 3 个方面。

1. 数据安全

合理的权限管理可以防止未经授权的访问和数据泄露。通过限制用户的访问权限，可以确保只有授权用户才能查看、修改或删除敏感数据。

2. 遵守合规要求

许多行业都有严格的法规和合规要求，要求企业对数据库进行安全管控。例如，金融行业需要遵守 PCI DSS 规范，美国医疗行业需要遵守 HIPAA 法规。合理的权限管理可以帮助企业满足这些合规要求，避免因不合规操作而引发的法律风险和罚款。

说明：PCI DSS 是 2004 年启动的一套安全标准；这些标准适用于任何接受、处理、存储或传输信用卡数据的企业。PCI DSS 现已成为加强支付卡数据安全和防止安全漏洞的一项全球公认标准。

HIPAA 法规是美国针对医疗保健行业的关键法律框架，旨在保护个人健康信息（PHI）的私密性和安全性。随访信息作为患者治疗过程的重要组成部分，包含了丰富的敏感数据，如病情进展、药物使用情况等。

3. 数据完整性

通过权限管理，可以限制用户对 MySQL 数据库的误操作，如执行高危命令。这有助于减少因误操作导致的系统崩溃或数据丢失等风险，从而维护系统的稳定性。

8.2 MySQL 8.0 的安全管理关键点

8.2.1 安全管理制度的执行和管理

对于 MySQL 数据库而言，安全管理制度的执行和管理应当涉及访问控制、数据备份、加密措施、审计监控、安全培训、应急响应、定期检查和合规性等方面。这些方面共同构成了数据库安全管理制度的核心内容，并通过有效的执行和管理来确保数据库的安全性与稳定性。

1. 访问控制

访问控制是数据库安全的基础，其主要目标是限制用户对数据库的访问权限，防止未经授权的访问和数据泄露。采取最小分配原则。

- 建立严格的访问权限管理制度，确保只有授权人员才能访问数据库。
- 对不同等级的数据和系统资源进行分类，按需分配访问权限。
- 定期审查和更新访问权限，确保权限与工作岗位和工作职责相匹配。

2. 数据备份

数据备份是保证数据库安全的重要措施。通过定期备份数据，可以在数据发生损坏或丢失时，迅速恢复数据，避免损失。

- 制定完备的数据备份和恢复计划，以防数据丢失或损坏。
- 定期进行备份数据的恢复演练，并确保备份数据的可读性和可用性。
- 在数据库发生故障时，应迅速采取措施恢复数据，并查明故障原因。

3. 加密措施

在数据库安全管理制度中，应考虑使用加密技术对重要数据进行加密，确保数据传输过程中不会被窃听和篡改。

- 在存储数据中，使用加密算法保存数据。确保数据在存储过程中的安全性。
- 在传输过程中，使用加密技术传输数据，保证数据在传输过程中的保密性。
- 在数据库内部，对敏感数据采取加密算法，即在直接查询时，无法读取，必须通过解密算法才可以读取。

4. 审计监控

为了及时发现数据库的安全隐患和风险，应建立审计监控系统，发现并消除潜在的安全隐患。

- 通过实时记录和分析数据库的访问行为、系统状态等数据，发现异常行为，提示潜在的安全风险。
- 审计监控系统还可以为后续的违规行为的调查提供有力的支持。

5. 安全培训

数据库管理员应学习数据库安全方面的知识，加强安全意识，提高对数据库安全的重视程度。
- 应定期为数据库管理员提供安全培训，使其了解数据库安全管理制度的相关规定和操作规范。
- 避免因操作失误引起的安全问题。
- 数据库管理员也应该了解如何应对常见的安全威胁，如网络钓鱼、恶意软件等。

6. 应急响应

为了应对突发的安全事件，应建立应急预案，确保数据库的安全稳定运行。
- 应包括事件报告、事件评估、事件处理和事件总结等环节。
- 在发生安全事件时，应迅速报告相关事件，并进行评估和处理。
- 应根据事件的实际情况，总结经验教训，完善数据库安全管理制度。

7. 定期检查

为了确保数据库安全管理制度的有效执行，应定期对数据库安全状态进行检查。
- 应清理过期的数据，修复已知的安全漏洞，确保数据库系统的安全性。
- 应定期对数据库服务器进行硬件和软件的维护与升级，保证其稳定性和性能。

8. 合规性

在制定和执行数据库安全管理制度时，应确保遵守相关法律法规和政策要求。同时，还应关注 MySQL 数据库的国内外相关法规和开源协议的更新。及时调整数据库安全管理制度以符合最新的法规要求。

9. 其他

除此之外，要设立专门的数据库安全管理团队，负责制定、实施和维护数据库安全管理制度，对数据库的访问和使用进行实时监控，发现并纠正违规行为。

8.2.2 建立数据库审计制度

审核数据库活动是加强数据库安全性的重要组成部分。权威统计显示，80%以上的安全问题源于企业内部。其中绝大部分问题源于权限、脚本的审核缺失或过于宽泛。数据库审计的主要目的和价值在于对内部进行审核。这包括识别漏洞，如弱密码登录凭据、过多的用户和组权限以及未修补的数据库。攻击者会利用这些漏洞来实现权限提升、SQL 注入和 DoS 攻击。因此，出于安全性和合规性原因，对重要数据库活动的审计需求上升。此外，为运维数据库的每个环节都记录日志，可以确保数据管理的合规性和可追溯性。

MySQL 数据库审计一般包含以下内容。

1. 系统审计

检查 MySQL 所在的系统级别，包括操作系统、启动项、用户权限、服务安全和硬件配置等。

2. 数据审计

检查数据库的访问权限、数据库的用户名和密码、数据库的角色和权限设置等的安全性，以及具体的数据库安全控制措施。

3. 应用审计

检查应用程序的安全设置（如安全补丁、漏洞修复等），包括代码安全检查和数据访问控制。

4. 网络审计

检查网络设置是否安全，包括防火墙、VPN 设置、IP 白名单、网络隔离、网络流量监控、网络设备和安全策略等。

5. 操作日志审计

日志内容主要是与数据库系统有关的活动，包括数据库对象的创建、修改和删除、用户的登录和退出，以及数据库的访问等。

MySQL 的社区版本是不带审计插件（Audit Plugin）的。可以使用 MySQL 的 binlog 日志和 general 日志中的操作内容进行审计。binlog 日志记录了数据库上的所有改变，但不会记录用户登录信息。general 日志不仅记录了登录信息、操作命令，甚至还记录了错误的登录信息。但它们都过于详细，无法控制具体的审计项，同时占用极大的空间，影响数据库性能。所以目前 MySQL 开源版本无法做到真正意义上的审计，可以结合第三方插件实现（在开源方面，目前有 MariaDB 审计插件、McAfee 审计插件）。

8.2.3 敏感数据加密

敏感数据是指那些一旦泄露可能会给社会或个人带来严重危害的数据。这包括个人隐私数据，如姓名、身份证号码、电话号码、银行账号、密码、教育背景等，同时也包括企业或社会机构不宜公开的数据，如企业的经营状况、企业的网络结构等。

数据加密是指对数据库中的数据进行加密处理，使其变得不易被窃听、破解、篡改等方式访问、读取或修改等。对数据加密可以确保数据的机密性、完整性和可靠性，特别是对于敏感数据的加密，非常有必要。常见的加密技术有 AES（对称加密）、DES（非对称加密）。

- 对称加密。该加密算法使用相同的密钥进行加密和解密。对称加密算法是快速且易于使用的，但需要在安全的环境中存储密钥，以避免密钥泄露。
- 非对称加密。该加密算法使用不同的密钥进行加密和解密。一个密钥用于加密数据，而另一个密钥用于解密数据。非对称加密算法更加安全，因为公钥用于加密数据。私钥用于解密数据。但这种方法需要更多的计算资源，并且私钥需要严格保管。

1. MySQL 加密技术的使用场景

MySQL 加密技术在多种场景中都发挥着重要作用，以下是其主要的使用场景。

- 保护用户密码。如果将用户密码加密后存储在数据库中，那么，即使数据库遭受攻击，用户密码也能得到保护，不会被轻易泄露。
- 数据传输加密。在进行网络通信时，使用 SSL/TLS 协议对数据进行加密，这样可以确保在网络传输过程中的数据始终处于加密状态，即使被黑客拦截，也无法直接读取其内容，从而有效保护数据的安全性，防止非授权用户的搭线窃听和非法入网，以及数据在传输过程中被窃取或篡改。
- 数据加密保存。将个人信息、银行信息、企业信息等敏感数据使用加密算法和加密密钥转换为密文形式后保存，这样即使数据被未授权的人员获取，也无法直接阅读其内容，确保数据的机密性。
- 备份数据加密。对备份数据进行加密处理，以保护备份数据的安全，这样即使备份数据被盗或被未经授权的人访问，也无法读取其中的内容，确保数据的完整性和安全性。

2. MySQL 提供的加密技术

MySQL 提供了多种加密技术来保证数据的安全性和完整性。以下是一些 MySQL 提供的加密技术。

- AES 加密算法：MySQL 支持 AES（Advanced Encryption Standard）加密算法，这是一种广泛使用的对称加密算法。MySQL 提供了 AES_ENCRYPT( ) 和 AES_DECRYPT( ) 函数，允许用户在存储或检索数据之前与之后对数据进行加密和解密。
- SHA-2 哈希算法：MySQL 支持 SHA-2（Secure Hash Algorithm 2）系列哈希函数，其中 SHA-256 是 SHA-2 家族中常用的一种。SHA-2 是一种非对称加密算法，生成的是一个固定长度的哈希值，通常用于验证数据的完整性和是否被篡改。
- SSL/TLS 传输加密：MySQL 支持使用 SSL/TLS 协议来加密客户端和服务器之间的通信。这可以保护数据在传输过程中的安全，防止数据泄露或被篡改。
- 数据备份加密：虽然 MySQL 本身不提供直接的数据备份加密功能，但可以在备份过程中使用第三方工具或自定义脚本来加密备份文件。这可以通过加密文件系统、使用加密的压缩工具或专门的数据库备份软件来实现。

8.3 MySQL 8.0 的安全管理实践

8.3.1 密码插件的使用

密码插件在数据库系统中至关重要，因为密码校验规则是防止未经授权访问的关键。可以说，密码是数据库系统的第一道防线，对于确保数据库免受恶意攻击至关重要。从 MySQL 5.6 版本到 MySQL 8.0 版本，随着版本迭代和新安全加密技术的发展，MySQL 提供了多种身份验证插件，不断加强这一层防护。

表 8-1 所示为目前 MySQL 8.0 版本支持的密码插件列表。

表 8-1　MySQL 8.0 密码插件

插件名	支持版本	检查说明
mysql_native_password	ALL	基于本机密码哈希方法实现身份验证
sha256_password	ALL	基于本机使用 SHA-256 密码哈希执行身份验证
caching_sha2_password	ALL	实现 SHA-256 身份验证（如 sha256_password），但会在服务器端使用缓存
mysql_clear_password	ALL	客户端身份验证插件，可以让客户端将密码以明文形式发送到服务器，而不需要哈希或加密。这个插件内置在 MySQL 客户端库中
PAM Pluggable Authentication	MySQL Enterprise	使用 PAM（Pluggable Authentication Modules）来认证 MySQL 用户。外部身份验证，代理用户支持
Windows Pluggable Authentication	ALL	在 Windows 上使用本地 Windows 服务对客户端连接进行身份验证
LDAP Pluggable Authentication	MySQL Enterprise	使用 LDAP（轻量级目录访问协议）通过访问目录服务（如 X.500）来认证 MySQL 用户

（续）

插 件 名	支持版本	检查说明
Kerberos Pluggable Authentication	MySQL Enterprise	使用 Kerberos 进行身份验证。在 MySQL 8.0.26 及更高版本中可用 Windows 和 Linux 环境
No-Login Pluggable Authentication	ALL	mysql_no_login 服务器端身份验证插件阻止所有客户端连接到任何使用它的账户
Socket Peer-Credential Pluggable Authentication	ALL	服务器端 auth_socket 身份验证插件对通过 UNIX 套接字文件从本地主机连接的客户机进行身份验证
FIDO Pluggable Authentication	MySQL Enterprise	MySQL 8.0.27 及更高版本中可用，FIDO 身份验证允许使用智能卡、安全密钥和生物识别阅读器等设备进行身份验证
Test Pluggable Authentication	ALL	MySQL 包含一个测试插件，用于检查账户凭据，并将成功或失败记录到服务器错误日志中

下面介绍目前 MySQL 8.0 版本中常用的两种密码插件：mysql_native_password 和 SHA-256。

1. mysql_native_password 插件

MySQL 的 mysql_native_password 插件是一个用于本地身份验证的插件，它在 MySQL 5.7 及之后的版本中都被广泛使用。这个插件提供了一种相对安全的方式来验证用户密码。

在早期的 MySQL 版本，如 MySQL 4.1 之前的版本中，采用的是 mysql_old_password 认证方式。这种认证方式使用简单的密码哈希方法进行身份验证。随着安全性的需求增加，MySQL 后续版本逐渐淘汰了这种旧的身份验证方式。mysql_native_password 方式相对于 mysql_old_password 方式，在密码加密方面有所加强。

图 8-2 所示为 MySQL 8.0 版本之前 mysql_old_password 和 mysql_native_password 两种密码认证方式的示意图。在 mysql_old_password 方式中，客户端（Client）和服务端（Server）以 8 字节随机串加密密码（现已弃用）。在 mysql_native_password 方式中，使用 20 字节的随机串来加密密码，提供了更高的安全性。

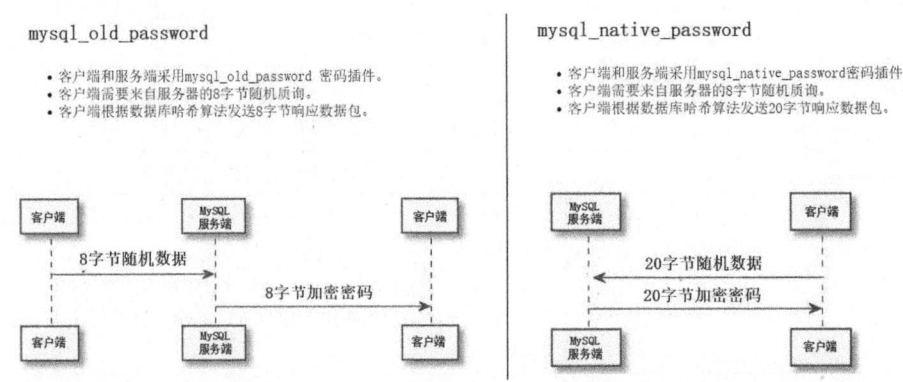

图 8-2　MySQL mysql_old_password 和 mysql_native_password 插件实现方式

在 MySQL 里使用的 SHA-1 加密方式如下。

```
SHA1( password ) XOR
SHA1( "20-bytes random data from server"<concat> SHA1( SHA1( password ) ) )
SHA1( SHA1( password ) )
```

- SHA-1 哈希：密码通过 SHA-1 算法进行哈希。SHA-1 是一种安全哈希算法，它会产生一个 160 位的哈希值。
- XOR 运算：服务器会生成一个 20 字节的随机数据。这个随机数据与上述 SHA-1 哈希结果进行 XOR（异或）运算。XOR 是一种位运算，它会比较两个对应位，如果两个位不同，则返回 1，如果相同，则返回 0。这步操作增加了密码的随机性和安全性。
- 双重 SHA-1 哈希：原始密码还会经过两次 SHA-1 哈希运算，即密码先经过 SHA-1 哈希，再将结果再次哈希。这样做可以增加哈希的复杂性和安全性。

mysql_native_password 密码插件在使用双重 SHA-1 加密后的值会被存储在 mysql.user 表的 authentication_string 中。如果两个用户使用了相同的密码，并且都是使用 mysql_native_password 插件进行加密的，那么这两个用户在 mysql.user 表中的 authentication_string 字段的值将会是相同的，这是因为相同的输入（即相同的密码）经过相同的加密流程会产生相同的输出。两个用户密码相同的命令行示例如下。

```
mysql> CREATE USER 'nativeuser1'@'localhost'
         IDENTIFIED WITH mysql_native_password BY '123456';

mysql> CREATE USER 'nativeuser2'@'localhost'
         IDENTIFIED WITH mysql_native_password BY '123456';
mysql> SELECT user,plugin,authentication_string
         FROM mysql.user WHERE user like '%nativeuser%';
+-------------+-----------------------+-------------------------------------------+
|user         |plugin                 |authentication_string                      |
+-------------+-----------------------+-------------------------------------------+
|nativeuser1  |mysql_native_password  |*6BB4837EB74329105EE4568DDA7DC67ED2CA2AD9  |
|nativeuser2  |mysql_native_password  |*6BB4837EB74329105EE4568DDA7DC67ED2CA2AD9  |
+-------------+-----------------------+-------------------------------------------+
2 rows in set (0.00 sec)
```

说明：美国国家标准与技术研究院（National Institute of Standards and Technology，NIST）建议 SHA-1 哈希算法的用户尽快迁移到更安全的 SHA-2 和 SHA-3 算法（最迟不晚于 2030 年 12 月 31 日）。一些大型科技公司，如 Meta、谷歌、微软和 Mozilla 等，在 2015 年时就差不多完成了对 SHA-1 算法的升级改造。2017 年，所有的主流浏览器不再将基于 SHA-1 的数字证书视为安全证书。对于 SHA-1 和其他哈希算法（如 MD5），可以很容易地预先计算好常见词语的哈希结果词典，利用诸如字典破解（Dictionary Attack）和暴力破解（Brute Force Attack）方式，可以匹配到密码，并且很多案例已证明非常容易破解。

2. SHA-256 插件

MySQL 为了避免用户使用相同或弱密码，提供了更安全的认证方法，这些方法不仅包括密码策略的实施，还包括使用强密码哈希算法，如 SHA-256。MySQL 8.0 及以后的版本中都提供了两种新的认证插件，它们利用 SHA-256 算法来增强密码的安全性。

说明：SHA-256（Secure Hash Algorithm 256bit）是一种密码学哈希函数，属于 SHA-2（安全哈希算法 2 系列）哈希函数家族的一部分。该函数能够将任意长度的数据转换为固定长度的唯一哈希值，这个哈希值的长度是 256 位，即 32 字节。

下面介绍 SHA-256 插件中的两种密码插件。

（1）sha256_password

使用 SHA-256 算法的密码哈希插件执行身份验证。该插件会对用户账户密码进行 SHA-256 哈希处理，并可能采用多轮哈希或"加盐"技术来增强安全性。与本地身份验证相比，这种加密方法更加强大，因此很难通过暴力破解方式来破解。

sha256_password 插件存在于 MySQL 服务器和客户端的形式如下。

- 服务器插件内置于服务器中，不需要显式加载，也不能通过卸载来禁用它。
- 客户端插件被内置到 libmysqlclient 客户端库中，对任何连接到 libmysqlclient 的程序都可用。

sha256_password 密码插件可与 RSA 公钥结合使用。对于使用 sha256_password 插件和 RSA 公钥进行密码交换的账户连接，服务器会根据需要向客户端发送 RSA 公钥。

- 当客户端使用 sha256_password 插件连接到服务器时，密码本身不会直接以明文形式发送。密码的传输安全性取决于是否使用了安全连接（如 TLS 加密）或 RSA 加密。
- 如果连接是安全的（例如，使用了 TLS 加密），那么密码将以加密形式发送，并且由于连接的安全性，密码在传输过程中无法被窥探。在这种情况下，不需要使用 RSA 密钥对进行额外的加密。
- sha256_password 插件通常不将共享内存连接视为安全的连接类型，即使共享内存传输在默认情况下是安全的。这是因为共享内存连接可能会受到其他安全因素的影响，如操作系统的访问控制和权限设置。
- 如果未使用安全连接（如 TLS 加密），并且 RSA 加密也不可用，则连接尝试会失败。这是因为在没有加密的情况下发送密码将暴露密码的明文形式，从而存在安全风险。
- 客户端用户可以通过两种方式获取 RSA 公钥：一种是数据库管理员可以提供公钥文件的副本，供客户端用户使用；另一种是客户端用户可以通过其他安全方式连接到服务器，并从服务器中获取 RSA 公钥。
- 数据库管理员可以提供 RSA 公钥文件的副本给客户端用户。
- 如果客户端用户能够通过其他方式连接到 MySQL 服务，则可以使用 SHOW STATUS LIKE 'Rsa_public_key' 语句来检索服务器的 RSA 公钥。客户端用户应将返回的键值保存在文件中，以便之后用于密码交换过程。

表 8-2 所示为 sha256_password 密码插件设置的参数列表。

表 8-2　sha256_password 密码插件参数

参　　数	说　　明
sha256_password_auto_generate_RSA_keys	MySQL 服务器在启动时自动生成 RSA 密钥对。这些密钥对用于 sha256_password 身份验证插件的密码交换
caching_sha2_password_auto_generate_rsa_keys	MySQL 服务器会在启动时自动为 caching_sha2_password 插件生成 RSA 密钥对
sha256_password_private_key_path	用于指定 sha256_password 插件使用的私钥文件的路径
sha256_password_public_key_path	用于指定 sha256_password 插件使用的公钥文件的路径

sha256_password 密码插件密钥的设置示例如下。

```
[mysqld]
sha256_password_private_key_path=myprivkey.pem
sha256_password_public_key_path=mypubkey.pem
```

```
# 通过命令行方式查看公钥信息
mysql> SHOW STATUS LIKE 'Rsa_public_key';
+-----------------+------------------------------------------------------------+
|Variable_name    |Value                                                       |
+-----------------+------------------------------------------------------------+
|Rsa_public_key   |-----BEGIN PUBLIC KEY-----
    MIIBIjANBgkqhkiG9w0BAQEFAAOCAQ8AMIIBCgKCAQEAx+VWXiKhYLjYPmSYn5AK
    jlyo258sjlNFTgvJk4A244newoNNCxAuVV5KR+zZgXCEdquz7Va4tFYxDqMK1WkS
    UKryGingqRxDRrl9zVQDh2Fo5zH42fehND/euDmvVU2xsoklYrLJf4MqjNpqfe3B
    HK0Obh4UOxQzwxJGnKksp3unM4JZW+snw07v+O9vhkE2/wRTvFCEOO1Ye25tuDqj
    q4GBjZIvP6oqjpxN/WqSP9hkf2geYKOrWlCjNxJIuR1fVdlUVTG6WZPnkw86sFLB
    beMUGkpZr3NTyIEyyaxzOOusZrboQ4q687ZrqIZt0nFHpJ8lip0Kf/uwM4CQzFg8
    YwIDAQAB-----END PUBLIC KEY-----                                           |
+-----------------+------------------------------------------------------------+
```

说明：对于密钥文件，使用官方提供的 mysql_ssl_rsa_setup 工具生成所需的 RSA 密钥对和证书。在认证过程中，SHA-256 哈希在安全连接和多轮哈希转换上需要更长的时间，这样会对性能产生一定的影响。

（2）caching_sha2_password

从 MySQL 8.0.3 版本开始，引入了一个新的身份验证插件 caching_sha2_password。在 MySQL 8.0.4 中，caching_sha2_password 成为默认的身份验证插件。这个插件是 sha256_password 身份验证插件功能的超集，它实现了 SHA-256 身份验证，同时在服务器端使用缓存来提高性能，并具有更广泛的适用性。由于 caching_sha2_password 既解决了安全性问题又解决了性能问题，因此不推荐使用 sha256_password。

使用 caching_sha2_password 进行身份验证的客户端必须使用安全连接（通过 TCP 使用 TLS/SSL 凭证、UNIX 套接字文件或共享内存进行连接），或者使用 RSA 密钥对支持密码交换的未加密连接。同时，服务器端使用 sha2_cache_cleaner 审计插件作为助手来进行密码缓存管理。sha2_cache_cleaner 和 caching_sha2_password 一样，都是内置的，不需要安装。

SHA_256 加密方式如下所示。

```
Scramble-XOR(SHA256(password),
SHA256(SHA256(SHA256(password)),Nonce))
Hash entry - account_name -> SHA256(SHA256(user_password))
```

- Scramble-XOR(SHA256(password),(SHA256(SHA256(SHA256(password)),Nonce))使用 Scramble-XOR 函数将密码进行 SHA-256 加密后，与客户端提供的密码（未经哈希）进行 XOR 操作，从而得到一个混淆的密码。通过两次 SHA-256 算法处理之后，再次利用随机值进行第三次 SHA-256 处理，这个过程涉及一个名为 Nonce 的随机值（每次连接时都会生成一个新的 Nonce）。
- SHA256(SHA256(user_password))：用户的密码首先通过 SHA-256 算法进行哈希，然后，这个哈希值再次通过 SHA-256 算法哈希，最终存储哈希值到用户表。

两个相同用户密码产生不同哈希值的命令行示例如下。

```
mysql> CREATE USER 'cachsha256user1'@'localhost'  IDENTIFIED WITH
caching_sha2_password BY '123456';
Query OK, 0 rows affected (0.01 sec)
```

```
mysql>CREATE USER 'cachsha256user2'@'localhost' IDENTIFIED WITH
caching_sha2_password BY '123456';
Query OK, 0 rows affected (0.01 sec)

mysql> SELECT user,plugin, authentication_string
       FROM mysql.user WHERE userLIKE 'cachsha256user%'\G
*************************** 1.row ***************************
                    user: cachsha256user1
                    host: localhost
                  plugin: caching_sha2_password
   authentication_string: $A $005 $04}
        SP5FJhpg4GTPTFKvaVuZFrZo9W/8bbP4czF8KP1nTyzBNxU1AD
*************************** 2.row ***************************
                    user: cachsha256user2
                    host: localhost
                  plugin: caching_sha2_password
   authentication_string: $A $005 $-nS)Il0`,Ab#
           ~XLrhOHJx0vAvyASvf1vHQrkkB7D7Y74lwjmATctcQ1U5
2 rows in set (0.01 sec)
```

从上述结果中可以发现，使用 caching_sha2_password 密码插件可以避免因为使用相同密码而产生的相同的哈希值问题，从而提高了安全性。表 8-3 所示为在 caching_sha2_password 机制下，MySQL 的 mysql.user 表中 authentication_string 字段的哈希值分割列表。

表 8-3　MySQL 8.0 版本中 caching_sha2_password 分割显示内容

内　容	字　节　数	说　　明
哈希算法	2 字节	目前仅为 $A，表示 SHA-256 算法
哈希轮转次数	4 字节	目前仅为 $005，表示 5×1000 = 5000 次
盐（salt）	21 字节	用于解决相同密码产生相同哈希值问题
哈希值	43 字节	更复杂的哈希值

虽然密码哈希值是根据一定的规律生成的，但这些规律并不足以泄露原始密码的信息。相比 sha256_password，caching_sha2_password 插件具有以下优势。

- 服务器端的内存缓存使得重新连接的用户能够更快地重新认证之前的连接。当然，对于第一次连接，缓存中是没有用户数据的。
- 无论 MySQL 连接到哪个 SSL 库，caching_sha2_password 都可以进行基于 RSA 的密码交换。
- 该插件还支持使用 UNIX 套接字文件和共享内存协议（此处可能需要补充具体支持的协议名称或功能描述）。

caching_sha2_password 密码插件可以与 SSL 配合使用，以提高数据传输的安全性。配置方式如下所示。

```
[mysqld]
caching_sha2_password_private_key_path=myprivkey.pem
caching_sha2_password_public_key_path=mypubkey.pem
caching_sha2_password_digest_rounds=10000
```

说明：上述配置中的 caching_sha2_password_private_key_path 和 caching_sha2_password_public_ key_path 分别指定了私钥与公钥文件的路径,而 caching_sha2_password_digest_rounds 参数(默认值为 5000)用于指定 caching_sha2_password 认证插件密码存储的哈希轮转次数,并控制密码哈希的复杂性,增加暴力破解的难度。这种方式会存在一定的性能损失。

在复制环节中,可以使用基于 RSA 密钥的密码交换连接到主机。所谓的 RSA 公开密钥密码体制,就是使用不同的加密密钥与解密密钥。RSA 密钥特别适合使用 caching_sha2_password 或 sha256_password 插件进行身份验证的场景。

- CHANGE MASTER TO 语句有一个子句,用于指定 RSA 公钥信息。
- 对于 MGR 集群,参数 group_replication_recovery_public_key_path 和 group_replication_recovery_get_public_key 的作用相同。
- SHOW SLAVE STATUS 语句和 performance_schema 下的 replication_connection_configuration 表可以显示复制从机的 RSA 公钥信息。

由于 caching_sha2_password 插件使用了缓存机制,因此能够实现快速的身份验证。然而,缓存清理操作将影响后续客户端连接的身份验证。以下是几种导致缓存清理的情况。

- 新创建账号,然后删除并重新创建。这种情况下,用户的缓存信息将从内存中移除。
- 更改用户密码。更改密码后,该用户的缓存信息将从内存中删除。
- 使用 RENAME USER 命令重命名用户。重命名用户时,其缓存信息将从内存中删除。
- 执行 FLUSH PRIVILEGES 命令。该命令会删除所有缓存的用户信息。
- 服务重新启动。当 MySQL 服务重新启动时,所有缓存的用户信息将被清除。

8.3.2 数据加密功能的使用

在当今信息化社会中,数据安全的重要性不言而喻。对于数据库而言,加密不仅用于隐藏敏感信息,还可以确保存储的数据紧凑且不浪费空间。因此,数据加密是一项一举多得的功能。

在 MySQL 方面,其提供的数据加密方式如下。

- MySQL Enterprise 版本提供了 Data Masking、Transparent Data Encryption 和 openssl_udf 等安全机制,以满足企业的安全需求。
- 对于 MySQL 的 GA 版本,它提供了加密函数来处理敏感数据。其中包括 MD5( )、SHA1( )、SHA2( )、STATEMENT_DIGEST( ) 和 STATEMENT_DIGEST_TEXT( ) 等函数。

表 8-4 所示为 MySQL 8.0 GA 版本提供的加密函数列表。

表 8-4 MySQL 8.0 GA 版本提供的加密函数

函 数	描 述
AES_ENCRYPT( )	使用 AES 加密
AES_DECRYPT( )	使用 AES 解密
MD5( )	计算 MD5 校验值
RANDOM_BYTES( )	返回随机字节量
SHA1( )	计算 SHA-1 的 160 位校验值
SHA2( )	计算 SHA-2 的 256 位校验值
STATEMENT_DIGEST( )	计算语句摘要哈希值
STATEMENT_DIGEST_TEXT( )	转换规范化语句摘要

下面具体介绍上述提到的 MySQL 的加密方式。

1. MD5 加密

MD5 算法是一种哈希算法，也是常用的方式，但这一算法是不可逆的。也就是说，通过哈希算法得到的数据，无法经过任何算法还原回原始数据。在底层处理中，数据会通过 MD5 或 SHA-1 等哈希算法，返回相应的十六进制字符串表示。为了提高存储效率和比较速度，这些十六进制字符串通常会被转换为二进制形式并存储在数据库中。需要注意的是，每两个十六进制数字表示一个字节的二进制数据，因此十六进制字符串的长度决定了所需的字节数。

- MD5 校验值需要 16 字节（因为 MD5 哈希是 128 位，而每字节是 8 位，所以需要 128/8＝16 字节）。
- SHA-1 校验值需要 20 字节（因为 SHA-1 哈希是 160 位，所以需要 160/8＝20 字节）。
- SHA-2 校验值取决于具体的 SHA-2 变体，如 SHA-224、SHA-256、SHA-384 和 SHA-512，它们分别需要 28、32、48 和 64 字节。

在使用场景方面，MD5 最常用于加密用户密码，它可以将用户输入的明文密码转换成一个 128 位的哈希值，这个哈希值可以用来验证用户输入的密码是否正确，而不必将用户的密码明文存储在服务器上，从而保护用户的密码安全。此外，MD5 密码还可以用于文件完整性检查，即检查文件是否被篡改。

在 MySQL 中，实现 MD5 数据加密方式的命令行示例如下。

```
mysql> CREATE TABLE md5_tbl (md5_val_char CHAR(32),md5_val_bin BINARY(16));
Query OK, 0 rows affected (0.02 sec)
# 使用 MD5()函数
mysql> INSERT INTO md5_tbl (md5_val_char) VALUES(MD5('abcdef'));
Query OK, 1 row affected (0.01 sec)
mysql> INSERT INTO md5_tbl (md5_val_bin ) VALUES(UNHEX(MD5('abcdef')));
Query OK, 1 row affected (0.00 sec)
# 查看加密数据
mysql> SELECT * FROM md5_tbl;
+----------------------------------+----------------------------------+
|md5_val_char                      |md5_val_bin                       |
+----------------------------------+----------------------------------+
|e80b5017098950fc58aad83c8c14978e  |NULL                              |
|NULL                              |0xE80B5017098950FC58AAD83C8C14978E|
+----------------------------------+----------------------------------+
```

说明： 除非使用 SSL 连接，否则作为加密函数参数提供的密码或其他敏感值将以明文形式发送到 MySQL 服务端。此外，这些值会出现在所写入的任何 MySQL 日志中。

虽然 MD5 加密方式在实际应用中很实用，但仍然存在被破解的风险。一种常见的破解方法是将提交的 MD5 哈希值与预先存储的哈希值进行比对。这种方法可能包括字典破解和暴力破解，尽管这些方法相对耗时且效率较低。为了提高破解效率，攻击者还可能采用更高级的技术，如查表法（通过预先计算的哈希值表来快速查找匹配项）、逆向查表法（根据哈希值反向查找原始数据）以及彩虹表（一种结合了时间和空间优化的预计算哈希值表）。这些高级技术虽然提高了破解效率，但仍然需要大量的计算资源和时间。

2. AES 加密

AES 是一种广泛采用的国际加密标准，由 NIST（美国国家标准及技术协会）制定。MySQL 提供了支持这个标准的函数,即 AES_ENCRYPT 和 AES_DECRYPT,分别用于对数据进行加密和解密。这些函数通常使用固定的加密模式,如 ECB 或 CBC,而不是通过一个单独的参数来指定。密钥长度可以是 128 位、192 位或 256 位,具体取决于 MySQL 的版本和配置。

在 MySQL 中使用 AES 加密方式时需要注意以下 3 点。

- 在复制集群中,使用 AES_DECRYPT 的语句对基于 STATEMENT 语句的复制是不安全的。
- 从 MySQL 8.0.30 版本开始,这些函数支持使用密钥派生函数(KDF)利用从 key_str 中传递的信息创建一个加密的强密钥。派生密钥用于加密和解密数据,它保留在 MySQL 实例中,用户无法访问。强烈建议使用 KDF,因为它提供了比指定自己的预制密钥或在使用函数时通过更简单的方法派生密钥更好的安全性。
- AES_ENCRYPT 和 AES_DECRYPT 函数允许使用控制块加密模式。

AES 加密算法的模式由参数 block_encryption_mode 控制,其默认值为 aes-128-ecb,表示使用 128 位密钥长度和 ECB 模式进行加密。允许的 keylen 值有 128、192 和 256,允许的模式值有 ECB、CBC、CFB1、CFB8、CFB128 和 OFB。

在 MySQL 中,查看和设置 AES 加密算法模式的命令行示例如下。

```
mysql> SHOW VARIABLES LIKE '%block_encryption_mode%';
+-----------------------+------------+
| Variable_name         | Value      |
+-----------------------+------------+
| block_encryption_mode | aes-128-ecb|
+-----------------------+------------+
1 row in set (0.00 sec)

# 设置 AES 加密函数使用 256 位的密钥长度和 CBC 模式
mysql> SET GLOBAL block_encrypto_mode='aes-256-cbc';
```

在 MySQL 中,AES 函数的语法如下所示。

```
AES_DECRYPT(crypt_str,key_str[,init_vector][,kdf_name][,salt][,info |iterations])
AES_ENCRYPT(str,key_str[,init_vector][,kdf_name][,salt][,info |iterations])
```

具体参数说明如下。

- str:用于指定纯字符串。
- crypt_str:加密字符串。
- key_str:用于指定用于加密字符串的字符串。因为经过加密和压缩的返回结果为二进制字符,所以建议配置为 VARBINARY 或 BLOB 二进制字符串数据类型的列,防止因字符集转换导致插入失败。
- init_vector:初始向量,用于块加密的模式(block_encryption_mode)。
- kdf_name:KDF 的名称,用于根据传入 key_str 的输入密钥材料和 KDF 的其他参数创建密钥。支持 HKDF(OpenSSL 1.1.0. 中的 HKDF)和 pbkdf2_hmac(OpenSSL 1.0.2 中的 PBKDF2)。
- salt:哈希值存储在数据库中依然是不够安全的。此时可采取 salt 方式,就是随机生成一个字符串。将"盐"与原始密码连接(concat)在一起(放在前面或后面都可以)。对密码可以这样处理,其他情况就不行。
- iterations:PBKDF2 在生成密钥时使用的迭代计数初始向量,用于块加密的模式。计数越高,对暴力攻击的抵抗力就越强。

在 MySQL 中,AES_ENCRYPT 和 AES_DECRYPT 函数通常与 INSERT 和 UPDATE 语句一起使用,存储或修改在数据库中的加密数据。AES 加密和解密的命令行示例如下。

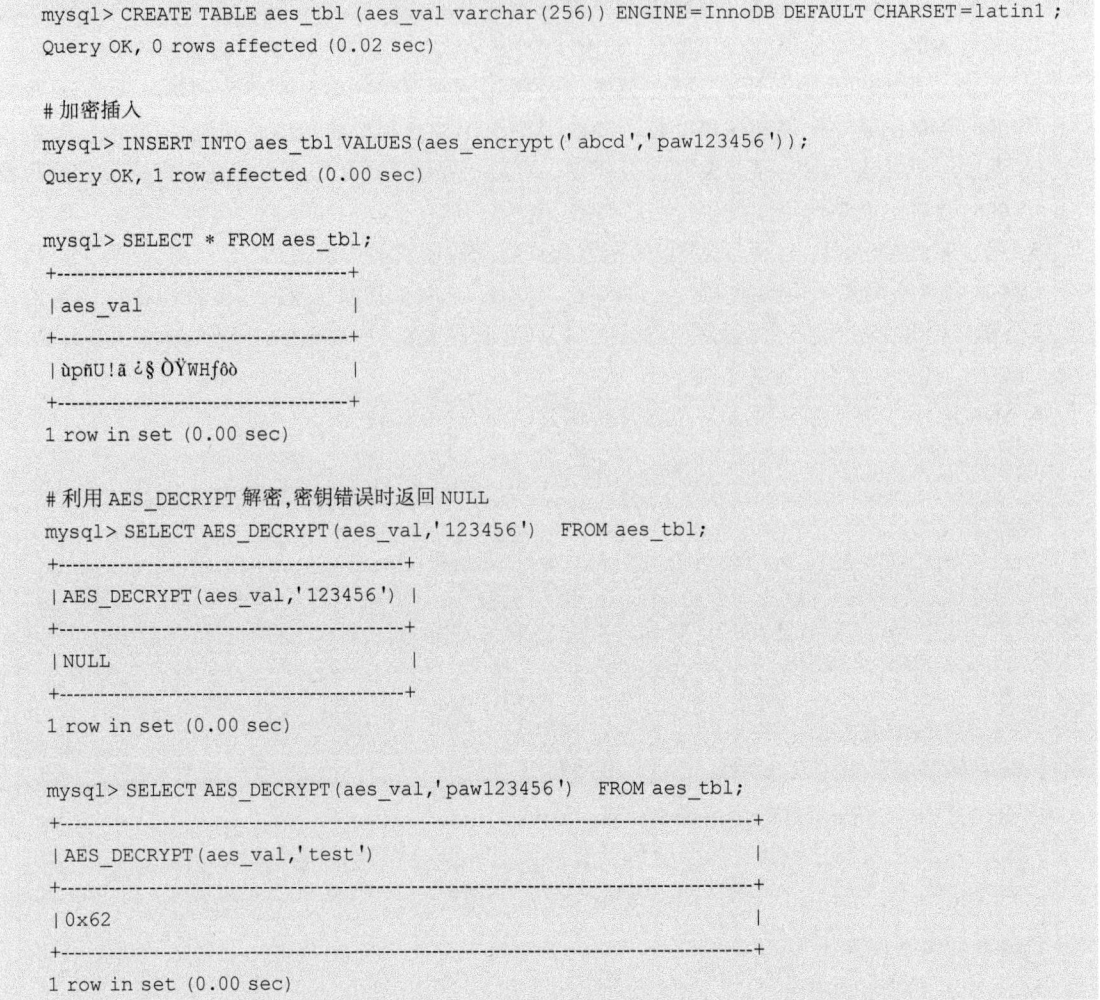

在实际应用中,密钥的管理至关重要,因为泄露密钥将直接导致加密数据的安全性受到威胁。同时,加密和解密操作可能会对性能产生影响,特别是在处理大量数据时。

3. RANDOM_BYTES 随机加密

在 MySQL 中,RANDOM_BYTES 函数用于生成加密安全的随机字节序列。这个函数返回指定长度的随机字节的二进制字符串,这些字节是通过 SSL 库的随机数生成器生成的。RANDOM_BYTES 函数指定长度(Len)允许的值范围为 1~1024。如果长度超出此范围,则将发生错误。如果长度为 NULL,则函数返回 NULL。

在 MySQL 中使用 RANDOM_BYTES 函数时,需要注意以下几点。

- RANDOM_BYTES 函数可用于为 AES_DECRYPT 和 AES_ENCRYPT 函数提供初始加密值。在这些上下文中使用时,长度必须至少为 16。虽然允许更大的值,但超过 16 字节的部分将被忽略。
- RANDOM_BYTES 生成的随机值结果是不确定的。因此,使用此函数的语句在基于语句的复制环境中可能是不安全的。
- 如果从 MySQL 客户端中调用 RANDOM_BYTES，则返回的二进制字符串将使用十六进制表示法显示。

在 MySQL 中，RANDOM_BYTES 函数的使用示例如下。

```
#设置密码 key
mysql> SET @init_vector = RANDOM_BYTES(16);
Query OK, 0 rows affected (0.00 sec)


mysql> SET @key = SHA2('secret key', 224);
Query OK, 0 rows affected (0.00 sec)
#加密
mysql> SET @crypto = AES_ENCRYPT('abcd', @key, @init_vector);
Query OK, 0 rows affected, 2 warnings (0.00 sec)
#查看
mysql> SELECT @crypto;
+--------------------------------+
| @crypto                        |
+--------------------------------+
| ?¯ rZªun XP ÝdG2               |
+--------------------------------+
1 row in set (0.00 sec)
#解密
mysql> SELECT AES_DECRYPT(@crypto, @key, @init_vector);
+------------------------------------------------+
| AES_DECRYPT(@crypto, @key, @salt)              |
+------------------------------------------------+
| abcd                                           |
+------------------------------------------------+
1 row in set, 1 warning (0.00 sec)
```

在 MySQL 中，由于 RANDOM_BYTES 生成的是随机值，因此它可能不适合所有基于语句的复制环境。如果主服务器和从服务器生成的随机值不同，则可能会导致数据不一致。RANDOM_BYTES 生成的随机值应该被视为敏感信息，并且在应用程序中应该妥善管理，避免泄露。

4. SHA-1 加密

SHA（Secure Hash Algorithm，安全哈希算法）是一个密码哈希函数家族，这些函数可以生成一个固定大小的哈希值，该值对于输入数据是唯一的。SHA 家族包括多种变体，其中 SHA-1 是最早的，也是最知名的版本之一。SHA-1 产生一个 160 位的哈希值，可以视为在密码学上比 MD5 更安全。

在 MySQL 中，使用 SHA1 函数进行数据加密的命令行示例如下。

```
mysql> SELECT SHA1('abc');
+------------------------------------------+
| SHA1('abc')                              |
+------------------------------------------+
| a9993e364706816aba3e25717850c26c9cd0d89d |
+------------------------------------------+
1 row in set (0.00 sec)
```

尽管 SHA-1 在过去被广泛使用并且目前仍然在一些场景中被接受,但它已经不再被认为是安全的,特别是在密码学应用中。这主要是因为已经存在实际的方法来找到 SHA-1 中的碰撞攻击,这意味着在某些情况下,攻击者可以生成两个不同的输入值,它们产生相同的 SHA-1 哈希值。这种能力可能会破坏某些依赖于哈希值唯一性的安全系统。

5. SHA-2 加密

SHA-2(Secure Hash Algorithm 2)是指 SHA-2 系列哈希函数,它们都是由美国国家安全局(NSA)设计的,并且由 NIST 发布以作为 SHA-1 的继任者。SHA-2 系列提供了更高的安全性和更长的哈希值长度,包括 SHA-224、SHA-256、SHA-384 和 SHA-512。

在 MySQL 中,使用 SHA2 函数进行数据加密的命令行示例如下。

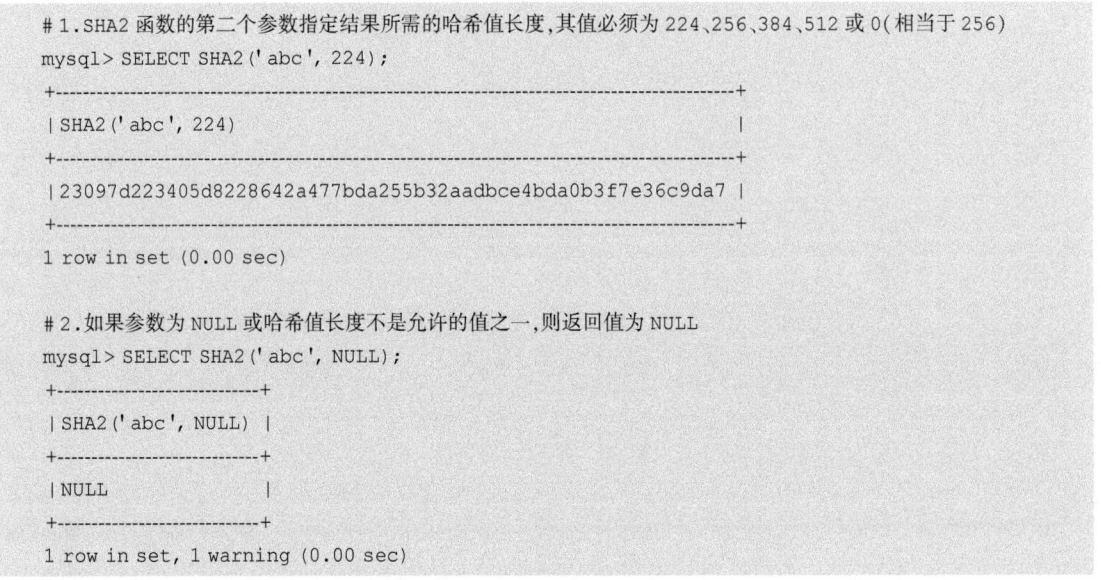

说明: 仅当 MySQL 配置为 SSL 时,SHA2 函数才起作用。

SHA2 函数的设计旨在防止碰撞攻击(即找到两个不同的输入值,产生相同的哈希值)和预图像攻击(即找到产生特定哈希值的输入)。由于 SHA-2 提供了比 SHA-1 更长的哈希值长度和更复杂的算法设计,因此它被广泛认为在密码学上更安全。

6. STATEMENT_DIGEST 加密

在 MySQL 中,STATEMENT_DIGEST(语句摘要)用于记录、摘要和统计 SQL 语句的执行情况。当 MySQL 服务器执行 SQL 语句时,它会将 SQL 语句通过一个哈希转换功能转换为一个简短的摘要。这样,相同的 SQL 语句(即使参数值不同)也会被归类到同一个摘要中。这种哈希转换对于监控和优化数据库性能非常有用,因为它允许查看哪些 SQL 语句频繁执行,以及它们各自消耗了多少资源,同时也保证了这些 SQL 语句不显示具体内容。

语句摘要提供的两个函数 STATEMENT_DIGEST 和 STATEMENT_DIGEST_TEXT,分别用于对 SQL 语句进行哈希加密和解密。下面介绍语句摘要函数。

(1)STATEMENT_DIGEST 函数

STATEMENT_DIGEST 函数接受一个 SQL 语句作为输入,并返回一个哈希值。这个哈希值是通过去除 SQL 语句中的条件和变量值后,对剩余的语句进行哈希处理得到的。这样,即使两个 SQL 语句在逻辑上相同但变量值不同,它们也会得到相同的哈希值。

STATEMENT_DIGEST 摘要加密函数的命令行示例如下。

```
# ID=1 的条件
mysql> SELECT STATEMENT_DIGEST("SELECT * FROM test WHERE ID=1;");
+--------------------------------------------------------------------+
| STATEMENT_DIGEST("SELECT * FROM test WHERE ID=1;")                 |
+--------------------------------------------------------------------+
| c4af45cba988541e319888c1bd5d79db763895a68320e5cce65e3b7e8547f919   |
+--------------------------------------------------------------------+
1 row in set (0.00 sec)
# ID=5 的条件
mysql> SELECT STATEMENT_DIGEST("SELECT * FROM test WHERE ID=5;");
+--------------------------------------------------------------------+
| STATEMENT_DIGEST("SELECT * FROM test WHERE ID=5;")                 |
+--------------------------------------------------------------------+
| c4af45cba988541e319888c1bd5d79db763895a68320e5cce65e3b7e8547f919   |
+--------------------------------------------------------------------+
1 row in set (0.00 sec)
```

从上述最终结果中可以发现，得到了一个哈希值"c4af45cba988541e319888c1bd5d79db763895 a68320e5cce65e3b7e8547f919"。

STATEMENT_DIGEST 摘要加密函数也支持调用存储过程和其他函数。具体命令行示例如下。

```
# 存储过程的摘要
mysql> SELECT STATEMENT_DIGEST('CALL abc()');
+--------------------------------------------------------------------+
| STATEMENT_DIGEST('CALL abc()')                                     |
+--------------------------------------------------------------------+
| b356ecfb586e0a761b9794bcb6d2e4395d40416f3110e688084bd67e87d27ae9   |
+--------------------------------------------------------------------+
1 row in set (0.00 sec)
# 自定义函数和系统函数的摘要
mysql> SELECT STATEMENT_DIGEST('SELECT abc()');
+--------------------------------------------------------------------+
| STATEMENT_DIGEST('SELECT abc()')                                   |
+--------------------------------------------------------------------+
| 0af0d93da64f52e740489a7bcd409469011ba78238c29facf60763264b64727e   |
+--------------------------------------------------------------------+
1 row in set (0.00 sec)

mysql> SELECT STATEMENT_DIGEST('SELECT NOW()');
+--------------------------------------------------------------------+
| STATEMENT_DIGEST('SELECT NOW()')                                   |
+--------------------------------------------------------------------+
| 2fad55a506c3342fe82629447a3412f5dfb1aaaa20b13be45f1b9c42175da923   |
+--------------------------------------------------------------------+
1 row in set (0.00 sec)
```

(2) STATEMENT_DIGEST_TEXT 函数

STATEMENT_DIGEST_TEXT 函数返回给定 SQL 语句的规范化摘要，该摘要以字符串形式表示。在这个摘要中，语句中的具体值和变量会被去掉，并替换成问号(?)符号，从而生成一个类似于通用语句的表示。摘要转化命令行示例如下。

```
mysql> select STATEMENT_DIGEST_TEXT("SELECT * FROM test WHERE ID=5;");
+--------------------------------------------------------+
| STATEMENT_DIGEST_TEXT("SELECT * FROM test WHERE ID=5;") |
+--------------------------------------------------------+
| SELECT * FROM `test` WHERE `ID` = ? ;                   |
+--------------------------------------------------------+
1 row in set (0.00 sec)
```

使用 STATEMENT_DIGEST 加密时需要注意以下两点。

- STATEMENT_DIGEST 函数转换的结果集受限于参数 max_digest_length 设置的最大长度。max_digest_length 的默认值是 1024 字节（在 MySQL 5.7 及更高版本中）。如果一个 SQL 语句的摘要文本超过了这个长度，那么它将被截断以适应这个限制。这可能导致一些较长的 SQL 语句的摘要信息不完整，从而影响通过摘要来识别和分析性能瓶颈的准确性。
- 对于 STATEMENT_DIGEST 函数，MySQL 5.7 版本中无此函数，只有 MySQL 8.0 版本中才存在。其实，MySQL 5.7 和 MySQL 8.0 使用的算法也不同。图 8-3 所示为源码 sql\sql_digest.h 里 DIGEST_HASH_TO_STRING_LENGTH 函数定义内容，包括每个函数使用的位数（bit）、处理方式和版本支持说明。在 MySQL 5.7 中使用 128bit 加密方式，从 MySQL 8.0 开始，使用 256bit 加密方式。

7. 总结

网络时代，信息极为丰富，但同时会造成信息泛滥，导致安全问题严重。因此，数据库的安全加密解密机制变得不可或缺。然而，值得注意的是，加密解密操作也会消耗一定的性能，如使用 AES 加密解密可能会导致 10%~25% 的性能损失。因此，在实际应用中，需要根据具体情况合理选择加密方式，以平衡安全性和性能。

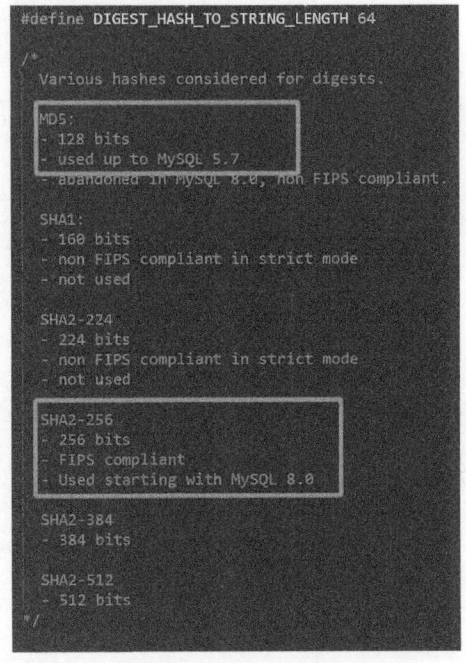

图 8-3　MySQL 8.0 DIGEST_HASH_TO_STRING_LENGTH 函数

8.3.3　SSL 安全的设置

在 MySQL 中，SSL 可以用于加密客户端与服务器之间的通信，以防止数据被窃取或篡改。它提供以下安全服务。

- 客户机和服务器之间的信息保密：通过使用公开密钥和对称密钥技术，SSL 可以确保客户端和服务器之间传输的数据是加密的，从而防止数据被窃取。
- 信息完整性：SSL 不仅提供加密，还提供了消息认证码（MAC），用于验证数据的完整性。
- 双向认证：SSL 支持双向认证，这意味着客户端和服务器都可以验证对方的身份。
- 安全服务对终端用户尽可能透明：终端用户不需要了解或参与加密和身份验证的过程，只需要使用标准的 MySQL 客户端连接到 MySQL 服务器。所有的加密和身份验证工作都是由 SSL 库在底层自动完成的。

1. SSL 介绍

SSL（Secure Socket Layer，安全套接字层）及其继任者传输层安全（Transport Layer Security，TLS）协议是为网络通信提供安全及数据完整性的一种安全协议。TLS 和 SSL 在传输层与应用层之间对网络连接进行加密。图 8-4 展示了有无 SSL/TLS 的对比情况。SSL/TLS 的加入确保了数据的完整性和实现了身份验证，从而大大增强了网络通信的安全性。

SSL 通信需要使用 SSL 安全证书。SSL 安全证书是一种在客户端和服务器之间建立安全连接的电子文档。它由权威的证书颁发机构（CA）颁发，包含了公钥、证书持有者的身份信息和 CA 的签名。公钥用于加密数据，而 CA 的签名则用于验证证书的真实性和有效性。客户端在建立 SSL 连接时会验证服务器的证书，以确保与正确的服务器进行通信，并且通信内容会被加密保护。

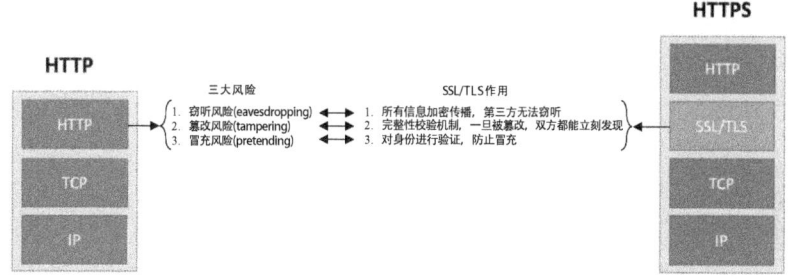

图 8-4　有无 SSL/TLS 的对比

说明：X.509 也是一种证书格式。对 X.509 证书来说，认证者总是 CA 或由 CA 指定的人。一份 X.509 证书是一些标准字段的集合，这些字段包含有关用户或设备及其相应公钥的信息。

OpenSSL 相当于 SSL 的一个实现。如果把 SSL 规范看成接口，那么 OpenSSL 则是接口的实现。

2. MySQL 中 SSL 使用方式

MySQL 支持 SSL 功能，用于客户端到服务器端的认证。SSL 的主要作用包括：提供对用户和服务器的身份验证；对传输的数据进行加密，以保护数据的隐私性；确保数据在传输过程中不被篡改，即保证数据的完整性。TLS 是 SSL 的升级版，提供了更为强大的安全保护。图 8-5 所示为 TLS 加密方式。

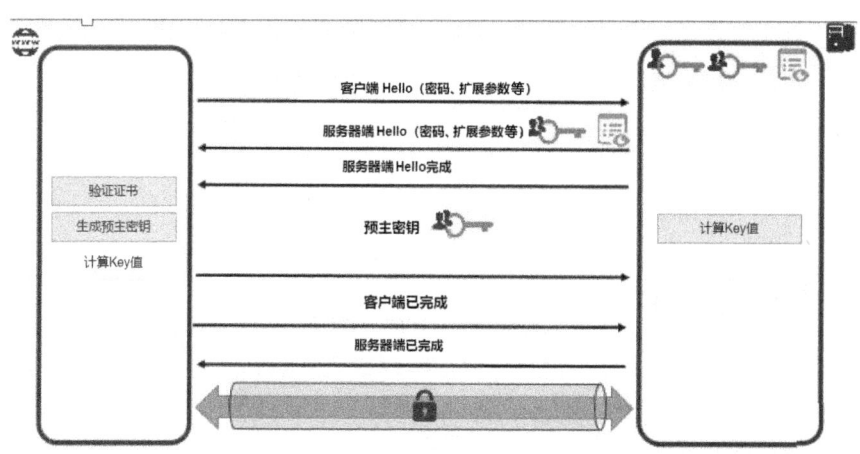

图 8-5　TLS 加密方式

实际上，目前 MySQL 使用的就是 TLS 协议，而不是单纯的 SSL 协议，因为 TLS 是 SSL 的后续版本，提供了更好的安全性。当在 MySQL 中启用 SSL 时，可以通过执行 STATUS 命令来查看 SSL 的状态和详细信息，如下所示。

```
mysql>STATUS
mysql  Ver 8.0.34 for Linux on x86_64 (MySQL Community Server - GPL)

Connection id:    22
Current database:
Current user:     ssluser@iZuf6178v14ipc59jbbpfnZ
SSL:              Cipher in use is TLS_AES_256_GCM_SHA384
...
```

上述信息中，显示 SSL 信息为 TLS_AES_256_GCM_SHA384，这意味着用户是通过 TLS 协议（具体来说，是通过 TLS 中使用 AES-256 加密和 GCM 模式的 SHA-384 哈希函数）连接到 MySQL 服务的。这条信息表明 SSL/TLS 加密已经被激活，并且正在使用特定的加密套件来保护连接。如果不是 SSL 协议，就会显示 "Not in use"。

MySQL 中 SSL 和 RSA 文件具有以下特征。

- SSL 和 RSA 密钥的大小为 2048 位。
- SSL CA 证书是自签名的。
- 通过 sha256WithRSAEncryption 签名算法，使用 CA 证书和密钥对 SSL 服务器端与客户端证书进行签名。
- SSL 证书使用以下通用名称（CN）值以及相应的证书类型（CA、服务器端和客户端）：

```
ca.pem:          MySQL_Server_suffix_Auto_Generated_CA_Certificate
server-cert.pem: MySQL_Server_suffix_Auto_Generated_Server_Certificate
client-cert.pem: MySQL_Server_suffix_Auto_Generated_Client_Certificate
```

扩展名的值基于 MySQL 的版本号。由官方工具 mysql_ssl_rsa_setup 执行生成的文件，也可以使用 suffix 选项显式指定扩展名。对于服务器生成的文件，如果生成的 CN 值超过 64 个字符，则省略名称中的_suffix 部分。

- SSL 文件中的"国家"、"州或省"（ST）、"组织"（O）、"组织单位名称"（OU）和"电子邮件地址"为空。
- 由服务器或 mysql_ssl_rsa_setup 创建的 SSL 文件自生成之日起 10 年内有效。
- RSA 文件不会过期。
- 对于每个证书/密钥对，SSL 文件具有不同的序列号（1 用于 CA，2 用于服务器端，3 用于客户端）。
- 由服务器自动创建的文件由运行服务器的账户拥有。使用 mysql_ssl_rsa_setup 创建的文件由调用该程序的用户拥有。可以通过操作系统的 chown 命令行更改文件的拥有者。
- 在 UNIX 和类 UNIX 系统上，证书文件的文件访问模式是 644（即世界可读），密钥文件的文件访问模式是 600（即仅由运行服务器的账户访问）。

(1) SSL 创建方式

SSL/TLS 证书和密钥可以通过 OpenSSL 命令手动生成，而 MySQL 提供了 mysql_ssl_rsa_setup 工具，该工具可用于自动创建这些证书和密钥。在执行 mysql_ssl_rsa_setup 命令时，通常不需要先停止 MySQL 实例。但是，在应用新的 SSL/TLS 证书和密钥到 MySQL 配置之后，需要重启 MySQL

服务以使更改生效。以下是创建证书的步骤。

```
# 执行 mysql_ssl_rsa_setup,指定 MySQL 的数据目录
shell $> mysql_ssl_rsa_setup --datadir=/opt/data8.0 --verbose
2023-08-01 10:15:22 [NOTE]    Destination directory: /opt/data8.0
2023-08-01 10:15:22 [NOTE]    Executing:openssl version
OpenSSL 1.1.1k  FIPS 25 Mar 2021
2023-08-01 10:15:22 [NOTE]    Executing:openssl req -newkey rsa:2048 -days 3650 -nodes -keyout ca-key.pem -subj /CN=MySQL_Server_8.0.34_Auto_Generated_CA_Certificate -out ca-req.pem && openssl rsa -in ca-key.pem -out ca-key.pem
Ignoring -days; not generating a certificate
Generating a RSA private key
...............................................................................
.....+++++
.........................................................................+++++
writing new private key to 'ca-key.pem'
-----
writing RSA key
2023-08-01 10:15:22 [NOTE]    Executing:openssl x509 -sha256 -days 3650 -extfile cav3.ext -set_serial 1 -req -in ca-req.pem -signkey ca-key.pem -out ca.pem
Signature ok
subject=CN = MySQL_Server_8.0.34_Auto_Generated_CA_Certificate
Getting Private key
...
..................................+++++
.......................................+++++
writing new private key to 'server-key.pem'
-----
writing RSA key
2023-08-01 10:15:22 [NOTE]    Executing:openssl x509 -sha256 -days 3650 -extfile certv3.ext -set_serial 2 -req -in server-req.pem -CA ca.pem -CAkey ca-key.pem -out server-cert.pem
Signature ok
subject=CN = MySQL_Server_8.0.34_Auto_Generated_Server_Certificate
Getting CA Private Key
...
..+++++
...............................................................................
....................+++++
writing new private key to 'client-key.pem'
-----
writing RSA key
2023-08-01 10:15:23 [NOTE]    Executing:openssl x509 -sha256 -days 3650 -extfile certv3.ext -set_serial 3 -req -in client-req.pem -CA ca.pem -CAkey ca-key.pem -out client-cert.pem
Signature ok
subject=CN = MySQL_Server_8.0.34_Auto_Generated_Client_Certificate
Getting CA Private Key
2023-08-01 10:15:23 [NOTE]    Executing:openssl verify -CAfile ca.pem server-cert.pem client-cert.pem
server-cert.pem: OK
client-cert.pem: OK
```

```
2023-08-01 10:15:23 [NOTE]    Executing:openssl genrsa  -out private_key.pem 2048
Generating RSA private key, 2048 bit long modulus (2 primes)
.............................+++++
.+++++
e is 65537 (0x010001)
2023-08-01 10:15:23 [NOTE]    Executing:openssl rsa -in private_key.pem -pubout -out public_key.pem
writing RSA key
2023-08-01 10:15:23 [NOTE]    Success!
```

执行完成，自动生成 SSL/TLS 证书和密钥对应的 8 个以 pem 为扩展名的文件，它们通常位于 MySQL 的数据目录中。以 pem 为扩展名的文件如下所示。

```
shell $> ll data/mysql/*.pem
-rw------- 1 mysql mysql 1676 Dec  6 11:11 ca-key.pem
-rw-r--r-- 1 mysql mysql 1112 Dec  6 11:11 ca.pem
-rw-r--r-- 1 mysql mysql 1112 Dec  6 11:11 client-cert.pem
-rw------- 1 mysql mysql 1676 Dec  6 11:11 client-key.pem
-rw-r--r-- 1 mysql mysql 1680 Sep 19 23:35 private_key.pem
-rw-r--r-- 1 mysql mysql  452 Sep 19 23:35 public_key.pem
-rw-r--r-- 1 mysql mysql 1112 Dec  6 11:11 server-cert.pem
-rw------- 1 mysql mysql 1680 Dec  6 11:11 server-key.pem
```

表 8-5 所示为 MySQL 8.0 版本中生成的以 pem 为扩展名的证书文件及其说明。

表 8-5　MySQL 8.0 版本中生成的以 pem 为扩展名的证书文件及其说明

文　件	说　明	文　件	说　明
ca.pem	CA 证书	server-cert.pem	服务器端使用的证书
ca-key.pem	CA 证书密钥	server-key.pem	服务器端使用的密钥
client-cert.pem	客户端使用的证书	private_key.pem	私钥
client-key.pem	客户端使用的密钥	public_key.pem	公钥

要查看 SSL 证书的内容，包括其有效日期范围，可以使用 Linux 中的 openssl 命令。使用 openssl 命令查看证书内容的命令行示例如下。

```
shell $>openssl x509 -text -in client-cert.pem
Certificate:
    Data:
        Version: 3 (0x2)
        Serial Number: 3 (0x3)
        Signature Algorithm: sha256WithRSAEncryption
        Issuer: CN = MySQL_Server_8.0.34_Auto_Generated_CA_Certificate
        Validity
            Not Before: Dec 6 03:11:19 2022 GMT      # 开始时间
            Not After: Dec 3 03:11:19 2032 GMT       # 结束时间
        Subject: CN = MySQL_Server_8.0.34_Auto_Generated_Client_Certificate
        Subject Public Key Info:
            Public Key Algorithm:rsaEncryption
                RSA Public-Key: (2048 bit)
```

```
            Modulus:
                00:c9:ea:ec:b1:9a:88:99:33:2f:e9:17:cc:d2:4a:
                91:9d:b4:ee:66:19:7a:b1:fb:ca:ae:4a:d5:41:c6:
                a4:eb
            Exponent: 65537 (0x10001)
        X509v3 extensions:
            X509v3 Basic Constraints: critical
                CA:FALSE
    Signature Algorithm: sha256WithRSAEncryption
         6f:09:b9:7d:d7:83:6c:72:3f:c4:8a:17:2a:f6:c7:0d:4a:22:
         a5:2f:1a:e4
-----BEGIN CERTIFICATE-----
MIIDBzCCAe+gAwIBAgIBAzANBgkqhkiG9w0BAQsFADA8MTowOAYDVQQDDDFNeVNR
eTixiNaMvkBXxDUky4xWZ00XU76ei6pWE8h1wP6OfUpw8NFZ7Wt4/dvRD2AtFxDH
PBOi6ppH/KUvGuQ=
-----END CERTIFICATE-----
```

也可以通过 MySQL 的命令行来检查 SSL 证书的相关信息，包括证书的过期时间。查看 SSL 证书的命令行示例如下。

```
mysql>SHOW STATUS LIKE 'Ssl_server_not%';
+-----------------------+----------------------------+
|Variable_name          |Value                       |
+-----------------------+----------------------------+
|Ssl_server_not_after   |Dec  3 03:11:19 2032 GMT    |
|Ssl_server_not_before  |Dec  6 03:11:19 2022 GMT    |
+-----------------------+----------------------------+
2 rows in set (0.00 sec)
```

最后在 MySQL 的 my.cnf 配置文件中设置 SSL 信息，需要指定 SSL 证书、私钥和证书的路径。在 my.cnf 中设置 SSL 示例如下。

```
[mysqld]
ssl_ca=ca.pem
ssl_cert=server-cert.pem
ssl_key=server-key.pem
require_secure_transport=ON
```

说明：require_secure_transport 参数指定客户端与服务器端的连接是否需要使用某种形式的安全传输。启用此变量后，服务器端仅允许使用 SSL/TLS 加密的 TCP/IP 连接，或者使用套接字文件（在 UNIX 上）或共享内存（在 Windows 上）的连接。此功能补充了每个账户的 SSL 要求，这些要求优先。

（2）密钥文件的访问权限

在 SSL/TLS 通信中，私钥是高度敏感的，应该受到严格的保护，通常只由服务器读取。公钥则可以被公开分发，用于客户端验证服务器的身份。

在 MySQL 的上下文中，当配置 SSL 时，需要确保 MySQL 服务器有权访问私钥文件，而客户端则只需要通过公钥和 CA 证书（如果使用了证书链验证）来建立安全的连接。

为了确保私钥的安全，应该采取相应措施，如设置文件权限，即确保私钥文件（如 private_key.pem、public_key.pem）的权限设置为只允许 MySQL 服务进程读取。使用 chmod 命令行设置权限的示例如下。

```
# 设置私钥文件权限为只读,并且只有文件所有者可以读取它
shell $> chmod 400 private_key.pem
shell $> chmod 444 public_key.pem
```

(3) 客户端重用 SSL 会话

从 MySQL 8.0.29 版本开始,SSL 会话重用功能被引入,以优化 SSL/TLS 连接的性能。此功能允许客户端和 MySQL 服务器重用先前建立的 SSL 会话,从而避免每次新连接时都进行完整的 SSL/TLS 握手过程。这可以显著减少连接延迟,提高应用程序的响应速度,特别是在需要频繁建立新连接的环境中。

SSL 会话重用功能通过客户端和服务器端的 SSL 库来实现。在 MySQL 中,服务器端的 SSL 会话信息会缓存在内存中,只要会话缓存未过期且服务器的配置允许会话重用,客户端就可以请求恢复先前的 SSL 会话。要启用 MySQL 的 SSL 会话重用功能,需要在 MySQL 的配置文件中(如 my.cnf 或 my.ini)设置相关的 SSL 选项,并确保服务器和客户端都支持此功能。设置 SSL 会话重用功能的命令行示例如下。

```
# ssl_session_cache 开启和重用有效时间
mysql> SHOW VARIABLES LIKE '%ssl_session%';
+--------------------------+-------+
|Variable_name             |Value  |
+--------------------------+-------+
|ssl_session_cache_mode    |ON     |
|ssl_session_cache_timeout |300    |
+--------------------------+-------+
2 rows in set (0.01 sec)
# 通过状态统计信息查看 ssl_session 状态
mysql> SHOW STATUS LIKE 'Ssl_session%';
+-----------------------------+--------+
|Variable_name                |Value   |
+-----------------------------+--------+
|Ssl_session_cache_hits       |0       |
|Ssl_session_cache_misses     |0       |
|Ssl_session_cache_mode       |SERVER  |
|Ssl_session_cache_overflows  |0       |
|Ssl_session_cache_size       |128     |
|Ssl_session_cache_timeout    |300     |
|Ssl_session_cache_timeouts   |0       |
|Ssl_sessions_reused          |0       |
+-----------------------------+--------+
8 rows in set (0.00 sec)
```

在某些场景下,希望调试或监控 SSL 会话的创建和重用情况。这时,可以使用 MySQL 提供的 ssl_session_data_print 函数。这个函数可以在 SSL 会话创建或重用时被调用,以输出会话数据的详细信息。这对于诊断 SSL 相关问题或理解会话数据如何在服务器和客户端之间传递都是非常有用的。调用 ssl_session_data_print 函数的命令行示例如下。

```
# 存储当前活动会话数据的文件的路径
mysql>ssl_session_data_print ~/private-dir/session.txt

# 调用任何 MySQL 客户端来建立到同一服务器的新的加密连接,重用 SSL 会话数据
shell $> mysql -u admin -p --ssl-session-data=~/private-dir/session.txt
```

3. 客户端使用 SSL 连接 MySQL

当 MySQL 数据库配置 SSL 之后，客户端应用程序也需要相应地配置 SSL，以便能够安全地连接到数据库服务器。以下是一些常见的客户端工具和应用配置 SSL 的方法。

（1）客户端配置禁用 SSL

如果在连接 MySQL 数据库时不想使用 SSL 加密，则可以在连接字符串或连接配置中禁用 SSL。在 MySQL 命令行客户端中，可以使用--ssl＝0、--skip-ssl 或--ssl-mode＝DISABLED 选项来禁用 SSL 连接。禁用 SSL 的命令行示例如下。

```
shell $> mysql -uroot -p****** --ssl=0
shell $> mysql -uroot -p****** --skip-ssl
shell $> mysql -uroot -p****** --ssl-mode=DISABLED
```

（2）强制用户使用 SSL

在 MySQL 中，强制用户通过 SSL 连接到数据库服务器可以确保数据在传输过程中的安全性。对于特定用户，使用 GRANT 语句时要指定 REQUIRE SSL 选项。强制用户使用 SSL 访问的命令行示例如下。

```
#新建用户,强制指定使用SSL
mysql> GRANT SELECT ON *.* to 'dba'@'%' IDENTIFIED BY 'xxx' REQUIRE SSL;
#修改用户
mysql> ALTER USER 'dba'@'%' REQUIRE SSL;
mysql> FLUSH PRIVILEGES;

#使用SSL方式访问数据库
shell $> mysql -ussluser -p -P3306 -h192.168.1.1
        --ssl-ca=/opt/ca.pem
        --ssl-key=/opt/client-key.pem
    --ssl-cert=/opt/client-cert.pem
```

（3）通过驱动程序配置 SSL

对于使用 JDBC 连接到 MySQL 数据库的应用，需要在 JDBC 连接字符串中包含 SSL 相关的参数。这些参数将告诉 JDBC 驱动程序如何与 MySQL 服务器的 SSL 配置进行交互。配置 JDBC 连接以使用 SSL 的示例如下。

```
driverClassName=com.mysql.cj.jdbc.Driver
"jdbc:mysql://hostname:port/dbname? useSSL=true&verifyServerCertificate=false&requireSSL=true&sslCertificate=/path/to/client-cert.pem&sslKey=/path/to/client-key.pem";
```

在上述示例中，JDBC 连接字符串包含以下 SSL 相关的参数。

- useSSL＝true：启用 SSL 连接。
- verifyServerCertificate＝false：不验证服务器证书。在生产环境中，应该设置为 true 并配置正确的 CA 证书，以确保服务器的身份是安全的。设置为 false 将跳过证书验证，这通常只用于测试环境。
- requireSSL＝true：要求 SSL 连接。如果服务器端配置为要求 SSL 连接，而客户端没有设置这个参数，则连接将失败。
- sslCertificate：客户端证书文件的路径。
- sslKey：客户端私钥文件的路径。

在某些 JDBC 驱动程序版本中，sslCertificate 和 sslKey 参数可能不被支持。如果驱动程序不支持这些参数，则需要在驱动程序的配置或属性中设置它们，或者在连接之后通过 SSLParameters 对象进行设置。确保 MySQL JDBC 驱动程序版本支持 SSL，并且已经包含必要的证书和密钥库。如果使用的是较旧的驱动程序版本，则可能需要升级以支持 SSL 连接。在生产环境中，建议设置 verifyServerCertificate=true 来验证服务器证书，以确保连接的安全性。同时，需要确保客户端证书和私钥是安全的，并且只有授权用户才可以访问它们。

（4）通过第三方管理客户端工具配置 SSL

为了确保数据传输的安全性，一些常用的 MySQL 第三方管理客户端工具在连接数据库时需要配置 SSL 信息。为此，需要生成以下 3 个文件：ca.pem、client-key.pem 和 client-cert.pem。这些文件分别用于验证服务器的身份、存储客户端的私钥，以及提供客户端的公钥证书。图 8-6 所示为 Navicat 配置 SSL 方式，首先选中"使用 SSL"复选框，然后把 ca.pem、client-key.pem 和 client-cert.pem 文件保存到客户端工具环境中以引用。

图 8-6　Navicat 配置 SSL 方式

（5）使用 caching_sha2_password 插件

在 MySQL 8.0 中，如果服务器端配置为使用 caching_sha2_password 插件，那么客户端在连接时需要能够检索并验证服务器的公钥。对于使用 SSL 连接的客户端，还需要提供 SSL 证书和密钥文件来加密通信。

使用将 MySQL 客户端连接到 MySQL 服务器的 caching_sha2_password 插件的命令行示例如下。

```
shell $>mysql -udbadmin -p******  -h192.168.1.1 -P3380 \
        --server-public-key-path=/opt/data/public_key.pem \
          --allow-public-key-retrieval=true
```

说明：在上述示例中，参数 --allow-public-key-retrieval=true 指定允许客户端在连接时检索服务器的公钥；参数 --server-public-key-path 指定服务器公钥文件的路径。因为 --allow-public-key-retrieval 允许客户端从服务器检索公钥，这可能会增加安全风险，所以，只有在信任服务器的情况下，才应使用此参数。

4. 启用 SSL 对性能的影响

启用 SSL 会对 MySQL 的性能产生一定的影响，尤其是在建立连接时。这是因为 SSL/TLS 握手过程需要额外的计算资源，包括加密、解密、证书验证等步骤。因此，在启用 SSL 后，数据库的整体 TPS 和 QPS 均有所下降，具体下降的比例取决于多种因素，如服务器的硬件性能、网络带宽、SSL 配置等。

因为目前压力测试工具 sysbench 没有提供直接支持 SSL/TLS 密钥和证书的选项，所以选择使用 mysqlslap 工具来进行 SSL 验证测试。使用 mysqlslap 进行压力测试的命令行示例如下。

```
shell $>mysqlslap -ussluser -p123456 -P3380 -h192.168.244.129 \
         --ssl-ca=/opt/ca.pem \
         --ssl-key=/opt/client-key.pem \
         --ssl-cert=/opt/client-cert.pem  \
         --create-schema=test -a -c 38 -i 500
```

在测试场景中，MySQL 数据库的整体 TPS/QPS 平均降低了 10%～20%，对性能的影响比较大。对于使用短连接的应用程序，可能产生更大的性能损耗。如果使用连接池或长连接，则对性能影响较小。

5. 总结

SSL 能有效预防多种攻击，提高一定的安全性。在传输过程中，如果黑客截获加密数据，那么在没有私有解密密钥的情况下无法读取或使用它。在国内的 MySQL 使用场景中，SSL 的应用相对较少，主要是因为数据库和应用通常部署在同一内部网络中，并且外围防火墙也提供了良好的隔离性。

- 在 MySQL 5.7 及更高版本（包括 MySQL 8.0）中，默认是开启 SSL 连接的。如果强制用户使用 SSL，那么应用配置中需要明确指定 SSL 相关参数，否则程序会报错。
- 在 SSL 加密的情况下，像 tcpdump 这样的网络抓包工具无法捕获或解析加密的数据包，在排查网络问题上加大了难度。
- 虽然 SSL 提高了安全性，但可能会使 TPS/QPS 降低 10%～20%，因此在选择使用 SSL 时需要谨慎权衡安全性与性能。对于非常敏感的核心数据，尽管 SSL 可能导致一定的性能损失，但它提供了更高的数据安全性。

8.3.4 用户数据库访问权限的设置

数据库系统中存储着大量重要数据和各类敏感信息，同时，它也是多个不同实体（包括开发人员、数据库管理员、应用等）共享的数据服务中心。因此，需要特定的账号来访问和操作对应的数据。通过访问控制机制来降低未经授权访问数据、资源和系统的风险。

数据库权限系统是安全体系的重要组成部分。完善的权限管理机制可以有效阻断恶意攻击，防止数据被篡改和隐私泄露。

1. MySQL 权限体系

目前，MySQL 8.0 版本的权限控制主要包括基于用户分配具体权限访问和基于角色的权限访问。这种灵活的权限机制大大提高了系统的安全性。然而，尽管提供了这些方法，但仍需要进行规范化的设置以确保安全。MySQL 数据库的权限体系通过制定不同的权限，为数据库对象和用户赋予相应的访问控制权限，从而满足绝大部分场景的权限分配需求。

（1）权限对象

MySQL 的权限体系可以实现精细的对象控制，遵循最小授权原则。图 8-7 所示为 MySQL 中的权限操作对象。对象的权限可以分为多个级别，包括数据库级别的权限（如 CREATE、DROP、ALTER）、表级别的权限（如 SELECT、INSERT、UPDATE、DELETE）、列级别的权限（针对表的特定列进行权限控制）、视图级别的权限（针对视图的查询权限）以及存储过程和函数的权限（如 EXECUTE、ALTER ROUTINE、CREATE ROUTINE 等）。这些权限级别使得管理员能够灵活地分配权限，以满足不同的安全需求。

MySQL 权限系统中各个对象的作用介绍如下。

- 库级别授权。对于大多数线上应用来说，它们通常只关心自己的数据，并不关心其他应用的数据。因此，为特定的应用分配特定的 SCHEMA 是一个很好的做法。这样做的好处是可以保持隔离性。每个应用都有其自己的数据空间，与其他应用的数据互不干扰。
- 表级别授权。当一个大型应用下有多个子应用，并且这些子应用由不同的团队开发时，为每个子应用分配其所需的表权限是很有必要的。这样做可以防止一个子应用无意中访问或

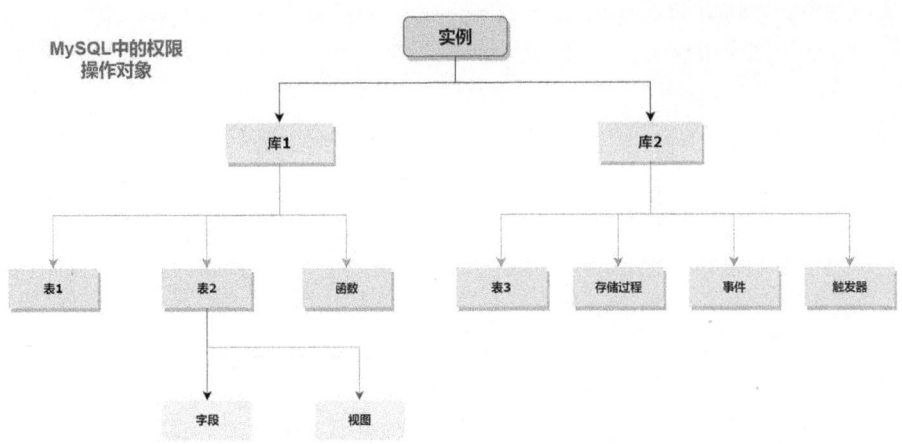

图 8-7　MySQL 中的权限操作对象

修改另一个子应用的数据。
- 列级别授权。当某个应用只需要表中某些列的权限时，列级别的授权就非常有用。例如，用户表的居住信息和密码信息可能由不同的应用修改。这时为特定应用的数据库用户只授予其所需的列权限。
- 其他数据库对象（视图、存储过程、触发器、函数和事件）授权。这些数据库对象允许封装复杂的 SQL 逻辑。当不希望应用直接访问或修改表时，可以通过这些对象来提供数据或服务。

（2）权限数据

MySQL 将权限信息存储在 mysql 系统库的授予表中。相关权限信息主要存储在 mysql.user、mysql.db、mysql.servers、mysql.table_priv、mysql.column_priv、mysql.proxies_priv 和 mysql.global_grants 表中。

权限表本身的信息数据量小，而且访问又比较频繁，所以 MySQL 在启动时就会将所有的权限信息都加载到内存中保存，并根据授予表的内存副本做出访问控制决策。所以，在手动修改权限相关的表后，需要通过执行 FLUSH PRIVILEGES 命令重新加载 MySQL 的权限信息。

图 8-8 所示为 MySQL 权限系统的 8 个关键系统表，它们共同维持着数据库的权限体系。这些表包括 user、servers、global_grants、db、tables_priv、columns_priv、procs_priv、proxies_priv。每个表都有其特定的用途，从全局权限控制到数据库、表、列和存储过程的细粒度权限控制，共同构成了 MySQL 强大且灵活的权限管理机制。

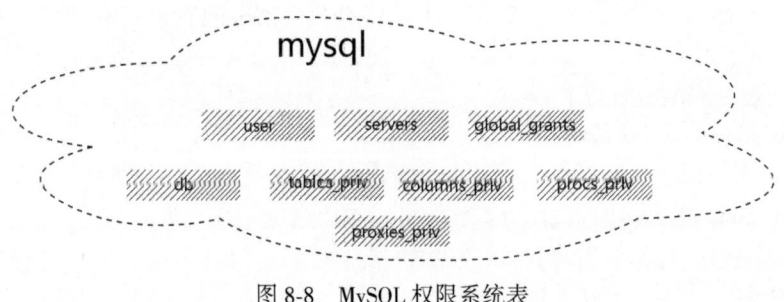

图 8-8　MySQL 权限系统表

下面介绍 MySQL 权限系统中部分表的作用。
- user 表：对于允许的连接，用户表中授予的任何权限都表示用户的静态全局权限。此表中授予的任何权限都适用于 MySQL 服务器上的所有数据库。
- global_grants 表：列出了当前为用户账户分配的动态全局权限。对于每一行，user 和 host 列共同确定哪个用户在哪个主机上具有在 privilege_type 列中命名的特权。
- db 表：确定哪些用户可以从哪些主机访问哪些数据库。Select_priv、Insert_priv 等列确定允许的操作。在数据库级别授予的特权适用于数据库和数据库中的所有对象，如表、视图和存储的程序。
- tables_priv 和 columns_priv 表：它们类似于 db 表，但有更细粒度，即分别应用于表和列级别。
- procs_priv 表：适用于存储例程（存储过程和存储函数）。它确定哪些用户可以对特定的存储过程和存储函数执行哪些操作。
- proxies_priv 表：确定哪些用户可以充当其他用户的代理，以及用户是否可以向其他用户授予代理权限。

（3）具体权限

在 MySQL 的用户表（mysql.user）中，权限相关的信息通常包含在各种列中。这些列大致可以分为几类，包括用户识别、权限控制、安全选项和资源限制等。

查看用户表的权限控制的命令行示例如下。

```
mysql> SELECT * FROM user limit 1\G;
*************************** 1.row ***************************
                  Host: %
                  User: dbadmin
            Select_priv: Y
            Insert_priv: Y
            Update_priv: Y
            Delete_priv: Y
            Create_priv: Y
              Drop_priv: Y
            Reload_priv: Y
          Shutdown_priv: Y
           Process_priv: Y
              File_priv: Y
             Grant_priv: N
        References_priv: Y
             Index_priv: Y
             Alter_priv: Y
           Show_db_priv: Y
             Super_priv: Y
   Create_tmp_table_priv: Y
        Lock_tables_priv: Y
           Execute_priv: Y
         Repl_slave_priv: Y
        Repl_client_priv: Y
         Create_view_priv: Y
           Show_view_priv: Y
```

```
            Create_routine_priv: Y
             Alter_routine_priv: Y
               Create_user_priv: Y
                     Event_priv: Y
                   Trigger_priv: Y
         Create_tablespace_priv: Y
               Create_role_priv: Y
                 Drop_role_priv: Y
...
```

表 8-6 所示为 MySQL 8.0 版本用户表权限控制字段及其说明。

表 8-6　MySQL 8.0 版本用户表权限控制字段及其说明

字 段 名	说　　明
Select_priv	通过 SELECT 命令查询数据
Insert_priv	通过 INSERT 命令插入数据
Update_priv	通过 UPDATE 命令修改现有数据
Delete_priv	通过 DELETE 命令删除现有数据
Create_priv	创建新的数据库和表
Drop_priv	删除现有数据库和表
Reload_priv	执行刷新和重新加载 MySQL 所用的各种内部缓存的特定命令，包括日志、权限、主机、查询和表
Shutdown_priv	关闭 MySQL 服务器。在将此权限提供给 root 账户以外的任何用户时，都应当非常谨慎
Process_priv	通过 SHOW PROCESSLIST 命令查看其他用户的进程
File_priv	执行 SELECT INTO OUTFILE 和 LOAD DATA INFILE 命令
Grant_priv	将自己的权限再授予其他用户
References_priv	创建外键约束
Index_priv	对索引进行增、删、查
Alter_priv	重命名和修改表结构
Show_db_priv	查看服务器上所有数据库的名字，包括用户拥有足够访问权限的数据库
Super_priv	超级管理员权限，如通过 KILL 命令删除用户进程；使用 SET GLOBAL 命令修改全局 MySQL 变量，或执行关于复制和日志的各种命令
Create_tmp_table_priv	创建临时表
Lock_tables_priv	使用 LOCK TABLES 命令阻止对表的访问或修改
Execute_priv	执行存储过程
Repl_slave_priv	读取用于维护复制数据库环境的二进制日志文件
Repl_client_priv	确定复制从服务器和主服务器的位置
Create_view_priv	创建视图
Show_view_priv	查看视图

(续)

字 段 名	说 明
Create_routine_priv	更改或放弃存储过程和存储函数
Alter_routine_priv	修改或删除存储过程及存储函数
Create_user_priv	用于创建新的 MySQL 账户
Event_priv	创建、修改和删除事件
Trigger_priv	创建和删除触发器
Create_tablespace_priv	创建表空间
Create_role_priv	创建角色
Drop_role_priv	删除角色

对于配置 MySQL 中 user 表的权限控制字段，MySQL 8.0 提供了相应的权限名称和权限对象。表 8-7 所示为 MySQL 8.0 版本的 user 表中权限控制字段与权限名称、权限对象的对应关系。

表 8-7　MySQL 8.0 版本的 user 表中权限控制字段与权限名称、权限对象的对应关系

权 限 名 称	对应 user 表中的权限控制字段	权 限 对 象	具 体 说 明
ALTER	Alter_priv	表	授予用户可以使用 ALTER TABLE 语句修改特定数据库中所有数据表的权限
ALTER ROUTINE	Alter_routine_priv	存储过程和函数	授予用户可以更新和删除数据库中已有的存储过程与存储函数的权限
CREATE	Create_priv	数据库、表、索引	授权用户可以使用 CREATE TABLE 语句在特定数据库中创建新表的权限
CREATE ROLE	Create_role_priv	服务	授权用户可以使用 CREATE ROLE 创建角色权限
CREATE ROUTINE	Create_routine_priv	存储过程和函数	授予用户可以为特定的数据库创建存储过程和存储函数的权限
CREATE TABLESPACE	Create_tablespace_priv	服务	授予用户可以创建表空间的权限
CREATE TEMPORARY TABLES	Create_tmp_table_priv	表	授予用户可以在特定数据库中创建临时表的权限
CREATE USER	Create_user_priv	服务	授予用户可以创建新用户的权限
CREATE VIEW	Create_view_priv	视图	授予用户可以在特定数据库中创建新的视图的权限
DELETE	Delete_priv	表	授予用户可以使用 DELETE 语句删除特定数据库中所有表的数据行的权限
DROP	Drop_priv	数据库、表、视图	授予用户可以删除特定数据库中所有表和视图的权限
DROP ROLE	Drop_role_priv	服务	授予用户可以删除角色的权限
EVENT	Event_priv	数据库	授予用户可以创建事件的权限
EXECUTE	Execute_priv	存储过程和函数	授予用户可以调用特定数据库的存储过程和存储函数的权限
FILE	File_priv	服务器本地文件访问	授予用户对文件加载处理等权限

(续)

权限名称	对应 user 表中的权限控制字段	权限对象	具体说明
GRANT OPTION	Grant_priv	数据库、表或存储过程和函数	授予用户赋予其他用户自己所拥有的权限的权限
INDEX	Index_priv	表	授予用户可以在特定数据库中的所有数据表上定义和删除索引的权限
INSERT	Insert_priv	表、列	授予用户可以使用 INSERT 语句向特定数据库中所有表添加数据行的权限
LOCK TABLES	Lock_tables_priv	数据库	授予用户可以锁定特定数据库的已有数据表的权限
PROCESS	Process_priv	服务	授予用户执行进程的权限
REFERENCES	References_priv	数据库、表	授予用户可以创建指向特定数据库中的表外键的权限
RELOAD	Reload_priv	服务	授予用户重新加载的权限
REPLICATION CLIENT	Repl_client_priv	服务	授予用户复制客户端的权限
REPLICATION SLAVE	Repl_slave_priv	服务	授予用户复制从设备的权限
SELECT	Select_priv	表、列	授予用户可以使用 SELECT 语句访问特定数据库中所有表和视图的权限
SHOW DATABASES	Show_db_priv	服务	授予用户执行显示数据库列表的权限
SHOW VIEW	Show_view_priv	视图	授予用户可以查看特定数据库中已有视图的视图定义的权限
SHUTDOWN	Shutdown_priv	服务	授予用户关闭服务的权限
SUPER	Super_priv	服务	授予用户超级用户的权限
TRIGGER	Trigger_priv	表	授予用户触发器管理的权限
UPDATE	Update_priv	表、列	授予用户可以使用 DELETE 语句删除特定数据库中所有表的数据行的权限

除此之外，MySQL 还存在管理数据库运维的权限。表 8-8 所示为 MySQL 8.0 版本中管理数据库运维权限与管理权限对象之间的关系。

表 8-8　MySQL 8.0 管理数据库运维权限对应表

管理数据库运维权限名称	管理权限对象	具体说明
APPLICATION_PASSWORD_ADMIN	双重密码	双密码功能
AUDIT_ABORT_EXEMPT	审计过滤	控制企业版 audit_log 插件，审核日志筛选器的查询内容
AUDIT_ADMIN	审计	启用审核日志配置
AUTHENTICATION_POLICY_ADMIN	认证	authentication_polICY 系统变量对如何使用 CREATE USER 和 ALTER USER 语句中与身份验证相关的子句设置了某些约束
BACKUP_ADMIN	备份	允许执行 LOCK INSTANCE FOR BACKUP 语句并访问
BINLOG_ADMIN	备份和复制	使用 PURGE BINARY LOGS 和 BINLOG 语句。启用二进制日志控制

(续)

管理数据库运维权限名称	管理权限对象	具 体 说 明
BINLOG_ENCRYPTION_ADMIN	备份和复制	二进制日志文件和中继日志文件的加密
CLONE_ADMIN	克隆	启用 CLONE 语句
CONNECTION_ADMIN	服务	使用 KILL 语句或 mysqladmin KILL 命令来终止属于其他账户的线程
ENCRYPTION_KEY_ADMIN	服务	InnoDB 加密密钥轮换
FIREWALL_ADMIN	防火墙	此权限表示由 MYSQL_FIREWALL 插件定义用户的防火墙规则
FIREWALL_EXEMPT	防火墙	此权限表示 MYSQL_FIREWALL 插件不受防火墙限制
FIREWALL_USER	防火墙	此权限表示 MYSQL_FIREWALL 插件更新自己的防火墙规则
FLUSH_OPTIMIZER_COSTS	服务	启用 FLUSH OPTIMIZER_COSTS 语句
FLUSH_STATUS	服务	启用 FLUSH STATUS 语句
FLUSH_TABLES	服务	启用 FLUSH TABLES 语句
FLUSH_USER_ RESOURCES	服务	启用 FLUSH USER_RESOURCES 语句
GROUP_REPLICATION_ADMIN	复制	启动和停止组复制
GROUP_REPLICATION_STREAM	复制	允许使用用户账户来建立组复制的组通信连接
INNODB_REDO_LOG_ARCHIVE	Redo 日志归档	使账户能够激活和停用 Redo 日志存档
INNODB_REDO_LOG_ENABLE	Redo 日志	启用或禁用 Redo 日志
MASKING_DICTIONARIES_ADMIN	服务管理员	企业版本数据屏蔽和去标识功能
NDB_STORED_USER	NDB 集群	允许共享和同步用户或角色及其权限
PASSWORDLESS_USER_ADMIN	认证	用于无密码用户账户
PERSIST_RO_VARIABLES_ADMIN	服务	持久性 PERSIST 命令权限
REPLICATION_APPLIER	复制通道用户	复制通道权限
REPLICATION_SLAVE_ADMIN	复制	连接到复制源服务器，使用启动和停止复制命令
RESOURCE_GROUP_ADMIN	资源池	资源组管理，包括创建、更改和删除资源组，以及向资源组分配线程和语句的权限
RESOURCE_GROUP_USER	资源	允许将线程和语句分配给资源组
ROLE_ADMIN	服务	角色管理员权限
SENSITIVE_VARIABLES_OBSERVER	服务	查看性能架构表 global_variables、session_variables、variables_by_thread 和 persisted_variables 中敏感系统变量的值，发送 SELECT 语句以返回其值，并在连接的会话跟踪器中跟踪对其的更改
SESSION_VARIABLES_ADMIN	服务	允许用户设置会话值
SET_USER_ID	服务	允许在执行视图或存储的程序时设置有效的授权 ID
SHOW_ROUTINE	服务	使用户能够访问所有存储例程（存储过程和存储函数）的定义和属性
SKIP_QUERY_REWRITE	服务	查询不受"重写器"插件重写的约束权限
SYSTEM_USER	服务	系统用户
SYSTEM_VARIABLES_ADMIN	服务	运行时启用系统变量更改

(续)

管理数据库运维权限名称	管理权限对象	具体说明
TABLE_ENCRYPTION_ADMIN	服务	表空间定义加密
TP_CONNECTION_ADMIN	线程池	允许使用特权连接方式连接到服务器。当达到 thread_pool_max_transactions_limit 定义的限制时,不允许进行新连接
VERSION_TOKEN_ADMIN	服务	version_kens 插件是否启动
XA_RECOVER_ADMIN	服务	启用 XA RECOVER 语句

这些不同的权限设置的主要目的是确保只有合适的人才能够访问和操作特定的数据。这样做能够防止不合理的、非法的数据查询和操作。

(4)授权管理

在 MySQL 中,权限管理是通过 GRANT 和 REVOKE 命令实现授权的,同时更新内存结构中的权限信息。MySQL 8.0 版本数据库授权系统的实现,分为用户定义权限和每个授权级别单独的授权表。用户定义是通过' user '@' host '的方式,其中' user '是数据库用户名,' host '是用户连接数据库服务器的主机名或 IP 地址。用户的权限基于' user '@' host '这种组合来确定,而不是单独基于用户名。此外,每个用户都可以被赋予不同的权限集,这些权限集定义了用户可以对数据库进行哪些操作。

MySQL 的授权管理主要涉及 3 个基本的命令行工具:GRANT、REVOKE 和 SHOW GRANTS。这些命令用于管理数据库、表、列以及存储过程和函数的权限。执行这些命令的用户需要有足够的权限。对于授权和撤销权限,建议操作完执行重新加载命令(FLUSH PRIVILEGES)。

以下是这些命令的基本语法和用法。

1)GRANT(赋予权限)。

在 MySQL 中,GRANT 是赋予权限命令,需要管理员权限。在 MySQL 的早期版本中,GRANT 语句可以用来创建用户并同时授予权限,但从 MySQL 8.0 开始,创建用户和授权被分离为两个独立的语句。GRANT 的基本语法如下所示。

```
GRANT
priv_type [(column_list)]
    [,priv_type [(column_list)]] ...
    ON [object_type]priv_level
    TO user_or_role [, user_or_role] ...
    [WITH GRANT OPTION]
    [AS user
        [WITH ROLE
            DEFAULT
          | NONE
          | ALL
          | ALL EXCEPT role [, role] ...
          | role [, role] ...
        ]
    ]
}
```

通过指定要授予的权限类型、特定列对象类型、对象名称,给不同的用户和不同来源的主机赋予权限。使用 GRANT 授权的命令行示例如下。

```sql
# 指定所有权限和所有库
mysql> GRANT ALL ON *.* TO 'tuser'@'localhost';

# 只用权限 SELECT 和 INSERT 面向所有库
mysql> GRANT SELECT, INSERT ON *.* TO 'tuser'@'localhost';

# 指定库
mysql> GRANT ALL ON db1.* TO 'tuser'@'localhost';

# 指定表
mysql> GRANT SELECT, INSERT, UPDATE ON db1.table1 TO 'tuser'@'localhost';

# 指定列
mysql> GRANT SELECT (col1), INSERT (col1, col2) ON db3.table1 TO 'tuser'@'localhost';

# 对用户 tuser 赋予存储过程的创建和执行权限
mysql> GRANT CREATE ROUTINE ON mydb.* TO 'tuser'@'localhost';
mysql> GRANT EXECUTE ON PROCEDURE mydb.myproc TO 'tuser'@'localhost';

# 对用户 tuser 赋予代理权限
mysql> GRANT PROXY ON 'proxyuser'@'localhost' TO 'tuser'@'localhost';

# 对用户 tester1 与 tester2 分别赋予角色 role1 和 role2
mysql> GRANT 'role1', 'role2' TO 'tester1'@'localhost', 'tester2'@'localhost';
```

2）REVOKE（权限回收）。

在 MySQL 中，REVOKE 语句用于从用户或角色中移除（回收）先前授予的权限。与 GRANT 语句一样，执行 REVOKE 语句的用户必须拥有相应的管理员权限。REVOKE 语句的基本语法如下。

```
REVOKE [IF EXISTS]
    priv_type [(column_list)]
      [,priv_type [(column_list)]] ...
    ON [object_type]priv_level
    FROM user_or_role [, user_or_role] ...
    [IGNORE UNKNOWN USER]
```

通过指定要回收的权限类型、特定列、对象类型、对象名称，对不同的用户和不同来源的主机撤销权限。使用 REVOKE 撤销权限的命令行示例如下。

```sql
# 撤销所有库的 INSERT 权限
mysql> REVOKE INSERT ON *.* FROM 'tuser'@'localhost';
# 撤销角色权限
mysql> REVOKE 'role1', 'role2' FROM 'tester1'@'localhost', 'tester2'@'localhost';
# 撤销对某个库的 SELECT 权限
mysql> REVOKE SELECT ON db1.* FROM 'tuser'@'localhost';

# 结合 IF EXISTS 使用。如果目标用户或角色存在,但由于某些原因没有发现分配给目标的特权或角色,则会发出警告,而不是错误
mysql> REVOKE IF EXISTS SELECT ON db1.table FROM 'tuser'@'127.0.0.1';
# 结合 IGNORE UNKNOWN USER 使用,则该语句会对语句中指定但未找到的任何目标用户或角色发出警告
mysql> REVOKE SELECT ON db1.table FROM 'tuser'@'127.0.0.1' IGNORE UNKNOWN USER;
```

3）SHOW GRANTS（查看权限）

在 MySQL 中，SHOW GRANTS 语句用于列出特定用户或角色所持有的所有权限。这个命令对于了解用户或角色的权限集合非常有用，特别是在进行权限管理或调试时。SHOW GRANTS 的基本语法如下。

```
SHOW GRANTS
    [FOR user_or_role
        [USING role [, role] ...]]
```

通过指定不同的用户和不同来源的主机来查看权限。使用 SHOW GRANTS 查看权限的命令行示例如下。

```
# 显示指定用户权限
mysql> SHOW GRANTS FOR 'tuser'@'localhost';
+-----------------------------------------------------------------+
| Grants for jeffrey@localhost                                    |
+-----------------------------------------------------------------+
| GRANT USAGE ON *.* TO `tester`@`localhost`                      |
| GRANT SELECT, INSERT, UPDATE ON `db1`.* TO `tester`@`localhost` |
+-----------------------------------------------------------------+
# 显示当前登录的用户的所有权限
mysql> SHOW GRANTS;
mysql> SHOW GRANTS FOR CURRENT_USER;
mysql> SHOW GRANTS FOR CURRENT_USER();
```

在返回的结果中，权限通常以 GRANT 语句的形式展示，这可以帮助了解如何授予这些权限，并可以在需要时通过复制这些 GRANT 语句来授予其他用户相似的权限。

2. MySQL 中权限设计

在 MySQL 中，面对众多的权限，应该如何设计运维权限体系呢？推荐采用最小授权法则。这意味着，在列、表、库级别上，应该授予应用可以运行的最小授权，以避免潜在的危险。特别是对于 DDL（如 CREATE、ALTER、DROP 等）操作，执行权限应该仅限于专职人员。可以按照不同的抽象角色来设置权限，如开发人员、开发 DBA、管理 DBA。这样做可以避免在开发和运维过程中，因疏忽或错误调用导致的潜在问题。

下面对于不同的角色，设置不同的权限。

（1）开发人员

开发人员主要负责数据库程序的连接和开发工作，对数据库的数据拥有增删查改（DML）的权限，但没有权限进行数据结构变更。根据需求，可以将其执行的操作分为以下几类。

- 由于报表用户主要执行查询操作，因此只需要授予 SELECT 权限。
- 由于数据导入专用的用户主要把数据导入到数据库中，因此应授予 INSERT 权限。
- 对于使用读写分离架构的应用，提供读写分离用户。对于执行读请求的用户，只需要授予 SELECT 权限。这有助于防止在开发过程中因错误调用导致的数据修改问题。对于执行写请求的用户，应授予 INSERT、UPDATE 和 DELETE 权限。

（2）开发 DBA

开发 DBA 是协助开发人员的数据库管理员，其主要工作职责包括数据库表结构的变更以及部分数据的一般变更，如数据的批量操作、存储过程的创建等。此外，开发 DBA 通常具有对数据库结构的变更权限。对于开发 DBA 的操作，需要注意以下内容。

- 在生产环境中进行一般的 DDL（库表结构）变更时，需要特别小心，因为这些变更直接关联数据的安全性和业务的连续性。错误的操作有时可能会导致业务中断。因此，对于执行这些变更的权限，必须有严格的限制。
- 数据批量操作（如导入和导出）可能会与正常的线上操作产生资源争抢，从而影响生产环境的性能。同时，数据的批量操作经常涉及数据库服务器文件的上传权限，如果控制不当，可能会导致数据库的安全受到威胁。为了平衡这种影响，通常情况下，对于普通的批量操作，可以采用资源限制的方式，以避免对生产数据库造成过大的压力。然而，这需要在确保不影响生产环境的前提下，与数据批量操作的速度进行合理的平衡，以避免数据导入时间过长。

（3）管理 DBA

管理 DBA（Management Database Administrator）在数据库管理中扮演着至关重要的角色，工作主要集中在确保数据库实例的稳定运行、数据安全，以及高效管理数据库资源上。以下是管理 DBA 的主要职责的介绍。

- 备份恢复操作。定期执行数据库的备份操作，确保在数据丢失或损坏时能够迅速恢复。同时，设计和实施备份策略，包括物理备份（直接备份数据文件、日志文件等）和逻辑备份（如使用 mysqldump 等工具导出 SQL 语句）。在需要时执行数据恢复操作，确保业务的连续性和数据的完整性。
- 复制操作。复制操作是 MySQL 中用于管理和执行与数据库复制相关的任务的关键部分。在 MySQL 的主从复制结构中，复制尤其重要，因为它负责管理和维护主从之间的复制关系，从而确保数据的可靠性和一致性。
- 日常操作。在日常的数据库操作中，存在一些由于其潜在影响和风险而不适合由开发 DBA 直接执行的操作。这些操作通常涉及数据库的重大变更，可能会对系统性能、数据完整性和业务连续性产生重大影响。因此，这些操作需要由更高级别的 DBA 或数据库管理团队进行审核和批准。具体来说，影响数据库运行和性能的操作、大批量数据的导入导出，以及 DDL（数据定义语言）命令，都需要在 DBA 的审核之后才能执行。

3. 总结

数据库权限体系在数据库管理体系中是一个非常重要的环节。但在实际中，常常有被忽略的防护漏洞，如误删数据、删库或执行高危命令等，这时，一个完善的权限机制就显得尤为重要。MySQL 权限管理的两个准则如下。

- 权限最小化原则：在满足业务需求的基础上，仅为所需用户配置必要的、最少的权限。
- 对于可能影响全局的操作，需要由专业人员进行审核和执行。

第 9 章 MySQL 8.0 架构设计与应用开发

理解 MySQL 8.0 的架构设计与应用开发对于构建高效、稳定、可扩展的应用程序至关重要。这不仅有助于提高应用程序的性能和用户体验，还有助于降低企业成本和增强系统的灵活性。本章将围绕 MySQL 8.0 的架构设计与应用开发进行详细讲解。

9.1 MySQL 8.0 架构设计

数据库架构设计是对数据库管理系统内部元素及其关系进行理性化和条理化的过程。它反映了设计者对于如何有效组织、存储和管理数据的理解，并基于这种理解来构建出满足特定业务需求的数据库系统框架。这个框架旨在提供高效、可靠的数据访问和操作机制，确保数据的完整性、安全性和可扩展性。

9.1.1 架构设计的原则

针对 MySQL 架构设计，结合多年的实战经验，建议遵循以下设计原则。

- 先进性。要求技术架构业内领先、技术成熟，以保证平台具备不断发展和扩充的空间。同时，应立足高起点规划、高标准建设，充分把握业务发展趋势，系统上线后能够满足未来 3~5 年的业务发展需求。例如，选择哪种数据库引擎、存储方案等。
- 开放性。相关的组件应尽可能选择开放、开源技术，便于对技术架构进行系统了解、深入掌握，最终实现对技术架构的完全自主可控。
- 经济性。通过合理的技术成本，完成 MySQL 平台的整体建设。在各组件中，优先考虑开源的技术组件，从而降低平台的经济成本。
- 高性能。具备高效、快速的数据处理能力，确保在多任务大数据量情况下仍能快速、高效、准确处理。
- 高扩展。总体架构和软件体系结构要有可扩展性，要充分考虑到未来业务的发展带来的数据规模的扩大、管理需求的变化以及系统保障级别的提高，方便对新需求的扩展和支持，软件体系结构不依赖于硬件设备。
- 易维护。平台应具备自监控性，能够监控平台运行情况，在平台出现异常情况下进行告警。在平台出现异常时，能够通过图形化界面或日志，快速、方便地定位出错误位置、原因，方便平台维护。
- 开放兼容。平台需要兼容各种硬件架构和各种数据库环境，采用开放的技术架构，能够适应存量和增量的数据库环境要求。

9.1.2 架构设计实践1：读写分离方案

MySQL 的读写分离方案旨在将数据库的读操作和写操作分配给不同的 MySQL 服务器，从而优化数据库的性能和提高可靠性。这种策略的主要目的是分散主数据库的负载，有效避免单点故障。当主数据库面临巨大的压力，无法单独应对所有业务请求时，通过读写分离可以显著提升数据库的整体性能。特别是对于高访问量的业务场景，MySQL 读写分离显得尤为重要。同时，结合分库分表方案，不仅可以突破 MySQL 的单点性能瓶颈，还能确保系统的整体性能和稳定性，进一步提升系统的可扩展性和可维护性。

1. 读写分离基本原则

MySQL 的读写分离基本原则是让主数据库（Master，又称主库）处理事务性操作（如写操作），而从数据库（Slave，又称从库）处理非事务性操作（如读操作）。由于 MySQL 主从复制通常采用逻辑复制，这可能会导致从数据库的数据同步相对于主数据库有一定的延迟。因此，在设计和实施读写分离时，需要注意以下两点。

- 延迟敏感业务：对于需要实时或接近实时数据的业务，应该避免从从数据库读取数据，因为从数据库的数据可能不是最新的。如果需要从从数据库读取，则应该实施主从验证机制来确保数据的一致性。
- 报表和统计类查询：这类查询通常对数据的实时性要求不高，而且处理的数据量很大，因此非常适合在从数据库上执行。这样既可以减轻主数据库的负担，又可以利用从数据库的资源优势。

2. 读写分离实现方式

在 MySQL 中，常见的实现读写分离的方式有以下 4 种。
- 代码层实现逻辑 1：应用程序代码直接处理读写分离逻辑，通过解析 SQL 语句或者识别特定的业务逻辑，将读请求和写请求分别发送到不同的数据库实例上。
- 代码层实现逻辑 2：这种方式涉及主库和从库之间的 GTID（全局事务标识符）同步验证。主库获取其最新的 GTID，并将其发送到从库进行验证。如果从库已经应用了该 GTID，则读取操作可以在从库执行。
- 基于类似 ProxySQL、MyCat 的中间件实现读写分离：使用像 ProxySQL 或 MyCat 这样的数据库代理或中间件来处理读写分离。中间件可以根据配置规则，将读请求路由到从库，将写请求路由到主库。
- 多副本写入方式：在这种架构中，写操作被复制到多个从库上，而不仅仅是主库。这可以提高系统的可用性和容错能力。不管读写那个节点，数据都能保证一致。

无论采用何种方式，都必须严格遵守读写账号的分离原则，即将写入操作（Writehost）和读取操作（Readhost）的账号分开。这一原则需要在应用端或中间件层面进行控制，以确保系统的安全性和稳定性。通过这种方式，实现了双保险机制。值得注意的是，Writehost 账号应该始终绑定在一个特定的节点上，以确保数据的一致性和完整性。

3. 读写分离中间件介绍

中间件位于应用系统和数据库软件之间，用于在 MySQL 场景中实现读写分离。根据配置信息，它能够将请求分发到不同的数据库节点，从而实现负载均衡、故障转移和读写分离。图 9-1

所示为目前常见的 MySQL 读写分离中间件。这些中间件可以根据其来源和性质分为国内开源、国内商业、国外开源和国外商业 4 种类型。

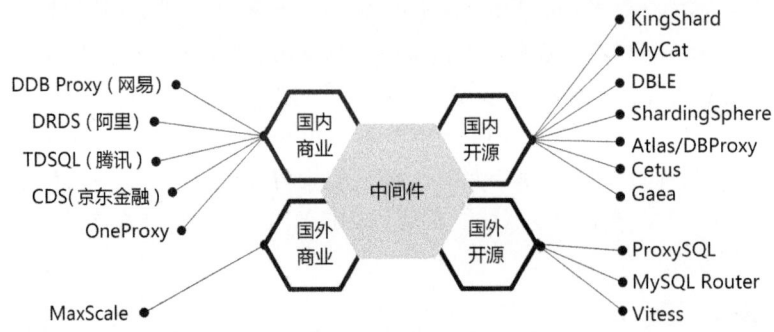

图 9-1　数据库读写分离中间件

下面介绍一些常用的读写分离中间件。

（1）MyCat（开源）

MyCat 提供了许多数据库中间件的功能和特性。

- 应用端连接 MyCat 服务器：应用端通过连接到 MyCat 服务器，可以间接地访问后端的数据库。MyCat 作为一个代理，负责接收和处理应用的数据库请求。
- 支持读写分离：MyCat 支持读写分离功能，它可以将读请求和写请求分别路由到不同的数据库服务器，以提高性能和可扩展性。这种策略对于大型分布式数据库系统来说非常有用。
- 支持多种分库分表算法：MyCat 支持多种分库分表算法，如 PartitionByFileMap、PartitionByLong、ModPartition、MurmurPartition、RoundRobinPartition、HashPartition、RangePartition、JoinPartition 和 AutoPartition 等。这些算法可以帮助用户根据业务需求和数据特性来选择合适的分片策略。
- MyCat 配置文件中包含 MySQL 路由相关信息：MyCat 的配置文件中包含 MySQL 路由的相关信息，这些信息用于定义如何路由到不同的数据库和表。通过配置路由规则，MyCat 可以实现灵活的数据库访问控制。
- 支持多种数据库：MyCat 支持多种数据库，包括但不限于 MySQL、PostgreSQL、Oracle 和 SQL Server。这使得 MyCat 可以方便地与各种数据库集成，满足用户的多样化需求。
- 性能损耗为 20%~50%：MyCat 作为一个代理服务器，会增加一定的性能损耗，根据官方数据和用户反馈，这种损耗通常在 20%~50% 之间。这意味着在使用 MyCat 时，需要权衡其带来的便利性和可能带来的性能影响。

MyCat 的整体设计思路和实现都很好，是一个相对成熟的解决方案。

（2）ShardingSphere（开源）

ShardingSphere 是一套开源的分布式数据库中间件解决方案，该项目由当当网捐入 Apache，并在京东数科逐渐发展壮大，成为业界首个 Apache 分布式数据库中间件项目。ShardingSphere 包含 3 个核心组件：Sharding-JDBC、Sharding-Proxy 和计划中的 Sharding-Sidecar。这个生态圈的目标是提供高性能、易用、易扩展、易维护的解决方案，特别适用于诸如 Java 同构、异构语言、容器、云原生等应用场景。

- Sharding-JDBC：它是一个轻量级的 Java 框架，在 Java 的 JDBC 层提供额外服务。它使用客户端直连数据库，以 JAR 包形式提供服务，无须额外部署和依赖，可以理解为增强版的 JDBC 驱动程序，完全兼容 JDBC 和各种 ORM 框架。支持任意实现 JDBC 规范的数据库，如 MySQL、Oracle、SQL Server 和 PostgreSQL 等。实现了分库分表、读写分离等核心功能，性能损耗约为 7%。
- Sharding-Proxy：它类似于 MyCat，是一个透明化的数据库代理端，提供封装了数据库二进制协议的服务端版本，支持对异构语言的操作。目前主要提供 MySQL 版本，实现了读写分离、分库分表等功能，性能损耗约为 20%。
- Sharding-Sidecar：它定位为 Kubernetes 或 Mesos 的云原生数据库代理。这个组件以 DaemonSet 的形式代理所有对数据库的访问，通过无中心、零侵入的方案提供与数据库交互的啮合层，即 Database Mesh，也可称为数据网格。

Sharding-JDBC 和 Sharding-Proxy 在开源中间件领域都是非常热门和广泛使用的读写分离中间件，它们各自在功能和适用场景上都有其独特之处。Sharding-JDBC 由于其轻量级和与 JDBC 的高度集成，通常被嵌入到应用程序中以提供透明的分库分表及读写分离功能。而 Sharding-Proxy 则作为一个独立的代理服务，支持对 SQL 的解析、改写和请求的转发，无须修改应用程序代码即可实现数据分片和读写分离。

ShardingSphere 作为一个开源项目，其社区活跃，不断有新的功能和优化被加入到项目中，以满足用户日益增长的需求。

（3）Cetus（开源）

Cetus 是一个由 C 语言开发的关系数据库 MySQL 的中间件，它基于 MySQL Proxy 开发，并且分为读写分离版和分库版两个版本。Cetus 被认为是 MySQL Router 的前身。尽管 Cetus 的社区活动不多，但由于其稳定性和功能性，一些企业仍然选择使用它来处理 MySQL 数据库的读写分离需求。

（4）MaxScale（闭源）

MaxScale 是由 MariaDB 公司开发的一款中间件，它位于应用系统和数据库软件之间，主要用于在 MySQL 或 MariaDB 场景中实现读写分离和读负载均衡。通过智能路由和数据转换功能，MaxScale 可以自动选择最佳的数据源，并根据预定义的策略（如读写分离）将请求转发到适当的服务器，以优化数据库负载。此外，MaxScale 还提供了数据过滤与转换、监控与日志、安全增强、高可用性和模块化设计等一系列高级功能，使得数据库管理变得更加便捷和安全。

尽管 MaxScale 是开放源代码的，但它并不遵循传统的 Open Source 协议。相反，它使用了特有的授权协议：商业源代码许可证（Business Source License，BSL）协议。根据 BSL 协议，用户在非生产环境下可以无限制地使用 MaxScale，但在生产环境中，如果后端实例超过 3 个，就必须购买商业授权。

在选择使用 MaxScale 时，务必了解其授权协议和相关限制，以确保它符合项目需求和使用场景。

（5）Vitess（开源）

Vitess 是一个复杂且功能强大的开源中间件，由 YouTube 在生产环境中使用。它旨在解决传统数据库扩展的难题，尤其是针对 MySQL 数据库的水平扩展。与以往中间件不同，它需要适配 Vitess 的分片、读写分离等特性，并且需要调整数据库连接池管理、事务处理策略以及查询优化等方面的代码。它支持读写分离、分库分表、故障切换和数据备份。

由于 Vitess 的复杂性和高度可配置性，使用它可能需要一些专门的知识和经验。在决定使用

Vitess 之前,建议仔细评估其是否适合具体需求和环境,并考虑进行充分的测试和验证。

(6) MySQL Router(开源)

MySQL Router 是 MySQL 官方推荐的读写分离中间件,是 MySQL Proxy 的替代方案。

- 读写分离:MySQL Router 支持读写分离,能够将读请求和写请求分别路由到不同的数据库服务器,以提高性能和可扩展性。
- 无法动态更改配置:MySQL Router 的一个限制是它不支持动态更改配置。一旦配置了 MySQL Router,如果需要更改配置,通常需要停止并重新启动服务来使更改生效。
- 在 MGR 中充当代理:MySQL Router 可以与 MySQL Group Replication(MGR)结合使用,充当代理角色。在这种情况下,它可以帮助管理数据库连接的路由,实现高可用性,并在主节点发生故障时自动切换到新的主节点。

由于 MGR 集群的这些优势以及 MySQL Router 在管理和优化这种集群方面的能力,使得 MySQL Router 在 MGR 集群部署中受到热捧。

(7) ProxySQL(开源)

ProxySQL 是 Percona 开发的一个开源软件,使用 C++编写,旨在为 MySQL 和 MariaDB 提供高性能、轻量级的代理服务。它允许用户实现读写分离、分片和故障转移等功能,是许多企业和组织在选择 MySQL 数据库中间件时的首选。

- 读写分离与语句级路由规则:ProxySQL 支持基于规则的读写分离,数据库管理员可以根据定义的规则将读请求和写请求路由到不同的数据库服务器。此外,ProxySQL 还提供了语句级的路由规则,允许更细粒度的控制,比如根据特定的 SQL 语句将请求路由到特定的服务器。
- Sharding 支持:Sharding 是将数据分布到多个数据库服务器上的过程,以提高性能和可扩展性。ProxySQL 支持简单地使用正则表达式来实现分片策略,允许用户定义数据分片的规则,并根据这些规则将请求路由到正确的分片。
- 动态更改配置:ProxySQL 的一个显著优点是它支持动态更改配置。数据库管理员可以在运行时添加、修改或删除规则,而无须停止和重新启动 ProxySQL 服务。这种动态性使得管理更加灵活,可以快速响应数据库架构的变化。
- 性能要求:由于所有的 SQL 请求都需要经过 ProxySQL 服务器进行路由和处理,因此对 CPU 和内存有一定的要求。在选择部署 ProxySQL 时,需要考虑服务器的硬件规格和性能,以确保其能够高效处理来自应用程序的数据库请求。
- 请求 SQL 都要经过 ProxySQL 服务器,对 CPU 和内存有一定的要求。
- 在 MGR 中充当代理:ProxySQL 也可以与 MGR 结合使用,作为代理服务器。在这种情况下,ProxySQL 可以帮助管理客户端与 MGR 集群之间的连接,实现读写分离、故障转移和负载均衡等功能。

虽然 ProxySQL 提供了丰富的功能和灵活性,但在实际部署时,还需要考虑其他因素,如网络延迟、安全性、监控和日志记录等。

4. 总结

MySQL 主从数据冗余架构能够很好地支持读写分离方案。但在使用时,应该考虑到延迟的敏感性。如果存在延迟,则应该将读请求放到从库执行;如果没有延迟或延迟在可接受范围内,则读请求也可以考虑在主库执行。

在中间件方面,从当前的社区活跃度和关注度来看,Sharding-JDBC 的发展趋势非常好。此外,轻量级的中间件(如 ProxySQL 和 MySQL Router)在社区活跃度和使用场景上也表现出色。上述这 3 个中间件是比较推荐的。

引入读写分离中间件,对于系统架构的复杂度来说,应该在可控范围之内。因此,在选择读写分离方案时,应结合业务模型进行合理选择。

9.1.3 架构设计实践 2:库内分库分表方案

对于轻量级数据库而言,总是存在一定的性能瓶颈。这些瓶颈通常与硬件资源有关,如 I/O、CPU、网络和内存等。当硬件资源受到一定的限制,无法进一步提升或无法提供更多的资源时,就需要从数据库设计方面进行优化,以达到预期的性能目标。

下面介绍 MySQL 库内分库分表的评判标准,以及库内分库分表方案。

1. 库内分库分表评判标准

MySQL 库内分库分表的评判标准是一个综合考虑的过程,需要根据实际业务需求、系统负载、数据安全性等因素进行权衡和决策。在进行 MySQL 库内分库分表的设计时,主要考虑如下 4 个方面,包括数据量、应用瓶颈、安全性和可用性。下面将展开介绍。

(1)数据量

图 9-2 所示为 MySQL 官方给出的 InnoDB 表空间的容量。单个表空间可以达到 TB 级别,因此,肯定能够处理亿行数据。

Table 15.31 InnoDB Maximum Tablespace Size

InnoDB Page Size	Maximum Tablespace Size
4KB	16TB
8KB	32TB
16KB	64TB
32KB	128TB
64KB	256TB

图 9-2 MySQL 数据量限制

但实际上,当单表数据量达到千万行级别或更高时,查询性能可能会受到影响。这通常是因为大量的数据导致 I/O 操作变得更为频繁,索引的维护成本增加,查询优化变得更加困难。MySQL 的 InnoDB 存储引擎使用 B+树作为其索引结构,并且是聚簇索引,意味着表中的数据记录本身就是按照主键顺序存储的。B+树是一种自平衡树,新增节点和删除节点后,都会按照规则进行平衡以保持 B+树的平衡。如果是 3 层的 B+树索引,则最多进行 3 次磁盘 I/O 操作。这在索引设计良好,或者采用覆盖索引的情况下,可以显著提高查询性能。3 层高的 B+树计算层高示例如下。

例如,InnoDB 存储引擎使用的默认页的大小是 16384B(16KB)。如果仅从简化的角度考虑,并且忽略页内其他必要的开销,一个 BIGINT 类型列通常占据 8B,指针又占用 6B。理论上,如果假设每行仅包含一个 BIGINT 值和一个指针,并且忽略所有其他开销,那么一个页中存放 16384/(8+6)≈1170 个指针。

那么可以算出一棵高度为 2 的 B+树,能存放 1170×16=18720 条这样的数据记录。

同理,可以算出一个高度为 3 的 B+树,可以存放 1170×1170×16=21902400 条这样的记录。

在机械硬盘时代,由于 I/O 性能的限制,B+树的深度通常建议不要超过 3 层,以保证数据库性能。然而,在固态硬盘时代,由于 I/O 性能的大幅提升,多几次 I/O 操作通常不会对性能产生显著影响。根据普遍的实践经验,当单表数据行数在 5000 万~8000 万之间时,数据库仍能保持较好的性能。这样的数据量通常对应着大约 10GB 的实际数据文件大小。

那么,数据量是否还可以继续提升?答案是肯定的。虽然数据量可以继续增加,但是,维持高性能变得更具挑战性。当数据行数超过 1 亿之后,性能下降会变得非常明显,这通常是 MySQL 的底层实现上的性能瓶颈造成的。这时,就需要考虑进行分库分表,以分散单表的数据量。

（2）应用瓶颈

当应用在数据处理方面遇到性能瓶颈，无法通过简单地提升硬件资源来满足不断增长的业务需求时，就需要考虑在数据库层面进行分库分表。

- 对于 TPS/QPS 要求比较高的情况，高 TPS/QPS 会对单表的处理能力产生一定的影响。一般来说，在 MySQL 的高配置环境下，TPS 普遍能维持在 3000 以内，QPS 普遍能维持在 8 万以内。超过上述范围，就无法满足性能要求了。
- 当某张表被频繁访问时，就会发现内存中的命中率非常低，处理能力逐步衰减，特别是在并发度比较大的情况下。

为了有效判断何时采取分库分表措施，可以结合应用处理能力的延迟反馈与数据库指标（如命中率），以及慢查询日志的积累情况。这些信息都可以在 MySQL 数据库中获取到，具体方式如下。

1）状态值：innodb_buffer_pool_read_requests 表示读请求的总次数，innodb_buffer_pool_reads 表示缓冲池从物理磁盘中读取数据的请求总次数。

2）缓存命中率=(innodb_buffer_pool_reads/innodb_buffer_pool_read_requests) ×100 %。

3）慢查询日志：通过对比慢查询日志记录，判断数据量是否有明显变化、执行效率是否变低等。

除此之外，添加多个应用节点或引入缓存机制（如 Redis），可以实现类似于读写分离的效果。这可以在一定程度上缓解数据库压力，但从根本上来说，这些问题并不能完全解决，原因如下。

- 多个应用节点最终还是会汇集到数据库上，数据库的压力仍然会上升，问题依然存在。
- 缓存机制可以有效解决查询性能问题，但对于增加、删除、修改等写操作，缓存并不能提供直接的解决方案。
- 虽然读写分离可以分散读写操作，但由于数据不是实时同步，因此存在一定的数据延迟。

（3）安全性和可用性

因为在设计方面需要考虑诸多系统类型的契合，并且库表之间存在一定的耦合性，所以普遍选择使用一个数据库。然而，在这种情况下，每个业务的数据量、访问量都不同，如果一个业务导致数据库出现问题，就可能会牵连其他业务。为了避免这种情况，每个业务使用独立的数据库，这样即使一个数据库出现问题，也不会影响其他业务。每个数据库只承担一部分业务，这样可以提高整体的可用性和安全性。这就像人们常说的：“不要把鸡蛋放在一个篮子里”。如图 9-3 所示，这种策略有助于分散风险并保持系统的稳定运行。

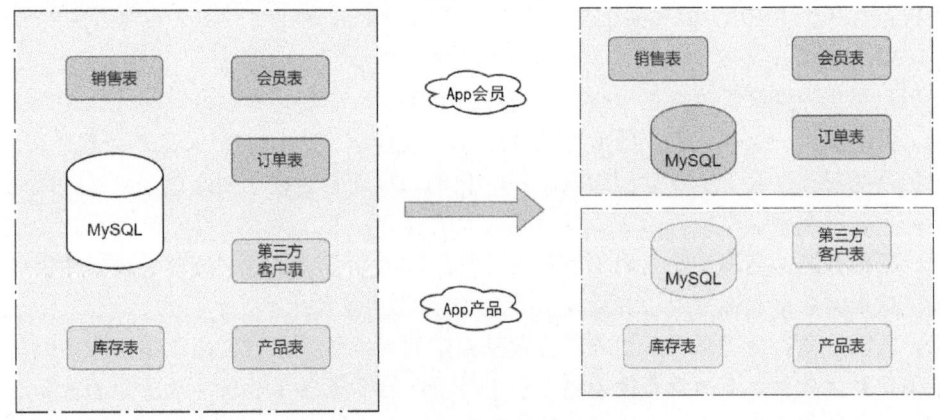

图 9-3　MySQL 库内实现分库分表

对于共用一个 MySQL 数据库实例的情况,虽然账号隔离是一种可行的方案,但这只是表面上的隔离,真正的问题在于资源隔离。然而,目前 MySQL 在资源隔离方面的支持并不完善。虽然 MySQL 8.0 引入了一些新功能,如资源组和连接内存限制,但这些功能仍然不足以实现全面的资源隔离。此外,MySQL 的底层架构是单进程多线程方式,这意味着对线程资源的控制仍然有限。因此,当考虑公用数据库实例时,需要谨慎评估资源隔离的需求和可行性,以确保不同业务或用户组之间的资源使用不会相互干扰。

2. 库内分库分表方案

库内分库分表的基本思路是将原有的单表数据分散到多个数据库或表的集合中,这样可以分散读写压力,从而提升系统的并发处理能力。常见的分库分表策略包括水平拆分和垂直拆分。水平拆分是将同一表中的不同行按照某种规则分散到不同的表中,而垂直拆分则是将同一表中的不同列分散到不同的表中。

在实施分库分表的过程中,必须充分考虑数据的完整性和一致性,同时还需要关注跨库表查询的优化问题,以确保分库分表后的系统能够稳定且高效运行。

下面介绍常用的库内分库和分表方案。

(1) 分库方案

在某些业务场景中,为了方便统一管理和节省资源,多个业务系统可能会共用一个 MySQL 数据库资源。然而,如果其中某个系统给数据库服务带来巨大的负载,那么它可能会影响所有业务系统。为了解决这个问题,可以采用"分库"的方式,即每个服务使用独立的 MySQL 数据库。这样,服务之间就不会相互竞争资源,从而提升了整体服务的性能。如图 9-4 所示,单个数据库能够支撑的并发量是有限的,通过将数据库拆分成多个独立的库,可以避免服务间的资源竞争,并进一步提高服务的性能。

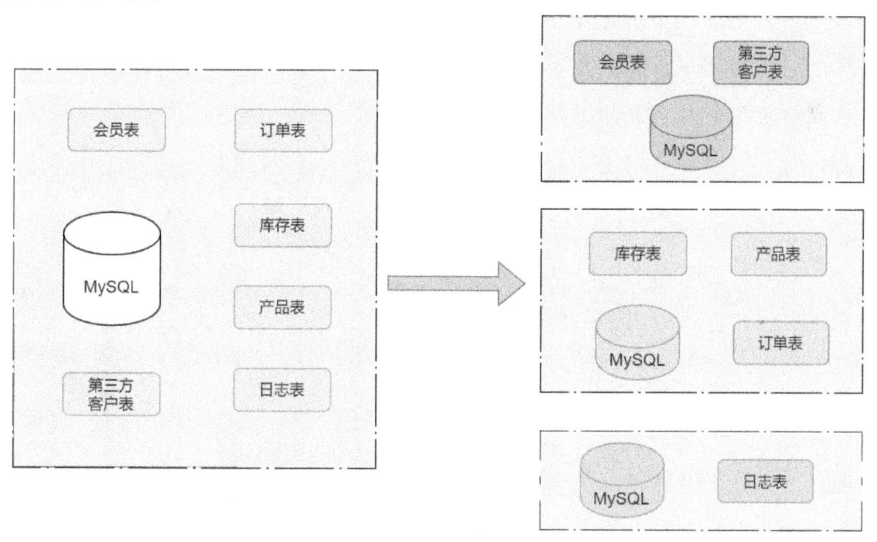

图 9-4　MySQL 库内分库方案

(2) 分表方案

分表方案与分库方案是解决不同数据库问题的两种策略。分库方案旨在分散单个数据库实例的处理压力,以应对高负载;而分表方案则专注于解决单表数据量过大的问题,以提升查询效率和管理能力。分表方案包括:水平拆分、垂直拆分和表拆分。

1）水平拆分和垂直拆分。这两种方式都能将一个大表拆分为多个小表,以减少不必要的数据访问,提升单表处理能力,进而加快数据库整体处理速度。
- 水平拆分：基于数据进行划分,拆分后的表结构相同,但数据内容不同。
- 垂直拆分：基于表或字段进行划分,拆分后的表结构不同。

图 9-5 所示为用户表的水平拆分和垂直拆分示意图。

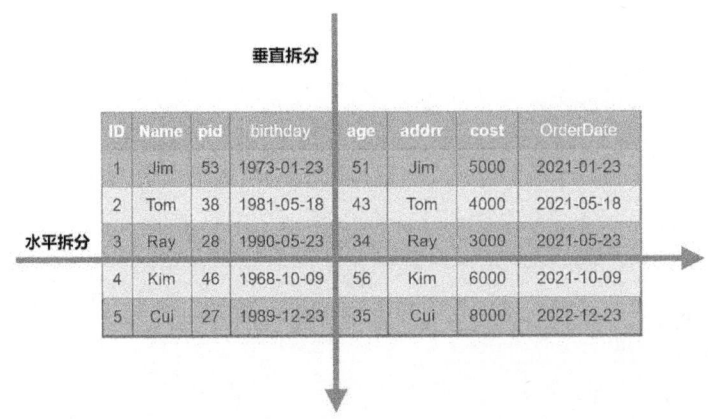

图 9-5　MySQL 分表中的水平拆分和垂直拆分

2）表拆分。冷热数据分离是一种常用的数据库优化手段,目的是将频繁访问的"热"数据和较少访问的"冷"数据分开存储,以提升数据库的性能。按照数据访问的规律进行拆分,可以更有效地利用存储和计算资源,从而使单表的处理能力得到提升,加快整个数据库的处理速度。

在实现冷热数据分离时,可以采用多种表拆分方法。
- 取模方式表：这是一种常见的水平拆分策略,可以通过哈希取模、范围划分、基于 ID 的奇偶性或除以某个特定值后取余数等方式来实现。这种方法可以将数据均匀分布到多个表中,从而分散单个表的负载压力。
- 每日表：按照数据的日期属性进行拆分,如只存储当天的数据。这种方式适用于那些数据量随时间增长非常迅速的场景。通过每日归档,可以保持单表的数据量在一个可管理的范围内。
- 每月表：与每日表类似,但是按照月份来归档数据。这种方式适用于数据增长速度稍慢,但仍需要定期归档的场景。
- 历史表：对于超过一定时间（如一年或几年）的历史数据,可以将其迁移到专门的历史表中。这样做既可以保持当前活动表的数据量较小,提高查询效率,又可以长期保存历史数据以备查询和分析之需。

图 9-6 所示为表拆分示例,即订单表按照每日表方法,每天进行拆分。当过期数据不用的时候,可以快速删除历史日表。这些拆分策略可以根据具体的业务需求和数据特点来选择与组合使用,以达到最佳的数据库性能和资源管理效果。

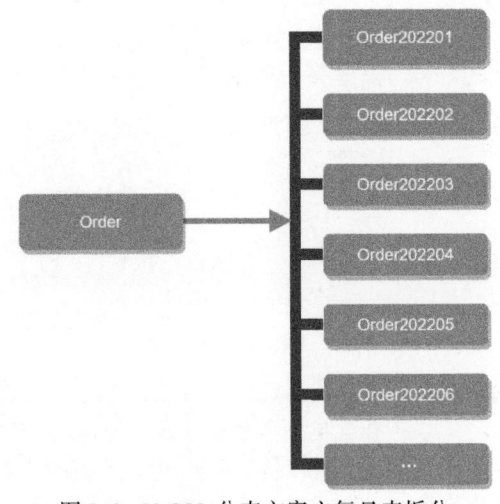

图 9-6　MySQL 分表方案之每日表拆分

3. 总结

对于库内分库分表，首先需要了解所使用的数据库的处理能力以及潜在的瓶颈，然后才能进行合理的拆分。拆分规则的设计应尽量简洁，只要能满足业务需求即可。

库内分库分表可以解决许多问题，但同时也给系统带来了不少复杂性工作，例如：
- 跨库关联查询。
- 聚合操作的实现。
- 定期维护和数据迁移。
- 单表数据量过大带来的管理挑战。

因此，在决定使用中间件分库分表之前，应该避免"过度设计"和"过早优化"。首先，应尽力采取其他可行的优化手段，如升级硬件、优化网络、实施读写分离、索引优化和库内分库分表等。当这些手段都已达到其使用极限时，再考虑通过中间件分库分表策略。

9.2 MySQL 8.0 应用开发

9.2.1 MySQL 8.0 应用开发的概念

数据库最基本的服务就是提供数据存储功能。以结构化的方式将数据存储在表中，为用户的数据管理和查询提供了便利。数据库应用开发则是指基于数据库系统提供的功能，实现一系列流程操作的过程。这些操作可以帮助企业更好地管理和优化业务流程，提高生产力和效率，从而更好地适应数字化时代的变化。

在进行 MySQL 数据库应用开发时，需要掌握一些基础知识，包括数据库的基本概念、关系数据库的原理和常用操作、SQL 的基本语法，以及数据库设计和优化等。

1. 数据库的基本概念

数据库是按照数据结构来组织、存储和管理数据的仓库，是一个在计算机中长期存储、有组织、可共享的数据集合。相比 Oracle、DB2 等数据库，MySQL 被视为轻量级数据库。它的特点在于简单、轻量，适合保存一定量的数据，但可能不适合保存大批量或复杂类型的数据。

2. 关系数据库的原理和常用操作

关系数据库基于关系模型，其中一个关键特点是支持事务，可以满足原子性、一致性、隔离性、持久性等事务属性。MySQL 具备这些特性，同时支持多种基本数据类型（如文本、数字、日期等）和复杂的数据类型（如数组、JSON 等）。MySQL 还支持多种功能，如视图、存储过程、触发器和事件等。在 MySQL 中，可以使用 SQL 进行数据的查询、插入、更新和删除等操作。常用的 SQL 语句包括 SELECT、INSERT、UPDATE 和 DELETE 等。

3. SQL 的基本语法

SQL（Structured Query Language，结构化查询语言）是用于管理和操作关系数据库的标准编程语言。SQL 包括数据定义语言（DDL）、数据操纵语言（DML）和数据控制语言（DCL）等部分。DDL 用于创建和修改数据库对象，如表、索引、视图和触发器等。DML 用于对数据库中的数据进行操作，包括查询、插入、更新和删除等。而 DCL 则用于管理数据库的访问权限和保证其安全性。MySQL 同样支持这些基础 SQL 语句操作。

4. 数据库设计和优化

数据库设计（Database Design）是指对于一个给定的应用环境，构造最优的数据库模式，建立数据库及其应用系统，使之能够有效地存储数据，满足各种用户的应用需求（信息要求和处理要求）和业务规则，设计出适合的数据库结构和关系模式。在 MySQL 数据库设计过程中，需要考虑实体和实体之间的关系、属性的定义和约束、索引的选择和设计等因素。MySQL 数据库优化是指通过优化数据库结构、查询语句和索引等手段，提高数据库的性能和效率。在 MySQL 中，常用的数据库优化技术包括合理使用索引、优化查询语句、分析和调整数据库参数等。

5. 总结

上述这些基础知识是 MySQL 数据库应用开发的基石。它们不仅帮助开发者更好地进行数据库的设计、开发和管理工作，还提供了解决问题和优化性能的思路，从而确保应用程序的稳定性和高效性。

9.2.2 MySQL 8.0 常用的开发规范

MySQL 数据库开发规范化是一个旨在消除设计和使用不良数据库缺陷的过程，它对于确保 MySQL 数据库的稳定性和可靠性至关重要。规范化实践在这个过程中扮演着非常重要的角色。

1）选择 InnoDB 存储引擎。使用 InnoDB 存储引擎是因为它支持事务处理、行级锁定和外键约束，这些都是提高数据库性能和保证数据完整性的重要特性。

2）字符集统一（UTF-8 和 utf8mb4）。使用 UTF-8 或 utf8mb4 字符集可以确保数据库能够存储任何 Unicode 字符，这对于支持多语言环境和特殊字符非常重要。与 utf8mb4 相比，UTF-8 可以存储更多的字符，包括 emoji 表情等。

3）数据库中库名、表名规范。统一使用小写字母和下画线可以确保数据库对象名称的一致性，并且符合 MySQL 的命名规范。限制名称长度（如 32 个字符）有助于避免某些潜在的问题，以便于管理和维护。

4）单表设计：MySQL 数据库单实例的表数目应限制在 2000 个以下，单表的列数目应限制在 30 个以下。所有表都应添加注释以提高可读性。建议将单表的数据量控制在 3000 万条以内，以避免性能问题。限制单实例表数目和单表列数目是为了优化查询性能与提升数据库管理的效率。控制单表数据量则是基于性能考虑，因为过多的数据可能会导致查询效率下降，以及备份和恢复时间增加等问题。

5）表必须有主键。主键是唯一标识表中每一行的约束，它可以提高查询效率并确保数据的唯一性。尽量避免使用 VARCHAR 类型作为主键，因为字符类型的主键在性能上可能不如整数类型的主键。

6）禁止使用外键。如果有外键完整性约束，则需要应用程序控制。外键检查需要在数据库层面进行，这可能会增加数据库操作的性能开销，特别是在高并发或大数据量的场景下。每次插入、更新或删除记录时，数据库都需要检查外键约束，这可能会导致操作速度变慢。

7）字段设置默认值。建议为所有数据库字段设置非空（NOT NULL）约束，并为它们提供默认值。这有助于确保数据的完整性和一致性。当查询包含 NULL 值的列时，MySQL 可能无法有效地使用索引，从而导致全表扫描，影响查询性能。聚合函数（如 COUNT( )、SUM( ) 和 AVG( ) 等）在处理包含 NULL 值的列时，通常会忽略 NULL 值。

8）字段使用的原则如下。

- 字段数据类型取值范围越小越好。使用合适的数据类型可以优化数据库的性能和存储效率。
- 不同表里相同内容的字段应该采用统一的名称和类型，这有助于保持数据模型的清晰和一致。
- 禁止使用 TEXT、BLOB 等类型大字段。这些类型用于存储大量文本或二进制数据，可能会导致性能问题，特别是在进行复杂查询或索引时。此外，它们也可能增加数据库的备份和恢复时间。

9）索引规范如下。
- 单表索引不允许超过 5 个。过多的索引会导致数据插入、更新和删除的性能下降，因为每次操作都需要更新索引。此外，索引本身也会占用额外的存储空间。
- 禁止在更新十分频繁、区分度不高的属性上建立索引。这样的索引不仅不能提高查询性能，反而会增加更新的开销，因为每次更新都需要重新构建索引。
- 建立组合索引，必须把区分度高的字段放在前面。在组合索引中，MySQL 会按照索引列的顺序进行匹配，因此将区分度高的字段放在前面可以更快地过滤出需要的数据，提高查询性能。

10）禁止使用 MySQL 高级功能，如存储过程、触发器、函数、视图、事件等。这些功能在某些场景下可以极大地提高数据库操作的灵活性和效率，但也可能带来额外的复杂性、维护成本和性能问题。

11）不建议使用分区表。MySQL 的分区表在实际应用中性能表现不佳，且管理和维护成本较高。

12）操作规范如下。
- 禁止使用 SELECT *，只获取必要的字段。SELECT * 会检索表中的所有字段，这会增加网络传输的数据量、消耗更多的内存和处理资源。
- 禁止使用 INSERT INTO t_××× VALUES(×××)。这种方式不够灵活，而且当表结构变更时，代码可能需要相应调整。
- 禁止使用属性隐式转换。隐式转换可能会导致性能下降和预期外的行为。
- 禁止负向查询，以及%开头的模糊查询。负向查询（如 NOT LIKE）和%开头的模糊查询通常效率较低，因为它们需要全表扫描。
- 禁止使用跨库查询，禁止大表使用 JOIN 查询，禁止大表使用子查询。跨库查询会增加网络延迟和复杂性；大表上的 JOIN 和子查询可能导致性能问题。
- 禁止核心业务流程 SQL 包含复杂查询。复杂查询可能会影响核心业务流程的性能和稳定性。
- 无论是使用 SELECT，还是使用"破坏力"极大的 UPDATE 和 DELETE 语句，一定要检查 WHERE 条件判断的完整性。不完整的 WHERE 条件可能导致表锁操作，影响整体性能。

13）尽量避免负载高的操作。大的 SQL 语句、大事务和批量操作会消耗大量的数据库资源，影响性能。

14）安全性规范如下。
- 应用不可用 root 账号访问。使用 root 账号进行应用访问会带来严重的安全风险。应为应用创建专门的数据库账号，并分配适当的权限。
- 遵守应用账号最小权限（SELECT、INSERT、UPDATE 和 DELETE）原则。根据应用的实际需要，只授予必要的权限，限制应用账号的权限可以减少潜在的安全风险。

9.3 MySQL 8.0 应用开发实践

9.3.1 时间类型的设置

在 MySQL 中，正确的时间字段设置可以确保数据库中的时间戳信息准确反映数据的实际时间，这对于确保数据的完整性和准确性至关重要。在许多业务场景中，时间是一个重要的维度。通过正确设置时间字段，可以支持各种业务逻辑，如时间范围的筛选、时间段的统计和分析等，从而满足业务需求、提升用户体验和推动业务发展。

下面将介绍在 MySQL 开发中如何正确使用这些时间字段来满足不同的业务需求。

1. 时间类型的字段

在 MySQL 中，有多种关于时间的类型可以选择，以满足不同的场景需求。表 9-1 所示为 MySQL 时间类型字段列表，以及每种类型的简要描述。

表 9-1 MySQL 支持的时间类型

类型	大小	范围	格式	用途
DATE	3 字节	1000-01-01 ~ 9999-12-31	YYYY-MM-DD	日期值
TIME	3 字节	-838：59：59 ~ 838：59：59	HH：MM：SS	时间值或持续时间
YEAR	1 字节	1901 ~ 2155	YYYY	年份值
DATETIME	8 字节	1000-01-01 00：00：00 ~ 9999-12-31 23：59：59	YYYY-MM-DD HH：MM：SS	混合日期和时间值
TIMESTAMP	4 字节	1970-01-01 00：00：01 UTC ~ 2038-01-19 03：14：07 UTC		时间戳数值

在选择这些数据类型时，重要的是要考虑哪些场景适合哪些时间类型。

- DATE：时间信息不重要或已知时使用，如用于生日、节日、纪念日等只需要日期的场景。
- TIME：在只需要时间数据时使用，如会议时间、工作时间等。例如，用于记录事件或活动开始和结束的时间。
- YEAR：在时间精度只需要到年份时使用，如出生年份、毕业年份，以及统计或报告按年份分组的数据、年度销售报告等。
- DATETIME：用于记录事件发生的确切时间，如订单创建时间、用户登录时间等。
- TIMESTAMP：用于记录数据创建和修改的时间戳。当需要自动跟踪记录的变化（如博客文章的发布和最后修改时间）时，需要使用它。

2. 影响时间类型参数配置

在 MySQL 应用开发中，TIMESTAMP 时间类型使用场景较多，但由于其特性和行为，它也是容易出现问题的一个数据类型。time_zone、SQL_MODE 和 explicit_defaults_for_timestamp 这 3 个参数在时间处理上各自扮演着不同的角色，并且它们之间的相互作用也会影响 TIMESTAMP 列的行为。

（1）TIMESTAMP 受到 time_zone 参数的影响

在使用 MySQL 的 TIMESTAMP 时间类型时，需要注意其与时区相关的特殊行为。

TIMESTAMP 时间类型在 MySQL 中的行为与其他日期和时间类型（如 DATETIME）不同，主要体现在其对时区的处理上。当存储一个 TIMESTAMP 值时，MySQL 会将其从当前会话的时区转换为UTC（协调世界时）进行存储。同样，当检索一个TIMESTAMP值时，MySQL会将其从UTC转换回当前会话的时区。图9-7所示为参数time_zone在使用SYSTEM和UTC时区时，查询显示的时间不一样的示例。

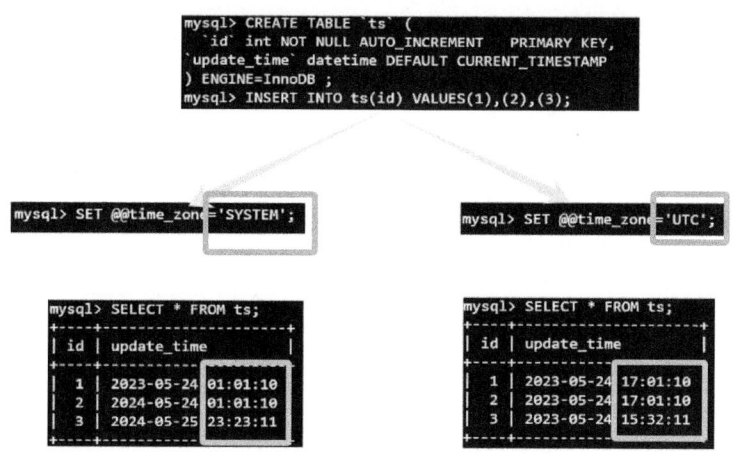

图9-7　MySQL TIMESTAMP时间类型对时区处理的示例

当应用程序需要考虑多个时区时，time_zone参数就变得非常有用。例如，一个全球性的公司可能需要根据用户的地理位置显示相应时区的时间。

（2）TIMESTAMP受到SQL_MODE参数的影响

在MySQL中，SQL_MODE参数用于定义MySQL应该如何处理SQL语句的执行和数据的校验。SQL_MODE的设置可以影响多个方面，包括数据类型、数据完整性、错误处理等。SQL_MODE参数设置不同同样也会影响TIMESTAMP值。影响TIMESTAMP的命令行示例如下。

```
# 创建一张TIMESTAMP的测试表
mysql> CREATE TABLE ts (
        id INTEGER NOT NULL AUTO_INCREMENT PRIMARY KEY,
        col TIMESTAMP NOT NULL
    ) AUTO_INCREMENT = 1;
# 查看当前SQL_MODE参数设置
mysql> SHOW VARIABLES LIKE '%sql_mode%';
+---------------------+--------------------------------+
|Variable_name        |Value                           |
+---------------------+--------------------------------+
|sql_mode             |STRICT_TRANS_TABLES             |
+---------------------+--------------------------------+

# 插入不在TIMESTAMP类型取值范围内的时间数据,会报错
mysql>INSERT INTO ts (col) VALUES ('1969-01-01 01:01:10');
ERROR 1292 (22007): Incorrect datetime value:'1969-01-01 01:01:10' for column 'col' at row 1

# 清空SQL_MODE值
mysql> SET sql_mode="";
Query OK, 0 rows affected (0.00 sec)
# 查看当前SQL_MODE
mysql> SHOW VARIABLES LIKE '%sql_mode%';
```

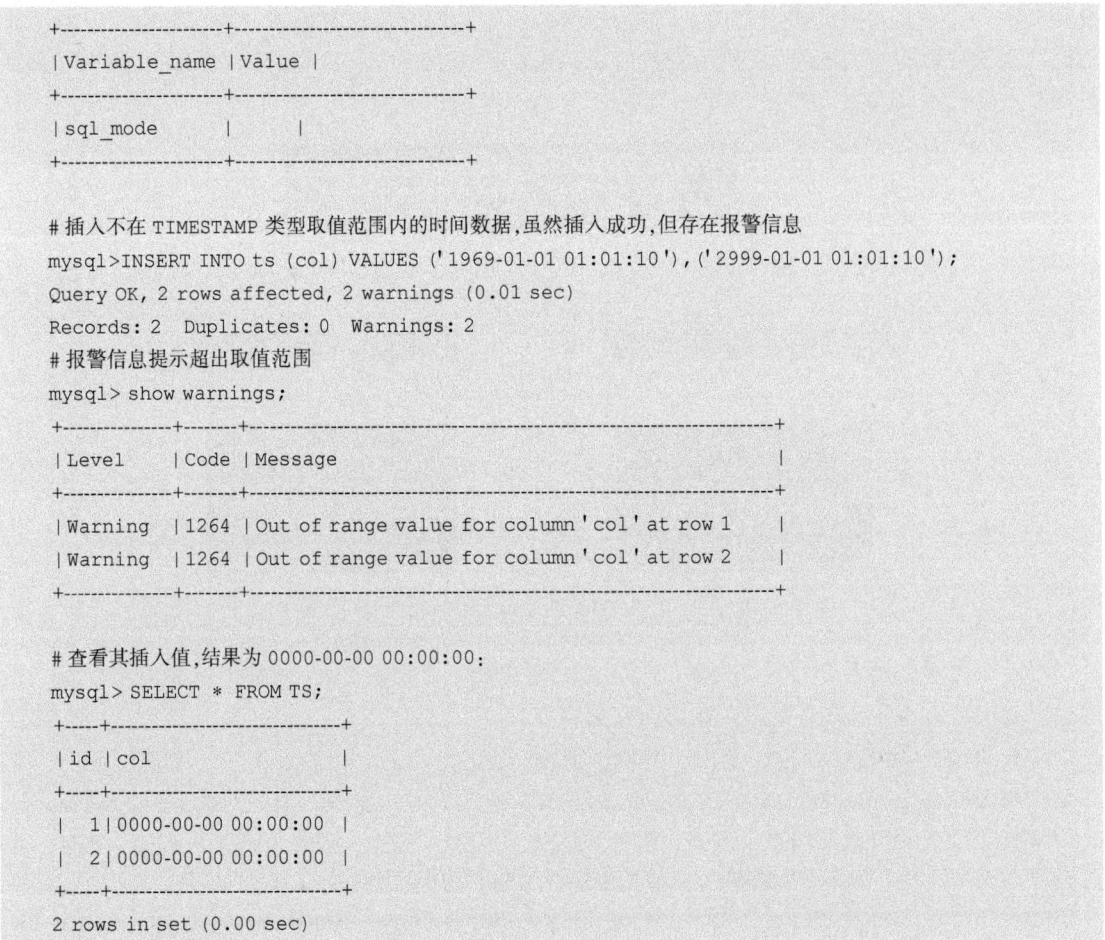

说明：当 SQL_MODE 包含 STRICT_TRANS_TABLES 时，对于 TIMESTAMP 列，插入数据，MySQL 通常会返回一个错误，提示数据不符合列的定义。在不包含 STRICT_TRANS_TABLES 时，如果插入的值超出了其范围，则 MySQL 会尝试调整该值。它不会返回一个错误，而是将超出范围的值转换为 "0000-00-00 00∶00∶00"。

（3）TIMESTAMP 受到 explicit_defaults_for_timestamp 参数的影响

explicit_defaults_for_timestamp 参数的作用是改变 TIMESTAMP 列在没有显式指定默认值时的行为。当 explicit_defaults_for_timestamp 设置为 ON（启用）时，如果 TIMESTAMP 列被声明为 NOT NULL 并且没有显式地设置默认值，那么试图插入 NULL 值将会导致错误。当设置为 OFF 时，就会按照定义方式处理。

图 9-8 所示为 explicit_defaults_for_timestamp 影响数据插入场景。当设置为 ON 的时候，提示无法插入 NULL 值；当设置为 OFF 时，直接使用 CURRENT_TIMESTAMP 当前时间，插入到表中。

但这种行为是为了增加数据的一致性和可预测性，因为它要求开发者明确地为 TIMESTAMP 列提供值，而不是依赖于可能不是预期的隐式行为。

3. 总结

在 MySQL 中，时间的设置和使用涉及多个方面，包括时间类型、系统参数和系统时区。对于开发人员来说，理解这些概念并正确地设置和使用它们是非常重要的。

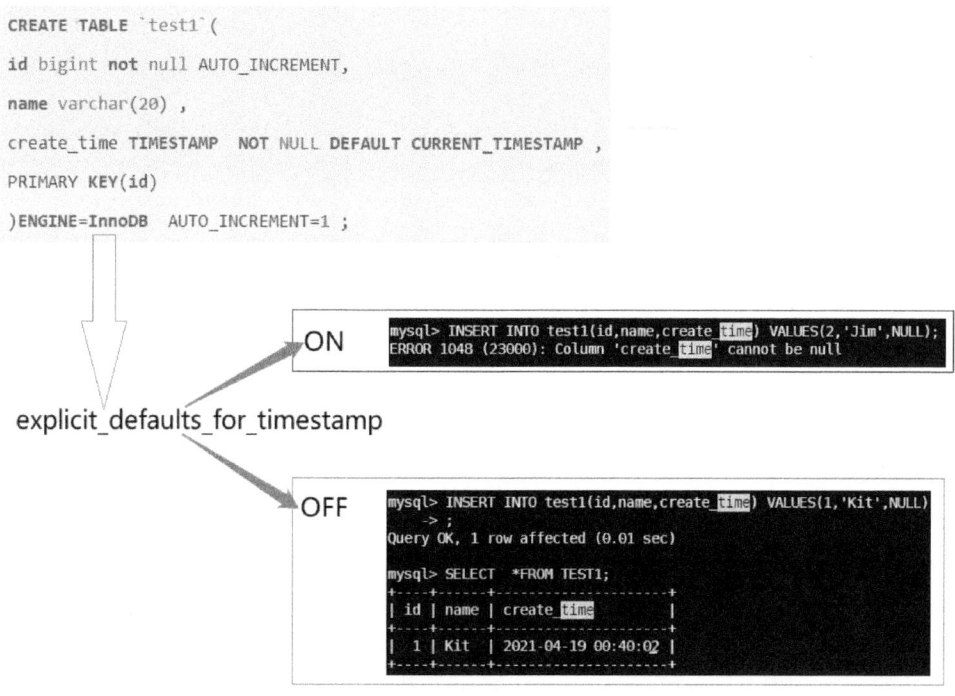

图 9-8　MySQL explicit_defaults_for_timestamp 影响数据插入场景

开发人员应该：
- 根据实际需求选择合适的时间类型。
- 谨慎更改 SQL_MODE 的设置，避免引入潜在的问题。
- 明确设置 explicit_defaults_for_timestamp，以确保 TIMESTAMP 列的默认行为符合预期。
- 根据应用程序的需求设置系统时区，以确保时间的正确显示和处理。

9.3.2　BIT 数据类型的使用

MySQL 中的 BIT 数据类型是一个非常特殊且有用的数据类型，专门用于存储位值。它被归纳为 Numeric Data 类型，并且支持 MyISAM、MEMORY、InnoDB、NDB 引擎表。由于 BIT 数据类型只能包含 0 和 1，因此它非常适合表示逻辑状态、开关、权限，或者任何需要二选一的情况。

BIT 数据类型的特性如下。
- 存储：BIT（M）类型允许存储 M 位值，其中 M 的取值范围为 1~64。例如，BIT（1）可以存储一个位（0 或 1），而 BIT（64）可以存储一个 64 位的值。
- 存储大小：尽管 BIT 数据类型可以存储多达 64 位，但它实际上并不总是占用 64 个单独的字节。相反，它通常根据所需的位数动态分配空间，以节省存储空间。
- 默认值：如果不为 BIT 列指定默认值，则它的默认值将为 b'0'。
- 显示：当查询 BIT 列时，它将显示为二进制字符串，如 b'101'。
- 填充：如果给长度小于 M 位的 BIT（M）列赋值，则该值将在左侧用 0 填充。例如，对于 BIT（6）列，赋值 b'101'实际上等同于 b'000101'。
- 在数据类型中，BIT 应该占据的空间最小。BIT（M）类型允许存储 M 位值。M 取值范围为 1~64。NDB 集群中所有 BIT 列的最大总和不能超过 4096 位。

BIT 数据类型适用场景如下。
- 权限管理：存储用户的权限，每个权限对应一个位。例如，一个 8 位的 BIT 列可以表示 8 种不同的权限。
- 状态跟踪：表示对象的各种状态。例如，订单的状态（待处理、处理中、已完成等）。
- 二进制操作：对数据进行位级别的操作，如按位与、按位或等。
- 数据压缩：在某些情况下，使用 BIT 数据类型可以以更高效的方式存储数据。
- 编码：某些数据或配置信息可以以二进制形式存储，节省空间。

下面介绍为出勤率设计的应用开发示例。

1. 数据查询

出勤率就是一个典型的 BIT 数据类型使用示例。可以使用 BIT 数据类型来存储每天的出席情况，其中 1 表示出席，0 表示缺席。这种数据类型与出勤率的二元性质（出席和缺席）非常匹配。模拟数据和使用 BIT 数据类型显示学生出勤信息的命令行示例如下。

```
# 创建出勤率表
mysql> CREATE TABLE attendance (
    emp_no CHAR(3),
    emp_name CHAR(50),
    attend BIT(5),class INT,KEY `idx_bit` (`attend`)
);
# 使用 b'val'的编写方式，Val 是使用 0 和 1 编写的一周出勤率信息,使用二进制值
mysql> INSERT INTO attendance (emp_no, emp_name, attend, class)VALUES
('001','Jim',b'11111',5),('002','Kim',b'11000',5),('003','Cui',b'00111',5),
('004','King',b'11101',5),('005','Wang',b'101',5),('006','Chen',NULL,5),
('007','Piao',0,5),('008','Hu',1,5);

# 查看出勤率表信息
mysql> SELECT emp_no,emp_name,attend,class FROM attendance;
+--------+----------+----------+-------+
| emp_no | emp_name | attend   | class |
+--------+----------+----------+-------+
| 001    | Jim      | 0x1F     |     5 |
| 002    | Kim      | 0x18     |     5 |
| 003    | Cui      | 0x07     |     5 |
| 004    | King     | 0x1D     |     5 |
| 005    | Wang     | 0x05     |     5 |
| 006    | Chen     | NULL     |     5 |
| 007    | Piao     | 0x00     |     5 |
| 008    | Hu       | 0x01     |     5 |
+--------+----------+----------+-------+
8 rows in set (0.00 sec)

# 以 BIN 方式显示出勤率信息
mysql> SELECT emp_no,emp_name,BIN(attend),class FROM attendance;
+--------+----------+-------------+-------+
| emp_no | emp_name | BIN(attend) | class |
+--------+----------+-------------+-------+
| 001    | Jim      | 11111       |     5 |
```

```
| 002       | Kim       | 11000                 | 5       |
| 003       | Cui       | 111                   | 5       |
| 004       | King      | 11101                 | 5       |
| 005       | Wang      | 101                   | 5       |
| 006       | Chen      | NULL                  | 5       |
| 007       | Piao      | 0                     | 5       |
| 008       | Hu        | 1                     | 5       |
+-----------+-----------+-----------------------+---------+
8 rows in set (0.00 sec)

# 出勤率信息使用 LPAD 函数左填充 0
mysql> SELECT emp_no,emp_name,LPAD(BIN(attend),5,0),class FROM attendance;
+--------+----------+-----------------------+-------+
| emp_no | emp_name | LPAD(BIN(attend),5,0) | class |
+--------+----------+-----------------------+-------+
| 001    | Jim      | 11111                 | 5     |
| 002    | Kim      | 11000                 | 5     |
| 003    | Cui      | 00111                 | 5     |
| 004    | King     | 11101                 | 5     |
| 005    | Wang     | 00101                 | 5     |
| 006    | Chen     | NULL                  | 5     |
| 007    | Piao     | 00000                 | 5     |
| 008    | Hu       | 00001                 | 5     |
+--------+----------+-----------------------+-------+
8 rows in set (0.00 sec)
```

说明：可以使用位值函数，并且可以通过十进制、二进制或任何其他数据转换函数检索位值。使用 LPAD 和 BIN 函数以适当的格式检索数据。

当查询某个学生的出勤率时，对应 WHERE 条件 BIT 字段必须是 b'val' 或 INT 类型。查询命令行示例如下。

```
#1.采用 INT 类型数字进行查询
mysql> SELECT emp_no,emp_name,BIN(attend),HEX(attend),attend,class
       FROM attendance where attend=7;
+--------+----------+-------------+-------------+--------+-------+
| emp_no | emp_name | BIN(attend) | HEX(attend) | attend | class |
+--------+----------+-------------+-------------+--------+-------+
| 003    | Cui      | 111         | 7           | 0x07   | 5     |
+--------+----------+-------------+-------------+--------+-------+

#2.采用位置进行查询
mysql> SELECT emp_no,emp_name,BIN(attend),HEX(attend),attend,class
       FROM attendance where attend=b'111';
+--------+----------+-------------+-------------+--------+-------+
| emp_no | emp_name | BIN(attend) | HEX(attend) | attend | class |
+--------+----------+-------------+-------------+--------+-------+
| 003    | Cui      | 111         | 7           | 0x07   | 5     |
+--------+----------+-------------+-------------+--------+-------+
```

```
#3.以 string 类型进行查询:无信息
mysql> SELECT emp_no,emp_name,BIN(attend),HEX(attend),attend,class
       FROM attendance where attend='7';
Empty set (0.00 sec)

#4.IN 语句部分失效(字符串)
mysql> SELECT emp_no,emp_name,BIN(attend),HEX(attend),attend,class
       FROM attendance WHERE attend in('7',5);
+--------+----------+-------------+-------------+--------+-------+
| emp_no | emp_name | BIN(attend) | HEX(attend) | attend | class |
+--------+----------+-------------+-------------+--------+-------+
| 003    | Cui      | 111         | 7           | 0x07   |   5   |
+--------+----------+-------------+-------------+--------+-------+

#5.可以采用 NULL 值查询
mysql> SELECT emp_no,emp_name,BIN(attend),HEX(attend),attend,class
       FROM attendance WHERE attend IS NULL;
+--------+----------+-------------+-------------+--------+-------+
| emp_no | emp_name | BIN(attend) | HEX(attend) | attend | class |
+--------+----------+-------------+-------------+--------+-------+
| 006    | Chen     | NULL        | NULL        | NULL   |   5   |
+--------+----------+-------------+-------------+--------+-------+
1 row in set (0.00 sec)
```

说明：BIT 字段只有是整数类型或 BIT 类型才能匹配。在数字范围内，b'val'与整数类型等价。例如，上述例子中的 7 和 b'111'等价。对于 NULL 值查询，BIT 类型依然等于 NULL 值。

2. BIT 数据类型字段索引

在 MySQL 中，是否能够有效使用索引（BIT 字段的索引）取决于查询的具体条件和索引的创建方式。对于 BIT 字段，可以创建索引来加速查询，但是否能够利用这个索引取决于查询的写法。具体命令行示例如下。

```
#1.数字类型,可以使用索引
mysql> EXPLAIN SELECT emp_no,emp_name,BIN(attend),HEX(attend),attend,class
       FROM attendance WHERE attend=7 \G
*************************** 1.row ***************************
           id: 1
  select_type: SIMPLE
        table: attendance
   partitions: NULL
         type: ref
possible_keys: idx_bit
          key: idx_bit
      key_len: 2
          ref: const
         rows: 1
     filtered: 100.00
        Extra: NULL
1 row in set, 1 warning (0.00 sec)
```

#2.采用位置进行查询,可以使用索引
```
mysql> EXPLAIN SELECT emp_no,emp_name,BIN(attend),HEX(attend),attend,class
    FROM attendance WHERE attend=b'111'\G
*************************** 1.row ***************************
           id: 1
  select_type: SIMPLE
        table: attendance
   partitions: NULL
         type: ref
possible_keys: idx_bit
          key: idx_bit
      key_len: 2
          ref: const
         rows: 1
     filtered: 100.00
        Extra: NULL
1 row in set, 1 warning (0.00 sec)
```

#3.使用范围进行查询,可以使用索引
```
mysql> EXPLAIN SELECT emp_no,emp_name,BIN(attend),HEX(attend),attend,class
    FROM attendance WHERE attend>17\G
*************************** 1.row ***************************
           id: 1
  select_type: SIMPLE
        table: attendance
   partitions: NULL
         type: range
possible_keys: idx_bit
          key: idx_bit
      key_len: 2
          ref: NULL
         rows: 3
     filtered: 100.00
        Extra: Using index condition
1 row in set, 1 warning (0.00 sec)
```

#4.使用二进制字符串b' '方式,也可以使用索引
```
mysql> EXPLAIN SELECT emp_no,emp_name,BIN(attend),HEX(attend),attend,class
    FROM attendance WHERE attend>b'111'\G
*************************** 1.row ***************************
           id: 1
  select_type: SIMPLE
        table: attendance
   partitions: NULL
         type: range
possible_keys: idx_bit
          key: idx_bit
      key_len: 2
          ref: NULL
```

```
         rows: 3
     filtered: 100.00
        Extra: Using index condition
1 row in set, 1 warning (0.00 sec)

#5.采用 string 类型字段,不能使用索引
mysql> EXPLAIN SELECT emp_no,emp_name,BIN(attend),HEX(attend),attend,class
        FROM attendance WHERE attend='7'\G
*************************** 1.row ***************************
           id: 1
  select_type: SIMPLE
        table: NULL
   partitions: NULL
         type: NULL
possible_keys: NULL
          key: NULL
      key_len: NULL
          ref: NULL
         rows: NULL
     filtered: NULL
        Extra: no matching row in const table
1 row in set, 1 warning (0.00 sec)
```

说明：从上述验证中可以发现，BIT 字段类型索引只接受 INT 和 b'val'方式。

3. 位函数和运算符

在 MySQL 8.0 中，扩展了对位函数和运算符的支持，允许它们接受二进制字符串类型（BINARY、VARBINARY 和 BLOB）的参数。这使得可以在更大的数据上执行位操作，而不仅仅是 BIT 数据类型的列，而且能够产生大于 64 位的返回值，更容易执行位操作。表 9-2 所示为位函数和运算符。

表 9-2　MySQL 位函数和运算符

名　称	描　述
&	按位与
>>	右移
<<	左移
^	按位异或
BIT_COUNT( )	返回设置的位数
|	按位或
~	逐位反转

注意：在 MySQL 8.0 版本中，对二进制字符串参数的位操作可能会产生与 MySQL 5.7 版本不同的结果。所以，MySQL 5.7 版本和 MySQL 8.0 版本之间可能存在不兼容的信息。

以下是 MySQL 8.0 开发应用中用到的位函数和运算符示例。

【例 9-1】 BIT_COUNT 函数返回一个无符号的 64 位整数，命令行示例如下。

```
# BIT_COUNT()
mysql> SELECT emp_no,emp_name,BIN(attend),HEX(attend),attend,
```

```
class,bit_count(attend) FROM attendance;
+--------+----------+-------------+-------------+--------+-------+-------------------+
| emp_no | emp_name | BIN(attend) | HEX(attend) | attend | class | bit_count(attend) |
+--------+----------+-------------+-------------+--------+-------+-------------------+
| 001    | Jim      | 11111       | 1F          | 0x1F   | 5     |                 5 |
| 002    | Kim      | 11000       | 18          | 0x18   | 5     |                 2 |
| 003    | Cui      | 111         | 7           | 0x07   | 5     |                 3 |
| 004    | King     | 11101       | 1D          | 0x1D   | 5     |                 4 |
| 005    | Wang     | 101         | 5           | 0x05   | 5     |                 2 |
| 006    | Chen     | NULL        | NULL        | NULL   | 5     |              NULL |
| 007    | Piao     | 0           | 0           | 0x00   | 5     |                 0 |
| 008    | Hu       | 1           | 1           | 0x01   | 5     |                 1 |
+--------+----------+-------------+-------------+--------+-------+-------------------+
8 rows in set (0.00 sec)
```

从上述结果中可以了解到，BIT_COUNT 函数对于统计某些状态或条件的数量很有用。

【例 9-2】MySQL 8.0 扩展了位操作。对于 UUID 和 IPv6 数据的转换，其中包含的 "-" "::" 等符号需要特殊处理。具体命令行示例如下。

```
# 对于 UUID 和 IPv6 数据的转换,需要特殊处理"-" "::"等符号
mysql> SELECT HEX(UUID_TO_BIN('6ccd780c-baba-1026-9564-5b8c656024db'));
+----------------------------------------------------------+
| HEX(UUID_TO_BIN('6ccd780c-baba-1026-9564-5b8c656024db')) |
+----------------------------------------------------------+
| 6CCD780CBABA102695645B8C656024DB                         |
+----------------------------------------------------------+

# IPv6
mysql> SELECT HEX(INET6_ATON('fe80::219:d1ff:fe91:1a72'));
+---------------------------------------------+
| HEX(INET6_ATON('fe80::219:d1ff:fe91:1a72')) |
+---------------------------------------------+
| FE800000000000000219D1FFFE911A72            |
+---------------------------------------------+
```

这些二进制值很容易通过位操作进行处理，如从 UUID 值中提取时间戳或从 IPv6 地址中提取网络和主机部分信息。这对于在应用开发中获取时间戳信息和进行数据库中的 IP 地址管理都非常有用。

4. 总结

在 MySQL 8.0 开发应用中，BIT 数据类型可以有效地存储和处理二进制信息。通过结合适当的索引策略，可以提高数据库的性能和效率。然而，在选择使用 BIT 数据类型时，还需要考虑具体的需求和数据模式，以确保选择最适合当前应用场景的数据类型。

9.3.3 INSERT INTO 语句的使用

INSERT INTO 语句是最常用的插入数据的语句之一。在 MySQL 中，除 INSERT INTO 语句本身以外，它还有其他多种变体用于插入数据。下面介绍 MySQL 中常用的 4 种插入数据的语句。

1. INSERT INTO 可选参数

在 MySQL 中，基础插入语句 INSERT INTO 支持多种可选参数，这些参数可以影响插入操作的行为。这些可选参数关键字包括 DELAYED、LOW_PRIORITY、HIGH_PRIORITY 和 IGNORE。需要注意的是，这些参数并不是在所有 MySQL 版本和配置中都可用，而且它们的行为也可能因 MySQL 版本的不同而有所变化。

- DELAYED 关键字在 MySQL 的 INSERT INTO 语句中用于尝试延迟地插入数据到表中。当使用这个关键字时，MySQL 会立刻返回一个标识，告知上层程序数据已经"预备"插入。实际上，数据会在表没有其他线程使用时被插入，因此真实的插入时间是不确定的。这种写法对数据的安全性没有保障，因为插入操作可能会在任何时候执行。在 MySQL 5.6 版本中，延迟插入和替换是不推荐的，因为它们的行为可能不如预期。在 MySQL 5.7 版本和 MySQL 8.0 版本中，DELAYED 关键字不再被支持。MySQL 服务会识别但忽略它，将插入处理为非延迟插入，并生成一个警告（er_warn_legacy_syntax_convert），告知用户"不再支持延迟插入"。这意味着任何尝试使用 DELAYED 关键字的 INSERT 语句都会被当作普通的 INSERT 语句来处理。为了确保数据的一致性和安全性，建议避免使用 DELAYED 关键字，并寻找其他方法来实现延迟插入的效果。

- LOW_PRIORITY 关键字在 MySQL 的 INSERT INTO 语句中用于尝试延迟插入的执行，直到没有其他客户端从表中读取数据为止。这包括在现有客户端正在读取时以及 INSERT LOW_PRIORITY 语句正在等待开始读取的其他客户端。因此，对于发出 INSERT LOW_PRIORITY 语句的客户端，可能需要等待很长时间，尤其是在高并发读写的环境下。注意，LOW_PRIORITY 只影响那些仅使用表级锁的存储引擎，如 MyISAM、内存（MEMORY）和合并（MERGE）表。对于使用行级锁的存储引擎（如 InnoDB），LOW_PRIORITY 可能不会产生预期的效果。在应用开发时，需要理解 LOW_PRIORITY 的行为和限制，以确保数据的正确性和性能。在需要保证数据实时性的情况下，可能需要考虑其他并发控制策略或调整应用的设计。

- HIGH_PRIORITY 关键字用于 INSERT INTO 语句，以尝试给予该插入操作更高的优先级。如果服务器以 low_priority_updates 选项启动，那么 HIGH_PRIORITY 会覆盖这一设置，使得当前的插入操作具有更高的优先级，这可能会导致其他同时进行的插入操作被取消或延迟。当 low_priority_updates 参数设置为 1 时，它会使得所有的插入、更新、删除和锁表写语句等待，直到受影响的表上没有未决的选择或锁表读取为止。这意味着，在这些操作执行时，如果有其他查询正在读取或锁定这些表，那么这些更新操作将会等待。HIGH_PRIORITY 和 LOW_PRIORITY 选项只影响那些使用表级锁的存储引擎，如 MyISAM、MEMORY 和 MERGE。

- IGNORE 关键字用于 INSERT INTO 语句，如果在尝试插入新数据时，由于主键或唯一键的冲突而导致错误，那么系统将忽略这次插入操作，而不是返回一个错误。这意味着，如果表中已经存在与新数据在主键或唯一键上相同的记录，那么新数据将不会被插入，但也不会阻止其他可能的插入操作。

对于大多数应用开发来说，如果没有特殊的需求，使用最基本的 INSERT INTO 语句就足够了。这样可以确保代码的可移植性和稳定性，减少因 MySQL 版本升级或配置变更而引入的潜在问题。

2. INSERT INTO...SELECT

在 MySQL 中，INSERT INTO...SELECT 语句用于从一个或多个表中选择数据，并将这些数据插入到另一个表中。这是一种非常有用的方法，可以用来复制数据、转换数据或结合多个表的数据。

INSERT INTO...SELECT 语句是在一个事务中执行的。这意味着，如果在这个过程中出现任何错误，那么整个操作（即整个事务）都会被回滚，数据库将保持一致性。图 9-9 所示为对 INSERT INTO...SELECT 语句记录 binlog 内容的解析，整个解析包含在一个事务内容里进行处理。

当从同一个表中选择和插入时，MySQL 创建一个内部临时表来保存 SELECT 中的行，然后将这些行插入到目标表中。如果 temp_table 临时表在同一个 INSERT INTO... SELECT语句中被引用了两次，就会导致错误。

INSERT INTO...SELECT 语句特别适用于批量数据迁移和转换，同时通过条件语句（如 WHERE、JOIN 等）来筛选、转换和组合数据，然后将其插入到目标表中。然而，大量的 INSERT 操作可能会消耗大量的系统资源（如 CPU、内存和磁盘空间），从而影响数据库的性能。因此，在开发过程中使用该语句时，建议先在测试环境中进行充分的性能测试和验证。

图 9-9　binlog 解析的 INSERT INTO... SELECT 语句

3. REPLACE INTO

REPLACE INTO 是 INSERT INTO 的加强版，两者在功能上有相似之处，但处理已存在记录时的行为是不同的。以下是 REPLACE INTO 的具体行为。

步骤 1：插入尝试。当使用 REPLACE INTO 语句时，MySQL 会尝试将新记录插入到表中。

步骤 2：在插入之前，MySQL 会检查新记录的主键或唯一键值是否已存在于表中。这是通过比较新记录的主键或唯一键值与表中现有记录的主键或唯一键值来实现的。

步骤 3：删除与插入。如果找到具有相同主键或唯一键的现有记录，那么 MySQL 会执行两个操作：首先删除该现有记录，然后插入新记录。这意味着新记录会替换具有相同主键或唯一键的旧记录。如果没有找到具有相同主键或唯一键的现有记录，那么 REPLACE INTO 会像 INSERT INTO 一样直接插入新记录。

步骤 4：权限要求。要使用 REPLACE INTO 语句，用户必须拥有对目标表的 INSERT、UPDATE 和 DELETE 权限。这是因为 REPLACE INTO 实际上是结合了 INSERT、UPDATE 和 DELETE 操作的组合体。

步骤 5：表结构要求。为了使 REPLACE INTO 能够正常工作，目标表必须有一个定义好的主键或唯一键。如果没有主键或唯一键，那么 REPLACE INTO 的行为将与 INSERT INTO 相同，并且不会删除任何现有记录。这可能导致表中出现重复的数据行，因为没有唯一的标准来确定哪些行是"重复"的。

当表中同时存在主键和唯一键时，REPLACE INTO 的行为可能会变得更为复杂，因为它需要同时考虑主键和唯一键的唯一性约束。这些行为可以通过解析 binlog 文件得到验证。图 9-10 所示为解析 binlog 得到的 REPLACE INTO 的行为。

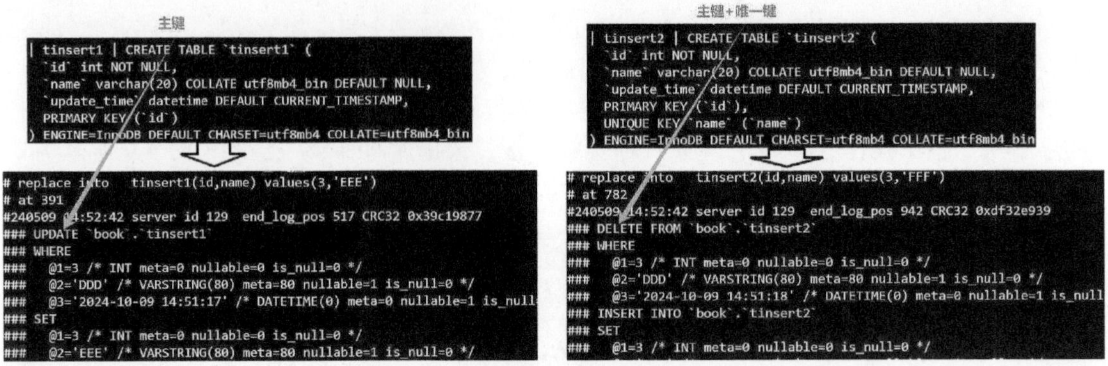

图 9-10 binlog 解析的 REPLACE INTO 语句

总结的规律如下。

- 在主键存在时，进行 UPDATE 操作。
- 在主键和唯一键同时存在时，进行 DELETE+INSERT 操作。
- 当同一个数据在所有行都一样时，REPLACE INTO 就不会进行更新操作。

此外，在开发过程中使用 REPLACE INTO 语句可能触发其他数据库行为，如外键约束、触发器或级联删除等。在使用 REPLACE INTO 语句之前，理解这些潜在的影响并相应地规划数据库操作是非常重要的。

4. INSERT INTO...ON DUPLICATE KEY UPDATE

在开发过程中，有时由于业务需求，可能需要先根据某一字段值查询数据库中是否有记录，如果有记录，则更新，如果没有记录，则插入。这时，可以使用 ON DUPLICATE KEY UPDATE 语句。判断记录是否存在是基于主键或唯一键的。如果出现重复值（即主键或唯一键冲突），则会对旧行进行更新。当主键和唯一键同时存在时，MySQL 会基于主键进行判断。`INSERT INTO ... ON DUPLICATE KEY UPDATE` 是 MySQL 中的一个非常有用的语句，它允许在插入新记录时检查是否存在具有相同主键或唯一键的记录。如果存在，则执行更新操作；如果不存在，则执行插入操作。图 9-11 所示为 INSERT INTO ... ON DUPLICATE KEY UPDATE 在主键和唯一键影响下不同的处理方式。

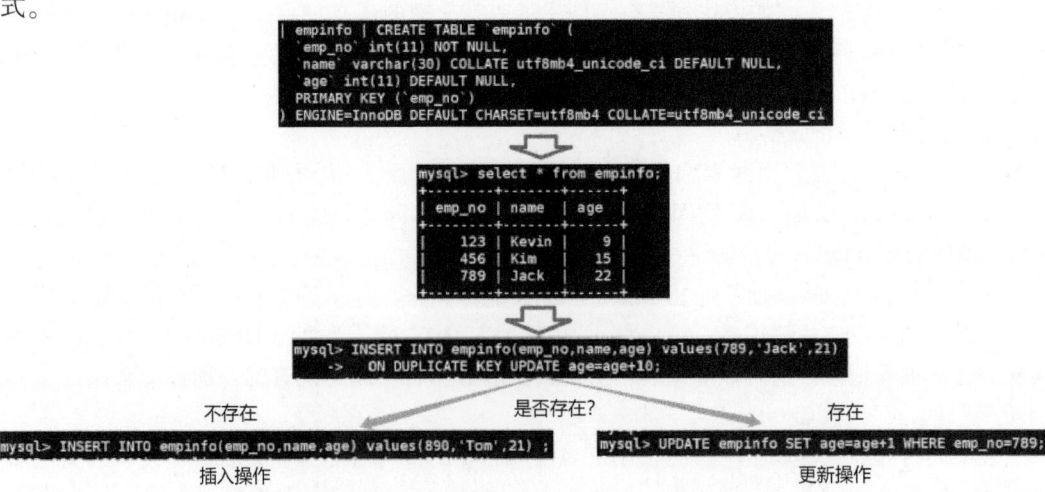

图 9-11 ON DUPLICATE KEY UPDATE 操作

MySQL 中的 INSERT INTO...ON DUPLICATE KEY UPDATE 语句减少了两个单独操作（先检查记录是否存在，再决定是插入还是更新）执行的需求。这是一个原子操作，在一个语句中完成这两个操作既可以提高效率，又可以减少由于多个操作同时尝试修改同一行数据而产生的锁争用。但是这个语句依赖于唯一键或主键来确定是否应该插入新行或更新现有行。如果表中没有适当的唯一键或主键，那么这个语句就无法使用。虽然这个语句可以提高效率，但如果表非常大且索引不是最优的，那么性能可能会受到影响。在开发过程中，应该根据具体的应用场景和需求来权衡使用。

5. 总结

在 MySQL 中，除了基本的 INSERT INTO 语句以外，还有 INSERT INTO...SELECT、REPLACE INTO 和 INSERT INTO...ON DUPLICATE KEY UPDATE 等变体，理解这些变体的用法对于有效开发非常重要。

- INSERT INTO...SELECT 是一种方便地在表之间迁移数据的方式。使用带索引的字段进行条件和排序限制可以提高性能。然而，如果 SELECT 查询返回的数据量非常大，则可能会形成一个大事务，这可能会影响整体性能。因此，在实际使用中需要根据具体情况进行权衡和优化。
- REPLACE INTO 在功能上相当于 DELETE+INSERT 的组合，但作为一个原子操作，它提供了一种方便的方式来替换或插入记录。然而，需要注意的是，REPLACE INTO 可能会导致索引页分裂，这可能会影响查询性能。
- INSERT INTO...ON DUPLICATE KEY UPDATE 语句允许在存在唯一键或主键冲突时更新记录。即使表定义有多个唯一键或主键，该语句也仍然是安全的，因为它会根据提供的键值来判断是执行插入还是更新操作。然而，表结构和索引设计仍然会影响该语句的性能。

9.3.4 分区表的使用

MySQL 数据库中的数据以文件形式存储在磁盘上，这些文件通常是以 .ibd 为扩展名的 InnoDB 表空间文件。当单张表的数据量变得非常大时，执行查找操作可能会变得缓慢。以 Linux 系统下的 vim 命令为例，尝试打开一个 30GB 大小的文件可能会遭遇性能问题，导致打开失败或需要很长时间来加载。

为了解决 MySQL 中大表带来的性能问题，一种常见的做法是在物理层面将表文件分割成多个较小的部分或块。通过分区（Partitioning）或分片（Sharding）等技术，可以将数据分散到不同的物理存储区域。这样，当需要查找某条数据时，MySQL 数据库只需要在包含该数据的分区或分片上进行搜索，而不是扫描整个表，从而大大提高检索速度。

1. 表分区定义

在 MySQL 中，表分区是指根据一定规则，将数据库中的一张表分解成多个更小、更容易管理的部分。如图 9-12 所示，虽然从逻辑上来看只有一张表（订单表），但是底层实际上是由 3 个分割的物理文件组成的。

在 MySQL 中，分区的物理文件名通常包含表名、分区键值和分隔符。分隔符用于区分表名、分区键值和文件扩展名。MySQL 允许自定义分隔符，但默认的分隔符通常是#。图 9-13 所示为 MySQL 表分区的物理文件，其中分区和子分区的文件名使用分隔符"#"创建。如果有一个名为 employees_sub 的表，并且按照范围对该表进行了分区，那么每个分区的物理文件名类似于 employees_sub#p#p0#sp#p0sp0.ibd、employees_sub#p#p0#sp#p0sp1.ibd、employees_sub#p#p1#sp#p1sp0.ibd 等。

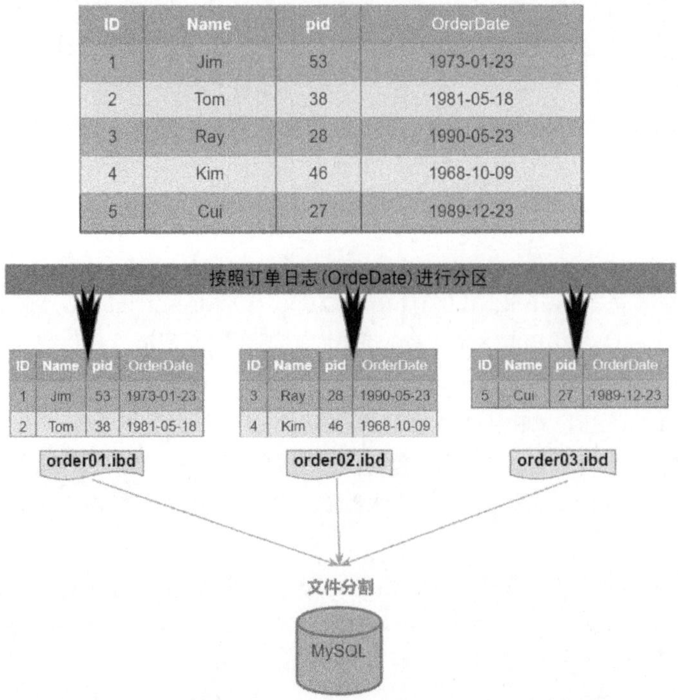

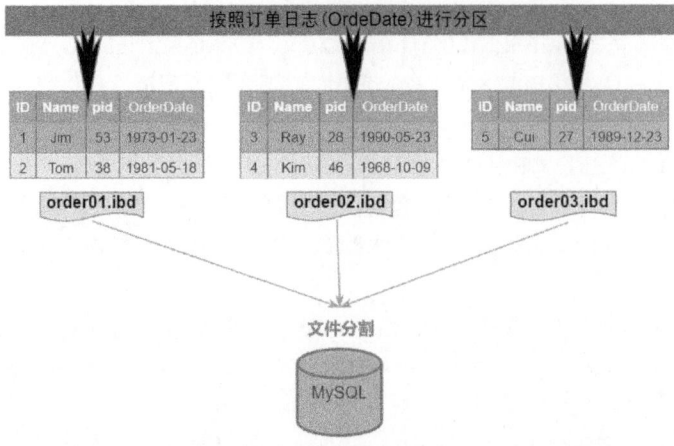

图 9-12　MySQL 中表分区实现

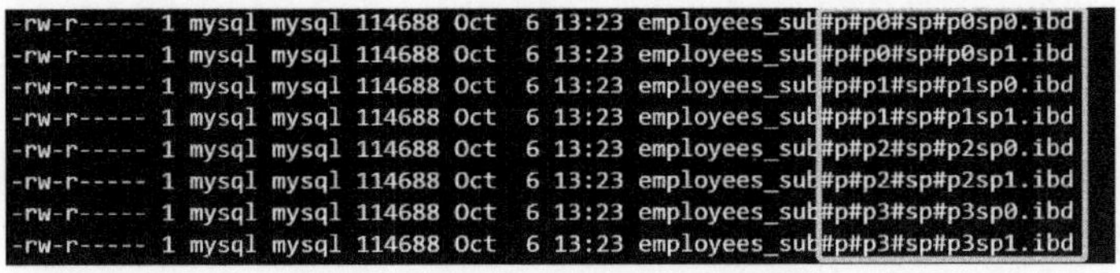

图 9-13　MySQL 表分区物理文件

在 MySQL 中，在下列场景中，分区可以起到非常大的作用。
- 当表非常大以致无法全部放在内存中，或者只在表的最后部分有热点数据，其他都是历史数据时，分区非常有用。
- 分区表的数据更容易维护，如批量删除大量数据时可以使用清除整个分区的方式。另外，还可以对一个独立分区进行优化、检查、修复等操作。
- 分区表的数据可以分布在不同的物理设备上，从而高效地利用多个硬件设备。
- 可以使用分区表来避免某些特殊的瓶颈，如 InnoDB 的单个索引的互斥访问、Ext3 文件系统的 inode 锁竞争等。
- 如果有需要，则可以备份或恢复独立的分区，这在处理非常大的数据集的场景下效果非常好。
- 分区可以优化查询，当 WHERE 子句中包含分区列时，可以只使用必要的分区来提高查询效率。同时，在涉及 SUM( ) 和 COUNT( ) 这类聚合函数的查询时，可以在每个分区上并行处理，最终只需要汇总所有分区得到的结果。

2. 分区类型

在 MySQL 8.0 版本中，支持 6 种分区类型，包括 RANGE（范围）分区、LIST（列表）分区、HASH（哈希）分区、KEY（关键字）分区、COLUMNS（列）分区以及子分区（复合分区）。下面具体介绍这 6 种分区类型。

(1) RANGE 分区

RANGE 分区是基于列值的连续范围来划分数据的。每个分区包含一定范围的值，范围之间不重叠。例如，可以根据日期列来分区，使得每个月的数据存储在不同的分区中。RANGE 分区特别适用于那些可以预测数据范围并且查询经常基于范围过滤的场景。RANGE 类型分区表创建命令行示例如下。

```
# 整型范围分区
mysql>CREATE TABLE employees (
    id INT NOT NULL,
    job_code INT NOT NULL,
    store_id INT NOT NULL
)PARTITION BY RANGE (store_id) (
    PARTITION p0 VALUES LESS THAN (6),
    PARTITION p1 VALUES LESS THAN (11),
    PARTITION p2 VALUES LESS THAN (16),
    PARTITION p3 VALUES LESS THAN MAXVALUE
);
# 时间范围分区
mysql>CREATE TABLE quarterly_report_status (
    report_id INT NOT NULL,
    report_status VARCHAR(20) NOT NULL,
    report_updated TIMESTAMP NOT NULL DEFAULT CURRENT_TIMESTAMP ON UPDATE CURRENT_TIMESTAMP
)PARTITION BY RANGE ( UNIX_TIMESTAMP(report_updated) ) (
    PARTITION p0 VALUES LESS THAN ( UNIX_TIMESTAMP('2021-01-01 00:00:00') ),
    PARTITION p8 VALUES LESS THAN ( UNIX_TIMESTAMP('2021-10-01 00:00:00') ),
    PARTITION p9 VALUES LESS THAN (MAXVALUE)
);
```

(2) LIST 分区

LIST 分区类似于 RANGE 分区，但它是基于列值匹配预定义列表中的值来进行分区的。每个分区包含列表中的一个或多个值。LIST 分区适用于那些数据值可以预知并且查询经常基于这些特定值的场景。LIST 类型分区表创建命令行示例如下。

```
mysql>CREATE TABLE employees (
    id INT NOT NULL,
    job_code INT,
    store_id INT
)PARTITION BY LIST(store_id) (
    PARTITION pNorth VALUES IN (3,5,6,9,17),
    PARTITION pEast VALUES IN (1,2,10,11,19,20),
    PARTITION pWest VALUES IN (4,12,13,14,18),
    PARTITION pCentral VALUES IN (7,8,15,16)
);
```

(3) HASH 分区

HASH 分区基于用户定义的表达式的返回值进行分区。MySQL 服务提供哈希函数，该函数将表达式的结果映射到分区。HASH 分区适用于确保数据均匀分布到各个分区的场景，从而提高查询性能。它特别适用于那些无法预测数据分布或无法有效使用 RANGE 或 LIST 分区的场景。哈希类型分区表创建命令行示例如下。

```
mysql>CREATE TABLE employees (
    id INT NOT NULL,
    job_code INT,
    store_id INT
)PARTITION BY HASH(store_id)PARTITIONS 4;
# 函数处理
mysql>CREATE TABLE employees (
    id INT NOT NULL,
    hired DATE NOT NULL DEFAULT '1970-01-01',
    job_code INT,
    store_id INT
)PARTITION BY HASH( YEAR(hired) )PARTITIONS 4;
# 线性哈希
mysql>CREATE TABLE employees (
    id INT NOT NULL,
    hired DATE NOT NULL DEFAULT '1970-01-01',
    job_code INT,
    store_id INT
)PARTITION BY LINEAR HASH( YEAR(hired) )PARTITIONS 4;
```

说明：MySQL 也支持线性哈希（LINEAR HASH），与常规哈希不同的是，线性哈希使用线性的 2 次幂算法，而常规哈希使用哈希函数值的模，也就是通过 2 的取模方式。例如，数字 137 的二进制表示为 10001001，将这种二进制表示写成 2 次幂的和的形式，令次幂高的排在前面，可得到如下表达式：$137 = 2^7 + 2^3 + 2^0$。

线性哈希分区的优缺点如下。
- 优点。分区的添加、删除、合并和拆分速度都要快得多，这在处理包含大量数据的表时非常有用。
- 缺点。与使用常规哈希分区获得的数据分布相比，数据不太可能在分区之间均匀分布。

(4) KEY 分区

KEY 分区类似于 HASH 分区，但它使用 MySQL 服务器提供的哈希函数基于列的值进行分区。KEY 分区要求分区列必须是整数类型。它适用于那些基于唯一键或主键进行分区的场景，因为这些键通常具有均匀的分布。KEY 分区表创建命令行示例如下。

```
mysql>CREATE TABLE tm1 (
        s1 CHAR(32)
)PARTITION BY KEY(s1)PARTITIONS 10;
```

说明：物理文件会生成 10 个 tm1 的分区文件（tm1#p#p0.ibd～tm1#p#p9.ibd）。

(5) COLUMNS 分区

COLUMNS 分区是 RANGE 和 LIST 分区的扩展，它允许根据多个列的值进行分区。这使得可以基于复合条件来定义分区键，而不仅仅是单一列的值。COLUMNS 分区在 MySQL 8.0 及更高版本中可用，它提供了更大的灵活性，特别适用于那些需要基于多个列进行分区的场景。COLUMNS 分区表创建命令行示例如下。

```
mysql>CREATE TABLE prt_column_range (
        a INT,
        b INT,
        c CHAR(3),
        d INT
    )
    PARTITION BY RANGE COLUMNS(a,d,c) (
        PARTITION p0 VALUES LESS THAN (5,10,'ggg'),
        PARTITION p1 VALUES LESS THAN (10,20,'mmm'),
        PARTITION p2 VALUES LESS THAN (15,30,'sss'),
        PARTITION p3 VALUES LESS THAN (MAXVALUE,MAXVALUE,MAXVALUE)
);
```

说明：KEY 分区键支持 TINYINT、SMALLINT、MEDIUMINT、INT（INTEGER）、BIGINT、DATE、DATETIME、CHAR、VARCHAR、BINARY 和 VARBINARY；KEY 分区键不支持 TEXT、BLOB、DECIMAL 和 FLOAT。

（6）子分区（复合分区）

子分区（也称为复合分区）允许在主分区的基础上进一步划分数据。每个主分区可以包含一个或多个子分区，子分区可以使用与主分区相同的分区类型（RANGE、LIST、HASH、KEY 或 COLUMNS）。子分区可以提高管理的灵活性。例如，可以在一个主分区内使用不同的子分区策略来优化查询性能或简化数据管理。

例如，ts 表有 3 个 RANGE 分区。每个分区（p0、p1 和 p2）被进一步划分为 3 个子分区。实际上，整个表被划分为 3×3＝9 个分区。

范围分区+哈希子分区表创建命令行示例如下。

```
# 范围分区+哈希子分区：
mysql>CREATE TABLE ts1 (id INT, purchased DATE)
        PARTITION BY RANGE( YEAR(purchased) )
        SUBPARTITION BY HASH( TO_DAYS(purchased) )
        SUBPARTITIONS 3 (
            PARTITION p0 VALUES LESS THAN (1990),
            PARTITION p1 VALUES LESS THAN (2000),
            PARTITION p2 VALUES LESS THAN MAXVALUE
);

# LIST 分区 + KEY 子分区
mysql>CREATE TABLE ts2(
        id INT NOT NULL,
        store_id INT
)PARTITION BY LIST(store_id)
        SUBPARTITION BY KEY( store_id )
        SUBPARTITIONS 2 (
        PARTITION pNorth VALUES IN (3,5,6,9,17),
        PARTITION pEast VALUES IN (1,2,10,11,19,20),
        PARTITION pWest VALUES IN (4,12,13,14,18),
        PARTITION pCentral VALUES IN (7,8,15,16)
);
```

图 9-14 所示为上述子分区创建命令在底层生成的物理文件。ts1 表会生成 9 个物理分区文件，p0、p1 和 p2 各 3 个文件。ts2 表生成 8 个物理分区文件，KEY 值 pNorth、pEast、pWest 和 pCentral 各两个文件。

说明： 只能对 RANGE 和 LIST 分区划分子分区，不能对 HASH 和 KEY 分区划分子分区。

在选择分区策略时，应该根据应用程序的需求、数据库的工作负载以及数据的特性来决定。合理的分区策略可以提高查询性能、简化数据管理和备份过程。同时，也应该考虑分区对系统维护、备份恢复以及数据迁移等方面的影响。

```
Table ts1:
-rw-r----- 1 mysql mysql 114688 May  4 21:48 ts1#p#p0#sp#p0sp0.ibd
-rw-r----- 1 mysql mysql 114688 May  4 21:48 ts1#p#p0#sp#p0sp1.ibd
-rw-r----- 1 mysql mysql 114688 May  4 21:48 ts1#p#p0#sp#p0sp2.ibd
-rw-r----- 1 mysql mysql 114688 May  4 21:48 ts1#p#p1#sp#p1sp0.ibd
-rw-r----- 1 mysql mysql 114688 May  4 21:48 ts1#p#p1#sp#p1sp1.ibd
-rw-r----- 1 mysql mysql 114688 May  4 21:48 ts1#p#p1#sp#p1sp2.ibd
-rw-r----- 1 mysql mysql 114688 May  4 21:48 ts1#p#p2#sp#p2sp0.ibd
-rw-r----- 1 mysql mysql 114688 May  4 21:48 ts1#p#p2#sp#p2sp1.ibd
-rw-r----- 1 mysql mysql 114688 May  4 21:48 ts1#p#p2#sp#p2sp2.ibd

Table ts2:
-rw-r----- 1 mysql mysql 114688 May  4 21:48 ts2#p#pcentral#sp#pcentralsp0.ibd
-rw-r----- 1 mysql mysql 114688 May  4 21:48 ts2#p#pcentral#sp#pcentralsp1.ibd
-rw-r----- 1 mysql mysql 114688 May  4 21:48 ts2#p#peast#sp#peastsp0.ibd
-rw-r----- 1 mysql mysql 114688 May  4 21:48 ts2#p#peast#sp#peastsp1.ibd
-rw-r----- 1 mysql mysql 114688 May  4 21:48 ts2#p#pnorth#sp#pnorthsp0.ibd
-rw-r----- 1 mysql mysql 114688 May  4 21:48 ts2#p#pnorth#sp#pnorthsp1.ibd
-rw-r----- 1 mysql mysql 114688 May  4 21:48 ts2#p#pwest#sp#pwestsp0.ibd
-rw-r----- 1 mysql mysql 114688 May  4 21:48 ts2#p#pwest#sp#pwestsp1.ibd
```

图 9-14　MySQL 子分区生成文件

3. 分区表操作支持的 SQL 语句

MySQL 中的分区表支持大部分的通用 SQL 语句，如 SELECT、INSERT、UPDATE、DELETE 等。然而，由于分区表在物理存储层面上的特殊性，因此一些 SQL 操作在分区表上可能会有一些特定的行为或限制。下面介绍在分区表上的 SQL 操作。

（1）通过 SELECT 语句指定分区

通过 SELECT 语句指定分区，可以实现对特定分区的访问。这样可以对特定分区进行查询和统计。指定分区的命令行示例如下。

```
mysql> SELECT * FROM ts PARTITION (p1,p2);
```

（2）通过 DROP PARTITION 语句删除分区

在 MySQL 中，DROP PARTITION 语句可用于删除一个或多个 RANGE 或 LIST 分区，从而删除这些分区中的数据并移除相应的分区文件。这个操作会改变表的结构，因为它减少了表中的分区数量。需要注意的是，DROP PARTITION 不适用于使用 NDB 存储引擎的表，并且不支持 IF EXISTS 子句来避免当分区不存在时的错误。删除分区表的命令行示例如下。

```
#示例分区表 t1
mysql>CREATE TABLE t1 (
        id INT,
        year_col INT
)PARTITION BY RANGE (year_col) (
        PARTITION p0 VALUES LESS THAN (1991),
        PARTITION p1 VALUES LESS THAN (1995),
        PARTITION p2 VALUES LESS THAN (1999)
```

```
);
# 通过 DROP PARTITION 语句删除 p0 和 p1 分区
ALTER TABLE t1 DROP PARTITION p0, p1;
```

（3）通过 TRUNCATE PARTITION 语句删除分区中的所有行

在 MySQL 中，TRUNCATE PARTITION 语句用于删除一个分区中的所有行，而不删除分区本身。这意味着分区结构保持不变，只是分区内的数据被清空。与 DROP PARTITION 不同，TRUNCATE PARTITION 不会改变表的结构，也不会删除分区文件。TRUNCATE PARTITION 语句命令行示例如下。

```
mysql>ALTER TABLE t1 TRUNCATE PARTITION p0;
mysql>ALTER TABLE t1 TRUNCATE PARTITION p1, p2;
mysql>ALTER TABLE t1 TRUNCATE PARTITION ALL;    # 所有表分区
# 需要验证是否删除了行,因为有时候会存在不准确情况
mysql> SELECT TABLE_SCHEMA, TABLE_NAME, PARTITION_NAME, TABLE_ROWS,
       AVG_ROW_LENGTH, DATA_LENGTH
    FROM INFORMATION_SCHEMA.PARTITIONS;
```

TRUNCATE PARTITION 语句同样不适用于使用 NDB 存储引擎的表。此外，这个操作也是不可逆的，一旦执行，分区内的数据将永久丢失，因此在执行前务必做好数据备份。此外，TRUNCATE PARTITION 也不支持 IF EXISTS 子句，所以如果指定的分区不存在，将会返回一个错误。

（4）通过 COALESCE PARTITION 减少分区数量

COALESCE PARTITION 是 MySQL 中用于合并相邻分区的命令，通常用于减少按 HASH 或 KEY 分区的表的分区数量。这个命令合并相邻的分区，从而减少表中的分区总数。注意，COALESCE PARTITION 并不适用于 RANGE 或 LIST 分区类型的表。COALESCE PARTITION 命令行示例如下。

```
mysql>CREATE TABLE t2 (
    name VARCHAR (30),
    started DATE
)PARTITION BY HASH( YEAR(started) )PARTITIONS 6;

# 要将 t2 使用的分区数量从 6 减少到 4,可以使用如下命令
mysql>ALTER TABLE t2 COALESCE PARTITION 4;
```

合并分区可能会导致数据的重新分布和整理，这可能需要一些时间，并且可能影响正在进行的查询。COALESCE PARTITION 也不支持 IF EXISTS 子句。

（5）通过 REORGANIZE PARTITION 重组分区

REORGANIZE PARTITION 用于添加、删除、合并或拆分分区，而无须重建整个表。这通常涉及创建一个新的分区表，将旧表的数据复制到新表，然后删除旧表。这个过程并不直接通过单个分区操作完成，需要所有分区重新定义。REORGANIZE PARTITION 命令行示例如下。

```
mysql>CREATE TABLE members (
    id INT,
    create_dt DATE
)PARTITION BY RANGE( YEAR(create_dt) ) (
    PARTITION p0 VALUES LESS THAN (1980),
    PARTITION p1 VALUES LESS THAN (2020),
    PARTITION p2 VALUES LESS THAN (2021)
```

```
);

# 将 p0 分区分成 n0 和 n1 两个分区
mysql> REORGANIZE PARTITION p0 INTO (
           PARTITION n0 VALUES LESS THAN (1970),
           PARTITION n1 VALUES LESS THAN (1980)
);
```

在执行 REORGANIZE PARTITION 操作时，MySQL 会尝试保持数据的完整性，并且不会丢失数据。然而，这个过程可能需要相当长的时间，特别是当表中的数据量非常大时。此外，这个过程可能会中断对旧表的访问，因此最好在低峰时段执行，并且确保有足够的磁盘空间来容纳两个表的数据。

(6) REMOVE PARTITIONING

在 MySQL 中，REMOVE PARTITIONING 不是一个直接可用的删除表的分区的命令。实际上，如果想删除一个表的分区，就需要重新组织表的结构，这通常涉及创建一个新的非分区表，将旧分区表的数据复制到新表，然后删除旧表。这个过程并不直接删除分区，而是删除了表的分区结构，并保留所有数据在一个非分区表中。REMOVE PARTITIONING 命令行示例如下。

```
mysql>CREATE TABLE e (
        id INT NOT NULL,
        fname VARCHAR(30),
        lname VARCHAR(30)
    )
    PARTITION BY RANGE (id) (
        PARTITION p0 VALUES LESS THAN (50),
        PARTITION p1 VALUES LESS THAN (100),
        PARTITION p2 VALUES LESS THAN (150),
        PARTITION p3 VALUES LESS THAN (MAXVALUE)
);

mysql>INSERT INTO e VALUES (1669, "Jim", "Smith"), (337, "Mary", "Jones"), (16, "Frank", "White"), (2005, "Linda", "Black");

mysql>ALTER TABLE e REMOVE PARTITIONING;

mysql>SHOW CREATE TABLE e;
    CREATE TABLE `e` (
       `id` int NOT NULL,
       `fname` varchar(30) COLLATE utf8mb4_bin DEFAULT NULL,
       `lname` varchar(30) COLLATE utf8mb4_bin DEFAULT NULL
    ) ENGINE=InnoDB;
```

这个过程可能需要相当长的时间，并且会消耗大量的磁盘空间，因为同时在维护两个表。此外，复制数据期间可能会影响旧表的性能。在执行这些操作之前，务必备份旧表的数据，以防在复制或删除过程中出现问题。如果只是想要简化表的结构，而不是完全删除分区，则还可以考虑使用 TRUNCATE PARTITION 语句来清空一个或多个分区的内容，而不是删除分区本身。这会保留分区结构，但删除分区中的所有数据。

（7）EXCHANGE PARTITION

EXCHANGE PARTITION 用于将分区表中的一个分区（或子分区）与非分区表之间的数据进行交换。可以将分区中的数据移动到非分区表中，或者将非分区表中的数据移动到分区表中。此操作要求分区表和非分区表具有相同的表结构，但非分区表不能包含分区。EXCHANGE PARTITION 命令行示例如下。

```
mysql>ALTER TABLE e EXCHANGE PARTITION p3 WITH TABLE e2;
```

如果分区存在问题并需要重新分区，或者需要清理旧数据，那么使用 EXCHANGE PARTITION 是一个很好的方案。

（8）其他分区命令

除了上述常用的分区命令以外，还有一些不太常用但同样重要的分区命令。表 9-3 列出了这些不常用的分区命令。当需要执行特定功能时，这些命令将非常有用。

表 9-3 MySQL 其他分区命令

分区命令	说 明
REBUILD PARTITION	重建分区。这与删除存储在分区中的所有记录，然后重新插入它们具有相同的效果
OPTIMIZE PARTITION	优化分区。若在一个分区中删除了大量的行，那么优化分区会回收任何未使用的空间，并整理分区数据文件的碎片
ANALYZE PARTITION	用于更新分区或子分区的统计信息
REPAIR PARTITION	修复损坏的分区
CHECK PARTITION	检查分区

说明：InnoDB 存储引擎目前不支持分区优化。ALTER TABLE ... OPTIMIZE PARTITION 会重新构建和分析整个表。

虽然分区表支持大部分的 SQL 语句，但在使用时仍然需要考虑其特殊性和限制，以确保能够有效地利用分区带来的性能和管理优势。

4. 分区限制

MySQL 对分区表有一些限制，了解这些限制对于有效地使用 MySQL 分区表至关重要。在设计分区策略时，务必考虑这些限制，并确保它们符合应用开发需求。下面介绍限制内容。

（1）分区表不允许使用的对象和内容

分区表在 MySQL 中有一些结构和使用上的限制。分区表不允许使用以下对象和内容。

- 分区表不支持直接在存储过程或函数中作为操作对象。同时，与分区表相关的操作不支持使用可加载的函数或插件。
- 在分区表的操作中，不允许声明变量或用户变量。
- 对于使用 InnoDB 存储引擎的分区表，不支持外键约束。同时，分区表不支持 FULLTEXT 索引，这意味着不能在分区表上创建全文搜索索引。在键分区中，不支持列索引的前缀。
- MySQL 不允许对临时表进行分区。
- 分区键必须是整数列或能够解析为整数的表达式。不允许使用 ENUM 列作为分区键的表达式。

（2）SQL 模式对分区表的影响

对于 SQL 模式对分区表的影响，当创建分区表时，该表的行为是基于创建时的 SQL 模式来确定的。如果在创建分区表后更改了 SQL 模式,那么这些表的行为可能会发生重大变化,这可能导致数据损坏或丢失。因此,强烈建议在创建分区表之后不要更改服务器的 SQL 模式。其次,当使用分区表进行复制时,主节点和从节点上的 SQL 模式必须保持一致。如果主节点和从节点上的 SQL 模式不同,就可能会导致复制过程中出现问题。

图 9-15 所示为在 RANGE 分区下根据位于给定范围内的列值将行分配给分区。当 SQL 模式设置为 NO_UNSIGNED_SUBTRACTION 时,MySQL 不允许对无符号整数(UNSIGNED)和有符号整数(SIGNED)进行减法运算。如果查询中涉及这样的运算,并且结果可能导致无符号整数变成负数,那么 MySQL 将返回一个错误。

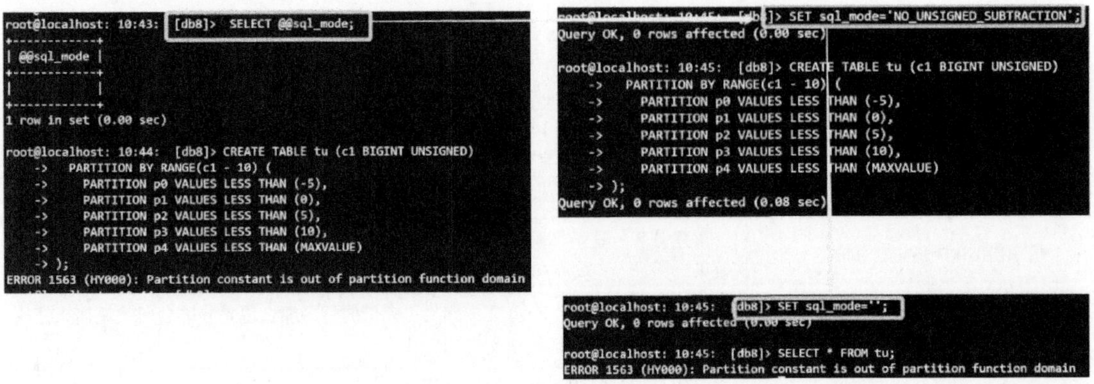

图 9-15 MySQL 中 SQL 模式对分区表的影响

当使用 MySQL 分区表时,应谨慎处理服务器 SQL 模式,并在创建分区表后避免更改它。同时,确保在复制过程中主节点和从节点使用相同的 SQL 模式,以避免潜在的问题。

(3)分区键、主键和唯一键的关系

分区表上的主键和唯一键必须包含分区键。这是因为分区键用于将数据分布到不同的分区中,从而确保每个分区内的数据唯一性。同样,分区表达式中使用的所有列都必须是表可能拥有的每个唯一键的一部分。换句话说,如果分区表达式涉及多个列,那么这些列的组合也必须是唯一键的一部分,以确保分区内部的数据唯一性。

如果在创建分区表时,表的主键或唯一键约束中不包含分区键,分区键部分在表的主键或唯一键约束中未找到,那么 MySQL 将返回相应的错误信息,提示无法创建分区表。以下是创建分区表过程中,唯一键不包含分区键时可能出现的错误信息示例。

```
mysql>CREATE TABLE t1 (
    col1 INT NOT NULL,
    col2 DATE NOT NULL,
    col3 INT NOT NULL,
    col4 INT NOT NULL,
    UNIQUE KEY (col1, col2)
)PARTITION BY HASH(col3)PARTITIONS 4;
ERROR 1503 (HY000): A PRIMARY KEY must include all columns in the table's
partitioning function (prefixed columns are not considered)
```

如果表没有定义主键或唯一键,那么分区键和分区表达式的上述限制不适用。在这种情况下,只要列的类型与分区类型兼容,就可以在分区表达式中使用表中的任何一列或多列。需要注意的是,一旦分区表被创建,后续向该表添加唯一键将受到限制。具体来说,如果尝试添加一个唯一键,而这个键不包含分区表达式中使用的所有列,那么将会导致错误。以下是一个尝试这样做时出现的错误信息示例。

```
mysql>CREATE TABLE t_no_pk (c1 INT, c2 INT)
        PARTITION BY RANGE(c1) (
            PARTITION p0 VALUES LESS THAN (10),
            PARTITION p1 VALUES LESS THAN (20),
            PARTITION p2 VALUES LESS THAN (30),
            PARTITION p3 VALUES LESS THAN (40)
);

mysql> ALTER TABLE t_no_pk ADD PRIMARY KEY(c2);
ERROR 1503 (HY000): A PRIMARY KEY must include all columns in the table's
partitioning function (prefixed columns are not considered).
```

在分区键创建之后,不要尝试去替换和更改主键与唯一键,因为分区键决定了数据在分区表中的分布方式。更改分区键可能会导致数据不一致和完整性问题,需要谨慎处理。

(4) 存储引擎对分区的限制

在 MySQL 8.0 中,InnoDB 存储引擎是默认的存储引擎,并且它支持本地分区。InnoDB 分区表在物理存储级别上提供了数据的划分,这对于管理和查询大量数据特别有用。通过使用分区,可以将数据分散到多个物理子表中,但逻辑上它们仍然表现为一个表。

与此同时,MyISAM 存储引擎在 MySQL 8.0 中仍然可用,但它不支持分区。如果尝试在 MyISAM 表上创建分区,那么将会收到一个错误,指出该存储引擎不支持分区。

NDB 存储引擎是 MySQL Cluster 的一部分,它也支持分区。NDB 的分区限制包括要求表具有显式的主键,并且表分区表达式中列出的所有列都必须是主键的一部分。这是一个强制性要求,因为 NDB 使用主键来确保数据的唯一性和分布的均匀性。在 MySQL 8.0 版本中,NDB 表的每个分区最大容量为 16GB,但在 MySQL 8.0 中,这个限制已经增加到 128TB,这是一个显著的提升,允许 NDB 存储引擎处理更大规模的数据集。

当在 MySQL 中开发使用分区时,了解不同存储引擎的分区支持和限制是非常重要的,这有助于确保数据库设计的有效性和性能。

(5) 升级分区表

在 MySQL 中,分区表的升级通常涉及表结构的更改或数据库版本的升级。对于使用 KEY 分区且基于非 InnoDB 存储引擎的表,从 MySQL 5.7 升级到 MySQL 8.0 版本时,需要确认兼容性。这是因为从 MySQL 5.7 开始,官方推荐使用 InnoDB 存储引擎,并且某些分区类型和功能可能仅在 InnoDB 上得到支持。

如图 9-16 所示,使用非 InnoDB 存储引擎的分区表不能从 MySQL 5.7 版本及以前版本升级到 MySQL 8.0 及以上版本。此时,必须转换成 InnoDB 存储引擎表,使用 ALTER TABLE ... ENGINE 语句变更表引擎。

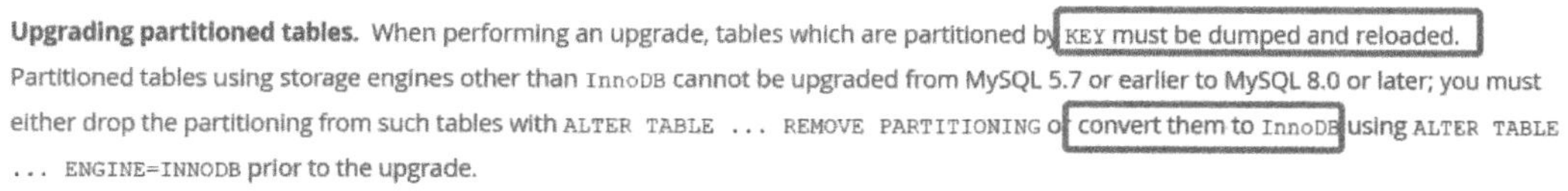

Upgrading partitioned tables. When performing an upgrade, tables which are partitioned by KEY must be dumped and reloaded. Partitioned tables using storage engines other than InnoDB cannot be upgraded from MySQL 5.7 or earlier to MySQL 8.0 or later; you must either drop the partitioning from such tables with ALTER TABLE ... REMOVE PARTITIONING or convert them to InnoDB using ALTER TABLE ... ENGINE=INNODB prior to the upgrade.

图 9-16 MySQL 分区表升级

随着 MySQL 版本的升级，还需要注意新版本可能对分区行为做出的改变，确保分区表在新的版本中仍然能够正常工作。

（6）分区表对 NULL 数据的处理

MySQL 中的分区不会禁止将 NULL 作为分区表达式的值，无论它是列值还是用户提供的表达式的值。尽管允许使用 NULL 作为表达式的值，但必须记住 NULL 不是一个数字，这一点很重要。MySQL 的分区将 NULL 视为小于任何非 NULL 值。但这些 NULL 值会根据分区类型的不同而被特别处理。

- RANGE 分区：当分区表达式的值为 NULL 时，MySQL 会将该行放入最低范围的分区中，因为 NULL 被视为小于任何非 NULL 值。
- LIST 分区：如果 LIST 分区的定义中没有显式包括 NULL 作为一个值，那么任何导致分区表达式结果为 NULL 的插入操作都会被拒绝。LIST 分区要求分区表达式的结果必须严格匹配定义中的某个值。
- HASH 和 KEY 分区：对于 HASH 和 KEY 分区，如果分区表达式的结果为 NULL，那么 MySQL 会将这个 NULL 值视为 0 来处理。这意味着 NULL 值会基于 0 的哈希值来确定分区。

在实际开发应用中，当使用分区表时，应该考虑到数据的实际情况，包括是否可能包含 NULL 值，以及这些 NULL 值应该如何处理。如果不考虑 NULL 值的处理方式，则可能会导致数据分布不均匀，或者在某些情况下，甚至无法插入数据。

（7）SQL 语句对分区表的影响

SQL 语句对分区表的影响程度主要取决于语句的类型和分区表的配置。以下是一些常见 SQL 语句对分区表可能产生的影响。

- 分区表 DDL 的操作实现依赖于文件系统操作，这意味着这些操作的速度受到文件系统类型和特征、磁盘速度、交换空间、操作系统的文件处理效率、MySQL 服务器选项和与文件处理相关的变量等因素的影响。特别是，应该确保启用了 large_files_support，并正确设置了 open_files_limit。启用 innodb_file_per_table 可以使涉及 InnoDB 表的分区和重分区操作更加高效。
- 对于 MySQL 来说，优化 SQL 查询的基本原则是确保查询能够高效地使用索引。对于分区表，优化的关键在于避免扫描不包含匹配值的分区，因为与扫描整个表的所有分区相比，只扫描包含匹配行的分区可以大大减少查询所需的时间和资源。
- 对于 DML 语句（如 SELECT、DELETE 和 UPDATE），MySQL 会根据分区键的值确定需要访问的分区，并仅锁定和操作这些分区，而不是整个表。在执行这些操作时，相关的表文件句柄可能会被缓存在内存中以提高性能。然而，如果同时打开的文件句柄过多，则可能会耗尽系统资源，影响性能。因此，建议单个表不应有过多的分区，以避免过多的文件句柄操作。具体的分区数量限制可能需要根据硬件和 MySQL 的配置来确定。建议分区文件不要超过 3000 个。
- 在对分区表进行范围查询时，执行计划必走第一个分区，因为第一个分区文件里包含所有基础信息，所以建议把第一个分区作为空的分区。图 9-17 所示为在 create_at 大于某个时间的时候执行计划扫描第一个分区和第四个分区。

在应用开发中，SQL 语句对分区表的影响主要体现在查询性能、数据修改效率以及特定操作的限制上。为了充分利用分区带来的性能优势，需要合理设计分区策略，选择合适的分区键，并考虑分区带来的限制和约束。

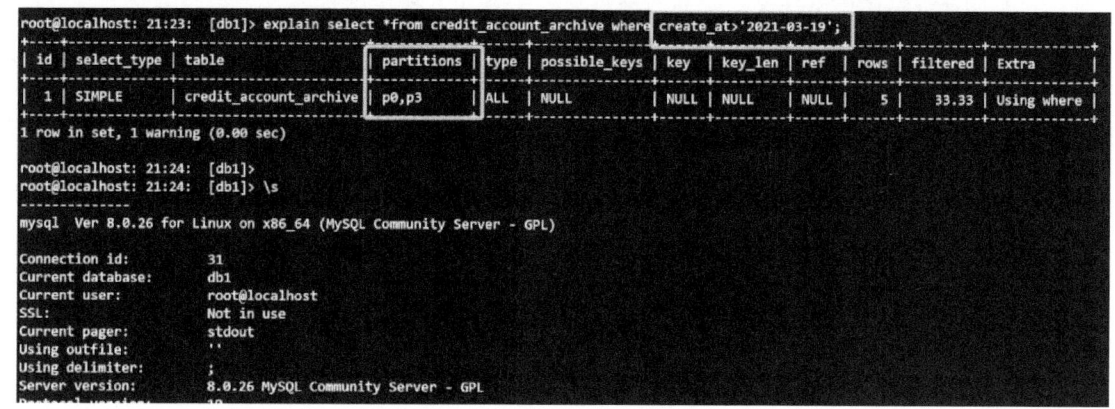

图 9-17 MySQL 分区表查询

5. 总结

想要使用 MySQL 分区表，需要仔细规划和优化，根据实际需求选择合适的分区键、类型和策略，并且监控分区表的使用情况并维护好，以确保它们能够满足业务需求并提高性能。分区表使用注意事项如下。

- 分区键的重要性。选择合适的分区键对于分区表的性能至关重要。如果查询条件不包含分区键，那么查询可能需要扫描所有分区，从而无法充分利用分区带来的性能优势。
- 数据分布不均匀。如果数据在分区之间分布不均匀，某些分区可能比其他分区大得多或小得多，就可能导致某些查询性能下降。因此，需要根据数据的实际访问模式和数据量来选择合适的分区策略。
- 文件描述符的限制。当分区表第一次被访问时，无论该次访问需要操作多少个分区，都需要访问该分区表上所有分区。当分区表上分区数量较大时，可能会因为打开文件数量超过参数 open_file_limit 限制而出错（操作系统支持 65535 个文件）。
- 对分区表的 DDL 操作需要锁定整个表，这会影响所有分区的操作。因此，在对分区表进行 DDL 操作时需要谨慎，并确保在低峰时段进行。
- 分区表的数量不宜过多。最大分区数目建议不要超过 1024 个，对于在整张表上分区的表，不易一次性创建太多分区表。例如，可以实现每 7 天、10 天或一个月创建一次分区表的方式。
- 分区表的清理。定期清理分区表中的数据是很重要的，以确保表的大小和性能得到维护。但是，在清理数据之前，应该确保没有正在进行的查询或事务，以避免数据不一致或锁定问题。

9.3.5 全文索引的使用

全文索引是搜索引擎中的关键技术，它可以显著提高搜索速度和效率。在构建全文索引时，首先定义一个词库，然后遍历文章，查找每个词条（term）出现的频率和位置。接着，将这些频率和位置信息按照词库的顺序进行归纳与组织，从而形成一个以词库为目录的索引。这样，当需要查找某个词时，搜索引擎可以快速定位到该词在文章中出现的位置。

1. 全文索引介绍

在 MySQL 数据库中，当需要在文本字段中进行高效的模糊搜索时，通常会使用全文索引。全文索引是为这种特定场景设计的,它允许数据库在大量文本数据中快速找到包含特定关键词的记录。

MySQL 中的全文索引表与普通的表不同,使用全文索引的表在底层文件系统中会生成以 fts 开头的索引文件。这些文件的存在表明表上已创建全文索引。查看全文索引的命令行示例如下。

```
# 创建表和全文索引
mysql> CREATE TABLE my_stopwords(value VARCHAR(30)) ENGINE = INNODB;
mysql> CREATE FULLTEXT INDEX idx ON table_name(`columns`);

# 通过系统表查询全文索引
mysql> SELECT * FROM INFORMATION_SCHEMA.TABLESPACES_EXTENSIONS WHERE
TABLESPACE_NAME LIKE 'db7/fts%';
+---------------------------------------------------+------------------+
| TABLESPACE_NAME                                   | ENGINE_ATTRIBUTE |
+---------------------------------------------------+------------------+
| db7/fts_00000000000005c4_00000000000002fa_index_1 | NULL             |
| db7/fts_00000000000005c4_00000000000002fa_index_2 | NULL             |
| db7/fts_00000000000005c4_00000000000002fa_index_3 | NULL             |
| db7/fts_00000000000005c4_00000000000002fa_index_4 | NULL             |
| db7/fts_00000000000005c4_00000000000002fa_index_5 | NULL             |
| db7/fts_00000000000005c4_00000000000002fa_index_6 | NULL             |
| db7/fts_00000000000005c4_being_deleted            | NULL             |
| db7/fts_00000000000005c4_being_deleted_cache      | NULL             |
| db7/fts_00000000000005c4_config                   | NULL             |
| db7/fts_00000000000005c4_deleted                  | NULL             |
| db7/fts_00000000000005c4_deleted_cache            | NULL             |
+---------------------------------------------------+------------------+
```

2. 全文索引实现方式和限制

全文索引的工作原理是主要基于倒排索引(Inverted Index)和自然语言处理技术。下面介绍 MySQL 中全文索引实现方式和限制(以 InnoDB 存储引擎为例)。

(1)全文索引设计之倒排索引

全文索引采用了倒排索引设计。倒排索引存储单词列表,对于每个单词,存储该单词出现的文档列表。为了支持邻近搜索,每个单词的位置信息也以字节偏移量的形式存储。这里提到了倒排索引。倒排索引是根据关键词反向得到该关键词的其他所有信息,如该关键词所在的文件、在文件里出现的次数和行数等,这些信息就是用户查找该关键词时要用的信息。

可用不同的数字索引不同的句子,如以下 3 句在文本中是按照 0、1、2 的顺序排列的。

```
0: "I love you"
1: "I love you too"
2: "I dislike you"
```

如果要用单词作为索引,而将句子的位置作为被索引的元素,那么索引就发生了倒置,如下所示。

```
"I": {0,1,2}
"love": {0, 1}
    "you": {0,1,2}"dislike": {2}
```

如果要检索 "I dislike you"，那么可以这样计算：{0, 1, 2} ∩ {0, 1, 2} ∩ {2}。这样就比较好理解了。

（2）全文索引对应的 DOC_ID 和 FTS_DOC_ID 列

InnoDB 存储引擎使用一个称为 DOC_ID 的唯一文档标识符来将全文索引中的单词映射到该单词出现的文档记录中。映射需要索引表上的 FTS_DOC_ID 列。如果没有定义 FTS_DOC_ID 列，那么 InnoDB 存储引擎表会在创建全文索引时自动添加一个隐藏的 FTS_DOC_ID 列。创建 FTS_DOC_ID 列的示例如下。

```
# 创建表
mysql> CREATE TABLE opening_lines (
        id INT UNSIGNED AUTO_INCREMENT NOT NULL PRIMARY KEY,
        opening_line TEXT(500),
        author VARCHAR(200),
        title VARCHAR(200)
        ) ENGINE=InnoDB;
Query OK, 0 rows affected (0.03 sec)
# 创建全文索引
mysql> CREATE FULLTEXT INDEX idx ON opening_lines(opening_line);
Query OK, 0 rows affected, 1 warning (0.26 sec)
Records: 0  Duplicates: 0  Warnings: 1
# 通过 Warnings 信息,查看 FTS_DOC_ID 列添加信息
mysql>SHOW WARNINGS;
+---------+------+------------------------------------------------------+
| Level   | Code | Message                                              |
+---------+------+------------------------------------------------------+
| Warning |  124 | InnoDB rebuilding table to add column FTS_DOC_ID     |
+---------+------+------------------------------------------------------+
1 row in set (0.00 sec)
```

在 InnoDB 存储引擎表创建全文索引时，InnoDB 存储引擎会自动为表添加一个隐藏的 FTS_DOC_ID 列，并在该列上创建一个唯一键（通常名为 FTS_DOC_ID_INDEX）。如果用户想在表中自定义 FTS_DOC_ID 列，那么该列必须定义为 BIGINT UNSIGNED NOT NULL 类型，并且列名必须严格写为 FTS_DOC_ID（全大写）。用户还需要确保 FTS_DOC_ID 列中的值是唯一的，并且不会重用或包含空值，以确保全文索引的正确性和性能。创建 FTS_DOC_ID 列和 FTS_DOC_ID_INDEX 索引的命令行示例如下。

```
# 创建表时指定 FTS_DOC_ID 列
mysql> CREATE TABLE `opening_lines` (
  `FTS_DOC_ID` bigint unsigned NOT NULL AUTO_INCREMENT,
  `opening_line` text COLLATE utf8mb4_bin,
  `author` varchar(200) COLLATE utf8mb4_bin DEFAULT NULL,
  `title` varchar(200) COLLATE utf8mb4_bin DEFAULT NULL,
  PRIMARY KEY (`FTS_DOC_ID`)
) ENGINE=InnoDB;
# 创建 FTS_DOC_ID_INDEX 索引
mysql>CREATE UNIQUE INDEX FTS_DOC_ID_INDEX on opening_lines(FTS_DOC_ID);
```

除此之外，FTS_DOC_ID_INDEX 不能定义为降序索引，因为 InnoDB 存储引擎的 SQL 解析器不使用降序索引。为了避免重新构建表，在删除全文索引时需要保留 FTS_DOC_ID 列。

（3）全文索引缓存

当插入文档时，InnoDB 存储引擎会对其进行标记，并从文档中提取单词和其他相关数据以构建全文索引。这个过程可能会涉及多个小的数据修改操作，尤其是在处理大型文档时。为了减少这些操作对系统性能的影响，InnoDB 存储引擎使用了一个称为全文索引缓存的内存结构来临时存储这些待写入的数据。这个缓存会在适当的时候（例如，当缓存已满或系统需要释放内存时）批量将这些数据刷新到磁盘上的辅助索引表中。查询全文索引缓存的命令行示例如下。

```
mysql> SELECT * FROM INFORMATION_SCHEMA.INNODB_FT_INDEX_CACHE;
```

图 9-18 所示为全文索引表对应生成的物理文件和缓存文件。

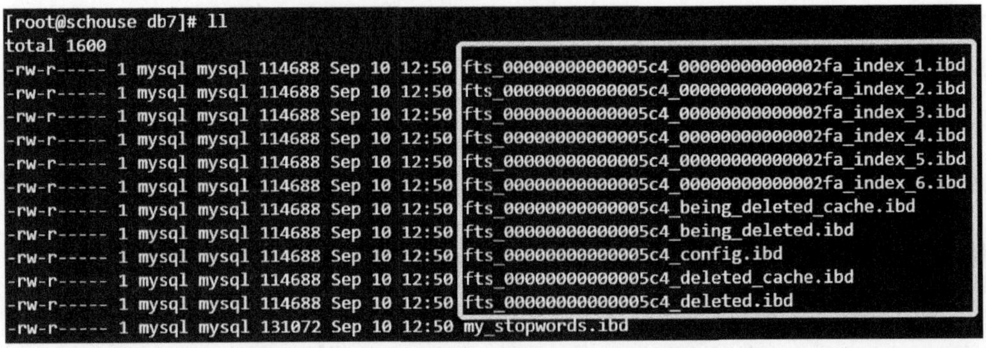

```
[root@schouse db7]# ll
total 1600
-rw-r----- 1 mysql mysql 114688 Sep 10 12:50 fts_00000000000005c4_00000000000002fa_index_1.ibd
-rw-r----- 1 mysql mysql 114688 Sep 10 12:50 fts_00000000000005c4_00000000000002fa_index_2.ibd
-rw-r----- 1 mysql mysql 114688 Sep 10 12:50 fts_00000000000005c4_00000000000002fa_index_3.ibd
-rw-r----- 1 mysql mysql 114688 Sep 10 12:50 fts_00000000000005c4_00000000000002fa_index_4.ibd
-rw-r----- 1 mysql mysql 114688 Sep 10 12:50 fts_00000000000005c4_00000000000002fa_index_5.ibd
-rw-r----- 1 mysql mysql 114688 Sep 10 12:50 fts_00000000000005c4_00000000000002fa_index_6.ibd
-rw-r----- 1 mysql mysql 114688 Sep 10 12:50 fts_00000000000005c4_being_deleted_cache.ibd
-rw-r----- 1 mysql mysql 114688 Sep 10 12:50 fts_00000000000005c4_being_deleted.ibd
-rw-r----- 1 mysql mysql 114688 Sep 10 12:50 fts_00000000000005c4_config.ibd
-rw-r----- 1 mysql mysql 114688 Sep 10 12:50 fts_00000000000005c4_deleted_cache.ibd
-rw-r----- 1 mysql mysql 114688 Sep 10 12:50 fts_00000000000005c4_deleted.ibd
-rw-r----- 1 mysql mysql 131072 Sep 10 12:50 my_stopwords.ibd
```

图 9-18　MySQL 全文索引物理文件和缓存文件

这些全文索引物理文件的说明如下。

- InnoDB 存储引擎使用以 fts_为前缀和以 index_#为后缀（其中#是数字）的辅助索引表来存储全文索引数据。前 6 个索引表（如 index_1 ~ index_6）共同组成了倒排索引，用于快速定位包含特定单词的文档。这些索引表根据单词的逆序和位置信息进行排序与分区，以优化搜索性能。InnoDB 支持并行索引创建，实际使用的线程数由 innodb_ft_sort_pll_degree 配置参数决定。查看当前配置的线程数的命令行示例如下。

```
mysql>SHOW VARIABLES LIKE 'innodb_ft_sort_pll_degree';
+---------------------------+-------+
|Variable_name              |Value  |
+---------------------------+-------+
|innodb_ft_sort_pll_degree  |2      |
+---------------------------+-------+
1 row in set (0.00 sec)
```

- fts_*_deleted 和 fts_deleted_cache：包含已删除但其数据尚未从全文索引中删除的文档的文档 ID（DOC_ID）。fts_deleted_cache 表是 fts_deleted 表的内存版本。
- fts_*_being_deleted 和 fts_being_deleted_cache：包含被删除文档的文档 ID（DOC_ID），这些文档的数据目前正在从全文索引中删除。fts_being_deleted_cache 表是 fts_being_deleted 表的内存版本。
- fts_*_config：存储关于全文索引的内部状态的信息。更重要的是，它存储 FTS_SYNCED_DOC_ID，用于标识已解析并刷新到磁盘的文档。在崩溃后恢复的场景下，FTS_SYNCED_DOC_ID 值用于标识尚未刷新到磁盘的文档，以便可以重新解析文档并将其添加回全文索引缓存。查询系统内存表信息的命令行示例如下。

```
mysql> SELECT * FROM INFORMATION_SCHEMA.INNODB_FT_CONFIG;
```

缓存和批处理行为避免了对辅助索引表的频繁更新，但这可能会在繁忙的插入和更新期间导致并发访问问题。批处理技术还避免了对同一个单词的多次插入，并尽量减少重复条目。不是单独刷新每个单词，而是将相同单词的插入合并并作为单个条目刷新到磁盘，从而在保持辅助索引表尽可能小的同时提高插入效率。表 9-4 所示为全文索引缓存刷新参数。

表 9-4　MySQL 全文索引缓存刷新参数

缓存刷新参数	说明
innodb_ft_cache_size	用于配置全文索引缓存的大小（以每个表为基础），这将影响全文索引缓存刷新的频率
innodb_ft_total_cache_size	为给定实例中的所有表定义一个全局全文索引缓存大小限制

全文索引缓存存储与辅助索引表相同的信息。但是，全文索引缓存只缓存最近插入行的标记化数据。在查询时，已经刷新到磁盘（到辅助索引表）的数据不会返回到全文索引缓存中。直接查询辅助索引表中的数据，并在返回之前将辅助索引表的结果与来自全文索引缓存的结果合并。

（4）全文索引删除处理

当从具有全文索引的表中删除记录时，会涉及特殊的处理机制。直接删除这些记录可能会导致辅助索引表中产生大量的小规模删除操作，这可能会增加对这些表的并发访问时的锁争用。为了避免这种情况，InnoDB 采用了以下策略。

- 当从索引表中删除一条记录时，不会立即从全文索引中移除该记录的索引项。相反，已删除文档的 DOC_ID 会被记录在一个特殊的 FTS_DELETED 表中。
- 在执行全文搜索查询时，InnoDB 会使用 FTS_DELETED 表中的信息来过滤掉那些已经标记为删除的文档，从而确保查询结果的准确性。

这种设计有以下几个优点。

- 删除操作更快：由于不需要立即从全文索引中移除记录的索引项，因此通常可以更快地完成删除操作。
- 减少锁争用：通过减少直接对辅助索引表的删除操作，可减少并发访问时的锁争用，从而提高系统的并发性能。
- 索引大小延迟减少：需要注意的是，尽管记录已被标记为删除，但相应的索引项仍然保留在全文索引中，这意味着索引的大小不会立即减少。

如果清理这些已标记为删除的索引项并减少索引的大小，则可以执行以下操作。

- 在索引表上设置 innodb_optimize_fulltext_only 配置选项为 ON。
- 执行 OPTIMIZE TABLE 命令来重建全文索引。这将触发 InnoDB 重新构建全文索引，同时移除那些在 FTS_DELETED 表中记录的已删除文档的索引项。

通过这种方式，可以平衡删除操作和索引维护之间的性能影响，确保全文搜索功能的效率和准确性。

（5）全文索引事务处理

全文索引在 InnoDB 存储引擎中具有特殊的事务处理特征，这主要与其使用的缓存和批处理行为有关。以下是关于 InnoDB 全文索引事务处理的一些要点。

- 事务提交时的处理：当在 InnoDB 表上进行更新或插入操作时，这些更改并不会立即反映到全文索引上。相反，这些更改会在事务提交时进行处理。这意味着，只有在事务成功提交后，全文索引才会更新，从而确保数据的完整性和一致性。
- 可见性：由于全文索引的更新是在事务提交时进行的，因此，在事务提交之前，全文搜索是无法看到这些未提交的数据的。这确保了全文搜索返回的结果都是已经提交并且一致的数据。
- 批量操作与实时查询：在批量操作的情况下，由于全文索引的更新是在事务提交时进行的，因此可能会出现实时查询无法立即看到新插入或更新的数据的情况。这是因为在事务提交之前，这些更改对全文索引是不可见的。这种行为确保了最终一致性，但可能牺牲了实时性。
- 性能优化：InnoDB 通过使用缓存和批处理来优化全文索引的性能。例如，它使用 FTS Index Cache（全文检索索引缓存）来提高全文检索的性能。这个缓存结构是一个红黑树，根据(word、ilist)进行排序。当有新数据插入时，数据首先会更新到这个缓存中，而不是直接更新到全文索引中。然后，InnoDB 会批量地将这些更改应用到全文索引中，从而提高了性能。

InnoDB 全文索引的事务处理特征确保了数据的一致性和完整性，同时也通过缓存和批处理优化了性能。

（6）全文索引的限制

以下是关于全文索引的限制。
- 全文索引主要用于 InnoDB 和 MyISAM 存储引擎的表。对于 InnoDB，从 MySQL 5.6 版本开始支持全文索引。全文索引只能为 CHAR、VARCHAR 或 TEXT 列创建。
- MySQL 提供了一个内置的全文 ngram 解析器，该解析器支持中文、日文和韩文（CJK）字符；还提供了一个可安装的 MeCab 日文全文解析器插件，但它是外部插件，不是 MySQL 内置的。
- 创建表时使用 CREATE TABLE 语句直接定义 FULLTEXT 索引。也可以使用 ALTER TABLE 或 CREATE INDEX 语句稍后为现有表添加全文索引。
- 对于非常大的数据集，先将数据加载到没有全文索引的表中，再创建索引，会比在加载数据的同时创建索引更快。
- 分区表在 MySQL 中不支持全文搜索。

3. 全文索引的 3 种类型查询方式

（1）自然语言搜索

自然语言搜索（Natural Language Search）将搜索字符串解释为人类语言中的短语（自由文本中的短语）。除了双引号（"）字符以外，没有特殊操作符。下面介绍停用词设置和自然语言搜索方式。

1）停用词介绍如下。

停用词（stopword）是指在信息检索中，为了提高搜索效率而自动过滤掉的一些常用词或短语。这些词由于出现频率非常高，几乎在每篇文档或每个句子中都会出现，因此它们对于确定文档或句子的主题内容并没有太大帮助。在搜索引擎、文本挖掘、自然语言处理等领域，停用词通常会被自动忽略，以提高搜索的准确性和效率。例如，"在""里面""也""的""它""为"等都是停用词。在 INFORMATION_SCHEMA 库下的 INNODB_FT_DEFAULT_STOPWORD 表中定义了停用词。查看停用词信息的命令行示例如下。

```
mysql> SELECT * FROM INFORMATION_SCHEMA.INNODB_FT_DEFAULT_STOPWORD;
+----------+
| value    |
+----------+
| a        |
| about    |
| an       |
| are      |
...
| who      |
| will     |
| the      |
| www      |
+----------+
36 rows in set (0.00 sec)
```

也可以自定义停用词,这样自主性比较高。设置停用词信息的命令行示例如下。

```
mysql> CREATE TABLE my_stopwords(value VARCHAR(30)) ENGINE = INNODB;
mysql> INSERT INTO my_stopwords(value) VALUES ('Ishmael');
mysql> CREATE FULLTEXT INDEX idx ON my_stopwords(`value`);
mysql> SET GLOBAL innodb_ft_server_stopword_table = 'db7/my_stopwords';
```

不同的应用场景可能需要不同的停用词列表。有些情况下,某些停用词可能对于确定文本的主题或意图仍然很重要,因此在某些特定的自然语言处理任务中,可能会根据具体需求调整或扩展停用词列表。

2)自然语言搜索介绍如下。

在 MySQL 中,自然语言搜索是通过 MATCH ... AGAINST 语法来实现的,并且可以通过 IN NATURAL LANGUAGE MODE 修饰符来指定搜索模式。这种搜索模式对于执行基于全文的搜索特别有用。自然语言搜索的命令行示例如下。

```
# 创建表和模拟数据
mysql> CREATE TABLE articles (
         id INT UNSIGNED AUTO_INCREMENT NOT NULL PRIMARY KEY,
         title VARCHAR(200),
         body TEXT,
         FULLTEXT (title,body)
       ) ENGINE=InnoDB;
Query OK, 0 rows affected (0.08 sec)

mysql> INSERT INTO articles (title,body) VALUES
       ('MySQL Tutorial','DBMS stands for DataBase ...'),
       ('How To Use MySQL Well','After you went through a ...'),
       ('Optimizing MySQL','In this tutorial, we show ...'),
       ('1001 MySQL Tricks','1.Never runmysqld as root.2....'),
       ('MySQL vs.YourSQL','In the following database comparison ...'),
       ('MySQL Security','When configured properly, MySQL ...');
# 使用 MATCH 和 AGAINST 函数查询
mysql> SELECT * FROM articles
       WHERE MATCH (title,body)
       AGAINST ('database' IN NATURAL LANGUAGE MODE);
```

在 MATCH 函数中指定了想要搜索的列名。通常，这些列应该包含 FULLTEXT 索引，以便能够执行全文搜索。在 AGAINST 函数中指定要搜索的查询字符串。这种搜索模式通常比简单的 LIKE 查询更加智能，因为它不仅基于字面匹配，还考虑了词汇之间的关系和权重。

（2）布尔搜索

布尔搜索（Boolean Search）是一种使用特殊查询语言规则的搜索方法，允许用户通过包含操作符来精确定义他们的搜索需求。字符串包含要搜索的单词。它还可以包含一些操作符，这些操作符指定了一些要求，如某个单词必须在匹配的行中出现或不出现，或者该单词的权重应该高于或低于通常值。布尔搜索的命令行示例如下。

```
mysql> SELECT * FROM articles WHERE MATCH (title,body)
       AGAINST ('+MySQL -YourSQL' IN BOOLEAN MODE);
```

布尔搜索的基本操作符介绍如下。

- AND：这个操作符要求搜索的字符串中同时包含两个或多个关键词。例如，搜索"apple AND juice"时将只返回同时包含"apple"和"juice"的文本。
- NOT：这个操作符用于排除某个关键词。例如，搜索"apple NOT juice"时将返回包含"apple"但不包含"juice"的文本。
- 除此之外，还支持更高级的操作符，如>、<、（）、~、*等条件符号。

通过组合这些操作符和关键词，用户可以构建出高度精确和详细的搜索查询，从而更准确地找到他们需要的信息。

（3）查询扩展搜索

查询扩展搜索（Query Expansion Search）是一种改进的自然语言搜索技术，它试图通过自动扩展原始查询的词汇来提高搜索的精度和召回率。在查询扩展搜索中，首先执行基于用户提供的初始查询字符串的自然语言搜索。然后，分析返回的最相关文档，提取这些文档中的关键词，并将这些关键词添加到原始查询字符串中。最后，使用扩展后的查询字符串重新执行搜索，返回的结果通常会更加精确和相关。查询扩展搜索的命令行示例如下。

```
# 创建表和模拟数据
mysql>CREATE TABLE articles01 (
          id INT UNSIGNED AUTO_INCREMENT NOT NULL PRIMARY KEY,
          title VARCHAR(200),
          FULLTEXT (title)
     ) ENGINE=InnoDB;
mysql>INSERT INTO articles01 (title) VALUES('MySQL Tutorial'),
        ('Optimizing Tutorial'),('analyze db');

# 分为两个阶段。第一阶段:第一条 SQL 语句直接返回 MySQL 包含的数据
mysql> SELECT * FROM articles
       WHERE MATCH (title,body)
       AGAINST ('database' IN NATURAL LANGUAGE MODE);
+-----+--------------------------------+
| id  | title                          |
+-----+--------------------------------+
|   1 | MySQL Tutorial                 |
+-----+--------------------------------+
```

```
# 第二阶段;若 QUERY EXPANSION 是上面结果当中的词汇,则再次进行对应的查询
mysql> WHERE MATCH (title,body)
    AGAINST ('database' WITH QUERY EXPANSION);
+-----+--------------------------------+
|id   |title                           |
+-----+--------------------------------+
|   1 |MySQL Tutorial                  |
|   2 |Optimizing Tutorial             |
+-----+--------------------------------+
```

查询扩展搜索的实现方式就是使用结果集再次进行匹配操作。查询扩展搜索可以提高搜索性能,因为它不仅考虑了用户提供的初始关键词,还考虑了与这些关键词相关的其他词汇和上下文信息。这有助于克服用户查询时的词汇偏差,并捕捉到更多的相关信息。

4. 解析器

全文索引解析器是一种用于数据库全文搜索的重要工具,它能够将文本内容进行分词和索引,从而提高搜索的效率和准确性。全文解析提供了两种解析器。

(1) ngram 解析器

ngram 解析器是 MySQL 中用于全文搜索的一种特殊解析器,它专门设计用来处理那些不使用空格作为单词分隔符的语言,如中文、日文和韩文等。在这些语言中,单词之间没有明显的分隔符,因此传统的基于空格的分词方法无法有效工作。ngram 解析器通过创建文本中所有可能的连续字符组合(称为 n-grams)来解决这个问题,如 abcd 字段,可以生成多个组合,如下所示。

```
n=1:'a','b','c','d'
n=2:'ab','bc','cd'
n=3:'abc','bcd'
n=4:'abcd'
```

在 MySQL 中,使用 ngram 解析器时,可以指定 ngram_token_size 的值,这决定了 n-grams 的长度。例如,如果 ngram_token_size 参数设置为 2,则解析器将为文本生成所有可能的双字符组合。例如,对于文本"abcd",拆分的单词为:"ab""bc"和"cd"。

对于中文文本,ngram 解析器会基于字符来生成 n-grams,这意味着它将处理每个汉字,而不是像英文那样基于单词。这对于在 MySQL 中实现中文全文搜索非常有用,因为它允许用户搜索文本中的任意字符组合,而不仅仅是完整的单词。例如,对于文本"生日快乐",就会拆分成"生日"和"快乐"两个分词。

(2) MeCab 全文解析器

MeCab 全文解析器是一个专门用于日文的全文解析器。MeCab 能够将文本序列标记为有意义的单词,这是它在处理日文时相比 ngram 解析器的优势之一。MeCab 索引通常更小,这意味着它占用的存储空间更少,同时全文搜索通常也更快,因为索引更小意味着搜索时所需处理的数据量更小。然而,与 ngram 全文解析器相比,MeCab 全文解析器的一个缺点是它需要对文档进行标记,这个过程可能需要更长的时间。这意味着在使用 MeCab 进行全文搜索时,初始的索引创建或更新可能会比使用 ngram 解析器更耗时。

MeCab 全文解析器配置方式如下。

```
[mysqld]
loose-mecab-rc-file=MYSQL_HOME/lib/mecab/etc/mecabrc
innodb_ft_min_token_size=1
```

在决定是否使用 MeCab 或 ngram 解析器时，需要权衡一些因素。如果需要更快的全文搜索速度和较小的索引大小，同时不介意初始标记过程的耗时，那么 MeCab 可能是一个好选择。如果更看重初始标记的速度和灵活性（例如，需要支持多种语言或处理不使用空格分隔符的语言），那么 ngram 解析器可能更适合。

5. 总结

MySQL 的全文索引主要基于词频来判断搜索结果的相关性，但也会考虑其他因素，如文档长度和词的权重。与 B+树索引不同，全文索引不会记录匹配的词在字符串中的位置。全文索引的性能通常与数据量和内存使用有关，当全文索引能够完全加载到内存中时，性能最佳。然而，当索引过大不能全部读入内存时，性能可能会受到影响。

9.3.6 自增键的设计

在 MySQL 表设计中，数字类型的自增键（AUTO_INCREMENT）是为了满足递增规则而设计的，尤其是在没有自然主键的情况下。自增键的存在是为了确保每行数据都有一个唯一的标识符，并且这个标识符是递增的。由于自增键的递增特性，它可以让主键索引保持顺序插入，这有助于减少页分裂（page split），从而使索引结构更加紧凑。

在使用 MySQL 时，需要区分 Sequence 和自增键。MariaDB 和 Oracle 等数据库系统提供了 Sequence 的功能，MySQL 本身并没有内建的 Sequence 机制。自增键和 Sequence 都用于生成唯一数字标识符，但自增键只支持简单的向上累加，而 Sequence 则提供了更多的灵活性和配置选项。

在 MySQL 中，自增键的实现相对简单，只需要在创建表时为相应的列指定 AUTO_INCREMENT 属性。这样，当向表中插入新行时，MySQL 会自动为这个列生成一个唯一的、递增的数字。

1. 自增键类型和取值范围

MySQL 的自增键支持整数类型。可以使用 UNSIGNED 属性来允许更大的范围。表 9-5 所示为所有整数类型的取值范围。

表 9-5 MySQL 整数类型取值范围

类型	字节数	最小值	最大值	最大整数（UNSIGNED）
TINYINT	1	-128	127	255
SMALLINT	2	-32768	32767	65535
MEDIUMINT	3	-8388608	8388607	16777215
INT	4	-2147483648	2147483647	4294967295
BIGINT	8	$-2^{63}-1$	$2^{63}-1$	$2^{64}-1$

选择适当的整数类型和是否使用 UNSIGNED 属性都是非常重要的。如果知道某个字段的值永远不会是负数（例如，用户 ID、订单编号等），那么使用 UNSIGNED 属性可以允许更大的取值范围，同时节省一半的存储空间。然而，如果字段的值有可能是负数，那么应该避免使用 UNSIGNED 属性，并确保选择足够大的整数类型来容纳可能的负值。

2. 自增键数据插入方式

在涉及数据插入而不指定自增键的值时，MySQL 会自动为这个字段分配一个唯一的、递增的数值。这个过程确保了即使在高并发的插入操作中，每条记录的自增键值也是唯一和递增的，从而避免了重复和冲突。

表9-6所示为自增键在处理不同插入模式时的3种方式。

表9-6 MySQL自增键数据插入方式

插入方式	说明	举例
简单插入	可以预先确定要插入的行数的语句（在语句最初处理时），不包含INSERT ... ON DUPLICATE KEY UPDATE	INSERT INTO Table1（field1）values（'test'）
批量插入	对于事先不知道要插入的行数（以及所需的自动递增值的数量）的语句，InnoDB存储引擎会在处理每一行时为AUTO_INCREMENT列一次赋一个新值。例如，INSERT ... SELECT、REPLACE ... SELECT和LOAD DATA	INSERT INTO Table2（field1，field2，...）SELECT value1，value2，... FROM Table1
混合模式插入	一种情形是为一些（但不是所有）新行指定自动增加的值。另一种情形是INSERT ... ON DUPLICATE KEY UPDATE，在最坏的情况下，实际上是一个INSERT，后面跟着一个UPDATE，其中为AUTO_INCREMENT列分配的值在更新阶段可能使用，也可能不使用	INSERT INTO Table2（field1，field2）VALUES（1,'a'），（NULL,'b'），（5,'c'），（NULL,'d'）

对于所有插入方式，自增键确保了每条记录都有一个唯一的标识符，这有助于维护数据的完整性和一致性。至于选择哪种插入方式，取决于具体的应用场景和需求。在实际应用中，应该根据数据量、性能要求、并发需求等因素来做出决策。

3. 自增锁模式

在MySQL的InnoDB存储引擎中，对于自增键数据的插入，使用了特殊的锁机制来保证唯一性和连续性。不过，需要注意的是，InnoDB对于自增锁的实现细节在不同版本中可能有所不同。在较早的版本中，自增锁是在语句级别上的，这意味着一个语句在获取自增锁后，会持有这个锁，直到该语句执行结束为止。然而，在后续版本中，InnoDB引入了一些改进，使得自增锁在获取后可以更早地释放，以提高并发插入的性能。

在MySQL中，关于自增锁的配置，通过innodb_autoinc_lock_mode参数来控制，该参数有3个值：传统模式（traditional）、连续模式（consecutive）和交叉模式（interleaved）。

（1）传统模式：innodb_autoinc_lock_mode=0

在这种锁模式下，所有"类INSERT"语句（如INSERT、REPLACE等）都会获得一个特殊的表级AUTO-INC锁，用于插入具有AUTO_INCREMENT列的表。这个锁通常保持到语句结束，而不是事务结束。这是为了确保自动递增值在一个给定的INSERT语句序列中是以可预见和可重复的顺序分配的，并确保连续的自动递增值。

- 传统的锁模式选项提供向后兼容性。传统的锁模式是为了与早期版本的MySQL或InnoDB兼容而保留的。对于那些依赖于早期行为的应用程序，使用传统锁模式可以确保它们正常运行。
- 所有"类INSERT"语句都会获得一个特殊的表级AUTO-INC锁，表级锁一直保持到语句结束。当一个INSERT语句执行时，它会获取一个表级锁来保护AUTO_INCREMENT计数器，防止其他事务同时修改它。这个锁会在INSERT语句执行完毕（而不是事务结束）后释放。
- 通过表级锁，InnoDB可以确保INSERT语句序列中的自增值是以可预测和可重复的顺序分配的。
- 当多个事务同时执行插入语句时，这些表级锁限制了并发性和可伸缩性。由于每个 INSERT 语句都需要获取表级锁，因此在高并发环境下，多个事务争用锁资源可能导致性能瓶颈。这限制了系统的并发处理能力和可扩展性。

（2）连续模式：innodb_autoinc_lock_mode=1

连续模式是 InnoDB 存储引擎中用于控制 AUTO_INCREMENT 锁行为的一种模式。在这种模式下，InnoDB 存储引擎会优化 AUTO_INCREMENT 值的分配，以提高并发插入的性能。

- 提高可伸缩性：连续的 AUTO_INCREMENT 值会在每个 INSERT 语句执行后立即分配，而不是在整个事务结束时。这意味着在单个事务中执行多个 INSERT 语句时，不需要持续持有锁，从而提高了并发插入的性能。
- 基于语句的复制安全：在这种模式下，主节点上的 AUTO_INCREMENT 值分配与从节点上的复制操作是兼容的。即使从节点使用基于语句的复制（Statement-Based Replication，SBR），也能正确地复制 AUTO_INCREMENT 值，因为每个 INSERT 语句在主节点上都会分配一个新的 AUTO_INCREMENT 值。

表 9-7 所示为连续模式下对数据插入方式的影响。

表 9-7 MySQL 连续模式下对数据插入方式的影响

插入方式	对插入方式的影响
简单插入	提前获知要插入的行数，避免表级 AUTO-INC 锁
批量插入	使用特殊的表级 AUTO-INC 锁，并将其保持到语句结束
混合模式插入	InnoDB 存储引擎会分配比要插入的行数更多的自动增量值。注意，所有自动分配的值都是连续生成的

虽然连续模式可以提高并发插入的性能，但它并不总是最佳选择。在某些情况下，如当需要确保事务中所有 INSERT 语句的 AUTO_INCREMENT 值都是连续的时，就需要使用其他锁模式。

（3）交叉模式：innodb_autoinc_lock_mode=2

在交叉模式下，InnoDB 不再使用表级 AUTO-INC 锁来为每个 INSERT 语句单独分配自增值。相反，它允许多个语句并发执行，从而提高了并发性能。

以下是交叉模式的主要特点。

- 无表级 AUTO-INC 锁：在交叉模式下，没有单个 INSERT 语句会持有表级锁来分配自增值。这意味着多个 INSERT 语句可以同时执行，无须等待其他语句完成。
- 唯一且单调递增：尽管多个语句可以同时生成自增值，但 InnoDB 保证了这些值在全局范围内是唯一的，并且是单调递增的。这意味着不会出现两个行具有相同自增值的情况，并且随着时间的推移，自增值会不断增大。
- 值分配交叉进行：由于多个语句可以同时生成自增值，因此值的分配不是连续的。换句话说，一个给定语句插入的行所生成的自增值可能与前一个或后一个语句生成的值之间存在间隔。
- 可伸缩性提高：由于不再需要等待表级锁，因此交叉模式可以显著提高并发插入操作的性能，从而增强数据库的可伸缩性。
- 可能的非连续性：虽然自增值是唯一且单调递增的，但它们可能不是连续的。这对于某些应用程序来说可能不是问题，但对于依赖连续自增值的应用逻辑来说可能需要额外的考虑。

选择交叉模式还是连续模式或传统模式，取决于业务需求。如果并发插入性能是关键，且应用程序不需要自增值连续，那么交叉模式可能是一个好选择。然而，如果连续性是一个需求，那么需要使用连续模式或传统模式。

4. 自增键注意事项

在使用自增键时，需要注意以下几点事项。

（1）不同版本中默认设置不同

innodb_autoinc_lock_mode 的默认设置在不同版本的 MySQL 中有所不同。在 MySQL 8.0 版本之前，连续模式是默认的 AUTO_INCREMENT 锁模式。这意味着，在大多数情况下，INSERT 语句会获得连续的自增值，并且锁会在语句执行后立即释放，以提高并发性能。在 MySQL 8.0 版本中，默认设置更改为交叉模式。这个模式允许多个 INSERT 语句同时生成自增值，这些值在全局范围内是唯一且单调递增的，但可能不是连续的。这种设置旨在进一步提高并发性能，特别是在高并发环境中。

这些默认设置可能会根据 MySQL 的具体配置和部署环境有所不同。在调整这些设置之前，需要评估应用程序的需求和性能要求。

（2）索引必不可少

对于 AUTO_INCREMENT 列，通常建议将其定义为主键（PRIMARY KEY）或唯一键（UNIQUE）。这样做有以下几个原因。

- 保证唯一性：AUTO_INCREMENT 列必须保证唯一性，这样每次插入新记录时，都会自动分配一个新的、唯一的值。将 AUTO_INCREMENT 列设置为主键或唯一键可以确保这一点。
- 性能优化：索引可以提高数据检索的速度。在 InnoDB 中，AUTO_INCREMENT 列的索引允许数据库系统快速地查找下一个可用的自增值，这在进行插入操作时尤为重要。
- 避免重复值：将 AUTO_INCREMENT 列定义为主键或唯一键可以防止插入重复的值。如果没有这样的约束，那么插入重复的自增值将会导致错误。
- 系统内部要求：InnoDB 存储引擎需要 AUTO_INCREMENT 列有索引，以便能够高效地管理和分配自增值。

在创建表命令行里将 auto 字段定义为索引的示例如下。

```
mysql> DROP TABLE IF EXISTS t_user;CREATE TABLE `t_user` (
     `id` int unsigned zerofill NOT NULL AUTO_INCREMENT,
     `name`varchar(30) NOT NULL,
  KEY `idx_id` (`id`)
) ENGINE=InnoDB ;
```

（3）自增序列和步长参数设置

在 MySQL 中，可以为整个实例设置全局的自增序列和步长参数。这些参数会影响整个 MySQL 实例中所有使用 AUTO_INCREMENT 列的表。表 9-8 所示为自增字段自增序列和步长参数说明列表。

表 9-8 MySQL 自增序列和步长参数说明

参数	范围	说明
auto_increment_offset	动态变量	表示自增字段的序列的初始值。取值范围是 1~65535
auto_increment_increment	动态变量	表示自增字段步长，即每次递增的量，其默认值是 1，取值范围是 1~65535

说明：如果这两个参数（auto_increment_offset 和 auto_increment_increment）的值都设置为大于 65535，则将导致它们的值被设置为 65535；如果小于 1，则将导致它们的值被设置为 1。

一旦设置全局的 AUTO_INCREMENT_INCREMENT，就将影响所有新创建的表，但不会影响已经存在的表，除非这些表也明确地修改了它们的自增步长设置。另外，需要注意的是，尽管可以设置全局的 AUTO_INCREMENT_INCREMENT，但每个表仍然可以通过 AUTO_INCREMENT_OFFSET 属性来定义自己的起始自增值。这意味着，即使全局步长被设置为 2，也可以为特定的表设置一个为 10 的起始值，并且每次插入时自增值增加 1（尽管全局步长是 2）。

（4）复制中自增字段处理方式

MySQL 复制环节中，记录的 binlog 语句有 3 种模式：STATEMENT、ROW 和 MIXED。不同模式下对自增字段处理的方式不一样。具体方式如下。

- 在进行基于 STATEMENT 模式的复制，从节点没有人为写入的情况下，innodb_autoinc_lock_mode 设置为 1，能确保生成的自增值和主节点是一致的。如果 innodb_autoinc_lock_mode = 2，则不能确保副本上的自动增量值与源上的相同。
- 如果使用基于 ROW 或 MIXED 模式的复制，那么所有自动增量锁模式都是安全的，因为基于行的复制对 SQL 语句的执行顺序不敏感（MIXED 模式对任何基于语句的复制不安全的语句使用基于行的复制）。
- 在 STATEMENT 模式下，主节点记录的是 SQL 语句本身，而不是行级的更改。当从节点执行这些 SQL 语句时，它会自己生成自增 ID。因此，主节点上的自增 ID 可能会不同，尤其是当从节点上有其他写入操作时。
- 在 ROW 模式下，主节点记录的是行级的更改，而不是 SQL 语句。因此，当从节点执行这些更改时，它会使用主节点上生成的自增 ID，而不是自己生成。
- 在 MIXED 模式下，它通常根据 SQL 语句的类型决定使用哪种复制模式。对于能够安全使用 STATEMENT 模式的语句，以及可能导致产生不一致的语句，都会使用 ROW 模式。

为了确保主从节点上的自增 ID 一致，特别是在高并发环境下，建议使用 ROW 模式。但是，需要注意的是，ROW 模式可能会产生更大的 binlog，并可能增加从节点的处理负担。因此，在选择复制模式和自增锁模式时，需要根据具体的业务需求和系统性能进行权衡。

（5）自动递增连续性

在 MySQL 中，一旦为 AUTO_INCREMENT 列生成一个值，这个值就被"消耗"了，即使后续的 INSERT 操作由于某种原因（如事务回滚）没有完成，这个值也不会被重新利用。这意味着，在自增序列中，可能会出现"间隙"（gap），即不是所有可能的自增值都被实际使用。

这种情况在以下几种场景中表现得特别明显。

- 事务回滚：如果一个事务尝试插入一行并生成了一个自增值，但该事务最终回滚，那么这个自增值仍然会被"消耗"，不会被重新用于后续的插入操作。
- 复制延迟：在复制环境中，如果从节点因为某些原因落后于主节点，则它可能会跳过一些自增值。当从节点最终追上并应用这些被跳过的自增值时，新的数据仍然会生成新的自增值，导致序列中出现间隙。
- 手动干预：如果数据库管理员手动干预了自增序列（例如，通过 ALTER TABLE 语句重置 AUTO_INCREMENT 的值），那么之前的自增值仍然会被视为已使用，不会被重新利用。

图 9-19 所示为自增键事务回滚场景。在这 3 个窗口中，会话 1 和会话 2 提交事务成功，会话 3 回滚，最终导致自增字段不连续。

图 9-19　自增键事务回滚场景

（6）自增键值重复利用问题

在 MySQL 5.7 及更早版本中，自动增量计数器存储在内存中，而不是磁盘上。当 MySQL 服务重启后，InnoDB 通常从上次关闭时的 AUTO_INCREMENT 值开始，或者在没有任何先前的值时从 1 开始。如果自增最大值之前被删除，那么再次插入数据时，自增字段可能会从较小的数字开始，导致重复使用。

在 MySQL 8.0 版本中，这种行为得到了改进。当前最大的自动增量计数器值被写入 Redo log，并且每次改变时都会保存到数据字典的检查点中。这些更改确保了在服务器重新启动后，当前最大自动增量计数器值保持不变。在崩溃恢复的服务重启过程中，InnoDB 使用存储在数据字典中的当前最大自动增量值来初始化内存中的自动增量计数器，并扫描 Redo log 中自上次检查点以来写入的自动增量计数器值。如果 Redo log 中的值大于内存中计数器的值，则应用 Redo log 中的值。然而，如果出现 Redo log 丢失或损坏的情况，则问题可能还会存在。

（7）自增键管理和维护问题

在管理和维护自增键时，需要注意下列一些关键事项。

- TRUNCATE 和 DELETE 的区别。TRUNCATE 命令会删除表中的所有记录，并且重置自增字段的计数器以回到初始值。DELETE 命令删除表中的记录，但不会重置自增字段的计数器。
- 在复制集群中，为了避免自增 ID 冲突，每个节点通常会有不同的自增步长（auto_increment_increment）。这样，即使两个节点同时插入数据，它们的自增 ID 也不会重叠。步长通常是根据集群配置来设定的，并且在集群运行过程中保持不变，除非人为地进行更改。但在某些情况下，可能存在 bug 或配置错误，导致步长不正确，从而引发 ID 冲突。
- 补齐宽度值 ZEROFILL 的用途。对于自增字段来说，如果设置了 ZEROFILL，那么当自增数值没有达到显示宽度时，MySQL 会在前面填充零。在实际场景中，ZEROFILL 的用途有限，因为它可能会导致混淆（例如，001 和 1 看起来不同，但在数值上是相等的）。此外，ZEROFILL 会影响性能，因为它需要额外的存储空间，并且在进行数值计算时可能会导致类型转换。
- 自增主键作为业务主键。是否使用自增主键作为业务主键取决于具体业务需求。对于一些没有特定业务主键需求的场景（如日志数据），添加自增主键可以提高复制和运维效率。然而，如果业务逻辑依赖于自增主键的连续性，或者自增主键本身没有业务意义，那么将自增主键作为业务主键可能是不合适的。这可能导致数据冲突和其他问题，尤其是在处理关联表时。
- 自增键耗尽的问题。如果一个 AUTO_INCREMENT 整数列的值达到了其数据类型的上限（例如，对于 UNSIGNED INT，上限是 4294967295），则后续的 INSERT 操作将失败，并返回一个错误。为了避免这种情况，需要定期监控自增字段的使用情况。
- 通过查询系统表，可以监控自增字段的当前值和使用情况。这对于预测自增键何时耗尽非常有用，还可及时采取行动进行调整。查询自增键使用情况的命令行示例如下。

```
mysql> SELECT infotb.TABLE_SCHEMA, infotb.TABLE_NAME,
infotb.AUTO_INCREMENT,infocl.COLUMN_TYPE FROM information_schema.TABLES as
infotb INNER JOIN information_schema.COLUMNS infocl ON infotb.TABLE_SCHEMA =
infocl.TABLE_SCHEMA AND infotb.TABLE_NAME = infocl.TABLE_NAME AND
infocl.EXTRA='auto_increment';
```

5. 总结

在 MySQL 应用开发中，使用自增键能够自动为每行数据提供一个唯一的标识符，从而简化了数据插入过程，保证了数据的唯一性，并提高了查询效率。然而，为了充分利用自增列的这些优点并避免潜在的问题，如 ID 冲突和 ID 溢出，需要仔细考虑其用途、数据类型、初始值和增量等因素。此外，在并发插入、复制和集群等场景中，还需要特别关注 ID 的生成和分配，以确保数据的完整性和应用的性能。

9.3.7 外键的设计

外键（Foreign Key）在数据库中扮演着关键角色，可用于维护数据的一致性和完整性，确保关联表之间的数据关系得到正确管理。尽管 MySQL 支持外键约束，但在实际应用中，开发者会选择在应用程序层面实现关联关系，以避免外键可能带来的性能影响和复杂性。

1. 外键特点

在 MySQL 中，外键是用于建立和加强两个表之间数据连接的一列或多列。它表示一个表中的一个字段被另一个表中的一个字段引用。外键约束主要用来维护两个表之间数据的一致性，以保证参照完整性。在外键使用上，需要注意以下几点。
- 在 MySQL 中，InnoDB 和 NDB Cluster 存储引擎支持外键约束。lower_case_table_names 系统变量会影响表名的大小写敏感性，这可能会影响外键约束的创建，因为外键约束要求父表和子表的名称大小写匹配。

- 父表和子表（即被引用表和引用表）必须使用相同的存储引擎。这些表不能是临时表，因为临时表不支持外键约束。在创建外键约束时，用户需要对父表上的列拥有 REFERENCES 权限。
- 外键列和它所引用的列（即父表中的列）必须具有相同或兼容的数据类型。具体来说，包括数据类型、精度、长度和符号。对于字符集和排序规则，非二进制字符串列必须匹配，以确保字符比较的一致性。
- 在引用表中，外键列必须是被索引的。这个索引不必是唯一的，但必须是存在的，以便数据库可以快速检查外键约束。对于 NDB 存储引擎，被引用的列（即父表中的列）必须有一个显式的唯一键或主键。
- 外键列上的索引不支持前缀索引，这意味着不能使用列的部分值作为索引。BLOB 和 TEXT 类型的列不能包含在外键定义中，因为这些数据类型可能包含大量的数据，不适合用作外键。
- InnoDB 存储引擎不支持在分区表上使用外键约束。
- 外键约束不能引用虚拟生成的列。

在 MySQL 外键的使用中，包含多种级联（Reference）行为，语法如下所示。

```
[CONSTRAINT [symbol]] FOREIGN KEY
[index_name] (col_name, ...)
REFERENCES tbl_name (col_name,...)
[ON DELETE reference_option]
[ON UPDATE reference_option]
reference_option:
    RESTRICT | CASCADE | SET NULL | NO ACTION | SET DEFAULT
```

- 包含三个 Action 直接引用，以及 DELETE、UPDATE 对应操作。可以指定当父表（参考表）中的记录被删除或更新时，子表中相应记录的行为。
- RESTRICT：如果子表中存在依赖于父表的数据，则使用 RESTRICT 会阻止对父表的删除或更新操作，以确保数据的一致性。
- CASCADE：删除或更新父表中的行，并自动删除或更新子表中匹配的行。CASCADE 操作确保了数据的级联一致性。
- SET NULL：如果父表中的数据被删除或更新，则子表中的外键列将被设置为 NULL。
- NO ACTION：它的行为等同于 RESTRICT。这意味着，如果子表中有依赖于父表的数据，则不允许对父表进行删除或更新操作。
- SET DEFAULT：这个动作被 MySQL 解析器识别，但是 InnoDB 和 NDB 都拒绝包含 ON DELETE SET DEFAULT 或 ON UPDATE SET DEFAULT 子句的表定义。

说明：对于未指定的 ON DELETE 或 ON UPDATE，默认操作总是 NO ACTION。InnoDB 存储引擎使用深度优先搜索算法对对应外键约束的索引记录执行级联操作。

外键的级联行为是数据库设计中非常关键的一部分，它可帮助确保数据的完整性和一致性。

2. 外键锁机制

MySQL 中的外键锁是根据元数据锁扩展到外键约束相关的表。对于外键约束检查，在相关表上使用共享只读锁（LOCK TABLES READ）方式。对于级联更新，在操作涉及的相关表上使用无共享的写锁（LOCK TABLES WRITE）方式。当存在外键约束时，每次都要通过扫描来判断记录是否合格。所以，一旦关联字段操作，就容易出现死锁现象。

下面介绍两个由外键导致的 SQL 语句执行失败示例。

【例 9-3】 外键锁最终导致死锁的命令行示例如下。

```
# 表结构
mysql>CREATE TABLE categories(
    cat_id int not null auto_increment primary key,
    cat_name varchar(255) not null
) ENGINE=InnoDB;
mysql>CREATE TABLE products(
    prd_id int not null auto_increment primary key,
    prd_name varchar(355) not null,
    prd_price decimal,
    cat_id int not null,
    FOREIGN KEY fk_cat(cat_id) REFERENCES categories(cat_id)
    ON UPDATE CASCADE ON DELETE RESTRICT
) ENGINE=InnoDB;
# 插入数据
mysql>insert into categories(cat_id,cat_name) values(1,'水果'),(2,'蔬菜');
mysql>insert into products(prd_id,prd_name,prd_price,cat_id)
values(1,'苹果',9,1),(2,'菠萝',10,1),(3,'白菜',8,2),(4,'可乐',7,3);
```

图 9-20 所示为外键死锁现象。会话 1 对父表（categories）插入数据但未提交，子表（products）更新等待。会话 2 对父表进行更新但未提交，子表插入等待。会话 2 的事务因外键导致超时退出。

图 9-20　MySQL 外键死锁模拟

查看死锁状态信息的命令行示例如下。

```
# 查看外键死锁信息
mysql> SHOW ENGINE INNODB STATUS \G;
------------------------
LATEST FOREIGN KEY ERROR
------------------------
2023-05-31 17:22:52 140626893833984 Transaction:
TRANSACTION 9011856, ACTIVE 0 sec updating or deleting
mysql tables in use 1, locked 1
4 lock struct(s), heap size 1128, 2 row lock(s), undo log entries 1
MySQL thread id 23, OS thread handle 140626893833984, query id 168 localhost root updating
update categories set cat_id=3 where cat_id=2
Foreign key constraint fails for table `dbdemo`.`products`:
,
  CONSTRAINT `products_ibfk_1` FOREIGN KEY (`cat_id`) REFERENCES `categories` (`cat_id`)
Trying to update in parent table, in index PRIMARYtuple:
DATA TUPLE: 4 fields;
 0: len 4; hex 80000002;asc;;
 1: len 6; hex 000000898290;asc;;
 2: len 7; hex 01000001970151;asc Q;;
 3: len 6; hex e894ace88f9c;asc;;
But in child table `dbdemo`.`products`, in index fk_cat, there is a record:
PHYSICAL RECORD: n_fields 2; compact format; info bits 0
 0: len 4; hex 80000002;asc;;
 1: len 4; hex 80000003;asc;;
```

【例9-4】DDL操作会受外键约束，导致失败。TRUNCATE失败的命令行示例如下。

```
mysql > TRUNCATE TABLE products;
Query OK, 0 rows affected (0.03 sec)

mysql > TRUNCATE TABLE categories;
ERROR 1701 (42000): Cannot truncate a table referenced in a foreign key constraint (`dbdemo`.`products`, CONSTRAINT `products_ibfk_1`)

mysql> DROP TABLE categories;
ERROR 3730 (HY000): Cannot drop table 'categories' referenced by a foreign key constraint 'products_ibfk_1' on table 'products'.
```

说明：MySQL外键约束检查是即时检查的，对每一行都会运行外键检查。对于上述情况或用LOAD DATA导入数据，在检查外键约束上往往消耗大量时间。这时，通过SET foreign_key_checks=0（默认值是1），可以灵活处理，忽略外键检查。

外键的数据其实在子表中，是实际保存的数据，一旦父表的外键更新，子表就会随之更新。

3. 总结

外键在数据库中的主要功能是保证数据的完整性和一致性。尽管外键的使用可以带来这些好处，但它也会对数据库的性能和管理带来一定的影响。具体来说，外键可能会导致额外的资源消耗，因为当涉及外键字段的增、删、更新操作时，数据库需要进行额外的检查。此外，外键的维护可能需要重新调整索引，从而可能增加上锁的时间和死锁的风险。

综合上述这些因素，不推荐在MySQL里使用外键，建议用代码逻辑来保证外键相关关系的正确性和一致性。

9.3.8 表主键的设计

MySQL中的主键具有至关重要的作用。首先，主键是数据库中唯一标识每条记录的键。它确保了数据的唯一性和准确性，避免了重复记录的出现。其次，主键在数据检索、更新和删除操作中发挥着关键作用，因为它们通常是数据库索引的基础。通过使用主键，数据库可以快速定位到特定的记录，提高查询效率。此外，主键还参与了数据库的约束和关系维护，确保了数据的完整性和一致性。

1. 官方对主键的说明

主键（Primary Key）是一个或一组列，其值能够唯一地标识表中的每一行。一个表只能有一个主键，并且主键列中的值必须是唯一的，不能为NULL。在MySQL的InnoDB存储引擎表中，每一行都应该有可以唯一标识自己的一列。当没有设置主键时，MySQL本身会生成隐藏的列作为主键列。

- 当在表上定义一个主键时，InnoDB使用它作为聚集索引。如果没有逻辑唯一的非空列或列集，则建议添加一个新的自动递增列，其值将自动填充。
- 如果没有为表定义一个主键，则MySQL定位第一个UNIQUE索引，所有的键列都不是NULL，InnoDB使用它作为聚集索引。
- 如果表中没有PRIMARY KEY或合适的UNIQUE索引，那么InnoDB内部会再分配一个6字节的ID字段作为隐藏的聚集索引，名为GEN_CLUST_INDEX。行是以隐藏ID为主键进行排序的。因此，在插入新行时，单调递增，按行ID排序的行在物理上是按照插入顺序排列的。

说明：聚集索引=主键，非聚集索引=二级索引。

2. 表结构和SQL语句中的主键的意义

在MySQL的InnoDB存储引擎中，主键是表数据组织的核心。数据记录按照主键值的顺序存储在B+树索引的叶子节点上。每当有新记录插入，InnoDB都会根据其主键值计算出应该插入的叶子节点和具体位置。当页面空间使用率接近装载因子（默认为15/16，即93.75%）时，InnoDB会分配一个新的页面来存储更多的数据。这种基于主键的存储方式有助于提高数据检索和插入的效率。

下面介绍MySQL中主键的作用。

在MySQL中，对创建的表结构的普通查询的命令行示例如下。

```
# 创建一张表
mysql> CREATE TABLE users(
    id INT NOT NULL,
    Name VARCHAR(20) NOT NULL,
    city VARCHAR(20) NOT NULL,
    PRIMARY KEY(id)
);
mysql> INSERT INTO users(id,name,city) VALUES(3,'Tom','北京'),(5,'Kim','上海'),(9,'Jim','北京'),
(12,'Ku','上海'),(18,'Bill','上海');

# 新建一个以Name字段为索引名的二级索引,进行查询
mysql>ALTER TABLE users ADD INDEX index_name(Name);
mysql>SELECT * FROM users WHERE Name='Bill';
```

将上面创建的表中的数据构建成B+树结构如图9-21所示。主键是表数据的根基，叶子节点包含整行的数据。

第 9 章 MySQL 8.0 架构设计与应用开发

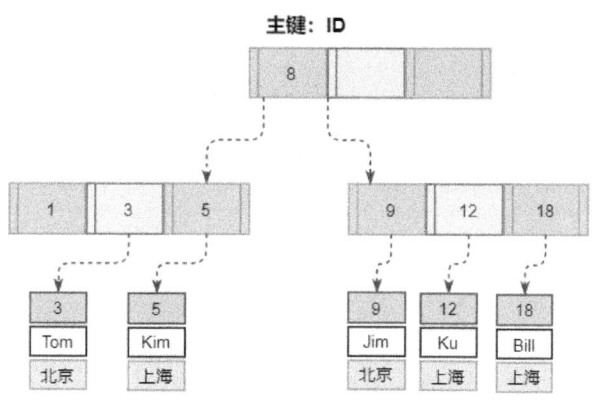

图 9-21　MySQL B+树结构

在上面创建的表中，以 Name 字段作为条件进行查询。图 9-22 所示为在二级索引下的回表操作。首先通过二级索引 Name 查找 Bill 的主键，然后通过主键定位对应的所有数据。

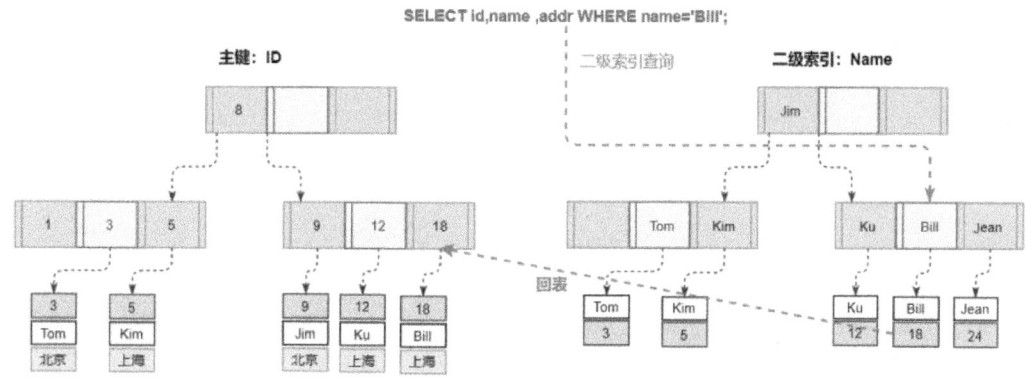

图 9-22　MySQL B+树回表操作

说明：二级索引是指定字段与主键的映射，主键长度越小，普通索引的叶子节点就越小，二级索引占用的空间也就越小，所以要避免使用过长的字段作为主键。

在 MySQL 的 InnoDB 存储引擎中，主键是表数据的根基，它决定了数据在磁盘上的物理存储顺序，并且每个表都有一个主键。B+树结构使得数据行按照主键的顺序存储在叶子节点上。这意味着，通过主键可以高效地检索到对应的数据行。

除此之外，InnoDB 还支持二级索引。二级索引的叶子节点包含了对应主键值的引用，而不是实际的数据行。当通过二级索引查询数据时，InnoDB 会首先找到对应的主键值，然后通过主键检索到实际的数据行。

3. 高可用架构中主键的作用

MySQL 高可用复制是逻辑复制，虽然与主库执行语句有些差异，但等价于把 SQL 语句在另一个节点执行。

在主从复制模式下，主键的作用尤为重要。主节点执行的 SQL 语句，都需要在从节点上准确、高效地应用整个事件。当涉及数据变更，如 INSERT、UPDATE、DELETE 操作时，主键的存在是至关重要的。如果没有主键，那么 MySQL 将无法生成一个唯一的、能够准确标识每一行数据的标识符。这会导致在从节点上执行这些变更操作时，因为缺少一个可以快速定位特定行的键，所以需要执行全表扫描来定位需要更新的行。

全表扫描是一个资源密集型操作，特别是在大型表中，它会导致复制延迟显著增加。因为每个数据变更都需要扫描整个表来找到相应的行，所以这种延迟会累积起来。尤其是在高并发的环境中，这会导致从节点无法及时追赶主库的状态，从而进一步加大主从延迟。

图 9-23 所示为主从复制回放语句判断的过程，包括如何根据主键（或其他唯一键）来定位需要更新的行。

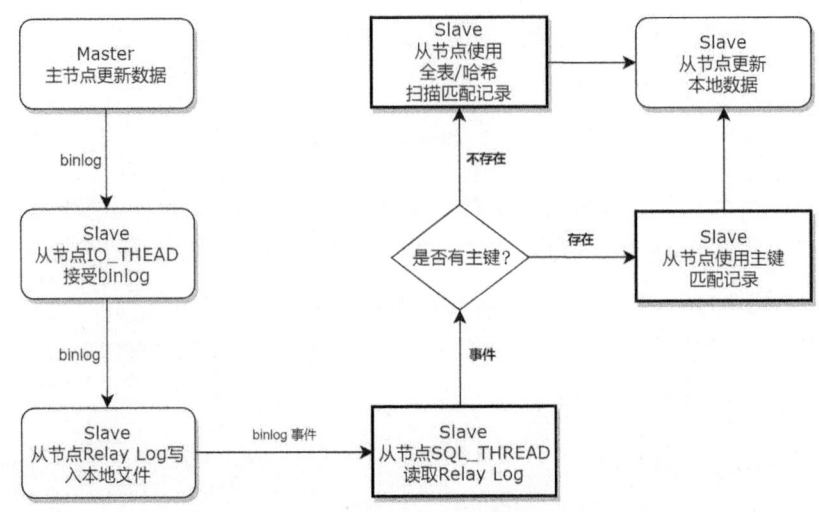

图 9-23　MySQL 复制回放语句流程

在另一个高可用架构 MGR 集群中，要求每个表都必须有一个定义好的主键，或者一个等效的主键，其中等效的主键是一个非空的唯一键。这样的键作为表中每一行的唯一标识符是必需的，这样 MGR 集群就可以通过准确地识别每个事务都修改了哪些行来确定哪些事务发生了冲突。

4. 怎样选择主键

在 MySQL 数据结构中，理想状态下的主键应该具备以下特点。这些特点有助于设计出一个既满足业务需求，又具有良好性能和数据完整性的数据库表结构。

- 不更新主键列的值。主键的值一旦分配给某个记录，就不应该更改。这有助于保持数据的完整性和一致性。
- 不重用主键列的值。主键必须是唯一的，确保在整个表中，主键列的值不重复。这为每行数据提供了唯一的标识。
- 不在主键列中使用可能会更改的值。理想的主键应该是稳定的，不会因为数据的变化而更改。
- 需要控制好字段长度，避免索引的频繁分裂和合并。较短的字段长度意味着更紧凑的索引结构，这有助于提高查询性能，并避免索引的频繁分裂和合并。

推荐使用的主键如下。

（1）自增主键

在无特殊需求情况下，对于 InnoDB 存储引擎，建议使用与业务无关的自增 ID 作为主键。

如果在表中使用自增主键，那么有利于插入性能的提高，即每次插入新的记录，记录就会顺序添加到当前索引节点的后续位置,当一页写满时,就会自动开辟一个新页,这样就会形成一个紧凑的索引结构,近似顺序填满。如图9-24所示,自增主键由于每次插入时不需要移动已有数据,因此效率很高,也不会在维护索引上增加很多开销。

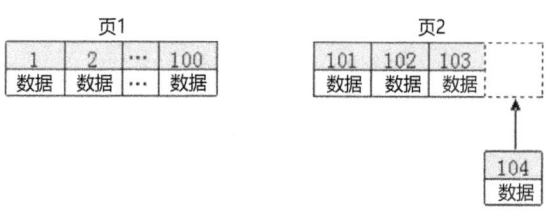

图 9-24　MySQL 自增主键索引构造

同时,自增主键设计(INT、BIGINT)可以减少二级索引的使用空间,提升二级索引的内存命中率。自增主键可以减少页中的碎片,提升空间和内存的使用率。

如果使用非自增主键(如果身份证号、机构唯一代码等),如图 9-25 所示,那么,由于每次插入主键的值近似于随机,因此每次新记录都要插入到现有索引页的中间某个位置。MySQL 为了将新记录插入到合适位置而不得不移动数据,甚至目标页面可能因为已经被回写到磁盘上而从缓存中清除,此时又要从磁盘上读回来,这增加了很多开销,同时频繁的移动、分页操作造成了大量的碎片,得到了不够紧凑的索引结构,后续不得不通过 OPTIMIZE TABLE 来重建表并优化填充页面。

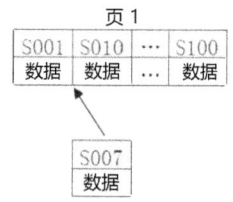

图 9-25　MySQL 非自增主键索引构造

自增主键的优点如下。

- 占用很小的数据存储空间。自增主键通常是整数类型,如 INT 或 BIGINT,它们在数据库中的存储效率很高,占用的空间相对较小。
- 性能较好。自增主键的插入性能通常较好,因为它使用顺序插入方式,不会导致页分裂(page split),从而减少了磁盘 I/O 操作。
- 容易缓存。由于自增主键值是基本可以保证顺序生成的,因此自增主键通常更容易被数据库和查询缓存所利用,这有助于提高查询效率。

自增主键的缺点如下。

- 如果存在大量数据,那么自增主键可能会达到其取值范围的上限。
- 可能存在不连续的情况。在 MySQL 5.7 版本中,当服务重新启动时,有可能出现自增序列丢失问题。虽然这通常不会对应用造成太大的影响,但在某些需要连续主键值的场景下可能会成为问题。
- 自增主键通常是自动生成的,没有实际的业务意义。这意味着它不能直接用于业务逻辑或数据整合。
- 在分布式存储或分库分表的场景下,自增主键可能会遇到合并冲突的问题,因为不同的节点或数据库可能会生成相同的自增主键值。
- 由于自增主键的生成规律是可预测的,因此可能会面临一些安全风险,如被恶意用户利用进行数据篡改或注入攻击。

(2)将 UUID 作为主键

UUID(Universally Unique Identifier)是在一定的范围内(从特定的名字空间到全球)唯一的机器生成的标识符。

- 利用一定的算法由机器生成是为了保证 UUID 的唯一性。UUID 的生成规范定义了网卡 MAC 地址、时间戳、名字空间(Namespace)、随机或伪随机数、时序等元素,以及从这些元素生成 UUID 的算法。
- 非人工指定，非人工识别。UUID 是不能人工指定的。UUID 的复杂性决定了"一般人"不能直接从一个 UUID 知道哪个对象和它关联。
- 在特定的范围内重复的可能性极小。UUID 的生成规范定义的算法的主要目的就是保证 UUID 的唯一性。但这个唯一性是有限的，只在特定的范围内才能得到保证。
- UUID 通常以 36B（含短横线）的字符串表示，示例如下：3F2504E0-4F89-11D3-9A0C-0305E82C3301。

说明：虽然 UUID 去除"-"之后是 32B，但对于 MySQL 的主键来说字段还是太长，并且字段基本无意义。

将 UUID 作为主键的优点如下。
- UUID 的设计目标就是确保在全球范围内生成的标识符是唯一的。使用 UUID 作为主键可以大大降低出现重复主键的风险。
- UUID 对于分布式系统来说是一个很好的选择，因为它可以在不同的节点上独立生成而不会发生冲突。这对于需要跨多个数据库服务或节点存储数据的系统来说特别有用。
- UUID 具有全局唯一性，当从不同的数据源或服务器合并数据时，不必担心主键冲突的问题。这使得数据整合过程更加简单和可靠。
- UUID 本身是安全的，因为它是随机生成的，很难从 UUID 中推断出任何有关数据的信息。

将 UUID 作为主键的缺点如下。
- 占用存储空间大：UUID 通常比传统的整数型主键长得多（例如，标准 UUID 的长度是 36B），这意味着它会占用更多的存储空间，并可能增加索引的大小，从而影响查询性能。
- 性能降低：由于 UUID 较长，索引可能会变得更大，因此可能会增加数据库在查找和检索数据时的负担。
- 无法缓存大量数据：由于 UUID 的随机性，因此缓存的效率可能会降低。
- 插入性能下降：由于 UUID 的随机性，新插入的记录可能会被放置在索引页的任意位置，因此可能导致页分裂的发生。页分裂是一个"昂贵"的操作，因为它需要重新分配页空间并可能触发索引的重建。
- 索引维护成本高：UUID 的随机性还会导致索引结构不够紧凑，从而产生大量的碎片。碎片化的索引会降低查询性能，因为数据库需要扫描更多的索引条目来找到相关的记录。

下面介绍如何通过 Snowflake 算法生成唯一 ID。

Snowflake 算法是一个用于生成分布式唯一 ID 的算法，它通常用于大型分布式系统中，如 Twitter。原始的 Snowflake 算法生成的 ID 是一个 64bit 的整数。

利用 Snowflake 算法生成唯一 ID 的优点如下。
- 全局唯一性：由于 Snowflake 算法结合了时间戳、工作机器 ID 和数据中心 ID 等多个因素，因此它能够保证生成的 ID 在全局范围内是唯一的。
- 有序性：由于时间戳是 ID 生成的一部分，因此生成的 ID 是有序的，这对于某些需要按时间排序的场景（如日志记录、事件追踪等）非常有用。
- 高性能：Snowflake 算法生成 ID 的过程不需要访问数据库或其他外部系统，因此具有非常高的性能。
- 灵活性：算法中的各个部分（如时间戳、工作机器 ID 等）可以根据实际需求进行调整和配置，提供了较大的灵活性。

利用Snowflake算法生成唯一ID的缺点如下。

- 生成速度限制：虽然理论上每毫秒可以生成4096个ID，但在高并发场景下，这个速度可能会成为瓶颈。尽管可以通过等待到下一个毫秒来解决这个问题，但在某些需要极高生成速度的场景下，这可能会成为问题。
- ID空间浪费：如果系统在某些时间段内没有生成ID，那么这些时间段的ID空间就会被浪费。这在高负载和低负载交替出现的场景中尤为明显。
- 系统时间依赖：Snowflake算法依赖于系统时间，如果系统时钟被回退，则将会导致ID重复。虽然系统时间可以向前调整而不受影响，但需要确保系统时钟的准确性和稳定性。
- 缺乏业务含义：与业务相关的UUID相比，Snowflake算法生成的ID通常没有具体的业务含义，这可能会在某些场景下（如需要人类可读或可解释的ID时）造成不便。

（3）业务含义的主键

业务含义的主键指的是那些与业务逻辑紧密相关，能够直接反映业务实体特征的主键。这样的主键不仅具有技术上的唯一性，还能够直观地表达数据的业务意义，使得数据库的设计和使用更加符合实际业务需求。例如，对于订单编号，可以使用"地区+UNIX_TIMESTAMP+产品编号+随机号"（如SH+1637416219+P23+05）的方式生成业务含义的主键。

也可以结合Snowflake算法生成唯一业务含义的ID。例如，Snowflake算法生成Long类型前缀UUID（占8个字节），之后可在生成的UUID后面加2个长度的类型的业务意义的字符串。如下设置：

- 业务UUID=Long类型占8个字节+2个英文字节（20个长度的UUID）。
- 业务UUID=UUID去除"-"+2个英文字节（32个长度的UUID）。

5. 总结

对于MySQL应用开发和业务使用场景来说，主键是非常重要的。无论是否需要，通常建议为表设置主键。选择合适的主键对于构建可扩展、松耦合和高可用的系统至关重要。以下是一些关于主键的建议。

- 可以使用自增主键，例如：ID BIGINT（20）UNSIGNED NOT NULL AUTO_INCREMENT。
- 直接使用UUID作为主键。
- 根据业务需要，自定义主键（例如结合Snowflake算法和业务的ID）。
- 使用具有业务含义的主键。

9.3.9 字符集的设计

MySQL的字符集从最初的Latin-1演变到UTF-8，再到utf8mb4，这一历程相当曲折。在实际应用开发中，从一个字符集迁移到另一个字符集时，经常会面临一些挑战和无奈。尽管并非没有解决方案，但这一过程通常较为烦琐。在MySQL 8.0版本中引入了一个新的字符集系列utf8mb4_0900_*，这对于需要全面支持Unicode标准的用户来说，字符集的设计非常重要。

说明：Latin-1主要支持西欧语言字符，UTF-8支持大部分Unicode字符，而utf8mb4则进一步扩展了对Unicode的支持，包括一些新的字符、生僻字和表情符号。

1. 字符集介绍

在MySQL中，字符集（Character Set）是用于定义和表示存储在数据库、表和列中的字符的编码集合。字符集的选择对于确保数据的正确存储和检索至关重要，特别是在处理多语言或特殊字符时。MySQL 的字符集可以在多个层次上进行设置。除了字符集以外，MySQL 还使用排序规则（Collation）来定义字符如何进行比较和排序。查看 MySQL 的字符集相关设置的命令行示例如下。

```
mysql> SHOW VARIABLES WHERE variable_name LIKE '%character%' OR variable_name LIKE '%collation%';
+--------------------------------+---------------------------------------+
| Variable_name                  | Value                                 |
+--------------------------------+---------------------------------------+
| character_set_client           | utf8mb4                               |
| character_set_connection       | utf8mb4                               |
| character_set_database         | utf8mb4                               |
| character_set_filesystem       | binary                                |
| character_set_results          | utf8mb4                               |
| character_set_server           | utf8mb4                               |
| character_set_system           | utf8                                  |
| character_sets_dir             | /opt/idc/mysql8.0.34/share/charsets/  |
| collation_connection           | utf8mb4_bin                           |
| collation_database             | utf8mb4_bin                           |
| collation_server               | utf8mb4_bin                           |
| default_collation_for_utf8mb4  | utf8mb4_0900_ai_ci                    |
+--------------------------------+---------------------------------------+
```

从上述返回结果中可以了解到，client、connection、database、results 等各个环节是环环相扣的。如果在这些环节中的任何一个环节的字符集不兼容，那么都可能导致出现乱码问题。

在 MySQL 8.0 版本中，utf8mb4_0900_ai_ci 是一个新的默认字符集和排序规则，它提供了对 Unicode 字符的全面支持，并且具有一些改进的特性。

以下是对字符集命名中的各个部分的详细解释。

- utf8mb4：这表示字符集使用 UTF-8 编码，并且支持最多 4 个字节的字符。这是为了能够存储更多的 Unicode 字符，包括 emoji、生僻字和其他一些特殊字符。utf8mb3 也是 Unicode 字符集的 UTF-8 编码，每个字符使用 1~3 个字节。（UTF-8 是 utf8mb3 的别名。）
- 0900：这指的是 Unicode Collation Algorithm（UCA）的版本号。UCA 是用于比较和排序 Unicode 字符的算法。0900 表示该字符集使用的是 2009 年发布的 UCA 版本，它提供了一些改进和修正，使得排序更准确。
- ai：这表示排序是重音不敏感的（Accent-Insensitive）。这意味着在排序和比较时，不会考虑字符的重音差异。例如，在法语中，带有重音的字母和不带重音的字母在排序时会被视为相同。
- ci：这表示排序是不区分大小写的（Case-Insensitive）。在排序和比较时，大写和小写字母被视为相同。例如，在比较时，"APPLE" 和 "apple" 会被认为是相同的。

除此之外，在 MySQL 配置中，常用的字符集还有 utf8mb4_bin、utf8mb4_general_ci 和 utf8mb4_unicode_ci。

- utf8mb4_bin：将字符串中每个字符用二进制数据编译存储，区分大小写。
- utf8mb4_general_ci：由于没有实现 Unicode 排序规则，因此在遇到某些特殊语言或字符集时，排序结果可能不一致。但在绝大多数情况下，这些特殊字符的顺序并不需要那么精确。
- utf8mb4_unicode_ci：基于标准的 Unicode 来排序和比较，能够在各种语言之间精确排序。

Unicode 排序规则为了能够处理特殊字符的情况,实现了略微复杂的排序算法。

2. 字符集设置

MySQL 的字符集设置涉及多个方面,并且需要按照特定的顺序进行配置,以确保字符集的一致性和数据的正确显示。

1) 在 MySQL 环境初始化时就需要设置好字符集。初始化时设置字符集的方式如下。

```
[mysqld]
character-set-server = utf8mb4
collation-server = utf8mb4_unicode_ci
character-set-client-handshake = FALSE
# 此处忽略客户端的字符集,使用 MySQL 服务的设置
```

说明:除此之外,配置文件中可以使用 init_connect='SET NAMES utf8mb4'方式对每个连接的客户端执行初始化配置,但对于数据库超级管理员就不会生效。

2) 在 MySQL 使用中,可以对库、表和列单独设置字符集。单独设置字符集的命令行示例如下。

```
# 更改数据库
ALTER DATABASE `db1` DEFAULT CHARACTER SET utf8mb4 COLLATE utf8mb4_unicode_ci
# 更改表
ALTER TABLE `t1` DEFAULT CHARACTER SET utf8mb4 COLLATE utf8mb4_unicode_ci
# 更改列字段
ALTER TABLE `t1` modify `name` varchar(80) CHARACTER SET utf8mb4 COLLATE utf8mb4_unicode_ci COMMENT '昵称';
```

3) 检查字符集设置是否合理,可以通过系统表进行统计。统计字符集的命令行示例如下。

```
mysql > SELECT b.SCHEMA_NAME, b.DEFAULT_CHARACTER_SET_NAME,
b.DEFAULT_COLLATION_NAME,a.TABLE_NAME,a.TABLE_COLLATION
FROM information_schema.SCHEMATA b left join information_schema.TABLES a
on b.SCHEMA_NAME = a.TABLE_SCHEMA
WHERE b.SCHEMA_NAME not in
('information_schema','mysql','performance_schema','sys')
ORDER BY TABLE_SCHEMA,TABLE_NAME ;
```

通过以上提到的多方面字符集设置和检查,可以确保 MySQL 数据库中的字符集得到统一设置。这种统一设置对于避免字符集不规范问题至关重要,因为它确保了数据库、数据表、数据列以及客户端连接在字符编码上的一致性。

3. 字符集对于数据库的影响

MySQL 中的字符集设置对数据库的性能和稳定性有着显著的影响。当字符集设置不正确或不一致时,可能会导致下列多种问题。

- 数据损坏和乱码:如果数据库、表或字段使用了不同的字符集,那么当数据在这些不同的存储层之间传输时,可能会发生字符集不匹配的问题,导致数据损坏或显示为乱码。
- 性能下降:使用不合适的字符集或排序规则可能会影响查询性能。例如,使用较大的字符集可能会增加存储和检索数据时的开销,而某些排序规则可能不如其他规则高效。
- 索引和搜索问题:如果不同字段使用了不同的字符集,那么索引可能无法正常工作,导致搜索效率降低或搜索结果不准确。
- 连接和交互问题:如果客户端和数据库服务器使用了不同的字符集,那么在数据传输过程

中可能会出现乱码或数据丢失的问题。
- 存储空间浪费：某些字符集可能比其他字符集占用更多的存储空间，如果选择了不合适的字符集，则可能会导致不必要的存储空间浪费。

下面以示例形式探讨字符集不一致导致的问题。

【例 9-5】 隐式转换导致无法匹配数据。

如图 9-26 所示，虽然查询字段上有索引，但执行计划没有使用索引，因为字符集使用的是隐式转换。

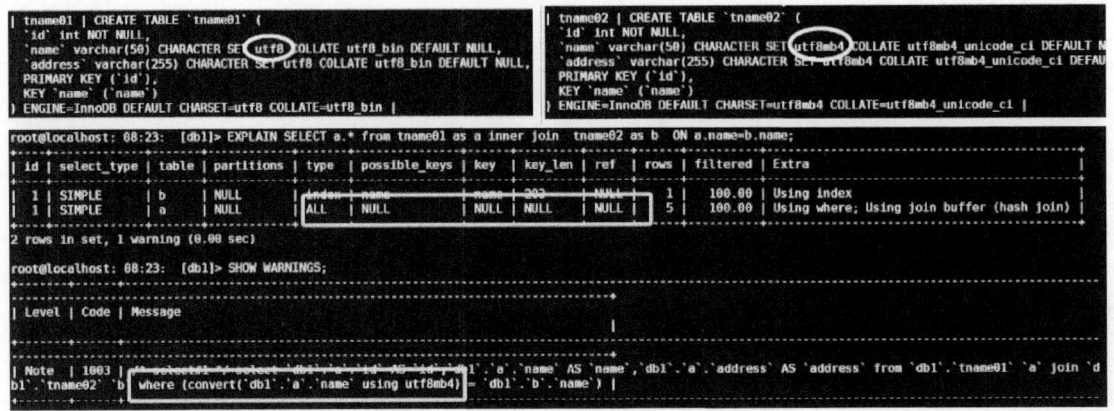

图 9-26　MySQL 字符集隐式转换

【例 9-6】 在数据后面尾随空格，导致查询结构错误。

在 MySQL 中，字符串值（CHAR、VARCHAR 和 TEXT）的比较与其他排序规则在尾随空格方面不同。图 9-27 所示为插入数据是 'a' 但查询数据为 'a ' （a 后有空格）时显示的结果。

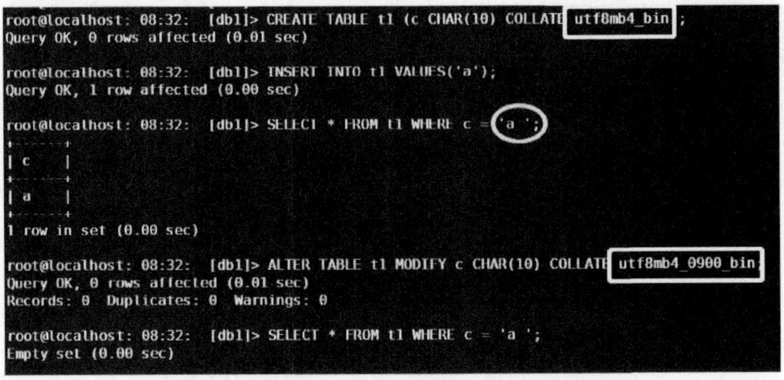

图 9-27　MySQL 字符集中空格对查询的影响

在 MySQL 中，对于字符集排序，系统表中记录的字符串末尾的空格也有对应的处理方式。查看字符集对应空格的处理方式的命令行示例如下。

```
mysql> SELECT COLLATION_NAME, PAD_ATTRIBUTE FROM INFORMATION_SCHEMA.COLLATIONS
    -> WHERE CHARACTER_SET_NAME = 'utf8mb4';
```

```
| COLLATION_NAME            | PAD_ATTRIBUTE |
+---------------------------+---------------+
| utf8mb4_general_ci        | PAD SPACE     |
| utf8mb4_bin               | PAD SPACE     |
| utf8mb4_unicode_ci        | PAD SPACE     |
| utf8mb4_icelandic_ci      | PAD SPACE     |
| utf8mb4_latvian_ci        | PAD SPACE     |
| ........                  |               |
| utf8mb4_0900_ai_ci        | NO PAD        |
| utf8mb4_de_pb_0900_ai_ci  | NO PAD        |
| utf8mb4_0900_bin          | NO PAD        |
+---------------------------+---------------+
```

说明：PAD_ATTRIBUTE 字段标识空格是否处理。在选择使用 utf8mb4_0900 字符集之后，空格就需要处理。

【例 9-7】数据的大小写敏感，导致查询结果不一致。

在 MySQL 中，除了参数 lower_case_table_names 以外，还提供了有效区分大小写的字符集。如图 9-28 所示，utf8mb4_bin 字符集可以区分大小写。因为使用 utf8mb4_bin 字符集，所以 WHERE 条件语句为' Jo%'和' JO%'时，都能准确匹配到对应的数据。

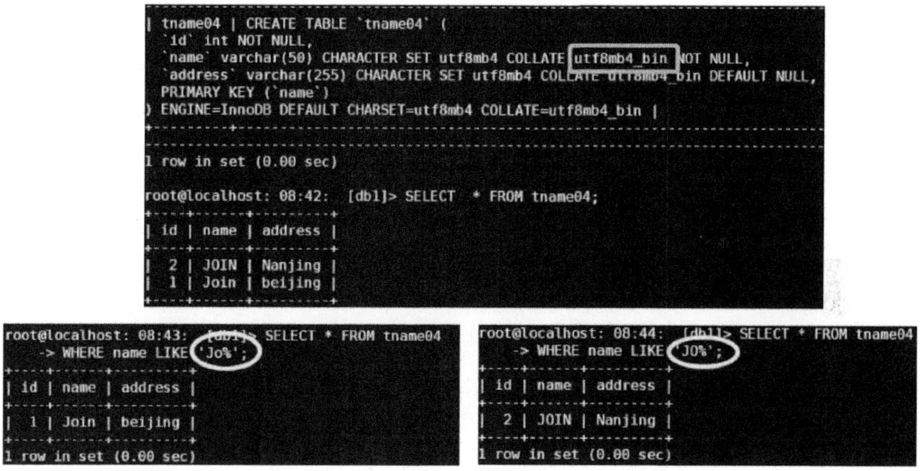

图 9-28　MySQL 使用 utf8mb4_bin 字符集查询对比

【例 9-8】表情符或生僻字导致无法插入。

图 9-29 所示为 MySQL 的表设置 utf8mb3 字符集，并插入表情符时，提示的错误信息 "Incorrect string value：'\xF0\x9F\x99\x82' for column 'name' at row 2"。对于不合法的字符串类型，需要把表改成 utf8mb4 字符集，同时在配置参数时设置 character-set-client-handshake 属性。这样就可以正常插入了。

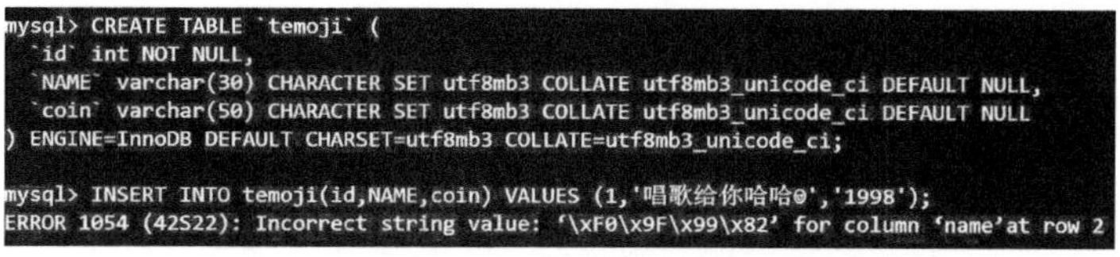

图 9-29　MySQL utf8mb3 下表情符插入数据失败

说明：对于很多应用驱动程序来说，没有 utf8mb4 字符集这样的说法，如 JDBC。

4. 总结

从 MySQL 的安装开始，就需要配置好字符集，并且数据库、表和字段的字符集需要与 MySQL 的配置文件中的设置保持一致。同时，MySQL 8.0 版本的 utf8mb4_0900 排序规则目前没有已知的 bug，但建议谨慎使用，推荐使用 utf8mb4_unicode_ci 排序规则。

9.3.10 MySQL 对 InnoDB 存储引擎、列、行格式的限制

在 MySQL 使用过程中，为了提高性能，以及更好地维护 B+树结构，对表的列数和行大小等做了限制。通常，当超过限制时，就会提示 ERROR 1118(42000)报错信息。

1. 列宽限制

当创建表超过列宽限制时，提示信息如下所示。

```
# 65535 列宽限制
mysql> CREATE TABLE t3(
       c1 VARCHAR(32765) NULL, c2 VARCHAR(32766) NULL
) ENGINE =MyISAM CHARACTER SET latin1;
ERROR 1118 (42000): Row size too large.The maximum row size for
the used table type, not counting BLOBs, is 65535.
This includes storage overhead, check the manual.
You have to change some columns to TEXT or BLOBs
```

说明：ERROR 1118（42000）：行数据最大长度太大。所使用的列类型（不包括 TEXT 或 BLOB）对应的行数据最大长度是 65535。必须将一些列更改为 TEXT 或 BLOB 类型。

2. 单行数据大小限制

innodb_page_size 参数是 MySQL InnoDB 存储引擎的基本存储单元，用于将数据存储到磁盘中。不同的页大小对 InnoDB 存储引擎的性能和存储空间占用有不同的影响。对于 4KB、8KB、16KB 和 32KB 的 innodb_page_size 设置，InnoDB 存储引擎将单行数据大小限制为略小于数据库页设置的一半。对于默认页（16KB 大小的页），限制为略低于 8KB。当创建表超过单行数据大小限制时，提示信息如下所示。

```
mysql> CREATE TABLE t4 (
       c1 CHAR(255),c2 CHAR(255),c3 CHAR(255),
       c4 CHAR(255),c5 CHAR(255),c6 CHAR(255),
       c7 CHAR(255),c8 CHAR(255),c9 CHAR(255),
       c10 CHAR(255),c11 CHAR(255),c12 CHAR(255),
       c13 CHAR(255),c14 CHAR(255),c15 CHAR(255),
       c16 CHAR(255),c17 CHAR(255),c18 CHAR(255),
       c19 CHAR(255),c20 CHAR(255),c21 CHAR(255),
       c22 CHAR(255),c23 CHAR(255),c24 CHAR(255),
       c25 CHAR(255),c26 CHAR(255),c27 CHAR(255),
       c28 CHAR(255),c29 CHAR(255),c30 CHAR(255),
       c31 CHAR(255),c32 CHAR(255),c33 CHAR(255)
) ENGINE=InnoDB ROW_FORMAT=DYNAMIC DEFAULT CHARSET latin1;
ERROR 1118 (42000): Row size too large (> 8126).Changing some columns to TEXT or BLOB may help.
In current row format, BLOB prefix of 0 bytes is stored inline.
```

说明：ERROR 1118 (42000)：行数据大小太大（> 8126）。将一些列更改为 TEXT 或 BLOB 类型可能会有所帮助。

从上述两个错误提示中可以看出，除 TEXT 或 BLOB 类型以外，列数据的长度总和不能超过 65535B，单行的实际数据保存记录不能超过 8126B。

相关官方说明如下。

- InnoDB 对表的最大行大小强制执行 65535B 的限制。表中 BLOB 和 TEXT 类型的列的总大小不计入此限制。
- InnoDB 表可以存储在行的主数据页中的最大数据量取决于 innodb_page_size 参数的值。单行在行的主数据页上最多可以消耗的数据大小是 innodb_page_size 参数值的一半。当默认值为 16KB 时，意味着单行在该行的主数据页上最多可以消耗 8KB 空间。但是，行的主数据页上的限制并不是行大小的绝对限制。

所有的 InnoDB 行格式都可以在溢出页中存储某些类型的数据，因此 InnoDB 表的最大行大小可以大于行的主数据页中可以存储的最大数据量。某些行格式可以在溢出页中存储比其他行格式更多的数据。

3. 列数限制

MySQL 对每个表都有最大列数为 4096 的硬性限制，但是对于给定的表，有效的最大值可能更小。确切的列数限制取决于以下 4 个因素。

- 表的最大行数据大小限制了列的数量（以及可能还有列的大小），因为所有列的总长度不能超过这个大小。
- 单个列的存储需求约束了符合给定最大行大小的列的数量。某些数据类型的存储需求取决于存储引擎、存储格式和字符集等因素。
- 存储引擎可能会施加限制表列数的附加限制。例如，InnoDB 存储引擎对每个表有 1017 个列的限制，但实际的限制可能更低，因为还需要考虑行的大小和其他因素。
- 功能键，如隐藏的虚拟生成列和存储列，也会对表的总列数限制产生影响，因为它们在表索引中被视为列。

4. 行数据大小限制

给定表的最大行大小由以下几个因素决定。

- MySQL 表的内部表示有一个最大行数据大小限制，通常为 65535B，即使存储引擎能够支持更大的行。对于 BLOB 和 TEXT 类型的列，它们只向行大小限制"贡献"了 9~12B，因为它们的内容与行的其余部分分开存储。
- 如果一行包含的变长列的数据超过了 InnoDB 存储引擎的最大行大小限制，那么 InnoDB 存储引擎会选择将超出限制的变长列数据作为外部页外存储。这样，行本身的大小就会符合 InnoDB 存储引擎的行大小限制。对于存储在页外的变长列，本地存储的数据量因行格式而异。
- 不同的存储格式使用不同数量的页头和尾数据，这会影响每行可用的实际存储量。
- InnoDB 和 MyISAM 等不同的存储引擎也有各自的行格式和限制。

5. InnoDB 存储引擎限制

在 MySQL 中，InnoDB 存储引擎有一些限制。了解这些限制有助于更好地使用 InnoDB 存储引擎存储数据。

- 一个表最多可以包含 1017 列。虚拟生成的列也包含在此限制中。
- 一个表最多可以包含 64 个辅助索引（包括主键）。
- 对于使用 DYNAMIC 或 COMPRESSED 行格式的 InnoDB 存储引擎表，索引键前缀长度限制为 3072B。对于使用 REDUNDANT 或 COMPACT 行格式的 InnoDB 存储引擎表，索引键前缀长度限制为 767B。
- 索引键的最大长度将按比例降低，基于 16KB 页大小对应的 3072B 的限制。
- 如果一行的长度小于半页，则该行的所有数据都存储在该页的本地部分。如果一行超过半页的大小，则会选择将可变长度列的数据存储到页外，直到该行的大小符合 InnoDB 的行大小限制。
- 尽管 InnoDB 内部支持大于 65535B 的行大小，但 MySQL 本身对所有列的组合大小施加了 65535B 的行大小限制。
- InnoDB 日志文件的最大组合大小为 512GB。
- 最小表空间大小略大于 10MB。表空间的最大大小取决于 InnoDB 页面的大小。例如，如果 InnoDB 页面大小为 16KB，则最大表空间大小约为 64TB。
- InnoDB 实例最多支持 2^{32}（4294967296）个表空间，其中一小部分表空间保留给 undo 和临时表。共享表空间最多支持 2^{32} 个表。

6. 行格式限制

表的行格式决定了行的物理存储方式，物理存储方式又会影响查询和 DML 操作的性能。由于单个磁盘页可以容纳更多行，因此查询和索引查找速度更快，缓冲池中所需的缓存更少，写入更新值所需的 I/O 操作也更少。

InnoDB 存储引擎支持 4 种行格式：REDUNDANT、COMPACT、DYNAMIC 和 COMPRESSED。

（1）REDUNDANT 行格式

使用 REDUNDANT 行格式的表将可变长度列值（VARCHAR、VARBINARY、BLOB 和 TEXT 类型）的前 768 个字节存储在 B 树节点内的索引记录中，其余的存储在溢出页上，大于或等于 768B 的固定长度列被编码为可变长度列，可以存储在页外。如果列的值为 768B 或更少，则不会使用溢出页，并且可能会节省 I/O，因为该值完全存储在 B 树节点中。这对于相对较短的 BLOB 列值很有效，但可能导致 B 树节点填充数据而不是键值，从而降低了效率。

（2）COMPACT 行格式

COMPACT 行格式是 REDUNDANT 的增强版，它减少了大约 20%的行存储空间，但代价是增加了某些操作的 CPU 使用量。工作负载受到 CPU 速度的限制，紧凑格式可能会慢一些。使用 COMPACT 行格式的表将可变长度列值（VARCHAR、VARBINARY、BLOB 和 TEXT 类型）的前 768 个字节存储在 B 树节点内的索引记录中，其余的存储在溢出页上。

（3）DYNAMIC 行格式

DYNAMIC 行格式提供与 COMPACT 行格式相同的存储特性，但增加了针对长可变长度列的增强存储功能，并支持大索引键前缀。

当用 ROW_FORMAT=DYNAMIC 创建表时，InnoDB 存储引擎可以完全在页外存储长可变长度的列值（对于 VARCHAR、VARBINARY、BLOB 和 TEXT 类型），索引记录只包含一个指向溢出页的 20B 指针。大于或等于 768B 的固定长度字段被编码为可变长度字段。

列是否存储在页外取决于页大小和行的总大小。当一行太长时，将选择最长的列作为页外存储，直到聚集索引记录适合 B 树页为止。小于或等于 40B 的 TEXT 和 BLOB 列按行存储。

在 DYNAMIC 行格式下，如果长数据值的一部分存储在页外，那么将整个值存储在页外通常是最有效的。使用 DYNAMIC 格式，较短的列可能保留在 B 树节点中，从而最大限度地减少给定行所需的溢出页数。DYNAMIC 行格式支持最多 3072B 的索引键前缀。

（4）COMPRESSED 行格式

COMPRESSED 行格式提供与 DYNAMIC 行格式相同的存储特性和功能，但增加了对表和索引数据压缩的支持。对于页外存储，COMPRESSED 行格式使用了与 DYNAMIC 行格式类似的内部细节，同时对表和索引数据进行了额外的存储与性能考虑，并使用了更小的页大小。对于 COMPRESSED 行格式，KEY_BLOCK_SIZE 选项控制在聚集索引中存储多少列数据，以及在溢出页上放置多少列数据。COMPRESSED 行格式支持最多 3072B 的索引键前缀。

使用 InnoDB 存储引擎的压缩特性，可以帮助提高原始性能和可伸缩性。压缩意味着更少的数据在磁盘和内存之间传输，并且占用更少的磁盘和内存空间。对于具有二级索引的表，好处会被放大，因为索引数据也被压缩了。压缩对于 SSD 存储设备尤为重要。当然，压缩也会消耗额外的 CPU 资源。

应用程序的总体性能、CPU 和 I/O 利用率以及磁盘文件的大小都是衡量应用程序压缩效率的有效指标。

7. 限制示例

在 MySQL 中，当尝试创建表或更改现有表的数据类型时，可能会遇到各种限制和提示信息。这些限制与字符集、数据类型、字段长度、行大小等多种因素有关。

下面提供两种常见的限制示例。

【例 9-9】 在 MySQL 中，当创建同长度但类型不同的字符串表时，有些创建语句会执行成功，有些则会失败。这取决于所使用的字符集，因为不同的字符集对存储空间有不同的需求。例如，UTF-8 字符集通常占用 2 个字节，utf8mb4 字符集占用 3 个字节，GBK 字符集占用 2 个字节，而 Latin-1 字符集则占用 1 个字节。图 9-30 所示为分别使用类型 CHAR 和 VARCHAR 创建表时提示的行大小限制信息。

```
root@localhost: 09:22:  [db9]> CREATE TABLE t_char4 (
    ->        c1 CHAR(255),c2 CHAR(255),c3 CHAR(255),
    ->        c4 CHAR(255),c5 CHAR(255),c6 CHAR(255),
    ->        c7 CHAR(255),c8 CHAR(255),c9 CHAR(255),
    ->        c10 CHAR(255),c11 CHAR(255),c12 CHAR(255),
    ->        c13 CHAR(255),c14 CHAR(255),c15 CHAR(255),
    ->        c16 CHAR(255),c17 CHAR(255),c18 CHAR(255),
    ->        c19 CHAR(255),c20 CHAR(255),c21 CHAR(255),
    ->        c22 CHAR(255),c23 CHAR(255),c24 CHAR(255),
    ->        c25 CHAR(255),c26 CHAR(255),c27 CHAR(255),
    ->        c28 CHAR(255),c29 CHAR(255),c30 CHAR(255),
    ->        c31 CHAR(255),c32 CHAR(255),c33 CHAR(255)
    -> ) ENGINE=InnoDB DEFAULT CHARSET=latin1 ROW_FORMAT=DYNAMIC;
ERROR 1118 (42000): Row size too large (> 8126). Changing some columns to TEXT or BLOB may help. In current row format
f 0 bytes is stored inline.
root@localhost: 09:22:  [db9]>
root@localhost: 09:22:  [db9]> CREATE TABLE `t_varchar4` (
    ->        c1 VARCHAR(255),c2 VARCHAR(255),c3 VARCHAR(255),
    ->        c4 VARCHAR(255),c5 VARCHAR(255),c6 VARCHAR(255),
    ->        c7 VARCHAR(255),c8 VARCHAR(255),c9 VARCHAR(255),
    ->        c10 VARCHAR(255),c11 VARCHAR(255),c12 VARCHAR(255),
    ->        c13 VARCHAR(255),c14 VARCHAR(255),c15 VARCHAR(255),
    ->        c16 VARCHAR(255),c17 VARCHAR(255),c18 VARCHAR(255),
    ->        c19 VARCHAR(255),c20 VARCHAR(255),c21 VARCHAR(255),
    ->        c22 VARCHAR(255),c23 VARCHAR(255),c24 VARCHAR(255),
    ->        c25 VARCHAR(255),c26 VARCHAR(255),c27 VARCHAR(255),
    ->        c28 VARCHAR(255),c29 VARCHAR(255),c30 VARCHAR(255),
    ->        c31 VARCHAR(255),c32 VARCHAR(255),c33 VARCHAR(255)
    -> ) ENGINE=InnoDB DEFAULT CHARSET=latin1 ROW_FORMAT=DYNAMIC;
Query OK, 0 rows affected (0.02 sec)
```

图 9-30　MySQL 的 CHAR 和 VARCHAR 类型列长度限制

【例 9-10】 在 MySQL 中，在同一张表中插入数据长度不一样时，有些插入语句会执行成功，有些则会失败。

对于插入数据长度小于 8126B 的场景（每列数据 240B×33 列 = 7920B），可以成功。相关插入命令行示例如下。

```
mysql>INSERT INTO t_varchar4 (c1,c2,c3,c4,c5,c6,c7,c8,c9,c10,c11,c12,c13,c14,c15,c16,
c17,c18,c19,c20,c21,c22,c23,c24,c25,c26,c27,c28,c29,c30,c31,c32,c33)
    values(repeat('a',240),repeat('a',240),repeat('a',240),repeat('a',240),repeat('a',240),
repeat('a',240),repeat('a',240),repeat('a',240),repeat('a',240),repeat('a',240),repeat('a',
240),repeat('a',240),repeat('a',240),repeat('a',240),repeat('a',240),repeat('a',240),repeat
('a',240),repeat('a',240),repeat('a',240),repeat('a',240),repeat('a',240),repeat('a',240),
repeat('a',240),repeat('a',240),repeat('a',240),repeat('a',240),repeat('a',240),repeat('a',
240),repeat('a',240),repeat('a',240),repeat('a',240),repeat('a',240),repeat('a',240));
Query OK, 1 row affected (0.01 sec)
```

对于插入数据长度大于 8126B 的场景（每列数据大小为 250B×33 列 = 8250B），则会失败。相关插入命令行示例如下。

```
mysql>INSERT INTO t_varchar4 (c1,c2,c3,c4,c5,c6,c7,c8,c9,c10,c11,c12,c13,c14,c15,c16,
c17,c18,c19,c20,c21,c22,c23,c24,c25,c26,c27,c28,c29,c30,c31,c32,c33)
    values(repeat('a',250),repeat('a',250),repeat('a',250),repeat('a',250),repeat('a',250),
repeat('a',250),repeat('a',250),repeat('a',250),repeat('a',250),repeat('a',250),repeat('a',
250),repeat('a',250),repeat('a',250),repeat('a',250),repeat('a',250),repeat('a',250),repeat
('a',250),repeat('a',250),repeat('a',250),repeat('a',250),repeat('a',250),repeat('a',250),
repeat('a',250),repeat('a',250),repeat('a',250),repeat('a',250),repeat('a',250),repeat('a',
250),repeat('a',250),repeat('a',250),repeat('a',250),repeat('a',250),repeat('a',250));
ERROR 1118 (42000): Row size too large (> 8126).Changing some columns to TEXT or BLOB may help.
In current row format, BLOB prefix of 0 bytes is stored inline.
```

因为 VARCHAR 类型长度是可以缩小的，所以创建时，列总字节不超过 65535B，表就会创建成功。但实际插入数据若超过 65535B（innodb_page_size 参数设置的半页长度），则还是会被限制插入。

8. 总结

在 MySQL 应用开发中，应该确保单个行的总大小不超过 innodb_page_size 参数值的一半，同时也不应超过 65535B 的限制。对于可能产生行溢出的列（如 VARCHAR、VARBINARY、BLOB 和 TEXT 类型），虽然其数据可以存储在页外，从而突破半页的限制，但每次跨页操作都会对性能产生不小的影响。

第 10 章
MySQL 8.0 云数据库建设

10.1 云数据库的概念和发展趋势

随着云计算技术的广泛应用和深入发展，云数据库逐步成为云计算领域新的方向和热点，目前已经得到各行各业的认可和重视。

10.1.1 云数据库的概念

云数据库是一种基于云计算平台的数据库服务，其核心在于将数据库的存储、计算和网络访问功能全部迁移到云计算环境中。其中 DBaaS（DataBase as a Service，数据库即服务）是云计算领域中的一种重要服务模式，它为企业提供了一种高效、灵活、成本效益高的数据库解决方案。Gartner 认为，DBaaS 是任何数据库管理系统（DBMS）或数据存储，设计为可伸缩、弹性、多租户的订阅服务，具有一定的自我管理能力，并由云服务提供商（CSP）或 CSP 基础设施上的第三方软件供应商销售和支持。

DBaaS 是一种云计算服务模型，为用户提供对数据库的某种形式的访问，而无须设置物理硬件、安装软件或配置性能。所有管理任务和维护都由服务提供商负责，所有用户或应用程序所有者需要做的就是使用数据库。当然，如果客户选择对数据库进行更多控制，则该选项可用，并且可能因提供商而异。

简而言之，DBaaS 是指将数据库的管理移到云计算环境的一种服务。云计算相关的实践证明，相比传统的数据库运行支撑环境，DBaaS 是一种属性更为优良的数据库相关资源、能力或者运行环境的供给与管理平台。

DBaaS 的引入，深刻改变了传统的资源部署、资源供给、资源运行和资源维护模式。

一方面，通过数据库资源池的合理构建，增强了数据库资源的弹性、共享性和可扩缩能力。另一方面，通过规范化、标准化、自动化、无差错的数据库软件部署方式，全面提高了数据库资源（能力或者运行环境）交付过程的效率、敏捷度、灵活性和可靠性。同时，依托 DBaaS 先进的运维理念和丰富的运维工具，大幅减少了运维人员各种烦琐、重复的日常运维工作，从根本上提升了数据库的日常运维效率和运维质量。

10.1.2 云数据库的发展趋势

纵观整个数据库产业，云数据库在最近几年逐渐成为业界焦点。Private Cloud dbPaaS 已经成为最受关注的技术领域之一,成为产业数字化、数据库基础设施数字化的基础。"

中国信通院在 2023 年发布的《数据库发展报告》中提及:"云数据库正在引领全球数据库市场稳步增长。根据 IDC 统计数据,在数据爆发式增长、数据复杂度提升的驱动下,2019 年全球 DBMS 市场规模高达 493 亿美元,同比增长 18.2%。2020 年,全球 DBMS 市场规模小幅降至 487 亿美元,主要原因是数据库市场受到疫情冲击,本地部署的数据库项目被推迟,同比下滑 6.2%,但同时云数据库仍保持 11.6% 的增速。长期来看,数据量高增叠加数据结构复杂度提升是长期驱动因素,预计云数据库将成为未来数据库主要部署方式。根据 IDC 预测,2024 年全球数据库市场规模将稳步增长至 739 亿美元,2020~2024 年 CAGR 将达 8.7%,其中云数据库市场规模将达到 404 亿美元,占比提升至 55%。"

云和恩墨公司创始人、国内知名数据库专家盖国强在多次数据库大会上,对当前的趋势发表看法,他表示 DBaaS 建设是大势所趋,但长期的行业趋势不是将所有数据库转移到公共云,而是将云体验转移到数据库上来,并认为 to B(to Business,面向企业)市场的认同成为数据库成败的关键,未来云上和云下的融合是大势所趋。如图 10-1 所示,云计算形态会走向混合云。

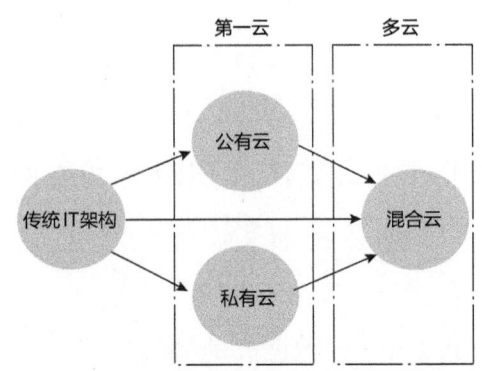

图 10-1 混合云是云时代的发展趋势

随着 DBaaS 技术的成熟,企业逐渐摆脱跟风的状态,云平台能力将逐渐演进为 PaaS 层能力,以构建适合企业或者行业特点的 DBaaS 平台。作者作为解决方案架构师和售前咨询顾问,常年工作在面向企业用户的一线,能够第一时间获取各个企业用户的需求。当前企业的云平台呈现不同的形态,有些企业希望部署到公有云,有些企业自建私有云,有些企业构建混合云,还有一些企业进行公有云下移,但总体上是向着混合云方向演进的。企业用户构建私有化 DBaaS,在获得云平台所具备的低成本、可用性、易用性、扩展性和并行处理等优势的同时,又能兼顾 IT 系统的自主性和安全性,因此构建适合企业特色的 DBaaS 平台是大势所趋,可把云的体验转移到数据库上来。

艾瑞咨询在近些年的中国数据库行业研究报告中也有专门章节阐述了这种趋势,报告中提到"部分企业出现反向迁移情况,混合部署成为未来常态。一方面,尽管上云是大势所趋,但是由于数据库基础软件的特性和公司战略考虑,在一定时间内,云数据库很难完全替代本地数据库,混合部署成为企业的必然选择。现阶段绝大部分企业都具有一定的 IT 基础,业务数据都存储在本地自建的数据库里,经过了几十年的积累,具有复杂和海量的特点。短时间内让企业放弃原本投入了大量成本的本地数据库,把海量复杂的数据全面迁移上云,是不现实且不划算的。另一方面,企业私有云部署成为当下的热门选择,公有云数据库市场增速放缓。当企业业务发展到一定规模时,对核心系统自主可控的要求会有相应提升,这一阶段的企业反而出现了反向迁移的现象,更多地考虑把部分业务数据从公有云迁移到私有云部署的环境里。"

10.1.3 云数据库面临的挑战

当前大部分企业的数据库云环境,还是采用传统"烟囱式"的建设模式,存在不规范、建设随意、效率低下、缺少管控等大量问题,对业务的安全、稳定、连续、高效运行带来极大挑战,制约企业业务发展。

作者结合多年的工作经验，总结了目前大部分企业数据平台环境中存在的主要问题，具体说明如下。

- 单点隐患问题：大部分数据库都没有实现本地高可用，一旦主机或者存储出现问题，业务的连续性就将受到很大的影响，甚至会出现数据丢失的重大风险。
- 数据库备份问题：大部分数据库没有备份，甚至日志没有开启归档模式，存在出现问题后直接导致数据丢失的风险。
- 数据库字符集问题：在建立数据库时没有规范和标准，导致使用的字符集不统一，甚至使用 GBK 和 UTF-8 两种字符集，这样会导致数据迁移时中文出现乱码。
- SQL 质量问题：存在不少低效的 SQL 语句，将直接影响数据库的运行性能。
- 操作系统及数据库版本问题：选择使用的操作系统版本多种多样，数据库版本不一致，甚至有些版本已经超过厂商支持期限，存在安全、性能、稳定性以及维护方面的问题。
- 权限问题：数据库中业务使用的是 DBA 的最高权限，能够操作所有数据库，这具有很大的隐患。权限管理应该规范，业务用户和 DBA 用户必须要分开。
- 系统负载不一，导致资源浪费：很多系统负载非常低，导致资源利用率较低。硬件资源浪费的主要原因是资源配置不合理。
- 系统无标准，运维压力大：新系统上线时缺乏数据架构的设计、规范和标准，造成"各自为政"的局面。数据库上线后缺少控制，导致后端数据库越来越多，运维压力越来越大。
- 业务运行速度越来越慢：系统上线后，没有进行有效的日常运维，导致系统中存在大量无效对象、重复数据、索引和视图，若不进行定期的清理和归档，就将导致系统越来越慢。
- 缺少专业化运维能力：若没有专业化的数据库运维团队，那么，当出现问题时就会常采用简单粗暴的"重启"方式来处理，缺少潜在风险的提前发现能力、疑难故障的快速处理能力以及极端情况的紧急救援能力。

近年来，随着数字化转型战略不断推进，组织的数据库类型和数量与日俱增，数据库技术已经从单一架构支持多类应用演变为多类架构支持多类应用，数据库管理进入多数据源、多云环境混合架构的时代。企业的数据库结构越来越复杂，云数据库既要满足企业数据库所需的安全、稳定、连续的稳态的要求，又要能够快速响应前端业务诉求、敏捷高效的敏态的诉求。传统的供给应对面对多元异构的管理难题具有局限性，企业建立 DBaaS 平台面临巨大的挑战，作者认为主要包括以下几方面的挑战。

- 数据库平台承载层面的挑战。数据库运行环境复杂，可能存在私有云、公有云、混合云，如何同时兼顾性能、容量、成本、扩展能力、可靠性等方面的要求是当前企业面临的巨大挑战，企业需要构建"高性能、高可靠、弹性扩展、云化共享"的数据库支撑架构环境，降低资源总投入成本，建设稳定高效的数据库基础架构环境。
- 数据库服务处理层面的挑战。企业大部分业务系统上线时，采用传统的"烟囱式"建设模式，每一套系统对应一套或多套数据库，而且数据库种类繁多，数据库版本也多种多样，很多版本甚至都已经超过原厂的支持期限，存在严重的安全隐患，企业需要给每套业务系统分配数据资源，带来不少管理问题和风险隐患，此时迫切需要建立标准、开放兼容的数据库资源池环境，同时满足企业存量和新增数据库环境需求。
- 数据库运维管理层面的挑战。随着数据库类型和数量的增加，人工运维难度越来越大，数据库资源申请变得缓慢，运维效率逐渐低下，难以实现对多元异构数据库资源的全生命周期、精细化、一体化、高效运维管理。想要降低运维难度、提高运维管理效率，迫切需要构建一体化运维管理和服务平台，以租户化形式给不同使用用户提供对应的数据库服务能力，以提升服务交付效率和运维效率。
- 数据库规范标准层面的挑战。当前大部分企业的数据库云环境，还是采用传统"烟囱式"的建设模式，以前端应用侧或者应用厂商的需求为主导，存在不规范、建设随意的特点，基本是一个业务一套系统，没有标准的安装配置管理，没有统一的数据库资源申请、使用、运维管理等方面的规范，特别是在当前国产化背景下，针对新引入的数据库管理系统，迫切需要建立适合企业特点的数据库标准规范体系。
- 数据库运维保障层面的挑战。随着企业业务系统上线规模不断扩大，平台系统运维压力也日渐凸显。目前各企业运维水平不一，特别是针对开源、国产数据库，运维技能薄弱。一些企业长期使用国外商业数据库技术栈，面对开源及国产数据库时的管理能力较低，学习周期长。想要实现出现问题时的快速故障处理，迫切需要建立专业的数据库运维保障团队。

10.2 MySQL 云数据库设计方法

在作者看来，针对包括 MySQL 数据库在内的任何一种云数据库，在设计的时候，需要遵循一些基本的设计原则。以下设计原则可供参考借鉴。

- 先进性原则：平台技术架构业内领先、技术成熟，以保证平台具备不断发展和扩充的空间。同时立足高起点规划、高标准建设，充分把握数字化发展趋势，系统上线后能够满足未来 3~5 年的业务发展需求。
- 开放性原则：平台相关的组件应尽可能选择开放、开源技术，便于企业对平台技术架构进行系统了解、深入掌握，最终实现对平台的完全自主可控。
- 经济性原则：平台通过合理的技术成本，完成平台的整体建设。在平台的各组件中，优先考虑开源的技术组件，从而降低平台的经济成本。
- 高可用原则：平台的架构需要具备高可用的能力，当在系统的任一组件出现异常时，通过主备或者集群等高可用技术方式，确保系统在短时间内进行恢复，保证业务持续处理。
- 高性能原则：平台具备高效快速的数据处理能力，确保在多任务大数据量情况下仍能快递、高效、准确地进行处理。
- 高扩展原则：平台总体架构和软件体系结构要有可扩展性，要充分考虑到未来业务的发展带来的数据规模的发展、管理需求的变化以及系统保障级别的提高，方便对新需求的扩展和支持，软件体系结构不依赖于硬件设备。
- 维护原则：平台具备自监控性，能够监控平台运行情况，具有平台出现异常时的告警能力。在平台出现异常时，能够通过图形化界面或日志，快捷、方便地定位出错误位置、原因，有利于平台维护。
- 开放兼容原则：平台需要兼容各种硬件架构、各种数据库环境，采用开放的技术架构，以及能够适应存量和增量的数据库环境要求。

遵循以上设计原则，可实现企业独特的"标准、自动、敏捷、自服务"的云数据库。云数据库设计应充分考虑企业实际情况，按照"分层解耦、异构兼容"等标准规范，提供硬件异构兼容和组件分层解耦的开放兼容能力，以及 DBaaS 平台调度和资源监控的统一管理能力。

整个 DBaaS 平台使用分层解耦的架构设计（见图 10-2）。

- 一是层次化基础架构承载层。为了适应各种数据库场景需求，满足业务发展需要，建议采用层次化设计,可以考虑按需规划高配区、中配区、低配区及敏捷区 4 种承载架构。对于整个基础架构承载层,不必关心底层硬件结构,它可以运行在多种 IaaS 平台(如 OpenStack、VMware)上,也可以直接运行在物理机上,也就是说,平台本身也支持部署在 x86、ARM 等型号芯片的机型上。当然,各企业可以结合自己的特点,按需选择适合自己的底层承载层架构。

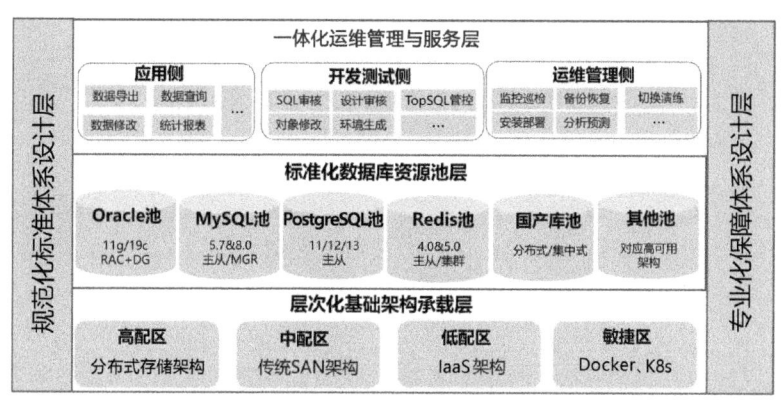

图 10-2 DBaaS 平台总体设计架构

- 二是标准化数据库资源池层。规划设计标准的适合自己当前现状和发展需求的标准化数据库池,包括 Oracle 池、MySQL 池、PostgreSQL 池、Redis 池、国产库池(集中式、分布式)、其他池等多个标准化数据库资源池,确保企业核心业务系统(兼顾存量和新增)数据平台的规范稳定运行,并分阶段完成各资源池的建设工作。当然,各企业应该结合当前数据库的现状和业务发展要求,按需构建符合企业自身特色的数据库池。
- 三是一体化运维管理与服务层。通过建设支持多中心多种数据库环境的统一数据库云化管理平台,以租户化的方式给应用侧、开发测试侧、运维管理侧提供不同的数据库服务能力和自助服务界面,实现数据库服务的自助申请、自助开通的能力。
- 四是规范化标准体系设计层。建设规范化的针对各种数据库的规范体系。通过将数据库领域专家多年积累的经验进行归纳、总结,结合企业的运维管理规范要求,建立适应企业发展的数据库标准规范体系。
- 五是专业化保障体系设计层。构建适合企业自身特点的运维保障体系。要建设专业化的运维保障团队,应该考虑建设进行日常运维和运行支撑的一线团队,以及提供兜底保障和持续赋能的二线团队或者外部专业团队,实现数据库全生命周期专业化运维。

以上架构设计方法适用于各种云数据库的设计,本书中,作者仅以 MySQL 云数据库为例,详细阐述这种设计方法。

10.2.1 层次化基础架构承载层设计

基础架构作为 MySQL 数据库运行的底座,直接关系到 MySQL 的运行是否安全、稳定、连续和高效,以及上层业务的连续性。良好的基础架构承载层设计,不仅能够满足各种业务场景的需求,还能够实现企业的降本增效的目标。

1. 常见承载层基础架构技术

数据库承载的基础架构承载层目前是非常成熟的技术,随着技术的逐步发展,出现了很多种基础架构,主要包括传统 SAN 架构、IaaS 虚拟化架构、容器化架构和专用数据库分布式存储架构等。

下面对上述几种架构进行简单说明。

(1) 传统 SAN 架构

存储区域网络(Storage Area Network,SAN)是一种在服务器和外部存储资源或独立的存储资源之间实现高速可靠访问的专用的高速网络。SAN 采用可扩展的网络拓扑结构连接服务器和存储设备,每个存储设备不隶属于任何一台服务器,所有的存储设备都可以在全部的网络服务器之间作为对等资源共享。目前常见的 SAN 有 FC SAN 和 IP SAN 两种类型,其中 FC SAN 通过光纤通道转发 SCSI 协议,IP SAN 通过 TCP/IP 通道转发 SCSI 协议。

图 10-3 所示为典型的 FC SAN 架构。

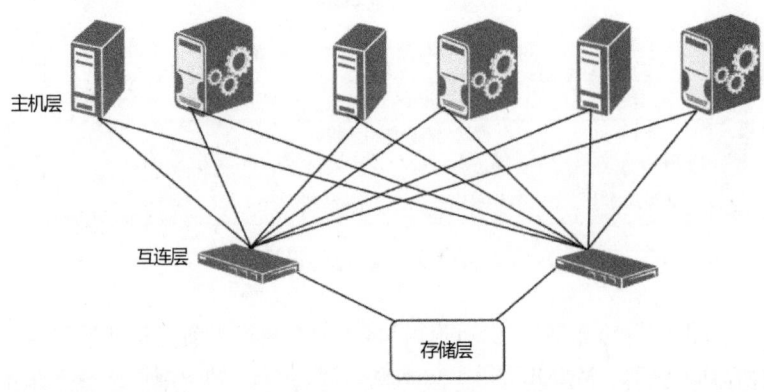

图 10-3 FC SAN 架构

这种架构主要包含以下几部分。

- 主机层:由两台或多台 PC 服务器组成,其上安装和运行 MySQL 数据库集群软件。
- 互连层:由两台 SAN 光纤交换机组成,通过它将上层服务器和下层存储互连起来,组成 SAN 网络。
- 存储层:用来集中存放数据。

(2) IaaS 虚拟化架构

虚拟化(Virtualization)是一种资源管理技术,它将计算机的各种实体资源,如服务器、网络、内存及存储等,予以抽象、转换后呈现出来,打破实体结构间的不可切割的障碍,使用户可以比原本的组态更好的方式来应用这些资源。虚拟化技术种类很多,如软件虚拟化、硬件虚拟化、内存虚拟化、网络虚拟化(虚拟 IP)、桌面虚拟化、服务虚拟化、虚拟机等,这里就不一一阐述了。

在目前市面上,比较常用的虚拟化架构形式包括基于传统 SAN 架构的虚拟化以及基于超融合基础架构的虚拟化。

1) 基于传统 SAN 架构的虚拟化。它基于传统的 SAN 架构,借助虚拟化软件实现整个基础架构的虚拟化。图 10-4 所示为典型的 VMware 加传统 SAN 架构的虚拟化

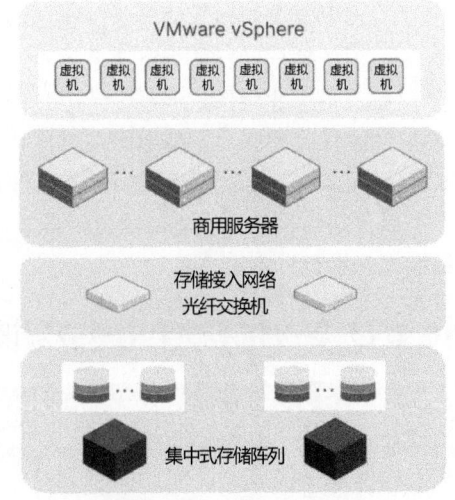

图 10-4 VMware 加传统 SAN 架构的虚拟化架构技术

架构技术。

2）基于超融合基础架构的虚拟化。超融合基础架构也称为超融合架构，是指在同一套单元设备（x86 服务器）中不仅具备计算、网络、存储和服务器虚拟化等资源与技术，还包括缓存加速、重复数据删除、在线数据压缩、备份软件、快照技术等元素，而多节点可以通过网络聚合起来，实现模块化的无缝横向扩展（scale-out），形成统一的资源池。图 10-5 所示为超融合基础架构的核心，首先是分布式存储技术对传统存储的替代，其他更多优势（如基于 x86 服务器构建、并发与易于扩展）都是这种替代带来的。典型的厂家有 SmartX、Nutanix、VMware。

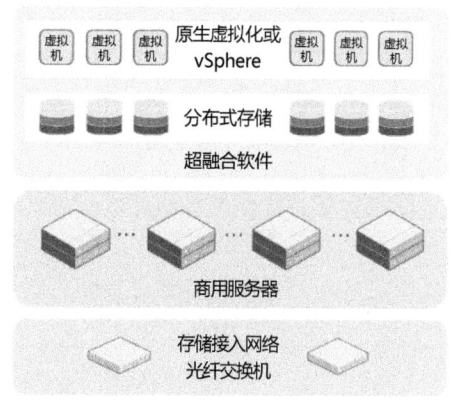

图 10-5　超融合基础架构

虚拟化的本质是资源的抽象化，要想资源充分利用，必须把资源最小单位化（池化），这样上层才能按需使用资源。虚拟化不但解放了操作系统，还解放了物理硬件，大大提高了资源的利用率。这种架构天然适用于低压力负载的应用场景，如 Web 应用、程序应用、低压力负载的数据库等。

（3）容器化架构

虚拟化技术通过 Hypervisor 首先让虚拟机与底层硬件解耦，实现了资源的有效利用和系统的弹性扩展能力，但是对于追求极致敏态的互联网场景，这还不够，因为虚拟机安装一整套操作系统软件这种方式还是"太重"了，所以就产生了一个新的基于操作系统的虚拟化技术，即容器技术。容器（Container）技术是一种更加轻量级的操作系统虚拟化技术，将应用程序及其运行依赖环境打包封装到标准化、强移植的镜像中，通过容器引擎提供进程隔离、资源可限制的运行环境，实现应用与操作系统平台及底层硬件的解耦，一次打包，随处运行。容器基于镜像运行，可部署在物理机或虚拟机上，通过容器引擎与容器编排调度平台实现容器化应用的生命周期管理。

表 10-1 所示为虚拟机和容器的主要区别。

表 10-1　虚拟机和容器的主要区别

特　　性	虚　拟　机	容　　器
隔离级别	操作系统级	进程级别
隔离策略	Hypervisor	CGroups
占用的系统资源	5%~15%	0~5%
启动时间	分钟级	秒级
镜像存储	GB 或 TB 级别	KB 或 MB 级别
集群规模	上百	上万
高可用策略	备份、容灾、迁移	弹性、负载、动态

（4）专用数据库分布式存储架构

最近几年，传统数据库架构进行了一些演化。随着 x86 技术的发展，特别是 Intel x86 CPU 性能的巨大进步（最新的 Intel x86 CPU 的性能基本接近 IBM 的 Power CPU）和 Linux 操作系统的迅猛发展（Linux 的稳定性、性能、扩展性和兼容性都已经超过传统的 UNIX 系统），在数据库领域，x86 服务器被普遍用于代替小型机。

很多厂商都推出了基于 x86 架构的高性能数据库产品，因此这里称这种架构为 x86+分布式存储架构。x86+分布式存储架构将计算层与存储层分开，实现功能的隔离，可根据需要灵活地对计算或者存储资源进行扩容，性能与稳定性方面互不干扰，提高了整体系统的可用性。

图 10-6 所示为专用数据库分布式存储架构。

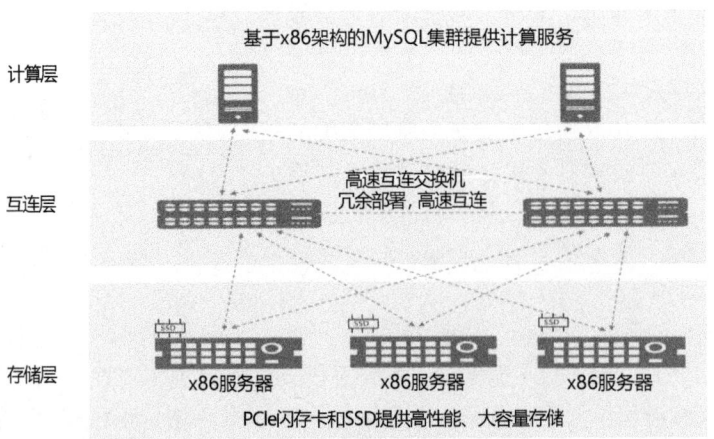

图 10-6　专用数据库分布式存储架构

专用数据库分布式存储架构从下到上分为 3 层：存储层、互连层和计算层。

存储层由若干个存储节点组成，每个存储节点都会配置 PCIe 闪存卡和 SSD 磁盘，形成一个独立的存储单元。多个存储节点组合在一起，形成一个存储池，性能和容量可按需线性扩展。

互连层采用分布式存储管理软件，通过两台高速互连交换机实现存储节点和计算节点间的高速互连，防止单点故障。

计算层由两个计算节点组成，其中部署数据库集群软件，通过增加计算节点，可以实现系统计算能力的线性扩展。计算节点通过多链路访问存储节点资源，单机故障不会影响数据库使用或导致数据丢失，同时可实现存储在线扩容和数据重分布。

这个架构的本质是采用软件定义的技术，通过高带宽低时延的高速互连交换机，将计算节点服务器和以闪存为核心的存储节点服务器进行软硬件配合，紧密耦合成高性能计算和存储资源池，实现了资源池化、资源弹性伸缩的能力，提高了数据库系统所需的极致性能、高可用性。

2. 承载层基础架构设计建议

在作者看来，以上 4 种架构，各有各的优势和适用场景，用户应该根据自身的业务特点，结合现状、投资预算等各个方面，进行层次化的设计，满足各个业务场景的需求。

根据作者多年咨询顾问经验，建议用户在设计承载层基础架构时，要充分考虑利旧和投资保护，除非有必要新建，否则尽可能少地去改变现有的架构。但是，规划合理的区域是很有必要的，所以强烈建议采用层次化设计，根据自身的业务特点规划几个不同的区域。

承载层基础架构技术不是本书的重点，作者不进行过多阐述，目前这种技术已经很成熟，市面上有很多成熟的解决方案。

10.2.2　标准化 MySQL 资源池设计

在基础架构已确定时，下一个关键步骤就是进行 MySQL 资源池的设计，这一层的设计的关键是标准化，其中包含 MySQL 数据库版本的标准化、高可用架构的标准化等方面，下面进行详细阐述。

1. MySQL 数据库版本的标准化

对于每个版本的特点，在本书前面的章节中已有涉及，这里不再重复阐述，重点说明如何实现版本的标准化。

对于 MySQL 数据库的版本选择，作者基于多年的实践经验，建议按照以下原则进行选择。

- 选择开源的社区版的稳定版本（GA 版本）。
- 选择发布 6 个月以上的 GA 版本。
- 要选择前后几个月没有大的 bug 修复的版本，而不要选择大量修复 bug 的集中版本。
- 要考虑开发人员开发程序时使用的版本是否兼容你选择的版本。
- 作为内部开发测试数据库环境，要运行大概 3~6 个月的时间。
- 企业非核心业务可采用数据库的新版本 GA 版本。
- 向 DBA 高手请教，或者在技术氛围好的群里和大家一起交流，使用真正的高手们用过的好用的 GA 版本产品。
- 不要盲目追求新版本，需要考虑现有环境的情况，以及尽量减少对生产环境的影响。

结合以上选择原则，作者建议，对于需要采用 MySQL 8.0 以下版本的系统，建议选择 5.7.30 以上版本，因为 MySQL 5.7 版本已经被广泛使用，一些问题可通过预先的告警或设置方式避免。而对于新增的业务系统，建议选择目前稳定可靠的 MySQL 8.0 版本。

2. MySQL 数据库高可用架构的标准化

高可用（High Availability）是系统架构设计中必须考虑的因素之一。它通常是指，通过设计减少系统不能提供服务的时间。如果一个系统能够不间断地提供服务，那么这个系统的可用性为 100%。如果一个系统每运行 100 个时间单位，就会出现 1 个时间单位无法提供服务的情况，那么该系统的可用性是 99%。目前大部分企业的高可用目标是 4 个 9，即 99.99%，也就是允许这个系统的年停机时间为 52.56 分钟。

MySQL 数据库常用的高可用架构介绍如下。

（1）MySQL 主从架构

主从复制是 MySQL 自带的功能，无须借助第三方工具，通过二进制日志将主库内容持续传输到从库上，然后由从库重做日志，从而使得从库和主库保持数据的一致。可从一个主库复制到一个或多个从库。MySQL 默认采用异步复制方式，从库可以复制主库中的所有或特定数据库，或者特定的表。

（2）MGR 高可用架构

MySQL Group Replication（MGR）是 MySQL 官方推出的一种基于 Paxos 协议的状态机复制机制，解决了传统的异步复制和半同步复制中数据一致性无法得到保证的问题。由若干个节点共同组成一个复制组，一个事务必须经过组内大多数节点（$N/2+1$）决议并通过，才能提交。

（3）MHA 高可用架构

MHA 架构是一种一主多从的数据库高可用解决方案。它的特点是在保证高可用自切换的前提下，最大限度地保证主从数据的一致性。在 MySQL 故障切换过程中，MHA 能做到在 0~30 秒之内自动完成数据库的故障切换操作，并且能做到故障自动检测和切换，扩展性较好，可靠性更高。

（4）Galera Cluster 架构

Galera Cluster 是 Codership 公司开发的一套免费开源的高可用方案（官网地址：http://galera-cluster.com），具有 multi-master 特性，支持多点写入。Galera Cluster 中的三个（或多个）节点都是

对等关系，每个节点均支持写入，集群内部会保证写入数据的一致性与完整性。

Galera Cluster 的实现方案有两种：PerconaXtrDB Cluster、MariaDB Galera Cluster。

以上只是列出了一些常用的高可用架构方案，具体的原理和复制机制已在本书第 3 章进行了详细阐述，这里不再重复阐述。

针对 MySQL 数据库，在进行架构的规划设计时，建议优先选择原生的主从复制高可用架构来搭建 MySQL 数据库资源池环境，可以满足大部分业务系统的高可用要求。随着 MGR 架构技术的愈发成熟，对于高可用要求更高的业务系统，可以考虑采用这种架构。

因为 MySQL 是开源产品，所以与 MySQL 数据库配套的产品五花八门，有些是开源的，有些是商业的，有些是过时陈旧的，不适用于目前主流数据库，等等。DBA 很难有时间和精力来完整测试和选择合适的开源软件构建 MySQL 数据库的运行环境。作为 DBA，或者传统企业的运维团队，可能没有足够的开发力量来整合开发并维护 MySQL 数据库运行所必需的整体环境。MySQL 数据库运行环境的组件，在必要的情况下需要进行二次开发或定制化。所以，为了更好地进行 MySQL 高可用架构的管理，就需要辅助使用一些自动化的高可用架构管理工具，实现 MySQL 资源池的管理，并借助专业的服务团队来提供支持保障。

10.2.3 一体化运维管理与服务设计

通过进行一体化运维管理与服务设计，建立适合企业自身特点的 MySQL 云数据库（DBaaS），能够实现 MySQL 数据库的标准化、自动化、集中化、智能化管理，大幅提升 MySQL 数据库服务的获得效率，集约化资源使用，提供 MySQL 主从高可用、MGR 高可用、MySQL 读写分离等服务，固化企业最佳实践。

1. 能力设计

通过 MySQL 云数据库的建设，企业在提高运维效率、降低运维难度的基础上，可最大程度消除运维工作对于运维人员的过度单一依赖，逐步削减组织的人力成本投入，同时，也使得现有运维人员能够从烦琐、复杂的日常运维工作中解放出来，适度减轻运维人员的运维压力，使运维人员能够将更多的时间和精力投入到业务模式与应用系统创新等更具价值的工作中。例如，开发工程师能够更加专注于业务逻辑，运维工程师能够更加专注于前置性风险管控，产品工程师能够更加专注于系统功能的设计与实现。

图 10-7 所示为针对 MySQL 云数据库的能力架构图。

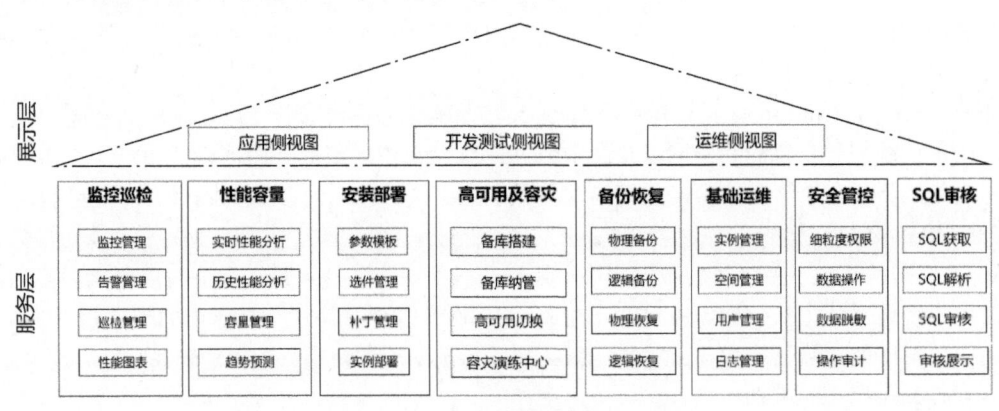

图 10-7 MySQL 云数据库的能力架构

在整个 MySQL 云数据库的能力架构中，不考虑底层的技术实现，在能力上至少包含以下两个层面。

（1）展示层

展示层是面对最终的使用用户的视图，应该包含应用侧视图、开发测试侧视图、运维侧视图。因为每一侧关注的点不同，所以 MySQL 云数据库应该进行合理的视图设计，实现易用、好用的目标。

1）应用侧视图：应用侧用户更关注的是数据相关的操作支持，如数据导出、数据查询、数据修改以及统计报表等，不会太过关注 MySQL 数据库本身。

2）开发测试侧视图：开发测试侧用户关注如何更加敏捷地支持他们的开发测试工作，如关注环境生成、对象修改、SQL 审核和模型审核等能力，对于 MySQL 数据库的深度运维能力，则不会过多关注。

3）运维侧视图：对于运维侧视图，就需要对 MySQL 实现全生命周期的运维管理，实现通过平台进行问题发现、问题定位和问题解决的目标，包括监控告警、性能分析及预测、容量分析及预测、安装部署、高可用管理、备份恢复等能力。

（2）服务层

服务层是整个 MySQL 云数据库中非常关键的一层，需要把对于 MySQL 资源池的运维管理的能力进行自动化实现。

1）监控巡检：对 MySQL 数据的关键指标进行采集，能够从不同的维度对数据库的运行指标及变化趋势进行分析，快速发现数据库的各种异常并实时告警。同时，能够通过对 MySQL 数据库运行状态数据的判断，以及指标关联分析，根据专家视角并结合算法知识库，全面检查数据库存在的健康隐患，帮助用户更好地了解数据库运行状态。

2）性能容量：指的是对 MySQL 数据库进行性能和容量的管理与分析，能够通过对数据库性能及容量指标和相关对象的深入分析，快速定位数据库性能和容量的问题。其中很关键的是需要基于算法建立数据库的性能和容量基线，这样会让平台更加智能。

3）安装部署：能够对 MySQL 数据库版本不同架构类型（单实例、主从架构、MGR 架构等）进行自动化安装部署，部署过程能够融合多种基于最佳实践经验的配置模板，实现一键安装部署 MySQL 数据库实例，实现标准化的数据库服务交付，最大程度地避免人为原因导致的数据库安装配置问题、安全性问题和性能问题。

4）高可用及容灾：实现对 MySQL 高可用架构的管理、复制状态的检测以及异常情况下的切换管理的能力。

5）备份恢复：实现对 MySQL 数据库进行物理备份、物理恢复、逻辑备份、逻辑恢复以及任意时间点的恢复的能力。

6）基础运维：实现对 MySQL 数据库实例的基础运维管理的能力，包括实例管理、参数管理、日志管理、用户管理和空间管理等。

7）安全管控：实现对 MySQL 数据库的 SQL 操作的管控，确保数据的安全，包括细粒度的权限控制、数据操作、数据脱敏、操作审计等能力。

8）SQL 审核：基于审核规则或者算法，对 SQL 语句进行质量的审核，找出可能会影响系统性能的问题 SQL 语句。

下面详细介绍对每个能力的关键实现。

2. 关键实现

（1）监控巡检

对于监控巡检，关键点包括：应该监控哪些体现 MySQL 数据库状态的指标？如何设置阈值？

如何设置采集频率？在平台实现中，主要是通过这些采集到的指标值进行展示、分析。

监控可分为以下几个层次。

- 应用层：包括响应时间、调用链、异常日志、SQL 异常、交易计数。
- 系统软件层：包括操作系统、数据库、缓存、应用服务器、云平台。
- 服务器硬件层：包括 CPU、硬盘、内存、网卡、温度、CMOS 电池。
- 网络层：包括设备状态、流量、健康情况、会话连接、防火墙。
- 基础设施层：包括电源、温度、湿度、线路、消防、安全。

对于 MySQL 监控，作者认为需要关注表 10-2 所示的指标。

表 10-2　MySQL 监控指标和告警阈值

监控指标	指标说明	告警阈值
Aborted Clients	因客户端没有正确关闭而被丢弃的连接的个数。数字增大意味着有客户端成功建立连接但很快就被断开或者终止。这一般发生在网络不稳定的环境中，主要原因如下。 • 客户端没有主动关闭 MySQL 连接（mysql_close） • wait_timeout 设置时间过短，导致连接丢失 • 客户端由于某些原因而连接丢失	建议：大于 1000 时，告警
Aborted Connects	试图连接 MySQL 服务器但没有成功的次数。当有大量请求连接不上 MySQL 的时候，数值激增，主要原因如下。 • 没有授权或者密码不对，一般错误日志中会有提示（access denied for 'user'@'host'） • 连接数已满（too many cononection） • 超过了连接时间限制。连接时间限制由参数 connect_timeout 控制，MySQL 默认设置为 10s。除非网络极端不好，一般不会超时	建议：大于 15 次时，告警
Active Transactions	正在执行的事务数量	建议：大于 512 时，告警
Binary Log Space	二进制日志的大小	建议：大于 300GB 时，告警
Com Delete	过去的 1 秒内执行删除命令的次数	—
Com Insert	过去的 1 秒内执行插入命令的次数	—
Com Select	过去的 1 秒内执行查询命令的次数	—
Com Update	过去的 1 秒内执行更新命令的次数	—
Connections	所有尝试连接到 MySQL 服务器的连接数，不管成功还是失败。所以这个数值的增量并不等于执行 show processlist 语句的结果数值，这一点需要注意	建议：大于 1500 时，告警
CreatedTmp Tables	MySQL 服务器在对 SQL 查询语句进行处理时在内存里创建的临时数据表的个数。如果该值太高，则唯一的解决办法是：优化查询语句	建议：大于 256 时，告警
Current Transactions	当前的事务（包括 not started、ACTIVE 等各种状态的事务）数量	建议：大于 3000 时，告警
Innodb Locked Tables	所有事务锁定的表的数量	建议：大于 256 时，告警
Innodb Lock Wait Secs	显示每秒处于锁等待的 InnoDB 事务总数。如果有一个非常大的值，则应该检查执行语句是否为大事务或语句之间是否存在锁冲突	建议：大于 5s 时，告警

(续)

监控指标	指标说明	告警阈值
Innodb Row Lock Time	该模板读取的 Innodb_row_lock_time 状态变量，表示 InnoDB 引擎在每次申请数据行锁定时等待的总时间（以 ms 为单位）	建议：大于 5s 时，告警
Innodb Row Lock Waits	读取的 Innodb_row_lock_waits 状态变量，表示 InnoDB 经过多长时间才获得一个行锁（以 ms 为单位）	建议：大于 5s 时，告警
Innodb Transactions	显示了 InnoDB 事务相关的信息	—
Locked Transactions	锁住的事务数量	建议：大于 1000 时，告警
Max Connections	允许同时保持在打开状态的客户连接的最大个数	建议：大于 2800 时，告警
Max Used Connections	同一时间并行连接的用户的最大连接数	建议：大于 1800 时，告警
MySQL running slave	复制运行状态	建议：等于 0 时，告警
Os Waits	操作系统等待时间	建议：大于 5s 时，告警
Query Time Count	查询响应时间，未开启	—
Questions	服务器收到的查询和命令的总数	—
Relay Log Space	中继日志的大小	—
Slave Lag	复制延迟	建议：大于 100 时，告警
Slave Open Temp Tables	从服务器中的 SQL 线程曾经打开的临时文件的个数	建议：大于 50 时，告警
Slave Retried Transactions	从服务器中的 SQL 线程重新尝试执行一个事务的次数	建议：大于 10 时，告警
Slave Running	从服务器的 I/O 线程和 SQL 线程是否在运行	建议：等于 0 时，告警
Slave Stopped	从服务器的 I/O 线程和 SQL 线程是否停止	建议：等于 0 时，告警
Slow Queries	慢查询的次数（执行时间超过 long_query_time 值）	建议：等于 50 时，告警
Table Locks Waited	显示有多少表被锁住并且导致服务级的锁等待（存储引擎级的锁，如 InnoDB 行级锁，不会使该变量值增加）。如果这个值比较高或者正在增加，那么表明存在严重的并发瓶颈	建议：大于 5s 时，告警
Threads Connected	正处于打开状态的连接的个数	建议：大于 3000 时，告警
Threads Running	正在运行的线程数	建议：大于 2000 时，告警
Total MemAlloc	总的内存分配数	—
Total number of mysqld processes	MySQL 总进程数	建议：大于 1800 时，告警
InnoDB Buffer Pool	• Pool Size：InnoDB 缓冲池的页数量，每页大小为 16KB。 • Database Pages：数据页大小。 • Free Pages：空闲页大小。 • Modified Pages："脏"数据页。如果脏数据页太多，则需要检查磁盘 I/O 状态	—
InnoDB Current Lock Waits	当前有多少事务正在等待获取锁资源	—
InnoDB I/O	• File Reads：显示每秒文件的读次数。 • File Writes：显示每秒文件的写次数。 • Log Writes：写日志的次数。 • FileFsyncs：调用 fsync() 函数的次数。与 innodb_flush_log_at_trx_commit 值的设置有关	—

（续）

监控指标	指标说明	告警阈值
InnoDB Insert Buffer	非聚集索引使用缓冲区的操作数。具体分为以下几个指标。 • Ibuf Inserts：插入的记录数。 • Ibuf Merged：合并的页的数量。 • Ibuf Merges：合并的次数	—
InnoDB Row Operations	以下几项表现了 InnoDB 内部的操作情况。 • Ibuf Inserts：插入的记录数。 • Row Read：读取行总数。 • Row Deleted：删除行总数。 • Row Updated：更新行总数。 • Row Inserted：插入行总数。	—
InnoDB Tables In Use	显示以下两个指标。 • InnoDB Tables In Use：打开表的次数。 • InnoDB Locked Tables：锁定表的次数	—
InnoDB Transactions Active/Locked	以下几项显示了处于不同状态的 InnoDB 事务的数量。 • Active Transactions：正在执行的事务数量。 • Locked Transactions：锁住的事务数量。 • Current Transactions：当前的事务数量（包括 not started、ACTIVE 等各种状态）。 • Read Views：打开一致性视图的次数	—
InnoDB Transactions	以下两项显示了 InnoDB 事务相关的信息。 • InnoDB Transactions：InnoDB 内部的事务总数。 • History List：Undo 空间记录的长度	—
MySQL Network Traffic	MySQL 网络传输数据量如下。 • Bytes Send：发送字节数。 • Bytes Received：收到字节数	—

图 10-8 所示为目前比较流行的开源监控框架 Prometheus。

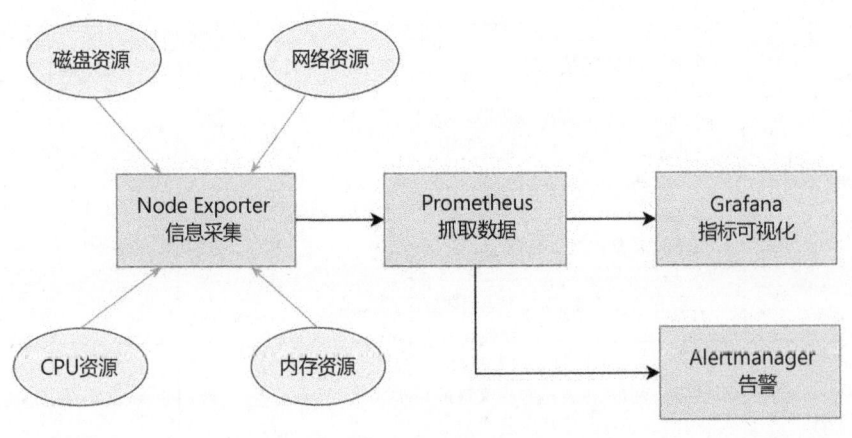

图 10-8　Prometheus 框架

对于监控项的采集，目前已有不少成熟的监控告警框架，比较典型的是 Prometheus。Prometheus 是一套开源的系统监控告警框架。它启发于 Google 的 borgmon 监控系统，由工作在 SoundCloud 的 Google 前员工在 2012 年创建，作为社区开源项目进行开发，并于 2015 年正式发布。2016 年，Prometheus 正式加入 Cloud Native Computing Foundation，成为当时受欢迎程度仅次于 Kubernetes 的项目。

Prometheus 的主要功能部件如下。
- Node Exporter：安装在监控目标上，为服务器提供数据抓取的接口，支持定制化。
- Prometheus：用于抓取数据，并存储到时序数据库中。
- Grafana：用于显示指标数据，实现可视化。
- Alertmanager：通用告警功能，用于实时处理警报。

完成采集后，下一个关键就是 Exporter。为 Prometheus 提供监控数据源的应用都可以称为 Exporter，如 Node Exporter 用来提供节点相关的资源使用状况，而 Prometheus 从这些不同的 Exporter 中获取监控数据，然后可以在诸如 Grafana 这样的可视化工具中进行结果的显示。

Exporter 根据来源可以分为社区提供的 Exporter 和自定义的 Exporter 两种。

很多软件都内嵌支持 Prometheus，如 Kubernetes、etcd，简单来说，这种类型的软件中不需要单独的 Exporter 来向 Prometheus 提供监控数据，这是它们本身的功能特性之一。当然，更多的情况则是通过独立运行的 Exporter 来提供监控数据，如 Node Exporter。操作系统本身由于不像 Kubernetes 那样提供对 Prometheus 的支持，所以需要单独运行 Node Exporter，用于提供节点自身的信息给 Prometheus 进行监控。

具体细节内容，读者可以去相关社区进行详细了解，建议有能力的读者可以基于这个框架进行 MySQL 数据库的更多细粒度的监控巡检管理的开发，如个性化监控展示、监控得分项设置、智能告警和智能巡检等相关内容。

如图 10-9 所示，以监控得分项设置为例，说明一个大致的设计思路。

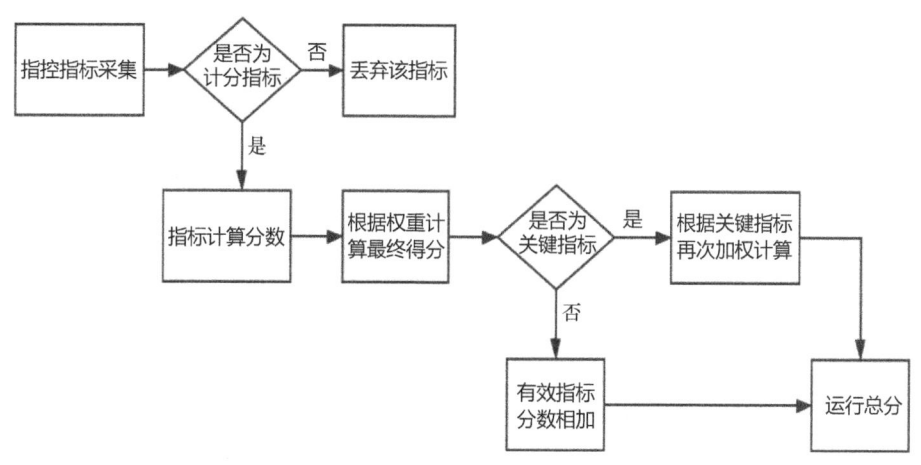

图 10-9　监控得分项设计思路

监控得分项关键设计内容说明如下。
- 传统的监控方法列举了海量数据，这些数据通常只有专业的数据库管理员才能准确解读。
- 从这些庞大的数据中快速提取出关键信息，对一般用户而言较为困难。
- 通过引入分数的计算方式，可以非常直观地展现当前的运行状态，使得监控结果更易于理解。
- 对于不同级别的重要指标，进行了独立的加权计算处理，以确保评估结果的准确性和公正性。

（2）性能容量

性能容量管理分为性能管理和容量管理两部分。

1）性能管理是指通过对数据库性能指标和相关对象的分析，快速定位数据库性能问题，预测数据库的性能趋势，快速定位问题和提前预防问题的发生。主要包含实时性能分析、历史性能分析和性能预测等关键能力。

- 实时性能分析是指通过实时监控以及告警信息，从数据库资源使用状况、事务运行及并发状况、I/O 及 CPU 使用情况、TOP SQL 等多个维度进行深入分析，快速定位发现影响性能问题的告警 SQL、TOP SQL、TOP Event、长事务、长会话和锁等性能数据并进行展示，能够从执行计划、对象信息展示、谓词使用信息、负载占比、性能数据详情和锁阻塞等方面进行性能分析，找出问题并给出优化的建议，以及一键处置的方案，如会话查杀。
- 历史性能分析是指通过存储和管理数据库历史运行数据，构建基于时间线的性能变化趋势，便于运维人员查看不同时间、不同维度的关键性能数据。用户可以基于时间段来进行针对性的性能分析，找出存在的隐患。
- 性能预测，首先要基于历史性能数据，生成性能基线，然后融合一定的算法进行性能走势的预测分析，便于提前进行性能问题的规避。例如，在业务高峰到来之前，提前进行计算能力的扩容。

2）容量管理是指通过对底层存储资源、数据库空间、数据库对象进行全生命周期管理，快速定位数据库空间异常问题，预测数据库的容量变化趋势，帮助用户减少前期硬件投入，在业务快速变化，数据库容量需求也发生变化的情况下，提前进行容量的扩缩。主要包含容量分析、容量管理和容量趋势预测等关键能力。

- 容量分析，通过对表空间、用户空间、存储对象、主机磁盘空间、归档日志等使用情况进行监控和信息获取，能够对容量的使用概况进行下钻分析，能够对容量的变化进行统计分析，找出容量的异常增长的问题。
- 容量管理，要实现对于容量的扩容、缩容的管理。
- 容量趋势预测，首先要基于历史的容量使用数据，生成容量基线，然后融合一定的算法进行容量使用走势的预测分析，如平台提供空间可用天数预测，便于提前进行容量问题的规避。例如，在业务高峰到来之前，提前进行容量的扩容。

（3）安装部署

安装部署是指针对不同 MySQL 数据库版本不同架构类型的自动化安装部署，部署过程能够融合多种基于最佳实践经验的配置模板，从而快速交付标准化的、符合最佳实践的数据库，需要能够支持各种 MySQL 数据库架构，包括单实例、主从集群和 MGR 集群，以及基于架构所必需的数据库中间件。数据库创建过程无须关注底层设施，即可获得稳定可靠、高性能的数据库服务。

安装部署主要包括参数模板、选件管理、补丁管理和实例部署等功能模块。

安装部署的前提是标准化，就是把安装部署过程中所需配置的参数通过参数模板来进行设定。表 10-3 所示为 MySQL 数据库在安装部署过程中常用的一些参数和建议值。

表 10-3　MySQL 核心参数建议

参　数　名	建议值	描　　述
innodb_buffer_pool_size	RAM 的 50%~70%	按照系统有效内存进行配置
server-id	IP 地址+随机值	可将 server-id 设置为 IP 地址+两位随机数所组成的数字，如 IP 地址为 192.168.1.23，则设置为 19216812367
binlog-format	Row	1）STATEMENT 模式（SBR）：每一条会修改数据的 SQL 语句会记录到 binlog 中。 2）ROW 模式（RBR）：不记录每条 SQL 语句的上下文信息，而是记录哪条数据被修改。 3）MIXED 模式（MBR）：以上两种模式的混合使用
lower_case_table_names	1	大小写区分
open-files-limit	635535	MySQL 打开文件的数量，与系统打开文件/etc/security/limits.d 配合使用
slow-query-log-file	1	开启慢查询日志
long_query_time	0.1	0.1s
expire_logs_days	7	binlog 日志过期时间
max_connections	2000	系统总的连接用户数
max_user_connections	1500	通过 API 连接的用户数
table_open_cache	10000	打开表的数量，取决于业务系统表
wait_timeout	28800	等待时间
max_allowed_packet	256MB	返回给客户端的网络包大小
innodb_data_file_path	ibdata1:1G:autoextend	数据设置范围
innodb_temp_data_file_path	ibtmp1:1G:autoextend:max:30G	临时表设置范围
innodb_flush_log_at_trx_commit	1	取值可以是 0、1 或 2。 设置为 1（默认）：每个事务提交时，重做日志会被物理写入磁盘。 设置为 2：每个事务提交时，重做日志会被写入操作系统的缓存。 设置为 0：事务提交时，日志不刷新到磁盘，只是每秒进行一次刷新
character-set-server	utf8	服务器字符集设计
innodb_file_per_table	1	独立表空间设置

说明：平台应该支持用户可以自行设置或者修改参数。

（4）高可用及容灾

高可用及容灾是指平台通过一键部署高可用 MySQL 数据库架构，实时探测和自动诊断故障，快速发现业务节点异常状态，在保证数据完整性、一致性的前提下，快速进行高可用切换，从而保证业务的连续性。MySQL 数据库基于成熟、稳定的数据库中间件（如 HAProxy、KeepAlive、ProxySQL、ShardingSphere 等），实现 MySQL 集群的自动高可用、负载均衡、读写分离、分库分表等高级特性。

高可用及容灾至少应该包括如下模块。

1）备库管理模块,能够支持新建备库、纳管备库、重建备库、删除备库等备库管理操作。

2）切换管理模块,能够支持自动切换、手动切换,自动切换时能够根据 RTO/RPO 条件判断选主库,支持设置从库切换权重等。

切换管理的关键技术主要包括下列几种。

- 前置状态检查,检查复制延迟、大事务、中间件状态等。
- 检测从库 I/O 线程,再次确认主库故障,避免误切。
- 进行切换设置,支持采取延迟切换、立即切换、不切换、数据完全同步后执行切换等多种切换策略。
- 根据配置切换优先级设置、从库延迟时选择新主库。
- 自动回调,修改中间件配置。
- 切换失败,暂停强制切换。

3）主从一致性检测模块。由于 MySQL 采用的是逻辑复制方式,有可能在运行时因误操作而造成主从数据不一致,因此需要进行主从一致性校验。

主从一致性校验的关键技术介绍如下。

- 支持从指定实例、指定库表进行主从一致性校验。
- 支持单次即时校验,也支持定时校验。

4）容灾演练中心模块。在很多行业(特别是金融行业)中,对容灾演练有很高的要求,平台应该设计容灾演练中心,用于进行容灾演练的统一管理。

(5) 备份恢复

针对 MySQL 数据库,通过云平台自动部署备份服务器,实现数据库的全量备份、增量备份、binlog 备份、逻辑备份和备份数据集校验,从而保证了备份数据的安全性、有效性、正确性,实现了备份数据的生命周期管理,减少了运维人员的工作量。通过恢复管理及自动恢复演练,可支持数据库异机和原机恢复,满足了高安全性、重要业务系统的数据保护及恢复到指定时间点的要求,减少了企业数据库的系统风险。

MySQL 数据库备份是为了发生意外时尽可能地恢复数据,没有恢复能力的备份是无意义的。MySQL 云数据库的自动化备份恢复主要包含以下内容。

- 通过定期恢复演练,确保备份的有效性。
- 由于不可能将所有备份都通过实际还原的方式进行校验,因此可使用文件对比方式(如 MD5 算法)进行每日基础校验。
- 备份异常时,需要有相应的处理机制来保证下一次备份能够正常进行。

(6) 基础运维

对于运维人员来说,对数据库进行日常运维是高频操作,其中包括实例的启停、实例状态查看、表空间管理、日志管理、参数管理和用户管理等。

对于 MySQL 的数据库实例日常运维管理,作者简单介绍一些常见的运维操作。

1）MySQL 权限管理。

MySQL 数据库的权限管理是日常运维中一项非常重要的基础性工作,建议进行权限的细粒度管控,不要授予过大的权限。

MySQL 账户权限信息被存储在内置 MySQL 数据库的 user、db、host、tables_priv、column_priv 以及 procs_priv 表中,MySQL 启动时服务器将这些数据库表内容读入内存。设置权限如下所示。

```
# 查看权限
mysql> show grants for 'jack'@'%';

# 授权:
mysql> grant all privileges on *.* to jack;

# 回收权限
mysql> revoke delete on *.* from 'jack'@'localhost';

# 为了满足安全策略需求,可能需要定期修改用户密码,修改命令如下
mysql> SET PASSWORD FOR 'root'@'localhost' = PASSWORD('123456');

# 由于 MySQL 数据库的用户、密码均存在于数据库的内置表中,因此可直接进行更新
mysql> update user set password = PASSWORD('123456') where user = 'root';

# 无论是权限变更还是用户密码的更改,完成之后建议进行权限的刷新
mysql> flush privileges;
```

2)错误日志检查。

MySQL 数据库的错误日志记录了 mysqld 启动和停止,以及服务器在运行过程中发生任何严重错误时的相关信息。可用 --log-error 选项指定路径。可以检查数据库运行过程中是否存在报错,如下所示。

```
shell $> tail -200 /var/log/mysql/mysqld.err
```

3)慢查询日志检查。

MySQL 的慢查询日志记录了业务系统运行时间较长的 SQL 语句,通过 slowlog 可以抓取慢 SQL 语句,同时还可以结合数据库监控、报错等相关日志进行问题的诊断与分析。

对于慢查询日志,建议进行定期检查,如下所示。

```
# 使用 mysqldumpslow 分析慢查询日志
shell $>mysqldumpslow /opt/mysql/data/mysqldb-slow.log
Reading mysql slow query log from /opt/mysql/data/mysqldb-slow.log
Count: 1  Time=0.00s (0s)  Lock=0.00s (0s)  Rows=0.0 (0), 0users@0hosts

# 也可以使用 pt-query-dest 工具进行分析。如下是一个简单的 Shell,可以通过 crontab 每天自动分析慢查询日
# 志,并将汇总的慢语句输出到日志
lowlog_path=/opt/mysql/data/mysqldb-slow.log
everyslow=/tmp/everydayslow
pt_digest=/usr/local/bin/pt-query-digest
start_string=$(grep `date --date="0 days ago" +%y%m%d` ${slowlog_path} |head -1)
start_pos=$(grep -n `date --date="0 days ago" +%y%m%d` ${slowlog_path} |head -1|awk -F:'{print $1}')
end_pos=$(grep -n `date --date="0 days ago" +%y%m%d` ${slowlog_path} |tail -1|awk -F:'{print $1}')
print_linecnt=$(expr ${end_pos} - ${start_pos})

if [ -z "${start_string}" ];
    then exit 1
else
grep -i "${start_string}" -A ${print_linecnt} ${slowlog_path} > ${everyslow}
    /usr/bin/perl ${pt_digest} ${everyslow} >/tmp/anaslowlog_`date +%Y-%m-%d`
fi
```

4）统计某个时间段内 SQL 语句执行情况。

对于数据库的日常运行情况，DBA 应该了如指掌。当有问题时，为了便于排查，需要统计数据库中 DML 语句的执行情况（分析 binlog 日志）。相关分析如下所示。

```
shell $> mysqlbinlog --no-defaults --base64-output=decode-rows-v -v log-bin.000057 | awk '/###/
{if($0~/UPDATE|INSERT|DELETE/)count[ $2" "$NF]++}END{for(i in count) print i,"\t",count[i]}'
|column -t |sort -k3nr
    INSERT `benchmarksql`.`test01` 27500
    INSERT `benchmarksql`.`test02` 2760
    UPDATE `benchmarksql`.`test03` 1107
    INSERT `benchmarksql`.`test04` 900
    ...
```

5）数据库运行情况详细检查。

检查执行的语句和底层 InnoDB 引擎运行情况，如下所示。

```
# 检查数据库运行状态
mysql>SHOW FULL PROCESSLIST;

# 查看 InnoDB 状态
mysql> SHOW ENGINE innodb status \G;
```

6）主从延迟检查。

对于 MySQL 主从环境，需要检查从库是否存在延迟，如果存在很大的延迟，则需要进行处理。相关查询如下所示。

```
# 通过检查 show slave status 命令的输出日志,判断 Seconds_Behind_Master 是否为 0。如果该值不为 0,则
说明从库存在延迟。一般来讲,如果该值超过 30 秒,则认为延迟较大,需要进行分析处理
mysql> show processlist; show slave status \ G;

# 也可使用 pt-slave-delay 工具进行检查,更加方便、直观
shell $> pt-slave-delay --delay=1m --interval=5s --run-time=10m u=delay_slave, p=delay_slave_
dba, h=14.17.119.220, P=6301
    # 说明：其中 delay_slave 为用户名，delay_slave_dba 为用户密码
```

（7）安全管控

安全管控是指在让所有用户通过云平台能够操作所有的 MySQL 数据库的同时，保证数据的安全和一致性，避免操作事故、安全事故等，实现事前的权限管控和身份认证，事中的访问控制和资产保护，事后的行为审计和溯源分析。

安全管控模块至少应该包含细粒度权限控制、数据操作、数据脱敏、行为审计等关键能力。

- 细粒度权限控制是指平台能够根据不同人员的使用性质从不同业务层面进行权限赋予，实现人员的权限分离和细粒度的管控。权限管控可细化到表数据操作的关键语句和关键字，严格控制数据操作的流程。
- 数据操作是指可针对数据库数据进行操作。在操作功能上，平台具备可视化语句执行界面，同时支持原生终端命令行，满足不同的应用场景。
- 数据脱敏是指平台能够集成脱敏规则，对数据库中的敏感数据进行脱敏保护，能够自适应规则、代码识别分析、SQL 识别分析。通过从产品底层 API 获取数据，取代传统的网络协议包解析，实现更准确的数据脱敏。
- 行为审计是指平台能够针对数据库的操作行为，进行操作中的同步监控，操作后的全方位

审计，包含系统管理类和安全行为的审计，并留存日志和记录，能够记录 SQL 语句执行人账号、IP 地址、语句、执行结果、耗时、影响行数等信息。它可动态地监测用户操作行为，并通过页面展示详细的行为记录和安全趋势。

（8）SQL 审核

SQL 审核由 SQL 获取、SQL 解析、SQL 审核和审核展示四大模块组成。通过对业务系统进行在线、离线 SQL 采集，平台将采集的所有 SQL 语句全部存储在资料库中，然后对采集的 SQL 语句通过算法或者审核规则进行分析，发现数据库设计和 SQL 编码中的问题，最终通过可视化方式将分析结果进行展示。

SQL 审核能力的设计，主要包括 SQL 解析、SQL 规则、SQL 优化三大关键引擎模块。

1）SQL 解析引擎。

SQL 解析引擎是一套能够理解 SQL 的引擎，可解决多种数据库 SQL 统一解析的问题，通过对静态 SQL 和动态 SQL 的解析，包括 DDL、DML、DQL、DCL 等语句，从多个维度进行 SQL 分析，并判断 SQL 语句执行目的，更加智能。

SQL 解析引擎基于 ISO/IEC 发布的国际标准，融合各主流数据库特征，对 MySQL 数据库 SQL 进行高效语义解析，生成增强特征语法树（Enhanced-AST）并与原生执行计划融合。

SQL 解析引擎参考 ANTLR 编写 SQL 词法/语法，构建抽象语法树（AST），接着针对 Oracle、MySQL 等数据库进行特征扩展，生成新的增强特征语法树，然后到目标数据库获取原生执行计划，将执行计划与增强特征语法树融合，形成新的结果集。此结果集可以用于各种 SQL 质量分析的场景，如专家规则式 SQL 分析、语义启发式 SQL 分析、机器学习式 SQL 分析等。

AST 是参考 ANTLR 解析引擎根据 SQL 语法构建的，整个语法树生成的逻辑包含两个重要的步骤：词法分析和语法分析。词法分析会读取 SQL 语法，然后把它们按照预定的规则合并成一个个标识（token）。同时，它会移除空白符、注释等。最后，整个 SQL 语法将被分割进一个 token 列表（或者称为一维数组）。语法分析会将词法分析得来的数组转换成树的形式。ANTLR 引擎内部是通过 DFA（Deterministic Finite Automaton，确定有穷自动机）算法来实现的，它通过事件和当前的状态得到下一个状态，即事件+状态=下一个状态，同时在 SQL 语法构建中，必须要减少运算，而 DFA 算法中几乎没有什么计算，有的只是状态的转换。最终达到语法与正则表达式贪婪匹配，因此 AST 构建过程可以看作正则匹配过程。完成 AST 的构建后，再针对 Oracle、MySQL、DB2、SQL Server、GaussDB、OceanBase、PostgreSQL 等数据库进行特征扩展，生成新的增强特征语法树。特征扩展的实现是到目标库提取 SQL 中用到对象的各种元数据，如表属性、列信息、索引信息、约束信息、统计信息等，将这些信息用为不同数据库设计的模型承载，然后和 AST 上的相关节点连接，最终形成了增强特征语法树。特征扩展后的增强语法树，信息更充分、全面，对后续的使用能够提供更强力的支撑。

SQL 语句"SELECT rule_id, rule_name FROM audit_rules WHERE id>10 AND level>5;"对应的增强特征语法树如图 10-10 所示。

由于 SQL 存在复杂的嵌套，以及各

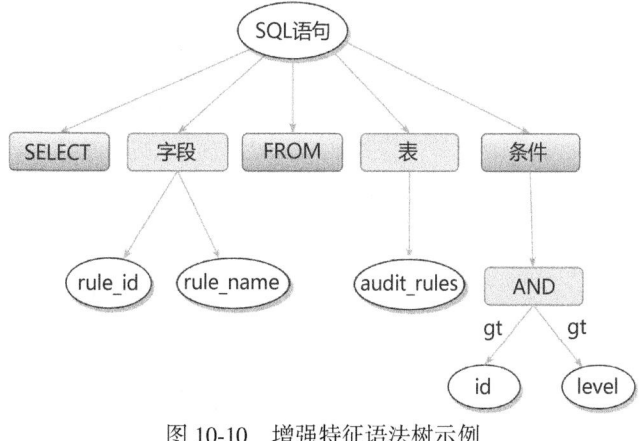

图 10-10 增强特征语法树示例

种函数和运算符，造成 AST 结构较为复杂，通常需要复杂的遍历，有时还要依赖数据字典，造成直接应用 AST 不方便，因此需要对 AST 进行改进。

2）SQL 规则引擎。

依靠专业数据库技术能力及多年对数据库的深入研究，将痛点需求与自身经验相融合，可形成一套融合了上千条专家经验的 SQL 规则引擎。它通过管理 SQL、表、索引、序列、执行计划、数据字典、性能数据，并与 SQL 引擎有机结合，可呈现出全面的审核结果、最优的修改建议，并最终形成最佳的实践方案和标准。

SQL 规则引擎的规则分成固定规则和自定义规则。

- 固定规则通过正则表达式或代码逻辑实现针对性的规则触发。在代码逻辑实现上，使用了多线程并发处理机制，提升了 SQL 规则引擎的审核速率。使用 Java 反射机制，能够很方便地创建灵活的代码，这些代码可以运行时再装配，不需要在组件之间进行源代码链接。
- 自定义规则可以让用户自行新增规则，通过自己编写的规则（正则表达式）实现规则触发。同时用户可以自行编辑自定义规则的适用 MySQL 数据库、影响程度（严重、警告、提示）、规则说明等信息。

3）SQL 优化引擎。

SQL 优化引擎，通过对 MySQL 数据库的绑定变量、绑定执行计划、调用对象类的规则和调用 MySQL 的优化包操作，完成了 SQL 语句的智能优化提示，方便开发人员通过优化引擎的优化建议自行完成 SQL 优化，大大提升了研发人员和 DBA 的工作效率。

基于专家经验的智能优化方案如下。

- 获取 SQL 执行数据。
- 获取 SQL 文本，先查询缓存，如果未查询到，则查询 AWR。
- 获取全部的执行计划。
- 分析 SQL 文本，得到表名、访问列、过滤列、关联列、分组列、排序列。
- 过滤全表扫描。
- 过滤隐式转换。
- 遍历全表扫描。
- 获取表统计信息、列统计信息、索引统计信息、索引列和索引状态。
- 判断是否绑定了 SQL Profile，如果已经绑定，就查找最优的执行计划并返回以结束优化流程；如果没有绑定，则会进入后续流程。
- 收集表统计信息，如果过滤列上有索引，则判断列上是否存在隐式转换，如果有，则建议改写 SQL 以避免隐式转换，如果没有，则遍历可选索引。
- 判断索引状态是否异常，如果异常，则建议重建索引。
- 判断索引统计信息是否异常，如果异常，则收集索引统计信息。
- 进行索引选择度判断，如果选择度好，则标记为可用索引。
- 判断索引是否可见，如果不可见，则将索引设为可见。
- 判断是否虚拟索引，如果是，则创建真实索引。

基于机器学习的智能优化方案如下。

- 查询特征表示。机器学习方法的性能在很大程度上取决于通过向量表示数据的形式或特征。本方案抽取用户查询、数据库元数据（表、属性、统计信息等）等特征，进行 One-hot 或 Word2Vec 等方式的特征表示。

- 查询解析。对 SQL 语句进行词法、语法分析，生成查询解析树，即抽象语法树（AST），并进行特征表征。
- 查询改写。查询改写分为基于规则的改写方法和基于学习的改写方法，基于规则的改写方法可以将人为确认的 SQL 优化类型进行特定语句的优化改写，而基于学习的改写方法则是通过机器学习和强化学习进行 SQL 语句优化选择。
- 查询优化。结合基数估计、成本估计、连接顺序的选择（采用 Tree-LSTM 模型结构），同时采用基于规则的优化器（RBO）和基于成本的优化器（CBO），研究基于机器学习和强化学习功能的优化查询。
- 机器学习模型管理器。研究机器学习模型库框架，并实现一个原型系统，支持 Spark MLlib、ML Algorithm、Stored Procedure、Spark UDF 等机器学习库和方法。

10.2.4 规范化 MySQL 标准体系设计

MySQL 标准体系是指将 MySQL 数据库领域专家多年积累的理论知识、实践案例、技能和经验进行归纳、总结，结合用户企业自身的运维管理规范要求，形成包含规划、建设、运维、优化提升 4 个阶段的全生命周期的 MySQL 数据库技术支撑能力，从而形成标准化的 MySQL 数据库技术支撑能力。

标准规范建议至少包含以下内容。

- 基础架构部署规范，包含命名规范、应用部署规范、数据库部署规范、数据存储规范（集中式、分布式或混合式）和网络规范等。
- 数据库资源池选型和管理规范。
- 数据库架构设计（表空间、表结构、索引）规范。
- SQL 开发及编写规范。
- SQL 质量管控规范。
- 数据库安全配置规范。
- 数据库运维管理规范。
- 数据库容灾备份规范。

下面以 MySQL 数据库部分常用的规范进行说明。

1）引擎选择：数据库服务器默认存储引擎一律使用 InnoDB（MySQL 5.7 以后的默认引擎），该引擎支持事务和行级锁，具有更好的恢复性，高并发下的性能更好，还会对多核、大内存、SSD、CPU 及内存缓存页优化，使得资源利用率更高。

2）字符集统一：要求新建 MySQL 数据库服务器默认字符集一律使用 UTF-8，必要时可申请使用 utf8mb4 字符集。

- UTF-8 字符集存储汉字占用 3 个字节，存储英文字符占用 1 个字节。
- UTF-8 统一且通用，不会在转码时出现乱码风险。
- 如果遇到 emoji 等表情符号的存储需求，可使用 utf8mb4 字符集。

3）数据库 Server Name：Server Name 是对拥有相同数据的一套主从实例的定义，目标是在同一 Server Name 下的实例可以相互切换。

Server Name 命名规范为：系统标识符（系统标识符要尽量体现系统特征且字符数不超过 6 个、不少于 3 个，并保证与已存在的 Server Name 无冲突）。

4）库名、表名和字段名规范如下

- 库名、表名、字段名必须使用小写字母并采用下画线分隔。
- 库名、表名、字段名禁止超过 32 个字符，须见名知义，禁止拼音与英文混用，禁止使用 MySQL 保留字。
- 临时库、临时表名必须以 tmp 为前缀并以日期为后缀。
- 备份库、备份表名必须以 bak 为前缀并以日期为后缀。

5）单表设计：本规范要求 MySQL 数据库单实例表数目必须小于 500，单表列数目必须小于 30。所有表都需要添加注释，单表数据量建议控制在 3000 万条以内。

6）主键：表必须有主键，如自增主键，禁止使用 VARCHAR 类型作为主键进行语句设计。
- 主键递增，数据行写入可以提高插入性能，可以避免页分裂，减少表碎片，提升空间和内存的使用效率。
- 主键要选择较短的数据类型，InnoDB 引擎普通索引都会保存主键的值，较短的数据类型可以有效减少索引的磁盘空间，提高索引的缓存效率。

7）外键：禁止使用外键。如果有外键完整性约束，则需要应用程序控制。

外键会导致表与表之间耦合，UPDATE 与 DELETE 操作都会涉及相关联的表，影响 SQL 的性能，甚至会造成死锁。高并发情况下容易造成数据库性能下降，大数据高并发业务场景中数据库的使用以性能优先。

8）表操作：对同一个表的多次 ALTER 操作必须合并为一次操作。

MySQL 对表的修改的绝大部分操作都需要锁表并重建表，而锁表则会对线上业务造成影响。为了减少这种影响，必须把对表的多次 ALTER 操作合并为一次操作。例如，要给表 t 增加一个字段 b，同时给已有的字段 aa 建立索引，通常的做法分为两步：先加字段，再加索引。

ALTER TABLE t ADD COLUMN b VARCHAR（10）；
ALTER TABLE t ADD INDEX idx_aa（aa）；

正确的做法是：

ALTER TABLE t ADD COLUMN b VARCHAR（10），ADD INDEX idx_aa（aa）；

9）字段默认值：最好将所有字段定义为 NOT NULL 并且提供默认值。
- 有 NULL 的列使索引、索引统计、值比较都更加复杂，对 MySQL 来说更难优化。
- NULL 类型在 MySQL 内部需要进行特殊处理，这增加了数据库处理记录的复杂性。在同等条件下，若表中有较多空字段，则数据库的处理性能会降低很多。
- NULL 值需要更多的存储空间，无论是表还是索引中，每行中有 NULL 的列都需要额外的空间来标识。
- 在对 NULL 处理时，只能采用 IS NULL 或 IS NOT NULL，而不能采用 =、in、<、<>、! =、NOT IN 这些操作符号。

10.2.5 专业化 MySQL 保障体系设计

为了保证 MySQL 数据库安全、稳定、连续、高效运行，需要建立专业化 MySQL 保障体系，该体系通过建立主动预防机制和快速排除机制，实现数据库全生命周期运维保障。

1）主动预防机制：通过定期检查、技术交流与培训、典型故障案例分析等主动性服务计划，将故障隐患提前查找，并消灭在萌芽之中。

2）快速排除机制：对于发生的故障（无论是软件、硬件故障，还是网络故障），能够通过电话、现场等方式综合响应，以最快的速度恢复系统，保证业务的稳健、安全运行。

作者认为，对于保障团队的建设，比较合适的方式是建立内外部团队相结合的专业保障团队。内部团队可以采用企业自主构建或者服务外包的方式组建，实现日常运行维护、故障响应、应急演练、技术支持等相关工作。建议内部团队按照高低搭配、技能互补、聚焦重点的方式来搭建。而外部团队应该聚焦在紧急和疑难故障处理、行业最佳实践的引入，以及团队能力培训和提升这些方面，可以由外部的专业服务厂商来提供。

以上观点仅是作者的一家之言，具体到各个企业，应该结合自身的实际情况，构建符合自身特点的 MySQL 保障体系。

10.3 MySQL 云数据库设计方案

DBaaS 是将数据库以云服务的方式提供，是以自助服务和便捷管理为导向的，可以对环境中的资源进行调配，以支持新老业务应用程序和操作型系统。

表 10-4 所示为业界存在的公有云 RDS 和私有云 DBaaS 两种形式的主要区别。

表 10-4 公有云 RDS 和私有云 DBaaS

	公有云 RDS	私有云 DBaaS
分享方式	跨区域、跨企业共享	本地数据中心，跨部门共享
网络	公网传输	数据中心内网
部署	容器、微服务	虚拟机、物理机、公有云、私有云

说明： RDS 是关系数据库服务（Relational Database Service）的简称，是一种即开即用、稳定可靠、可弹性伸缩的在线数据库服务。它具有多重安全防护措施和完善的性能监控体系，并提供专业的数据库备份、恢复及优化方案，使用户能专注于应用开发和业务发展。

云厂商的 MySQL 云数据库设计主要分成以下两种类别。

10.3.1 公有云 RDS：公有云厂商的云数据库设计

以华为云、阿里云、腾讯云、AWS、天翼云、移动云等公有云厂商和传统的 IaaS 云厂商为代表的云厂商有领先的云平台的优势，从原来提供 IaaS 层云服务器（ECS）能力，演进发展为提供 PaaS 层能力，而 RDS 是其中关键的一环。

在国外公有云为主流用云策略的背景下，根据不同的应用场景选择专用的数据库引擎，通过云化进行资源统一管理成为新兴的数据库使用方式。AWS 拥有超过十年的云数据库服务经验，基于其丰富的产品和配套工具，能够为用户提供从数据库迁移起全生命周期的云数据库管理服务（见图 10-11）。同时，亚马逊云数据库服务具有全托管的特性，基于 Serverless 和 AI 实现自部署、自伸缩、自修复、自优化，减轻了用户在数据库管理上的压力，最小化用户的运营负担。

以 Amazon RDS 为例，它兼容 Aurora、MySQL 等 6 种关系数据库引擎，让用户能够在云中轻松设置、操作和扩展关系数据库。如图 10-12 所示，Amazon RDS 可以自动执行耗时的管理任务（如硬件预置、数据库设置、修补和备份），持续提供高可用的数据库环境，从而让用户专注于应用程序，具备轻松管理、快速高效、高度扩展、持久稳定等优势。

对于公有云 MySQL RDS，通过底层高可用架构，结合本身 ECS 强大的高可用的能力，可以实现单中心或多中心的高可用架构。图 10-13 所示为一个典型的多中心的 MySQL RDS 高可用架构。

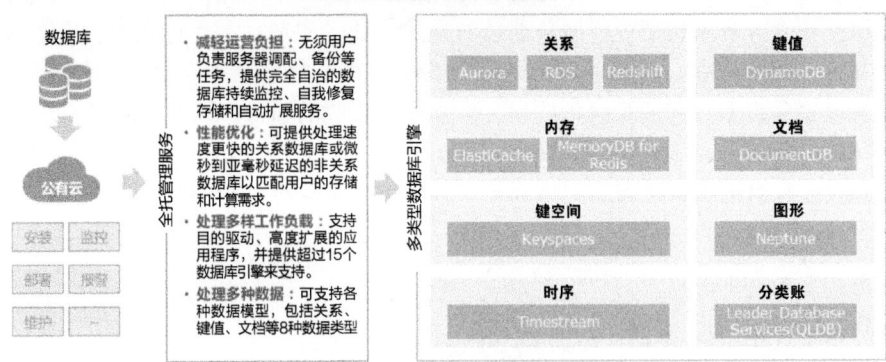

图 10-11　AWS 公有云数据库管理服务

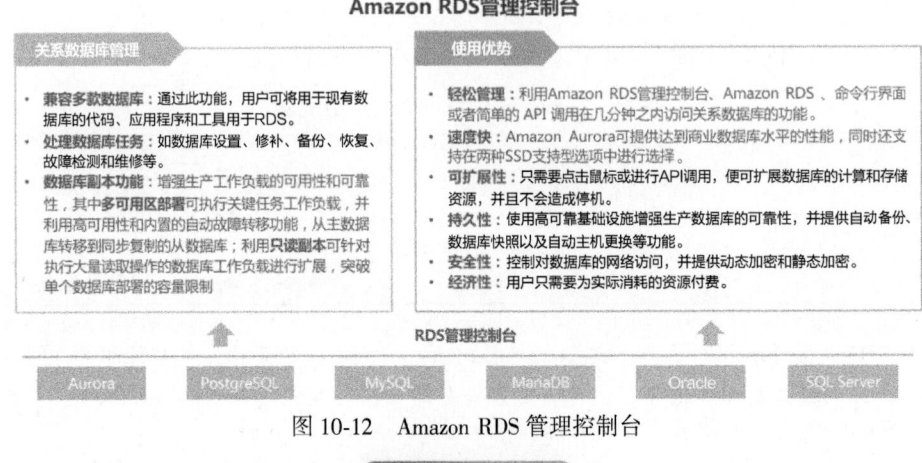

图 10-12　Amazon RDS 管理控制台

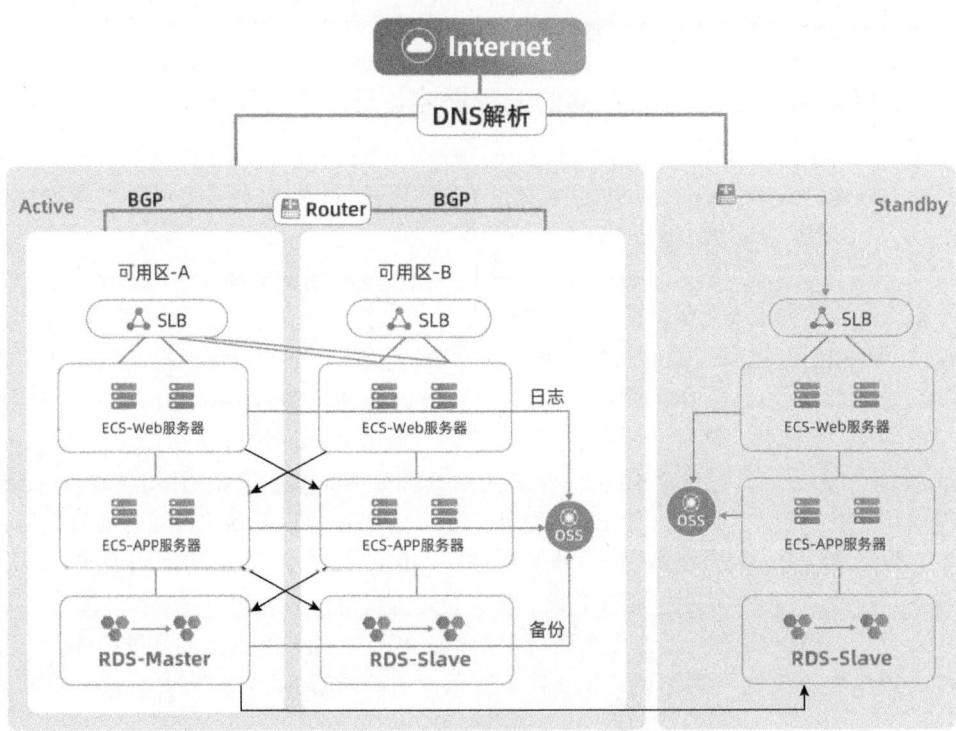

图 10-13　公有云 MySQL RDS 高可用架构

10.3.2　私有云 DBaaS：专业数据库服务厂商云数据库设计

以云和恩墨、爱可生为代表的专业数据库服务厂商结合多年积累的数据库服务经验和对客户需求的洞察，提出了以"数据库云管平台的本质是数据库管理经验的代码化"的理念来构建 MySQL 云数据库。云数据库以标准化、自动化、智能化的方式为企业提供 MySQL 数据库安装部署、监控巡检、性能容量管理、提高可用性、安全管控等全生命周期数据库管理服务，为企业打造面向未来、统一规范、安全稳定、弹性高效、能力增强的数据库基础环境。

作者以云和恩墨 zCloud 为例进行说明（来源：云和恩墨官网 www.enmotech.com 及白皮书等公开资料）。

云和恩墨 zCloud 数据库云管平台，提供端到端、一站式数据库服务支持解决方案。采用微服务架构，可以实现快速迭代更新、业务之间的低耦合，保证了平台运行的独立性、安全性和稳定性。zCloud 是自治智能的数据库云管平台，以智慧即服务（Waas，Wisdom as a Service）为产品理念，持续汇聚专家知识和经验，融合行业标准和最佳实践，提供云化自治的部署能力、智能巡检和诊断能力、知识即代码的沉淀能力，通过多元数据库统一纳管，实现服务化、自动化、智能化的数据库生命周期管理。

图 10-14 所示为 zCloud 平台架构。

图 10-14　zCloud 平台架构

如图 10-15 所示，对于 MySQL 云数据库，zCloud 基于成熟稳定的数据库中间件（HAProxy、KeepAlived、ProxySQL、ShardingSphere 等），满足 MySQL 集群的自动高可用、负载均衡、读写分离、分库分表等高级特性。MySQL 高可用管理，不仅支持自动高可用切换、手动切换、数据校验，还可以跨机房快速搭建。

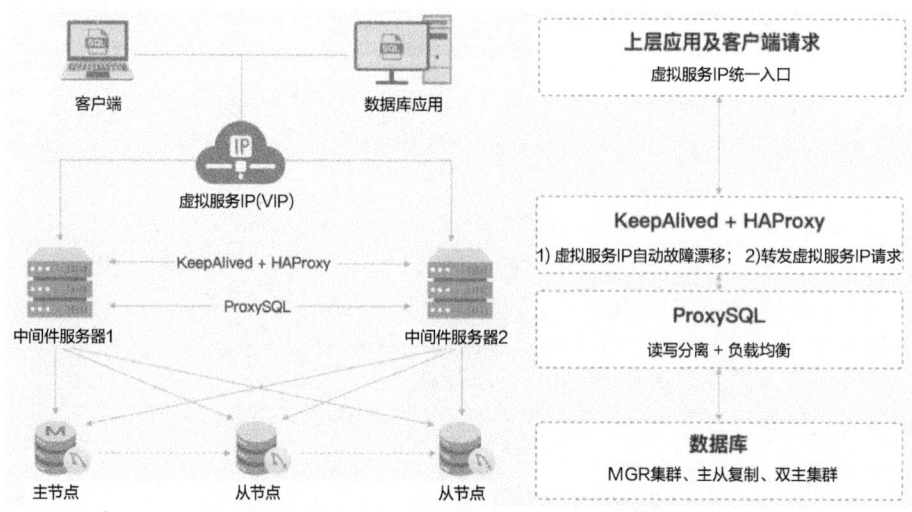

图 10-15　zCloud 平台 MySQL 高可用架构

目前市面上，还有一类厂商，它们借助底层虚拟化的云管的能力，或者容器化的能力，构建敏捷、易用的 MySQL 服务化能力。但是对于 MySQL 运维管理能力，它们相对专业数据库服务厂商还有一些差距，这里作者就不一一阐述了。